APPLIED ANATOMY AND PHYSIOLOGY OF THE SPEECH & HEARING MECHANISM

APPLIED ANATOMY AND PHYSIOLOGY OF THE SPEECH AND HEARING MECHANISM

By

HUGH E. BATEMAN, Ph.D.

Professor of Communicative Disorders
Assistant Dean, College of Liberal Arts
University of Mississippi
University, Mississippi

and

ROBERT M. MASON, Ph.D., D.M.D.

Professor and Chief of Orthodontics
Division of Plastic, Maxillofacial and Oral Surgery
Department of Surgery
Duke University Medical Center
Durham, North Carolina

CHARLES C THOMAS • PUBLISHER
Springfield • Illinois • U.S.A.

Published and Distributed Throughout the World by

CHARLES C THOMAS • PUBLISHER
2600 South First Street
Springfield, Illinois 62717

ISBN 0-398-04912-2

Library of Congress Catalog Card Number: 84-134

With THOMAS BOOKS *careful attention is given to all details of manufacturing and design. It is the Publisher's desire to present books that are satisfactory as to their physical qualities and artistic possibilities and appropriate for their particular use.* THOMAS BOOKS *will be true to those laws of quality that assure a good name and good will.*

Printed in the United States of America
Q-R-3

Library of Congress Cataloging in Publication Data

Bateman, Hugh E.
Applied anatomy and physiology of the speech & hearing mechanism.

Bibliography: p.
Includes index.
1. Speech. 2. Hearing. 3. Anatomy, Human--Pictorial works. I. Mason, Robert M., 1937-
II. Title.
QP306.B33 1984 612'.78 84-134
ISBN 0-398-04912-2

PREFACE

WE view speech, language, and hearing clinicians as important individuals on the health-care-delivery team. As such, speech and language pathologists and audiologists need to have a firm base in the interdisciplinary aspects of patient care. This book is designed for the student clinician as well as for the clinician in practice. While a major purpose of this effort is to present the basic anatomy and physiology of the speech and hearing mechanism, our feeling is that applications are needed for the most meaningful unfolding of information pertinent to scholarly activity and clinical work in speech and hearing.

Pictorial descriptions and clinical applications dominate this book with the idea that students ought to be reminded why they need to know the basic framework and function of the speech and hearing apparatus. Throughout the writing and selection of illustrations, we attempted to keep our intended readers in mind. These include beginning and advanced student clinicians in speech, language, and hearing; teachers of courses in introductory speech science and clinical applications; and practicing speech, language, and hearing clinicians who integrate thoroughly their patient management with professionals in medicine and dentistry.

The information is multilevel, permitting the reader to select introductory material to establish a visual-associated foundation of the structure and function of the speech and hearing apparatus or to gain an appreciation of important clinical concepts related to patient care. Chapters are designed to be used optionally in teaching introductory and advanced courses. The index is highly detailed for useful reference of terms in the text, including the terms found on the copious illustrations.

The authors are grateful to all publishers referenced in legends for permissions granted to reproduce their excellent illustrations. In addition to this acknowledgment, special appreciation is hereby accorded the following for courtesy use of illustrations: W. B. Saunders (Joyce Williams); CIBA (Shirley Kiefer); C. V. Mosby (Majorie McCoy); Abbott Laboratories (Gordon Stein); American Dental Association (Joan Glauberman); Springer-Verlag, Heidelberg; Charles C Thomas (Marion Taylor); Little, Brown and Company (Joyce Connell); Lea and Febiger (M. A. Dunleavy); Thieme-Stratton, Inc. (Jill Rudansky); Williams and Wilkins (Annette Boulmitis); Year Book Medical Publishers (Arlene LaMantia); Elsevier Publishers; Journal of Oral Myology;

British Association for the Advancement of Science; Josiah Macy, Jr., Foundation; and Federation of American Societies for Experimental Biology.

The long hours of typing, cutting, pasting, corresponding with publishers, keeping the authors organized, and a host of other time-consuming operations were dutifully and conscientiously performed by Patsy N. Bateman, who should be listed as a co-author.

H.E.B.
R.M.M.

CONTENTS

APPLIED ANATOMY AND PHYSIOLOGY OF THE SPEECH & HEARING MECHANISM

PART ONE: OVERVIEW OF THE SPEECH AND HEARING MECHANISM

Chapter 1

INTRODUCTION

OVERVIEW

WHEN a disorder of communication arises that is caused by an obvious malfunction of the anatomic system producing speech, much interest is generated, especially for the student of speech and hearing. To explain the *why* of such a hearing or speaking pathology necessitates a fundamental investigation and comprehension of the structure and function of the speech, language, and hearing mechanisms. Most students are unprepared to attempt any such investigation independently. The temptation is great for the student to look for a very superficial and general explanation of the makeup and operation of the communicative system.

Speech anatomy is a concrete discipline. It is an investigation into the machinery by which we live and communicate. Understanding the structure and function of the speaking apparatus opens up many new possible introductions to the appreciation of other vast fields such as medicine and the natural and behavioral sciences. But the application of speech anatomy in the clinical situation is the immediate benefit and utility for the speech clinician (Fig. 1-1).

Management procedures for the communicatively handicapped can only be generated and practiced appropriately if the student has considerable knowledge of the normal, intact speech and hearing mechanisms. Working with the communicatively handicapped without prerequisite knowledge of the normal can easily lead the clinician to develop strange concepts of the cause and remediation of speech and hearing problems. A graphic example of this may be found during the tragic period of the early 1840s when Dieffenbach introduced surgery on the tongue as a treatment of stuttering. He viewed stuttering as a convulsive condition that could be changed if the nerve supply to the tongue muscles were interrupted. Three methods of cutting into the tongue were recommended to interrupt nerve transmission to the tongue. A flood of painful and dangerous operations were loosed in Europe and America from the Dieffenbach example until more causes of death and failures brought disillusionment. (Remember, this was before anesthesia and antisepticism, so much pain and even

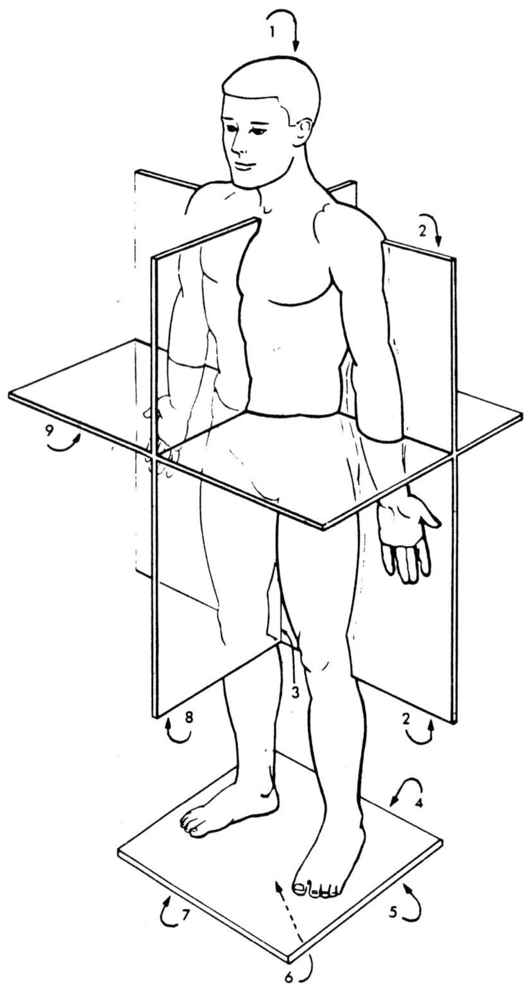

Key:

1. Cranial (superior)	6. Caudal (inferior)
2. Coronal plane	7. Anterior (ventral)
3. Medial	8. Mid-sagittal plane
4. Posterior (dorsal)	9. Transverse plane
5. Lateral aspect	

Figure 1-1. Common terms of position, direction, and function are widely used in anatomy and physiology. The planes of reference are an example of anatomy as a concrete discipline. (From Jacob, S.W., and Francone, C.A.: *Laboratory Manual of Structure and Function in Man*, 5th ed., 1982. Courtesy of W.B. Saunders Co., Philadelphia.)

death stemmed from this form of stuttering therapy.) Systematic study of anatomy, designed in approach for speech and hearing students, will, hopefully, equip students to develop sound assumptions underlying communicative disorders. Such an approach should also encourage a technically accurate exchange of patient treatment information with other professionals.

A basic purpose of this text concerns the usefulness of anatomy and physiology to the student of speech and hearing as s/he relates with the communicatively impaired. Its design may prevent incorrect interpretations, false theories, and unfounded courses of action pertaining to the physical makeup and operation of the speaking and listening machinery by leading the student into sound comprehension of the structural and functional dimensions of normality and abnormality. This should permit the clinician to formulate proper therapeutic responses for communicative handicaps, engendering clinical ability and confidence.

THE COMMUNICATION PROCESS

Communication may be thought of as an active and complex social *process*. It is extremely difficult to separate the parts of such a highly integrated process. Nonetheless, a study of the processes contributing to communication may best be understood if introduced as sequential steps and sometimes as separate components. The introduction in these first two chapters presents components of the communicative process in a rather discrete arrangement. As much as possible, an attempt is made periodically throughout the text to remind the student of the integrative nature of the total act and to foster an appreciation for the unity of the systems.

Speech Communication

Communication may be thought of as a process involving the sorting, selecting, and sending of symbols in such a way as to help a listener recreate in his own mind the meaning contained in the mind of the speaker. Speech may be viewed as a subdivision of communication. Defined, *speech is the communication of thought and emotion by means of an oral linguistic code designed to influence the behavior of others.*

Communication Models

When a television repairman opens the back of a television set, the first thing s/he looks for is a schematic or block diagram pertaining to the construction of the set. S/He knows that repair will be easier if the workings and design of the television are understood and appreciated prior to tinkering with the set. By like token, the benefits of putting an abstract concept such as speech into a model or block diagram enables the student to have another, closer look at the process. However, making models of complex abstract processes, where physical measurements are often highly variable and elusive, introduces a potential problem of oversimplification.

The use of verbal-pictorial models or diagrams may be especially helpful to beginning students of speech and hearing science in providing an idea of the entire complex scope of the communication process. The anatomy and physiology of speaking and hearing might then be appreciated in a perspective that can translate into practical clinical applications.

A number of theoretical models have been proposed to explain the speech communication process. Probably the best known is the servosystem approach of Fairbanks (1954). This early block diagram and theoretical explanation of its constituent parts has motivated many other scientists to refine and expand Fairbanks' concepts.

Before one person can communicate with another, according to the seven-sequence, verbal-pictorial model in Figure 1-2, the speaker's (1) *idea* (intentive behavior of the speaker) or thought must be (2) *transformed* into language symbols (encoding behavior of the speaker). Next, muscles stimulated by impulses from the brain (3) *convert* these language symbols into audible speech. The sounds are (4) *transmitted* as wave patterns (the message) in the air until they strike the eardrums of the listeners. In addition, the speaker may send many accompanying visual cues with facial expressions and gesturing with the hands. Note there is (5) *interfering noise* introduced into the signal as it is picked up by the receiver. The listener's ear (6) *decodes* the sound wave pattern (changing the air pressure waves into patterns of nerve impulses) and sends them to the brain. Hopefully, (7) an idea — the *inter-*

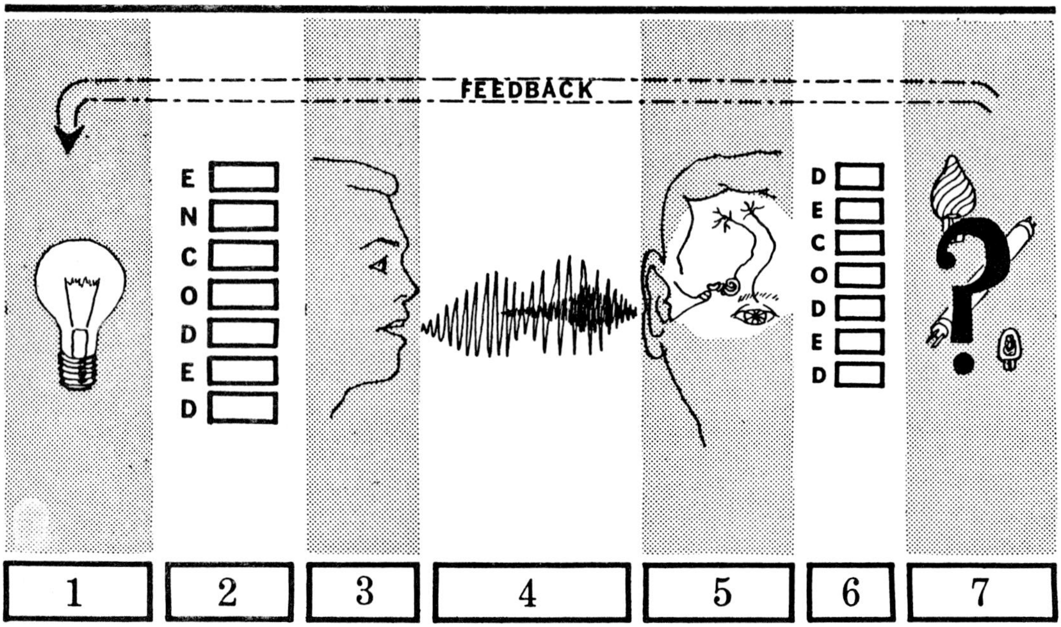

Figure 1-2. Communication model: (1) idea — intentive behavior of speaker; (2) encoding — changing idea into language symbols; (3) programming the speech articulators — converting language symbols into speech; (4) transmission — sound pressure waves; (5) interference — noise introduced to ear with signal; (6) decoding — changing sound pressure waves to nerve impulses; (7) reconstructed idea — listener interprets signal. Finally, the speaker observes the reaction (feedback) of the listener, which permits adaptation of subsequent ideas to enhance the effectiveness of further communications. (From Milton Valentine, Information theory and the psychology of speech. In Eisenson/Auer/Irwin (Eds.): *The Psychology of Communication*, 1963, p. 168. Reprinted by permission of Prentice-Hall, Inc., Englewood Cliffs, New Jersey.)

pretative behavior of the listener — similar to that which started the process is stimulated in the mind of the person to whom the communication was directed.

Our speaker can usually tell if the communication process was carried out effectively as s/he observes the listener's reaction, for there is often sufficient *feedback* from the listener's responses to judge if the idea that "sparked" in the receiver's mind was the appropriate one. Actually, the feedback phase simply reverses the entire seven-sequence process as the original sender becomes the receiver, then the sequence reverts back to the original process.

Of course, there must be sufficient similarity in background (education, culture, etc.) experiences to expect the communication process to work. Take the concept of "riding a bicycle up a hill" and imagine the differing concepts likely to be in the minds of two youngsters who have lived all their lives in their respective farm communities, but one lives in the great plains of Kansas and the other lives in the eastern mountains of Tennessee. Similarly, a phrase like "cold weather" is likely to be interpreted differently by native Wisconsinites as compared with native Louisianians.

Communicative Interference

A break or distortion anywhere along the chain of events between speaker and listener will result in the listener receiving a message unlike that intended by the speaker. Deficits in language resulting from traumas to those centers in the brain that maintain memory and process vocabulary and apply grammatical rules would be an example of a disorder of steps 1 and 2 (Fig. 1-2). Such language disorders may be found in patients following strokes; the terms *dysphasia* (disturbance in language) and *aphasia* (loss of language) are used to describe these deficits. It should be pointed out that even poor choice of language unassociated with disease or injury to the brain of the speaker may yield a breakdown in the complete communication sequence, since the listener would be confused and misread the intentive behavior of the speaker.

A disorder of articulation whereby speech sounds are distorted or omitted will interfere or miscue the listener, illustrating a malfunction in steps 3 and 4. Another example of a break in these steps would be vowel distortions heard from foreign individuals. This kind of distorted speech places a great handicap on the listener's ability to correctly code the off-target sound patterns and extract the intentive behavior of the speaker.

Even with a well-formed sound wave pattern, there is always some noise (step 5) in the environment that intrudes into the pattern as it reaches the hearer. Trying to have a conversation in a car on a busy highway with the windows open is usually devastating to the communication process.

Hearing loss would illustrate a breakdown in the chain of events for steps 5 and 6, beginning with the mechanical portion of the ear, especially the eardrum and three tiny bones. These structures must faithfully trace out the wave pattern of the compressed and rarefied air particles impinging upon them and transmit the waves to the fluid of the inner ear. In the inner ear, the waves are again transformed into nerve impulses and sent to the brain as a special pattern of electrical signals. (Remember that the origin of the speech process began with nerve impulses in the brain.) Ear infections or tumors on the auditory nerve would be examples of anomalies that would interfere with receiving and sending auditory information in steps 5 and 6. Loss of hair cells in the inner ear that transduce mechanical vibrations of the fluid into nerve impulses would break or interfere with the transmission as quickly in some cases as if the speaker omitted sibilants such as /s/ and /sh/. In young adults, the destruction of hair cells in the high frequencies may be due to excessive stimulation of the cells by listening to music set at very high loudness levels. Noisy factories for decades were the former biggest culprit purveying injurious sound on workers.

An inadequate vocabulary or language disturbance (dysphasia) will break the sequence in step 7 and result in incomplete communication. In some instances, the listener's attention is distracted to another idea at the critical time that the message arrived and the brain attempted to catalog the signal accurately.

But failure of the speaker to observe the listener's reaction (feedback) would break the chain or compound an already weakening communication process. Relegating the speech act to a one-way process is to introduce little more than a mechanical system. A radio or television set or tape player serves to illustrate speech without feedback. Remember, our definition calls for influencing the behavior of others; therefore, some feedback is essential to determine if anyone's behavior has been influenced.

Aspects of Speech Production

The vocal mechanism used in the production of speech involves organs of the body whose functions include a number of biological roles. The definition offered of speech does not imply a speaking apparatus apart from the basic structures of air and food intake. Nerves, muscles, organs, and bones involved in speech production have as their first responsibility those processes involved with the maintenance of life. These include respiration, chewing, and swallowing. Still, as we will mention later, the speech and listening mechanisms are most remarkable in design and function for communication purposes. The speculation is imposing for us to consider these highly developed structures as being especially specialized for speech. A general outline of the structures comprising the vocal tract may be seen in Figure 1-3.

Aspects of speech production include (1) an *energy source*: expiratory air; (2) a *vibrator*: the vocal folds; (3) *resonators*: oral, nasal, and pharyngeal cavities; and (4) the *modifiers*: the articulators. Although speech in each phase is produced by organs whose basic functional duty is not communication, special design features of these organs make speech a first-class citizen in the biological activities of man (Fig. 1-4).

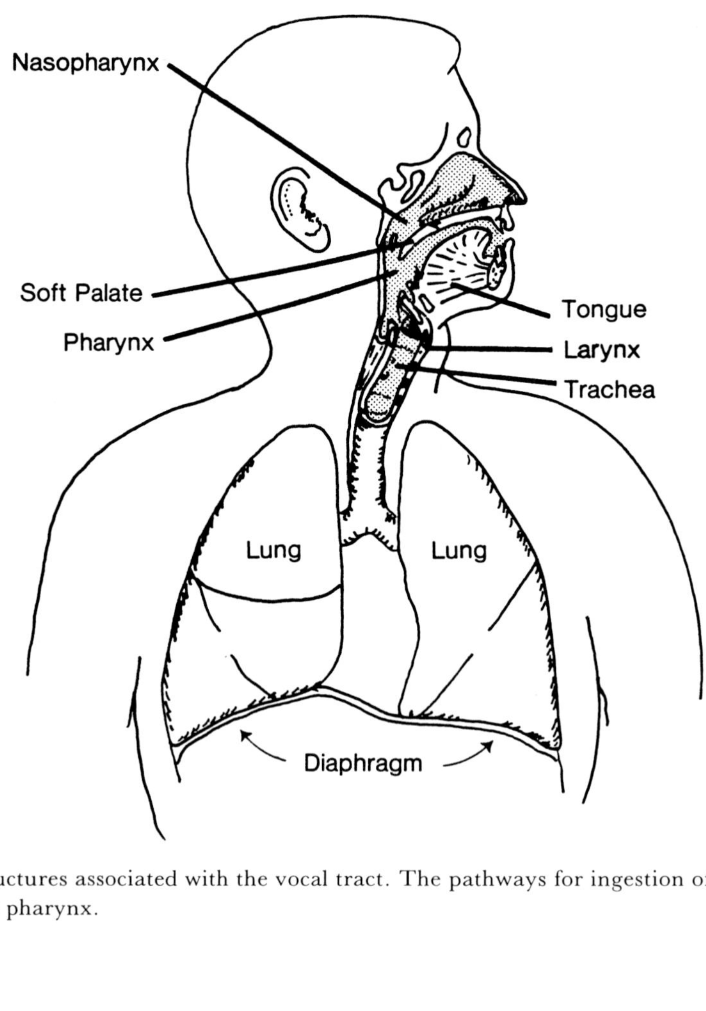

Figure 1-3. Structures associated with the vocal tract. The pathways for ingestion of food and respiration cross in the pharynx.

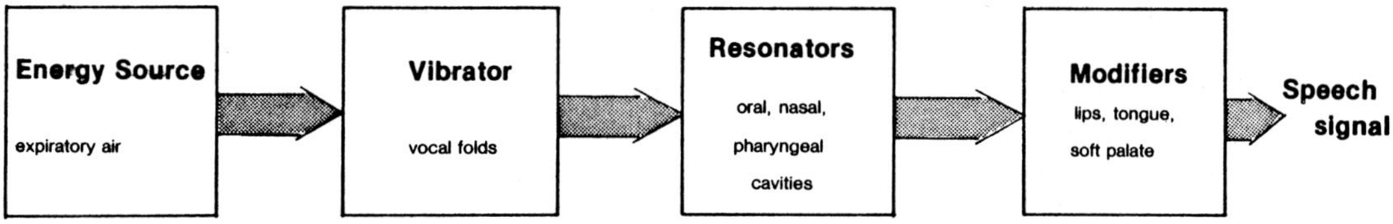

Figure 1-4. Aspects of speech production. While these are actually overlapping events, it is helpful to study each aspect separately.

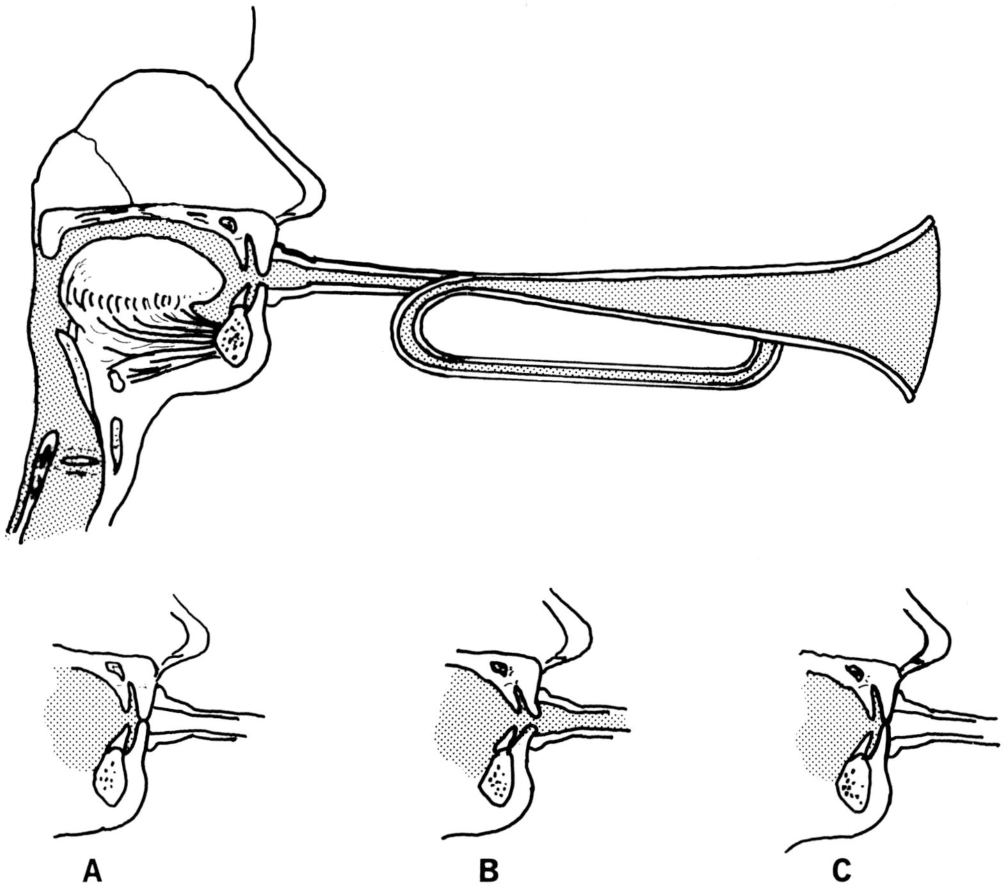

Figure 1-5. Brass musical instruments represent to a basic degree the manner we produce speech-sound production. The lips serve as a vibrator. They open and close (A, B, C) rapidly to produce a buzz-like sound source, which excites the resonating cavities of the instrument. By adding valves, as in a trumpet, the potential of musical articulation is vastly increased.

If we focus our attention on the anatomical machinery concerned with the production of speech sounds, we find that brass musical instruments are roughly analogous to the speech mechanism; both utilize expiration of air as an energy source. For a vibrator, the lips are pressed against a mouthpiece and are made to vibrate as air is forced through them. Resonation, which acts to alter sound, is achieved by the release of the vibrating lip sound into the coiled tubes of the instrument. The length of the column of air vibrating in the instrument is changed by means of valves or a sliding tube, which in turn causes the pitch to vary for musical articulation (Fig. 1-5).

Speech Monitoring

The speed by which the vocal mechanism produces speech immediately suggests a very complex and integrated behavior with the benefit of automatic neural control. Through *sound energy* by way of the ear, by *tactile* sensations created from contacts of the articulators, and by *muscle position* and muscle sense *receptors* (proprioception), the speaker's brain is given feedback during speech production. Adjustments in the speech musculature can be made in midstream to correct or alter patterns of articulation, voice quality, pitch, loudness, rhythm, or melody aspects (the *prosodic* features) of ongoing speech. The speaker monitors his speech for at least two basic requirements: (1) correct content (semantics) of words that convey the meaning intended and (2) correct speech sound pattern (articulation).

From the above, it is clear that speech automation truly involves elaborate neural control. This regulatory apparatus brings into focus a basis for explaining speech pathologies and their underlying etiologies. Faulty monitoring by the neural systems can yield faulty parameters in the speech product to the extent that, if adequate feedback controls are not established, correction of a deficit may be beyond reach.

Summary

Coupled with one another, each aspect of the speech mechanism presents a complex anatomic and physiologic apparatus. It is controlled with precision programming from the brain. (a) The expiratory muscles compel the lungs to send their inhaled quantities of air under pressure up the windpipe to operate the vocal folds in (b) a vibratory pattern for sound production. The sound stream (or air stream for whisper-like sounds) leaves the larynx and (c) resonates in the cavities of the throat, oral, and nasal chambers. At the same time, (d) the articulators of the oral cavity begin to halt, constrict, and redirect the sound stream so as to form meaningful

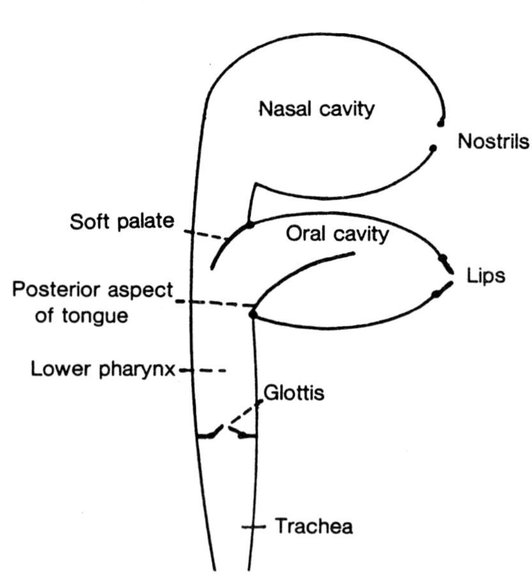

Figure 1-6. Schematic of the vocal tract cavities and points of constriction. A series of valves and cavities comprise the speech apparatus. Four valves are showing in the vocal tract; three of them associated with the oral cavity.

chunks of sound interpreted as words. All the while, the entire operation is watched over by the brain's neural sensory and speech perception units, checking for adequacy according to learned standards (Fig. 1-6).

There are actually two cardinal components in speech production. These were presented in the foregoing paragraph as (1) *a concept of neurophysiologic timing interactions* and (2) *a concept of pressure variations.* While neural activity is the initiator of speech as well as being a monitoring system, each speech sound involves an intricate distribution of pressures along the vocal tract. Further, each speech sound requires an adjustment of pressures along the vocal tract, in part because of the nature of the individual sound but also because of the influence of neighboring sounds.

BIBLIOGRAPHY

Brown, C.T., and Van Riper, C.: *Speech and Man.* Englewood Cliffs, NJ: P-H, 1966

Clevenger, T., Jr., and Mathews, J.: *The Speech Communication Process*, Glenview, IL: Scott F, 1971.

Denes, P.B., and Pinson, E.N.: *The Speech Chain: The Physics and Biology of Spoken Language.* Garden City, NY: Anch Doubleday, 1973.

Eisenson, J., Auer, J.J., and Irwin, J.V.: *The Psychology of Communication.* New York: Appleton, 1963.

Fairbanks, G.: Systematic research in experimental phonetics: 1. A theory of the speech mechanism as a servosystem. *J Speech Hear Disord, 19*:133-139, 1954.

Gray, G.W., and Wise, C.M.: *The Bases of Speech*, 3rd ed. New York, Har-Row, 1959.

Ross, R.S.: *Speech Communication: Fundamentals and Practice*, 5th ed. Englewood Cliffs, NJ: P-H, 1980.

Chapter 2

SUPERFICIAL ANATOMY

INTRODUCTION

THE surface anatomical features involved with verbal communication are not presented in the literature with the same dramatic thrust as are the imbedded structures responsible for speech. Yet from the view of surface anatomy, one can virtually outline the speaking mechanism and observe a host of activity.

A study of the human body is termed *surface anatomy*, which falls under the general heading of *gross anatomy*. Other aspects of anatomy provide the student with opportunity to study structure and form of the body from other approaches to suit the purpose of the inquiry. These other possible approaches are *microscopic, developmental, comparative*, and *applied anatomy*. If the purpose of the study is to understand the *function* of the body's structures, then the discipline is known as *physiology*. Therefore, by examining the structural makeup of the speech apparatus as well as the function of the involved structures, one is studying the anatomic and physiologic aspects, respectively. *Morphology* is a branch of biology concerned with form and structure without regard to function.

Anatomy is that body of knowledge pertaining to the structure of the body. A dictionary may state in a more formal manner that anatomy is the science of the structure of organisms and the relations of their various parts. Several different approaches to achieve specific objectives are used in the study of anatomy. Each approach may be called a subdivision of the field of anatomy.

Approaches to the Study of Anatomy

Gross Anatomy

Gross anatomy is a study of structures with the unaided eye or with low magnification. It was the only method available to the early anatomists. (a) *Regional anatomy*, a type of gross anatomy, is most frequently found in the curriculum of professional schools. In this method, which is sometimes known as *topographical anatomy*, portions of the body, such as the head and neck, thorax, abdomen, the extremities, etc., are studied. The selected division of the body is studied individually, along with the component parts of the region. (b) *Systematic anatomy* is another approach to the study of gross anatomy. In this method, the systems of the body are studied individually without regard to regional boundaries. The systems include:

The Skeletal System. The bones and cartilages, along with the tissues that bind these structures together forming joints, make up the skeletal system. Its roles are support, protection, movement, and blood formation. The study of this system is called *osteology*. The basic framework for support of the soft tissues functioning in the speech processes is considered under osteology (Fig. 2-1).

The Muscular System. This system is comprised of the structures capable of contraction and relaxation that produce movements of the body as a whole and of its parts. *Myology* is the study of the muscular system. Through muscular activity, movements are initiated to produce a sound stream of air, which is articulated into sequences of speech sounds by the articulators (Fig. 2-2).

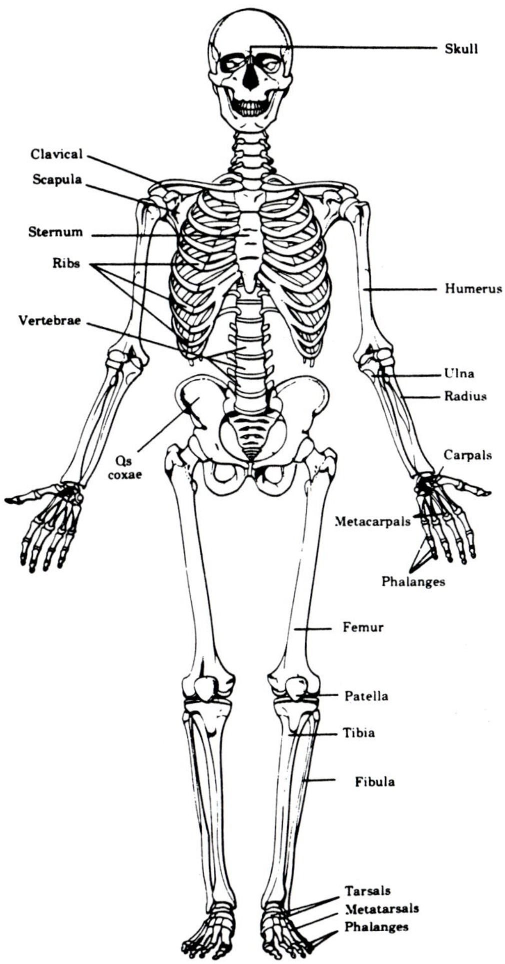

Figure 2-1. Skeletal system. (From Dawson, H.L.: *Basic Human Anatomy*, 2nd ed., 1974. *Courtesy of* Appleton-Century-Crofts, New York.)

Figure 2-2. Muscular system. (From Dawson, H.L.: *Basic Human Anatomy*, 2nd ed., 1974. Used by permission of Appleton-Century-Crofts, New York.)

Endocrine System. Numerous body functions are controlled and coordinated by the glands of this system. The glands secrete hormones into the blood for distribution to those tissues and organs that are influenced by these products. *Endocrinology* is the study of this system. A number of glandular secretions influence speech and hearing functions through systemic influence. Voice mutation during puberty is one such example.

The Integumentary System. This system is comprised of the skin and its derivatives, which are the hair, nails, and associated glands. Protection, reception of stimuli, excretion, temperature regulation, and a reservoir for food and water are functions of the integumentary system. It is studied under *dermatology*.

The Excretory System. The essential organs of excretion are the kidneys, but also included are the urinary bladder and associated ducts, which conduct, store, and eliminate urine produced by the kidneys. This system is also referred to as the *urinary* system.

The Digestive System. All the organs that act on ingested food, both mechanically and chemically, comprise the digestive system. The function of this system is to enable food to be absorbed into the blood or lymph for nutritive purposes. Structures of the mouth, which comprise the articulatory mechanism, are members of this system (Fig. 2-3).

Respiratory System. Although the lungs are the essential organs, this system includes the passageways served by the nose, pharynx, larynx, trachea, and bronchi. Absorption of oxygen and release of

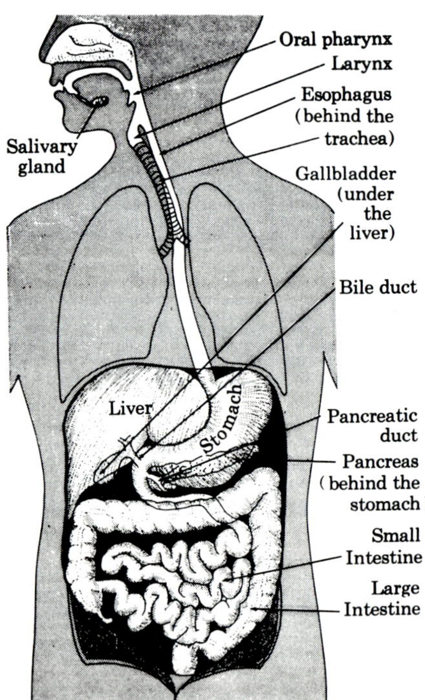

Figure 2-3. Digestive system. (From Dawson, H.L.: *Basic Human Anatomy*, 2nd ed., 1974. Used by permission of Appleton-Century-Crofts, New York.)

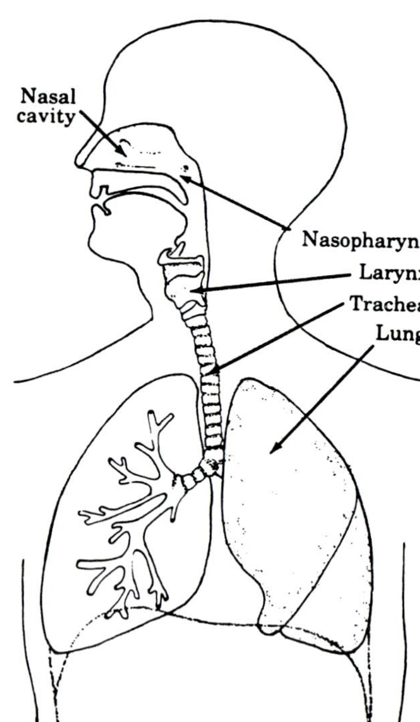

Figure 2-4. Respiratory system. (From Dawson, H.L.: *Basic Human Anatomy*, 2nd ed., 1974. Used by permission of Appleton-Century-Crofts, New York.)

carbon dioxide by the bloodstream takes place in the lungs, where a constant flow of air is maintained by this system. Other functions include voice production, olfaction, and loss of heat. In particular, speech and hearing clinicians are concerned about the process of controlled exhalation producing air energy for speech and singing, but structures pertaining to *phonation* (the larynx) and *resonation* (the pharyngeal, oral, and nasal cavities) are part of the respiratory tract (Fig. 2-4).

The Reproduction System. Both the male and female reproductive tracts are included in this system. The basic organs are the gonads: testes in the male and ovaries in the female. The reproductive tracts include other accessory organs designed to enable conception, embryonic development, and expulsion of the fetus at term.

The Circulatory System. The essential organs of this system are the heart, blood vessels, and lymphatics. They transport food and other materials to the cells and carry the cells' waste products away. The study of the circulatory system falls under *angiology* (Fig. 2-5).

The Nervous System. The essential parts of this system are the brain and spinal cord. When the peripheral nerves and ganglia, along with their autonomic connections, are added to the brain and spinal cord, the nervous system is formed. Intelligence, control and coordination of the body organs and systems, and response to the environment of the organ-

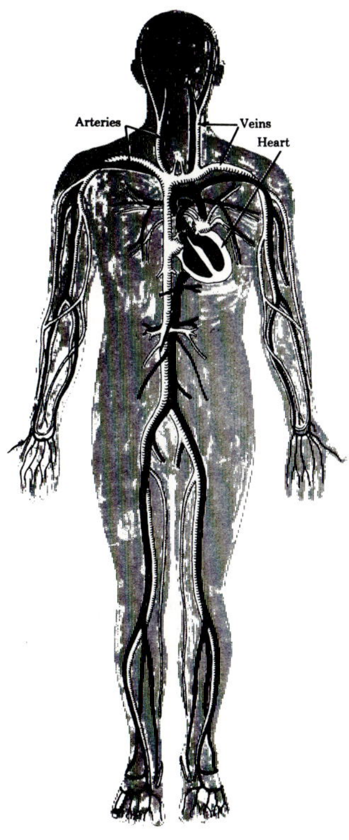

Figure 2-5. Circulatory system. (From Dawson, H.L.: *Basic Human Anatomy*, 2nd ed., 1974. Used by permission of Appleton-Century-Crofts, New York.)

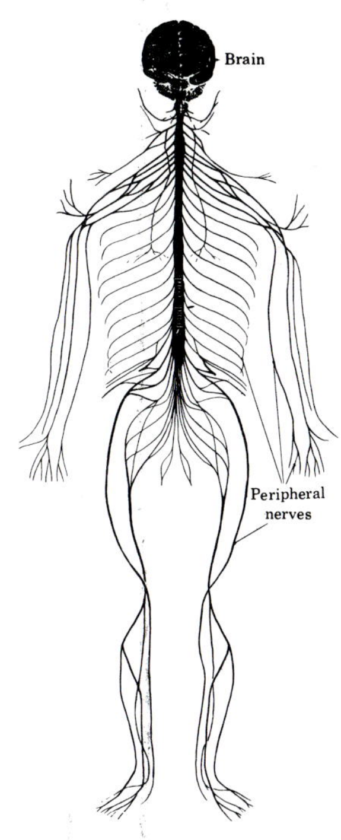

Figure 2-6. Nervous system. (From Dawson, H.L.: *Basic Human Anatomy*, 2nd ed., 1974. Used by permission of Appleton-Century-Crofts, New York.)

ism constitute the basic functions of this system. The study of the nervous system is called *neurology*. Under neurology, a study of those nerves, tracts, and centers concerned with hearing and language processing is included. Both the initiation and monitoring of speech are functions of the nervous system (Fig. 2-6).

The systematic approach is the method used most often in basic courses of anatomy and serves best here also. That the speech and hearing processes occur as a function of several systems readily indicates a highly complex interaction of many different structures.

Microscopic Anatomy

Microscopic anatomy is the study of the minute structures of the body. The principal areas are *cytology*, the study of cells; *histology*, the study of tissues which are derived from cells; and *organology*, the study of the groupings of tissues to form organs. In this subdivision of anatomy, the elemental structures of the body are considered.

Developmental Anatomy

Developmental anatomy is the study of development of the body from conception. Prenatal development includes that period of development beginning with conception and ending with birth. The first phase of the development period after fertilization is termed *embryology*. The period of developmental growth after birth is called *postnatal development*. Embryology of the oral, laryngeal, and aural aspects of the head and neck is of interest to students of speech and hearing science.

Comparative Anatomy

Comparative anatomy is the science of the structure of other animals and their relationship to man and to each other. *Phylogeny* is the study of the history of the race.

Applied Anatomy

Applied or *practical anatomy* is a special subdivision of anatomy. This branch includes surgical anatomy, pathologic anatomy, and radiologic anatomy. Clinical anatomy is applied anatomy, focusing on disabilities in the body's structure arising from development, injury, or disease process.

Physiology

The science of the function of the structures of the body is termed *physiology*. How the abdominal and thoracic regions generate air pressure for speech, how the larynx sets the air stream from the lungs into vibration, how the resonant chambers of the nasal, oral, and pharyngeal cavities modify the sound stream, how the structures of the oral cavity produce speech sounds, and how the ear changes airborne articulated speech into the language of the brain are all treated under speech and hearing physiology.

ANATOMICAL POSITION AND DIRECTION

To gain an orientation to the study of the visible body parts used in speech production, it is wise for one to begin with an introduction to the anatomical position and direction. Anatomists have agreed upon a specific body position in order to communicate effectively about the human body. The *anatomical position* has been selected to make terminology uniform for communication of such data. In this posi-

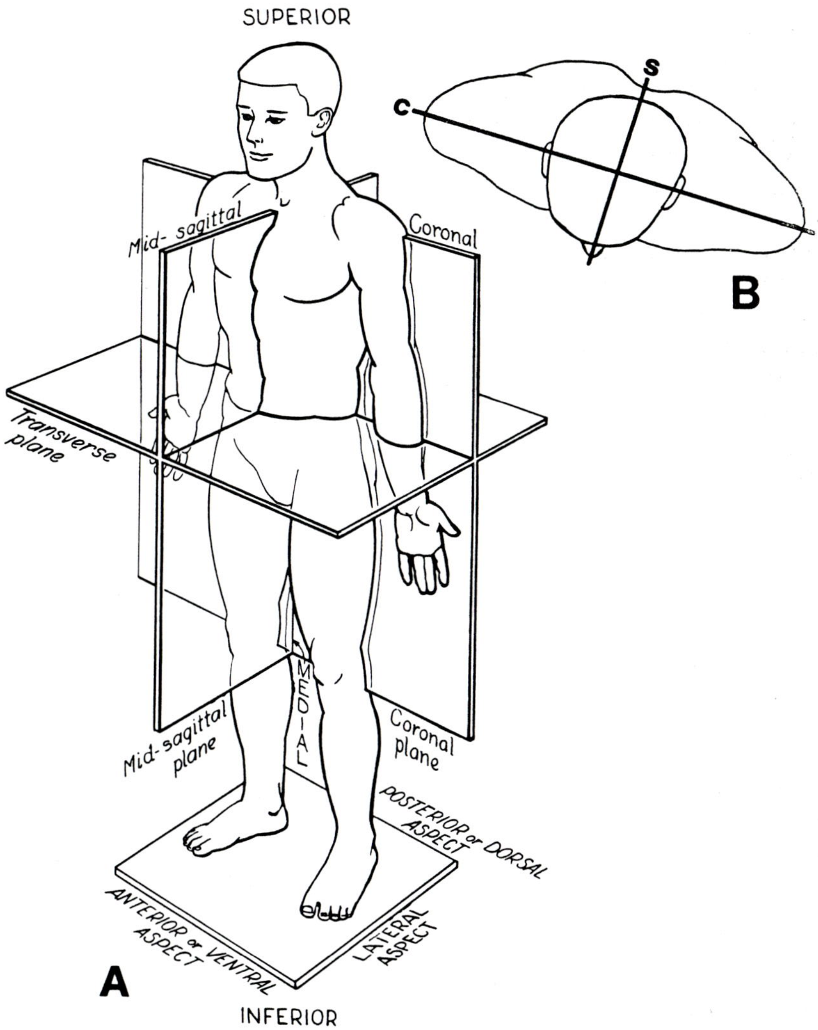

Figure 2-7. (A) Anatomical position and planes of reference. (From Jacob, S.W., and Francone, C.A.: *Structure and Function in Man*, 4th ed., 1978. Courtesy of W.B. Saunders Co., Philadelphia.) (B) Superior view. The coronal plane (c) and sagittal plane (s) are named after the sutures of the skull.

tion, the individual is standing erect, facing the observer; the arms are hanging at the sides with the palms forward (Fig. 2-7).

Planes of Reference

An aid to the understanding of human anatomy is the use of *planes of reference*, or *sections*, cut variously through the body. The types of cuts made in the body, or an organ that lies in a given plane, are given below.

The Medial Plane. This section is better known as *midsagittal*. It divides the body vertically into equal left and right portions.

The Sagittal Plane. Named after the sagittal suture of the skull, which runs in a front-to-back direction at the top of the head, this plane is sometimes synonymous with the medial plane. The sagittal plane is a vertical cut dividing the body into left and right portions. A *parasagittal plane* is any plane parallel to the medial plane.

The Frontal Plane. This plane is also known as a *coronal* section, named after the coronal suture of the skull, which courses from the peak of the skull downward on either side in front of the ears. It consists of a vertical cut, made at right angles to the medial plane, dividing the body into anterior and posterior portions.

The Transverse Plane. A horizontal cut made at a right angle to the axis of the body, dividing the body into upper and lower portions, characterizes this plane. The cut may be made at any level.

When the above planes of reference are used in relation to an organ of the body, the axis of the body is not considered; rather, the axis of the organ itself is used.

Some other traditional terms of direction used in describing the human body, and its organs (as illustrated in Fig. 2-8) are included in the following list.

Afferent and *efferent* are most commonly used in describing vessels or nerves and their pathways in relation to the central structure (e.g. the heart and the central nervous system). Afferent, meaning toward the central structure, and efferent, meaning away from the central structure (*see* Fig. 2-9).

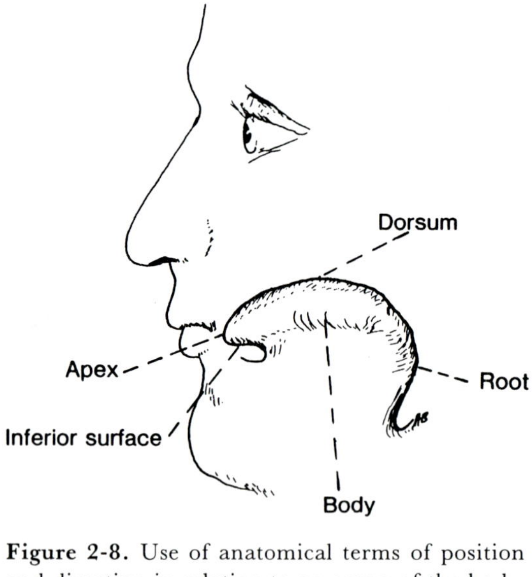

Figure 2-8. Use of anatomical terms of position and direction in relation to an organ of the body, the tongue.

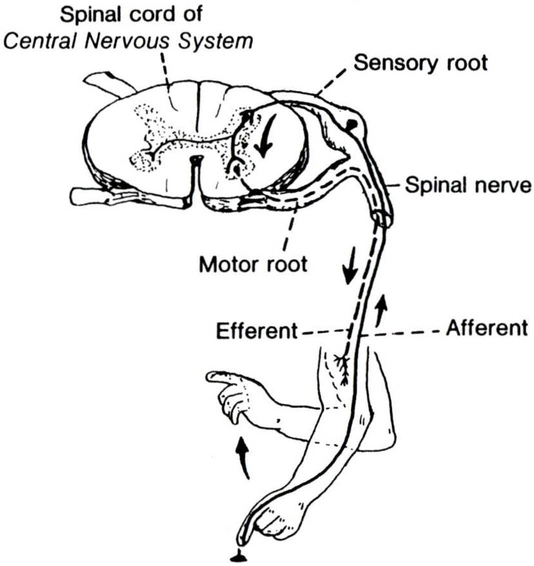

Figure 2-9. Use of anatomical terms in relation to a central structure of the body, and central nervous system. (Modified from Anderson, P.D.: *Clinical Anatomy and Physiology for Allied Health Sciences*, 1976. Courtesy of W.B. Saunders Co., Philadelphia.)

Anterior or *ventral* — toward the front.

Buccal — pertaining to the mouth or inner surfaces of the cheeks.

Cephalic — pertaining to the head.

Cephalocaudal — from the head toward the tail.

Cervical — pertaining to the neck.

Contralateral — opposite side.

Costal — pertaining to the ribs.

Cranial or *rostral* — toward the head.

Deep — away from the surface.

Inferior — toward the feet; under or lower.

Internal and *external* refer most often to the walls of body cavities and the hollow organs.

Ipsilateral — same side.

Lateral — away from the midline axis of the body or structure.

Medial — toward the midline axis of the body or structure.

Posterior or *dorsal* — toward the back.

Proximal and *distal* are most commonly used in reference to limbs.

Superficial — toward the surface.

Superior or *cranial* — toward the head; upper or higher.

SURFACE FEATURES OF THE ABDOMEN AND CHEST

The abdominal and chest region of the body are important in supplying the energy source (expiration) for speech production. Just as organ pipes are said to "speak" when provided with an ample wind supply from a blower, the speaking machinery also calls for a bellows or electric blower apparatus to generate a wind supply. For example, the heaving chests of sprinters competing in running events are highly noticeable. Particularly important are the differences in activity of the upper part of the body (the chest) and the lower part (the abdomen) during respiration. Yet during sleep, the stomach area appears to exhibit virtually all of the visible motion. In either case, what is being demonstrated is the result of the muscles of inhalation and exhalation (Fig. 2-10).

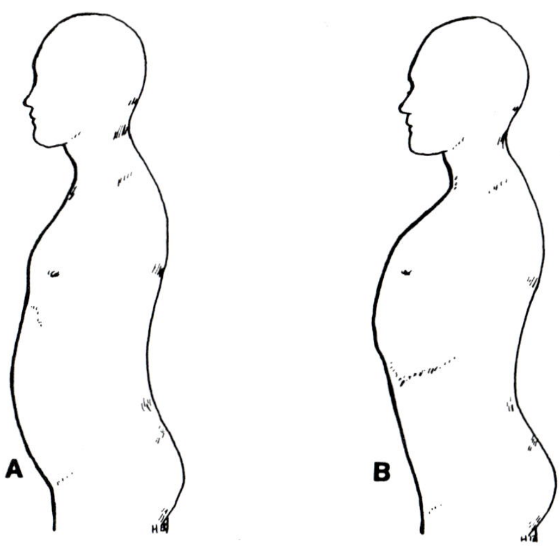

Figure 2-10. Two types of respiration. (A) Protrusion of the abdomen during inhalation due to descent of the diaphragm. Exhalation follows due to contraction of the abdominal wall musculature to force the viscera against the base of the lungs. (B) Chest expansion during inhalation. The diameter of the chest expands during inhalation and decreases for exhalation.

Often, singing teachers, speech teachers, and therapists have instructed their students to use the hands to feel their muscles in a given area of the abdominal region for more insight into proper breath control.

The Abdomen

The muscles of the abdominal wall may contribute to expiration by forcing the contents of the abdominal cavity against the underside of the diaphragm, thus compressing the lungs. During inspiration, they mostly relax and allow the contracting diaphragm to push the contents of the abdomen downward, causing a bulging appearance as the abdominal wall musculature relaxes and protrudes in a "pot belly" shape.

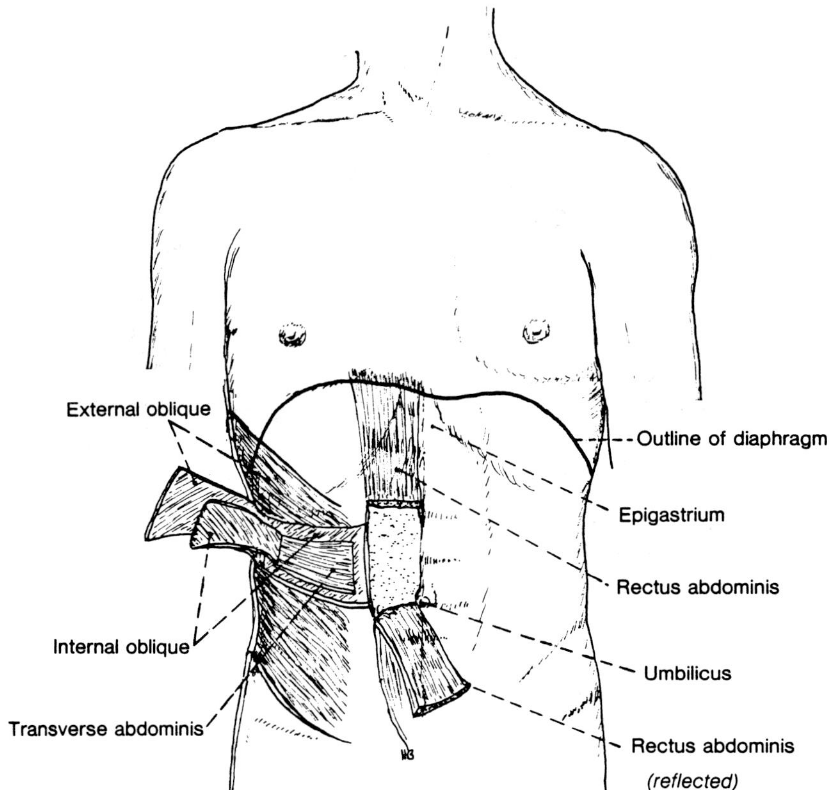

Figure 2-11. Muscles of the abdominal wall and position of the diaphragm. (Anderson, P.D.: *Clinical Anatomy and Physiology for Allied Health Sciences*, 1976. Courtesy of W.B. Saunders Co., Philadelphia.)

The upper central zone of the abdominal area, the epigastrium, encompasses the prominent notch of the lower central border of the rib cage, with the xiphoid process of the sternum at the angle of the notch. The xiphoid is easily felt and serves as an important landmark in x-ray studies of the chest. The term *midriff* refers to this part of the body also.

The diaphragm is a dome-shaped sheet of muscle and tendon (Fig. 2-11). It is located at about the upper margins of the hypochondriac and epigastric zones. Most important in inspiration, the diaphragm has been noted as the second most important muscle of the body, after the heart. The lungs rest on the superior surface of the diaphragm.

The Chest

Perhaps the most prominent feature of the chest area is the pectoralis major muscle (Fig. 2-12). In the female, the breast is attached to the covering of this muscle. Most body-building advertisements capitalize on the development of the pectoralis major to signal distinct masculinity. This superficial chest muscle arises from the medial half of the clavicle (collarbone), the sternum, the rib cartilages, and the abdomen's superficial tendinous sheet, the aponeurosis. After these multiorigins, the muscle fibers converge to attach on

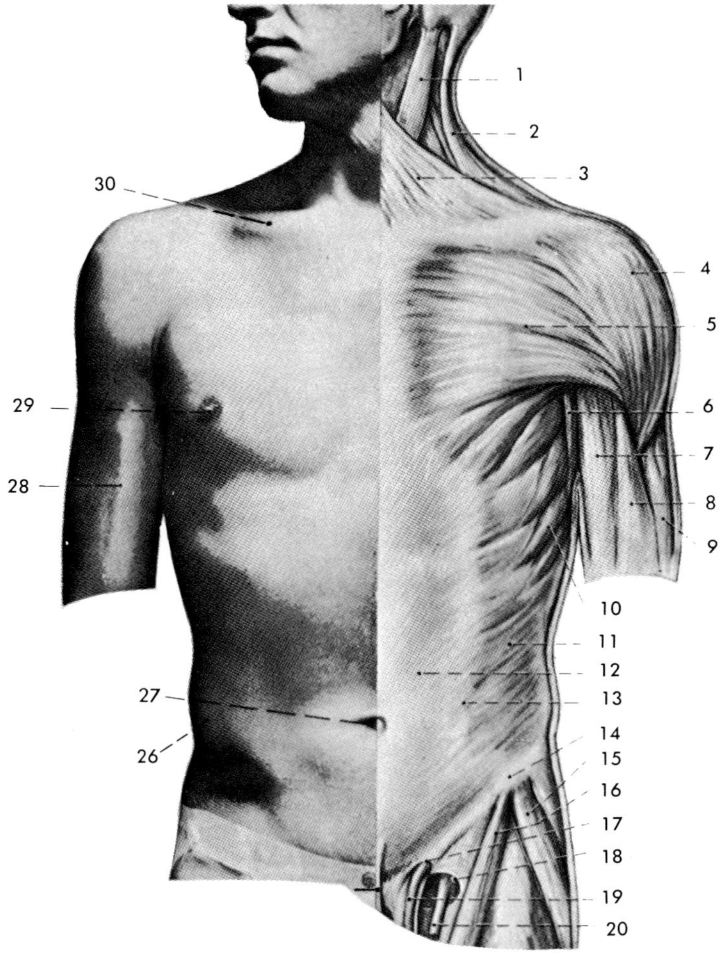

Key:

1. Sternocleidomastoid m.
2. Trapezius m.
3. Platysma m.
4. Deltoid m.
5. Pectoralis major m.
6. Latissimus dorsi m.
7. Short head biceps brachii m.
8. Longhead biceps brachii m.
9. Lateral head of triceps m.
10. Serratus anterior m.
11. External oblique m.
12. Fascia of rectus abdominis m.
13. Linea semilunaris
14. Anterior superior iliac spine
15. Tensor fascia lata m.
16. Sartorius m.
17. Superior inguinal ring
18. Saphenous opening (for veins of leg)
19. Spermatic cord
20. Great or long saphenous v.
26. Iliac crest
27. Umbilicus
28. Cephalic vein
29. Nipple
30. Clavicle

Figure 2-12. Surface features of the body; left half with skin removed, exposing the first muscle layer. (From Jacob, S.W., and Francone, C.A.: *Laboratory Manual of Structure and Function in Man*, 5th ed., 1982. Courtesy of W.B. Saunders Co., Philadelphia.)

the upper aspect of the humerus, or upper bone of the arm. A companion muscle to the pectoralis major, but not a muscle of the chest, is the deltoid muscle, which originates on the lateral half of the clavicle and also ends on the humerus.

Muscles of the rib cage are multilayered and arranged in overlapping patterns as found in the abdominal walls (*see* Fig. 2-11). The muscles interconnecting one rib to another are known as the *intercostals* (Fig. 2-13). During breathing, these muscles complement other respiratory muscles.

The superficial muscles of the thoracic walls may play important roles in respiration, especially as they influence speech, depending on the pattern of breathing being used. This is discussed in greater detail under the subject of the physiology of breathing in Chapter 6.

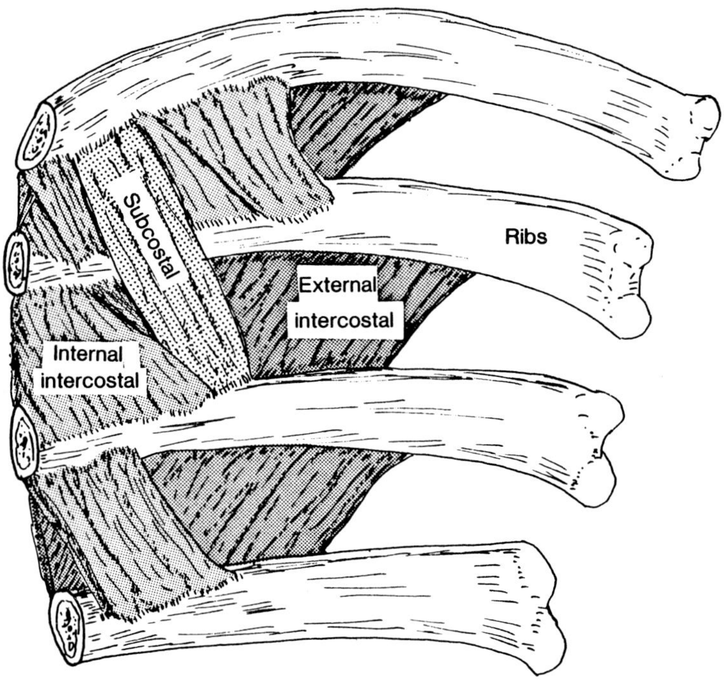

Figure 2-13. Inner surface of a portion of the rib cage showing muscles interconnecting the ribs.

SURFACE FEATURES OF THE HEAD AND NECK

The Neck

The neck offers several landmarks for identifying the underlying structures. On the anterior surface of the neck, the *laryngeal prominence* of the thyroid cartilage of the *larynx* is visible, particularly in certain male subjects (Fig. 2-14).

The larynx serves as a sound generator by converting the air stream from the lungs into a sound wave capable of exciting the auditory mechanism. This operation is designated as *phonation* or sound production. Air arrives at the lower en-

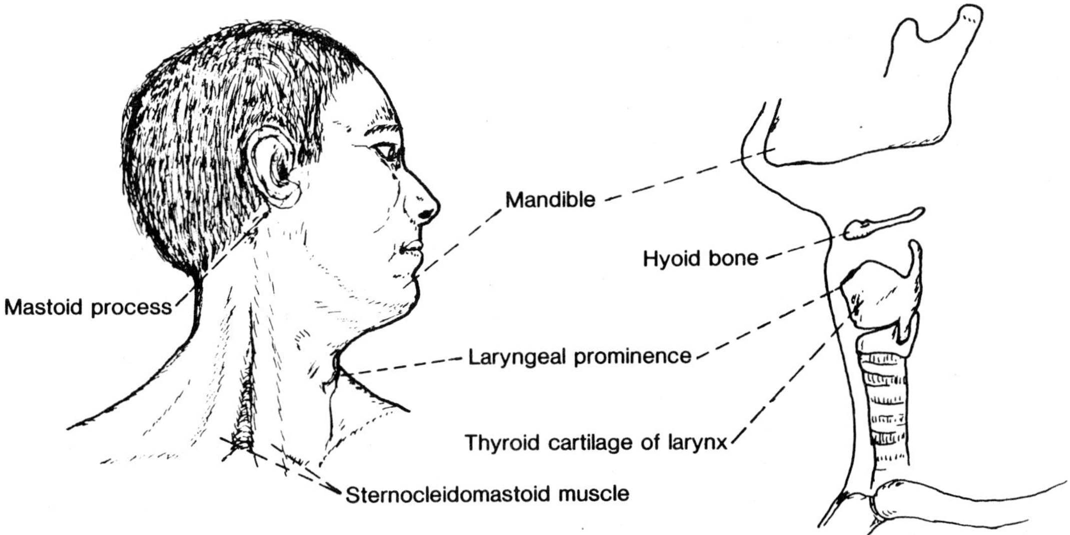

Figure 2-14. Anterolateral view of the neck and head (left). Some internal structures of the neck related to superficial landmarks (right).

trance of the larynx by way of the trachea or windpipe, which can be palpated (detected with the fingers) just a short distance below the thyroid cartilage and followed into the jugular notch, where the trachea disappears behind the sternum. In some thin subjects, the individual tracheal rings can be felt. While the thyroid gland covers the second, third, and fourth rings just below the larynx, approximately four other rings may be palpated above the sternum.

The upper limit of the larynx is the hyoid bone, a horseshoe-shaped structure without direct connections to any other bone. It is not difficult to locate the hyoid bone, for it is about 1 to 2 cm above the thyroid prominence. In addition, the hyoid bone may be located by inspecting the *angle of the chin* and anterior part of the neck while the head is tilted back slightly.

The sternocleidomastoid muscle is the most easily identified muscle on the surface of the neck. In the lower attachment, this muscle forms with its opposite mate a notch, the jugular notch, at the top of the sternum. A second group of fibers of this muscle take their origin on the clavicle near the head of the sternum. Hence, each sternocleidomastoid muscle has a split inferior end composed of a sternal head and a clavicular head. This strong muscle courses from the two heads as a

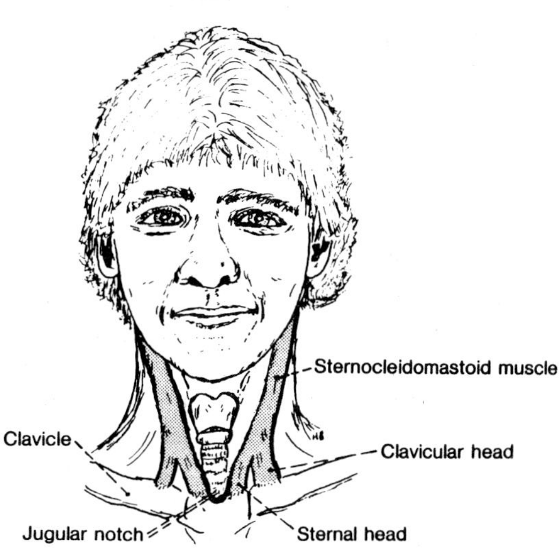

Figure 2-15. The jugular notch and adjacent structures. The trachea is behind the notch, and the sternum is at its base.

single, stout bundle of muscle fibers to attach onto the skull just below and back of the ear, an area known as the *mastoid process* of the temporal bone, which houses the auditory mechanism. The initial part of the name of the muscle depicts its fixed beginning, or *origin: sterno* for sternum and *cleido* for clavicle or collarbone. The latter part of the muscle's name describes its movable end, or *insertion: mastoid* for the nipple-shaped process of the temporal bone (Figs. 2-14 & 2-15).

By placing the fingertips on the sides of the larynx just at the anterior border of the sternocleidomastoideus, one can detect an arterial pulse as blood is pumped up toward the brain via the carotid artery in the neck.

A muscle covering the entire anterolateral neck region can be readily observed when it contracts. This muscle, the platysma, originates over a wide area of the upper front of the chest and travels up the neck as a superficial thin sheet of muscle to insert over the mandible or jaw bone. As with most muscles of the body, the platysma is paired. It is this muscle that assists in relieving the pressures of an uncomfortable collar or assists in opening the mouth.

The Head

Surface features of the face contribute to modifying the sound stream for speech sound production, called *articulation*. Also, the muscles of facial expression contribute to communication by providing emotional content, which complements articulation. Speech reading, often taught to the hearing impaired, depends on visual clues of the lips and facial expressions of the speaker to receive the message. Other fields such as drama and public address also emphasize communicative aspects of facial expression (Fig. 2-16).

The Lips

The lips are composed of muscles and glands, covered by skin on the outside and by mucous membrane on the inside. Known also as *labium superius oris* and *labium inferius oris*, for the upper and lower lips, respectively, the lips represent a highly variable valve for the anterior orifice of the mouth, with the point of contact of each lip during closure labeled as the *rima oris*. At the lateral margins of the lips is the *angle of the mouth*, or *angularis oris*. The midsection of the upper lip exhibits a slight prominence in most subjects, the *tubercle* of the upper lip (Fig. 2-17).

The *philtrum* is, in part, the indented median portion of the red zone of the upper lip. Altogether it is actually a shallow depression extending from the upper lip to the nose and forms, with the rounded parts of the lip, a bow, Cupid's bow. As the philtrum reaches the nose, it meets the *columella nasi*, the fleshy portion of the *nasal septum*, which is a column of tissue separating the nostrils (Fig. 2-18).

The *red* or *vermilion* zone of the lips is a characteristic feature of man only. In this red zone, the epithelium (skin covering the lips) is thin and contains a special substance in the layers of the epithelium. A rich vascular supply of the lips under the thin epithelium, seen as numerous capillaries carrying blood to the lips, provides the red color of the lips.

Movement by the principal muscle of the lip, the *orbicularis oris*, is closely followed by the skin of the lip and the mucous membrane on the inside. This occurs because the lip coverings are tightly bound to the connective tissue covering the orbicularis oris.

In youth, the lower lip has no boundary laterally. But with advancing age, a furrow appears, beginning at or close to the corner of the mouth. This groove, the *labiomarginal sulcus*, runs in a posteriorly convex arch toward the lower border of the mandible. Another groove, the *nasolabial groove*, is present in practically all persons. It forms an upper border, separating the cheek from the upper lip. The nasola-

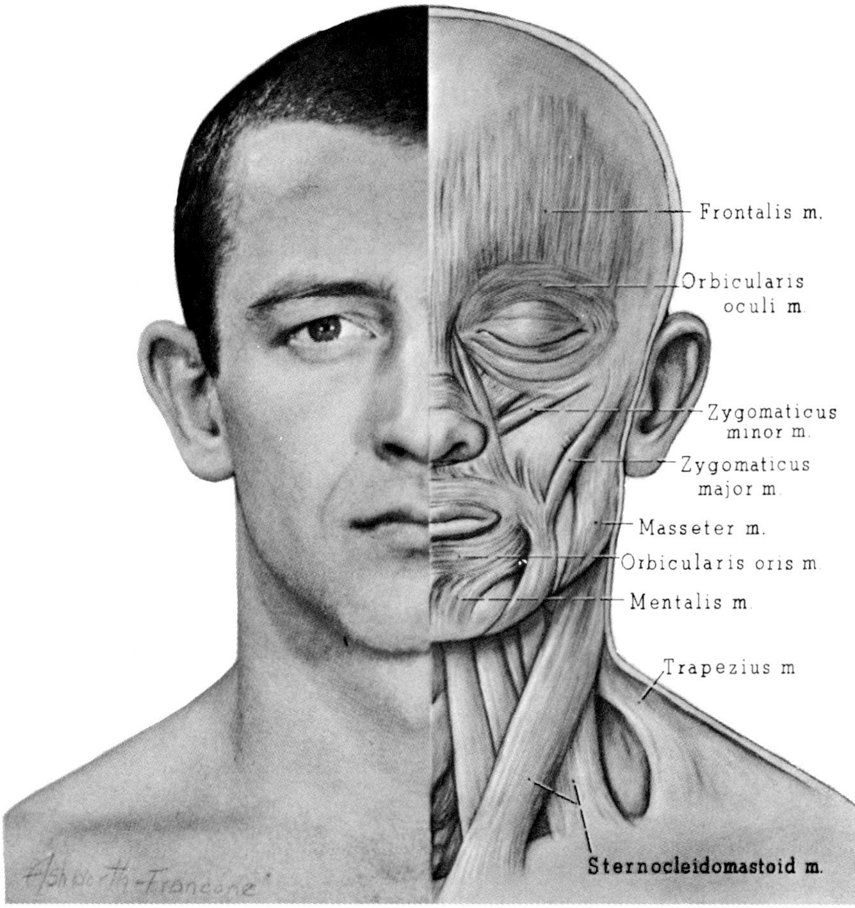

Figure 2-16. Muscles underlying the skin of the face, neck, and scalp. (From Jacob, S.W., and Francone, C.A.: *Elements of Anatomy and Physiology*, 1976. Courtesy of W.B. Saunders Co., Philadelphia.)

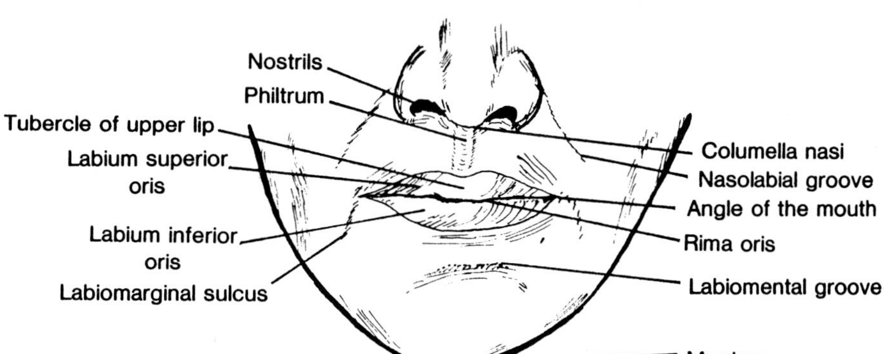

Figure 2-17. The lips and surrounding landmarks on an adult face.

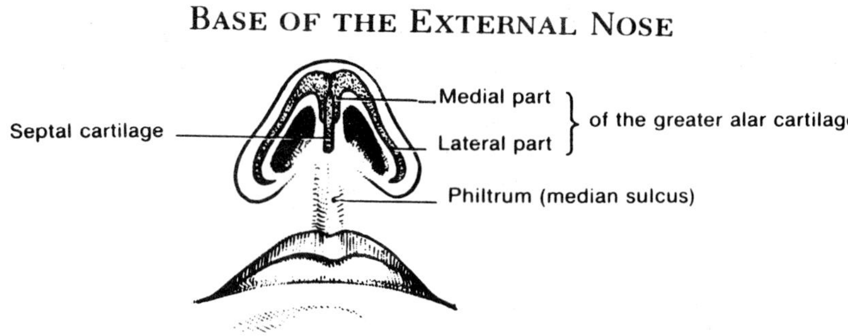

Figure 2-18. Inferior view of the nose. (From Bossy, Jean: *Atlas of Neuroanatomy and Special Sense Organs*, 1970. Courtesy of W.B. Saunders Co., Philadelphia.)

bial groove begins at the wing (ala) of the nose and courses downward and laterally to pass at some distance from the corner of the mouth.

Basically, the functions of the lips are of a biological nature. That is, the lips assist in the intake of solid and liquid foods into the oral cavity. The lips are also responsible for preventing the contents of the oral cavity from escaping during mastication and swallowing. They are very sensitive to touch and temperature, which may play a role in selective intake of food. Additionally, the lips are involved in speech production. Certain vowel sounds and especially the consonant sounds *b, p, f, v, m*, and *w* are produced with the aid of the lips. The sounds listed can be referred to as labial sounds.

Below the lips is the chin. The *point of the chin* is also known as the *mentum*, the Latin word for chin. Separating the chin and the lower lip is a somewhat sharp and deep groove, the *labiomental groove*. Age, fullness of the lower lip, and prominence of the bony and soft chin determine the depth of the labiomental groove.

The chin itself does not move in the male when the lower lip is depressed; at least this appears to be true of most subjects. This is because the skin of the face varies with sex, the male having the thicker and firmer skin. Mobility is thus restricted in the male, and the female is likely to show more of the upper teeth, including a portion of the gum (gingiva), during smiling.

The Cheeks

The cheeks, or *buccae*, serve as the lateral boundary of the oral cavity. Just under the skin are several muscles acting upon the lips at the angle of the mouth. But an inner layer of muscle, the buccinator muscle, forms the mobile part of the cheek with attachments to both upper and lower jawbones. The buccinator may be referred to as the trumpeter's muscle. It is the largest of the oral muscles and draws the cheek in upon contraction.

A peculiar body of fat tissue is found in the cheek and accounts for the appearance and fleshy feel of the cheeks. The variable *buccal fat pad* of Bichat makes the cheek feel especially pliable. In the infant, the buccal fat pad is extensive in volume and is said to aid the suckling movement. It has been termed the *suckling pad* as a result of this speculation.

The Nose

The *ala*, or wing of the nose, forms the lateral flared portion of the nose. The *nostrils* are positioned just under the ala on either side (Fig. 2-18). The alar wings

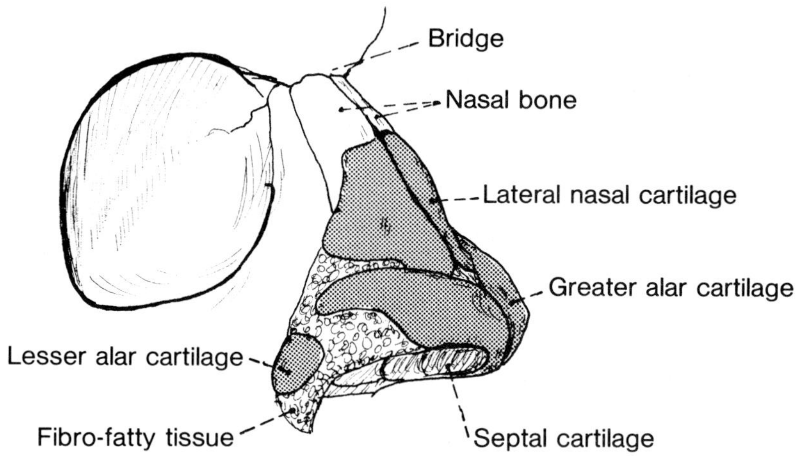

Figure 2-19. Framework of the nose showing the nasal bones and cartilages.

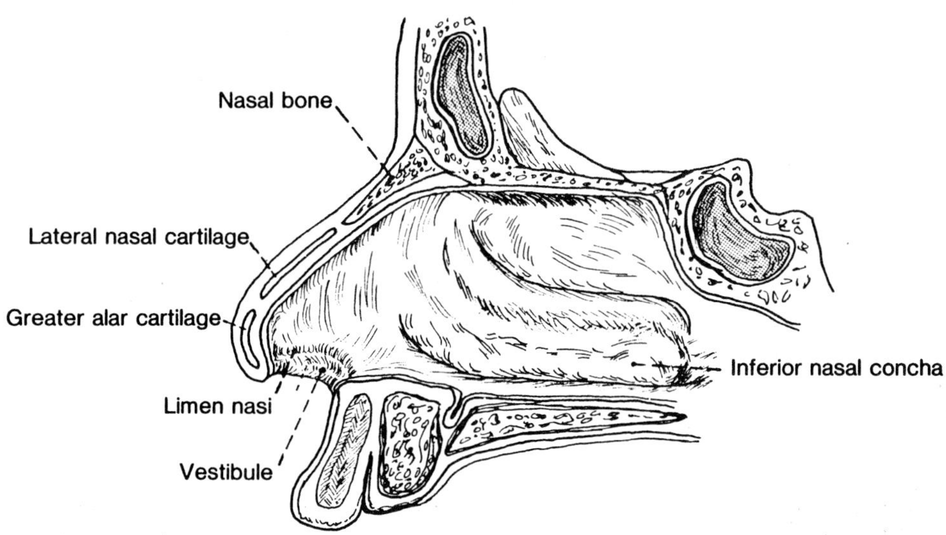

Figure 2-20. Diagram of a parasagittal section of the nose showing the right nasal cavity. The limen nasi, located about 1 cm behind the nostril opening of the nose, is also known as the threshold of the nose. This constricted passageway is designated as the liminal valve of the anterior nose.

narrow behind the flared lateral projections to provide a decreased diameter of the nostril opening. The most constricted portion of the nostril due to this anatomical arrangement is found about 1 cm behind the nostril opening. This constriction is referred to as the *liminal* valve, or the valve of Mink. Some patients who claim that they are not able to breathe freely through the nose experience a dramatic opening of the airway when the liminal valve is widened. This is often seen in aerodynamic assessments when nasal catheters of differing sizes are used to evaluate nasal airflow and resistance. Clinically, individuals who have either a very obtuse or acute nasolabial angle experience more difficulty in nasal respiratory competence that has been related in part to the angle created at the liminal

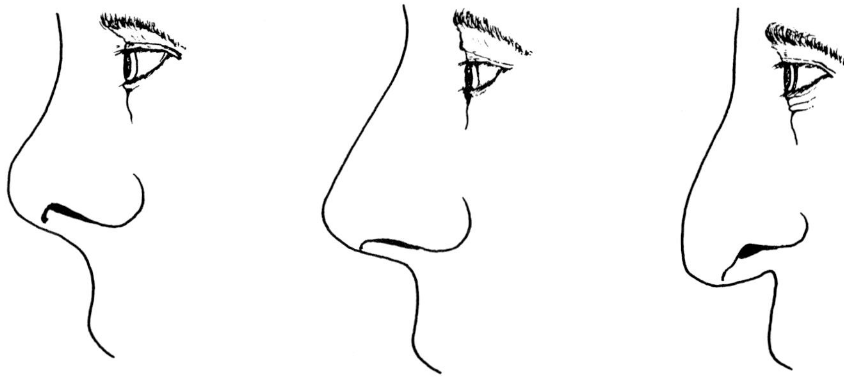

Figure 2-21. Various nose configurations.

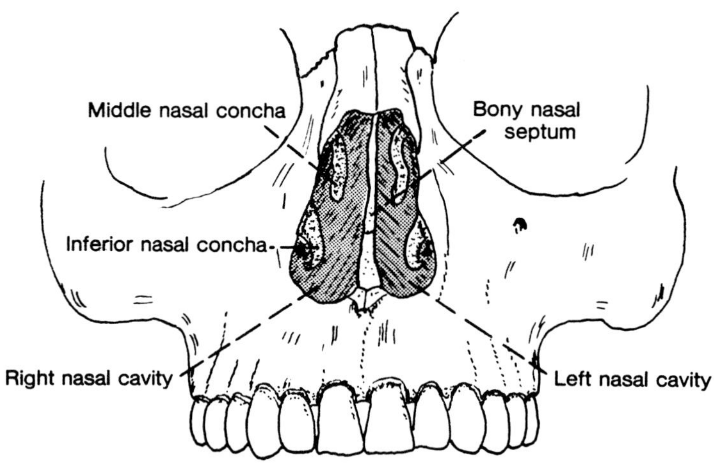

Figure 2-22. The bony nasal septum and nasal concha, anterior view.

valve by the shape of the external nose (Fig. 2-21). A framework of cartilage gives rigidity to the alae and other portions of the nose, especially the *apex* or tip of the nose. By feeling the side of the nose, one can locate near the *bridge* of the nose the bony foundation signaling the location for anchoring of the nasal cartilage to the skull (Fig. 2-19). The nasal bones are also found here at the bridge of the nose. Inside is the nasal septum, the partition separating the nasal cavity into halves. It may be seen just inside the nostrils, but it extends completely throughout the total nasal cavity. If the septum is deviated too much to one side, it may sufficiently obstruct the nasal cavity to prevent normal resonance of the sound stream. Consequently, the individual may sound as if he has a stopped nose, as in the case of a head cold, if the obstruction is at the posterior portion of the nasal cavity (Fig. 2-22).

The Mandible

The mandible lends itself to examination by palpation. Its horseshoe-shaped lower portion (the body or corpus) may be followed easily to the point of the chin. The posterior aspect begins at the *angle of the mandible*, where a sharp upward turn of a large process known as the *ramus* is located (Fig. 2-23).

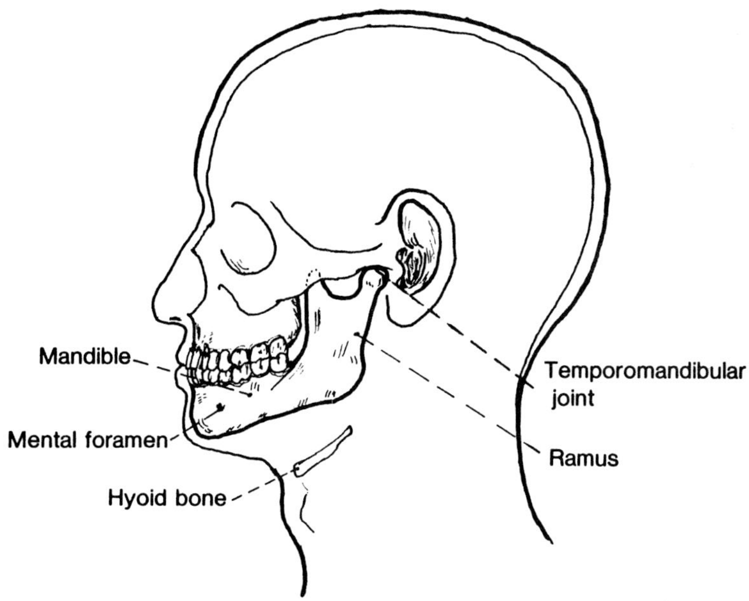

Figure 2-23. Lateral view of the mandible.

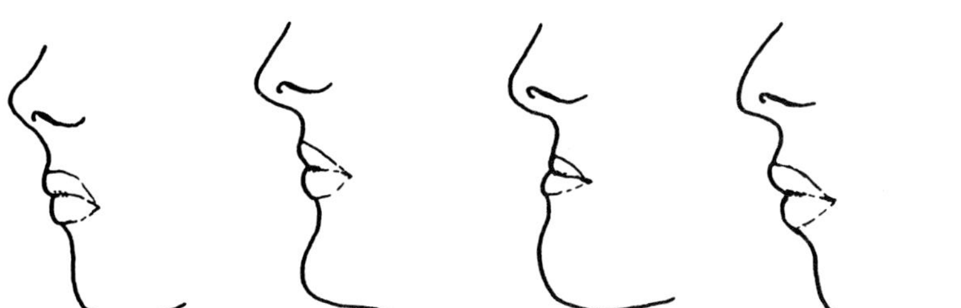

Figure 2-24. The shape and position of the mandible determines the configuration of the chin.

By placing the cap of a ball-point pen between the grinding teeth (the molars) and attempting to count aloud while holding the object in place, one can discover the relative importance of mandibular movement during speech production. Interestingly, persons who have their teeth wired together for weight reduction, a broken jaw, or other reasons have the capacity for speech which may sound quite normal. Mandibles may be too far forward or back in relation to the upper part of the face and the upper dental arch. A protruding mandible may position the tongue too far forward in the oral cavity, restricting its mobility and precision of articulating speech sounds. The appearance of the lower part of the face is directly related to the mandible (Fig. 2-24).

The Eyes

The eyes sit in special indented sockets of the skull for protection. Other protective features include the eyelashes, which protrude from both the upper and lower eyelids. Perhaps the eyebrow should be included in this frame of reference, although its role is probably not as great as the lashes. The eyebrows may be beneficial to some as they deflect rain or dust particles from the eye. However, dandruff from the eyebrow area may be an irritant to the eye.

Similar to the lips, the eyelids meet, forming lateral and medial *angles of the eye*. The normal width of the head is five eyes wide, and the eyes are one eye apart. Near the medial angle of the eye, small openings are present on the rims of the lids. These are the *puncta lacrimale*, which transport fluid from the eye into a duct system emptying into the nasal cavity. The *lacrimal gland*, located laterally above the top eyelid, may, under emotional stress and in response to irritation of the eye, accelerate secretions of fluid so that it overflows the eyelids creating tears (Fig. 2-25).

The whiteness over all but the cornea of the eyeball is due to the outer layer of the eyeball, the *sclera*. It is a tough covering of connective tissue. A specialized membrane known as the *conjunctiva* covers the anterior portion of the eyeball. Dilation of the minute vessels in the conjunctiva causes reddened or "bloodshot" eyes. Conjunctivitis or pinkeye is infection of the conjunctiva.

The transparent window of the eye is the *cornea*. Within the boundary of the cornea may be found the *iris*, a disk-like body giving color to the eye by its pigmentation. The *pupil* is the central hole in the iris that controls the amount of light passing into the interior of the eye. Through the pupil may be seen the *lens* of the eye, which is responsible for focusing the image on the retina (Fig. 2-26).

Without a doubt, the eye is the window of two important channels of communication: reading and writing. In addition, visual clues are picked up constantly during face-to-face conversations that may decidedly affect interpretation of the speech signal as well as listener reaction.

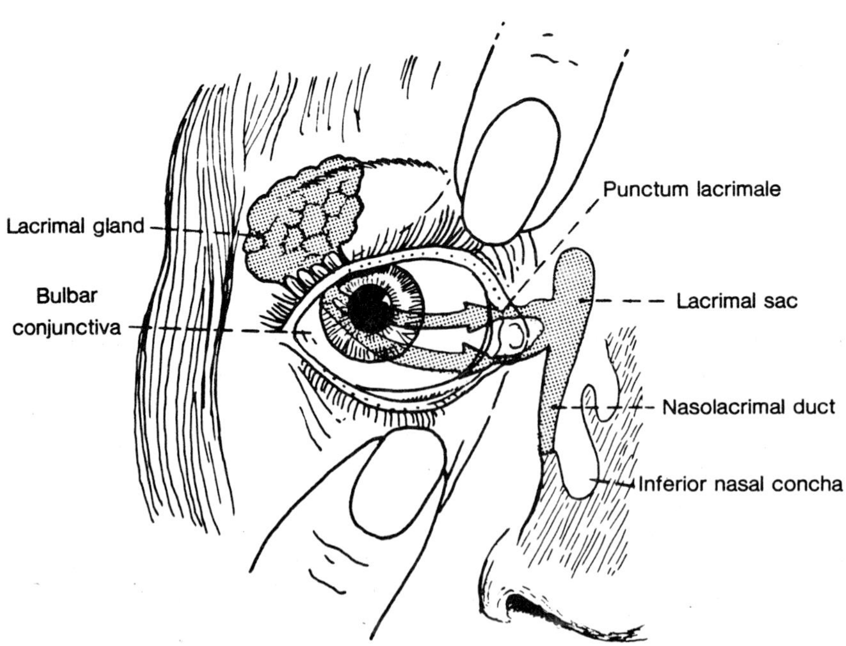

Figure 2-25. The lacrimal apparatus.

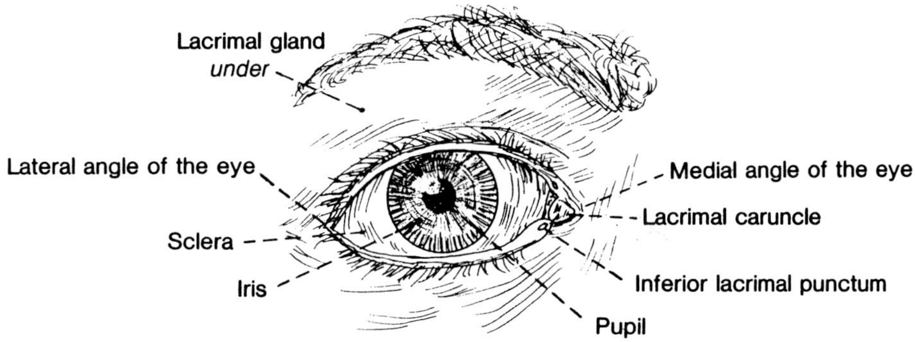

Figure 2-26. Anterior view of the eye

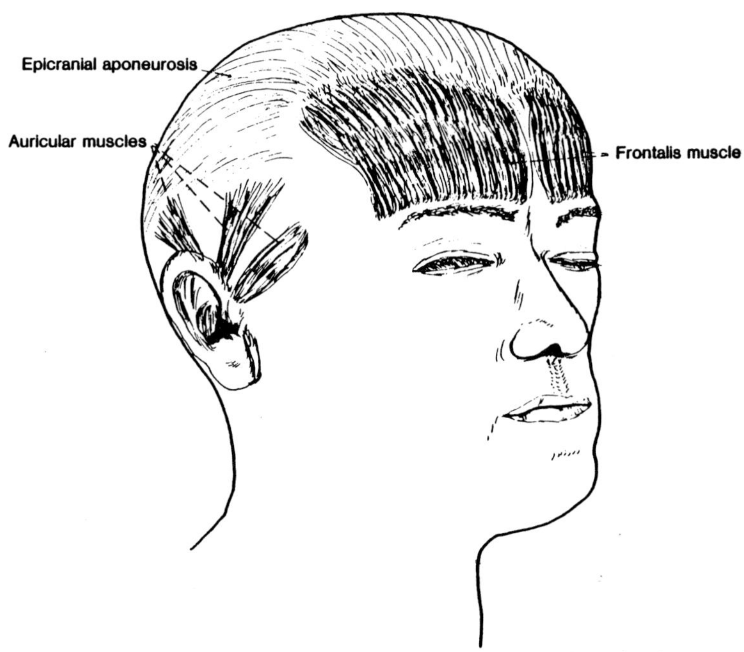

Figure 2-27. Diagrammatic representation of the muscles of the forehead and ear.

The eyes are indeed vital in their role of conveying emotional content during conversation. The essence of the credibility of what we say usually rests on the behavior of our eyes. For example, looking our listener in the eye without blinking adds intensity to the impact of our vocalization.

The Temporal Area and Scalp

Just lateral and inferior to the eye is the cheek or zygomatic bone, which is easily palpable. By following an extension of the cheek bone posteriorly, the *zygomatic arch* between the cheekbone and the temporal bone may be identified. The posterior limit of the arch is just above the external ear orifice. The only movable joint of the skull is located just under the posterior origin of the zygomatic arch. It is the *temporomandibular joint*. From the name of the joint, the bones contributing to this joint are identified. The action of the mandible can be felt by holding the fingers just in front of the external auditory canal orifice and moving the jaw in chewing motions (Fig. 2-23).

On top of the zygomatic arch between the ear and the eye, a depression, the *temporal fossa*, is located. An important muscle of the mandible, the *temporalis*, converges from the side of the scalp into the space between the arch and the skull to insert on the upper anterior portion of the mandibular ramus. In bald persons, the action of this muscle may be observed during mastication as the skin over the temporalis ripples upon alternate contraction and relaxation of this underlying muscle (Figs. 2-27 and 2-28).

The skin over the forehead may be wrinkled by action of the frontalis muscle, located under the skin. A large sheet of fascia, the *epicranial aponeurosis*, covers the scalp, serving as attachment for several scalp muscles including the frontalis. At the posterior area of the scalp, the occipitalis muscle assists in movement of the scalp. For practical purposes, it may be described in combination with the frontalis muscle and labeled as a single muscle, the *occipitofrontalis*, since the medial fibers of the two muscles are connected. Wrinkling of the forehead is accomplished by the action of the occipitofrontalis muscle.

The Ear

The external ear consists of the *auricle*, or *pinna*, which is unmistakably visible on the side of the head, and the *external acoustic meatus*, or canal. The auricle is able to make only feeble movements in response to action by three small rudimentary muscles attached to the skin and cartilage of the ear. These three muscles, the superior, anterior, and posterior auricularis muscles, are named from their positions around the external ear. All of these muscles take their origin on the fascia, the epicranial aponeurosis, around the auricle and insert in the cartilage at the base of the auricle (Figs. 2-27 and 2-28).

Directing the auricle toward a sound source for more acute hearing and more precise sound localization is not possible without moving the entire head. Many animals can accomplish this feat readily, but this is not a characteristic in man. No doubt everyone has observed a dog prick up its ears and listen to identify the source of a faint sound. While the auricle is the least important structure of the ear, it does serve to direct some sound waves into the external meatus in spite of the fact that there is no well-developed mechanism for active movement and control of the auricle. A parallel to the auricle's function is the cupping of the hands behind the ear to aid in a difficult listening situation, especially faint sound sources.

The most used clinical term to describe any malformation of the external ear is *microtia*. Literally, however, microtia is defective or incomplete development (hypoplasia) or lack of development (aplasia) of the external auricle with a blind or absent external acoustic canal. Another clinical term used to describe abnormalities of the auricle is *agenesis*: a congenital condition of marked deformity or absence.

The auricle is full size around eight years of age. Plastic surgery for protruding (lop) or deformed ears is usually not performed before this age.

Surrounding the rim of the auricle is a fold of tissue known as the *helix*. The inner rim that parallels the helix is the *anthelix*. It is this rim that is deficient or absent in individuals who have protruding or lop ears. The *lobule* is very familiar, for it is the part of the ear where earrings are affixed. The *concha* is the capacious cavity at the entrance to the external auditory meatus. Anteriorly positioned is the *tragus*, a small projection just ahead of the opening of the external canal.

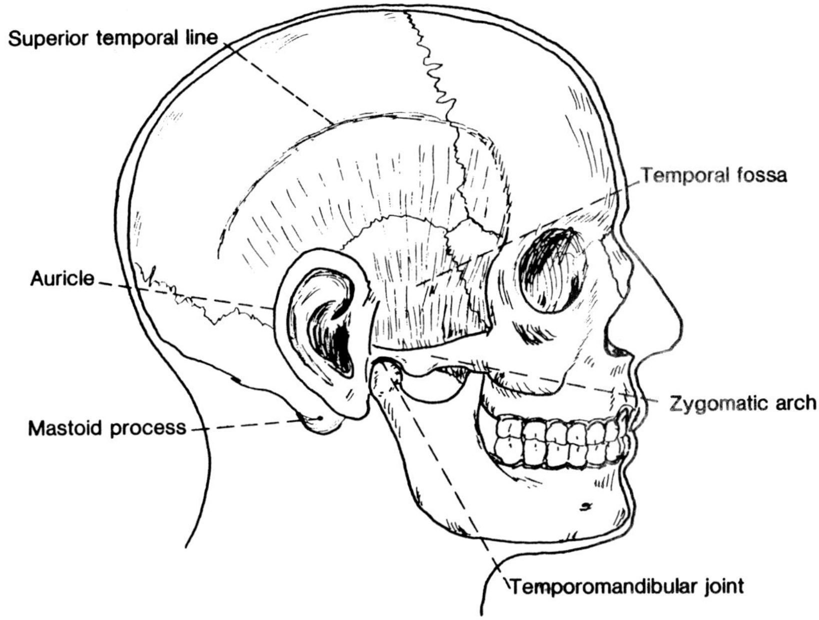

Figure 2-28. Lateral view of the skull; the external ear and temporal fossa.

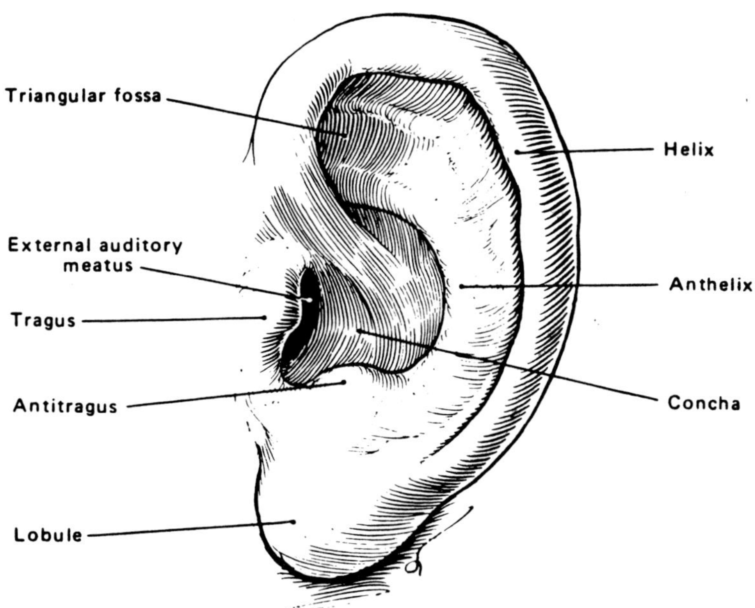

Figure 2-29. The auricle. (From *Dynamic Anatomy and Physiology*, 3rd ed., by L.L. Langley, I.R. Telford, and J.B. Christensen. Copyright© 1969. Used with the permission of McGraw-Hill Book Company.)

The external auditory meatus appears as a roughly cylindrical passage just over 0.5 cm in diameter. The passage ends about 2.54 cm deep in the temporal bone,

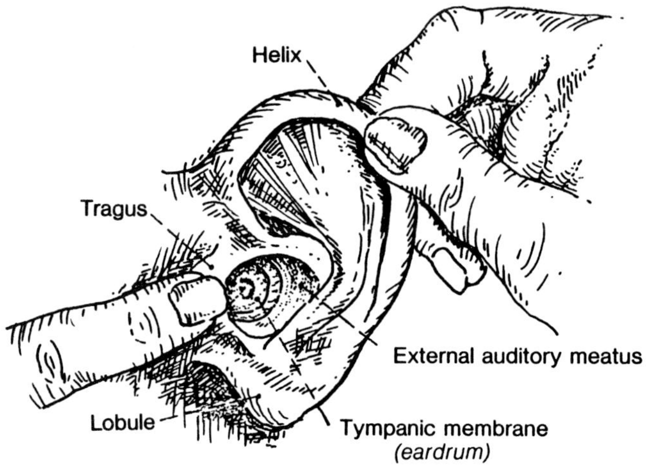

Figure 2-30. The left external ear. The index finger is shown retracting the tragus anteriorly while the helix is pulled upward and back. This results in straightening of the external auditory meatus and enables the examiner to view a portion of the tympanic membrane.

where the eardrum membrane or *tympanic membrane* caps the medial end of the external auditory meatus. In young persons, the external meatus is almost straight, permitting direct viewing of the tympanic membrane with ease. With age, the canal curves and requires more adjustments by the examiner to view the eardrum (Fig. 2-30).

For normal hearing, the external auditory meatus must be unstopped. Sometimes children will insert foreign objects and material into the meatus, resulting in reduced acuity of hearing. Wax secretions or *cerumen* from cells lining the external meatus may build up to obstruct the passage of sound through the canal in a way similar to plugging the ear with cotton. Occasionally, an insect may fly into the ear and cause a number of problems. But the insect will not find the ear a welcome haven, for ear wax is noxious to insects. External otitis (infection of the external ear canal) may also interfere with sound conduction into the middle ear as residue from the disorder accumulates.

Atresia of the external auditory canal indicates that the canal is absent, while stenosis denotes that the canal is occluded or closed down. An atretic ear canal is often found with a microtic auricle. In some cases, there are no functional middle-ear structures present. It is important to note that there is not a strong correlation be-

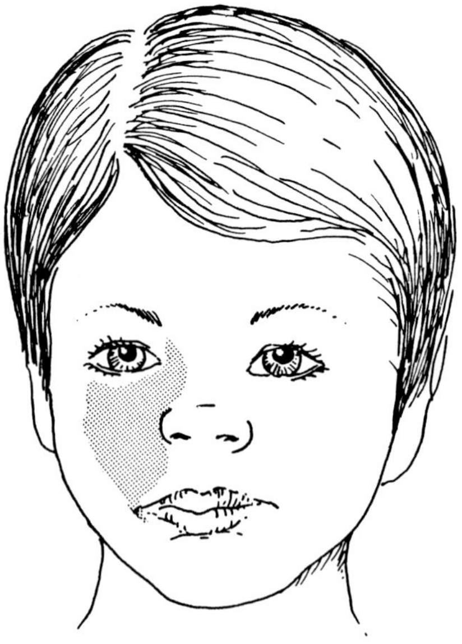

Figure 2-31. Port-wine hemangioma.

tween the severity of the external ear deformity and the status of the middle-ear contents.

Occasionally, the skin of the face will be marked with a large splotch of contrasting, dark color. Known as a port-wine hemangioma, this is a benign tumor made up of blood vessels. The tumor is nonpalpable (cannot be felt with the fingers). Enlarged red-blue capillaries in the skin of the face or upper trunk cause the port-wine mark, which is present at birth and remains throughout life. There are surgical techniques that reduce or eliminate this condition (Fig. 2-31).

BIBLIOGRAPHY

Elson, L.M.: *It's Your Body*. New York, McGraw, 1975.

Gray, H.: *The Anatomy of the Human Body* (Charles M. Goss, Ed.), 28th ed. Philadelphia, Lea and Febiger, 1973.

Jacob, S.W., and Francone, C.A.: *Elements of Anatomy and Physiology*. Philadelphia, Saunders, 1976.

Langley, L.L., Telford, I.R., and Christensen, J.B.: *Dynamic Anatomy and Physiology*, 4th ed. New York, McGraw, 1974.

Snell, R.S.: *Clinical Anatomy for Medical Students*, 2nd ed. Boston, Little, 1981.

Chapter 3

EMBRYOLOGY OF THE SPEECH AND HEARING MECHANISMS

INTRODUCTION

DEVELOPMENTAL **anatomy** was defined earlier as the study of an organism from its origin as a fertilized egg until birth. The prenatal period is synonymous with developmental anatomy but not with embryology. Broadly, **embryology** is concerned with the early developing stage of any organism, beginning with fertilization of the egg. In man, the embryo is usually considered to be the developing organism from conception until the *end* of the seventh week.

During this relatively short developmental period, the primordia (original or primitive) of all essential structures that concern speech and hearing clinicians, dentists, and otolaryngologists are laid down. The *fetal period* thus brings about only finer details in development. The embryo becomes a fetus at the beginning of the eighth week (Fig. 3-2).

Fetal anatomy, that period following the embryonic period, provides for further elaboration and growth of the organs and body systems. The fetal period is, in reality, a period of simple growth that, for some structures, may extend into the postnatal period (Fig. 3-3).

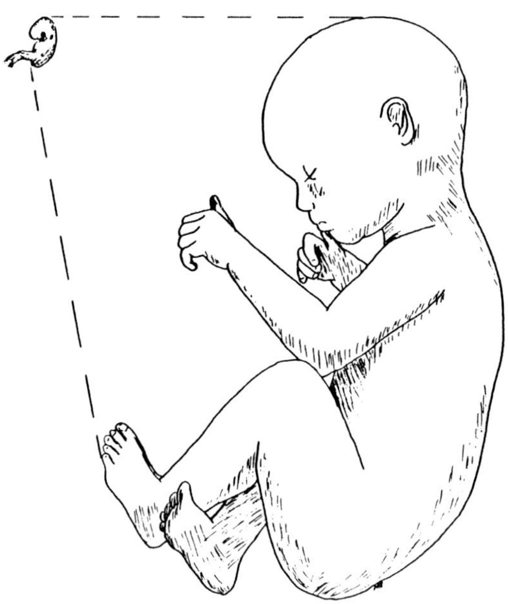

Figure 3-1. Fetal growth showing remarkable changes in size.

Teratology, the science of the study of monstrosities, is a branch of developmental anatomy concerned with abnormal development. How individual characteristics are passed from one generation to the

Figure 3-2. Fetal development indicating refinement of structures due to *stages* of growth.

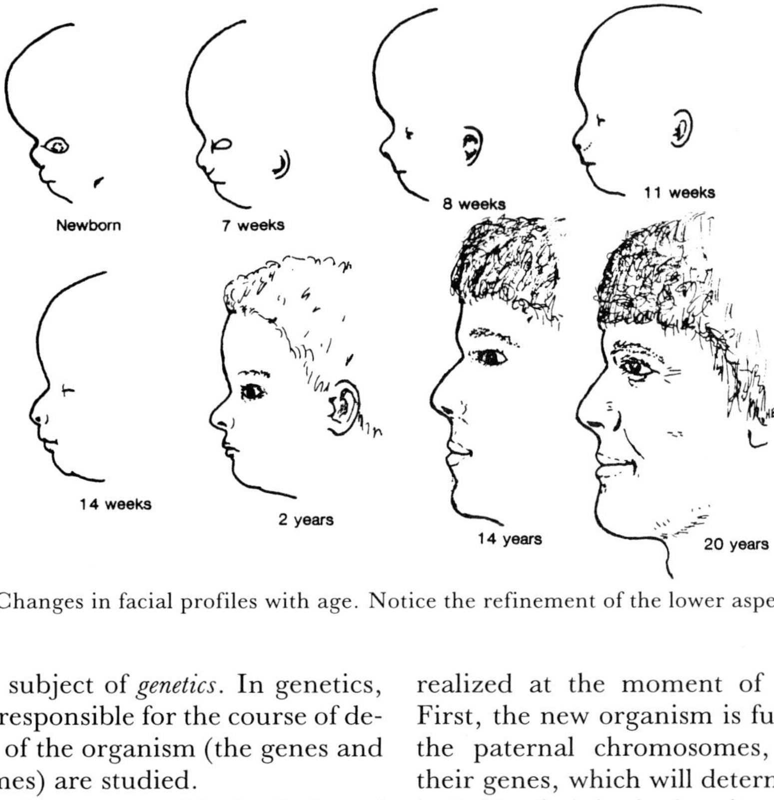

Figure 3-3. Changes in facial profiles with age. Notice the refinement of the lower aspect of the face.

next is the subject of *genetics*. In genetics, the agents responsible for the course of development of the organism (the genes and chromosomes) are studied.

Fertilization occurs with the fusion of the *ovum* and *sperm* to form the *zygote* (yoked, joined together). Two factors are realized at the moment of fertilization. First, the new organism is furnished with the paternal chromosomes, along with their genes, which will determine the new individual's inherited traits. Second, the ovum is activated to begin its development.

OVERVIEW: FIRST WEEK THROUGH SEVENTH WEEK

First Week of Development

From the time of fertilization to the end of the first week marks a period of cell division, multiplying greatly the size of the embryo. During this early period there is rapid growth from a single cell, the zygote, to a multicelled entity. *Implantation* of the zygote in the wall of the uterus takes place around day six (Fig. 3-4).

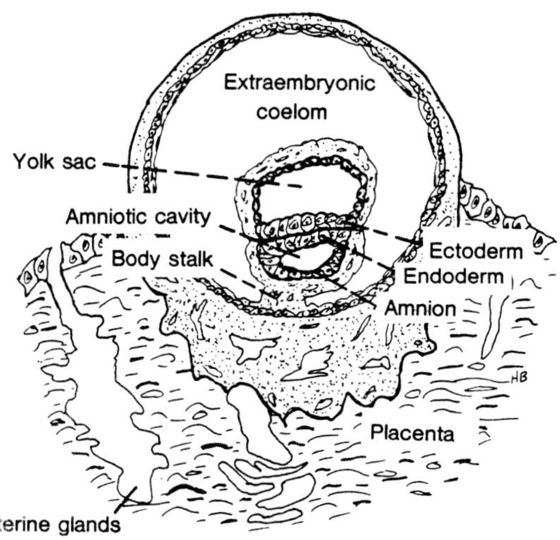

Figure 3-4. Schematic diagram of a cross section of an embryo in an early stage of development approximately twelve days after fertilization.

Second Week of Development

The second week marks further elaboration in implantation so that an early ecological rapport is established between the embryo and mother. The process of becoming even more deeply embedded in the uterine wall is apparent. By day twelve, maternal blood bathes a portion of the embryo to denote the first step in establishing placental circulation. Many changes are occurring simultaneously in the embryo itself. Cells are proliferating to form layers and leading to the formation of a two-layer embryonic disk.

Third Week of Development

Cessation of menstruation in the third week offers the mother the first symptom of pregnancy, although there are many factors other than pregnancy that may interrupt menstruation. Implantation is completed by day fifteen. The overall length of the developing embryo goes from about 0.2 mm to around 1.0-1.5 mm during week three. (A pencil is approximately 7 mm in diameter.) Additionally, the round, disk-shaped embryo enlarges into a pear shape because the cephalic end of the disk grows more rapidly.

Around day twenty, a developmental milestone arises in the cephalic region when a block of cells organizes together from adjacent masses of cells to form a pair of structures called *somites*. Eventually, there will be over forty pairs of somites marking the location of the *vertebral column* and from which will arise the bones and muscles of the back. Tissue overlaying the somites will give rise to the *nervous system* and *skin* in the region.

Rapid proliferation of embryonic tissue in the primitive head region of the embryo begins to produce a rounding prominence with a deep groove underneath, known as the *primary oral groove* or stomodeum (primitive mouth). This oral groove extends inward to halt in the blind cranial end of the foregut. A membrane at this blind end, the *buccopharyngeal* (mouth and pharynx) *membrane*, effectively separates the oral groove from the foregut (Fig. 3-5).

Structures in front (cephalic) of the buccopharyngeal membrane are destined to become the *mouth*, while the area on the other side of the membrane will form the *pharynx*. A series of arches, the *branchial* or *pharyngeal arches*, emerge below the primitive mouth and will contribute to the formation of most of the essential structures of the mouth and those structures in the vicinity of the mouth.

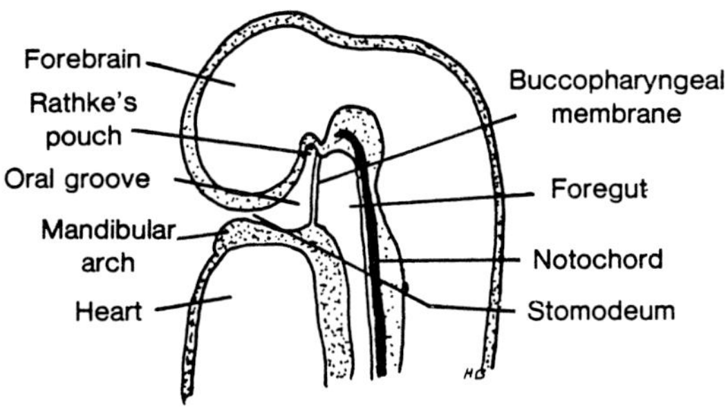

Figure 3-5. Diagrammatic representation of a median section through the head of an embryo in about the third week. The oral cavity and the foregut are separated by the buccopharyngeal membrane until the third week when the membrane ruptures. Rathke's pouch is the primordium of the anterior lobe of the pituitary gland. (From Arey, L.B.: *Developmental Anatomy*, 1965. Courtesy of W.B. Saunders Co., Philadelphia.)

The front of the future mouth is the developing heart with the site at which the liver will arise in front of the heart. Our embryo is in dire need of rearrangement to place the mouth between the brain and heart, rather than under the heart and above the brain. However, we need to wait only for week four to ascertain proper organization of these structures.

Fourth Week of Development

In three weeks, the embryo has increased in size by one million fold. A multicellular aggregate, the embryo now has discernible parts, which is an even more remarkable accomplishment in this short time frame. Still, as noted in week three, some major alterations in placement of structure are needed.

The disk-shaped embryo becomes a tubular, three-dimensional structure. Rapid growth of the embryo continues and is especially effective in elongating the embryo. That portion which will become the *brain* is more prolific and overrides the buccopharyngeal membrane to form the head-fold; similarly, a tail-fold is formed caudally. Through extension of the head-fold, while the heart and liver areas are held fixed, the anterior portion of the embryo is effectively rotated through an angle of about 180 degrees to reverse the structures in this region and to form a larger ventral surface beneath the head-fold (Fig. 3-6).

An umbilical cord is formed to serve better the nourishment needed by the enlarging embryo. During week four, additional somites are added until there are in excess of thirty pairs to enter week five.

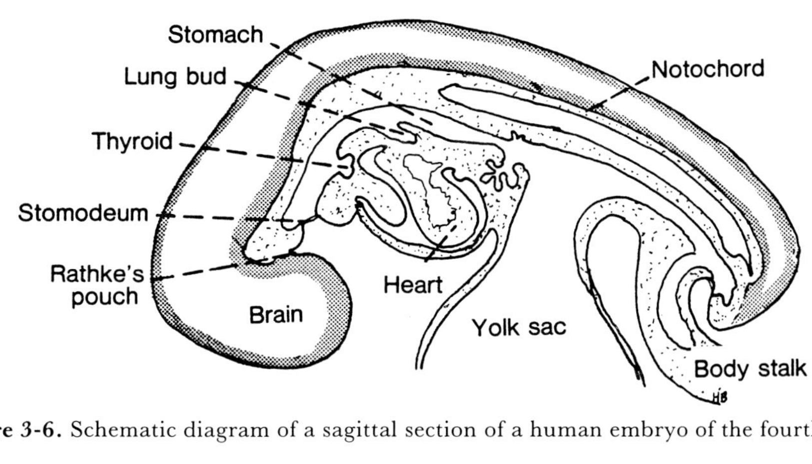

Figure 3-6. Schematic diagram of a sagittal section of a human embryo of the fourth week.

Late in week four, two small swellings appear on the lateral body walls at the level of the heart. These *arm-bud* swellings seem disproportionately low on the trunk of the embryo since the precociously developing head region dominates the body. *Leg buds* appear within one or two days later and always lag behind in development. In fact, the lower limbs take over two years to catch up with the upper limbs in terms of length.

Fifth Week of Development

The embryo has attained a length of around 6-7 mm by the beginning of the fifth week, which, as previously mentioned, is the diameter of a standard pencil. Already the heart has been functional for several days. Most organ systems are already present in some form, so week five will be largely a continuation of development of these structures. By the end of this week, the length of the embryo will be 11-14 mm or double its dimensions when compared with the first day of week five.

Only a few changes in body form take place in week five. These are increases in head size and in the structures that are on course to form the face. The limb buds also grow rapidly (Fig. 3-7).

Forty-two pairs of somites are evident by the end of this week, introducing a basic pattern of differentiation. Appearing in cranial to caudal numerical sequence and by body levels, the forty-two pairs are named: *occipital* (1-4), *cervical* (5-12), *thoracic* (13-24), *lumbar* (25-29), *sacral* (30-34), and *coccygeal* (35-42 +). Each somite is associated with the development of a single *spinal nerve*, at least for those somites in the regions giving rise to spinal nerves. As noted earlier, the somites are also responsible for the eventual development of the muscles and bones of the vertebral column.

Sixth Week of Development

Development during week six is rather modest in the shadow of the remarkable accomplishments during weeks four and five. Growth for the perfecting of the processes and organs laid down in earlier weeks is the major business in week six.

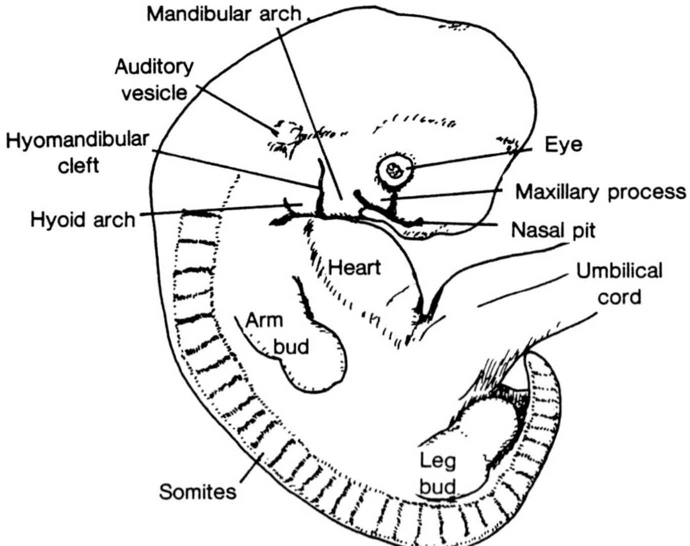

Figure 3-7. Lateral view of the embryo in its fifth week

The general configuration of the embryo in week six continues to be decidedly curved, although by week six and one-half the backbone region straightens. The limb buds advance considerably in this period, with the upper limb advancing ahead of the lower. Fingers appear by the end of the week, and indications for elbows and knees are evident (Fig. 3-8).

Figure 3-8. Embryo at six weeks.

Seventh Week of Development

By the end of the seventh week, differentiation and establishment of organs and organ systems are essentially complete, leaving the remaining prenatal period to development that focuses on growth and functional maturation. This is the final week of the embryological stage. The length of the embryo will be no more than 30 mm at the end of the seventh week — just over one inch.

The limbs elongate, with the fingers and toes following suit. Overall, the body contours become more rounded, especially the head.

EMBRYOLOGY OF SPEECH AND HEARING STRUCTURES

Now we will focus on the systems and structures more important to the nature of the text. Attempts to arrange development by *weeks* will be employed, but there is not complete agreement among writers for all of these dates. The reader is encouraged to compare the timing used here with other sources.

There are two phases that appear to occur in weeks five through eight of embryonic growth which are of special interest to speech clinicians. The first phase focuses on weeks five and six when the basic building blocks of the face arise, the communication of the *primitive oral cavity* with the *foregut* is established, and the early nasal ducts are formed. At the close of this phase the nasal cavities communicate extensively with the oral cavity, and the tongue is developing from the floor of the primitive oral cavity.

Weeks seven and eight are the second phase which brings about other critical changes which prestage the formation of complete facial and oral structures. In week seven the preparations for the division of the oral and nasal cavities are accomplished to culminate in week eight with the formation of the *secondary palate* (see Fig. 3-25).

Generally, one may presuppose that failures of development in the first phases, weeks five and six, are most likely responsible for clefts of the *lip* and *alveolar* (gum ridge) *process*. Growth failures in week eight (phase two) result in clefts of the *palates*. A pronounced growth failure during phase one could lead to disturbances of growth of the palates.

Nervous System

The first evidence of the differentiation of the nervous system appears at the end of the second week. In fact, the pace of growth of the structures belonging to this system seems to be an impatient pace. But much must be accomplished to develop a system as complex as the human nervous system (Fig. 3-9).

Fourth Week of Development

A tube, the *neural tube*, forms early in week four. Three primary brain vesicles (small fluid-filled bladder) are presented in this week. These are dilated segments at the cranial end of the neural tube. The *mesencephalon* or primitive *midbrain* is seen developing anteriorly about midway through the week. By the end of week four a development caudal to the primitive midbrain area denotes the junction of the *brain stem* and the *spinal cord*. Between the midbrain and brain stem/spinal cord junction is a region which gives rise to the *rhombencephalon* or primitive *hindbrain* (Fig. 3-10).

The apex of the neural tube, that portion anterior to the primitive midbrain region, becomes the *prosencephalon*, or primitive *forebrain*. Two outpocketings from the lateral walls of the primitive forebrain

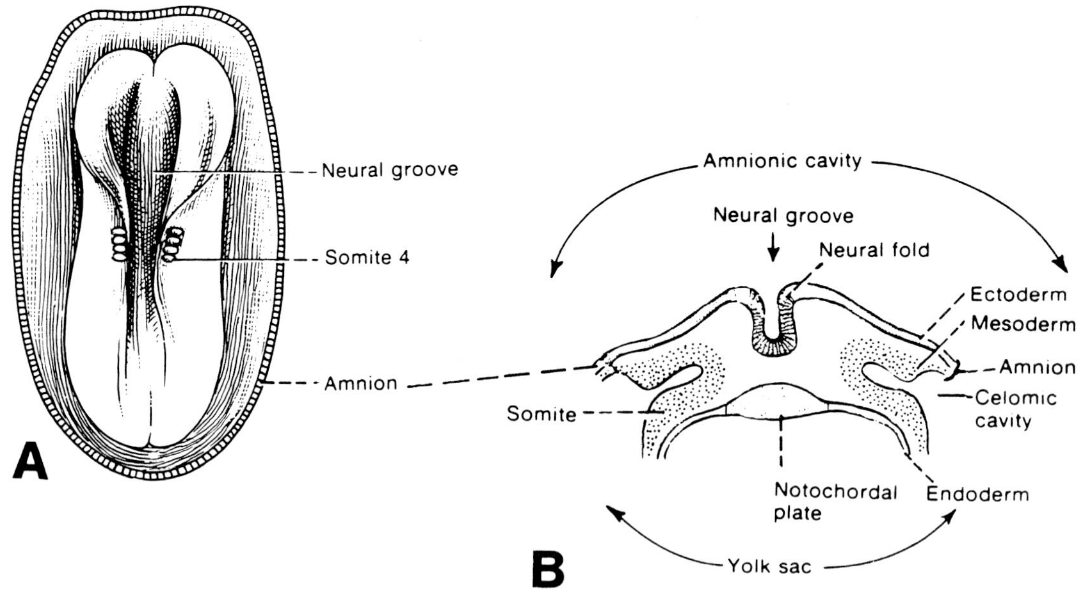

Figure 3-9. Early development of the nervous system. (A) The amnion is cut to expose the dorsal aspect of the embryo and the neural groove at about three weeks. (B) Schematic transverse section of a three-week-old embryo showing the neural groove. (From Bossy, Jean: *Atlas of Neuroanatomy and Special Sense Organs*, 1970. Courtesy of W.B. Saunders Co., Philadelphia.)

area arise to form the neural elements of the *eyes*.

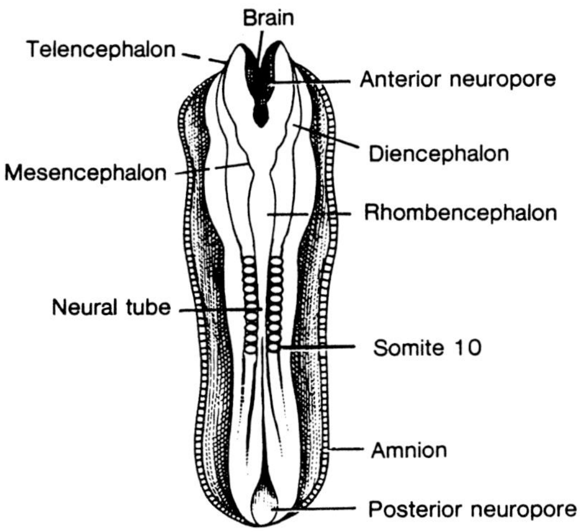

Figure 3-10. Dorsal aspect of human embryo in fourth week. The neural tube is beginning to form. (From Bossy, Jean: *Atlas of Neuroanatomy and Special Sense Organs*, 1970. Courtesy of W.B. Saunders Co., Philadelphia.)

The *spinal cord* develops from the caudal section of the neural tube. A longitudinal furrow in the caudal section of the neural tube divides the tube into dorsal and ventral regions which are destined to supply the *sensory* and *motor nerves* of the spinal cord.

Fifth Week of Development

The primary brain vesicles, which appeared in week four, elaborate in week five. As named in week four, they are, beginning at the cranial extreme, prosencephalon (primitive forebrain), mesencephalon (primitive midbrain), and rhombencephalon (primitive hindbrain) (Fig. 3-11).

All of the brain vesicles undergo rapid growth and development in week five, causing the brain region to be the most striking feature of the embryo. Together with the equally rapid developing face, the embryo acquires a disproportionate appearance – almost all head (Fig. 3-12).

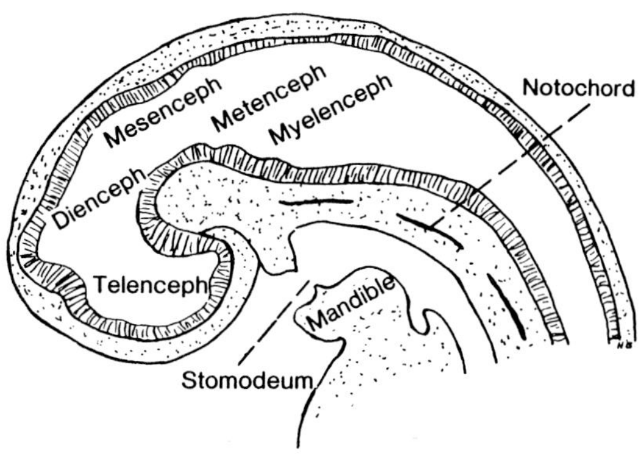

Figure 3-11. Diagrammatic sagittal section of a four-week-old embryo showing the neural and stomodeal portions of the head area.

The prosencephalon develops bilateral outpocketings from its lateral walls in midweek to lay down the beginnings of the *telencephalon* (primitive *cerebral hemispheres*) of the brain. But a portion of the primitive forebrain not used in forming the cerebral hemispheres contributes to the formation of the *diencephalon* (primitive *thalamic* structures). The rhombencephalon also subdivides to form two vesicles so that the combined number of brain vesicles comes to five in week five (Table 3-I).

Unlike the spinal nerves, the *cranial nerves* are not orderly and segmented; additionally, they are varied in their makeup of sensory and motor fibers. All spinal nerves exhibit a *sensory dorsal root* and a *motor ventral root* to supply rather well-defined regions of the body with afferent (sensory) and efferent (motor) nerve fibers.

Twelve pairs of nerves leave the brain straight forth; none utilize the spinal cord as a pathway. These are known as the *cranial nerves* and are essential to the higher

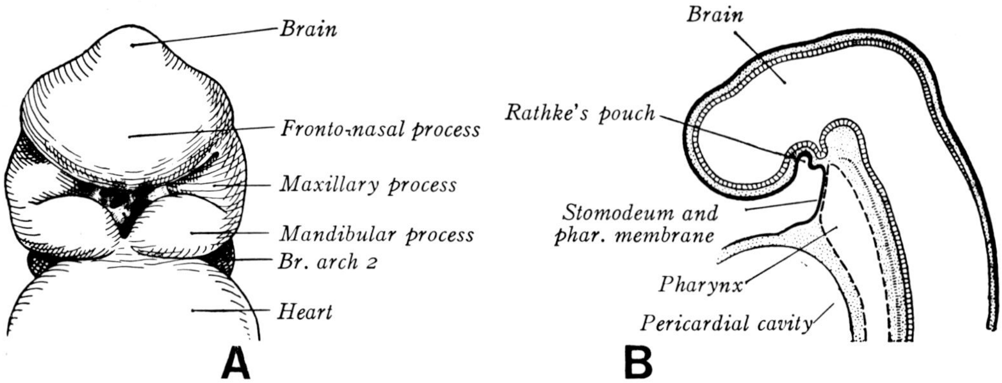

Figure 3-12. (A) Diagrammatic anterior view of the stomodeum (mouth) and its floor membrane. The boundaries of the stomodeum and a partly perforated buccopharyngeal membrane. (B) Sagittal schematic of A. (From Arey, L.B.: *Developmental Anatomy*, 1965. Courtesy of W.B. Saunders Co., Philadelphia.)

Table 3-I

DIFFERENTIATION OF BRAIN

Developing Brain	*Mature Brain*	
Prosencephalon	Gross Divisions	Subdivisions
	Telencephalon	Cerebrum, Basal ganglia
	Diencephalon	Thalamus, Hypothalamus, Epithalamus
Mesencephalon	Mesencephalon	Midbrain
Rhombencephalon	Metencephalon Myelencephalon	Cerebellum, Pons, Medulla

senses and control of speech, among other things. In a true sense, cranial nerves I and II (olfactory and optic) are processes of the brain, but they are traditionally included as cranial nerves. It should also be noted that cranial nerves V and XI (trigeminal and spinal accessory) have some spinal connections and interactions that relate these cranial nerves to the spinal cord.

Each of the cranial nerves is derived principally from structures of the primitive brain and the *branchial arches*. Roman numerals are assigned to each and may be used interchangeably with the names of the cranial nerves. (Fig. 3-13).

The three highly specialized cranial nerves of the special senses:

Olfactory (I)	Smell
Optic (II)	Vision
Vestibulo-cochlear (VIII)	Equilibrium and hearing

The somatic (body tissue) motor nerves:

Oculomotor (III)	5 muscles of the eyeball
Trochlear (IV)	1 muscle of the eyeball
Abducent (VI)	1 muscle of the eyeball
Hypoglossal (XII)	Muscles of the tongue

Spinal accessory (XI)	Muscles of the neck

The cranial nerves carrying both sensory and motor fibers:

Trigeminal (V)	Muscles of mastication sensory to face, mouth, head
Facial (VII)	Muscles of face and ear sensory to anterior two-thirds of tongue, glands
Glosso-pharyngeal (IX)	Muscles of pharynx, tongue sensory to posterior third of tongue, gland
Vagus (X)	Muscles of larynx, pharynx, viscera sensory to larynx and wide range of viscera

in the embryo. The maturity of the nervous system permits a generalized total body response to denote a primitive degree of integration of the nervous system components.

The *autonomic nervous system* (ANS), a visceral (important organs of the body) motor system, supplies the structures of the body *not* under voluntary control. These structures include smooth muscle, cardiac muscle, and glands, with most of these structures innervated (supplied with nerves) by fibers of the system having a dual nerve supply with opposing functions.

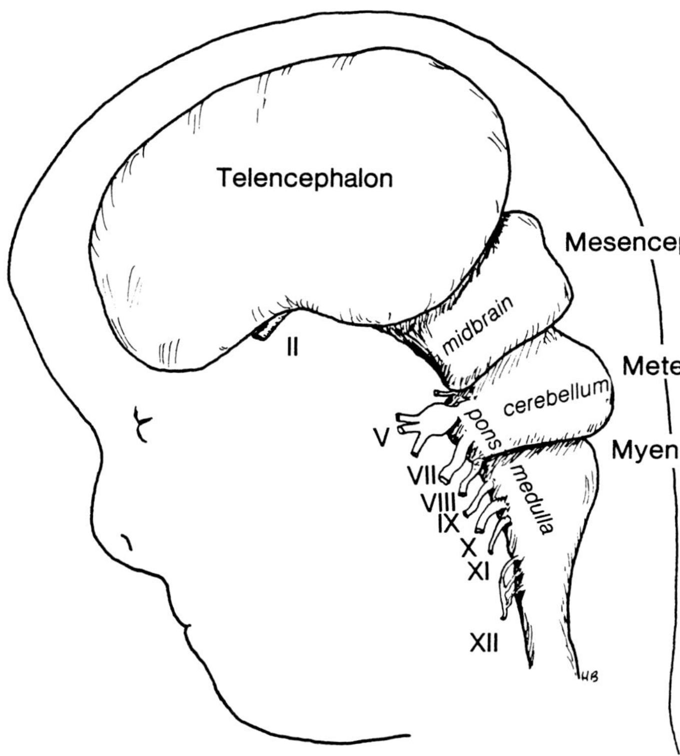

Figure 3-13. Divisions of the brain with arrangement of the cranial nerves illustrated.

Sixth Week of Development

A milestone occurs in week six in that *simple reflex responses* may be demonstrated

To provide this dual nerve supply, the ANS is divided into two major subdivisions, differing both in anatomy and function. These two subdivisions are the *sympathetic* and *parasympathetic* divisions. The sympathetic division supplies *stimulatory* fibers to organs and glands; these fibers arise from *thoracic* and *lumbar* segments of the spinal cord — the thoracolumbar division. Fibers arising from the *brain* and *sacral* part of the spinal cord — the craniosacral division of the ANS — supply organs and glands with *inhibiting* fibers and constitute the parasympathetic division. This basic plan of the ANS is established in week six.

Divisions of the brain continue to develop to establish this most complex organ of the body. The highest centers of the brain, the *cerebral hemispheres*, continue to expand. Among other developments are the laying down of the major tracts of the *brains stem*, which become distinguishable by the end of week six.

Seventh Week of Development

The *cerebral hemispheres* and other divisions of the brain continue to mature. At the end of week seven the spinal cord has grown caudally to the level of the fourth coccygeal segment. But the vertebrae will outgrow the spinal cord subsequently so that the cord ends in the upper lumbar region at maturity.

Pharynx, Mouth, Face

Condensations of tissue in the cephalic end of the foregut of the embryo — the region immediately behind the buccopharyngeal membrane — result in a series of six pairs of bars known as the *branchial* (resembling gills of fish) *arches* or *pharyngeal arches*. The first pair arises at the extreme of the cephalic portion of the primitive pharynx early in week four. The others appear sequentially until the sixth pair is present at the end of week four. Five of the six arches give rise to cartilaginous rods and branchial muscles. Arch number five is vestigial in man (Fig. 3-14).

Each pair of branchial arches has a *specific nerve* and *tissue* that develops into *specific muscles* and *cartilages*. Herein lies an important opportunity to simplify understanding of the highly complex adult anatomy. The muscles that develop from each arch associate directly with the bones arising from that arch and are innervated by the cranial nerve developed in the arch. This scheme of development has a logical, orderly, and readily recognizable rationale for the adult derivatives, which also has very important clinical applications (Fig. 3-15).

Fourth Week of Development

PHARYNX. Early in week four, shallow

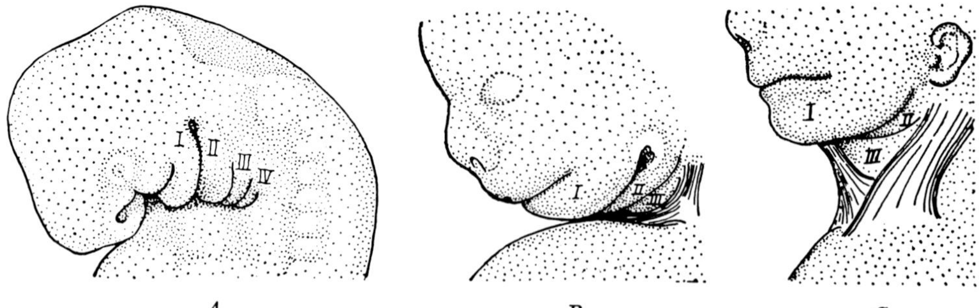

Figure 3-14. The branchial arches at three stages: (A) at 5 weeks; (B) at 7 weeks; (C) at 12 weeks. (From Arey, L.B.: *Developmental Anatomy*, 1965. Courtesy of W.B. Saunders Co., Philadelphia.)

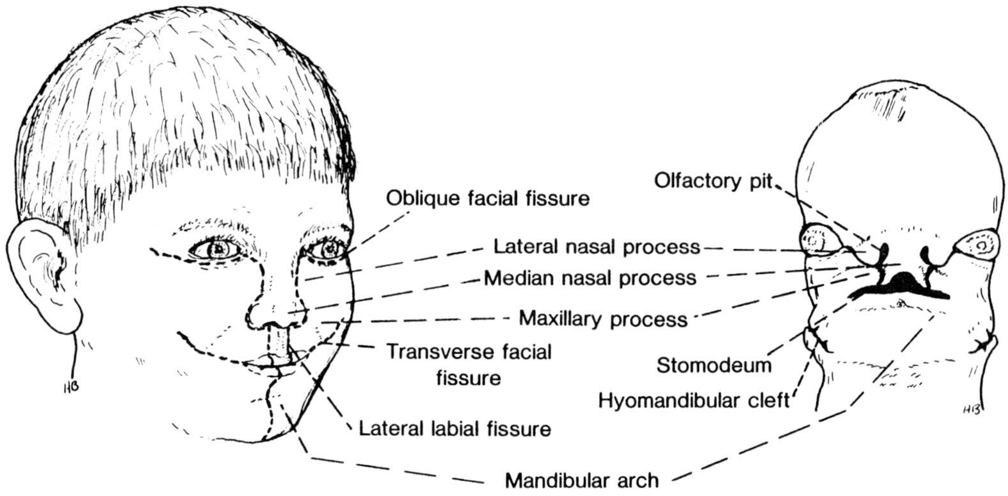

Figure 3-15. Development of the face from primitive structures of the head. The broken lines (left figure) indicate the correlate clefts on the face. The oblique facial fissure is the site of development of the nasolacrimal duct. Clefts of the upper lip are often found at the lateral labial fissures.

grooves appear bilaterally in the walls of the pharynx and separate with an interspace the successive *branchial arches*. There are a total of four pairs of *branchial grooves* or *clefts* that deepen to form the *branchial pouches* (Fig. 3-16).

The lateral portion of the branchial pouches develop a dorsal and ventral extension. Also, as the pouches expand, the connection between the pouch and the pharyngeal cavity becomes drawn out and narrow so that a duct-like structure is formed. The extensions and the duct, as well as each pouch, have different destinations.

The first pouch, which lies between the first or *mandibular arch* and second or *hyoid arch*, will eventually form the cavity of the middle and the *auditory (eustachian) tube*. The second pouch, which lies between the second and third arches, is destined to a limited role, contributing the *fossa* (trench or channel) for the *tonsils* in the oral cavity. The third pouch, which lies between the third and fourth arches, eventually separates itself from the phar-

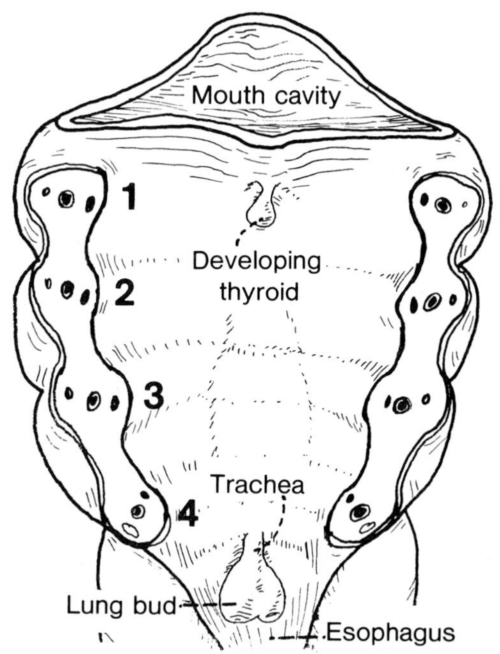

Figure 3-16. Diagrammatic frontal section of the mouth area showing the development of the pharynx and related structures. Notice the numbering of the branchial arches. Between the arches are pouches.

Table 3-II

DERIVATIVES OF THE PHARYNGEAL ARCHES

Arch	*Skeletal Structures*	*Muscles*	*Nerve*
I	Middle ear bones: *Malleus, incus*	Muscles of mastication; anterior belly of digastric, mylohyoid; Middle ear: tensor tympani; Soft palate: tensor veli palatini	Mandibular division of the trigeminal nerve (*Vth cranial nerve*)
II	Middle ear bone: *stapes; styloid process of temporal bone*; Hyoid bone: *lesser horn*, and *upper part of hyoid body*	Muscles of facial expression; posterior belly of digastric, stylohyoid; Middle ear: stapedius	Facial (*VIIth*) *cranial nerve*)
III	Hyoid bone: *greater horn* and *lower part of hyoid body*	Pharynx: stylopharyngeus, superior constrictor	Glossopharyngeal (*IXth cranial nerve*)
IV and V	Cartilages of the larynx: *thyroid, cricoid, arytenoid, corniculate*, and *cuneiform cartilages*	Pharyngeal and laryngeal muscles	Vagus (*Xth cranial nerve*)

ynx and forms two glands: parathyroids in the neck and thymus in the thorax. The fourth pouch, which lies posterior to the fourth arch, assists in forming the parathyroid gland (concerned chiefly with metabolism of calcium and phosphorus) (Table 3-II).

MOUTH. The first branchial arch is known as the *mandibular arch* because it eventually leads to the definitive *mandible*. A cartilaginous bar of this first arch is popularly termed *Meckel's cartilage*. Extensions from both sides of the *mandibular process* will later give rise to another important structure of the face, the *maxillary process*. The buccopharyngeal membrane is at the bottom of a depression, with the mandibular and maxillary processes at the borders of the membrane and an overhanging *head process* or *prominence* above. This depression is the *stomodeum*, or primitive mouth. Sometime during week four the buccopharyngeal membrane breaks down, establishing communication between the primitive mouth and pharynx (see Fig. 3-12).

The buccopharyngeal membrane in the primitive mouth ruptures because the structures around the stomodeum grow and enlarge at a rapid rate in contrast to this membrane which does not grow. While the membrane is near the entrance in the four-week embryo, it will be equivalent to the level of the tonsils in the adult. Therefore, a remarkable amount of growth must occur in front of the stomodeum.

TONGUE. The ventral ends of the first branchial arches meet in the midline ventrally. *Lingual* (tongue) *swellings* appear at the tips of the first arches relating to development of the tongue. Just behind the junction of the two lingual swellings, a midline swelling appears, known as the *tuberculum impar* (unpaired nodule or tubercle), which is also essential to the formation of the tongue. Posterior to the region of the second arches (hyoid arches), another swelling that is destined to assist in forming the tongue arises. It is known as the *hypobranchial eminence* (hypo = under) (Fig. 3-17).

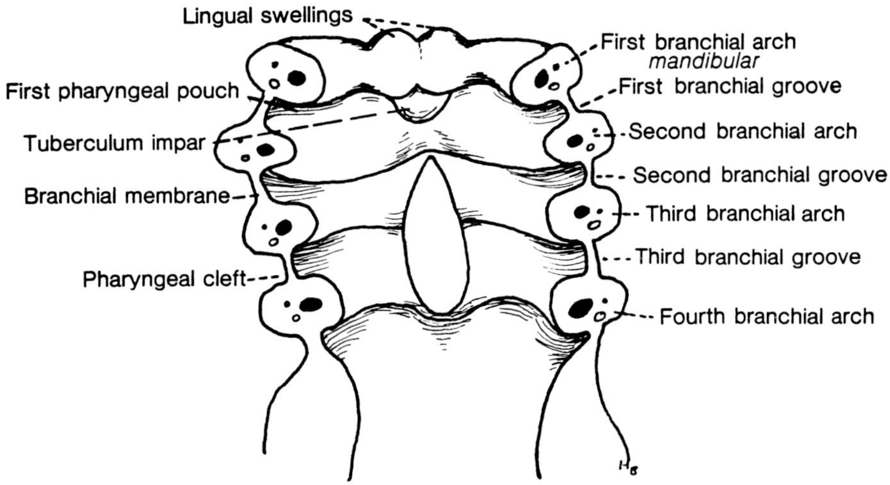

Figure 3-17. Diagrammatic frontal section of the pharyngeal area showing the primitive pharynx with the configuration of the branchial arches, branchial grooves (clefts), and pharyngeal pouches. Notice that the pouches are *internal* to the grooves.

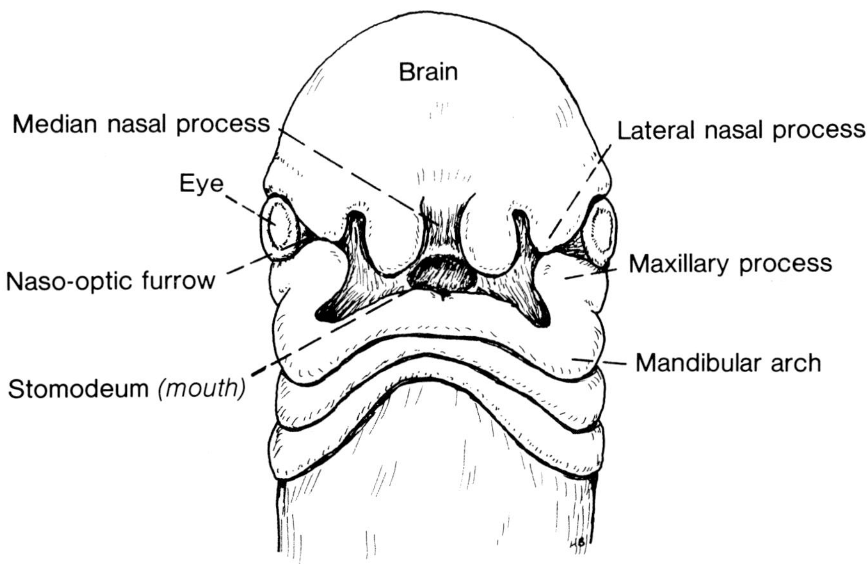

Figure 3-18. Facial region at about five weeks.

Fifth Week of Development

PHARYNX. Although there are numerous changes in the pharynx during week five, all of the changes may be viewed as progressive steps of development that began in previous weeks. Cranial nerve fibers invade the branchial arches, with the *trigeminal* (V) supplying the *mandibular arch*, the *facial* (VII) supplying the *second*, the *glossopharyngeal* (IX) supplying the *third*, and the vagus (X) supplying the remaining arches.

Further development in week five of the mouth and pharyngeal area focuses on the bars of tissue of the branchial arches. They give rise to muscles and cartilage in the facial and pharyngeal region. The muscles derived from the *mandibular arch* pertain to *mastication*, the *hyoid bone*, and the *middle ear*. The trigeminal (V) nerve innervates (supplies nerves) to the mandibular arch (Fig. 3-18).

The mandibular arch contributes to the formation of the *mandible* as might be expected, owing to its name. However, the *maxillary process* has a composite origin, with the mandibular arch contributing only a part. Located on the cephalolateral angles of the oral cavity, the maxillary processes grow medially; they are destined to become the lateral parts of the upper jaw.

The ventral end of the popularly named cartilage of the mandibular arch, Meckel's cartilage, turns into the mandible. Its dorsal portion gives rise to two bones of the middle ear (see Fig. 3-29).

The cartilage of the hyoid arch is *Reichert's cartilage*. Muscles of *facial expression*, two muscles of the hyoid bone, and the stapedius muscle of the middle ear arise from the second or *hyoid arch*. All of these structures are innervated by the facial (VII) nerve. The hyoid arch and the third arch form the hyoid bone. The hyoid arch contributes the third ossicle, the *stapes*, and gives rise to the *styloid* (long and pointed) *process* located beneath the skull, straddling the pharyngeal area. The styloid process is an important point of *origin* (fixed end of a muscle) for muscles associated with the hyoid bone, tongue, and pharynx — muscles vital to articulation and resonance.

MOUTH. In week four, lingual swellings appeared on the *ventral lips* of the first branchial arches. A slight elevation, the *tuberculum impar*, imposing itself on the floor of the pharynx between the paired lingual swellings, joins the arches. These three elevations fuse in week five and proceed forward into the mouth forming the body of the tongue. With the union of the two second arches, the root of the tongue is formed. Early on, this portion of the tongue is represented by a slight elevation, the *copula*.

Between the copula and the tuberculum impar lies a V-shaped groove. A foramen (opening or passage), the *foramen cecum*, is at the apex of the groove marking the site of the thyroid invagination (infolding of one part within another). Embryologically, the foramen cecum is a vestige of the thyroid gland that develops from the floor of the pharynx and is situated below the larynx at maturity (Fig. 3-19).

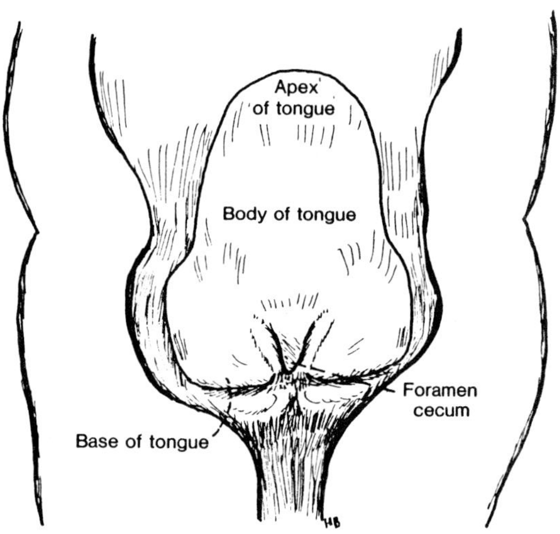

Figure 3-19. Diagrammatic representation of the developing tongue.

The nerve supply to the tongue reflects the genesis of the tongue. The *mucosa* of the body of the tongue contains sensory fibers from the trigeminal (tactile) and facial (taste) nerves that are derived, respectively, from the mandibular and hyoid arches. The *root* of the tongue is innervated by the nerve of the third branchial arch, the glossopharyngeal (IX). The vagus (X) nerve innervates the mucosa and taste buds in the vicinity of the epiglottis, implying that a small region of the fourth arch contributes to the tongue.

The main mass of the *extrinsic musculature* (muscles coming from parts outside the organ) of the tongue is not derived from the branchial arches. The hypoglossal (XII) nerve supplies motor fibers to these muscles, and neither it nor the muscles it serves arise from arches. Rather, they appear to develop from the transient occipital somites, but the origins are slurred (see week six).

Contributions are also made to the development of the tongue from other growth centers introduced in week four, namely, the hypobranchial eminence from the third and fourth arches.

FACE. The *nasal* or *olfactory placodes* appear on the ventrolateral surface of the head lateral to the frontal prominence, seem to sink beneath the surface, and proceed to form the nasal pits. They are bordered by vigorously growing horseshoe-shaped elevations, which deepen the depression in which each of the nasal placodes lies. The rim of each pit becomes elevated to form the *medial* and *lateral nasal folds* (see Fig. 3-18).

The medial limbs of these nasomedial folds or processes advance toward the midline to form the primordium that will differentiate into the middle part of the nose, the philtrum of the lips, the midpart (incisor) of the maxilla, and the small primary palate (see week six).

The frontal prominence dividing the nasomedial folds is actually little more than a segment of the head (brain) area. It is the upper boundary of the stomodeum or mouth.

Sixth Week of Development

PHARYNX. Growth is the primary consideration for the pharynx in week six. By midweek, the basic makeup of the pharynx is established. Already the glands derived from this area (the thyroid, parathyroid, and thymus) have separated. The primordium of the *pharyngeal musculature* becomes differentiated.

TONGUE. The tongue is mostly a muscular organ, so the development of the tongue in size and shape is dependent on the muscles of the tongue. It was speculated above that the muscles of the tongue are derived from the occipital somites (see weeks three and five) since the innervation of these muscles comes from the hypoglossal (XII) nerve. The twelfth nerve is associated with the occipital somites at development, so the assumption that the *intrinsic muscles* (situated entirely within the organ) of the tongue arise from the occipital somites appears valid.

The lingual swellings of the mandibular arch continue to enlarge, and the halves of the tongue are apparent. The tongue begins to be a recognizable, though primitive, organ.

PALATE FORMATION. Separation of the oral and nasal cavities is of special interest to speech clinicians. Week six provides the beginning staging of the formation of the definitive palate.

The palate develops from at least *three* directions of growth, converging to form a partition between the oral and nasal cavities. The anterior part of the palate will be considered first.

The *primary palate* is formed from the deeper portions of the *nasomedial processes* that make up the premaxilla. The configuration of the primary palate is triangular and is the lesser part of the palatal structures. Two blocks of tissue of the nasomedial processes give the characteristic wedge-shape to this midline structure. The primary palate is associated with the base of the developing *nasal septum*, the important junction for the parts of the palate (Fig. 3-20).

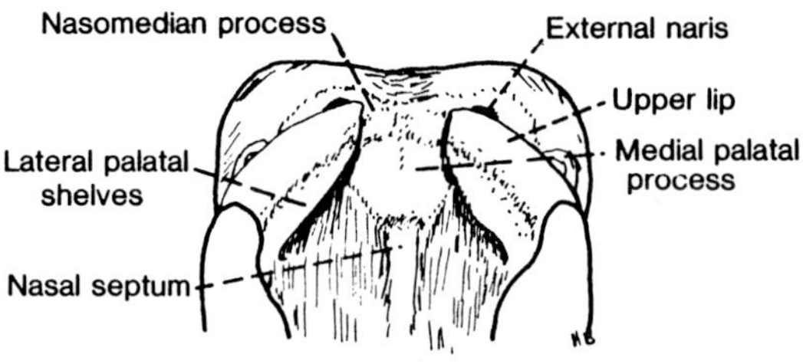

Figure 3-20. Diagrammatic representation of the early dévelopment of the palate, inferior view.

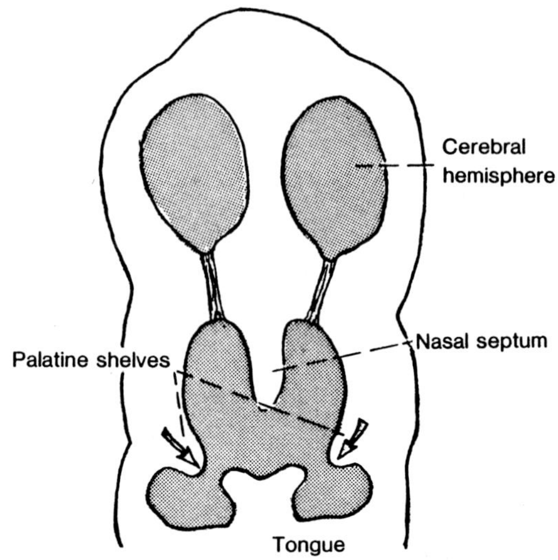

Figure 3-21. Schematic drawing showing palatal development. The arrows indicate the direction of growth of the palatal shelves.

Much uncertainty about the successive developing stages of the primary palate from its first appearance as a wedge-shaped mass in week six exists. It can be asserted that this block of tissue in the medial aspect of the mouth is the *primordium* (original or primitive) of the primary palate and that it stands as the roof of the anterior portion of the mouth. The upper lip and gum ridges are derived from this wedge-shaped mass and grow together in week six. Failure to unite results in a unilateral or bilateral cleft of the lip and can extend through the anterior aspect of the alveolar (gum ridge of the teeth) process.

The primordium of the *secondary palate*, which is the principal part of the complete palate, arises from the medial surface of the *maxillary processes*. These processes appear as shelf-like outgrowths of the maxillary processes on either side and will eventually form the horizontal hard and soft portions of the palate (Fig. 3-21).

At this time, the oral cavity is relatively small, cramping the space needed by the tongue and the developing *palatal shelves*. The tongue even extends into the nasal cavity and touches the *nasal septum*. Hence, the tongue rides high in the oral cavity interposed between the right and left shelves. To grow, these palatal processes must enlarge *downward* alongside the tongue. But the oral cavity is growing vigorously and increases greatly in vertical dimensions due to an enlarging mandible and other parts. The *nasal chambers* are still continuous with the oral cavity. There is sort of a race going on: the tongue is descending and the palatal shelves are growing downward. The tongue wins, freeing the shelves to turn horizontally and advance toward one another. Additionally, the walls of the oral and nasal chambers are expanding laterally and inferiorly, perhaps contributing to the upward swing of the palatal processes.

In week six and one-half, the nasal septum is developing downward from the *frontal prominence* of the head to mark the two nasal chambers. Only the small, anteriorly placed medial palatine process, the *primary palate*, lends a boundary between the nasal and oral cavities before formation of the *secondary palate*.

FACE. The maxillary swellings are enlarging and moving ventrally to merge with the nasomedial folds or processes; a

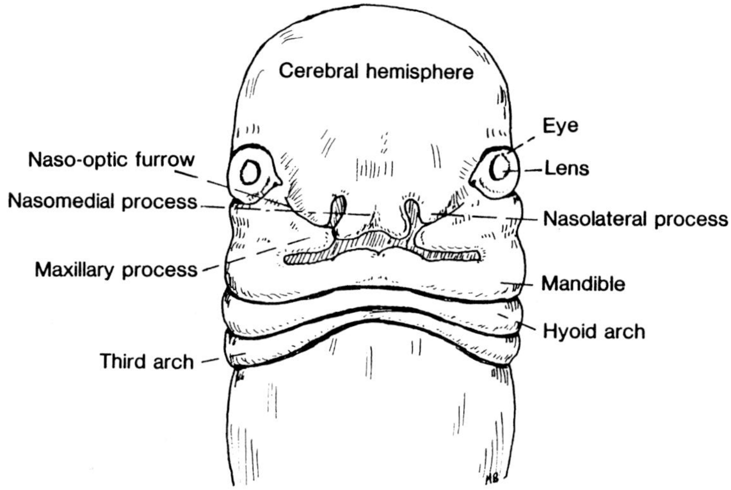

Figure 3-22. Development of the face at about the sixth week. The union of the maxillary process with the lateral and medial nasal processes from the cheek and upper lip.

closed, U-shaped arch is then formed. Parts contributed by the *nasomedial folds* of the intermaxillary segment (or *premaxilla*) are as follows: incisors, philtrum, tubercle of the lip, and primary palate. The canine, bicuspid, molar, as well as lateral lip parts, are derived from the *maxillary process* (Fig. 3-22).

Growth of the mandibular and maxillary processes permits them to fuse laterally in week six. They thus form the *angles* of the mouth, with the cheek area lateral to the point of fusion. The two *lips* also arise in this week. That thick, prominent central part of the upper lip that lies over the premaxilla is known as the *prolabium*. It is most prominent in infants. Formation of the lower lip arrives from the free edge of the mandibular processes, but the upper lip is more involved. The maxillary processes advance toward each other in the midline. First they fuse with the base of the lateral nasal processes and second with the premaxillary or nasomedial growth center.

From these fusion lines between the maxillary and nasal processes, potential developmental clefts are evident. Failure of either maxillary process to fuse properly with the medial nasal process will result in a cleft of the lip, the jaw underneath, or both (Fig. 3-23).

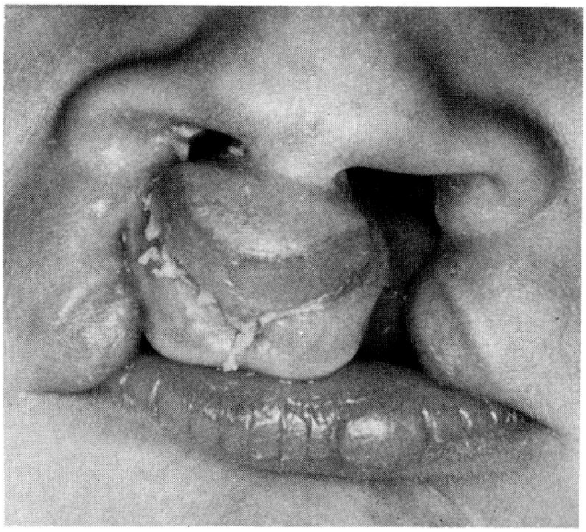

Figure 3-23. Bilateral cleft of the lip and jaw. The prolabium is prominent as the cleft extends through the premaxilla into the nasal cavities. The maxillary processes failed to fuse with the median nasal processes.

As noted earlier, the primary palate fills in a small, medial, triangular-shaped segment of the anterior palatal complex, the intermaxillary segment.

The medial and lateral nasal folds grow at a fast pace. Together with the frontal prominence, the two nasal folds form an elevation. By the end of week six, the nose has a tip and the lateral nasal folds give rise to the wings (alae) of the nose.

A number of noteworthy changes in the mouth region are apparent in week six. The lips become defined, primary bone-formation (ossification) centers are set up in the mandible and other facial structures, and the primary dental ledge or lamina (a thickened band of tissue along the upper and lower margins of the primitive gum ridges from which enamel organs will be developed) emerges. No time is wasted in utilizing the dental lamina, for the deciduous upper and lower canine teeth and first and second incisors begin to develop within the dental lamina by the end of week six. One other significant development is the appearance of the salivary glands over the course of week six.

Seventh Week of Development

PALATE. Although the destiny of the shelf-like processes was discussed in week six, they continue in week seven to extend inferiorly from the medial walls of each maxillary process. Separated by the body of the tongue, the palatal processes await a further lowering of the tongue. The mandibular arch grows, enabling the tongue to recede from the nasal septum and making it possible for the palatal processes to approach one another at the midline. At this point, the palatal shelves do not reach far enough to join each other even after the tongue no longer interferes with the midline fusion process. However, they grow rapidly so that by the end of week eight they have made contact. The nasal septum is also on a programmed growth track, proceeding inferiorly to meet the palatal processes at the midline (Fig. 3-24).

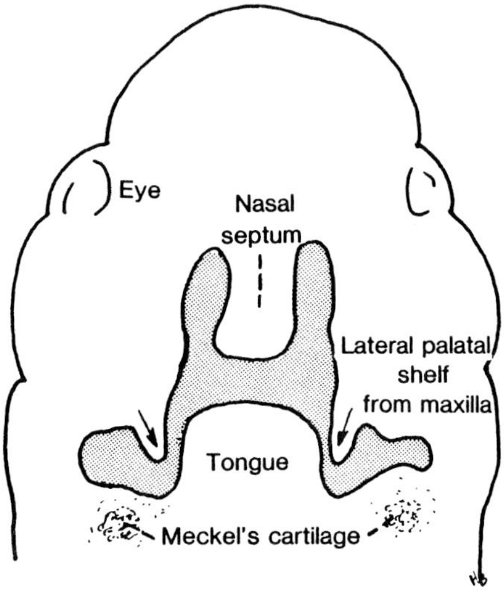

Figure 3-24. Schematic representation of the nasal septum, palate, and tongue, seventh week. The appearance of Meckel's cartilage (the basis for the mandible) is shown.

Subsequently, the palatal shelves meet and fuse with the small, triangular *premaxillary process*, or *primary palate*, anteriorly instead of with each other. These lateral palatine shelves fuse with each other posteriorly to the *primary palate*, proceeding toward the primordia of the *uvula*. They also fuse with the *nasal septum* above shortly after the tenth week. The completion of all fusions is not reached until around the twelfth week.

Together, the palatine shelves comprise the major part of hard palate. Since the process of fusing proceeds from the primary palatal area toward the uvula, the soft palate is the last to unite. It remains temporarily notched and represents an early developmental stage of the uvula. Occasionally, this bilateral origin of the uvula is clearly recognized in adults who exhibit a *bifid* (cleft into two parts) uvula. This can occur as an isolated, innocuous finding or can be related genetically to cleft lip or palate as has been shown in family pedigree studies (Fig. 3-25).

MOUTH. Meckel's cartilage, the principal support of the *mandibular arch*, appears in the latter part of week seven. Ossification begins to form the mandibular bone as the embryo passes into week eight. The oral cavity has deepened remarkably since week four.

By the end of week seven, a thickening of oral epithelium (covering of the internal and external body surfaces) can be identified more clearly on both the upper and lower jaw. This band of epithelial cells is the *dental ledge* that was described in week six.

Local *tooth buds* emerge in the dental ledge at points where a tooth is destined to be formed. The *enamel crown* of the tooth is derived from the cell masses of the dental ledge. Termed the *enamel organs*, those giving rise to the *deciduous teeth* are budded off from the dental ledge first. The cell clusters that give rise to the *permanent dentition* arise surprisingly early but lie dormant until the jaw has enlarged sufficiently to accommodate them.

Larynx

The entire respiratory system develops from a simple tube that arises as an outgrowth from the midventral portion of the lower primitive pharynx. The inlet to

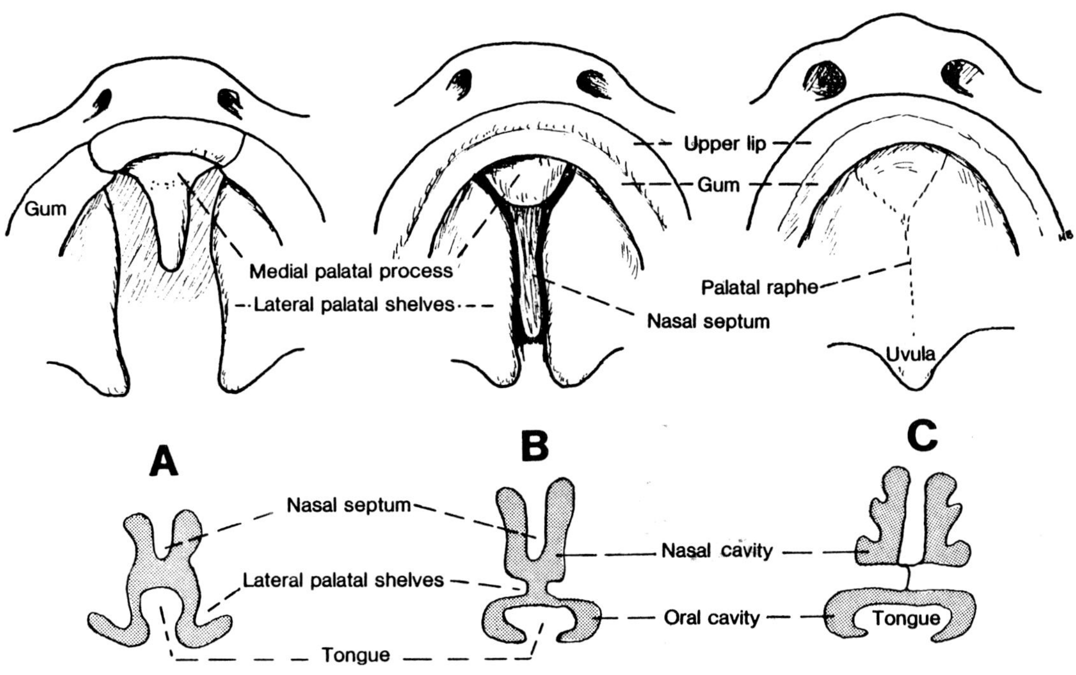

Figure 3-25. Growth and fusion of the palatal parts. Each stage of development is shown from an inferior view and a section below. (A) Upper jaw and roof of the mouth as seen from below in embryos about six weeks old. (B) Similar view of a seven-week-old embryo showing medial progression of the palatal shelves and descent of nasal septum. (C) Palate in the final trimester of pregnancy.

this tube is the larynx.

Fourth Week of Development

The first appearance of the respiratory system is in week four. About midway through this week, a longitudinal groove appears on the floor of the foregut just posterior to the *hypobranchial eminence*. This is the *laryngotracheal groove*, the forerunner of the respiratory system. The cephalic tip of the laryngotracheal groove is destined to become the larynx and is situated in the region of the sixth branchial arch. Several arches contribute to the formation of the larynx as described next.

No time is wasted in the development of the entrance to the larynx. The epiglottis, derived from the union of the third and fourth branchial arches, appears only as a transverse ridge immediately caudal to the tongue in week four. Placed at the opening of the larynx, the epiglottis serves as a protection against aspiration of food into the trachea in the postnatal period.

Fifth Week of Development

The fourth arch contributes a single muscle to the larynx, the *cricothyroid*, which is the "tensor" of the vocal folds. A framework cartilage (the large *thyroid cartilage*) arises from the fourth arch to provide the anterior attachment of the vocal folds.

Muscles controlling phonation and other valvular actions of the larynx arise from the sixth arch and are termed the *intrinsic muscles* of the larynx.

Sixth Week of Development

At six weeks, the epiglottis lies just below the root of the tongue. Inferiorly and laterally to the epiglottis are the emerging *arytenoid swellings*, which will be the *posterior attachments* of the vocal folds.

The *vocal folds* begin to develop in week six from a mass of cells located between the primitive thyroid cartilage and the arytenoid swellings. The cell mass atrophies (diminution in size) leaving a strand of cells that are destined to become the vocal folds. The folds arise from the fifth branchial arch.

Chrondrification (turning into cartilage) of the laryngeal, tracheal, and bronchial cartilages commences. Muscle fibers of the larynx, trachea, and bronchi begin differentiation in week six.

Seventh Week of Development

Although a complete ring in the fully developed *cricoid*, this framework cartilage of the larynx chrondrifies as separate parts. The primitive cricoid cartilage of the larynx chrondrifies anteriorly by week seven, but a soft tissue mass remains posteriorly. If cartilage fails to form in this posterior aspect, the infant may aspirate food due to lack of effective protective valve action in the young larynx. Like the posterior defect described above, a notch is also possible anteriorly if the halves of the cartilage fail to develop fully (Fig. 3-26).

By the end of week seven, the epiglottis and arytenoid swellings rest in a wider pharyngeal space, with the parts of the tongue well identified above in the broader primitive oral cavity. Because the arytenoid swellings grow toward the base of the tongue, they crowd against the epiglottis and form an interesting T-shaped slit. The top of the T is just below the epiglottis. The characteristic shape of the major laryngeal cartilages and the primordia of the laryngeal muscles are evident in week eight.

At this stage, a vigorous growth of laryngeal epithelium *occludes* (to close tight) the opening to the larynx. This may be nature's way to prevent amnionic fluid

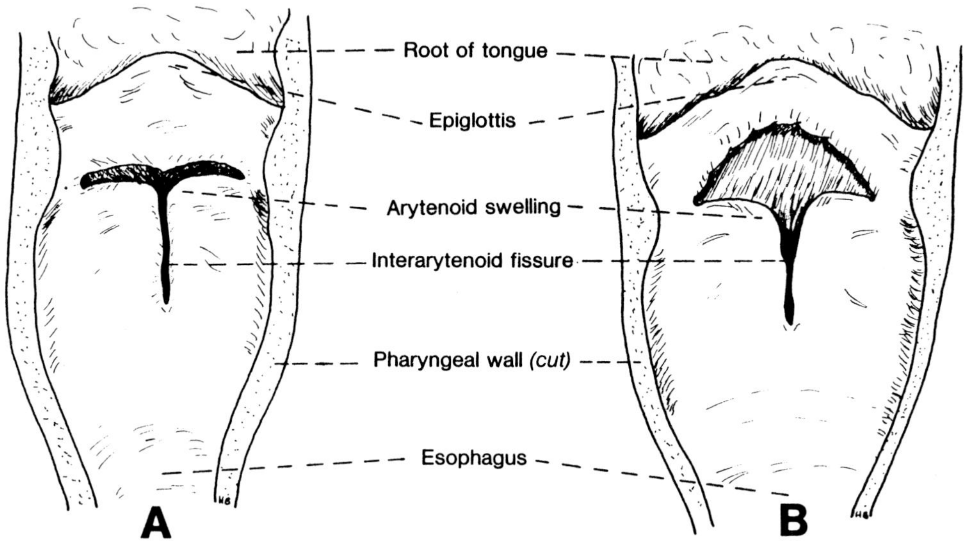

Figure 3-26. Diagrammatic representation of the developing larynx at eight weeks (A) and nine weeks (B). (From Arey, L.B.: *Developmental Anatomy*, 1965. Courtesy of W.B. Saunders Co., Philadelphia.)

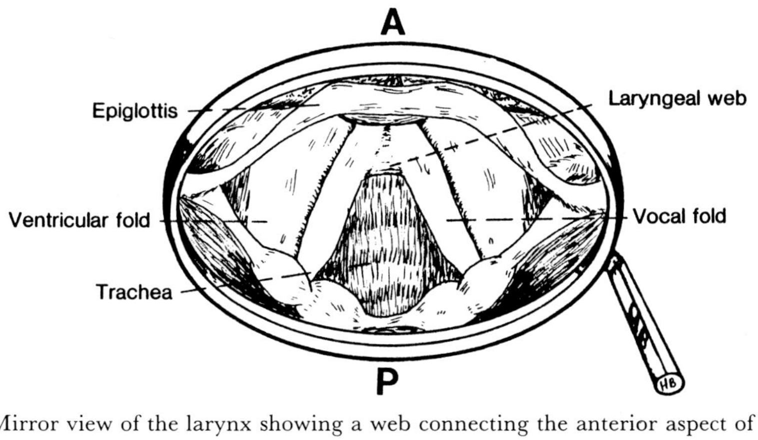

Figure 3-27. Mirror view of the larynx showing a web connecting the anterior aspect of the vocal folds. Failure of an early protective membrane to dissolve completely in the tenth week results in webbing of the folds. Complete webbing is incompatible with life and must be ruptured immediately at birth. A and P designate anterior and posterior aspects of the larynx.

from entering the developing respiratory tract. However, in the tenth week, the membrane begins to dissolve so that the opening is reestablished. Failure of this membrane to dissolve accounts for a variety of *congenital webs*, which might occur in between the margins of the vocal folds as well as webbing above or below the folds (Fig. 3-27).

During the time of occlusion by the membrane in week eight, a small cavity is formed in the larynx. This is the *laryngeal ventricle*, which will eventually separate the *ventricular* and *vocal folds*. On either side of this slit-like cleft, then, is the primordia of the ventricular and vocal folds, which appear as the membrane dissolves (Fig. 3-28).

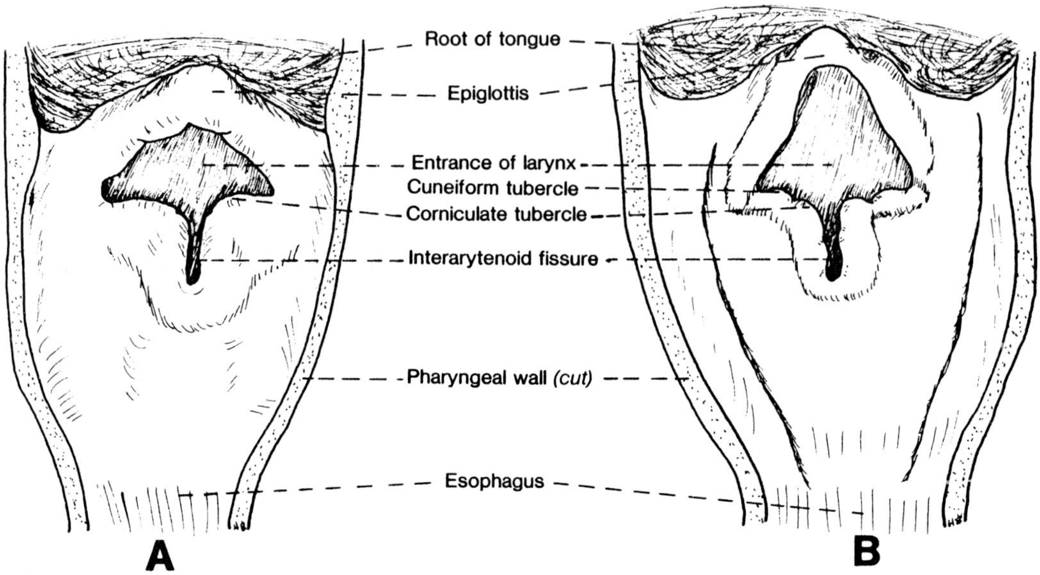

Figure 3-28. Diagrammatic representation of the larynx at ten weeks (A) and seven months (B). (Arey, L.B.: *Developmental Anatomy*, 1965. Courtesy of W.B. Saunders Co., Philadelphia.)

Ear

The ear is the first organ of special sense to be differentiated. Responses to sound have been demonstrated in an embryo not yet into week four of development (Johannson, Wedenberg, and Westin, 1964).

Fourth Week of Development

Early in week four, two thickenings appear on the enlarged cephalic end of the neural tube in the region of the primitive hindbrain. Known as the *auditory placodes* (plate-like), these swellings eventually give rise to the *inner ear*. Within a short time after appearance of the auditory placodes, they begin to sink below the surface and become shallow pits. These auditory pits sink rapidly so that by the end of week four they appear as dilated sacs, but growth of the sacs continues below the surface.

The first pharyngeal pouch contributes the middle-ear cavity or *tympanic cavity*. This cavity expands to surround the developing *ossicles*, or middle-ear bones. An extension of the tympanic cavity, the *auditory* or *eustachian tube*, also develops from the first pouch in week four. In a sense, the middle-ear cavity and the auditory tube form as a trumpet-like outpocketing of the pharynx.

The tympanic membrane or eardrum is a three-layer structure, deriving outside layers from the lining of the *external auditory meatus* and the *middle ear cavity*. The middle layer is presumably left over from the embryonic tissue that once plugged the ear canal.

Fifth Week of Development

The forerunner of the inner ear, which includes both the *cochlea* (organ of hearing) and the *semicircular canals* and associated structures (sense of equilibrium), appears as an *otic* (ear) *vesicle* in tissue below the surface of the embryo. Note that only the inner ear is derived from the sub-surface otic vesicle, since the middle-ear structures come from the first pharyngeal pouch. The external ear is derived from the first branchial groove or cleft.

Growth centers in both the first and second arches give rise to nodular elevations, which surround the opening of the first branchial cleft. The nodular elevations or hillocks (small prominences) will eventually fuse and form the *auricle* of the external ear. The opening of the first branchial cleft will become the *external auditory meatus*.

In addition to the contributions by the first arch to the external ear parts, the dorsal portion of the arch supplies the first two bones (the ossicles) of the middle ear. These are the *malleus* and *incus*. The malleus receives vibrations from the eardrum and transfers them to the incus. The third bone, the *stapes*, of the ossicles comes from the second arch and is responsible for transferring vibrations to the fluid-filled cavities of the cochlea in the developed ear. The tiny muscle attached to the stapes, termed the *stapedius*, is also derived from the second arch. Its role is to restrict movement of the stapes in response to loud sound to protect delicate hair cells in the cochlea (Fig. 3-29).

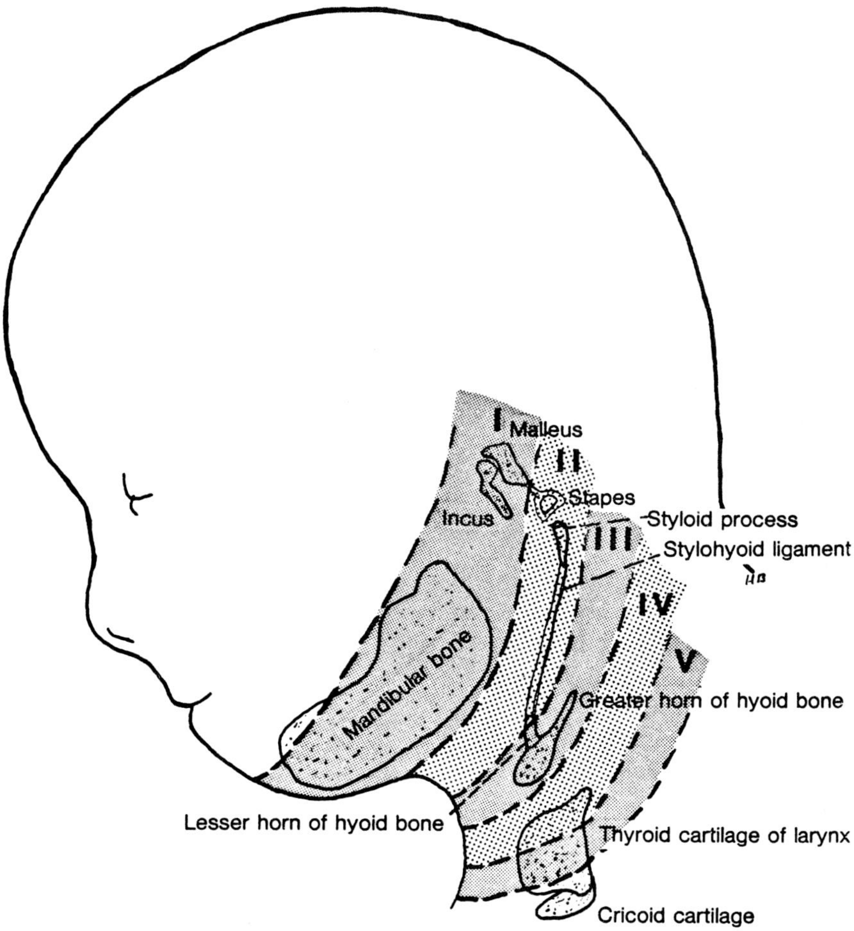

Figure 3-29. Schematic diagram showing the relations of the various structures to the branchial-arch territory from which these structures arise. The branchial arches appear in Roman numerals and are differentially shaded. Notice the contributions of the first and second arches to middle-ear structures.

Sixth Week of Development

The precartilage foundations for the walls of the external ear canal and the auricle are evident. The primordia of the *semicircular canals* (organ of equilibrium) and the *cochlea* (organ of hearing proper) continue to develop. Precartilage tissue, from which the middle-ear ossicles will be derived, develops. The auricle is still too low and too far ventral, but growth of the mandibular arch begins to displace the auricle dorsolaterally.

Seventh Week of Development

The cochlea coils about itself, twisting the *vestibulocochlear* (VIII) *nerve* with it for some two-and-one-half turns. The coiling is finished by the end of week seven. By the fourth fetal month, the inner ear has reached its full adult size (Fig. 3-30).

Figure 3-30. Schematic of the cochlea coiling about itself and twisting the vestibulocochlear nerve in the process. (From Daniloff, R., Schuckers, G., and Feth, L.: *The Physiology of Speech and Hearing: An Introduction*, 1980, p. 411. Reprinted by permission of Prentice-Hall, Inc., Englewood Cliffs, New Jersey.)

Before the tympanic cavity is formed, it is filled with embryonic tissue, which is subsequently resorbed. At birth, there continues to exist some residue of unresorbed embryonic connective tissue partially filling the tympanic space and damping, more or less, the free movement of the ossicles. Complete freedom of ossicular mobility is achieved within a few months after birth as the remnants of the loose tissue are absorbed.

It is interesting to note that the stapes is at first ring-shaped; but, with the disappearance of a temporary stapedial artery, it assumes the shape of a stirrup (see Fig. 3-29) with its footplate placed in the *oval window* of the otic capsule, the primitive inner ear (Fig. 3-31).

Progressive chrondrification leading to ossification of the ossicles occurs in this week. The ossicles are miniature in size with adult configuration. They will grow three times larger over the next three months and reach their full size by the seventh month.

Eye

Fourth Week of Development

As stated above, the eyes take their origins from a pair of lateral outgrowths of the primitive forebrain along with the developing brain structures. These outgrowths, the *optic vesicles* (small bladder with a liquid), form early in week four. They migrate toward the surface in the region of the primitive forebrain. Once in contact with the surface tissue, a thickening plate known as the *lens placode* (platelike) develops, which will become the lens of the eye. Thus, the optic vesicle and its lens placode prestage the formation of the complete eye (Fig. 3-32).

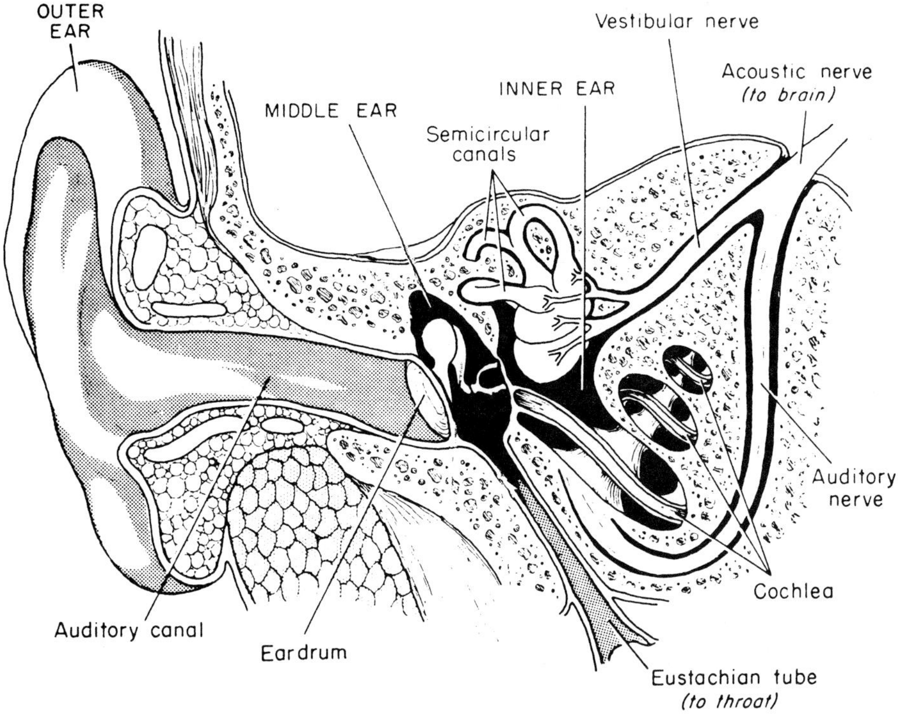

Figure 3-31. Divisions of the mature hearing apparatus. Three divisions of the ear are recognized: outer, middle, and inner. (From Brooks, Stewart M., and Paynton-Brooks, Natalie: *The Human Body,* ed. 2, St. Louis, 1980, The C. V. Mosby Co.)

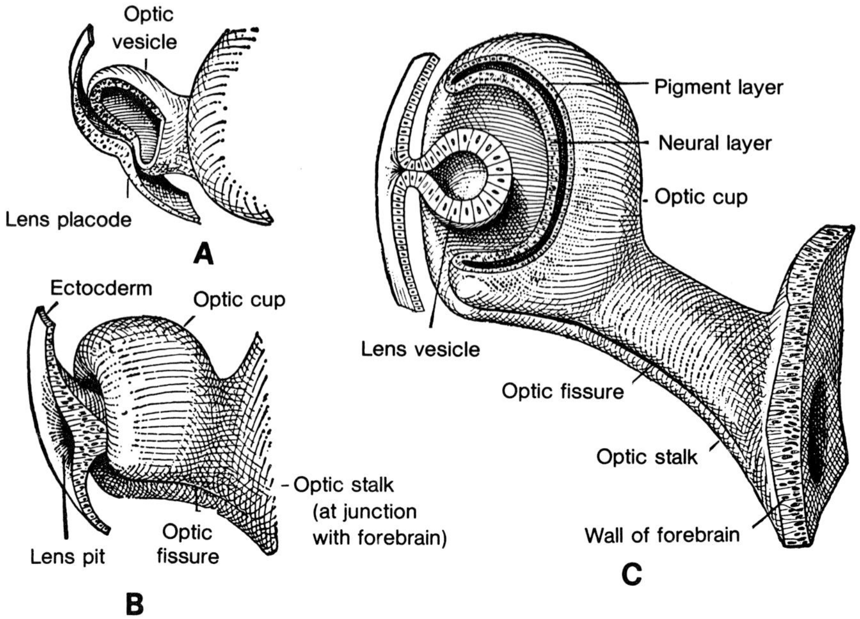

Figure 3-32. Various stages in the development of the eye. The lens is sectioned and the optic cup in A and C has been partly cut away. (From Arey, L.B.: *Developmental Anatomy*, 1965. Courtesy of W.B. Saunders Co., Philadelphia.)

Fifth Week of Development

First, the lens placode formed in week four sinks beneath the surface, forms a lens pit, and then becomes a lens vesicle. This process of sinking and forming a vesicle is termed *invagination* (infolding) and produces separation from the surface by midweek. The optic vesicle now lies in a depression termed the *optic cup*.

Growth continues on many fronts to develop parts of the eye. Of special interest is the formation of the primitive vitreous body in a cavity associated with the lens vesicle. The *vitreous body* is the glass-like gelatinous fluid filling the cavity of the eyeball.

Sixth Week of Development

The *retina* and the *optic nerve* fibers are differentiated and grow in week six. More specialized development of the parts of the eye, particularly in the region of the lens, is progressing. The muscles that will move the eyeball, the *extrinsic eye muscles*, are clearly apparent by the end of week six. Forward rotation of the eyes on the head brings them much closer to their proper positions on the face (Figs. 33 & 34).

Seventh Week of Development

Eyelids begin to grow in week seven. Continued differentiation of the eye structures is ongoing with maturation of the retina, lens, iris, cornea, and sclera. Nerve fibers (the optic fibers) advance to the brain from the eye.

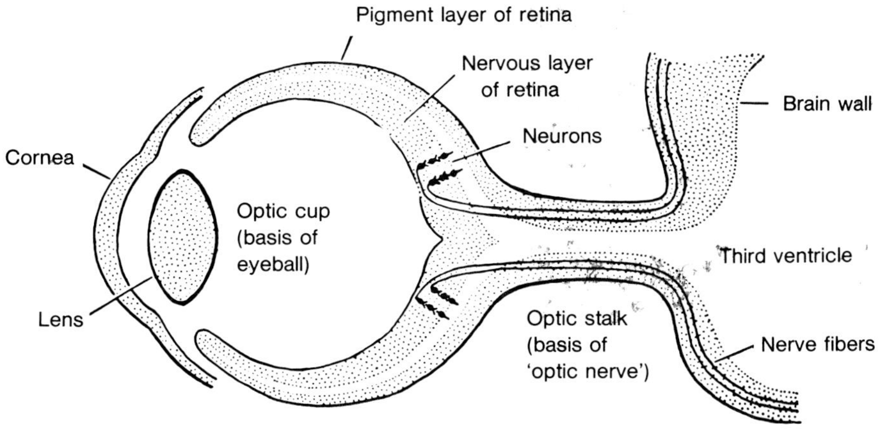

Figure 3-33. Schematic diagram showing the retina as an extended portion of the brain and the course of the optic nerve fibers through a connecting stalk (called the optic nerve) to reach the brain. It is apparent why some suggest that the optic tract, unlike other cranial nerves, is an extension of the brain. (From Arey, L.B.: *Developmental Anatomy*, 1965. Courtesy of W.B. Saunders Co., Philadelphia.

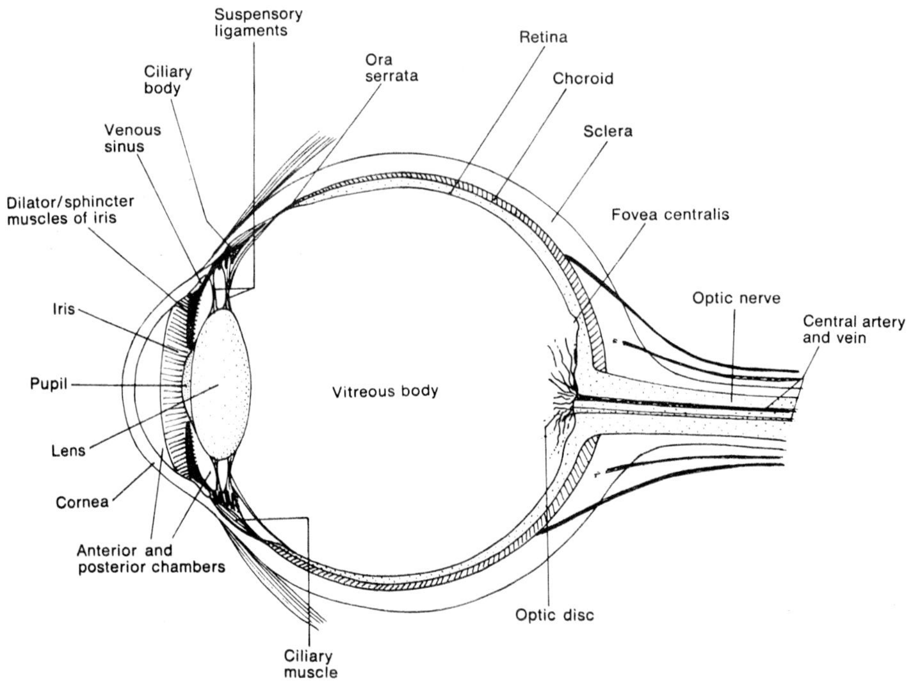

Figure 3-34. The eye at maturity. (From *It's Your Body* by L.M. Elson. Copyright© 1975. Used with the permission of McGraw-Hill Book Company.)

BIBLIOGRAPHY

Arey, L.B.: *Developmental Anatomy*, 7th ed. Philadelphia, Saunders, 1974.

Dorland's Illustrated Medical Dictionary, 25th ed. Philadelphia, Saunders, 1974.

Enlow, D.H.: *Handbook of Facial Growth*. Philadelphia, Saunders, 1975.

Gray, H.: *The Anatomy of the Human Body* (Charles M. Goss, Ed.), 28th ed. Philadelphia, Lea and Febiger, 1973.

Johannson, B., Wedenberg, E., and Westin, B.: Measurement of tone responses by the human foetus. *Acta Oto-laryngologica 57*:188-192, 1964.

Langley, L.L., Telford, I.R., and Christensen, J.B.: *Dynamic Anatomy and Physiology* 4th ed. New York, McGraw, 1974.

Myerson, M.C.: *The Human Larynx*. Springfield, IL: Thomas, 1964.

Orban, B.J.: *Oral Histology and Embryology*, 4th ed. St. Louis, C.V. Mosby, 1957.

Patten, B.M.: *Human Embryology*, 3rd ed. New York, McGraw, 1968.

Romanes, G.J. (Ed.): *Cunningham's Textbook of Anatomy*, 12th ed. New York, Oxford U Pr, 1981.

Thomas, J.B.: *Introduction to Human Embryology*. Philadelphia, Lea and Febiger, 1968.

Van Alyea, O.E.: *The Embryology of the Ear, Nose and Throat*. Chicago, American Academy of Ophthalmology and Otolaryngology, 1949.

PART TWO: THE VOCAL TRACT – LOWER ASPECTS

Chapter 4

FRAMEWORK FOR RESPIRATION

THE muscles and other soft tissues responsible for respiration functions are attached to a bony framework composed of the vertebral column, the rib cage, and the pelvic girdle. Most of the movement associated with respiration comes from the diaphragm and abdominal muscles without significant bone involvement. When active, the principal movement of the bony framework for respiration is carried out by the ribs in response to thoracic muscles (Fig. 4-1 & Table 4-I).

THE VERTEBRAL COLUMN

The segmented feature of the *vertebral column* permits movement of the *neck, thoracic*, and *lumbar regions*. Although the movement is only slight, it enables the skull to be turned and tilted to assist the senses of vision and learning in the communication process. This gently curved, rod-like structure extends from the base of the skull to the coccyx (tailbone). It serves as the axial support for the body. The vertebral column also provides protection for the spinal cord (Fig. 4-2).

The vertebral column, with a total of about thirty-three individual bones termed *vertebrae,* is divided into five segments: seven *cervical*, twelve *thoracic*, five *lumbar*, five *sacral*, and four *coccygeal* vertebrae. However, the total number may vary in some individuals. The vertebrae

Table 4-I
BONES PERTAINING TO RESPIRATION

Division	*No. in Subdivision*	*Total Number*
Vertebral Column		33
Cervical vertebrae	7	
Thoracic vertebrae	12	
Lumbar vertebrae	5	
Sacrum (fused vertebrae)	5	
Coccyx (variable)	4	
Rib Cage		25
Vertebrosternal (true) ribs	14	
Vertebrochondral (false) ribs	4-6	
Vertebral (floating) ribs	4-6	
Sternum	1	
Shoulder Girdle		4
Scapula	2	
Clavicle	2	
Pelvic Girdle		6
Ilium	2	
Ischium	2	
Pubis	2	

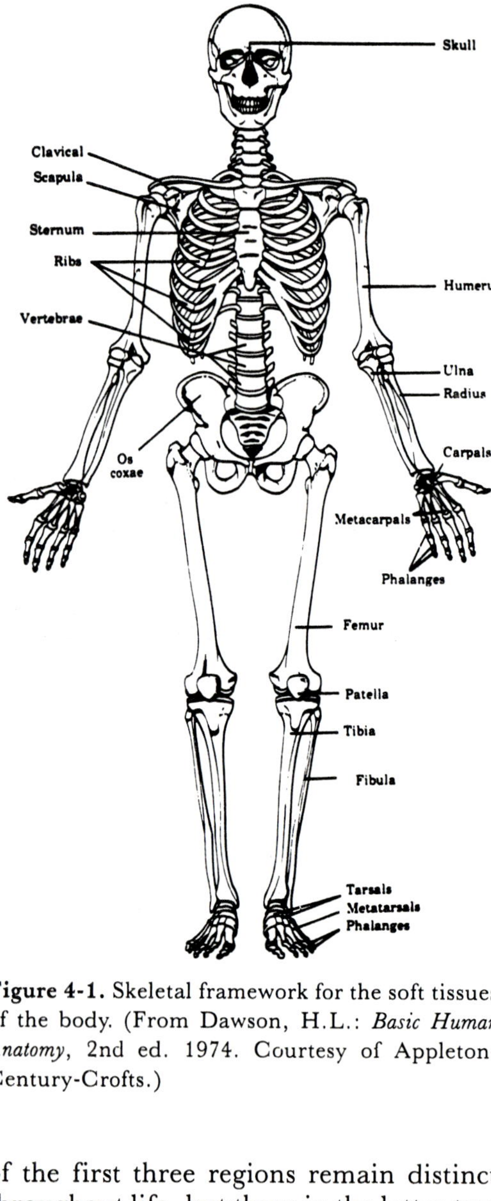

Figure 4-1. Skeletal framework for the soft tissues of the body. (From Dawson, H.L.: *Basic Human Anatomy*, 2nd ed. 1974. Courtesy of Appleton-Century-Crofts.)

of the first three regions remain distinct throughout life, but those in the latter two regions fuse. Thus, the adult typically has twenty-six distinct bones in the vertebral column.

Pads of fibrocartilage known as *intervertebral disks* are found separating the vertebrae above the sacrum. These disks are the shock absorbers, adding resilience and flexibility to the column. There are twenty-four intervertebral disks cushioning twenty-five vertebrae (Fig. 4-3).

A canal is created when the vertebrae are stacked in place because each vertebra has a foramen. The foramen of each vertebra, then, provide passage for the spinal cord.

The giraffe has the same number of cervical vertebrae as man except they are much larger, of course. Another point of interest is the method used by some natives to acquire long necks. As rings are added to the neck, the shoulder, especially the clavicle, is depressed. Due to the compression on the clavical caused by the neck rings, the neck becomes longer, but the cervical vertebrae do not expand.

Cervical Vertebrae

The *cervical vertebrae* comprise the neck bones. These seven vertebrae are the smallest of the bony spinal column and present oblong bodies.

The first two of the seven deserve special attention. The *atlas* supports the skull and is fitted to articulate with the occipital condyles of the occipital bone. It has no body per se and no spinous process. It is bounded by tubercles and also has a superior and inferior articular surface. The second cervical vertebra, the *axis*, is also an atypical vertebra. It contains a vertical projection known as the *dens* or *odontoid process*, which extends up through the ring of the atlas to make a pivot on which the atlas and head rotate. The atlas and axis are bound to each other and to the base of the skull by a system of ligamentous connections. Even so, it is possible for the odontoid process to be pushed into the foramen magnum during accidents to cause grave paralyses (Fig. 24-9, p.576).

The vertebral foramina of the cervical vertebrae are rather large in order to accommodate the enlarged upper portion of the tapered spinal cord. The spinous processes in the cervical region may vary

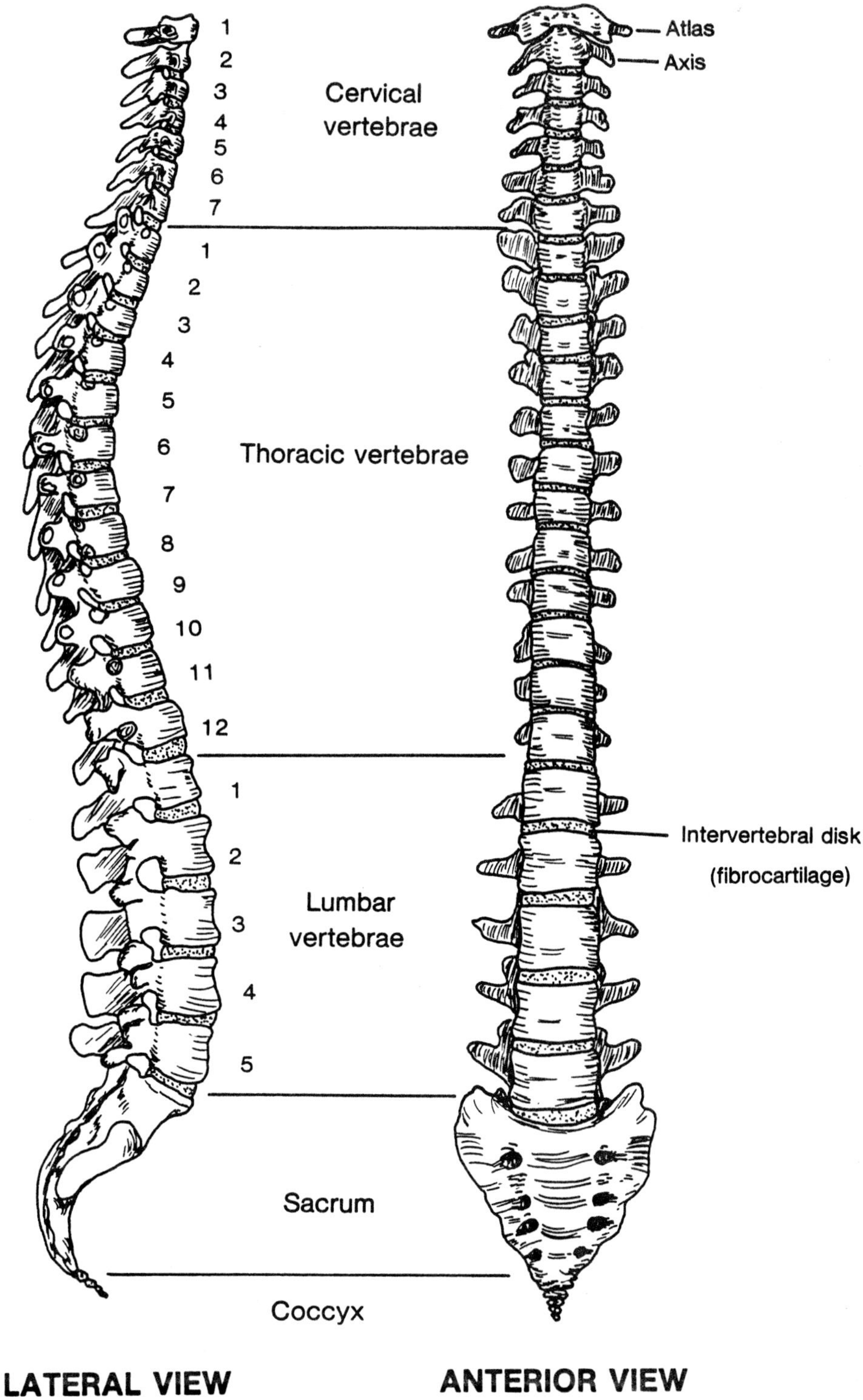

Figure 4-2. The vertebral column.

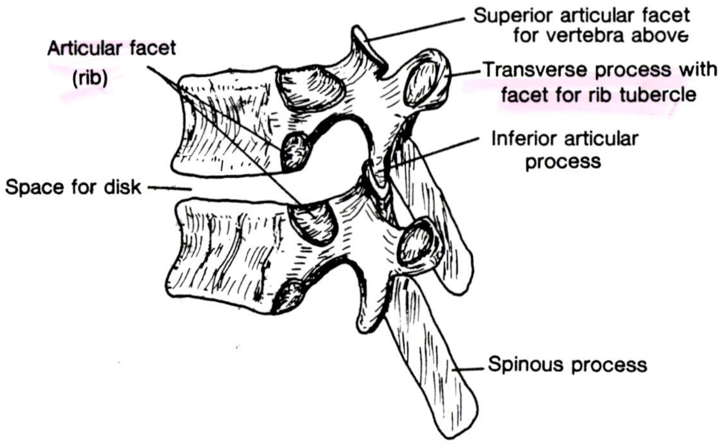

Figure 4-3. Lateral view of the thoracic vertebrae. A fibrocartilage disk separates the bodies of the vertebrae. Notice the design for articulation between the vertebrae.

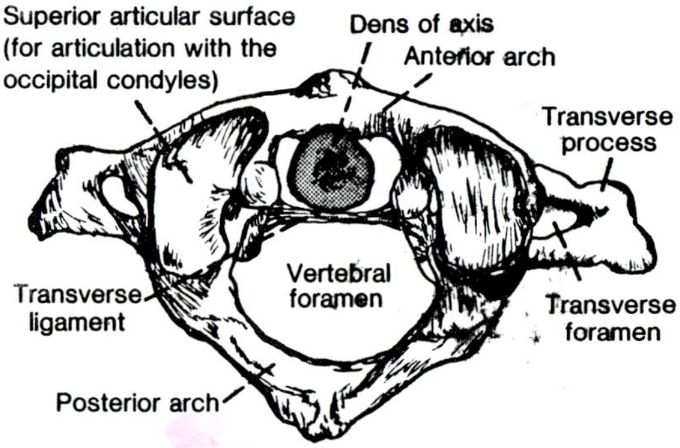

Figure 4-4. The atlas. The first cervical vertebra, the atlas, articulates with the skull. The large vertebral foramen transmits the spinal cord from the skull through the foramen magnum. Notice the dens of C2 shown in position with the transverse atlantal ligament. Such an arrangement permits turning of the head, the skull and atlas rotating about the dens of C2.

considerably in development from one vertebra to the next.

If the upper cervical vertebrae or cranial base are in malposition, they may distort the posterior pharyngeal wall soft tissue attachments and interfere with functional palatopharyngeal closure. Thus, posterior displacement of the cervical vertebrae could account for some hypernasality problems. From a lateral x-ray, it is apparent that the cervical column is off center with the skull, permitting the pharynx to assume a midaxial position (Fig. 4-6, also Fig. 24-8, p. 576).

Thoracic Vertebrae

The twelve thoracic vertebrae are typical vertebrae, with the exception of the added *costal facets*, also termed *demifacets*. There are three such facets on either side of each vertebra, two on the body and one on the transverse process, to articulate with the head of each pair of ribs. All except T11 and T12 carry a facet on the transverse process for the tubercle of the rib (Fig. 4-7).

Lumbar Vertebrae

The five vertebrae of the lumbar region are the largest and strongest. A number of powerful back muscles are attached to these vertebrae on the spinous processes.

Sacrum and Coccyx

The sacrum is situated between the two hip bones. This triangular region of

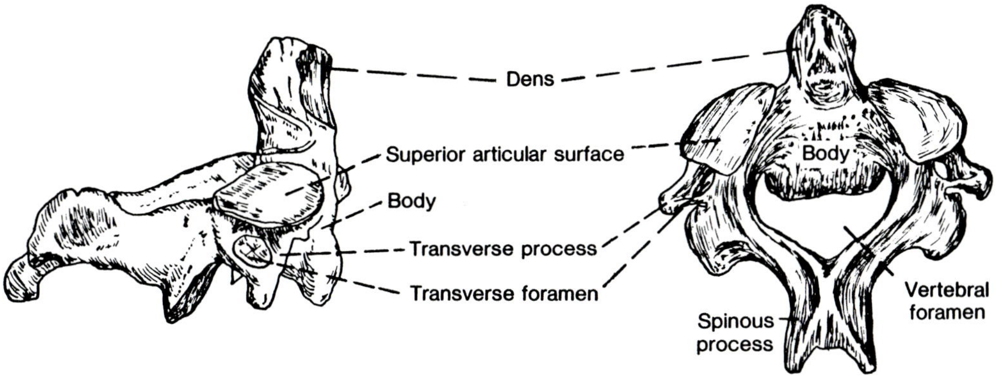

Figure 4-5. The second cervical vertebra, the axis, articulates with the atlas to permit rotation of the head.

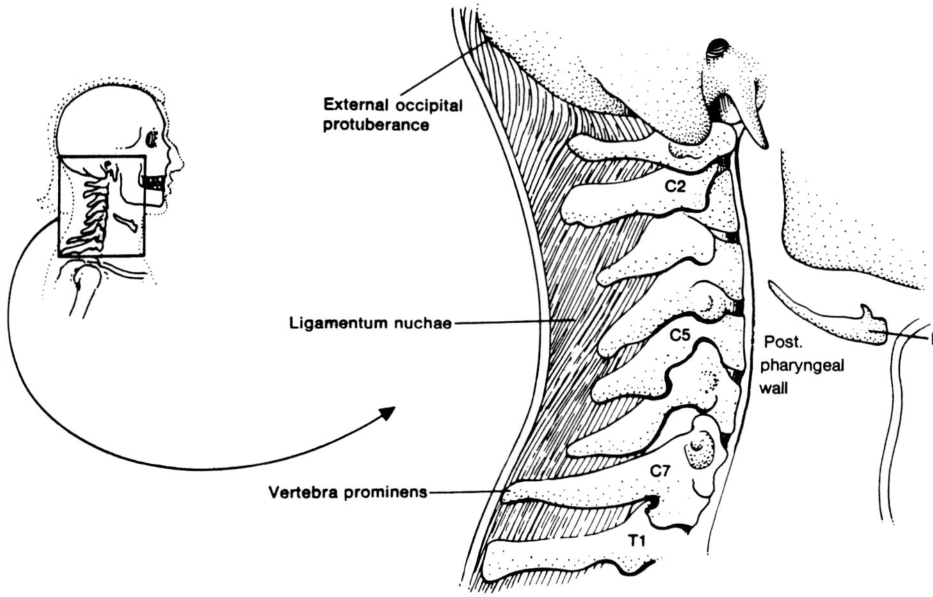

Figure 4-6. The cervical vertebrae of the neck. Malalignment of the upper cervical vertebrae may distort the position of the posterior wall of the pharynx and interfere with palatopharyngeal valving. The bodies of the cervical vertebrae are significantly smaller than those in the thoracic and lumbar regions. (From *It's Your Body* by L.M. Elson. Copyright© 1975. Used with the permission of McGraw-Hill Book Company.)

the vertebral column contains a cavity as a continuation of the spinal canal, although the spinal cord ends in the upper portion of the lumbar region. The sacrum articulates above with the fifth lumbar vertebra. With the iliac (hip) bones, the sacrum forms the bony pelvic girdle (Fig. 4-8).

The coccyx (like a cuckoo's beak) consists of a variable number of rudimentary vertebrae, usually three to five. It articulates with the tip of the sacrum, forming a slightly movable joint that serves to expand the size of the birth canal during delivery.

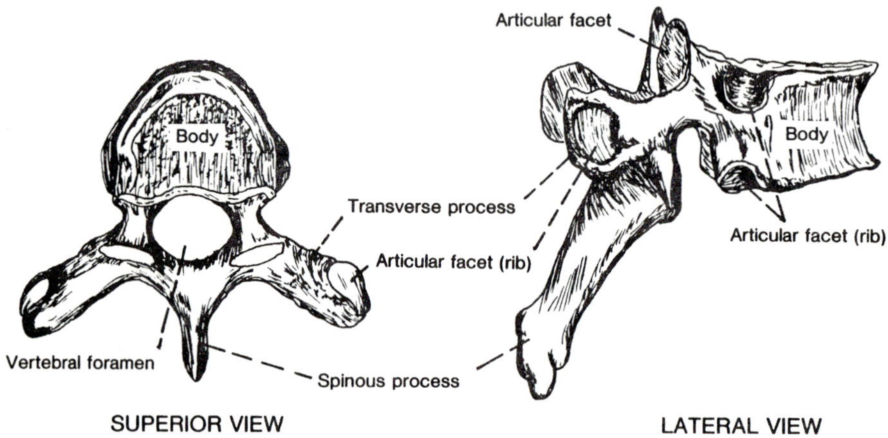

Figure 4-7. Thoracic vertebra.

Normal and Abnormal Conditions

The vertebral column appears straight from an anterior or posterior view. From a lateral view, it is either straight or slightly curved forward (convex) in the cervical area, concave (C-shaped) in the thoracic region, convex in the lumbar area, and concave in the sacral area. The vertebral column in the fetus has but a single anterior concave curve.

Speech pathologists often see clients with multiple handicaps that include vertebral column deviations, especially if the practice is in a medical setting.

Abnormal curves in the vertebral column are not rare. An abnormal lateral curve is known as scoliosis and may result from an imbalance in muscular support on one side of the body, diseased vertebrae, paralysis, or poor posture. Two other conditions interfere with normal alignment of the vertebral column. They are kyphosis or "hunchback," an abnormal curvature (increased convexity) in the thoracic region, and lordosis or "swayback," an abnormal curvature (increased concavity) in the lumbar region.

Disk pathology can result from injury, the condition being termed *herniated intervertebral disk*. Lay persons refer to it simply as a "slipped disk." A disk herniation in the lumbar region involves pinching of the sciatic nerve with pain in the leg, known as attacks of sciatica. Pressure on the roots of spinal nerves from the herniated portion can cause much pain, but most instances of backache in the older person are probably due to disk attrition caused by an insufficient blood supply (Fig. 4-9).

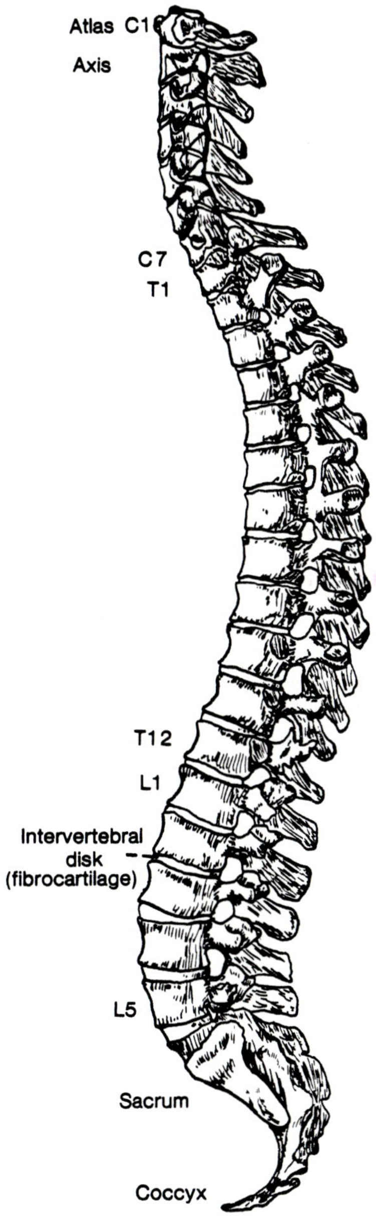

Figure 4-8. The vertebral column showing variations in the configuration of the vertebrae, especially in the lower region.

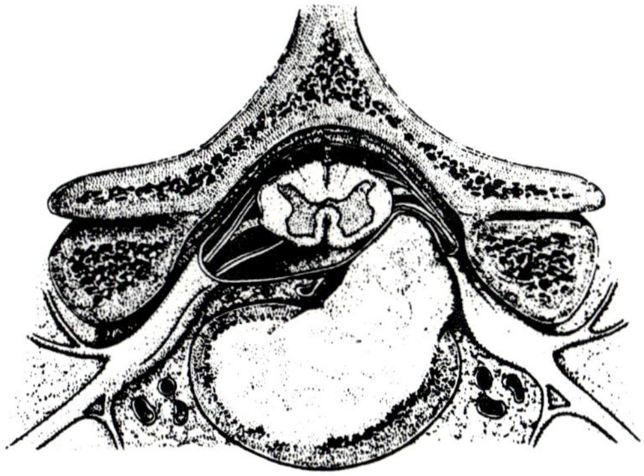

Figure 4-9. Example of compression trauma on a spinal nerve caused by herniation of an intervertebral disk. Depending on the level of the nerve compression, symptoms may include muscle weakness, pain, paresthesias, and change in reflexes. (From *Man, Microbes, and Matter* by B.C. Block. Copyright© 1975. Used with the permission of McGraw-Hill Book Company.)

FRAMEWORK FOR THORACIC AND ABDOMINAL CAVITIES

The Rib Cage

The *thoracic cage* is formed by the *sternum, ribs, costal cartilages*, and *thoracic vertebrae*. It is part of the axial skeleton. The shape is conical, beginning at the top with the narrow end and flaring to a broad base. In a lateral view, the rib cage slants forward and downward. In an anteroposterior dimension, the thoracic cage is flattened, forming a kidney shape when viewed from above. Its floor is the diaphragm. The lateral and posterior walls of the thoracic cage are formed by twelve ribs. Anteriorly, costal cartilages join all but the lowermost ribs to the sternum, or breastbone (Fig. 4-10).

The bony cage encloses and protects vital organs and structures of the chest cavity, including the lungs. Additionally, the thorax provides support for the bones of the shoulder girdle and upper extremities. Its movement plays an important role in respiration.

Sternum

The *sternum* presents three segments.

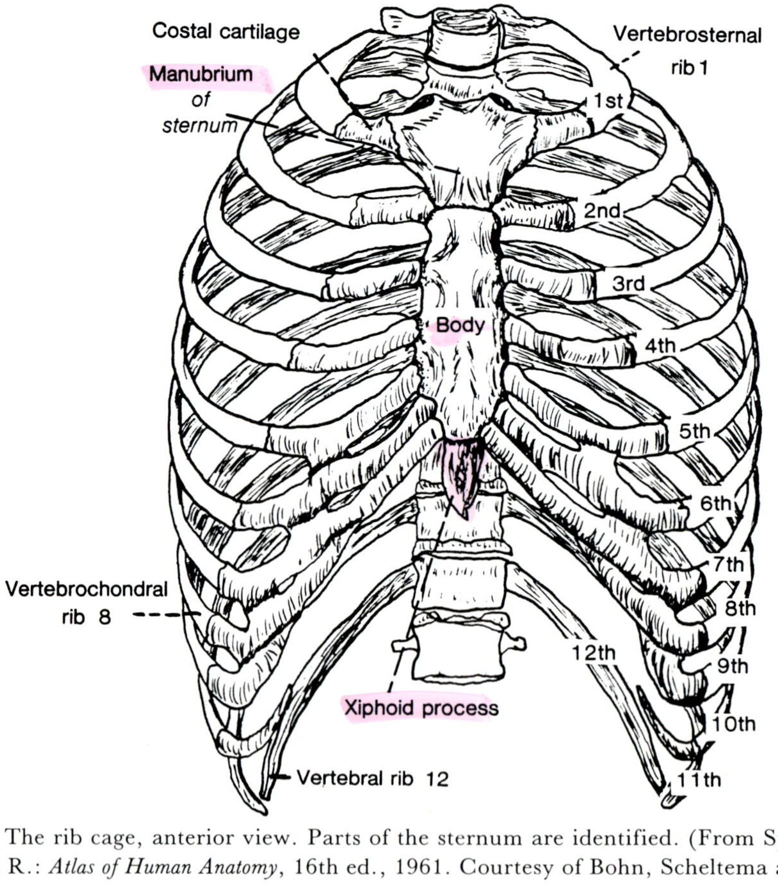

Figure 4-10. The rib cage, anterior view. Parts of the sternum are identified. (From Spalteholz, W., and Spanner, R.: *Atlas of Human Anatomy*, 16th ed., 1961. Courtesy of Bohn, Scheltema and Holkema, Amsterdam.)

The uppermost part is known as the *manubrium*, or handle-like structure. At its superior border, the jugular notch is located. The *clavicle* (collarbone) and the first rib articulate with the sternum; the rib contains a *costal cartilage* medially. The midportion of the *sternal body* is joined with the manubrium by a movable joint, permitting a hinge action. About 10 cm long, the body provides maximum protection for the heart. The sternal body is thus allowed to swing forward during each inspiratory movement. Along the lateral borders, six ribs are connected through costal cartilages. The *xiphoid* or *ensiform* (sword-shaped) process hangs inferiorly from the sternal body and may present a variety of forms: bifid (split by a cleft), curved, pointed, perforated, or deflected laterally. It gives attachment to the linea alba, the midline of the abdomen (Fig. 4-11).

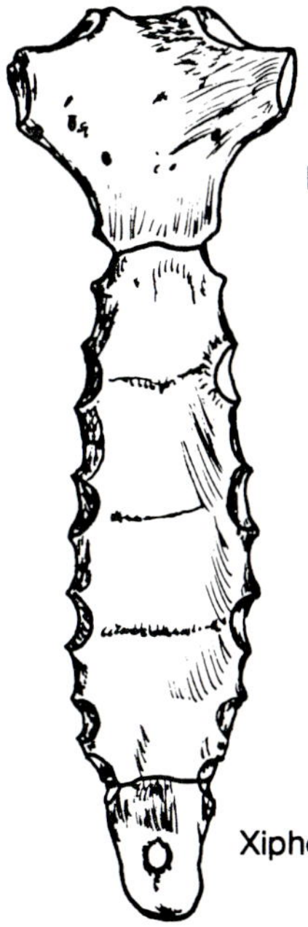

Figure 4-11. The sternum. The xiphoid process is perforated.

Ribs

The twelve pairs of ribs take their posterior origins on the vertebral column, being associated with the twelve thoracic vertebrae.

The first seven pairs are designated as *true ribs*, since they articulate through their costal cartilages to the sternum. The lower five pairs are termed *false ribs*, although they are attached to true ribs by costal cartilage. Nevertheless, they do not articulate directly with the sternum. Of the five false ribs, the last two or three (the 11th, 12th, and sometimes the 10th) end in the body wall without a firm attachment anteriorly. Embedded in the musculature of the abdominal wall, these are called *floating ribs* (Fig. 4-12).

The three kinds of ribs may be descriptively termed *vertebrosternal, vertebrochondral*, and *vertebral* to denote the attachment, or lack of attachment, of the ribs' two ends (Fig. 4-13).

The clavicles, although not a part of the thoracic cage proper, articulate with the manubrium medially. The lateral ends articulate with the *scapulae*, and the shoulder girdle is formed (Fig. 4-14).

Pelvis

At the base of the vertebral column is the pelvic girdle. It is a single bone in the adult. Originally it is composed of three pairs of bones, the *ilium, ischium*, and *pubis*. The terms are retained for convenience to describe regions of the adult pelvic girdle.

At the posterior aspect of the pelvic girdle is a gap, since the two iliac bones do not meet. This opening is completed by the sacrum in the articulated skeleton.

The female pelvis is adapted for childbearing. The ilia are less sloped and set farther apart. Also, the sacrum is less curved, increasing the distances between

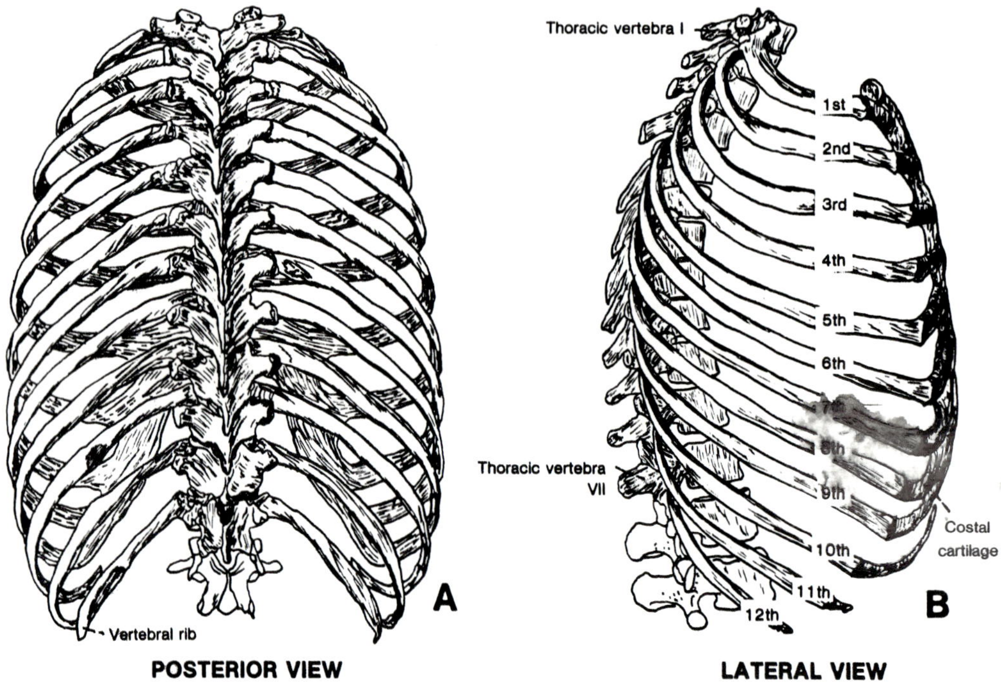

Figure 4-12. Two views of the rib cage. The downward sloping of the ribs represents a rest position with each hinging at the vertebral column. Lifting up on the ribs produces inhalation during chest breathing. (From Spalteholz, W., and Spanner, R.: *Atlas of Human Anatomy*, 16th ed., 1961. Courtesy of Bohn, Scheltema and Holkema, Amsterdam.

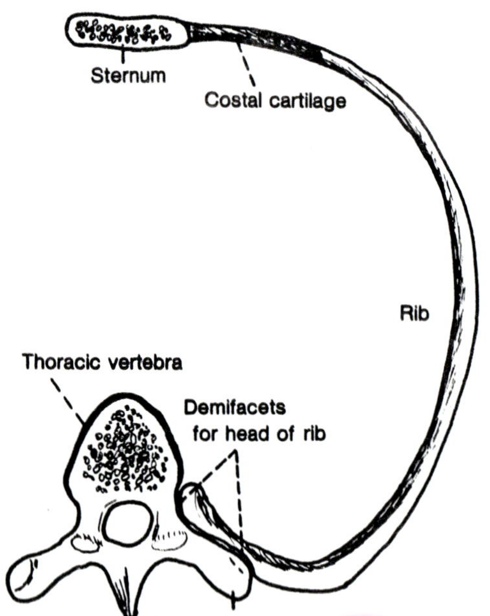

Figure 4-13. Anterior and posterior connections of a vertebrosternal rib.

the coccyx and symphysis (Fig. 4-15).

To the pelvic girdle, especially its superior borders, are attached a number of postural muscles. Additionally, muscles pertaining to respiration take their origin on the pelvic girdle.

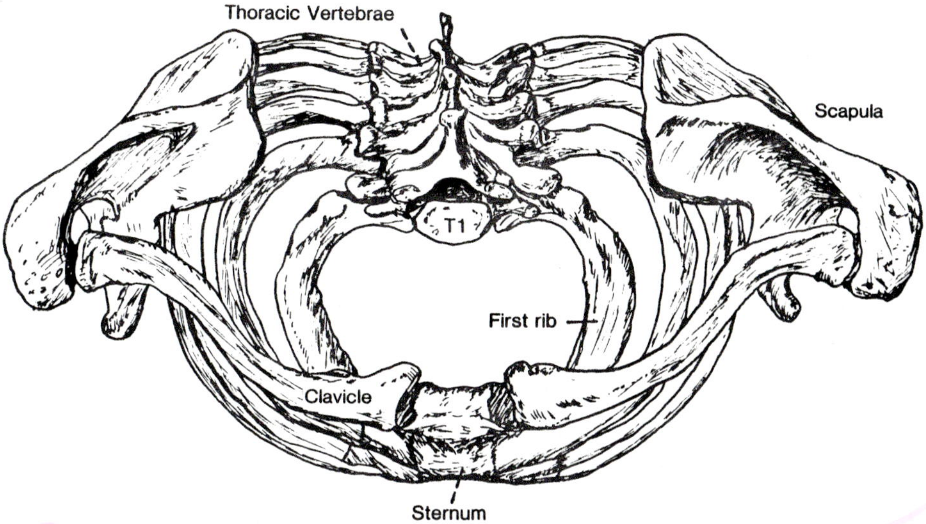

Figure 4-14. Superior view of the shoulder girdle and rib cage. (From Spalteholz, W., and Spanner, R.: *Atlas of Human Anatomy*, 16th ed., 1961. Courtesy of Bohn, Scheltema and Holkema, Amsterdam.)

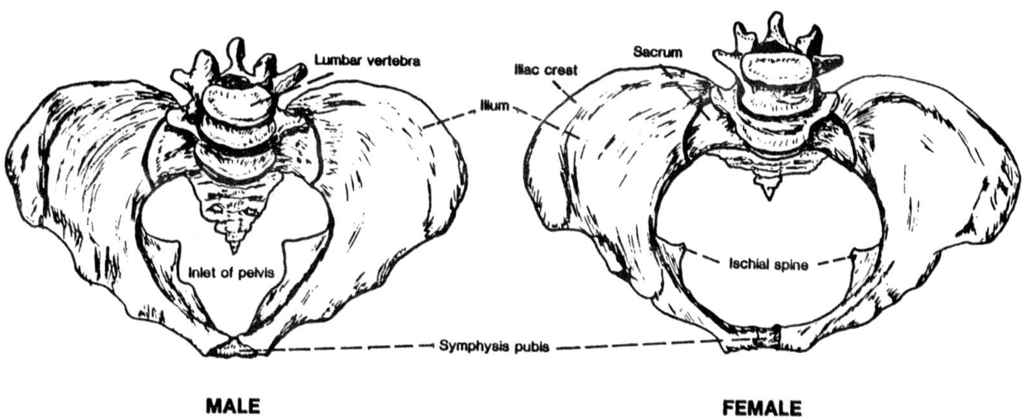

Figure 4-15. The pelvic girdle. Comparison of the male and female pelvic girdles. (From Spalteholz, W., and Spanner, R.: *Atlas of Human Anatomy*, 16th ed., 1961. Courtesy of Bohn, Scheltema and Holkema, Amsterdam.)

BIBLIOGRAPHY

Bruck-Kan, R.: *Introduction to Human Anatomy*. New York, Har-Row, 1979.

Gray, H.: *The Anatomy of the Human Body* (Charles M. Goss, Ed.), 28th ed. Philadelphia, Lea and Febiger, 1973.

Romanes, G.J. (Ed.): *Cunningham's Textbook of Anatomy*, 12th ed. New York, Oxford U Pr, 1981.

Woodburne, R.T.: *Essentials of Human Anatomy*, 6th ed. New York, Oxford U Pr, 1978.

Chapter 5

THE LUNGS AND TRACHEOBRONCHIAL TREE

RESPIRATION may be studied at the cellular level or in a general sense pertaining to breathing. Essentially, *respiration* concerns the exchange of gases: oxygen passes into the blood vessels within the lungs, and carbon dioxide leaves the blood vessels to be expelled from the lungs. *Breathing* refers to the mechanical process of moving air into and out of the lungs. It is this mechanical process that serves as the power source for generating sound in the upper aspects of the respiratory tract. Our attention here will first focus on those structures that provide the energy to drive the phonatory and articulatory apparatus.

THE LUNGS

The two lungs or pulmones are composed of tissue having a light, porous, and spongy texture. When handled in a cadaver specimen, a sound like that of crumpling tissue paper will be heard, owing to the fact that air remains in the lungs after removal from the thoracic cavity. The lungs, being highly elastic, appear small when viewed without supportive attachments (after removal from the thoracic cavity). Although the actual mass of tissue comprising the lung walls is quite small, they retain a relatively normal configuration outside the body due to the air remaining in the alveoli (Fig. 5-1).

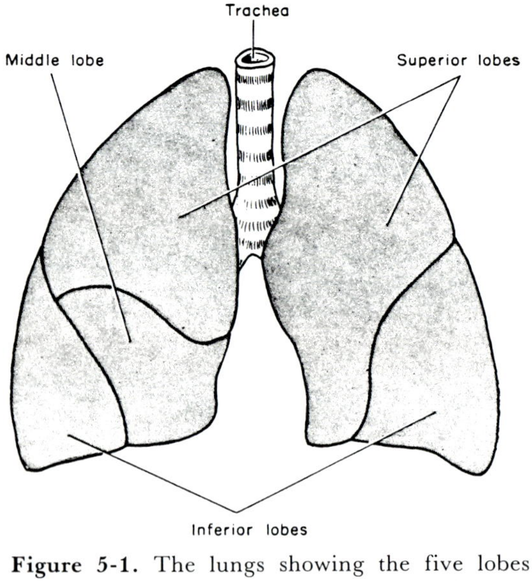

Figure 5-1. The lungs showing the five lobes (From Brooks, Stewart M.: *Basic Science and the Human Body.* St. Louis, 1975, The C. V. Mosby Co.)

The color of the lungs at birth is white to pinkish white. With age, they gradually turn a darker color, first a slate gray and then darker. This is due to contamination from the atmosphere which is inhaled and lodges in the lung tissue in increasing amounts. Male lungs are usually darker than female lungs.

The two lungs represent the basic organs of respiration, each occupying one side of the thoracic cavity. The *apex* of each lung points upward and fits into a space bounded by the upper ribs, the sternum, and the vertebrae. At the *base*, the lungs are concave and ride on the dome-shaped diaphragm. The general configuration of the lungs is similar to bisected cones (Fig. 5-2).

A space called the mediastinum separates the two lungs. Lying in it are the heart, large vessels, the esophagus, some nerves, and part of the network of branching tubes that carry air into and out of the lungs. Both lungs contribute to the *cardiac impression*, a deep cavity providing space for the heart between the lungs. The surface of each lung facing the mediastinum is termed the *mediastinal surface*. In the center of the mediastinal surface is an area called the root of the lung, or *hilum*, which admits the vascular and respiratory structures (Fig. 5-3).

The convex surface of the lung next to the ribs is known as the *costal* or rib surface and follows the rib cage back to the vertebral column. The costal surface area is the largest of the lung surfaces (Fig. 5-4).

On the costal surface of each lung are deep grooves or fissures that divide the lung into lobes (see Fig. 5-2). Beginning near the midbase of each lung and traveling in a superolateral direction, cutting deep into each lung, is the *oblique fissure*. An additional fissure, the *transverse fissure*, is present on the right lung. On the left lung, the oblique fissure divides the upper and lower lobes of the lung. The transverse fissure of the right lung adds a third lobe to this lung, the middle lobe, which separates the upper and lower lobes. There are, then, a total of five lobes.

Surrounding each lung is an important double-walled serous membrane known as the pleura. The outer layer lines the thoracic cavity and may be termed the *costal pleura*. The inner layer, or *pulmonary*

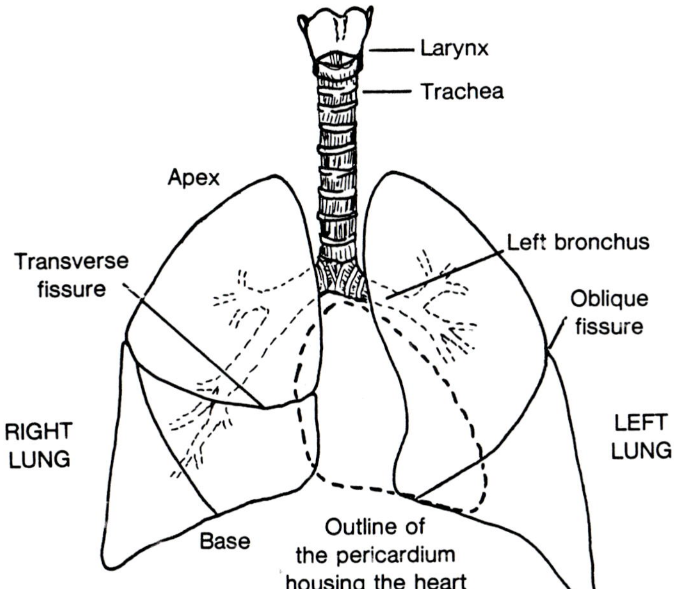

Figure 5-2. The tracheobronchial connections to the lungs. Structures associated with the lungs. Note the outline of the heart and the bronchi. The fissures of the lungs are also identified.

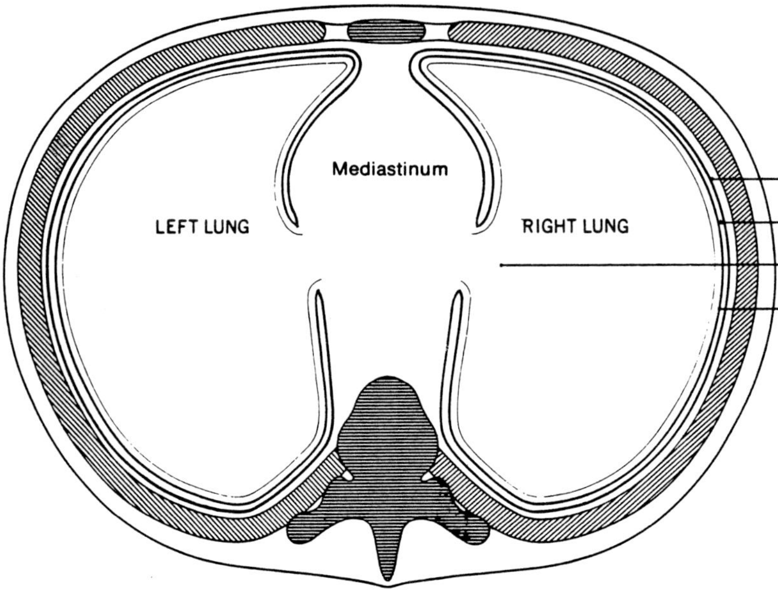

Figure 5-3. The linings of the lungs. At the root of the lung, the bronchus and blood vessels are admitted. (From *Man, Microbes, and Matter* by B.C. Block. Copyright© 1975. Used with the permission of McGraw-Hill Book Company.)

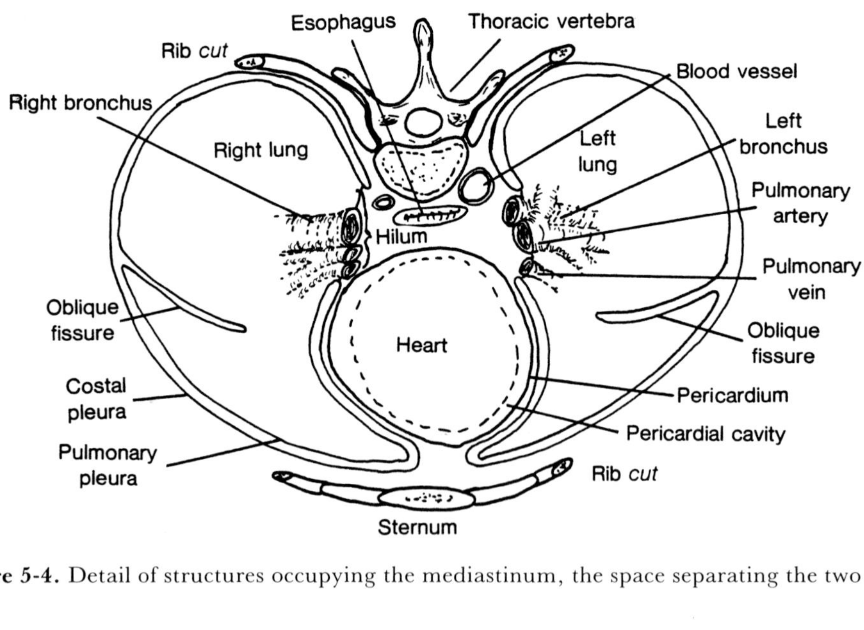

Figure 5-4. Detail of structures occupying the mediastinum, the space separating the two lungs.

pleura, adheres firmly to the lung. With each lung invested by its own pleura, a puncture through the chest cavity into one lung will result in a collapse of only the punctured lung (Fig. 5-5).

The pleural cavity is the potential, not

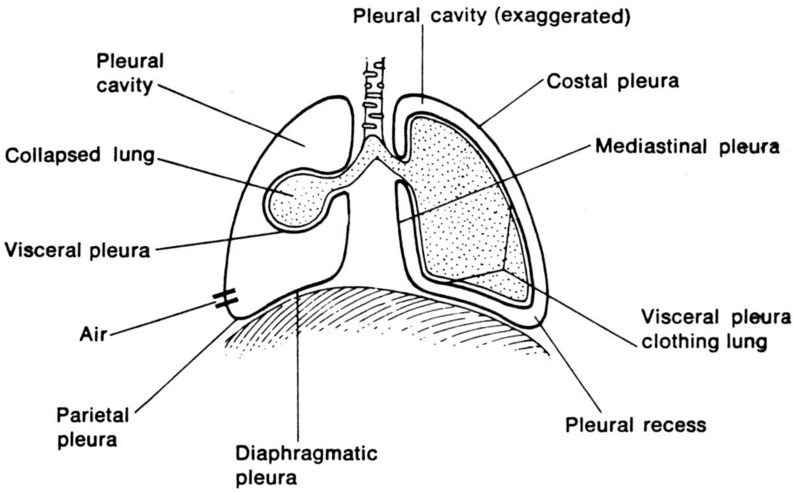

Figure 5-5. Collapse of a single lung. Puncture of one lung may not affect the other due to the separate double-wall pleura membranes investing each lung. (From *It's Your Body* by L.M. Elson. Copyright© 1975. Used with the permission of McGraw-Hill Book Company.)

actual, space between the double-walled membrane, or pleura. Serous fluid secreted by the cells of the serous membrane provides an antifriction lubricant, permitting the layers to move freely against each other with each phase of respiration.

Pleurisy is a familiar term referring to a disorder involving inflammation of the pleural membranes, also known as *pleuritis*. During the illness, an increase of fluid may pervade the pleural cavity, resulting in the layers adhering to each other. Each respiratory movement may be extremely painful, and in some cases, surgery may be required to separate the serous membranes.

A pneumothorax is the puncture of a lung's pleural lining, resulting in its collapse. Purposely creating a pneumothorax may be indicated as medical therapy for treatment of tuberculosis or similar illnesses. The collapsed lung apparently is able to benefit from the rest while the other lung takes on the total job of gaseous exchange, i.e. the exchange of carbon dioxide for oxygen. After the opening into the pleural cavity is closed, the collapsed lung slowly expands and returns to a functional state.

Atelectasis refers to collapse of the lung or a part of it. The condition may be congenital, arising from lack of inflation after lung development. If the lack of inflation is extensive, in the absence of successful intervention, the baby dies soon after birth. In other cases, the cause of collapse may be due to increased surface tension between the alveoli, the smallest compartments within the lungs. The normally moist alveoli would collapse due to surface tension of the moist film if it were not for surfactant, a substance which decreases surface tension. A small needle will float in a cup of tap water until a drop of liquid detergent, the surfactant, breaks up the strength of the surface tension. The delicate alveoli, then, do not collapse because of reduced surface tension. Sometimes babies, particularly those who are premature, are born without adequate quantities of surfactant. They experience respiratory distress that may prove fatal. This condition is known as *hyaline membrane disease*.

THE BRONCHIAL TREE

Entering each lung are many small branches of a larger trunk (Fig. 5-6). The major trunk is the *trachea*, which arises out of the thoracic cavity carrying air into and out of the thorax. Two primary divisions, the *primary bronchi*, arise from the lower end of the trachea. As the bronchi extend deeper into the lungs, they branch profusely, diminishing in diameter with each successive branching.

The primary bronchi have walls made up of a series of incomplete rings of cartilage, fibrous tissue, muscular fibers, and an internal lining of mucous membrane. The right bronchus is larger in diameter and straighter than its mate. However, the left bronchus is about twice as long, passing under the aortic arch, the large artery connecting the heart with the major arterial trunk of the body.

From the right primary bronchus, three *secondary bronchi* extend into the respective lobes of the right lung. The left primary bronchus divides into but two secondary bronchi. Then, each secondary bronchus proceeds to divide into smaller passages, the *tertiary bronchi*. Next, the *bronchioles* follow as branching of the tertiary bronchi.

After further branching, the bronchioles give rise to the *terminal bronchioles*. The ciliated epithelial lining of the respiratory tract ends in the terminal bronchioles, which is also the termination of the bronchial tree proper (Fig. 5-6).

Each terminal bronchiole divides repeatedly. Nonetheless, an important principle is observed. The cross-sectional area of any given subdivision is greater than the cross-sectional area of the parent division. This design leads to a tremendous functional capacity for gaseous exchange, which is a fundamental basis for the maintenance of life processes. From the terminal bronchioles, the very small *respiratory bronchioles* branch. They are usually less than 1 mm in diameter and contain no cartilage in their walls.

The final subdivision in the lower airways is found as the small resiratory bronchioles terminate with *alveolar ducts*, which open into minute *alveolar sacs* having small depressions called *aveoli*. The alveoli, like a complex maze with borders composed of short pots, each with its large mouth facing the passage running by it, are the locus of the end of the lower respiratory passage. The alveoli are like leaves on a tree. Viewed from the standpoint of the speech and hearing specialist, the alveoli represent, in one sense, the beginning of the vocal tract; it is from here that an air energy source emerges that can excite the vocal folds or otherwise eventually be changed into speech sounds by the articulators.

The elaborate capillary network surrounding each alveolus provides the mechanism for transference of oxygen to the red blood cells in exchange for carbon dioxide. The separating boundary between the capillary and alveolus is only one cell layer thick.

It is interesting to note that there are over 500 million alveoli in the lungs. Collectively, if laid out side by side, they would form a sheet equivalent to the amount of carpet necessary to cover the area of the floor of a single-car garage.

Except, perhaps, for scattered muscle fibers in the lung walls, the only significant muscle fibers associated with the lungs are smooth muscle fibers in the walls of the bronchioles. Their purpose is to alter the size of these tubes for regulation of the air volume. Except for the elastic recoil of lung tissue after inflation, the lungs are passive structures with complete dependence on surrounding musculature to "pump" the air in and out. Lung mus-

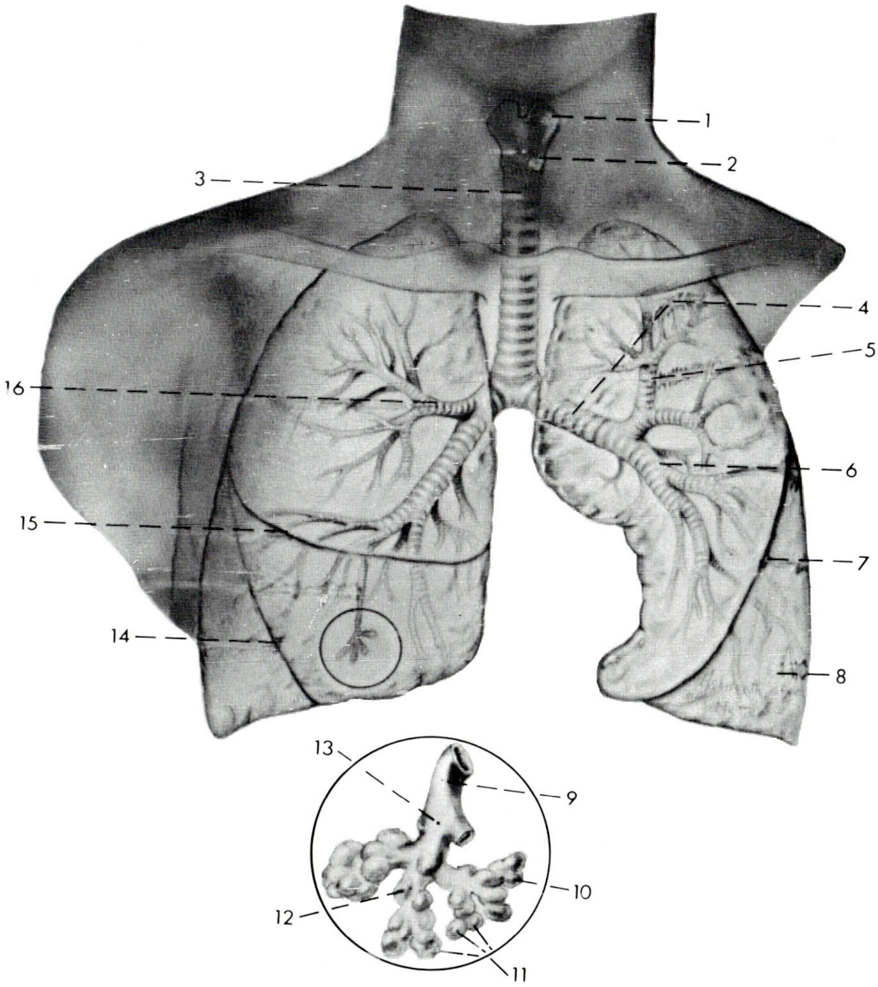

Key:

1. Thyroid cartilage
2. Cricoid cartilage
3. Trachea
4. Left main bronchus
5. Upper lobe bronchus
6. Lower lobe bronchus
7. Oblique fissue
8. Lower lobe
9. Respiratory bronchiole
10. Alveolar sac
11. Alveolus
12. Oblique fissure
13. Horizontal fissure
14. Right upper lobe bronchus

Figure 5-6. The lungs. (From Jacob, S.W., and Francone, C.A.: *Laboratory Manual of Structure and Function in Man*, 5th ed., 1982. Courtesy of W.B. Saunders Co., Philadelphia.)

cles are not responsible for respiration, although lung tissue may assist with expiration.

Whenever there is edema (swelling) in the bronchioles as well as bronchial spasms, the condition results in a marked decrease in the diameter of the bronchial branches. As the victim attempts to overcome the increased resistance to air flow, a dyspnea, or labored, apprehensive breathing, follows. The term designating this problem is *asthma* or, more accurately, *bronchial asthma*. The etiology may be an allergy and a psychological component, singly or in combination.

If some condition serves to block the bronchi, the result is collapse of the alveoli, since the air of the alveoli is absorbed and no new supply can enter. Some of these conditions include inflammation, tumors, foreign bodies, or external pressure caused by an abnormal process such as malignancy of the thyroid gland, which can constrict the trachea in a choking action.

Emphysema is a disorder characterized by the breakdown of the alveolar septa or partitions so that in the place of numerous small alveoli, there are fewer but larger ones. With the loss of each septum, blood vessels servicing the adjacent alveoli are lost. Consequently, the reduced surface area means less diffusion of respiratory gases and decreased vital capacity. This is caused by increased pulmonary resistance when air escapes into interalveolar spaces and is trapped there.

THE TRACHEA

The passage providing for transition of air into and out of the thoracic cavity is the *trachea* (see Fig. 5-6). It is less than 12.5 cm long in the adult and is about 2.5 cm in diameter. Essentially it is a muscular tube whose structural framework is composed of sixteen to twenty C-shaped cartilaginous rings. The framework holds the trachea open, enabling free passage of air through it. On the posterior wall, the trachealis muscle completes the tracheal tube, filling in the gap between the C-rings of cartilage.

From the primary bronchi, the trachea extends superiorly to the cricoid cartilage of the larynx. As previously mentioned, several tracheal rings may be felt, beginning in the depth of the jugular notch and following the trachea upwards.

The trachea is lined inside with a ciliated epithelial mucous membrane. A number of mucous glands secreting mucus, a slime, are located in the submucous layer. This lining moistens incoming air and moves particles out of the trachea with its sweeping cilia (Fig. 5-7).

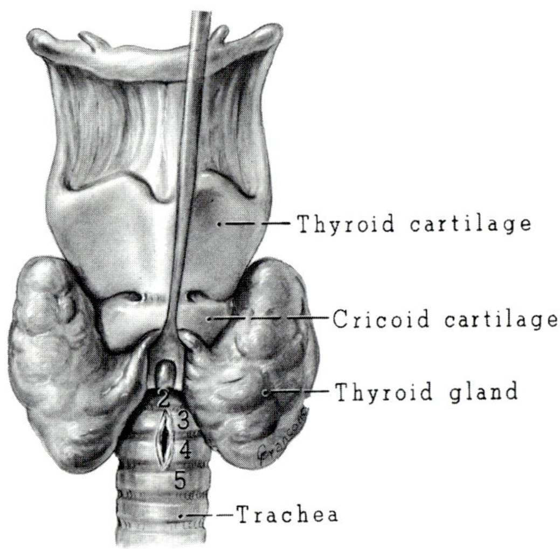

Figure 5-7. Site for incision for tracheotomy. The thyroid gland is retracted superiorly and the incision is made through tracheal cartilages 3-4. (From Jacob, S.W., and Francone, C.A.: *Elements of Anatomy and Physiology*, 1976. Courtesy of W.B. Saunders Co., Philadelphia.)

A tracheotomy is an incision of the trachea through the skin and muscles of the neck. The purpose is usually for exploration, taking a biopsy specimen, or removal of a tumor. Surgical creation of an opening into the trachea to establish an alternate breathing passage is termed a *tracheostomy*. A tracheostomy tube is inserted to facilitate respiration through the opening. During total removal of the larynx (laryngectomy), the tracheostomy is permanently created in the jugular notch (Fig. 5-8).

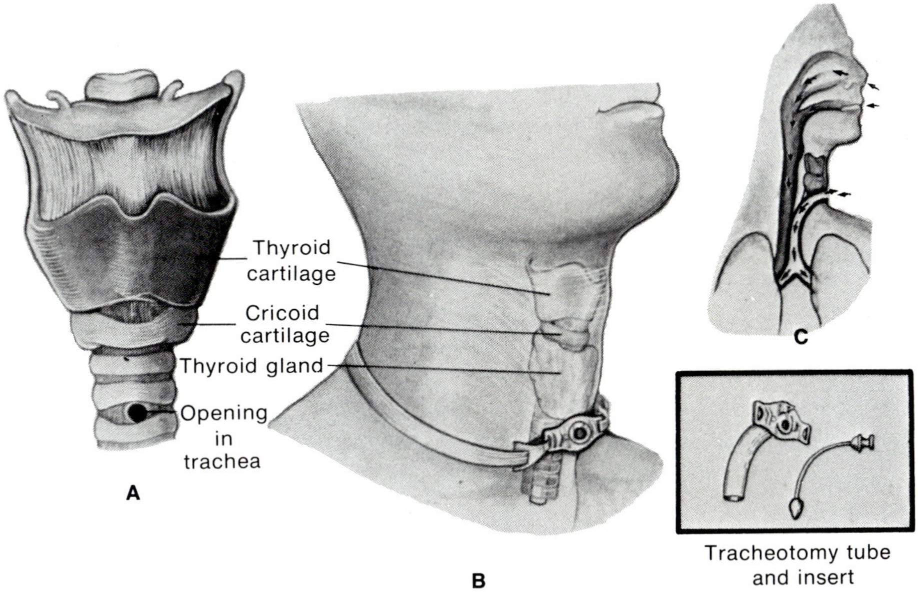

Figure 5-8. Tracheostomy. An opening (A) is made into the trachea (see Fig. 5-7) and a tube is inserted to allow the patient to breathe (B). The tube permits free access to outer air (C). (From Anderson, P.D.: *Clinical Anatomy and Physiology for Allied Health Sciences*, 1976. Courtesy of W.B. Saunders Co., Philadelphia.)

SUMMARY OF THE LOWER RESPIRATORY PATHWAY

If a gnome sets out to travel the length of the vocal tract, and if he begins in the most remote recesses of the lower vocal tract following the air pathway, his origin would be in the (1) alveoli of the lungs. From here, he moves out into the (2) alveolar sacs and then into the (3) alveolar ducts, which lead into the (4) respiratory bronchioles. The next corridor is in the lower bronchial branches, the (5) terminal bronchioles. Afterwards, the (6) bronchioles are entered, and they lead into the (7) tertiary bronchi, the third major branch of the inverted tree. From the tertiary bronchi, the still larger (8) secondary bronchi are encountered and lead into the (9) primary bronchus of the lung from which the journey began. Finally, the primary bronchus opens into the (10) trachea, the trunk of the tree. From this point, it is a single air duct, the exit of the thoracic cavity, opening into the larynx and the pharynx.

PURPOSES OF BREATHING

Vocalization is a secondary use of the lower respiratory tract. The body's need to expel carbon dioxide-laden air overrides any vocal purposes of respiration. Nevertheless, a cooperative and functional middle ground is readily achieved without reduced efficiency to either's claim on the breathing mechanism. Air is taken in for oxygen replenishment, and carbon dioxide is expelled as an outflowing breath during exhalation. Man employs this latter phase of the biological respiratory process for speech. While the timing of the inhalation and exhalation activities are compromised in favor of speech, the essential bodily needs of gaseous exchange are satisfied.

Without a doubt, sharing of respiratory processes for gaseous ventilation and speech does not imply a simplified muscular activity. Skillful coordination of the muscular control functions of inhalation and exhalation are requisites for healthy and efficient support of speech sound production. There is *no one way* to use the musculature to breathe for speech, yet there are inefficient patterns that provide poor phonatory-breath control.

At this point, something is known of the travel of air in the lower portion of the vocal tract. It should not be forgotten that this aspect of the speech machinery represents the reservoir for initiation of the breath stream, the energy source for operating the laryngeal and supralaryngeal generators to produce sound.

BIBLIOGRAPHY

Basmajian, J.V.: *Primary Anatomy*, 8th ed. Baltimore, Williams and Wilkins, 1982.

Crouch, J.E.: *Functional Human Anatomy*, 3rd ed. Philadelphia, Lea and Febiger, 1978.

Hard, A.: Respiration. In Hamlet, S. (Ed.): *Fundamentals of Hearing and Speech Science*. Columbus, OH: Collegiate Pub, 1973.

O'Rahilly, R.: *Basic Human Anatomy*. Philadelphia, Saunders, 1983.

Paff, G.H.: *Anatomy of the Head and Neck*. Philadelphia, Saunders, 1973.

Romanes, G.J. (Ed.): *Cunningham's Textbook of Anatomy*, 12th ed. New York, Oxford U Pr, 1981.

Scott, J.H., and Dixon, A.D.: *Anatomy for Students of Dentistry*, 4th ed. London, Churchill, 1978.

Snell, R.S.: *Clinical Anatomy for Medical Students*, 2nd ed. Boston, Little, 1981.

Spalteholz, W., and Spanner, R.: *Atlas of Human Anatomy*, 16th ed. Philadelphia, Davis, 1967.

Woodburne, R.T.: *Essentials of Human Anatomy*, 6th ed. New York, Oxford U Pr, 1978.

Chapter 6

RESPIRATION FOR SPEECH

BREATHING for speech requires a change in the normal pattern of respiration. Quiet breathing does not require the sophisticated muscular activity used to power the phonatory-articulatory modulators. More than the inherent elasticity of the expanded lungs is needed to collapse the lungs and to propel a pressure stream of air up through the vocal tract. For this, muscles of the abdominal walls and thoracic cage are employed.

The key factor in breathing for speech is controlled exhalation. Such control is taken for granted in most people, but many neurologically impaired individuals have great difficulty in regulating speech breathing. Occasionally also, individuals with normal neurologic systems exhibit poor breathing patterns in speech. While breathing strictly for intake of oxygen and discharge of carbon dioxide, the intake phase is about as long as the outgoing phase, even during vigorous activity such as running. During speaking or singing, the exhalation phase is quite long in proportion to the inhalation phase. Thus, the rather even cyclic pattern of breathing is interrupted whenever the expired air stream is required to serve as the power process for speech. In essence, the rate and pressure of the expired air is metered according to the demands of the vocal tract. The skeletal muscles of the abdominal and thoracic divisions of the body are responsible for this control.

Some of the musculature of the rib cage and the back is difficult to classify according to its role in respiration. If the basic principles of breathing are understood first, perhaps it will be less bewildering to consider those particular muscles lacking clear classification. Accordingly, the basic mechanics of breathing will be outlined first, followed by a description of muscle function.

RESPIRATORY CAPACITIES

The total volume of the fully expanded lungs for the average normal individual is about 6000 ml, or just over six liters of air (Fig. 6-1, Table I). *Vital capacity* is the maximum quantity of air that can be moved into or out of the lungs. This

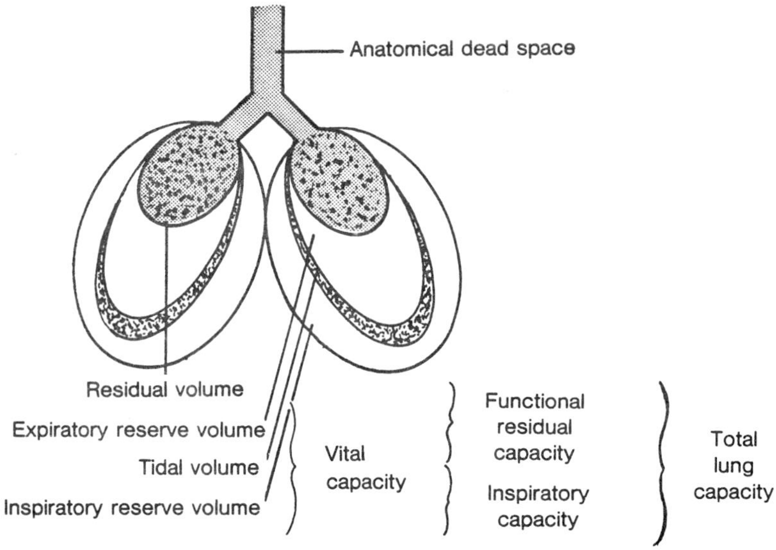

Figure 6-1. Schematic illustration of lung volumes and capacities.

figure averages about 4500 ml. It is less than the total lung capacity because there is always still some air in the lungs even after maximal expiration effort.

Thoracic activity is not sufficiently effective to result in complete collapse of the lungs during expiration. That volume remaining after maximal expiration is termed *residual volume*, which amounts to about 1500 ml (4500 + 1500 = 6000 ml) (Fig. 6-1).

The *tidal volume* is that quantity of air normally inhaled or exhaled with each breath while at rest. It amounts to about 500 ml on the average.

After normal expiration, an additional quantity of air can be forcibly expelled. This additional volume, about 1000 ml, is known as the *expiratory reserve volume*. The *inspiratory reserve volume* is similar. It is the quantity of air that can be inhaled after normal inspiration. This additional effort of inhalation will bring into the lungs about 3000 ml of air above the tidal volume inspired.

The *functional residual capacity* is obtained by adding the expiratory reserve volume and the residual volume (1000 + 1500 = 2500 ml). This capacity describes the volume of the lungs after normal expiration.

Since there is no provision for gaseous exchange above the terminal bronchioles, the volume of air contained in these passages is called the *anatomical dead space*. This quantity is, on the average, about 150 ml. Simple arithmetic indicates that only about 350 ml of air of the tidal volume reaches the alveoli during inspiration (500 - 150 = 350 ml). This concept can be made clearer by having a subject breathe through a tube, which will result in extension of the dead space and less effective gaseous exchange in the alveoli. By lengthening the tube until its volume exceeds the quantity of the tidal volume, there will be no renewal of air in the alveoli (Table 6-I).

In addition, *physiological dead space* is also present, hampering air exchange just as anatomical dead space interferes with respiration. Physiological dead space is that space within the lungs containing air of the same composition as room air just

Table 6-I

LUNG CAPACITIES AND RESPIRATORY AIR VOLUMES

Volume	*Quantity of Air*	*Description*
Total Lung Capacity (TLC)	6,000 cc	Total volume of air that the lungs can hold: TLC = VC + RV
Vital Capacity (VC)	4,500 cc	Maximum volume of air that can be exhaled after taking the deepest breath possible: VC = TV + IRV + ERV
Residual Volume (RV)	1,500 cc	Volume that remains in the lungs at all times
Tidal Volume (TV)	500 cc	Volume moved in or out of the lungs during quiet breathing
Inspiratory Reserve Volume (IRV)	3,000 cc	Volume that can be inhaled during forced breathing in addition to tidal volume
Expiratory Reserve Volume (ERV)	1,000 cc	Volume that can be exhaled during forced breathing in addition to tidal volume
Functional Residual Capacity	2,500 cc	Quantity of air left in lungs following normal expiration (ERV + RV)
Inspiratory Capacity	3,500 cc	Quantity of air that can be drawn into the lungs at the end of expiration (TV + IRV)
Anatomical Dead Space	150 cc	Volume of air contained in the respiratory passageways — nostrils down to terminal bronchioles — where no gaseous exchange occurs

prior to expiration. This means that nonfunctional alveoli may carry out little or no exchange of carbon dioxide for oxygen during the time the air was within the alveoli prior to the exhalation phase. Hence, the air that was taken into the alveoli is sent out during expiration without being adequately supplied with opportunity for air transference due to postural or other mechanical inefficiencies. Pulmonary diseases that interfere with activity of the alveoli result in the need for a greater tidal volume to provide sufficient air for basic respiratory purposes. The breathing rate of these individuals may be more than double as compared with persons having normal lungs.

BREATHING PHYSIOLOGY

The ventilation system of the body functions principally to renew the air in the lungs by exchanging oxygen for carbon dioxide in the alveoli. Breathing is the mechanical operation of moving air into and out of the lungs. At rest, the adult human breathes at the rate of sixteen to twenty times per minute. Children breathe more rapidly, and older adults breathe more slowly.

There are two phases in respiration that comprise each breathing cycle: inspiration and expiration. During inhalation the size of the thoracic cavity is increased, and during exhalation the size of the thoracic cavity is decreased. These two points are basic to respiration.

As a means of communicating information about breathing, terminology has been developed to describe various physiological properties of respiration. The following terminology is useful in appreciating breathing stages and in describing aberrant breathing patterns.

Inspiration

While at rest and during quiet breathing, an individual inhales just over a pint or one-half liter (500 ml) per inspiratory phase. This is termed *tidal air* (tidal volume refers to either the inspiratory or expiratory phase of quiet respiration). Unfilled space remains in the lungs after normal inhalation. An additional quantity can thus be inhaled at the end of the normal inspiratory phase. This added quantity is termed the *inspiratory reserve volume* and amounts to about 2½ to 3 liters (2,500 to 3,000 ml), or 5 to 6 pints of air. After vigorous exercise, the body requires an increased supply of oxygen and demonstrates a need to reduce the elevated quantity of carbon dioxide in the bloodstream. Naturally, the tidal volume is greater under these circumstances. At its maximum, the inspiratory capacity is placed at just over 3½ liters (3,500 ml), which is calculated by adding the tidal volume and the inspiratory reserve volume (see Table 6-I).

Sucking air into the lungs is accomplished by creating a partial vacuum in the thoracic cavity. Drawing serum into a hypodermic syringe illustrates how a partial vacuum pulls liquid or gas into an enclosure. The respiratory airways from the head to the lungs, including the bronchial tree, represent the needle-in-the-syringe analogy. The chest with the hollow lungs serves as the barrel or chamber of the syringe. Below, the diaphragm is the plunger, moving downward to increase the volume of the chest cavity, which draws air into the lungs.

The surface area of the dome-shaped portion between the abdominal and thoracic cavities, the diaphragm, is broad. If the diaphragm moves about a half inch or slightly more, as it typically does in quiet respiration, it will enlarge the thoracic capacity such to account for about 75 to 80 percent of the inspired volume. For increased inspiratory requirements, the diaphragm may descend about 3 inches.

As the diaphragm drops lower during contraction, intra-abdominal pressure rises. To relieve this contradicting pressure, the muscles of the abdominal wall make a reciprocal response by relaxing. This may continue up to a point. When the diaphragm descends to a level such that the abdominal muscles can relax no further and the liver, stomach, intestines, and other abdominal organs are compressed and are stable against the bony pelvic basin and backbone, additional lowering of the diaphragm is not possible. This can be observed by standing erect, inhaling deeply and concentrating on abdominal action while keeping thoracic movements to an absolute minimum. The anterior abdominal wall will protrude until a "bottoming effect" is felt.

If the diaphragm accounts for 75 to 80 percent of the work for inspiration, there must be another mechanism for further increasing the volume of the thoracic cavity in order to supply the 20 to 25 percent additional inspiration volume. That additional aspect of the inspiratory pump is served by rib movement. The ribs are hinged at their articulations with the backbone. The rest position is a forward and downward slant. Upon being raised by the associated musculature, the capacity of the thoracic cavity is increased, with the same effect as pulling on the plunger of the syringe. The lungs take in air and expand to fill the enlarged thorax. It should be apparent that inspiration is a function of both the diaphragm and rib cage musculature.

Forced inspiration brings a maximum increase in the capacity of the thorax. All muscles capable of raising the ribs are utilized, including the sternocleidomastoideus and other neck muscles. Muscles already involved become more active. By this action, the inspiratory capacity is reached.

A final statement on inspiration must be made for emphasis. The lungs and chest do not expand because air is entering during inspiration. In reality, it is the reverse. As the thoracic cavity is made larger by action of the diaphragm and rib cage muscles, intrapulmonic (within the lungs) pressure falls below atmospheric pressure. Atmospheric air moves into this low or negative pressure area, the lungs. Thus, the difference between atmospheric and intrapulmonic pressure is the force moving air into the lungs. All of this is possible by virtue of a fluid seal between the outer lung surface and the inner wall of the thoracic cavity, called the *pleural space* (Fig. 6-2).

Expiration

At the end of the inhalation phase, the lungs are stretched. Since the thoracic cavity capacity decreases during expiration, the intrapulmonic pressure rises above atmospheric pressure. This creates a pressure difference exactly opposite of that found during inspiration. As a result, air moves out of the lungs, just as serum leaves the hollow needle of a syringe during movement of the plunger down the barrel.

Expiration may be passive or forced. If *passive*, the contraction of the stretched lungs stays ahead of the collapsing movement of the thorax. This elastic recoil of the lungs, coupled with the relaxation of muscles of the rib cage, appears to be caused by the inhibition of the nerve impulses responsible for the inspiratory movements. *Quiet expiration* utilizes the economy of effort found in passive expiration and is characterized by increased tone of the anterior abdominal wall musculature to force the relaxing diaphragm back to its upward position. Some muscles may assist in this expiratory activity by pulling down on the ribs.

Forced expiration involves the active contraction of the anterior abdominal wall muscles and the quadratus lumborum, which pull down on the lowest ribs to force constriction of the rib cage. Other muscles may be involved in concert with those mentioned.

A normal expiration phase moves about a pint (½ liter or 500 ml) of air and, like a normal inspiration phase, is termed *tidal volume*. At the end of a normal expiration, it is possible to continue expelling air by a forceful effort. This additional quantity, about a liter, is known as the *expiratory reserve volume*. Fortunately, the lungs are not completely collapsed by activity of the

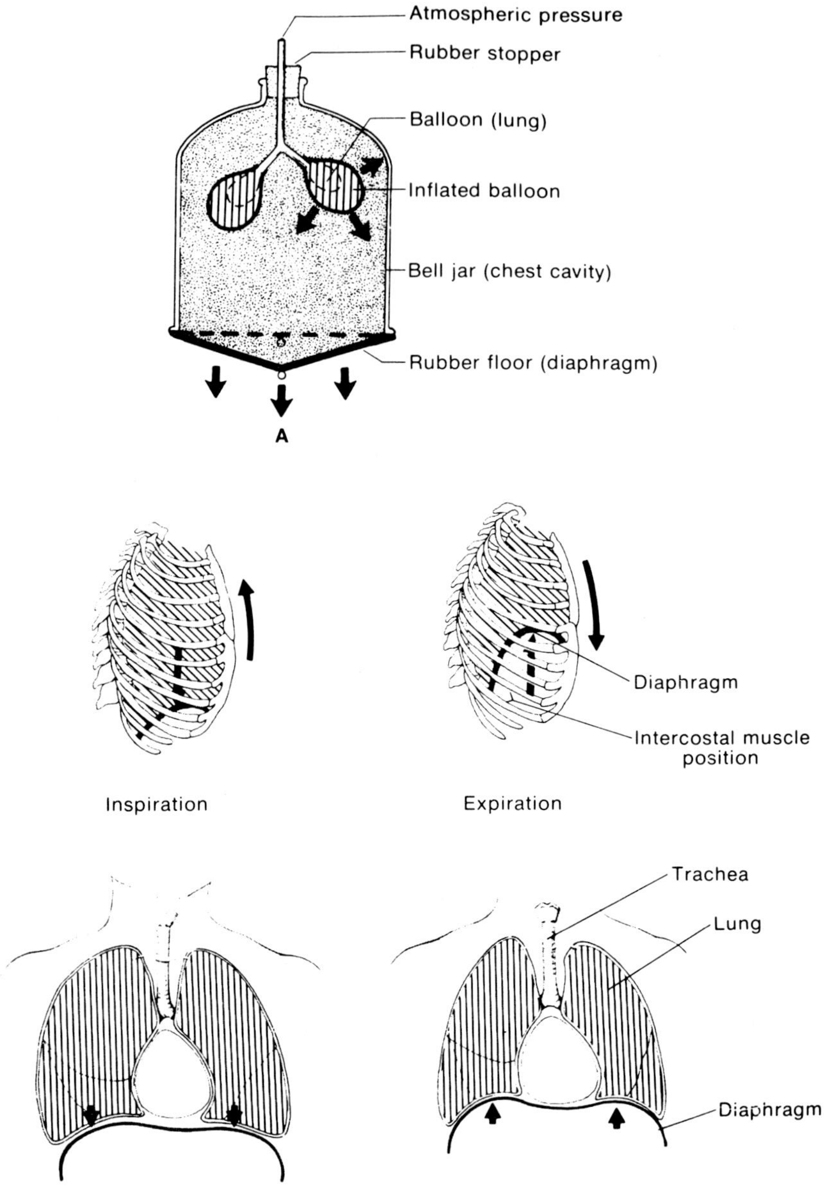

Figure 6-2. Principle for inhalation and exhalation. (From Anderson, P.D.: *Clinical Anatomy and Physiology for Allied Health Sciences*, 1976. Courtesy of W.B. Saunders Co., Philadelphia.)

muscles of exhalation. The air remaining in the lungs after maximum expiration amounts to about 1½ liters, or 1,500 ml, and is termed the *residual volume*. The functional residual capacity is that combined quantity of air in the residual and expiratory reserve volumes. By comparing the expiratory reserve volume with the inspiratory reserve volume, it is apparent that it is possible to inhale more deeply than to exhale. This should not be perplexing, since the two measurements are not taken one after the other; in other words, each is taken after the tidal volume, or after a normal inspiration or expiration phase.

The *functional residual volume* is made up of the expiratory reserve volume and the residual volume. Added together, these two latter volumes total about 2,500 ml of air left in the lungs (1,000 ml + 1,500 = 2,500 ml). The functional residual capacity is that quantity of air remaining in the lungs at the end of normal expiration.

Total lung capacity is about 6 liters of air for adult males and about 1 to 2 liters less in females. Vital capacity is the total volume of air that can be expelled with maximum expiration after maximum inspiration. This capacity also varies for males and females, being 3 to 5 liters in males and 2 to 3 liters in females.

MUSCLES OF RESPIRATION

Most muscles associated with respiration are active during portions of both inspiration and expiration. For example, some muscles of inhalation may act as checks against how rapidly the rib cage descends to a constricted state during expiration.

It is convenient to describe individual muscles according to primary function. Each muscle will be detailed according to the primary phase of breathing involved, with accessory muscles included.

Muscles of Inspiration

Quiet Inspiration

The diaphragm and the external intercostal muscles are the primary muscles involved during quiet inspiration. These muscles are usually sufficient for most conditions of quiet breathing. They may serve this function working as a team, or either one may be employed without loss of effectiveness under this mode of breathing.

Diaphragm

As previously described, this is a broad sheet of muscle and tendon separating the thoracic and abdominal cavities. On top of the diaphragm, the bases of the lungs ride, conforming to the curvature of the diaphragm. It presents two domes, a right and left. The dome is slightly higher on the right side, reaching the upper border of the fifth rib (Table 6-II).

The muscle fibers of the diaphragm take their origins from three parts around the circumference of the inside of the thoracic cavity. After arising from the circumference of the diaphragm, the fibers converge medially to insert into the *central tendon*. There are openings piercing the diaphragm for the passage of the esophagus, blood vessels, and nerves (Fig. 6-3).

Upon contraction, the central tendon is pulled downward. This increases the size of the thoracic cavity, since the bottom has, in effect, been lowered. In this role, it is a muscle of inspiration. Since the abdominal muscles relax and bulge in response to the lowering of the diaphragm, this has been termed *abdominal breathing*, but is well known as diaphragmatic breathing also. By contracting the abdominal muscles simultaneously with the diaphragm, high intra-abdominal pressures are possible. This may serve to brace the vertebral column as in weight lifting or contribute to abdominal viscera discharge of urine (micturition), evacuation of fecal material from the rectum

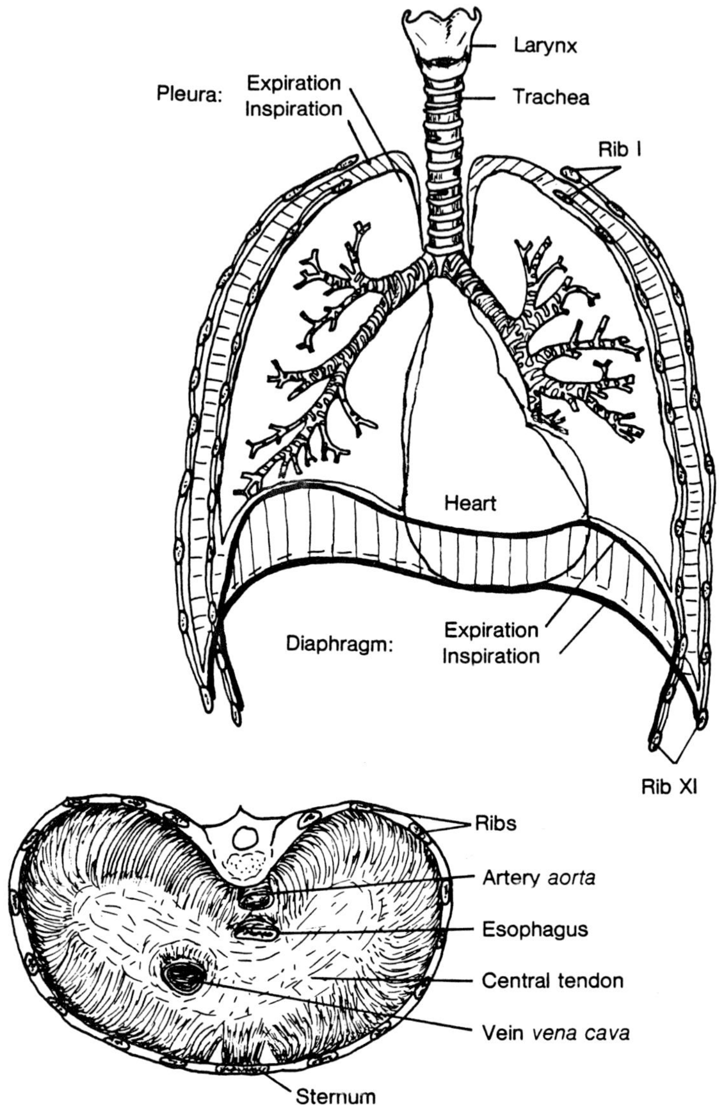

Figure 6-3. (Top) Frontal section of the thoracic wall, pleural cavity and diaphragm during expiration and inspiration. (Bottom) View of the diaphragm from above. The heart is located above the central tendon. (From Spalteholz, W., and Spanner, R.: *Atlas of Human Anatomy*, 16th ed., 1961. Courtesy of Bohn, Scheltema and Holkema, Amsterdam.)

(defecation), vomiting (emesis), or expelling a baby from the birth canal (parturition).

The diaphragm has been named as the second most important muscle of the body, the heart being first. The diaphragm can be viewed as the most important skeletal muscle in man. While its role is designated as inspiration, it is possible to use the diaphragm while in the reclining position as a check muscle of expiration. In this role, the abdominal viscera

Table 6-II

MUSCLES OF INSPIRATION

Muscle	*Origin*	*Insertion*	*Action*
Diaphragm	a. xiphoid process b. lower 6 ribs c. lumbar vertebrae	central tendon	lowers central tendon; essential muscle of inspiration
External intercostals	rib above	rib immediately below	pulls ribs together elevating rib cage
Scalenes	C2 – C6 vertebrae	first and second ribs	elevates first and second ribs
Sternocleidomastoid	sternum and clavicle	mastoid process	with head fixed, raises sternum
Subclavius	first rib	clavicle	with clavicle fixed, raises first rib
Pectoralis major	medial half of clavicle, sternum, and costal cartilage	humerus	with arm fixed, expands rib cage
Pectoralis minor	ribs 2-5	scapula	with shoulder fixed, elevates rib cage
Costal elevators	C7 – T11 vertebrae	rib just below vertebrae from which it takes its origin	raises ribs
Serratus posterior superior	C7 – T3 vertebrae	ribs 2-5	raises ribs
Serratus anterior	first 8 ribs	scapula	with shoulder fixed, raises ribs
Latissimus dorsi	iliac crest, T6 – T12 vertebrae, lumbar fascia, lower 3-4 ribs	humerus	with arm fixed, elevates rib cage
Trapezius	superior nuchal line, ligamentum nuchae, C7 – T12 vertebrae	scapula and clavicle	muscle of fixation to brace head and shoulder while other muscles of inspiration move the rib cage

are pulled back into place both by gravitational force and elastic recoil of tissues.

The phrenic nerve provides innervation to the diaphragm. It arises as a branch of the cervical plexus from spinal cord segment C3 to C5.

External Intercostals

Enlarging the thoracic cavity by elevating the rib wall, the external intercostals comprise the other principal musculature for inhalation. Still, the intercostals are overshadowed by the action of the diaphragm in quiet inspiratory movements. But a combination of diaphragmatic and costal inhalation is often present. It is possible, however, to use only the intercostals, to the exclusion of the diaphragm. Voluntary control is clearly present in these two means of enlarging the thoracic cavity. One can actually move the thorax and diaphragm in opposing directions during breathing.

The external intercostals are oblique bands of muscle fibers coursing from the edge of the rib above to the edge of the rib immediately below. In essence, they fill the space between the ribs, being assisted in this "filler function" by the internal intercostals. During inhalation, these intercostals stiffen the thoracic spaces between the ribs to prevent the tissue from being sucked inward or bulging outward (Fig. 6-4).

Contraction elevates the rib directly below, which has the effect of lifting the entire rib cage. The downward-sloping ribs hinge at the backbone, and the anterior-posterior diameter of the rib cage is increased. Costal breathing, then, alters the lateral dimensions of the thorax, while diaphragmatic breathing alters the vertical dimension, i.e. in the normal inhalation phase.

Perhaps it is appropriate to point out here that there is no one way to breathe. During inhalation, men tend to be diaphragmatic breathers, while women tend to be thoracic or costal breathers. Asking individuals to speak from the diaphragm as a technique to improve the speaking or singing voice is probably without any real evidential support from the standpoint of anatomy and physiology. For one, the diaphragm is sparsely innervated and an individual has no actual proprioceptive sense as to the position of the diaphragm in the body. What breathing exercises accomplish appears questionable. As will be seen later, coordination of the forced expiratory air stream with the phonatory mechanism is the crucial measure of the power process for speech.

Forced Inspiration

Deep inspiration may be thought of as an extension of quiet inspiration. All movements detailed under quiet inspiration are increased so that a more vigorous effort is made to draw air into the lungs.

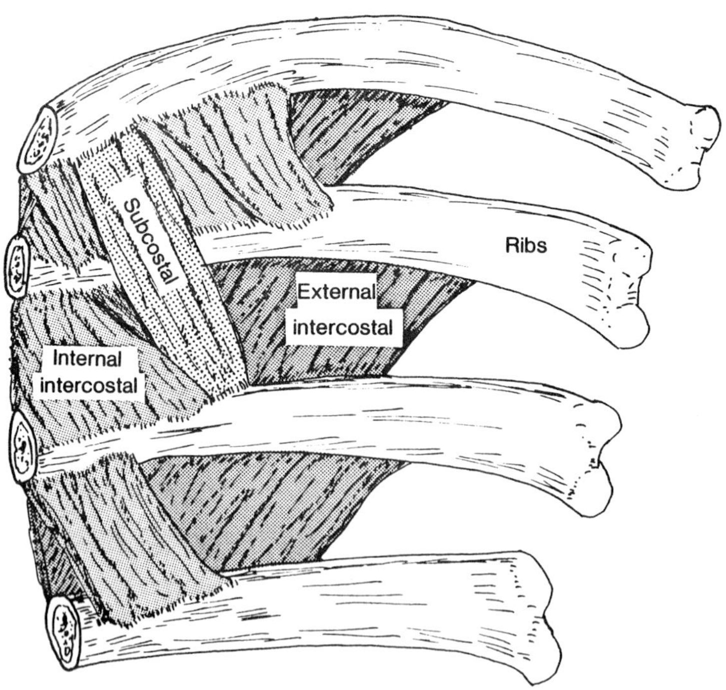

Figure 6-4. View of the internal surface of the ribs showing the intercostal muscles.

Forced inspiration should be thought of as being on a continuum from the quietest of inspiratory movements to maximum heaving, gasping actions of inspiration, as in cases of air starvation.

The depth of inspiration is determined by a basic biologic need for oxygen. To effect these various levels of forced inspiration, accessory muscles are brought into use. Essentially, they further expand the lateral, vertical, and anterior-posterior dimensions of the thoracic cavity. Each of these inspiratory accessory muscles is detailed below.

Scalenus

There are three sets of muscles, the *scaleni*, located in a lateral position in the neck. All originate on the cervical vertebrae and course to the rib cage at its top. Each group of these small muscles and the members of this group are named after their location. There is great variability among individuals in the origin and insertion of the scaleni.

The action of the anterior and medius scalenus increases the girth of the rib cage, reducing intrathoracic pressure to effect movement of air into the lungs to equalize air pressure inside and outside the thorax. Scalenus posterior elevates the second rib, which may assist in lifting the ribs below for increasing the capacity of the thoracic cavity.

Generally it is thought that the scaleni are active accessory muscles of inspiration, helping to raise the rib cage to its maximum position. This is particularly true in conditions of respiratory distress, when all muscles that affect the rib cage are brought into use. These muscles may be considered companions to the sternocleidomastoid, which is considered next (Fig. 6-5).

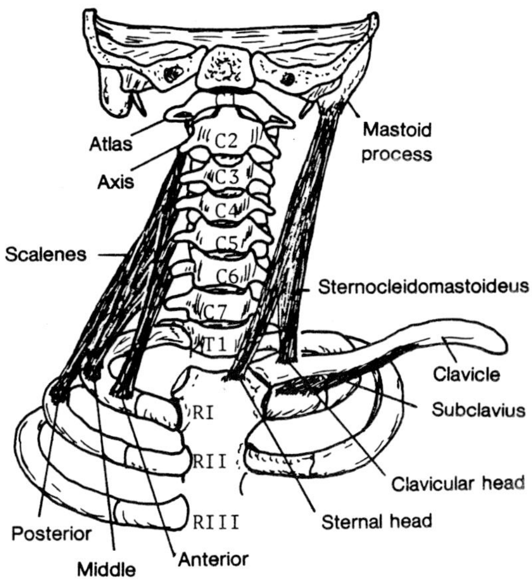

Figure 6-5. Three upper-placed muscles associated with respiration: scalenes, sternocleidomastoid, and subclavius.

Sternocleidomastoideus

This strong, superficial muscle of the neck is readily palpable from its double-headed origin on the sternum and clavicle up to the mastoid process of the temporal bone. It lies on the lateral portion of the neck, bilaterally.

The action of the sternocleidomastoid, both muscles contracting, raises the sternum and its attached ribs if the head and neck are fixated by other muscles. Otherwise, the sternocleidomastoid in unilateral contraction may serve the function of turning the head so that the eyes will look in the direction opposite the active muscle. This muscle is particularly significant in producing an elevated rib cage for inspiration, as in the clavicular type of breathing. The effect on the rib cage is to increase its A-P dimensions, measured from the sternum to the vertebral column. Again, with both muscles active and the rib cage fixed, the cervical vertebrae are flexed to bring the head forward and elevate the chin. This is probably the most frequent position of the head as one clasps his hands in back of his head and leans back in pensive thought, staring at the ceiling.

As the most important accessory muscle of inspiration, the sternocleidomastoid works in conjunction with the scaleni to move air into the lungs during deep inspiration. During distressful breathing, these muscles are observed contributing as inspiratory muscles. It should be pointed out that if the rib cage is raised by these neck muscles, the abdominal muscles cannot relax sufficiently to allow the descent of the contracting diaphragm. That is, the vertical distance between the sternum and pubic symphysis, for example, has been increased, causing the abdominal muscles to be stretched and disallowing them from bellying anteriorly as they normally do when the diaphragm is lowered and compresses the viscera of the abdomen. It is likely that insufficient quantities of air are inhaled (for speech or singing purposes) under such circumstances. Certain cases of cerebral palsy are plagued by prominent patterns of clavicular-type breathing, resulting in considerable fragmentation of the speech output from the standpoints of vocal quality, prosody, and rhythm. The problems are insufficient quantities of air inhaled for speech and excessive tension in the neck area resulting in poor phonatory-breath coordination.

Congenital shortening of the sternocleidomastoid is called *torticollis*, or wry neck. The muscle is injured at the time of birth and turns into a fibrous cord that is unable to lengthen as the neck grows. This situation is usually correctable to some extent with surgery.

Minor Muscles in Forced Inspiration

The next group of accessory muscles of inhalation assisting in producing forced inhalation involves several muscles of the back and thorax. These muscles, some of which are controversial concerning their role in breathing, coordinate and provide the vigorous actions during forced inspiration. These muscles pertain to rib cage movement in lateral and anteroposterior expansion. A number of these muscles depend upon fixation by other muscles in order that the pertinent muscles may act on the rib cage. Some of the muscles considered here are, at least at times, fixators for accessory inspiratory muscles.

Normal inhalation prior to exhalation for speech purposes does not depend upon this group of muscles unless perhaps the individual is speaking under stressful conditions, as may sometimes be observed in stutterers. These accessory muscles will be briefly discussed individually. It is noted that all are paired.

Subclavius

Figure 6-5 illustrates the position and attachments of this muscle. If the clavicle is fixed, the subclavius will raise the first rib and thereby increase the thoracic cavity slightly. In this aspect, the subclavius assists in inspiration. Otherwise, the clavicle will be pulled downward, drawing the shoulder forward upon contraction.

Pectoralis Major

Pectoralis major and latissimus dorsi have the same insertion, the humerus. Upon contraction, this muscle will pull the sternum and ribs upward if the arm is held firmly. This may be demonstrated by grasping onto the seat of one's chair and forcefully inhaling. Typically, the action is directed in the opposite direction, i.e. the arm is moved when not anchored by holding onto something. As the rib cage is moved, the thorax is increased in capacity (Fig. 6-6).

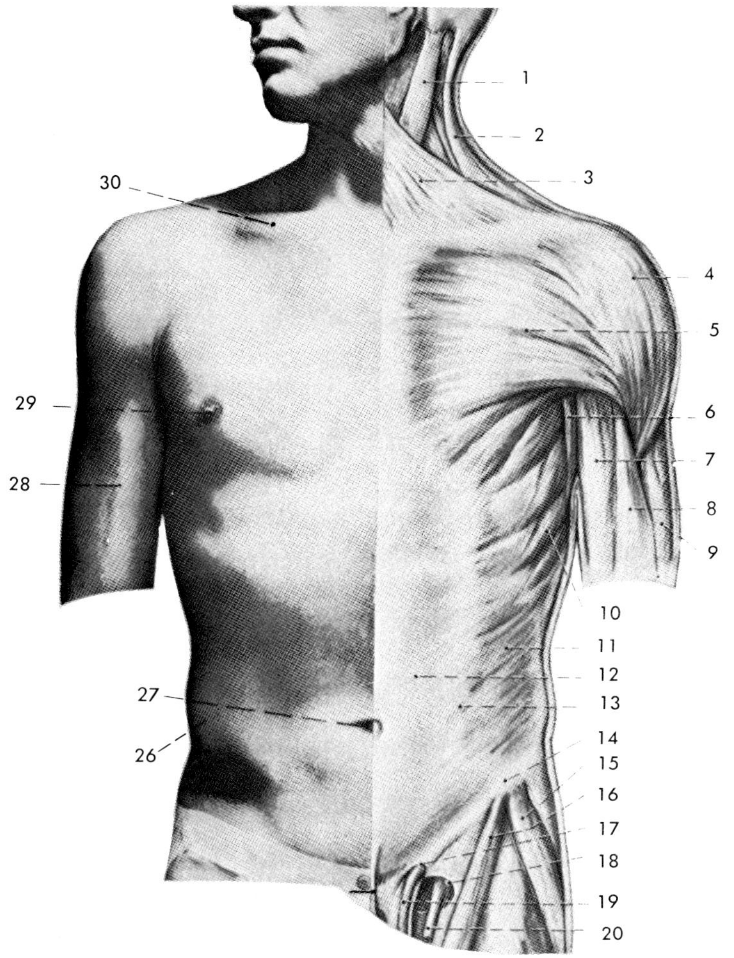

Key:

1. Sternocleidomastoid m.
2. Trapezius m.
3. Platysma m.
4. Deltoid m.
5. Pectoralis major m.
6. Latissimus dorsi m.
7. Short head biceps brachii m.
8. Longhead biceps brachii m.
9. Lateral head of triceps m.
10. Serratus anterior m.
11. External oblique m.
12. Fascia of rectus abdominis m.
13. Linea semilunaris
14. Anterior superior iliac spine
15. Tensor fascia lata m.
16. Sartorius m.
17. Superior inguinal ring
18. Saphenous opening (for veins of leg)
19. Spermatic cord
20. Great or long saphenous v.
26. Iliac crest
27. Umbilicus
28. Cephalic vein
29. Nipple
30. Clavicle

Figure 6-6. Superficial muscles of the neck, rib cage, and abdomen. (From Jacob, S.W., and Francone, C.A.: *Laboratory Manual of Structure and Function in Man*, 5th ed., 1982. Courtesy of W.B. Saunders Co., Philadelphia.)

Pectoralis Minor

This muscle lies under pectoralis major. With the scapula fixed by antagonistic back muscles, the pectoralis minor will elevate ribs 3 through 5.

Levatores Costarum (costal elevators)

There are twelve pairs of this relatively short muscle of the posterior aspect of the rib cage. These muscles raise the ribs upon contraction to enlarge the thoracic cavity. Spinal intercostal nerves innervate the costal elevators.

Serratus Posterior Superior

This muscle is found on the upper back (Fig. 6-11). Serratus posterior superior elevates the posterior portion of those ribs listed to increase the thoracic cavity.

Serratus Anterior

Serratus anterior draws the scapula forward; if the scapula is held in place by other muscles, the upper ribs will be

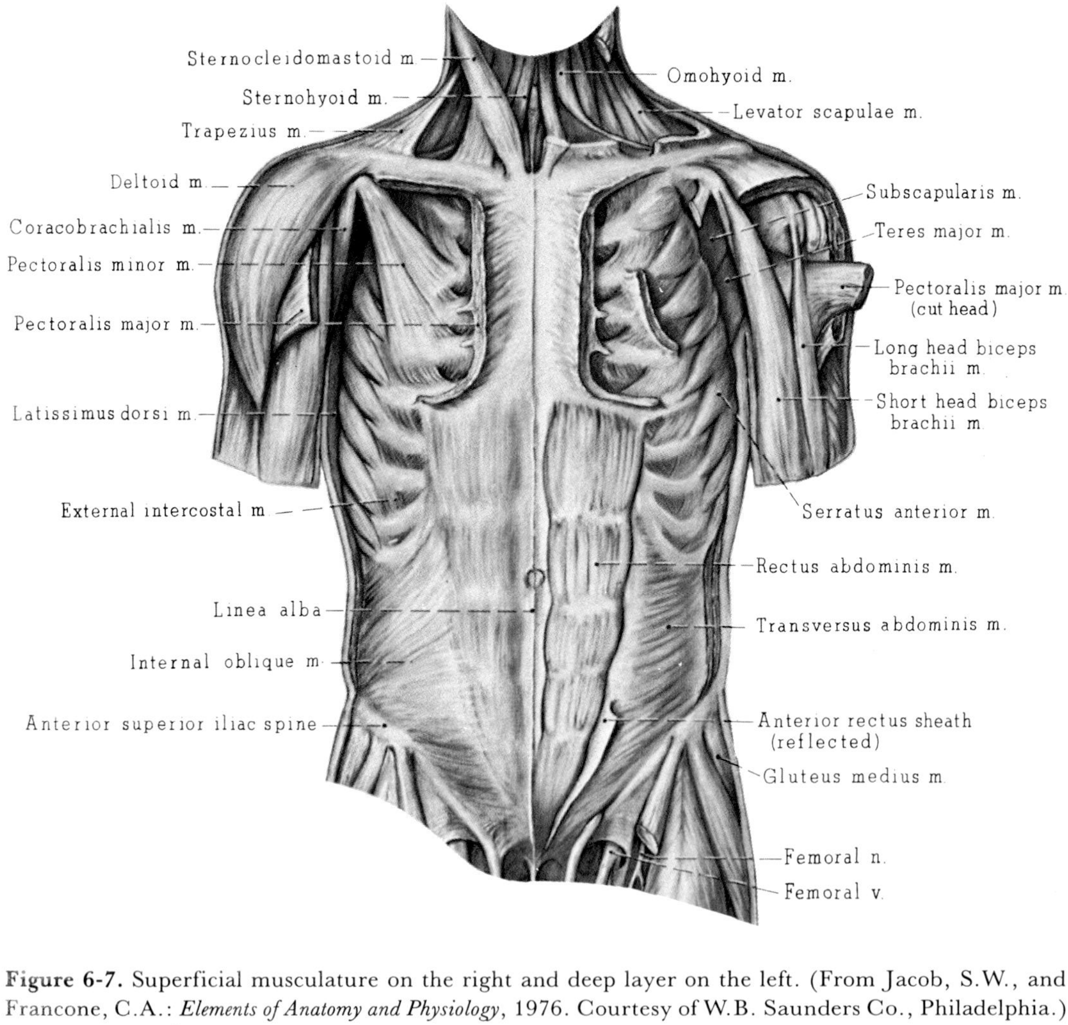

Figure 6-7. Superficial musculature on the right and deep layer on the left. (From Jacob, S.W., and Francone, C.A.: *Elements of Anatomy and Physiology*, 1976. Courtesy of W.B. Saunders Co., Philadelphia.)

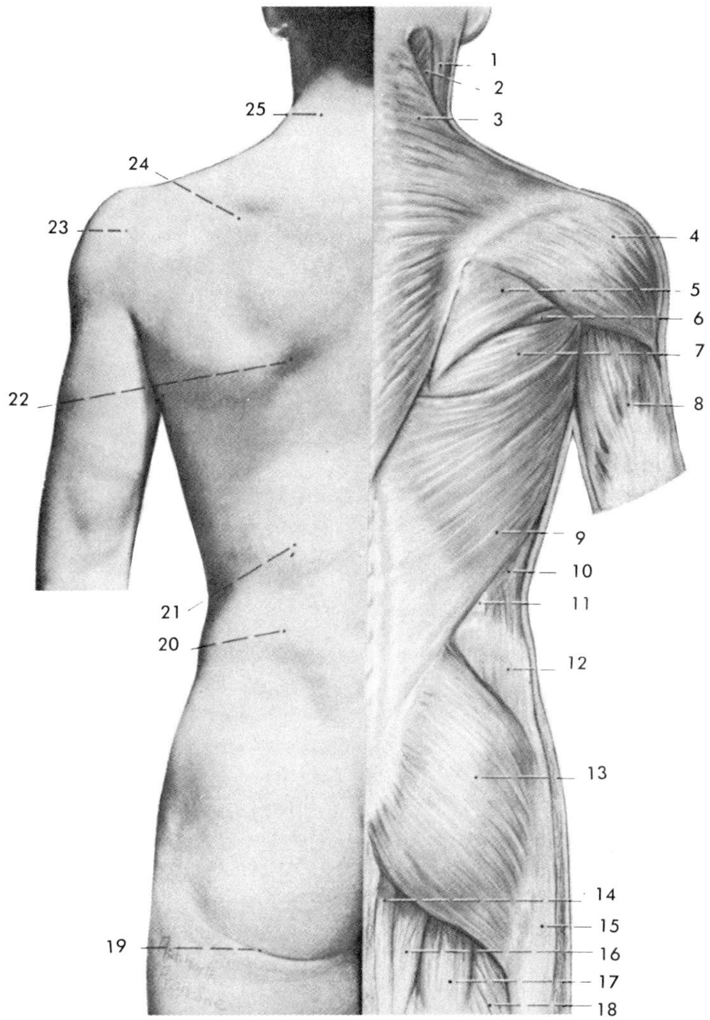

Key:

1. Sternocleidomastoid m.
2. Levator scapulae
3. Trapezius m.
4. Deltoid m.
5. Infraspinatus m.
6. Teres minor m.
7. Teres major m.
8. Triceps m.
9. Latissimus dorsi m.
10. External oblique m.
11. Lumbar triangle
12. Fascia of gluteus medius m.
13. Gluteus maximus m.
14. Gracilis m.
15. Iliotibial tract
16. Adductor magnus m.
17. Semitendinous m.
18. Biceps femoris m.
19. Fold of buttock
20. Iliac crest
21. Latissimus dorsi m.
22. Medial border of scapula and auscultatory triangle
23. Deltoid m.
24. Spine of scapula
25. Trapezius m.

Figure 6-8.-Posterior surface of the male with the outer layer of muscles exposed. (From Jacob, S.W., and Francone, C.A.: *Laboratory Manual of Structure and Function in Man*, 5th ed., 1982. Courtesy of W.B. Saunders Co., Philadelphia.)

raised. This latter action enlarges the thoracic cavity for inspiration purposes. This muscle is not given a role of much importance as an accessory muscle of inhalation (Fig. 6-7).

Latissimus Dorsi

Like other muscles that have been designated as accessory inspiratory muscles, latissimus dorsi is primarily responsible for other functions, namely, extending, adducting, and rotating the arm medially or drawing the shoulder back and down. If the humerus is braced, however, those fibers attached to the lower ribs will raise them to expand the thorax. As will be seen later, this muscle is also listed as a muscle of expiration (Fig. 6-8).

Trapezius

The trapezius is a very broad muscle of the upper part of the back. Upon contraction, the trapezius has as its main function the adduction, rotation, and elevation of the scapula or the drawing back of the head. Secondarily, it serves as a muscle of fixation to brace the head and shoulder, while the sternocleidomastoid and/or pectoralis muscles elevate the ribs during forced inspiration (Fig. 6-8).

Muscles of Expiration

The principal requirement for expiration is that there be sufficient pressure, quantity of air, and control if the expiratory air stream is to serve as an adequate energy source for speech or singing. In the expiratory phase, a number of the inspiratory muscles may act as check muscles for metering the breath stream according to needs for phonation and articulation. In essence, such a check action retards the rate of expiration as required to maintain a dynamic breath-pressure stream to produce in rapid succession the various combinations of high- and low-pressure speech sounds. But there are several important muscles directly involved in forced exhalation that have not been listed in this examination of the inspiratory musculature. They are presented next.

Quiet Expiration

A description of the mechanism for quiet or passive expiration was given earlier in this chapter. Essentially, this involves the elastic recoil of the tissues, including muscles as well as lung tissue, which has the effect of restoring the normal resting position of the thorax and abdomen. This compressive force on the lungs is sufficient to move air from them.

Forced Expiration

Innumerable gradations exist from quiet expiration to the point where maximum intrathoracic pressures are obtained. Ultimate expiratory pressures such as in sneezing and coughing are included in this section.

The abdominal musculature represents a significant role in expiration from quiet to forced conditions. The abdominal muscles may be designated as the most important muscles of forced expiration. There are four of these muscles and all are paired. All serve an identical function in forced expiration by compressing the abdominal contents to push the diaphragm against the base of the lungs.

Rectus Abdominis

Twin belt-like muscles on either side of the *linea alba* and the umbilicus run from the anterior portion of the pelvic girdle up to the large anterior notch of the rib cage. This pair of muscles is *rectus abdominis*. The origin is the crest of the pubic bone and the pubic symphysis. It courses vertically up until it reaches the xiphoid process and the seventh, sixth, and fifth costal cartilages, the insertion points (Fig. 6-9).

Three transverse fibrous bands characterize this muscle. Two are located at the umbilicus and at the tip of the xiphoid; the third is halfway between these two. These may be identified in muscular subjects tensing the anterior abdominal wall, as in sit-ups. The transverse bands are known as *tendinous intersections* or *inscriptions*. They do not break the vertical muscle fibers except in rare instances.

Also visible in muscular subjects is the lateral margin of rectus abdominis. It may be identified as a curved line along the mid and upper portion of the abdominal surface. This landmark is known as *linea semilunaris*. Linea alba divides the two recti muscles and is seen as an indentation between the two muscles during their contraction. It is a cord of fibrous tissue, with the umbilicus in its midpart.

A sheath completely encases this muscle. The sheath is not binding on the actions of the muscle, however, but does hold it in position. Formation of the sheath is from the aponeuroses of the three lateral abdominal muscles. Abdominal surgical incisions are readily tolerated if the cut is made in consideration of the aponeurosis and the margins of the abdominal muscles. Midline incisions, especially above the umbilicus, provide for rapid entrance into the abdominal cavity.

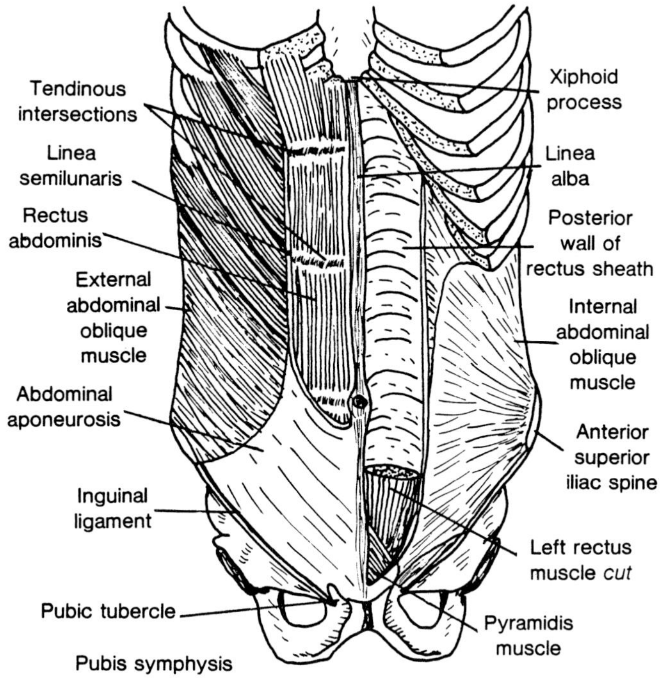

Figure 6-9. Anterior view of the abdominal wall. The rectus sheath has been partially removed to reveal the tendinous intersections of the right rectus abdominis muscle and the posterior wall of the left rectus sheath. The second layer of the abdominal musculature is illustrated by the internal abdominal oblique muscle. (From Snell, R. S.: *Clinical Anatomy for Medical Students*, 2nd ed., 1981. Courtesy of Little, Brown and Co., Boston.)

Table 6-III
MUSCLES OF EXPIRATION

Muscle	*Origin*	*Insertion*	*Action*
Rectus abdominis	xiphoid process, 5-7 costal cartilages	pubic crest, pubic symphysis	compresses abdominal wall, forcing diaphragm upward, or flexes vertebral column
External abdominal oblique	lower 8 ribs	xiphoid process, linea alba, pubic crest, iliac crest	compresses abdominal wall, forcing diaphragm upward
Internal abdominal oblique	inguinal ligament, lumbar fascia, iliac crest	lowest 3-4 costal cartilages, linea alba	compresses abdominal wall, forcing diaphragm upward
Transversus abdominis	inguinal ligament, iliac crest, lumbar fascia, lowest 6 costal cartilages	xiphoid process, linea alba, pubic symphysis	compresses abdominal wall, forcing diaphragm upward
Quadratus lumborum	iliolumbar ligament, iliac crest	12th rib, L1 – L4 vertebrae	pulls last rib down; fixes rib cage inferiorly
Subcostals	near angles on inner surface of ribs	2-3 ribs below origin	pulls ribs downward
Latissimus dorsi	iliac crest, T6 – T12 vertebrae, humerus, lumbar fascia, lower 3-4 ribs	humerus	with arm fixed and the whole muscle acting as a unit, constricts lower part of rib cage
Serratus posterior inferior	T11 – L2 vertebrae	last 4 ribs	pulls last 4 ribs downward
Internal intercostals	rib above	rib immediately below	working with other muscles, lower sections pull ribs downward
Transversus thoracis	lower third of sternum at inner surface	costal cartilage of ribs 2-6	pulls ribs downward and perhaps inward

The function of rectus abdominis, which is pertinent to expiration, is that of tensing the anterior wall of the abdomen. In this role, it assists in compressing the abdominal viscera, which moves the diaphragm cranialward and decreases the thoracic capacity. This assists also in micturition, defecation, emesis, and parturition. All of these acts pertain to compression of the abdominal contents. Other functions are postural in nature and pertain to flexing the vertebral column (Table 6-III).

Obliquus Externus Abdominis

The external oblique, unlike rectus abdominis, is broad and flat. As its name implies, it is the most superficial of the four abdominal muscles. It occupies the lateral and anterior portions of the abdomen and is reinforced by two other flat abdominal muscles underneath. The fibers of this muscle course at a downward slope.

The origin is from eight slips at the outside lower borders of the lower eight ribs. After fanning out, the insertion is made by means of a broad aponeurosis into the xiphoid process, linea alba, pubic crest, and anterior half of the iliac crest. The lower border between the iliac crest and the pubic tubercle of the aponeurosis

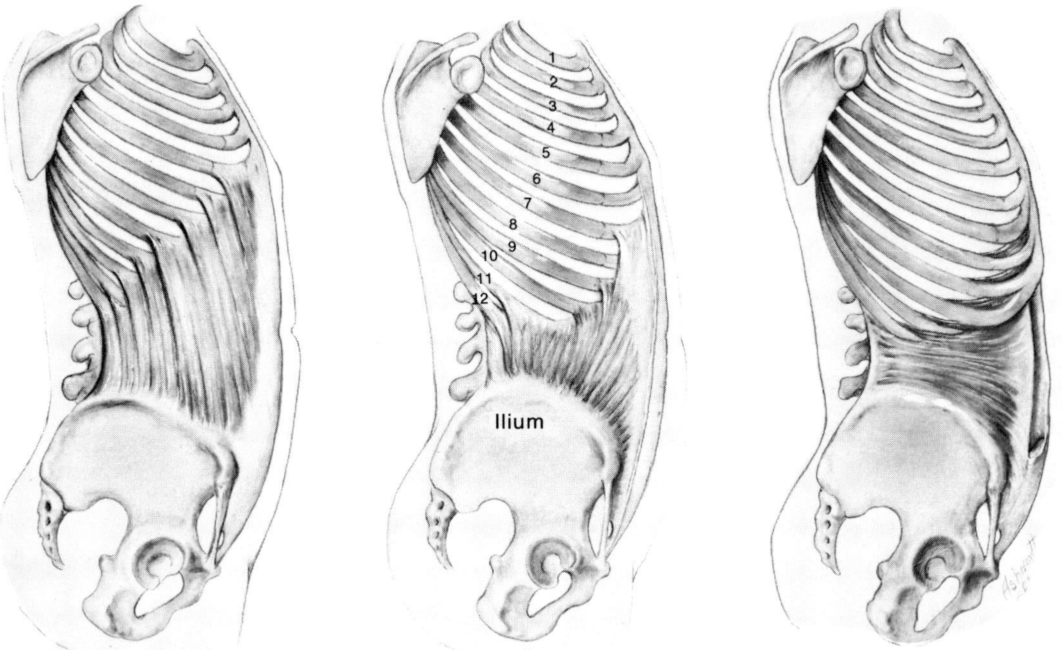

Figure 6-10. Three layers of abdominal musculature. (From Jacob, S.W., and Francone, C.A.: *Structure and Function in Man*, 4th ed., 1978. Courtesy of W.B. Saunders Co., Philadelphia.)

is folded backward on itself to form the inguinal ligament in the groin. At the pubic tubercle and medial to it, there is a break in the aponeurosis for passing the spermatic cord in males or the round ligament of the uterus in females.

The function of this abdominal muscle and its underlying two mates is identical to that of the rectus abdominis. Upon contraction, they constrict the abdomen, compressing its contents. In addition to forced expiration, this action assists in visceral functions outlined in the discussion of rectus abdominis. If the vertebral column is not held rigid, the column may be flexed as the muscles tense (Fig. 6-10).

Obliquus Internus Abdominis

This flat muscle of the abdominal wall is situated between two other such abdominal muscles, external oblique and transverse. It arises from the lateral portion of the inguinal ligament, the lumbar fascia, and the anterior two-thirds of the iliac crest. As indicated by its name, the fibers travel at an incline upward, an angle opposite that of external oblique.

The insertion is into the lowest three to four costal cartilages and the abdominal aponeurosis, which joins it to the linea alba and the other abdominal muscles. Actually, the three-layered abdominal aponeurosis is contributed by the respective three abdominal muscles on the anterolateral wall of the abdomen. And since the fibers of these aponeurotic sheets are like crosshatching, the membrane is strong. Through this wide aponeurosis, then, the muscles reach the midline cord, linea alba.

The internal abdominal oblique assists the other abdominal muscles in forced expiration and the other functions listed in the discussion of the external oblique.

Transversus Abdominis

The broad sheet of muscle fibers underneath the internal abdominal muscle travels in a horizontal plane. The transversus takes its origin from the lateral third of the inguinal ligament, the anteromedial aspect of the iliac crest, the lumbar fascia, and the lowest six costal cartilages. From this extensive origin, the fibers course toward the midline of the abdomen. Its aponeurosis permits insertion into the xiphoid process, the pubic symphysis, and the linea alba through the abdominal aponeurosis. The aponeurosis in question is formed by the three flat muscles of the anterolateral wall of the abdomen, as indicated above, and serves as the tendinous attachment.

As noted earlier, contraction constricts the abdominal viscera by tensing the abdominal wall. This may help move the diaphragm upward to compress the lungs for forced exhalation, or, if the diaphragm is held tense, it may assist in functions of micturition, defecation, emesis, and parturition.

Transversus Thoracis

This paired muscle of expiration is attached on the lower third of the inner surface of the sternum, which is its origin. The fibers, derived from the upper end of the rectus abdominis, slope upward and outward to attach by groups on the inner surface of each of five ribs (2 through 6), the insertion. The action of this muscle pulls the anterior aspect of the ribs downward and inward (toward the backbone), resulting in a decrease in the size of the thoracic cavity, including a reduction of its A-P dimensions. By constricting the rib cage, air is forcibly expelled. Intercostal nerve branches provide the neural supply for this muscle.

Quadratus Lumborum

Named after its quadrilateral shape, this muscle is considered a posterior muscle of the abdomen. The action of the quadratus lumborum, also a paired muscle, pulls the last rib down or fixes the twelfth rib during forced expiration. In pulling the last rib downward, there is direct constriction of the thoracic cage or assistance to other muscles of expiration in the fixation state. It also serves to fix the rib cage from below to enable other muscles to pull against this anchored aspect.

Subcostales

These muscles of the lower part of the posterior of the thorax originate from the inner surfaces near the angles of the ribs. They vary often in number. The course of the fibers is parallel to those of the internal intercostals. Upon contraction, the subcostals draw the ribs together, and with the twelfth rib fixed by quadratus lumborum, they pull the ribs down to constrict the rib cage and assist in forced expiration. They are innervated by the intercostal nerve branches.

Latissimus Dorsi

Arising from the lower portion of the back and inserting into the humerus, this broad triangular muscle of the back has been listed and described earlier as a muscle of inspiration. It is listed here as a muscle of expiration because the muscle may contract as a single sheet, which would constrict the inferior aspect of the rib cage. As a muscle of inspiration, those fibers arising on the lower three to four ribs contract to lift the lower ribs. Therefore, acting in part, latissimus dorsi may

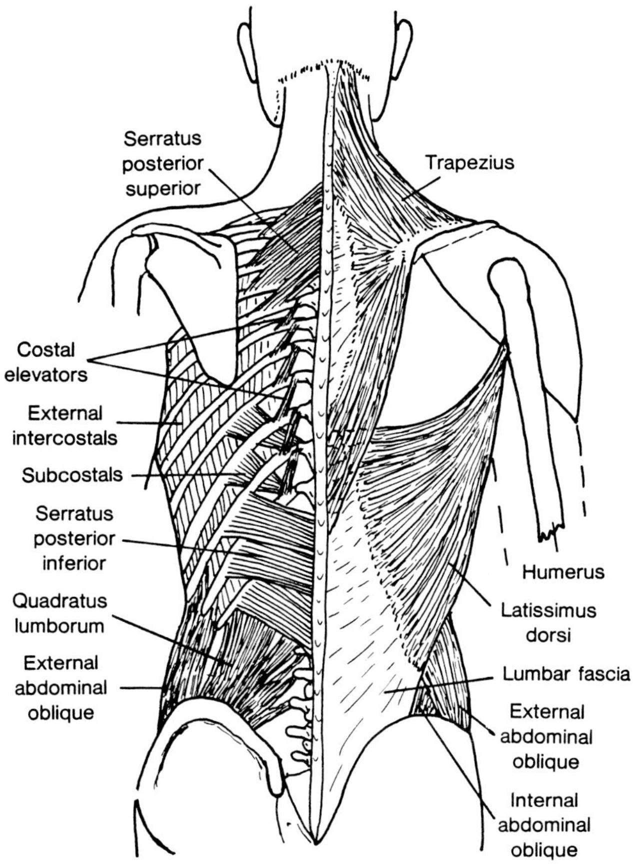

Figure 6-11. The superficial and deep muscles of the back involved in respiration. (From Snell, R.S.: *Clinical Anatomy for Medical Students*, 2nd ed., 1981. Courtesy of Little, Brown and Co., Boston.)

be considered a muscle of inspiration; acting as a unitary structure, it may serve as a muscle of forced expiration (Fig. 6-11).

Serratus Posterior Inferior

This muscle is more extensive than its superior mate, which has already been described as a muscle of inhalation. The location of serratus posterior inferior is on the back at the level of the thorax-lumbar junction. It arises from the thoracic vertebrae T11 and T12 and from the lumbar vertebrae L1 and L2. After passing at an angle upward and outward, it inserts into the inferior borders of the last four ribs.

Upon contraction, this muscle pulls the last four ribs downward and outward to constrict the rib cage at its base. This movement is antagonistic to the diaphragm, which pulls inward on these ribs.

Intercostales Interni

The internal intercostals (the *interni*) lie in the spaces between the ribs to help seal the openings. Each arises from the rib above and courses down and backward before inserting on the rib below. By contrast, it will be recalled that the external intercostals, classified as inspiratory muscles, course down and forward in front of

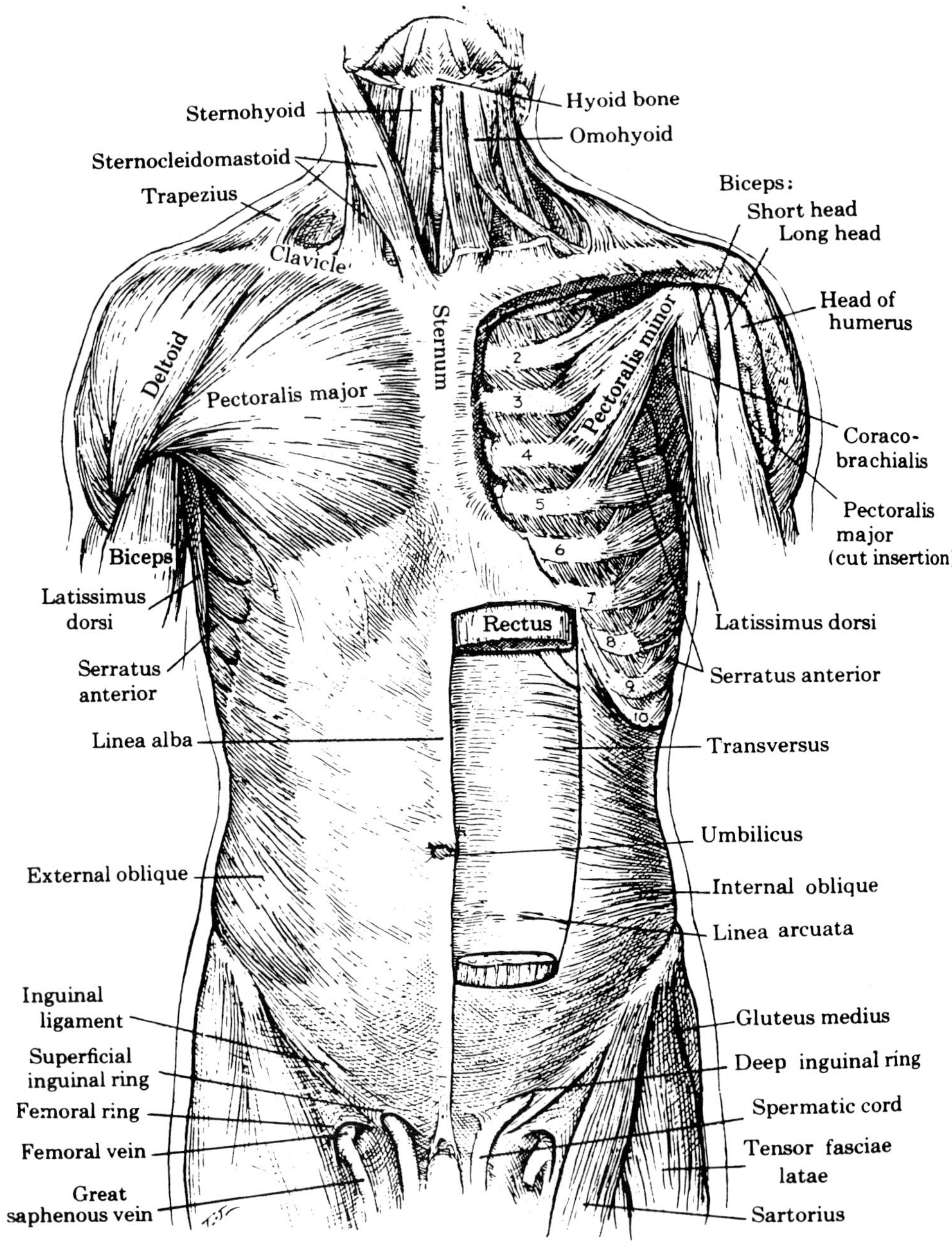

MUSCLES OF TRUNK. ANTERIOR VIEW

Figure 6-12. Musculature of the trunk, outer and deep layers. (From *Dorland's Illustrated Medical Dictionary*, 26th ed., 1981. Courtesy of W.B. Saunders Co., Philadelphia.

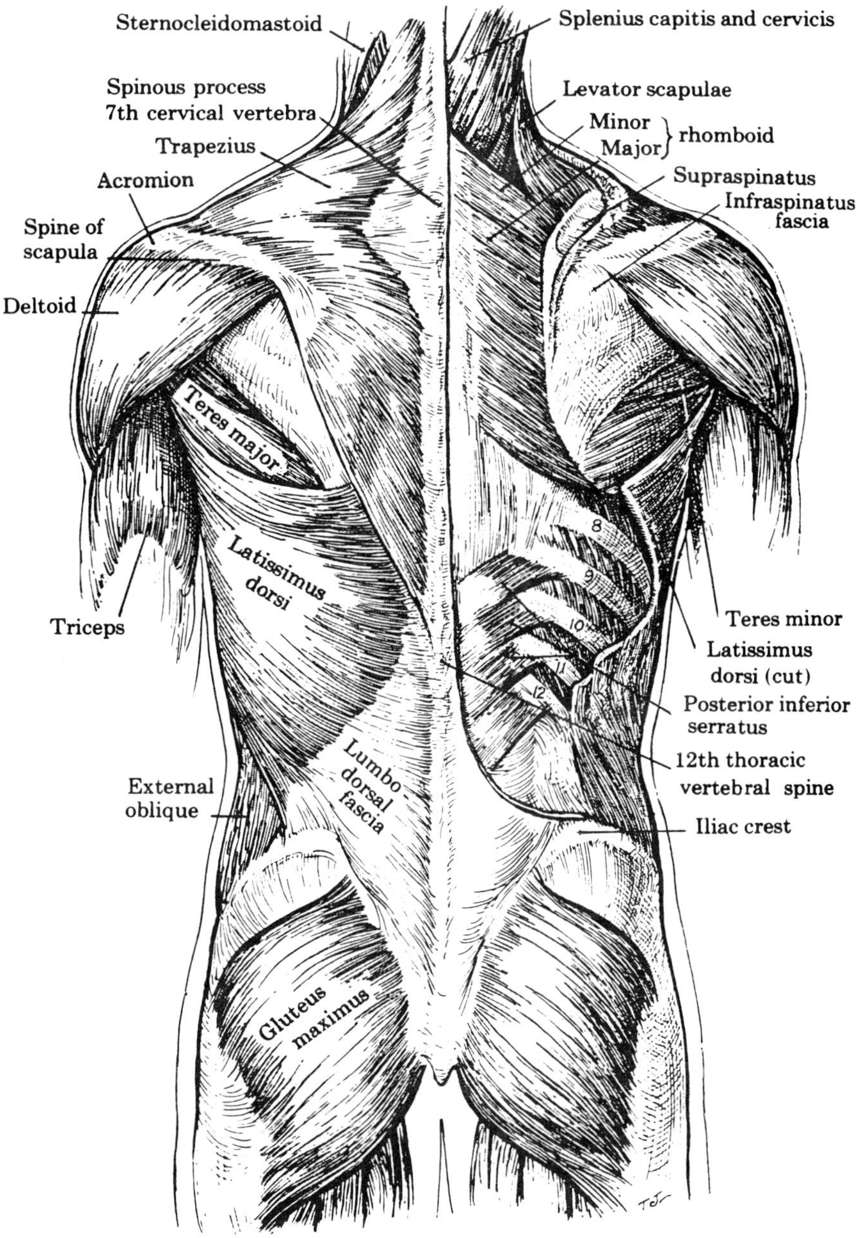

MUSCLES OF TRUNK, POSTERIOR VIEW

Figure 6-13. Musculature of the back, outer and deep layers. (From *Dorland's Illustrated Medical Dictionary*, 26th ed., 1981. Courtesy of W.B. Saunders Co., Philadelphia.)

the internal intercostals.

The internal intercostals, working with other muscles, draw the ribs downward in forced expiration. This collapsing movement of the rib cage forcibly expels air from the lungs. This pattern of action is most effective when quadratus lumborum fixes the twelfth rib and the interni then pull the ribs downward.

In summary, twelve muscles of *inspiration* (see Figs. 6-12 and 6-13) have been described. These included the following:

The diaphragm

Two neck muscles:

- scaleni (collectively)
- sternocleidomastoideus

Five muscles of the rib cage wall:

- subclavius
- serratus anterior
- intercostales externi
- pectoralis major
- pectoralis minor

Two muscles of the back of the rib cage:

- levatores costarum
- latissimus dorsi

Two muscles of the upper back shoulder area:

- serratus posterior superior
- trapezius

Ten muscles of *exhalation* (see Figs. 6-12 and 6-13) were listed and discussed as follows:

Three muscles of the rib cage:

- subcostales
- transversus thoracis
- intercostales interni

Four muscles of the abdominal walls:

- rectus abdominis
- obliquus internus abdominis
- obliquus externus abdominis
- transversus abdominis

Three muscles of the back:

- latissimus dorsi
- serratus posterior inferior
- quadratus lumborum

EXTRARESPIRATORY FUNCTIONS

There are several extrarespiratory functions involving air movement that are facilitated by the muscles of expiration and inspiration. The primary purpose in most of these activities is to move air through the upper respiratory tract without a basic regard for gaseous exchange. Nevertheless, the gaseous exchange rate remains essentially the same during respiration while speaking as during quiet respiration.

Phonation requires a specific volume and force of air movement through the larynx. Since all sounds of American speech are produced normally on expiration, whether voiced or unvoiced, the inhalation phase is for the purpose of taking in a supply of air for phonation. The vocal folds begin to close after the inspiration phase, just as the expiratory air begins moving through the laryngeal cavities. Upon complete adduction, the folds block the air flow until expiratory muscles create sufficient subglottal pressure to overcome the tense, elastic folds. The subsequent opening and closing in rapid succession creates a pulse train of air, a sound wave. The expiratory muscles contract progressively deeper to maintain the required breath energy for operating the glottal generator with a dwindling supply of air. Finally, a quick gulp of air through the mouth recharges the lungs for the next phase of speech.

Coughing and *sneezing* are two other activities utilizing air movement of an extrarespiratory nature. Both activities involve a sharp increase in intrapulmonary pressure so that air is expelled with great force through abducted vocal folds. The blast of air serves to clear the upper respiratory tract. Reflexive coughing may occur due to collection of material in the respiratory passageways that cannot be removed by the cilia. Sneezing, a type of upper respiratory tract cough, may be started by nasal cavity stimulants, mechanical or chemical. When considering the force of the breath stream during coughing and sneezing, it should be clear that the power and effectiveness of the expiratory muscles are more than adequate for providing sufficient minimum pressures for speech function, particularly loudness aspects of speech. It is said that during a cough, germs travel at the rate of 200 miles per hour for the first ten feet. (Remember to cover your mouth around others.)

Spasmodic contractions of the diaphragm produce hiccupping. The sound results from a sharp, involuntary inhalation, which pulls the vocal folds together following the Bernoulli principle and sets them into brief vibration. In essence, a reverse phonatory act is used in hiccupping and includes a modified inspiratory glottal stop. While the cause is unknown, it is well known that breathing in and out of a paper bag is one of a myriad of cure techniques. Elevation of the level of carbon dioxide in the bloodstream through repeatedly breathing the same air appears to alleviate this condition in most cases.

Yawning includes deep inspiration followed by deep exhalation, which, no doubt, has a refreshing effect on inadequately ventilated alveoli and perhaps on related vessels holding stagnant blood. How this reflexive air movement is triggered is not clearly known. It may be poorly oxygenated arterial blood or similar needs.

Blood circulation, an extrarespiratory function of a different nature from those above, is assisted by respiration. During the inspiratory phase of breathing, the thoracic cavity creates a suction effect due to increased volume. Air is sucked into the lungs as a result, but venous flow in the neck is mainly dependent upon this phenomenon of suction. Additionally, the root of the lung descends as the trachea moves down as much as two vertebrae during inhalation. Elongation of the bronchi and dilation result, assisting the pulmonary circulation.

Inspiratory snoring usually occurs as a type of palatal phonation. The air stream vibrates the soft palate against the pharyngeal wall through the Bernoulli principle.

MEASUREMENT OF RESPIRATORY FUNCTIONS

Spirometers are used to measure quickly the vital capacity. The procedure involves having the subject inhale maximally, followed by maximal exhalation into the instrument. The volume is read from a graph on the device. Since there are variations that must be considered that effect the vital capacity, i.e. age, sex, size, etc., a formula must be used to normalize the volume of each subject. A volume of 4.5 liters (just under 5 quarts) is average. Decreases in vital capacity often signal distress by the alveoli related to disease, which reduces the functioning lung tissue. Pneumonia, carcinoma, pulmonary congestion, and fibrosis are examples of such diseases. These can lead to dyspnea (labored breathing) and cyanosis. These conditions may gravely interfere with phonatory-breath activity for speech purposes (Fig. 6-14).

The electronic spirometer computes

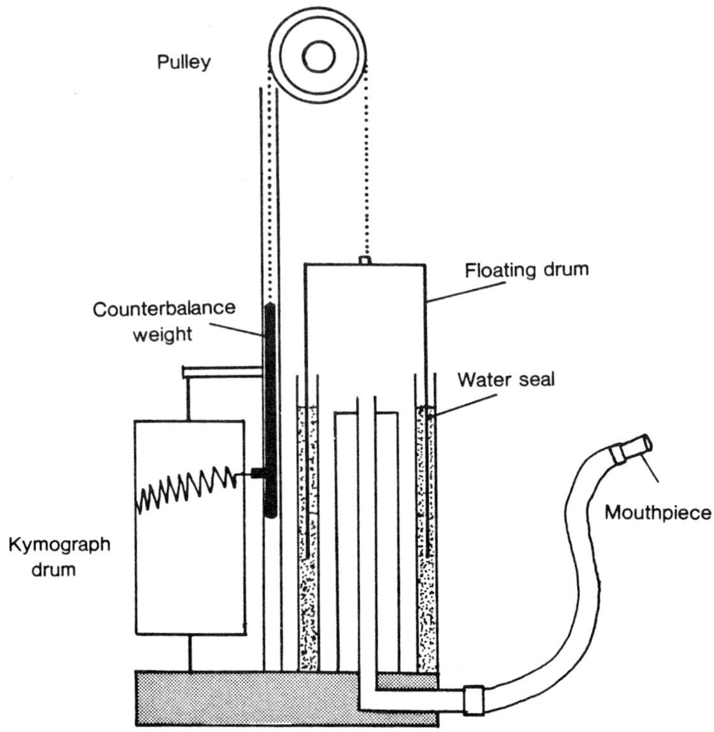

Figure 6-14. A spirometer. The spirometer is a volume recorder. A pen makes a downward stroke on the slowly moving drum of paper during expiration; during inspiration there is an upward excursion on the graph paper.

the results of breathing volumes and shows both a digital and graphic readout based on input data such as height and other factors. Normal lungs yield a steep curve as the patient exhales into the instrument. Pathologic lungs, asthmatic, emphysematous, cancerous, etc., reveal a slow rise and low peaking.

Oral manometers are used to estimate oral breath pressure. They can indicate the effectiveness of velopharyngeal closure and effort by the expiratory muscles to create oral breath forces.

Pneumographs are accordian-like rubber tubes containing a spiral spring for strapping around the thorax and/or abdomen. As the given section of the body expands and contracts, a small tube transfers the pressure changes of the pneumograph to recorders, where a pattern is traced out on moving paper. This is a highly visual method of recording breathing patterns in terms of rate and rhythm of inspiration and expiration, examining depth of breathing and synchrony of movement between the thorax and abdomen. Most have seen the lie detector machine being used in movies. This device utilizes a multichannel recorder of respiration and other parameters. The pneumograph is the transducer for the breathing parameter.

Electromyography involves recording the electrical currents from designated respiratory muscles to determine the degree to which certain muscles are active during speech and other modes of breathing. Amplifiers boost the weak signal from electrodes implanted in or over the areas of the muscle under study and pass the action current to paper recorders for analysis.

Intrathoracic and intratracheal pressures have been studied by interesting approaches. Passing a tube with a latex balloon on its end through the nose and down the esophagus allows the balloon to be inflated with a known amount of air and connected to a sensitive gauge and recorder. When the individual speaks, the changes in intrathoracic pressure caused by the expiratory muscles squeezing on the contents of the thorax reveal the amount of pressure involved.

The insertion of a hollow needle through the skin into the trachea between the larynx and jugular notch provides opportunity to take direct pressure readings of subglottal air pressure. The needle is connected by a small tube to elaborate recorders, where pressure changes are read and synchronized with speech production. Volunteers are used for such approaches.

A body plethysmograph requires that the subject sit in an enclosed chamber with only his head protruding from an airtight neck seal. Chamber pressures are recorded as the subject performs various breathing-related activities. By fitting an airtight dome over the subject's head while he sits encased in the chamber and providing an opening at the mouth, alveolar pressure during speech may be measured as a differential calibration procedure between the pressure-volume at the mouth and displacement of gas within and around the subject. Subglottal air pressures may thus be measured with much more efficiency than with the tracheal puncture technique described above.

Many of the techniques described above are inadequate for fulfilling the needs of researchers and clinicians in obtaining accurate, valid data regarding breathing processes. It is expected that new recording equipment or techniques will eventually replace some of the armamentaria currently in use.

BIBLIOGRAPHY

Anderson, P.D.: *Clinical Anatomy and Physiology for Allied Health Sciences*. Philadelphia, Saunders, 1976.

Langley, L.L., Telford, I.R., and Christensen, J.B.: *Dynamic Anatomy and Physiology*, 4th ed. New York, McGraw, 1974.

Lippold, O.: *Human Respiration*. San Francisco, Freeman, 1968.

Luchsinger, R., and Arnold, G.: *Voice-Speech-Language*. Belmont, CA: Wadsworth Pub, 1965.

Romanes, G.J. (Ed.): *Cunningham's Textbook of Anatomy*, 12th ed. New York, Oxford U Pr, 1981.

Spalteholz, W., and Spanner, R.: *Atlas of Human Anatomy*, 16th ed. Philadelphia, Davis, 1967.

Chapter 7

THE PHONATORY APPARATUS
Framework and Muscles

IN the functional design of the vocal tract, the act of phonation is coupled with the process of exhalation for speech. Phonatory processes are coupled with the remaining aspects of the speech act, resonation and articulation interactions. During speech, the larynx functions as the primary transducer for changing expiratory air into sound.

LARYNGEAL FUNCTIONS

The larynx connects the trachea with the pharynx or throat cavity. It has several important biological responsibilities. Through action of a valvular mechanism, the larynx serves to protect the lower respiratory passage by preventing entrance of foreign material, liquid or solid. Everyone has most likely inadvertently tested the reaction caused by a particle of food or liquid entering the larynx. Violent coughing ensues as a normal reflex for clearing the respiratory pathway of the intruding material (Fig. 7-1).

By inhaling deeply and using the larynx to hold the air captive in the lungs, the chest framework can be held rigid. This internal support appears especially important for activities that involve muscle tension on the outside of the body cavities as in lifting of heavy objects. The act of thoracic fixation is also important to facilitate activities demanding intra-abdominal pressures that assist the visceral musculature, such as expelling a baby from the birth canal or expelling the contents of the rectum.

While the larynx may be termed the *organ of voice*, it is first a vital section of the respiratory pathway. Nevertheless, the larynx is well suited as a volunteer generator. The vocal folds, formed by two shelves of muscle covered by a mucosal lining, are the principal structures of phonation. Constriction of the breath stream by this valvular structure enables the folds to produce segmental puffs of air from a continuous stream of outgoing air. As such, the vocal folds act as a *transducer*

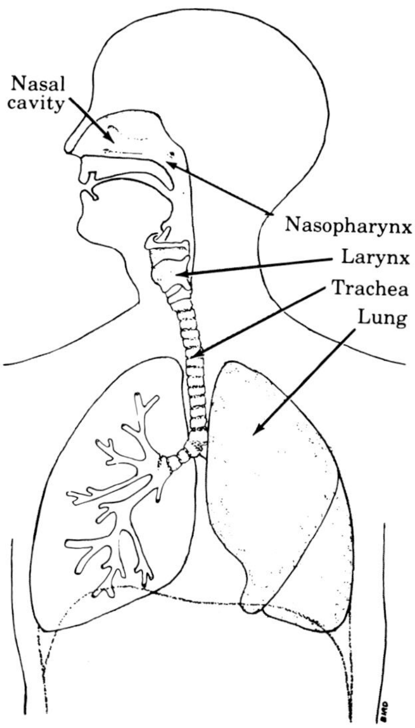

Figure 7-1. The respiratory system showing the position of the larynx in relation to the upper and lower aspects of the respiratory pathway. (From Dawson, H.L.: *Basic Human Anatomy*, 2nd ed., 1974. Courtesy of Appleton-Century-Crofts, New York.)

during vibration. The transduction of a direct current of air into an alternating current or a stream of segmented puffs of air produces a turbulence that is perceived as sound or voice. Hence, it is during those portions of the vibratory cycle when the vocal folds are open and a puff of air is released that sound is produced, rather than when the folds are approximated.

FRAMEWORK

The basic anatomical descriptions of the larynx were written many hundred years ago. These descriptions were based on a few cadaver specimens. Even modern anatomy texts fail to show proper regard for the tremendous variability in the form of individual larynges. While the delicate structures of the larynx may be most notable by virtue of individual variations, it is still convenient to describe an idealized anatomic form for the larynx. By virtue of the miniature nature of the laryngeal components and the intricate connections and interactions involved, a considerable amount of variation should be expected. In fact, some anatomists now contend that this is the rule rather than the exception.

The framework of the larynx consists of five major cartilages, positioned as if stacked on a base cartilage. Those unpaired are the thyroid, cricoid, and epiglottic cartilages. The paired cartilages are the arytenoid, corniculate, and cuneiform cartilages. The basic structure of the larynx is dependent upon two of the single cartilages, the thyroid and cricoid. All of the cartilages are dependent upon the extrinsic and intrinsic musculature and ligaments to hold them in position (Fig. 7-2).

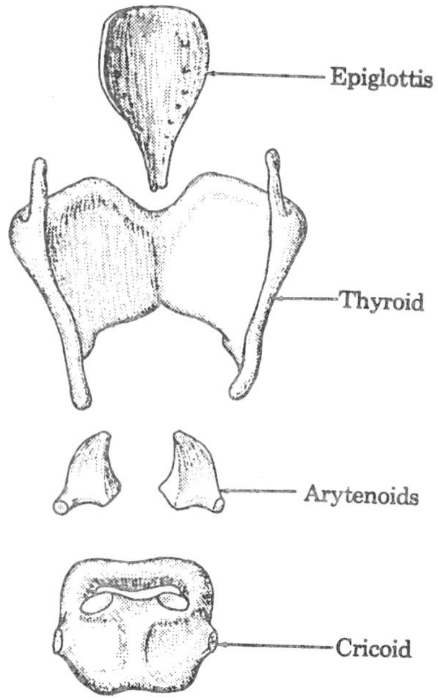

Figure 7-2. The cartilaginous structures comprising the skeleton of the larynx as viewed from behind. (From Dawson, H.L.: *Basic Human Anatomy*, 2nd ed., 1974. Courtesy of Appleton-Century-Crofts, New York.)

The approach taken in this section is to describe the typical configuration of the larynx while also recognizing certain structural variations that have been identified as being common findings in dissection studies.

Thyroid Cartilage

The thyroid cartilage is the largest and may be the most anatomically variable of the laryngeal cartilages. Its name is taken from the thyroid gland, located at its lower sides, which has the same basic configuration; it is angular toward the front and open behind. This cartilage consists of a pair of quadrilateral plates, termed the *thyroid laminae*. They meet anteriorly at almost a right angle in males but are somewhat rounded with a slightly wider angle in females. Many older anatomy texts state that the angle of the two thyroid laminae is around 90 degrees for males and 120 degrees for females. More recent texts indicate that these angles are excessive. It appears that the angle is on the order of 80 degrees for males and 90 degrees for females. On occasion, it is possible to find one lamina to be larger than the other (Fig. 7-3).

At the upper center of the laminae junction is the thyroid notch, a V-shaped upper central border. At the base of the thyroid notch is another landmark, the thyroid prominence. This prominence is also known as the laryngeal prominence or Adam's apple; it is particularly outstanding in the profile of the neck of many slender men and barely noticeable in females. It is easily palpated in both male and female adults (see Fig. 2-14).

Proceeding posteriorly, each lamina flares widely. Before ending in a free border at the back, it encloses the side of the larynx. Thus, the thyroid cartilage encloses the larynx anteriorly and laterally as a protective shield not unlike some of the more elaborate shields carried by spearmen under Prince Valiant in the days of King Arthur. Occasionally, a foramen or a few foramina are seen in one or both laminae. Blood vessels and nerves can, variably, enter or exit the larynx through this opening. Traditional anatomy texts ignore this rather common variation in the thyroid cartilage.

An oblique line occasionally extends downward and forward on the lateral outside surfaces. In other instances, there are superior and inferior tubercles joined by a ligamentous band. These permit attachment of a part of the inferior pharyngeal constrictor, the thyrohyoid, and the sternothyroid muscles (see the extremes of the oblique line, Fig. 7-27).

The laminae present a smooth inner surface. In addition to its closely applied covering, the perichondrium, it is covered

Figure 7-3. The thyroid cartilage of the larynx viewed from above showing, by a dotted line, the differences in the thyroid angle between an adult female (A) and male (B). The thyroid lamina, or lateral plates, meet at an angle of about 90 degrees for females and about 80 degrees for males.

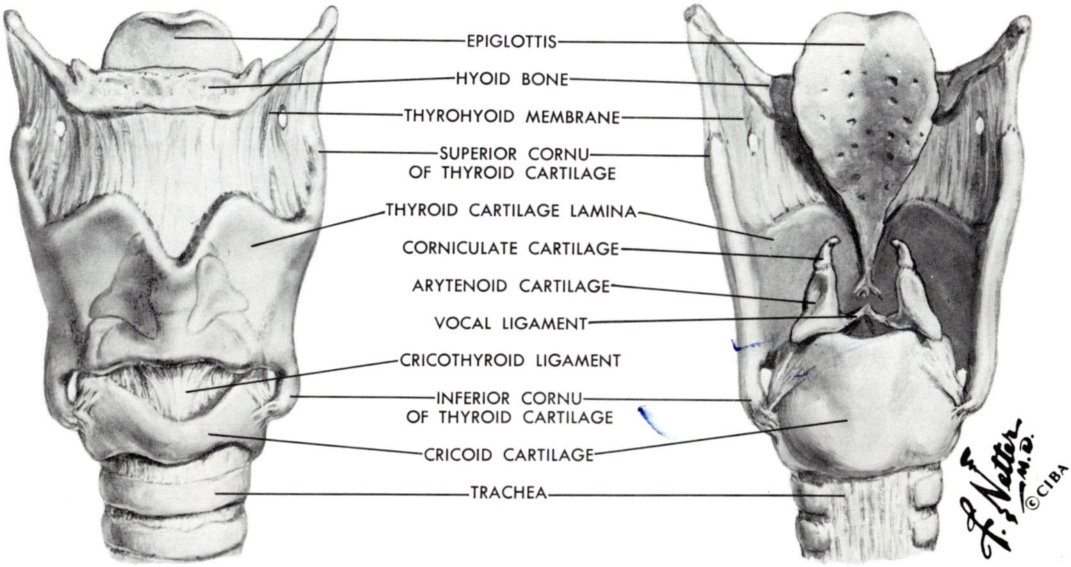

Figure 7-4. Framework cartilages of the larynx, the hyoid bone, and tracheal connection, anterior and posterior view. The V-shaped notch on the anterior aspect of the thyroid cartilage is easily felt (palpated) on the neck. (© Copyright 1964, CIBA Pharmaceutical Company, Division of CIBA-GEIGY Corporation. Reprinted with permission from CLINICAL SYMPOSIA illustrated by Frank H. Netter, M.D. All rights reserved.)

by mucous membrane. At the upper part of the junction of the two laminae, the stem of the epiglottis is fastened. The area immediately below serves as a place of attachment for the ligaments of the ventricular folds, closely followed by the ligaments and muscles of the vocal folds. In this narrow field also may be found the origin of a thyroepiglottic ligament and muscle fibers following this path upward to the epiglottis (Fig. 7-4).

The posterior border is a point of insertion of both the stylopharyngeal and palatopharyngeal muscles. Projecting superiorly from the dorsal border is the *superior cornu*, a long slender process which points toward the posterior limit of a like structure on the hyoid bone. The two are connected by the lateral thyroid ligament. In this ligament may be frequently found a small cartilaginous nodule called the *triticeal cartilage* (resembling a grain of wheat).

It may eventually ossify in some individuals. Projecting inferiorly is another process, the *inferior cornu*. This cornu is shorter than the superior process. The two inferior cornua have articular facets at their ends that rest upon a small facet found on the back portion of the cricoid cartilage. This is the only direct articulation by the thyroid cartilage, since all other connections are through ligaments or muscles (Fig. 7-5).

A number of structural anomalies may appear in the cornua. Sometimes one of the horns may be absent, bent forward, or vary considerably in height from specimen to specimen.

Behind the thyroid cartilage is the hypopharynx, with the fourth, fifth, and sixth cervical vertebrae of the spinal column behind. An important muscle attachment on the inner surface, lower border, and inferior cornu is that of the cricothyroid muscle, which is responsible for the rocking action between the thyroid cartilage and the cricoid cartilage below. As will be seen later, the cricothyroideus is a principal muscle of voice regulation.

Cricoid Cartilage

Lying directly below the thyroid cartilage and forming the base for the larynx is the cricoid cartilage. Encircling the lower cavity of the larynx, the cricoid is shaped liked a signet ring. It is low and narrow in front, the *arch portion*, and wide and thick in back, the *lamina portion*. The lower edge is horizontal and is attached to the first ring of the trachea below by the *cricotracheal ligament*. This ligament may rupture as a result of an excessive and abrupt increase in air pressure within the trachea

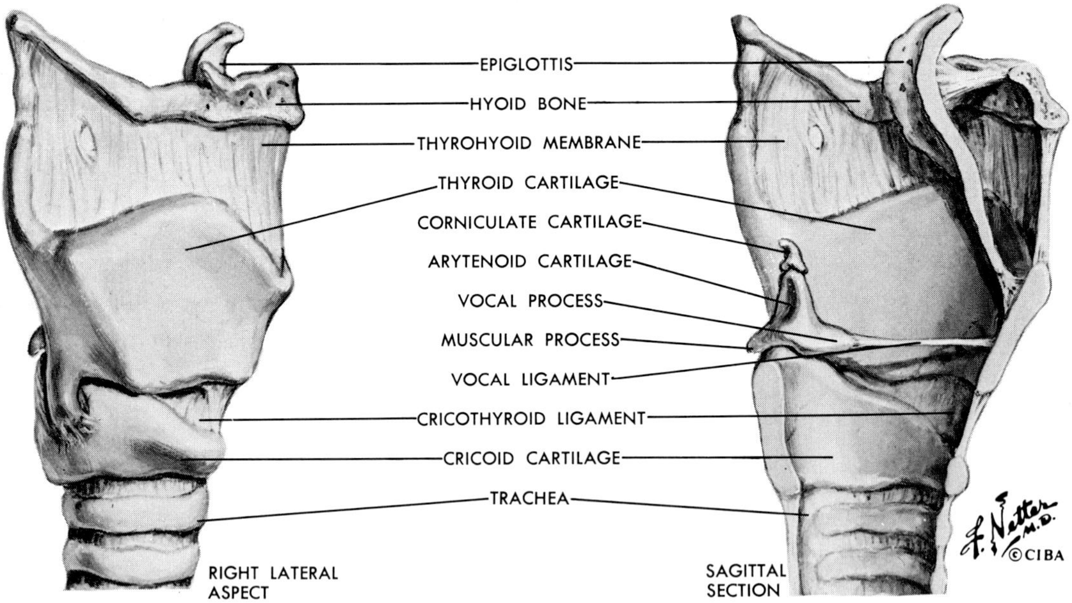

Figure 7-5. (Left) Lateral aspect of the framework cartilages of the larynx and their connections to the hyoid bone and trachea. The thyroid prominence, which is highly noticeable in some males, is at the base of the thyroid notch. (Right) Note the anterior attachment of the vocal ligament on the thyroid cartilage and the posterior attachment on the vocal process of the arytenoid. (© Copyright 1964, CIBA Pharmaceutical Company, Division of CIBA-GEIGY Corporation. Reprinted with permission from CLINICAL SYMPOSIA illustrated by Frank H. Netter, M.D. All rights reserved.)

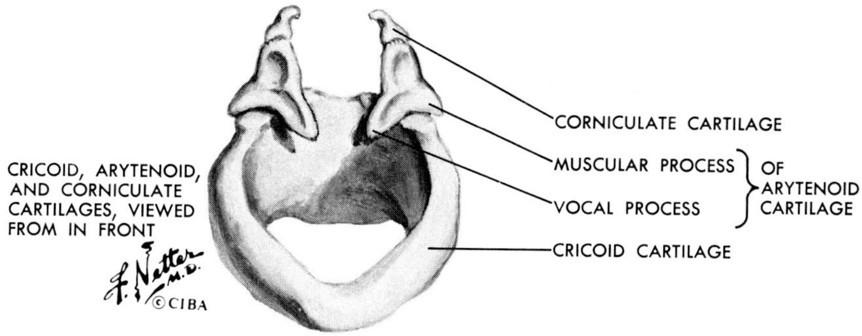

Figure 7-6. Cricoid cartilage viewed from the front with the arytenoids in position. The vocal ligament attaches to the vocal process. The corniculate cartilages sit atop the arytenoids. (© Copyright 1964, CIBA Pharmaceutical Company, Division of CIBA-GEIGY Corporation. Reprinted with permission from CLINICAL SYMPOSIA illustrated by Frank H. Netter, M.D. All rights reserved.)

while the vocal folds are closed. High intratracheal pressures from severe coughing and/or sneezing are the typical causes of such a rupture.

There are two small facets on the posterolateral surface to receive the inferior cornua of the thyroid cartilage. This articulation is a true synovial joint (freely movable) of the modified ball-and-socket type, permitting a rocking action of the cricoid cartilage on the thyroid cartilage. In addition, there is also a slight front-to-back sliding action at this joint. Ligaments hold the inferior cornua in place on the cricoid. The rocking action directly affects the length of the vocal folds and can be important in adjustments of vocal fold tension (Fig. 7-6).

The space at the front between the bottom of the thyroid and top of the cricoid arch is sealed medially by the triangularly shaped *cricothyroid ligament* or *membrane*, also termed the *conus elasticus* or *cricovocal membrane*. This membrane and ligament lie just under the skin of the neck; consequently, an emergency tracheotomy may be performed rapidly and without complications through the space between the thyroid and cricoid cartilages. Emergency airways created through the cricothyroid membranes are always temporary and used only until a tracheostomy, if needed, can be completed below the cricoid cartilage (Fig. 7-7).

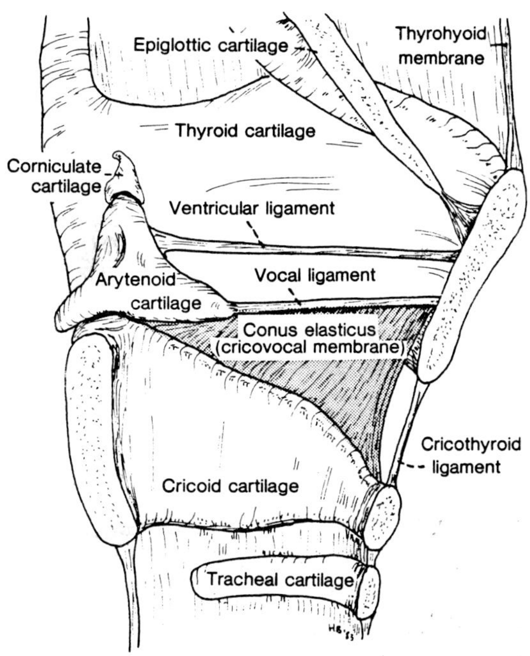

Figure 7-7. Sagittal aspect of the larynx showing the vocal and ventricular ligaments. Notice how the cricovocal membrane connects the border of the cricoid cartilage with the vocal ligament. Compare this view of conus elasticus with Figures 7-4 and 7-8. The space between the vocal and ventricular ligaments gives rise to a small laryngeal cavity in the soft tissue (see Fig. 7-13).

Arytenoid Cartilage

There are two arytenoid cartilages. The name means "shaped like a jug or pitcher." Each is a small pyramidal structure having a base, apex, and three surfaces. They are superimposed on the upper posterior border of the lamina of the cricoid cartilage. This cricoarytenoid articulation is a synovial joint of the modified ball-and-socket type, like that of the cricothyroid articulation. A ligament at the back, the posterior cricoarytenoid ligament, extends from the posterior part of the base of the arytenoid straight down to the cricoid lamina. Another connection between these two cartilages is contributed by the fibrous band, which forms a capsule at the joint. The capsular ligament encases a synovial membrane that provides lubrication of the joint. Each arytenoid cartilage is able to rotate around a vertical axis to some extent, to slide medially, and to rock forward and backward, although this is not agreed upon by all vocal anatomists, particularly when the anatomy of the joint is reported from fresh specimens.

There are two processes on the base to which intrinsic muscles are attached. The vocal process provides for the posterior attachment of the *vocal ligament* and the thyrovocal (medial) fibers of the vocal muscle. The muscular process is associated with the thyromuscular (lateral) portion of the vocal folds and the lateral and posterior cricoarytenoid muscle. These cause the arytenoid cartilage to rotate on its vertical axis, according to traditional views. But according to some, this results in sliding of the arytenoids medially or laterally. Nonetheless, the arytenoid is able to exhibit rocking motions on the cricoid.

Just a short distance above the vocal process, the posterior attachment of the *ventricular ligament* (also termed the *vestibular ligament*) is found. From around this ligament the ventricular folds arise, with a slightly crescent lower margin arching between the arytenoid cartilages and the anterior attachment on the inner surface of the thyroid cartilage (Fig. 7-8).

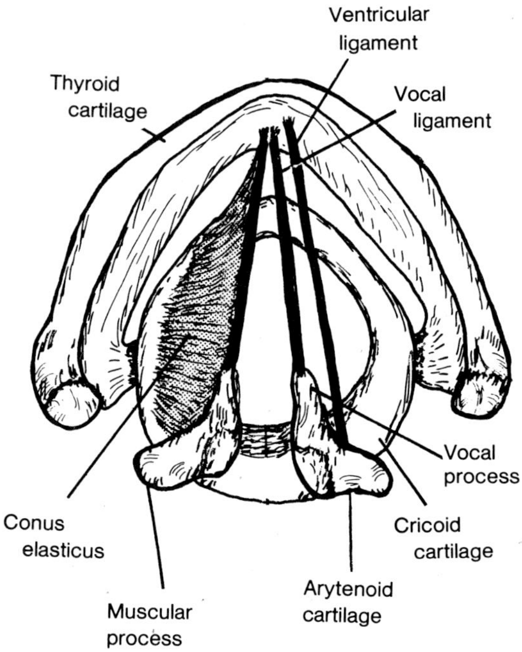

Figure 7-8. The vocal and ventricular ligaments viewed from above. Note that the attachments of the ventricular ligament are lateral (and above) the vocal ligament. Hence, the vocal folds appear to be only partially covered by the ventricular folds in a mirror view (see Fig. 7-9) of the larynx.

The apex of the arytenoid cartilage is pointed and is part of the support for the posterior part of the membrane sealing this part of the cavity of the larynx from the hypopharyngeal cavity.

The medial surface is smooth and covered by mucous membrane. This important surface makes up the lateral boundary of the posteriormost aspect of the glottis (see Fig. 7-7) and is sometimes referred to as the cartilaginous portion of

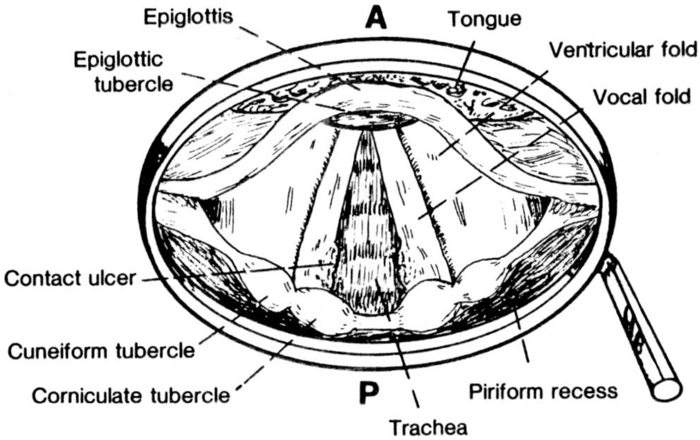

Figure 7-9. Mirror view of the larynx. A contact ulcer is shown at the posterior aspect of the vocal folds where the medial surfaces of the arytenoid cartilages come together. The forward portion of the vocal folds is membranous in contrast to the cartilaginous basement of the posterior aspect. Anterior (A) and posterior (P) aspects of the larynx are indicated.

the glottis. The glottis is identified as the space between the vocal folds. (Chapter 8 includes a detailed discussion of the glottis.)

If the two arytenoid cartilages are frequently brought together forcefully, the thin mucous membrane covering the cartilaginous glottis will be injured because of insufficient cushioning underneath. This condition may be described as a contact ulcer, and the ulceration may be unilateral or bilateral. Granulation (small fleshy masses forming in the wound) may accompany the ulceration. This is a very difficult voice disorder to treat because it appears to stem from personal adjustment problems. It is typically an adult male anomaly (Fig. 7-9).

Corniculate Cartilages

There is a tiny, horn-shaped cartilage perched on the apex of each arytenoid cartilage. They are most commonly known as the *corniculate cartilages* but are also termed the *cartilages of Santorini*. These small cartilaginous nodules sometimes fuse with the arytenoid cartilages; sometimes they are absent altogether. They are extensions of the arytenoid's support of the posterior part of the *aryepiglottic fold.*

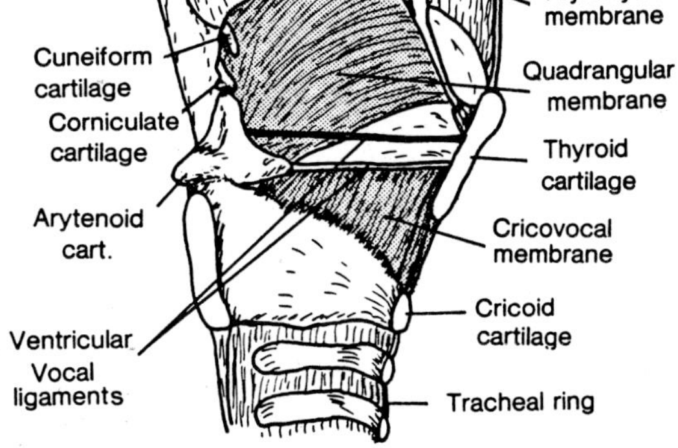

Figure 7-10. Sagittal section of the larynx showing the membranes of the larynx. The cuneiform cartilage acts as a stiffener in the posterior margin of the aryepiglottic fold, which follows the border of the quadrangular membrane.

When one views the larynx from above, the prominent enlargements on either side at the posterior margin of the aryepiglottic folds reveal the underlying corniculate cartilages (Fig. 7-10).

Cuneiform Cartilages

Also known as the *cartilage of Wrisberg*, the cuneiform cartilage is a small, rod-shaped cartilage embedded in either side of the aryepiglottic fold. When present, it is just anterior to the corniculate cartilage. It may reinforce the aryepiglottic fold but is commonly absent.

Epiglottic Cartilage

This large, variable, leaf-shaped cartilage rises from a stem attached on the inner front surface of the thyroid cartilage. It towers above all other laryngeal cartilages and is tied to the hyoid bone by the *hyoepiglottic fold*, which rises from below on the arytenoid cartilage. The upper border of the epiglottis is free. The *quadrilateral membrane* relates the epiglottis to the arytenoid cartilages and forms a foundation for the *aryepiglottic fold*.

The superior border of the epiglottis presents a variety of configurations as seen in laryngoscopic studies. The typical form is a gentle posterior arc, while an omega shape is also quite common. The posterior surface of the epiglottis (laryngeal surface) often presents a fairly prominent tubercle (see Fig. 7-9).

A clinical condition involving the epiglottic cartilage is laryngeal malacia. There is a softening of the cartilage, reducing its stiffness and causing the laryngeal aditus to be constricted. During attempts to produce phonation, the result is a laryngeal stridor (high-pitched, harsh respiratory sound heard during inspiration in acute laryngeal obstruction) (Fig. 7-11).

The ligaments connecting the tongue and epiglottis may be used to advantage in controlling the position of the epiglottis for examination of the vocal folds. During mirror examination of the larynx, the examiner grasps the tongue between his thumb and second finger and pulls it forward. This forward pull on the tongue moves the epiglottis forward with the tongue and uncovers the larynx for mirror viewing. The laryngeal mirror is inserted into the oropharynx to reflect the image of the vocal folds and related structures (see Chap. 25).

The epiglottis serves to deflect food and liquids being swallowed, so that nothing enters the laryngeal cavity. This

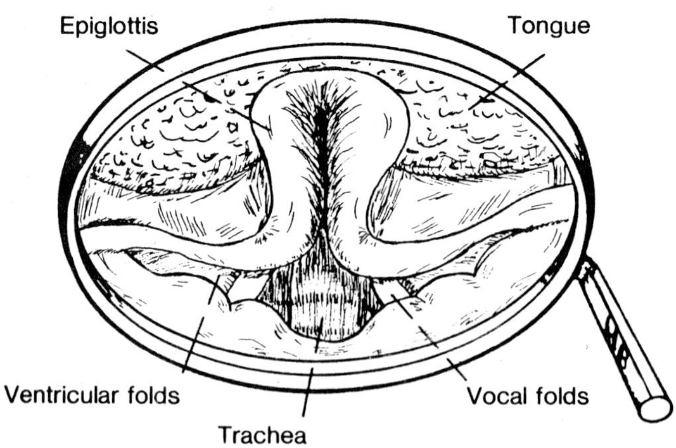

Figure 7-11. Collapse of the epiglottic cartilage due to softening of the cartilage (laryngeal malacia). The entrance into the larynx is restricted and a laryngeal stridor is heard (see text).

protective function involves sphincteric reduction of the opening into the larynx (laryngeal aditus) by contraction of the aryepiglottic folds. This action also draws the epiglottis down and back to shield the larynx. Additionally, the laryngeal elevator muscle system participates by tucking the larynx up and forward under the base of the tongue. The epiglottis is also able to influence resonance by its ability to manipulate the size of the opening between the laryngeal and pharyngeal cavities (Fig. 7-12). In the larynx, the laryngeal cavity or space, which is accessible through the pharynx as well as from the trachea, extends from the rim of the epiglottis downward to the cricoid cartilage level where the trachea unites with the lower laryngeal limit. The laryngeal space is divided into three compartments by the two sets of folds protruding from each side on the lateral walls of the larynx.

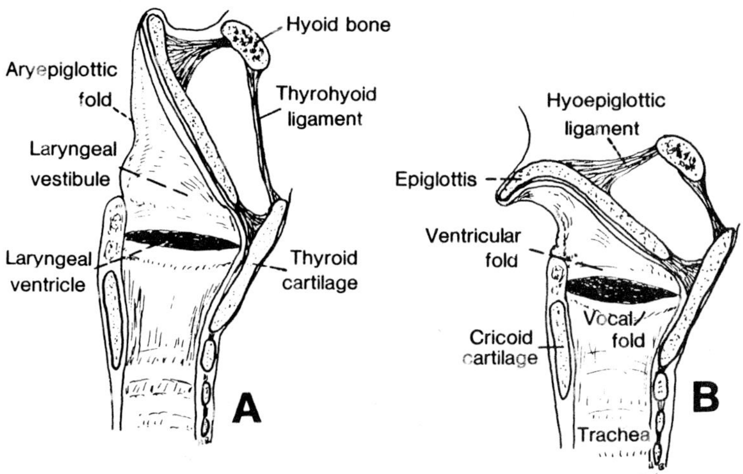

Figure 7-12. Action of the epiglottis in swallowing. (A) Normal position as in respiration and phonation. (B) Position during swallowing.

DIVISIONS OF THE LARYNGEAL CAVITY

Laryngeal Vestibule

Situated above the ventricular folds is the upper laryngeal space, having a triangular opening and known as the *laryngeal vestibule*. It is widest on the forward wall at the back of the tongue and slopes down toward the back wall. The anterior wall is formed by the *epiglottis*, a lid-like structure extending out of the laryngeal cavity. A lateral membrane rises to form the lateral boundaries of the vestibule. These lateral walls are known as the *aryepiglottic folds*. The aryepiglottic membrane forming the vestibule wall is descriptive of its location. It covers the aryepiglottic muscle, which extends between the arytenoid cartilages on the posterior margin of the larynx and the epiglottis.

The posterior wall of the *aditus*, another term for the entrance into the larynx, is formed by the upper parts of the arytenoid cartilages and is much lower than the anterior wall. The inferior limits of the vestibule are the ventricular folds.

Laryngeal Ventricle

Normally, the lower folds (the true vocal folds) extend farther into the laryngeal cavity than the ventricular folds. The true folds are separated from the ventricular folds by a small space. The ventricle between the two sets of folds represents the smallest of the three laryngeal compartments. This middle laryngeal cavity is known variously as the *laryngeal ventricle*, the *laryngeal sinus*, and the *ventricle of Morgagni* (1741) after the anatomist who provided the first detailed description (Fig. 7-13).

The upper and lower boundaries of the laryngeal ventricle, then, are the false and true vocal folds, respectively. The lateral limit is the mucous membrane covering the inner wall of the thyroarytenoid muscle. The membrane continues anteriorly, toward the thyroid prominence, and is prolonged upward into the *laryngeal sacculus* or appendix of the laryngeal ventricle. Numerous glands are found in the mucous membrane of the sacculus. Perhaps the laryngeal ventricle is a vestige of a resonating pouch, which is quite prominent in some species of monkeys and apes, for example, the orangutan (Fig. 7-14).

Inferior Laryngeal Ventricle

Below the true vocal folds lies the inferior laryngeal ventricle, also termed the *infraglottic space*. From the narrowest open-

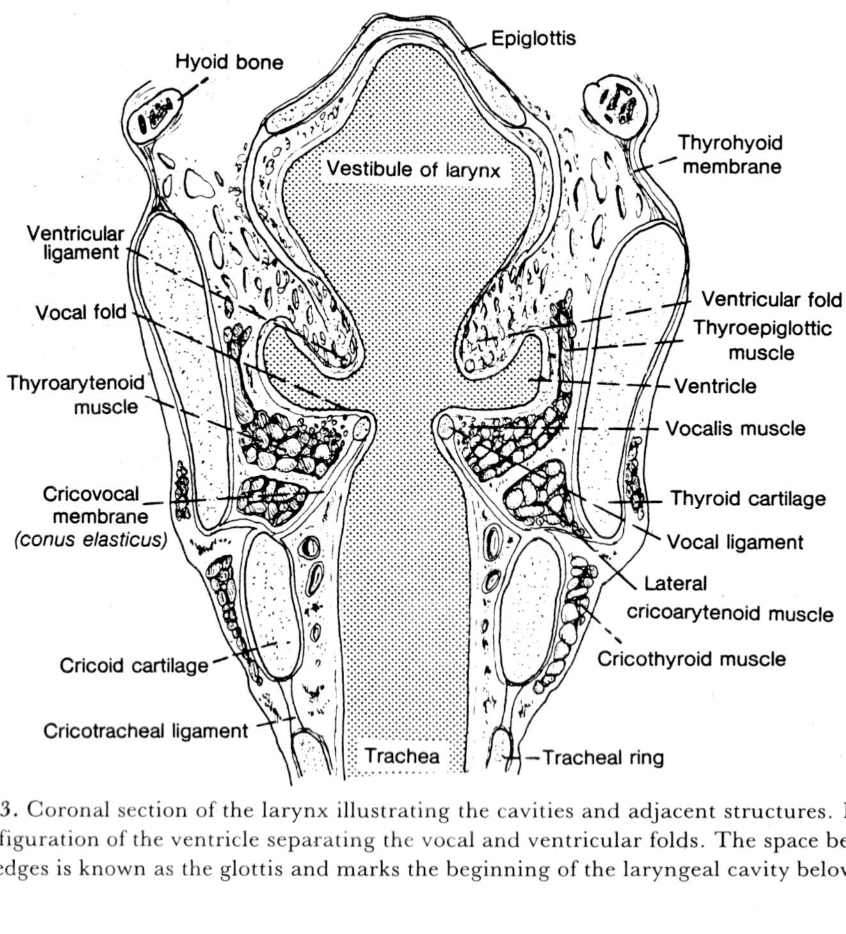

Figure 7-13. Coronal section of the larynx illustrating the cavities and adjacent structures. Notice the pocket configuration of the ventricle separating the vocal and ventricular folds. The space between the vocal fold edges is known as the glottis and marks the beginning of the laryngeal cavity below the glottis.

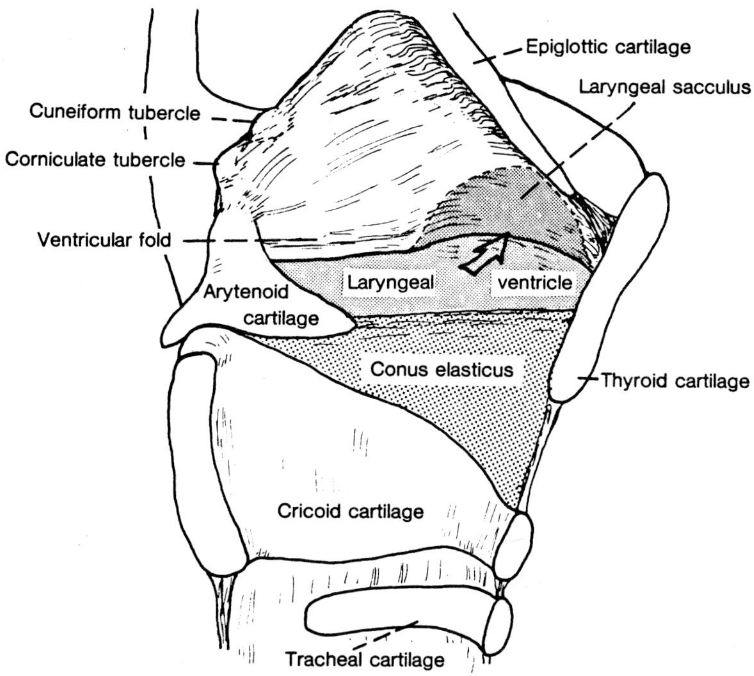

Figure 7-14. Sagittal section of the larynx showing the laryngeal sacculus as an outpouching of the laryngeal ventricle. This appendix of the laryngeal ventricle is a membranous sac. On its surface are the openings of 60-70 mucous glands, which are lodged in the submucous tissue.

ing between the true vocal folds in the whole of the cavity of the larynx, the portion below the true folds widens gradually to meet the tracheal lumen or opening. Viewed from a frontal section through the larynx, the inferior laryngeal space just below the vocal glottis is bell-shaped, as shown in Figure 7-13. Laterally and posteriorly, the cricoid cartilage forms the walls.

The anterior wall of the inferior laryngeal ventricle is a ligament connecting the cricoid cartilage with the thyroid cartilage immediately above, the *cricothyroid ligament* or *membrane*, referred to earlier as conus elasticus and the medial cricothyroid ligament. This ligament is superficially situated and is just below the level of the true folds, providing a favorable area for surgical entrance into the airway.

THE MUCOUS MEMBRANE OF THE LARYNX

The mucous membrane lining the laryngeal cavities is continuous up from the trachea. The cells comprising the tissue are columnar in shape with hair-like projections. The cilia, which are continuous with the tracheal lining, beat toward the pharynx to carry particulate matter out of the respiratory airway. However, at the margins of the vocal folds, the ciliated columnar tissue is replaced by squamous (pavement-like) epithelial tissue, which is structurally more suitable. Some authors indicate that the edges of the ventricular folds plus the upper posterior surface of

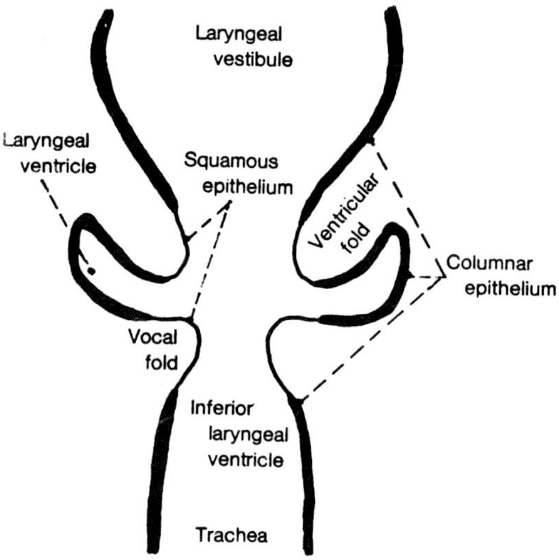

Figure 7-15. Schematic representation of the membranes covering the surfaces of the cavities of the larynx. Two types of epithelium are found forming the basement membrane within the larynx: stratified squamous epithelium and columnar ciliated epithelium.

the epiglottis are also covered by squamous epithelium (Fig. 7-15).

Over the vocal folds, the squamous is multilayered and tightly bound. The white appearance of the true vocal folds is due to lack of a rich vascular supply along the length of the mucosal surface of the folds.

In the area of the ventricular folds, especially in the aryepiglottic folds, the mucosa is loosely attached. In such places, fluid can accumulate and lead to significant swelling (edema) of the membrane.

The ventricular folds appear pink in color, in contrast to the whitish color of the true vocal folds just below. The ventricular folds also appear plump and have rounded edges. There are very few muscle fibers contained in these folds, which are made up principally of thick folds of the mucous membrane lining the larynx.

MUSCLES OF THE LARYNX

Movement required for changing the breath stream into a sound stream for speech purposes is dependent upon the muscle system of the larynx. Phonation, the act of modulating the breath stream to produce sound, is dependent on the laryngeal musculature but also on a dynamic breath stream.

The muscles of the larynx focus on the vocal folds. These are composed principally of muscles stretched horizontally in an A-P direction across the opening through the larynx, which serves to protect the respiratory pathway below as well as functioning for speech purposes.

Thus, regulation of the openings of the respiratory pathway is a function of laryngeal muscles. The valvular system includes the epiglottis, the ventricular folds, and the vocal folds, which serve as protective gates leading toward the lungs. Yet these variable openings are quite functional for speech purposes, especially the vocal folds. To understand the function of the larynx for speech purposes requires more than just a study of the vocal folds.

The muscles of the larynx are associated with its cartilaginous framework. Instead of the primary laryngeal muscles being lost in the mucous membrane of the laryngeal cavity, most have clear origins and insertions on the skeleton of the larynx.

Two primary sets of muscles are found attached to the larynx. Similar to those of the tongue, these muscles are classified as either *intrinsic* or *extrinsic* laryngeal muscles. The extrinsic muscles attach the larynx directly to the hyoid bone above the larynx, among other structures. Since the hyoid bone is tied to the larynx through ligaments and a membrane, any movement of the hyoid bone will likely move the larynx. The several muscles attached to the hyoid bone, then, should be considered along with the laryngeal extrinsic musculature. The intrinsic muscles regulate the valvular action of the laryngeal cavities; they are situated on or within the cartilaginous skeleton of the larynx (Table 7-I).

Intrinsic Muscles of the Larynx

The extrinsic laryngeal musculature connects the larynx to structures above and below, while the intrinsic muscles act upon the internal structures of the larynx. This latter set of muscles concerns itself with valvular adjustments within the larynx, especially the vocal folds.

The intrinsic muscles are short and thin and are found embedded within the mucous membrane and other membranes of the larynx. Altogether, these form the walls inside and partially outside of the larynx. The essential function of these muscles is to control the laryngeal inlet into the trachea, that is, to protect the lower respiratory airway from obstructions that might threaten life. The intrinsic muscles, by manipulating the air stream at the level of the rima glottidis (the space between the vocal folds, commonly called the glottis), produce sound for speech and singing purposes (Fig. 7-16).

The muscles of the larynx have traditionally been described as to specific functions they are said to perform. As such,

Table 7-I

SUMMARY OF THE INTRINSIC MUSCLES OF THE LARYNX

Muscle	*Origin*	*Insertion*	*Action*
Cricothyroid	anterolateral aspect on the arch of the cricoid cartilage	inner lower border of the lamina of thyroid cartilage, its inferior cornu	increases A-P dimension of the vocal folds; lengthens and tenses vocal folds
Thyroarytenoid	inner surface of the lamina of the thyroid cartilage near its angle	anterolateral base of the arytenoid cartilage	shortens vocal fold length; sphincter of larynx
Vocalis	inner surface of the lamina of thyroid cartilage at its angle	vocal process at the base of the arytenoid cartilage	shortens and increases tension of vocal folds
Lateral cricoarytenoid	upper margin on the lateral surface of the cricoid cartilage arch	muscular process at base of arytenoid cartilage	approximates vocal folds
Transverse arytenoid	one arytenoid cartilage	opposite arytenoid cartilage	approximates arytenoid cartilages, closing the glottis
Oblique arytenoid	muscular process of arytenoid cartilage	apex of opposite arytenoid cartilage	approximates arytenoid cartilages, closing aryepiglottic folds
Posterior cricoarytenoid	back of cricoid cartilage	muscular process of arytenoid cartilage	opens vocal folds
Aryepiglottic muscle	apex of arytenoid cartilage and extension upward of oblique arytenoid	epiglottis	closes laryngeal inlet

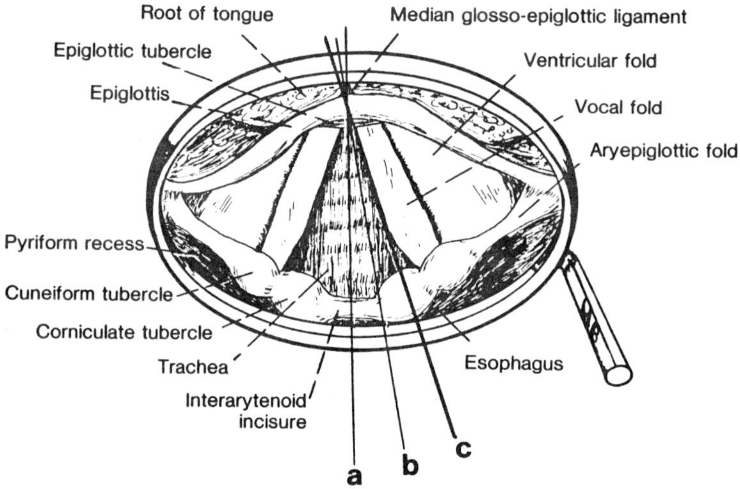

Figure 7-16. Mirror view of the larynx: (a) the vocal folds would contact each other for complete adduction as in phonation; (b) a paramedian position is shown; and (c) an abduction as during inspiration.

each muscle is considered in isolation. One can easily picture an anatomist dissecting out a particular muscle, tugging on it, and recording the movement involved. This is an oversimplified view of laryngeal muscle function that does not give due regard for the interactions of muscles and the subtle nuances that are involved in human laryngeal behaviors.

The description to follow will adopt the tradition of describing individual muscles and their expected primary activity. This approach, although simplistic, remains a convenient way of introducing each muscle to the neophyte student of laryngeal anatomy.

Each of the eight intrinsic muscles, except for the transverse arytenoideus, is paired.

Tensor Muscles

The three major muscles directly regulating the tension of the vocal folds are the cricothyroid, thyroarytenoid, and vocal muscles.

Cricothyroideus

Beginning from the outside of the cricoid cartilage, the fibers of the cricothyroideus travel upward in two large sections termed the *vertical* or *ventral part* and the *oblique part.* The vertically directed fibers insert into the lower inner surface of the thyroid cartilage. A few fibers are continuous with the vocal folds. The obliquely directed fibers insert into the anterior border of the inferior cornu of the thyroid cartilage (Fig. 7-17).

The oblique part of the cricothyroid muscle provides an anterior pull on the inferior cornu of the thyroid cartilage. As a result, there is a *translatory movement* of the thyroid cartilage. This action lengthens the vocal folds by increasing the A-P dimensions of the vocal ligament. The "visor angle" (see Fig. 7-19) is unaffected by the action of the oblique fibers. It is interesting to compare the action of the oblique part with the vertical part since the action of each elongates the vocal folds by pulling on the thyroid cartilage at different points along the inferior aspect of the cartilage. One (oblique part) involves

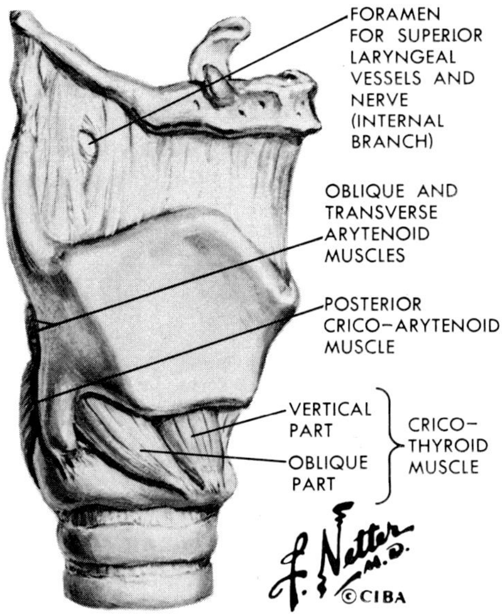

Figure 7-17. The cricothyroideus showing its vertical and oblique groups of fibers. (© Copyright 1964, CIBA Pharmaceutical Company, Division of CIBA-GEIGY Corporation. Reprinted with permission from CLINICAL SYMPOSIA illustrated by Frank H. Netter, M.D. All rights reserved.)

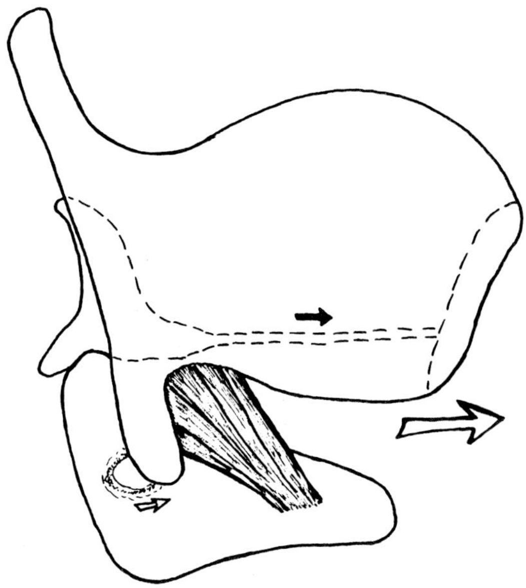

Figure 7-18. Cricothyroideus, oblique part. Because of the angular attachments of the oblique part, the thyroid cartilage is subject to a translatory movement as indicated by the arrows. The result is an elongation of the vocal ligament. The small arrow at the inferior cornu of the thyroid cartilage denotes the point of translation (subluxation, gliding) relative to the cricoid cartilage.

a sliding movement while the other (vertical part) relies on a hinge action (Fig. 7-18 and 7-19).

The action of this muscle is often described as a tensor of the vocal folds, but this is not necessarily accurate. It does increase the *length* of the vocal folds.

Upon contraction, the cricoid cartilage pivots at its joint with the inferior cornu of the thyroid cartilage and typically rocks upward, similar to the movement of the lower jaw when closing the mouth opening. However, if the cricoid is held stable by the extrinsic musculature with a corresponding relaxation of the thyroid fixators, then the thyroid will tilt forward to reduce the visor angle between these two cartilages. In either case, the arytenoid cartilages, riding on top of the posterior rim of the cricoid cartilage, are carried backward to increase the length of the vocal folds. The firmness of the folds at the time the cricothyroideus contracts will determine the relative tenseness of the folds (Fig. 7-19).

Essentially, then, the A-P length of the vocal folds is increased by the cricothyroideus. The vertical fibers of the cricothyroideus fibers pull the arch or anterior portion of the cricoid upward, while the oblique fibers pull the arch in a posterior direction. Recalling that the anterior attachments of the vocal folds are on the thyroid cartilage and the posterior attachments on the arytenoid cartilages, which are fixed on top of the posterior aspect of the cricoid, the cricothyroideus thus stretches the vocal ligament and folds between the two ends. The mesh pattern of

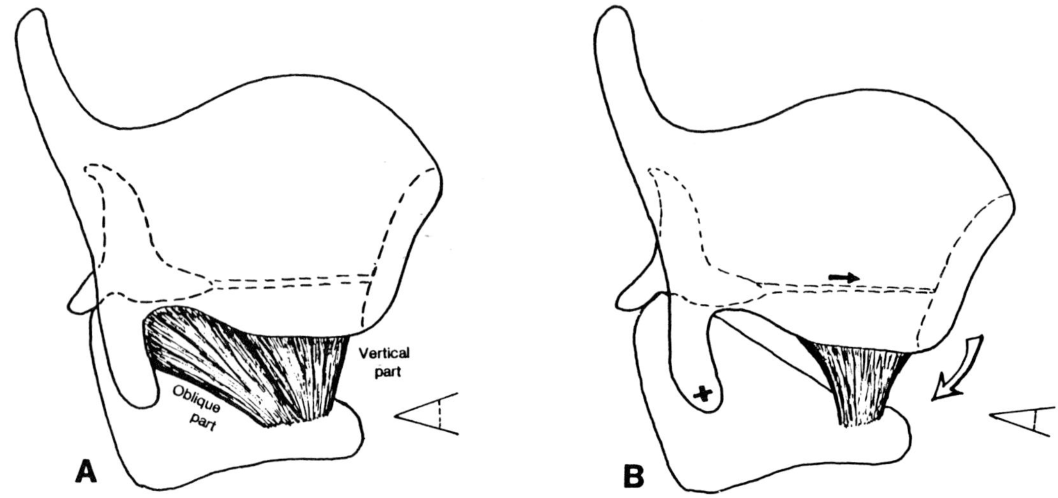

Figure 7-19. Changes in the visor angle as a function of the vertical part of the cricothyroideus. A rotation is produced by the pull of the vertical fibers, decreasing the distance between the thyroid and cricoid cartilages (A to B). Rotation of the thyroid cartilage stretches the vocal ligament.

the thyroarytenoid and vocal muscle fibers permits lengthening of the vocal folds *without* a corresponding increase in tension. In combination with the cricopharyngeus, the lowermost band of fibers of the inferior pharyngeal constrictor taking its origin on either side of the cricoid cartilage, the cricothyroideus is said to be active in producing low-pitched tones.

If the nerve supply to one of the two cricothyroid muscles is interrupted, the oblique portion will be unable to pull the cricoid cartilage posteriorly to lengthen the folds. Such a condition of unilateral paralysis will result in only one side of the cricoid being rotated backward toward the unaffected side. This results in a deviation of the glottis. When viewed through a laryngeal mirror, the glottis may appear twisted in an oblique fashion. A deviation of the posterior commissure pointing *away* from the healthy side and toward the side of paralysis is seen in unilateral conditions (Fig. 7-20). The oblique glottis to the laryngologist is a fundamental symptom of unilateral cricothyroid paralysis, especially if accompanied by anesthesia (loss of sensitivity) on the inner wall of the affected side of the larynx.

The external branch of the superior laryngeal nerves, which is derived from the vagus (X) cranial nerve, innervates this muscle and also supplies sensation from the wall within the larynx. All other intrinsic muscles are served by the recurrent laryngeal branch of the vagus.

When the recurrent laryngeal portion of the vagus nerve is paralyzed, some serious implications to voice and health can occur. Noteworthy among these is the finding of a hoarse voice of sudden onset with an accompanying complaint by the patient of listlessness. If examination of the larynx in this instance shows the left vocal fold to be paralyzed, the patient should be sent immediately to a hospital emergency room for a chest x-ray study. If cardiomegaly (enlarged heart) is found, this signals a problem of mitral valve insufficiency due to damage from a tumor or other cause associated with the left recurrent laryngeal nerve. Before reaching

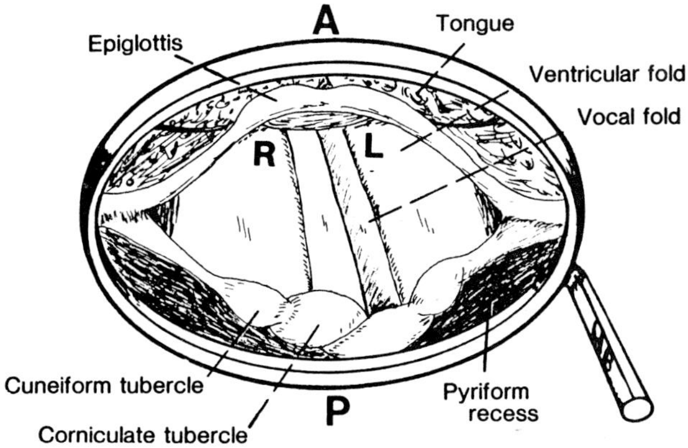

Figure 7-20. An oblique glottis due to unilateral paralysis of the cricothyroideus (left side) as viewed through a laryngeal mirror. The posterior (P) end of the glottis points *away* from the healthy side and toward the paralyzed side. The paretic fold is also lower than the unaffected fold.

the larynx, the recurrent laryngeal nerve, along with the main trunk of the vagus, winds through the upper thoracic cavity, while the main trunk of the vagus continues into the thoracic and abdominal cavities to supply many visceral organs. The left recurrent laryngeal nerve sends parasympathetic fibers to the mitral valve of the heart. Hence, hoarseness in the speaking voice can be the presenting symptom of mitral valve insufficiency. The speech clinician should be aware of this relationship and be prepared to screen laryngeal function in selected instances.

Thyroarytenoideus

This is a broad but thin muscle running parallel to the glottis. The thyroarytenoideus muscle is the essential muscle mass of the vocal folds. It takes its origin from the inner surface at the angle of the thyroid cartilage. Coursing straight back along with the vocal ligament, it inserts into the anterolateral base of the arytenoid cartilage.

Sometimes this muscle is listed as a relaxer of the vocal folds and sometimes as a tensor. By drawing the arytenoid cartilages forward, the thyroarytenoideus shortens and relaxes the vocal folds. To tense the folds, this muscle may act to increase the isometric tension in the folds to assist the vocalis muscle by acting as an antagonist to the cricothyroideus, which stretches the folds. The shortening of the folds would lower the pitch of the voice (Fig. 7-21).

A central band of fibers of the thyroarytenoideus, those that run alongside the vocal ligament, are attached posteriorly to the vocal process of the arytenoid cartilage. This medial band of fibers is known as the *vocalis*. Other segments of the thyroarytenoideus give rise to two additional muscles, the *thyroepiglottis* and *ventricularis*, which are presented in the following discussion; however, the student should remember that the bulk of this muscle is concerned with the vocal folds.

Thyroepiglotticus

The thyroarytenoid muscle has a significant portion of its fibers turned in a vertical direction. Some of these, namely, the *thyroepiglotticus*, end in the aryepiglottic fold, while others continue on to insert into the lateral border of the epiglottis,

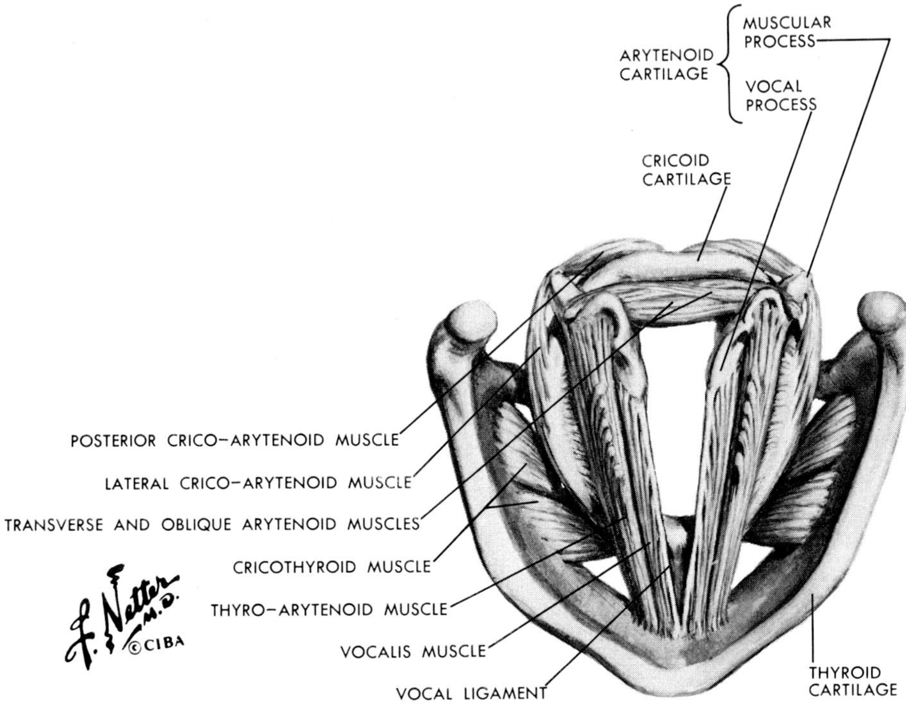

Figure 7-21. The intrinsic muscles of the larynx as viewed from above. Notice the two groups of fibers comprising each vocal fold. (© Copyright 1964, CIBA Pharmaceutical Company, Division of CIBA-GEIGY Corporation. Reprinted with permission from CLINICAL SYMPOSIA illustrated by Frank H. Netter, M.D. All rights reserved.)

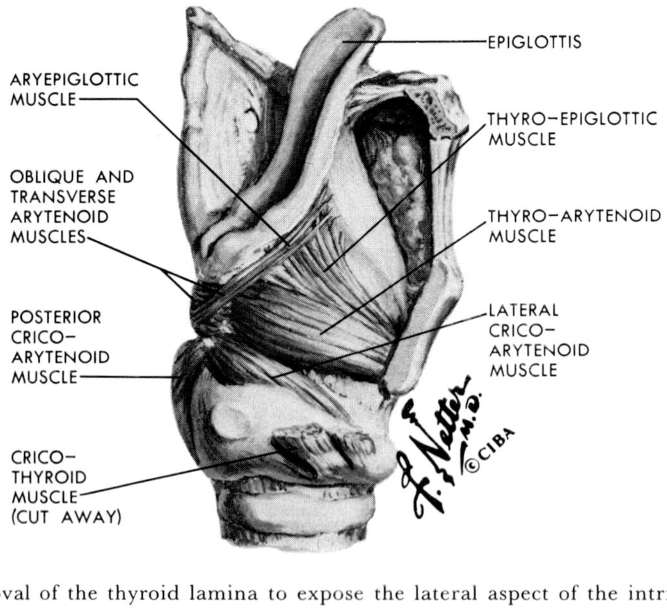

Figure 7-22. Removal of the thyroid lamina to expose the lateral aspect of the intrinsic muscles of the larynx. (© Copyright 1964, CIBA Pharmaceutical Company, Division of CIBA-GEIGY Corporation. Reprinted with permission from CLINICAL SYMPOSIA illustrated by Frank H. Netter, M.D. All rights reserved.)

meshing with fibers of the *aryepiglottic muscle*. These two muscles serve as sphincter muscles of the aditus, or entrance to the larynx. These muscles, the aryepiglotticus and the thyroepiglotticus, are active during swallowing. They are asssisted by the horizontal portion of the thyroarytenoid and transverse arytenoid in closing the entrance of the larynx for executing the second stage of deglutition (Fig. 7-22).

Ventricularis

Another segment of the thyroarytenoid muscle consists of muscle fibers within the ventricular fold. These few fibers are found along the wall of the laryngeal ventricle (the cavity between the ventricular fold or false vocal fold and the true vocal fold). These fibers, sometimes termed the *ventricular muscle*, arise from the lateral border of the arytenoid cartilage and end in the lateral margin of the epiglottis. The action of the ventricularis apparently affects the ventricular folds. If employed during phonation, the laryngeal ventricle would be erased as the ventricular fold collasped onto the vocal fold. Although the ventricular folds may not vibrate vigorously during phonation, they would interfere with vibration of the vocal folds. This disorder of voice production is termed the *ventricular voice*, or *dysphonia plicae ventricularis* in the medical literature (disturbance of the voice involving the ventricular folds; *plicae* = folds).

Vocalis

A number of anatomists do not regard the vocalis as being separate from the thyroarytenoideus. It is sometimes considered as part of the confluent muscle mass comprising the vocal folds. Certainly, gross dissection would support this view. Here, it is considered as a distinct muscle,

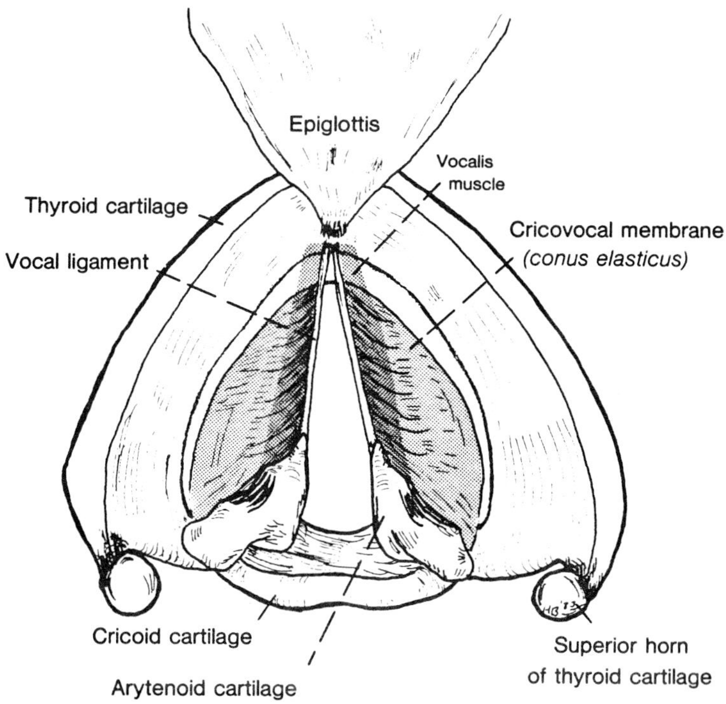

Figure 7-23. The vocal glottis as viewed from above schematically representing the cricovocal membrane and the vocalis muscle.

however, since a simplistic approach is followed for the sake of clarity in light of the introductory nature of this presentation.

The vocalis is a band of fibers that appears as the medial portion of the thyroarytenoid muscle. These fibers run from the angle of the thyroid lamina to the vocal process of the arytenoid cartilage posteriorly. It adheres to the vocal ligament located at its medial border and is covered by the cricovocal membrane, which is better known as the *conus elasticus*, or the elastic cone of the larynx. This is a thickened, fibrous, elastic membrane lying over the vocalis and serves as the basement for the mucosal or surface covering of this membrane. Some of the fibers of the vocalis take their origin from this membrane and its ligament, the vocal ligament (Figs. 7-23 & 7-13).

When contracting, the vocalis bunches up the conus elasticus and its mucosa, which would have the effect of thickening the vocal fold, an important factor for production of low-pitched tones. The vocalis is also responsible for increased tension of the vocal folds as with the thyroarytenoideus; however, the more delicate adjustments of the folds for pitch modulation are relegated primarily to the vocal muscle, with the other muscles serving important ancillary roles. The mucous membrane over the vocal muscle itself may be thought of as the principal vibrating part of the vocal generator.

The vocalis muscle is said to have a particularly rich nerve innervation. This would enable the brain to exert rather precise control over the vocalis, permitting specialized contractual activity in the muscle to achieve finely graded configurations needed for different aspects of phonation as in the control of voice quality. But much of the precise function of the vocalis muscle is only speculative until further research provides definitive information.

In summary, it should be noted that the thyroarytenoid muscle itself acts antagonistically to the cricothyroideus. That is, the thyroarytenoideus shortens the A-P length of the vocal folds. This muscle will, therefore, determine the degree of tension of the folds and is sometimes described as the internal tensor. In so doing, the thyroarytenoideus, along with its opposite mate, narrows the space between the vocal folds to close the glottis. Thus, together with its central muscle mass, the vocalis, the thyroarytenoideus is important in changing the shape and tension of the vocal folds to make fine adjustments for phonation. In addition, this muscle may cause the folds to bunch up, which increases the bulk of the vocal folds. Normally, this muscle is highly elastic. If weak, the folds may appear bowed, resulting in a disturbance of the vocal fold sound.

Adductor Muscles

There are three intrinsic laryngeal muscles involved specifically with closure of the vocal folds. They are the *lateral cricoarytenoid, oblique arytenoid*, and the *transverse arytenoid* muscles. One other muscle, the posterior cricoarytenoid, is responsible for opening the vocal folds. The adjusters of the glottis, then, will be considered in this group of intrinsic muscles (Fig. 7-24).

Cricoarytenoideus Lateralis

The lateral cricoarytenoid muscle arises from the upper border of the arch of the cricoid cartilage. It courses obliquely upward and backward and is inserted into the muscle process of the arytenoid cartilage.

The basic movement resulting from contraction of this muscle is adduction of the vocal process so as to approximate the

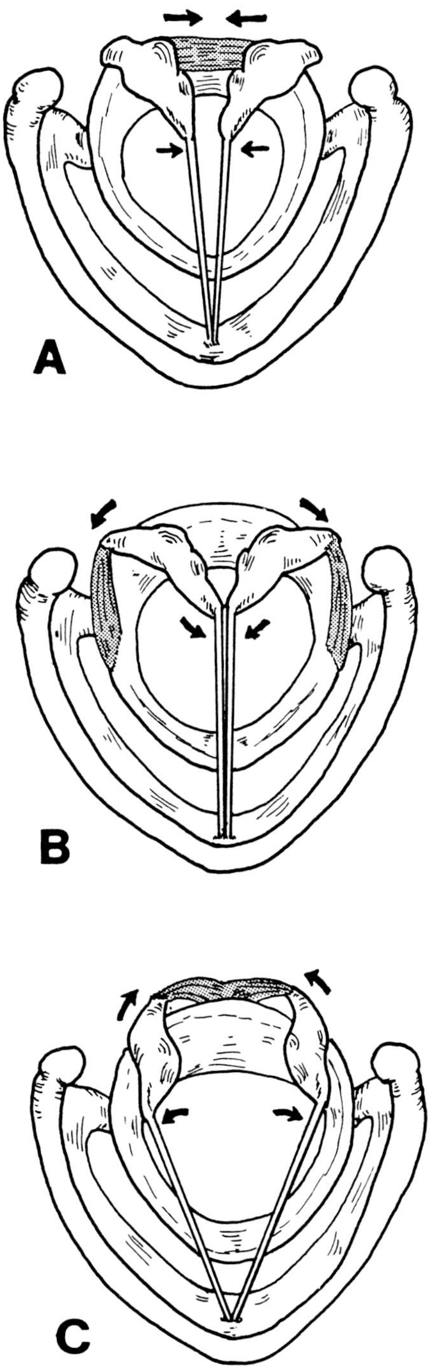

Figure 7-24. (A) Action of the transverse arytenoid muscle. (B) Action of the lateral cricoarytenoid muscles. (C) Action of the posterior cricoarytenoid muscles.

vocal folds and close the glottis. The arytenoid cartilage is caused to rotate, moving the vocal process medially, where it meets its mate from the opposite side. There is, additionally, some pull on the arytenoid cartilage in anterior and outward directions, depending on the location of the various fibers of this muscle. It should be mentioned that most but not all anatomists believe that the arytenoid cartilage is able to rotate on a vertical axis. The position of the glottis during a whisper indicates adduction is taking place (Fig. 7-25).

Arytenoideus Transversus

The posterior aspects of the arytenoid cartilages are joined by the transverse arytenoid muscle. Since both ends are equally movable, there is no distinct origin or insertion. Rather, both cartilages are pulled medially to adduct the vocal folds. This muscle appears to merit the designation of the true adductor of the arytenoids. Of course, approximation of the arytenoids at the midline requires a sliding motion on the part of the arytenoids. Perhaps it is appropriate to suggest that this interarytenoid muscle is the principal adductor of the vocal folds for phonation.

The transverse arytenoid is the only unpaired intrinsic muscle of the larynx. In addition to its function of approximating the vocal folds, it also functions as a structural boundary to close the posterior part of the glottis. Occasionally, this muscle and its close ancillary muscle, the oblique arytenoid, are included under one term, the *arytenoideus*.

Arytenoideus Obliquus

This small muscle, like the transverse arytenoid, interconnects the two arytenoid cartilages. It begins at the base of one arytenoid cartilage and ends on the

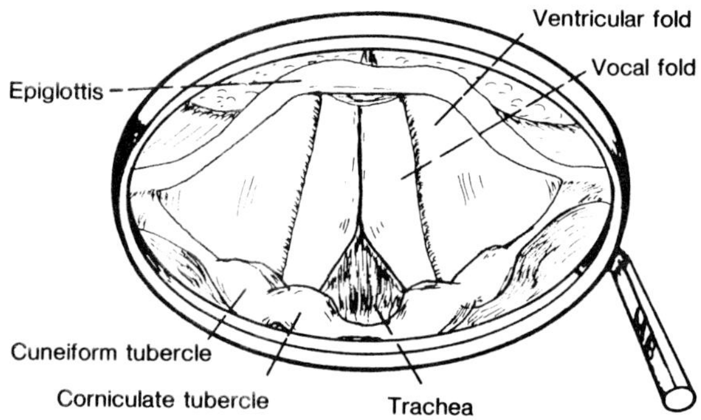

Figure 7-25. Shape of the glottis during a whisper. (Notice that a whisper may also be produced with the entire margins of the folds slightly abducted.)

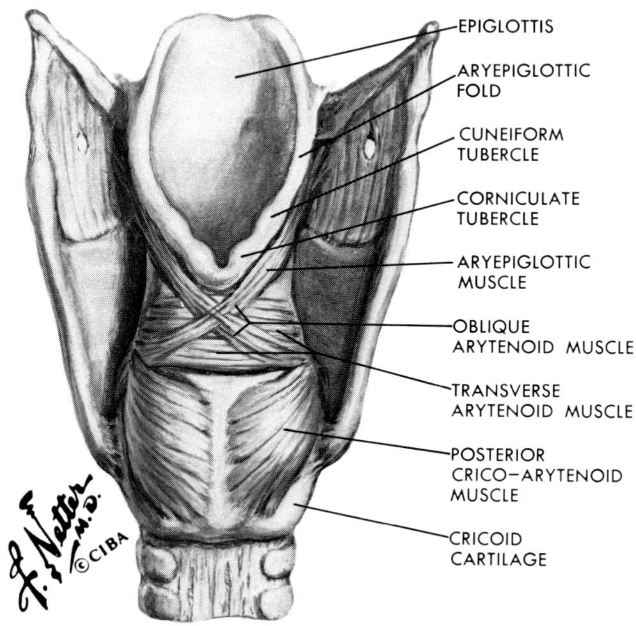

Figure 7-26. Posterior view of the larynx showing the intrinsic muscles. (© Copyright 1964, CIBA Pharmaceutical Company, Division of CIBA-GEIGY Corporation. Reprinted with permission from CLINICAL SYMPOSIA illustrated by Frank H. Netter, M.D. All rights reserved.)

apex of the other; therefore, there are two muscles. When viewed posteriorly, the two oblique muscles form an X as they cross each other over the transverse muscle. Instead of stopping at the apex of the arytenoid cartilage, some of the fibers continue around the lateral edge of the cartilage into the aryepiglottic fold. These fibers are regarded as part of a separate muscle, the aryepiglotticus, which is discussed later. Nonetheless, the oblique and the aryepiglottic are considered collectively as active agents in bringing the aryepiglottic folds together to guard the laryngeal inlet.

While the pair of oblique arytenoid muscles may be considered as adjustors of the sphincter at the entrance into the larynx (the aryepiglottic folds), the principal function is that of approximating the arytenoid cartilages, which close the glottis. This muscle is active in swallowing and may provide sensitive adjustments of the approximated folds during phonation (Fig. 7-26).

Abductor Muscle
Cricoarytenoideus Posterior

This muscle has been termed the *safety muscle of the larynx*, or the *laryngeal dilator*. It is the only muscle abducting the vocal folds that opens the airway of the larynx.

It arises from a broad area over the posterior outer surface of the cricoid cartilage. After taking its origin from the medial cricoid lamina, the posterior cricoarytenoid courses upward and lateralward to insert into the muscular process of the arytenoid cartilage. The muscle appears fan-shaped, with both vertical and oblique fibers present. The vertically directed fibers counter the pull of the thyroarytenoideus and keep the arytenoid cartilages from being pulled foward into the cavity of the larynx. The oblique fibers tend to level out at the top and become almost horizontal. These fibers pull the muscular process of the arytenoid cartilage medially, abducting the vocal folds by causing the arytenoids to rotate and slide outward laterally. The posterior cricoarytenoid muscle is antagonist to several others but principally to the lateral cricoarytenoid.

It has been noted that this muscle is responsible for the cessation of voicing during speech by abducting the folds for production of voiceless stops and fricatives in both medial and final positions. The cricoarytenoideus posterior is continuously active during respiration. During quiet breathing it keeps the folds out of the midline to permit air to be inhaled for deep or forced inhalation. This muscle strongly abducts the folds to achieve the laryngeal dilator maneuver (Fig. 7-24).

The Upper Sphincter of the Larynx
Aryepiglotticus

The aryepiglottic muscle is not a well-defined structure. These muscle fibers are embedded with the aryepiglottic fold surrounding the pear-shaped opening (aditus) into the laryngeal cavity. This may be referred to as the upper sphincter of the larynx, the lower sphincter being at the level of the glottis.

This muscle, which is not always present, is seen as a continuation of the oblique arytenoid muscles. Fibers also rise from the apical ends of the oblique arytenoid muscles and the apex of the arytenoid cartilages to travel upward and lateralward around the margins of the aditus. They insert into the quadrangular membrane (the membrane extending from the epiglottis to the arytenoid cartilage and covering the lateral wall of the laryngeal vestibule) and the lateral margin of the epiglottis. The aryepiglotticus serves to effect a sphincteric movement of the aryepiglottic folds to close the laryngeal entrance.

Extrinsic Muscles of the Larynx

There are two categories for the extrinsic laryngeal muscles, elevators of the larynx and depressors of the larynx. Most vivid of such activity is the movement of the larynx up and down during swallowing. Such movements during speaking or singing are often too slight to notice. Through the laryngeal elevators and depressors, the larynx may be rapidly raised and lowered during swallowing or adjusting the resonating characteristics of the pharynx during phonation.

The hyoid bone, located below the mandible, provides attachments for nine pairs of muscles. Through these muscles, the hyoid bone is related to the tongue, pharynx, temporal bone, mandible, scapula, larynx, and sternum. By means of the stylohyoid ligament, the hyoid bone is attached to the skull with more firmness than is provided by the muscles in this region. Since there are no articulations with any other bone, the hyoid bone is highly mobile. It forms a base for the tongue and is sometimes referred to as the *bone of the tongue*. Typically, the hyoid is viewed as belonging to the larynx due to suspensory ligaments and muscles tying the structures together.

Muscles moving the hyoid bone also move the larynx, since the two structures are connected by fibrous tissue plus muscles. Thus, if the hyoid bone is jerked upward, tilted forward, or pulled backward, the larynx is forced to follow. By action of the thyrohyoid muscle, the two structures are brought closer together, while the thyrohyoid membrane acts as a check ligament to limit the maximum distance of separation.

The extrinsic muscles of the larynx are discussed below under the headings denoting their influence on the larynx, i.e. elevators or depressors. Some of these are discussed in association with the tongue and mandible and are only listed here along with the extrinsic laryngeal musculature (Fig. 7-27).

Principal Elevators of the Larynx

These muscles are part of the suprahyoid complex and the tongue. The principle elevators include the digastric, mylohyoid, stylohyoid, geniohyoid; also, as supplemental elevators, the hyoglossus and genioglossus, and the thyrohyoid muscles. Many of these only weakly elevate the larynx when acting alone but effectively lift the larynx by acting in concert or in synergy.

There are four sets of suprahyoid muscles. They are the digastric, mylohyoid, geniohyoid, and stylohyoid. The classification of suprahyoids is given owing to their arrangement between the skull and mandible and the hyoid bone. Their function is to raise the hyoid bone, which in turn raises the larynx, if the mandible is fixed by action of the primary muscles of mastication. On the other hand, if the infrahyoid muscles are fixed (those below the hyoid bone), the mandible will be depressed upon contraction by the suprahyoid muscles attached to the mandible. As implied above, all the suprahyoid muscles are paired. Three cranial nerves, facial, trigeminal, and hypoglossal, innervate the suprahyoids.

The action of the suprahyoid muscles can be demonstrated by lightly grasping the hyoid bone at the angle of the neck with the thumb and index finger and then swallowing. The hyoid bone will cycle up and down with each swallow (Table 7-II).

Digastricus

The digastric muscle is a two-part muscle. It has an anterior belly relating the hyoid bone to the mandible and a posterior belly relating the hyoid bone to the skull through the mastoid process of the temporal bone. A strong, round tendon connects the anterior and posterior bellies. This intermediate tendon is attached to the hyoid bone by a fibrous loop, a pulley, similar to the small trochlea of the superior oblique muscle of the eye. Since this fascial loop of connective tissue is loose around the intermediate digastric tendon, the tendon is permitted to slide in this sling.

The posterior belly arises from the medial surface of the mastoid process. Its course is downward and forward, ending in the intermediate tendon. The anterior belly arises from a depression on the inner

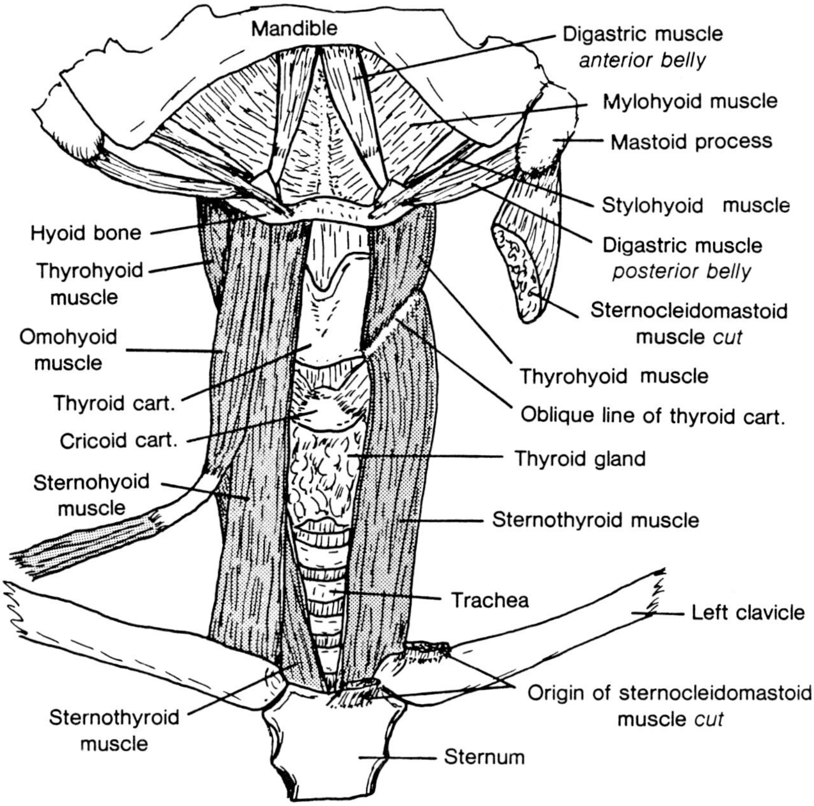

Figure 7-27. Extrinsic muscles of the larynx.

Table 7-II

PRINCIPAL ELEVATORS OF THE LARYNX

Muscle	*Origin*	*Insertion*	*Action*
Digastric	Mastoid notch	Mandible near mental symphysis	Raises hyoid bone and base of tongue
Mylohyoid	Mylohyoid line of mandible	Mylohyoid raphe and hyoid bone	Raises hyoid bone, floor of mouth, and tongue, or depresses mandible
Stylohyoid	Styloid process	Hyoid bone	Raises and retracts hyoid bone
Geniohyoid	Lower genial tubercle on mandible	Hyoid bone	Raises and protracts hyoid bone

side of the lower border of the mandible, in the vicinity of the mental symphysis and below the anterior point of the mylohyoid line of the mandible. Because this is potentially the most mobile end of the digastric muscle, it is designated as the insertion. From its anterior attachment, the anterior belly courses posteriorly and inferiorly to the intermediate tendon in the fascial pulley of the hyoid bone.

Upon contraction, the two parts of the digastric muscle, acting together, depress the chin and aid in opening the mouth when the infrahyoid muscles fix the hyoid bone. They elevate the hyoid bone when the mandible is fixed by the primary muscles of mastication, especially as during swallowing. The anterior belly draws the hyoid bone forward, and the posterior belly draws it backward.

Neural innervations to the anterior belly are by a branch of the trigeminal (V) cranial nerve. The posterior belly is innervated by a branch of the facial (VII) cranial nerve.

Mylohyoideus

The two mylohyoid muscles together make up what was once designated as the *oral diaphragm*, a term that has now passed out of general use. Structurally and functionally, this pair of muscles constitutes the floor of the oral cavity. A tendinous strip, the mylohyoid raphe at the midline, unites the two muscles to seal the floor of the oral cavity and support the tongue.

Each mylohyoid arises from a long ridge, the mylohyoid line, on the inner surface of the body of the mandible. The line extends from the mental symphysis anteriorly to the third molar posteriorly. From along this line, the fibers converge toward the mylohyoid raphe and the body of the hyoid bone. The fibers of each muscle are ordered in the shape of a triangle.

Branches of the trigeminal (V) cranial nerve supply the mylohyoid muscle. The action of this muscle is to raise the hyoid bone and tongue, as during initiation of the swallowing act. With the hyoid bone fixed, it aids in depressing the mandible. The anterior fibers, those meeting at the midline and inserting into the mylohyoid raphe, will, upon contraction, elevate the tongue in a fashion similar to the operation of a trampoline during its thrusting movment.

Geniohyoideus

A small cylindrical muscle arises above the anterior end of the mylohyoid line on the inner surface of the mandible and near the mental spines. It is the geniohyoid muscle. Coursing directly backward and slightly downward above the mylohyoid muscle, the geniohyoid inserts into the upper half of the hyoid body. During its course toward the hyoid bone, the fibers spread somewhat so that each geniohyoid muscle meets the other at the midline.

Upon contraction, the geniohyoid elevates and draws the hyoid bone forward, which in turn draws the tongue forward. If the hyoid bone is fixed by other muscles, the geniohyoid depresses the mandible. Fibers from the first cervical nerve innervate this muscle (Fig. 7-28).

Stylohyoideus

The stylohyoid muscle, unlike the other suprahyoid muscles, is not part of the submandibular musculature. It is a slender round muscle located above and in front of the posterior belly of the digastric muscle. The origin is the posterior surface of the styloid process of the temporal bone. From here it travels downward and forward along the upper border of the posterior belly of the digastric muscle. Then the stylohyoid divides to straddle the intermediate tendon of the digastric muscle at the junction of the greater horn and body of the hyoid bone.

The action of the stylohyoid is to draw the hyoid bone up and rearward during

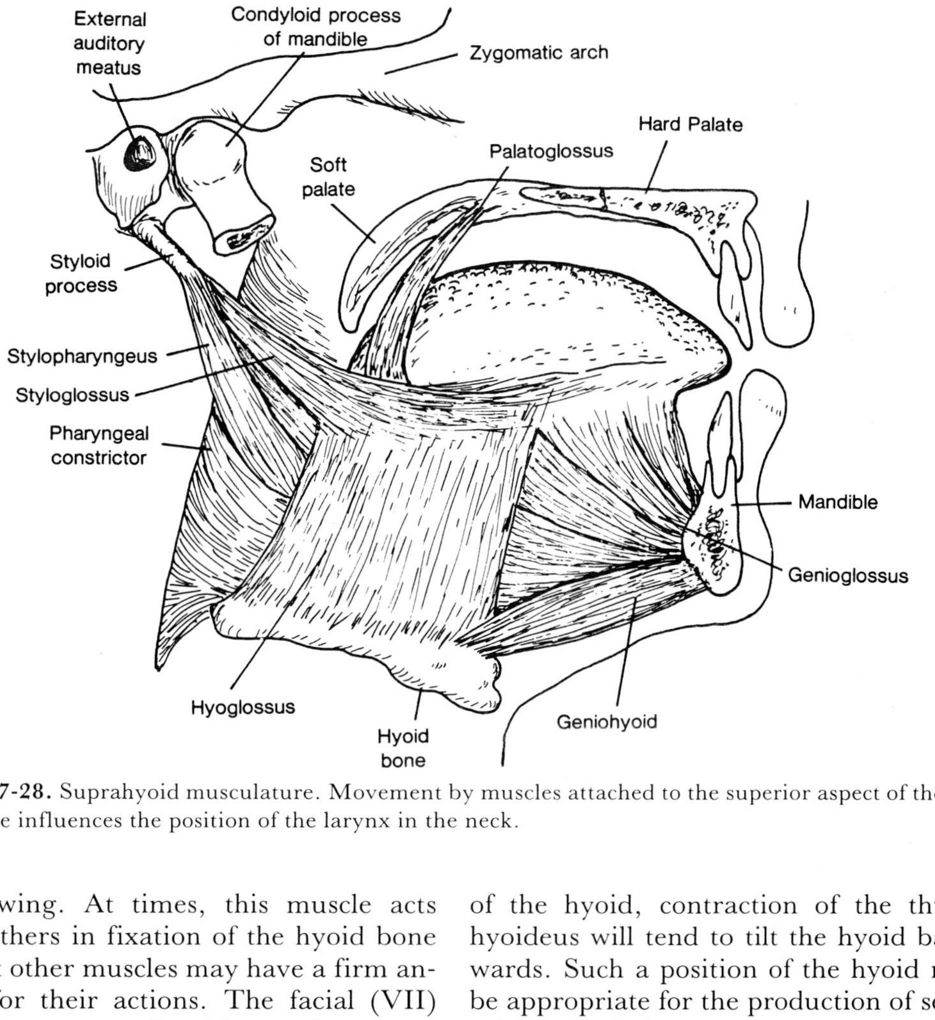

Figure 7-28. Suprahyoid musculature. Movement by muscles attached to the superior aspect of the hyoid bone influences the position of the larynx in the neck.

swallowing. At times, this muscle acts with others in fixation of the hyoid bone so that other muscles may have a firm anchor for their actions. The facial (VII) cranial nerve supplies this muscle.

Supplemental Elevators of the Larynx

Thyrohyoideus

This muscle shortens the distance between the larynx and the hyoid bone. If the hyoid bone is fixed, the action of the thyrohyoid lifts the larynx. If the hyoid bone is not held firm by the suprahyoid muscles, then the hyoid bone and the larynx will be depressed, causing the thyroideus to become a depressor of the larynx.

With its insertion in the greater cornu of the hyoid, contraction of the thyrohyoideus will tend to tilt the hyoid backwards. Such a position of the hyoid may be appropriate for the production of some velar articulations.

It is listed as a strap muscle of the neck, although it is perhaps the shortest of the extrinsic muscles. Due to its location below the hyoid bone, it is termed an *infrahyoid muscle*; however, the other infrahyoid muscles are depressors of the larynx.

Arising from the oblique line or ligament on the lamina of the thyroid cartilage, the thyrohyoideus courses vertically to insert into the lower border of the greater horn of the hyoid bone. A good case can be made to support the contention that the thyrohyoid and sternothy-

Table 7-III

SUPPLEMENTAL ELEVATORS OF THE LARYNX

Muscle	*Origin*	*Insertion*	*Action*
Thyrohyoid	Oblique line of the thyroid cartilage	Greater horn of the hyoid bone	Raises larynx or pulls hyoid bone down
Hyoglossus	Hyoid bone	Tongue	Can raise hyoid bone
Genioglossus	Upper genial tubercles	Hyoid bone (lower fibers)	Can protract and raise hyoid bone

roid are the same muscle, interrupted only by a ligament at the thyroid lamina. This is probably anatomically accurate in most specimens (Table 7-III).

The thyrohyoid is supplied from the ventral ramus of the first cervical nerve by fibers brought to the muscle through the hypoglossal nerve. The innervation is through the ansa cervicalis (ansa = looplike) nerve loop in the neck attached in front and above to the hypoglossal cranial nerve and behind to the upper cervical spinal nerves.

Hyoglossus

Regarded as a muscle of the tongue, the hyoglossus rises from the body and greater cornu of the hyoid bone. The fibers course up to the tongue, inserting into the tongue tip and interdigitating with other extrinsic tongue musculature in the root of the tongue.

When the tongue is fixed by its extrinsic muscles such as the palatoglossus, styloglossus, and genioglossus, contraction of the hyoglossus will aid in elevating the larynx. However, it should be noted that hyoglossus is a principal muscle of the tongue (see Chap. 12).

Genioglossus

Another muscle influencing both the larynx and tongue, but regarded as a major muscle of the tongue, is the genioglossus. This muscle forms most of the muscular core of the tongue (see Chap. 12).

This muscle takes its origin from the upper genial tubercles (mental spine) of the mandible. From this inside frontal position of the mandible, the fibers spread in a fan pattern upward to the tongue tip and horizontally and posteriorly to the body of the hyoid bone and interior surface of the epiglottis. It is these posteriorly directed fibers that are here regarded as possible elevators of the hyoid bone and larynx, but only if the mandible is held fixed while the genioglossus contracts.

Depressors of the Larynx

Under this category, there are only three muscles that oppose the elevators. Of course, this group of muscles is assisted by gravitational forces during upright postures. The actual position of the larynx in the neck is basically determined by the balance of tonus between the depressor and elevator musculature. The sternohyoid, omohyoid, and sternothyroid muscles are referred to as strap muscles of the neck because of their shape. All are paired (Table 7-IV).

General anatomy texts treat these as infrahyoid muscles, which they are. But they are logically treated under the larynx from a functional classification instead of

Table 7-IV

THE LARYNGEAL DEPRESSORS

Muscle	*Origin*	*Insertion*	*Action*
Sternothyroid	manubrium of sternum and first rib	oblique line of the thyroid cartilage	depresses the larynx
Sternohyoid	manubrium of sternum and clavicle	body of hyoid bone	depresses hyoid bone and larynx
Omohyoid	superior border of scapula	inferior border of the hyoid bone	depresses hyoid bone and larynx

a placement classification. Production of low tones may be aided by the depressors of the pharyngeal resonator. Some of the depressors are individually active in the pitch-changing mechanism of the larynx.

Nerve loops, the ansa cervicalis, from the ventral rami of the first three cervical nerves send fibers to these muscles.

Sternothyroideus

In production of pitches at the upper and lower end of the vocal scale, the sternothyroideus is active. Its role is antagonistic to the cricothyroid to permit fine adjustments of the vocal folds in the act of phonation. Like the other infrahyoid muscles, the sternothyroid muscle assists in positioning the larynx in the neck by achieving a balanced tonus with the suprahyoid muscles.

This muscle arises from the area of the manubrium of the sternum and (generally and with variation) from the cartilage of the first rib. It rises as a strap muscle, passing over the thyroid gland to insert into the oblique line of the thyroid cartilage. Some fibers may blend with the thyrohyoid muscle, which lies in front of the sternothyroideus (Fig. 7-29).

Sternohyoideus

Although not as wide as the sternothyroideus, the sternohyoid muscle has essentially the same origin and course. It arises in front of the sternothyroid on the manubrium of the sternum, plus the medial end of the clavicle. The origin of the large neck muscle, sternocleidomastoideus, lies anteriorly to the root of the sternohyoid. After passing up and leaning inward on its course, the sternohyoid meets the lower border of the hyoid bone for its point of insertion. As this pair of muscles rises on either side of the larynx, they may touch one another at their medial borders, giving the appearance of a wide band of muscle encompassing the anterior of the upper part of the larynx.

Upon contraction, it draws the hyoid bone inferiorly, which in turn depresses the larynx.

Omohyoideus

The central tendon of the omohyoideus makes this muscle parallel in structure to the digastricus. The omohyoid muscle begins at the upper border of the scapula, known to most as the *shoulder blade*. It rises upward on a course toward the lower aspect of the larynx. The median tendon is found just as the omohyoideus passes behind the sternocleidomastoideus, where the muscle changes direction. From

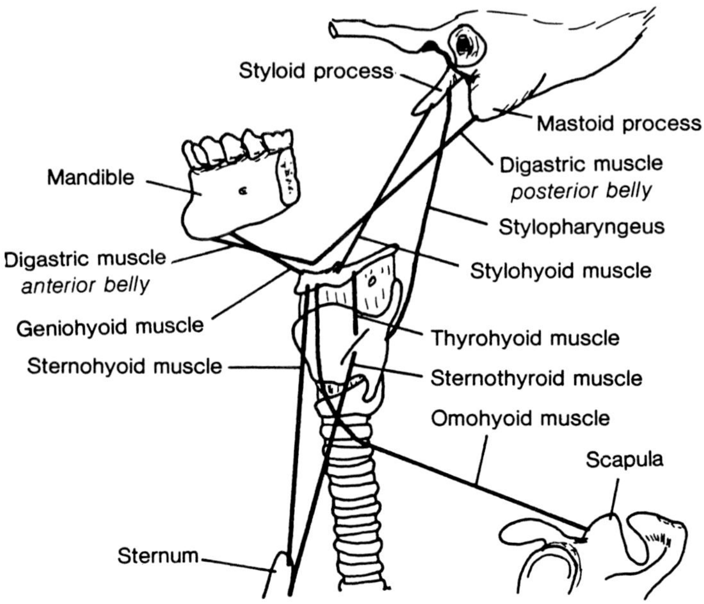

Figure 7-29. Schematic representation of the extrinsic muscles elevating and depressing the larynx.

this point on, the omohyoideus ascends straight upward, ending on the lower border of the body of the hyoid bone. This unusual angle in the muscle is held fixed by fascia of the neck region, which is attached to the clavicle below. A variation is that it may arise from the clavicle instead of the scapula.

The omohyoideus pulls the hyoid bone down upon contraction, depressing the larynx. The direction of pull is like that of the sternohyoid muscle. It also is capable of carrying the larynx backward or tilting it to one side if only one of the pair acts. In synergism with other muscles, the omohyoid has been shown to play a role in depressing the mandible.

The Suspensory System

The foregoing treatment of the extrinsic muscles of the larynx indicates a very complex mechanism for elevating and depressing the larynx. It should be pointed out that the picture is more complex than discussed in this section, for a group of elastic ligaments and membranes support the hyoid bone and the thyroid and cricoid cartilages. This suspensory system, together with the intrinsic muscles already presented and their physiology to be presented next, bind and support the larynx into a remarkable working unit.

The principal suspensory system is concerned with the suspension of the hyoid bone, allowing it considerable mobility. The major ligament is the stylohyoid, which suspends the hyoid from the base of the skull. Other ligamentous membranes connect the hyoid to the thyroid cartilage, the thyroid cartilage to the cricoid cartilage, and finally the cricoid cartilage to the upper tracheal ring.

The role of the extrinsic laryngeal musculature and the accompanying suspensory system on the action of the vocal folds during phonation and respiration is not well understood at present. The system often effectively shields its detailed and delicate actions from investigators.

BIBLIOGRAPHY

Dickson, D.R., and Maue, W.M.: *Human Vocal Anatomy*. Springfield, IL.: Thomas, 1977.

Gardner, E., Gray, D.J., and O'Rahilly, R.: *Anatomy: A Regional Study of the Human Structure*, 4th ed. Philadelphia, Saunders, 1975.

Grant, J.C.B.: *An Atlas of Anatomy*, 6th ed. Baltimore, Williams and Wilkins, 1972.

Gray, H.: *The Anatomy of the Human Body* (Charles M. Goss, Ed.), 28th ed. Philadelphia, Lea and Febiger, 1973.

Myerson, M.C.: *The Human Larynx*. Springfield, IL.: Thomas, 1964.

Neiman, G.S., Zemlin, W.R., and Meyerson, M.D.: Some Observations on Anatomy of the Human Larynx. Presented at annual convention of the American Speech and Hearing Association, New York, 1970.

Romanes, G.J.: *Cunningham's Manual of Practical Anatomy*, 14th ed. New York, Oxford U Pr, 1978, vol. 3.

Saunders, W.H.: *The Larynx*. Summit, NJ: CIBA, 1964.

Yokochi, C.: *Photographic Anatomy of the Human Body*, 2nd ed. Baltimore, Univ Park, 1978.

Chapter 8

LARYNGEAL PHYSIOLOGY

GENERAL INTRODUCTION

THE larynx is basically a constrictor mechanism situated between the trachea and the pharynx. For its size, the larynx is a highly versatile instrument. While the larynx is capable of highly versatile activities, it is common for most lay persons to think of the larynx only as an organ for production of speech and singing. But the essential laryngeal functions include (a) glottal closure for holding back air in the thorax for postural efforts (heavy lifting), (b) ventilation of the lungs (inspiration and expiration), (c) serving as a barrier and deflector of potential lung invasions from nongaseous material (food and drink), (d) sound production (phonation), and (e) clearing of the upper airway (coughing).

Laryngeal tones may be the mirror of us in several ways. From the characteristic vocal quality we use, our sex, to an extent our age, and our emotional state are revealed. There are all kinds of voices around us. We can pick out many that are unusually smooth and artistic as well as those which grate, squeak, or have so much breathiness that they are almost too weak to be heard. Those voices belonging to confident, self-assured individuals are often quickly, albeit often inaccurately, distinguished from those of shy, insecure persons.

In this chapter, we will examine the function of the larynx as a sound generator and how we use it to produce sounds to serve us on all occasions and to display the personality to which it is linked. Here, we will begin with a focus on the structures surrounding the laryngeal glottis, since it is this aspect of the cavernous larynx that enables our instrument of voice to serve its several roles so well.

That variable space between the margins of the true vocal folds is known as the *rima glottidis* (literally, the "crack of the glottis"), or simply the *glottis*. This landmark in the cavity of the larynx is a reference point to the passageways above and below the vocal folds. Hence, the area above the glottis is the supraglottic area and below it is the subglottic or infraglottic area (Fig. 8-1).

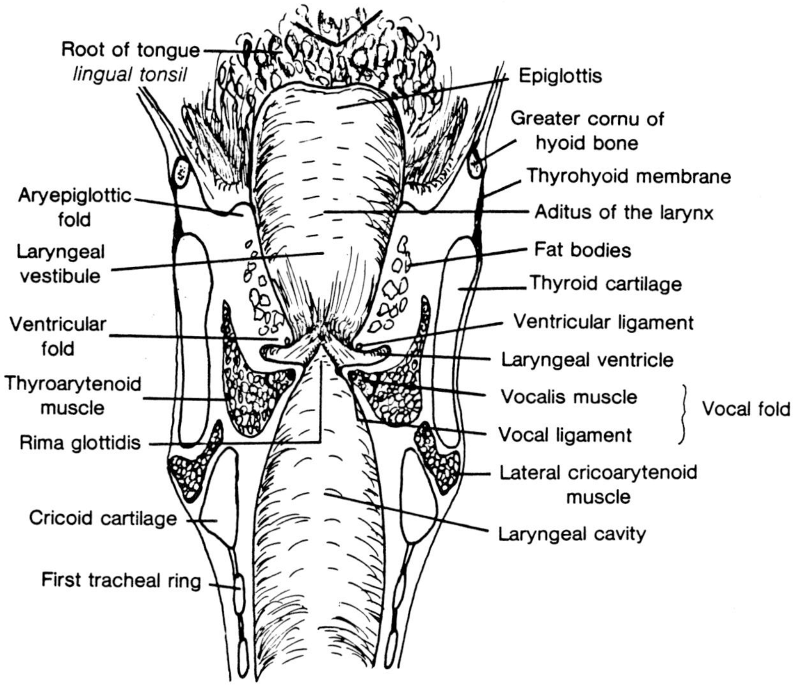

Figure 8-1. The laryngeal glottis as seen in a cross section from behind. The margins of the vocal folds form a V pointing anteriorly.

THE VOCAL FOLDS AND GLOTTIS

The shape of the glottis during quiet respiration is triangular, with the point at the front anterior to the neck. Termed the *vocal glottis*, the margins of the folds are formed by the vocal ligament and a mucous membrane lying over the ligament, which extends from the posteriorly placed arytenoid cartilage to the thyroid cartilage anteriorly.

The point of anterior attachment of the vocal ligament is relatively close to its mate, and the anteromost point of the glottis extending between the vocal ligaments is known as the *anterior commissure*. On the posterior attachment, the ligament ends on the vocal process of the arytenoid cartilage; this marks the posterior limit of the *membranous glottis*, or the principal vibrating portion. But the glottis continues posteriorly and passes medially to the arytenoid cartilages. This posterior portion is referred to as the *cartilaginous glottis*. It ends with the posterior limits of the arytenoids at the *posterior commissure* (Fig. 8-2).

In a nonphonatory state, such as quiet breathing, the length of the membranous glottis is about 15.5 mm (slightly greater than one-half inch) for males and 11.5 mm (just under one-half inch) for females. The length of the cartilaginous glottis is around 7.5 mm (slightly greater than one-fourth inch) in males and 5.5 mm (just under one-fourth inch) in females.

Since a part of the mucous membrane covering the cartilaginous glottis participates in phonation by vibrating along

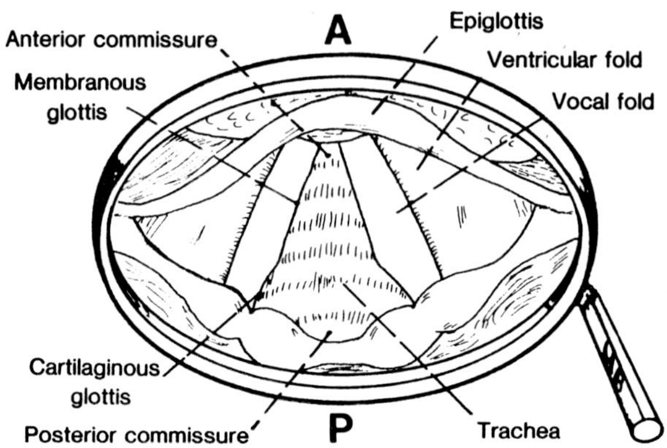

Figure 8-2. Mirror view of the larynx, with the anterior and posterior commissures of the glottal area indicated. The anterior (A) and posterior (P) aspects of the larynx are indicated.

with the membranous glottis, the measurement of the active portion of the vocal fold is lengthened. The A-P (anterior-posterior) measurement of the membranous glottis is maximal during at-rest breathing and is about three-fifths of the total glottal length.

The membranous glottis is over twice as long as the cartilaginous glottis. Because of these different lengths, the glottis may be roughly divided into thirds for a clinical application. Around two-thirds (really, three-fifths) of the glottis is membranous and about one-third is cartilaginous. Carefully note the location of the point on the glottis that represents the *junction* of the anterior third meeting the posterior two-thirds along the edge of the vocal folds. This site is almost always the location of vocal nodules arising from vocal abuse because it represents the area along the margin of the folds that is most active during vibrations (see Figs. 8-3 & 8-26).

Attached to the vocal ligament and sharing the same locus of attachments anteriorly and posteriorly is the *vocal musculature*. Those muscle fibers that are attached to the vocal ligament, and to the same anterior and posterior structures as this ligament, comprise the muscular vocal fold.

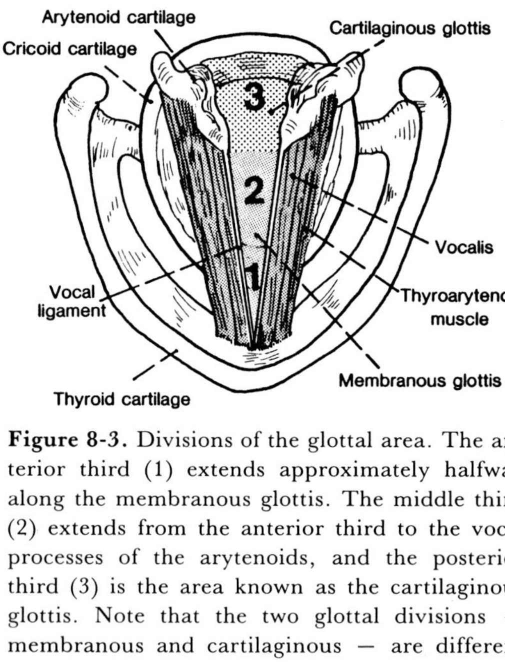

Figure 8-3. Divisions of the glottal area. The anterior third (1) extends approximately halfway along the membranous glottis. The middle third (2) extends from the anterior third to the vocal processes of the arytenoids, and the posterior third (3) is the area known as the cartilaginous glottis. Note that the two glottal divisions — membranous and cartilaginous — are differentially shaded.

The larynx, like other areas of the body with surfaces, is lined with epithelium (lining cells closely adhered to each other). Epithelium cells rest on supportive connective tissue, which contacts the epithelium at the basement membrane, as described microscopically. They perform a variety of activities and functions and are classified according to cell arrangement and shape. On the basis of cell shape, epithelium can be described as squamous (flattened), cuboidal (cube-shaped), or columnar (appearing in columns). The free surface of epithelial cells may contain cilia (motile hairs) or microvilli, which engage in absorptive activity.

The epithelial lining of the larynx, like that of many other structures, reflects its activities. Areas of stress and contact are lined with squamous epithelium, while other areas that are involved in moving mucous along the surface are lined with ciliated columnar cells (Fig. 8-4).

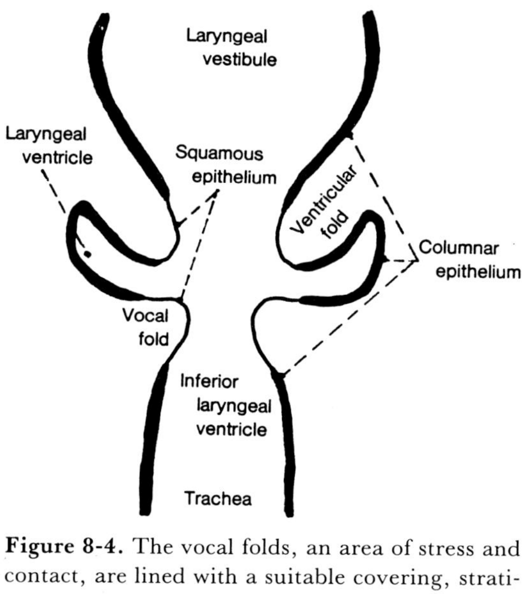

Figure 8-4. The vocal folds, an area of stress and contact, are lined with a suitable covering, stratified squamous epithelium. Other nonvalvular areas of the laryngeal cavities are lined with ciliated columnar epithelium designed for moving particulate material toward the pharynx.

That portion of the mucous membrane covering the vocal folds is stratified squamous epithelium with a pavement-like configuration of cell layers. The squamous epithelium is found only over the opposed surface of the folds, while the laryngeal cavities below and above the vocal folds are lined with ciliated columnar epithelium. The hair-like cilia sweep particulate material upward toward the pharynx.

The mucous membrane lining the laryngeal cavity adheres tightly to the basement membrane (conus elasticus) covering the vocal folds. Elsewhere, except for the epiglottis, the mucosal lining is only loosely adhered to the basement cell structures.

The mucous membrane forming the margins of the vocal glottis is pale due to its nonvascular attribute. Under lighting, such as that used during mirror examination of the folds, the folds appear glossy white. They are actually more a pale, slightly pink color. Near the anterior commissure, the color changes to yellowish owing to the elastic tissue in this area.

The laryngeal structures that control the size of the glottis play a prominent role in preventing aspiration of food, drink, or any foreign matter into the lungs. As part of the protective mechanism of the respiratory tract, the glottis is closed automatically by reflex, along with the epiglottis and other sphincteric structures above the vocal folds, when a foreign body enters the upper laryngeal area from the pharynx. The reflexive closure is so effective that in drowning, involuntary control keeps the glottis closed until death (Fig. 8-5).

During sound production, the larynx is capable of creating many variations in the shape of the glottis, thereby accounting for a wide range of complex sound wave forms. Closure of the glottis is affected by movement of the arytenoid cartilages toward midline (adduction) at the posterior aspect of the vocal glottis. The specific operation of the vocal folds in sound production is the topic of a later

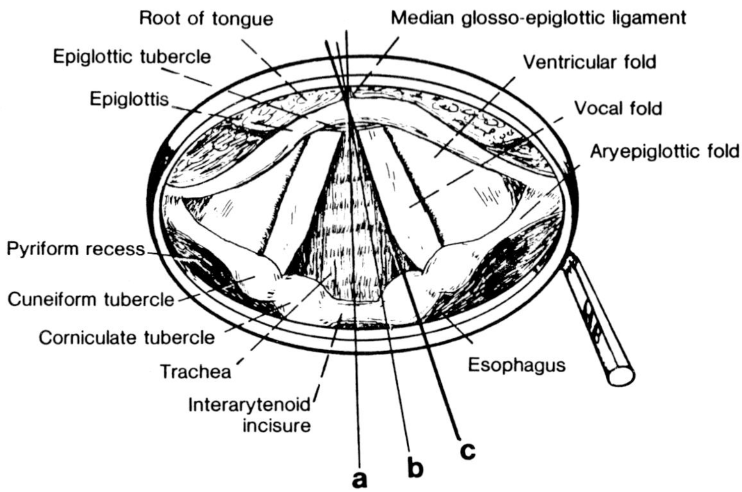

Figure 8-5. Positions of the vocal folds: (a) adduction, (b) paramedian, (c) abduction.

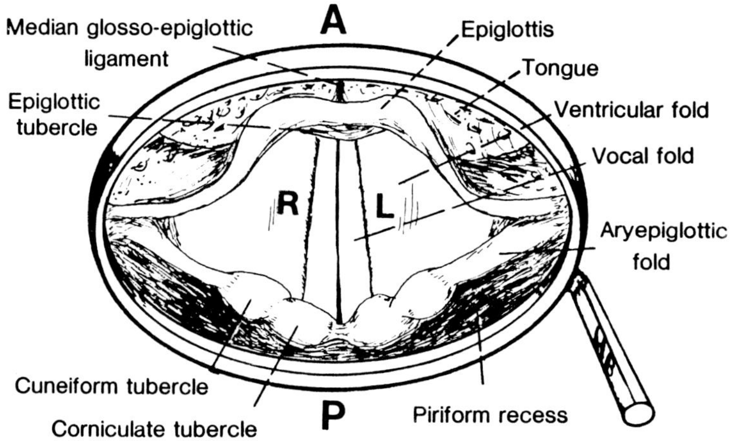

Figure 8-6. Mirror view of the vocal folds in adduction. Left and right sides of the larynx are indicated.

section in this chapter. However, it can be mentioned here that a sort of metered subglottic air pressure pushes through the closed glottis so as to cause a critical balance between the tension of the vocal folds and the subglottic air pressure. The rhythmic opening and closing of the vocal folds in the forced breath stream releases, with each opening of the folds, a pulse of air creating sound energy (Fig. 8-6).

THE VENTRICULAR FOLDS

The ventricular folds are not endowed with essential musculature for sound production purposes. They can be adducted through the sphincteric action of the laryngeal muscles, serving as an important part of the protective valvular apparatus.

During acts of swallowing and when containing excess intrapulmonary pressures to resist outside pressures, the ventricular folds may close as a complement to other parts of the laryngeal valve.

When the ventricular folds adduct, the result is usually a dysphonia (disorder of phonation). The chief symptom is hoarseness and low pitch with variations in severity. Termed *dysphonia plicae* (folds) *ventricularis*, hyperkinesia of the ventricular folds may be a functional disorder or secondary to organic disease of the vocal folds. Hyperkinesia of the ventricular folds may be elusive to observe, since the first inspection with the laryngeal mirror may reveal no abnormality whatsoever. But then suddenly the ventricular folds may sweep partially or even completely closed. The voice quality breaks like that of an adolescent (Fig. 8-7).

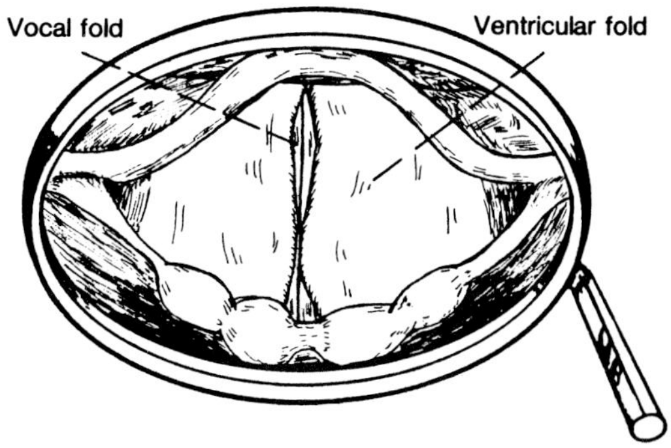

Figure 8-7. Hyperkinesia of the ventricular folds as viewed through a laryngeal mirror. Also termed *dysphonia plicae ventricularis*, this condition seems to be a functional disorder in some cases. Normally, the ventricular folds are positioned laterally to the vocal folds (see Fig. 8-6).

REFLEXIVE ACTIVITIES OF THE VOCAL FOLDS

Altogether, the human vocal mechanism can be viewed as a finely tuned musical instrument. It is the site of much tension buildup during emotional responses and is an important mode of expressing feelings in human interactions. Efficient production of voice gives promise of a vocal generator that will be of service, almost endlessly without failure. Implied in efficient voice production is (a) absence of strain, (b) coordination of the phonatory-breath stream, (c) soft onset of phonation, and (d) absence of constriction in the vocal tract above the glottis.

There are several interesting activities that may be classified as reflexive phonatory functions. Some of these promote efficient voice physiology and some are aversive to voice hygiene. Clinical applications are mentioned for each of the activities to emphasize the need by the voice clinician for having a good understanding of the range of vocal fold operations.

Laughing

Like artistic singing and yodeling, laughing phonation appears to be a natural use of the larynx in sound production. Laughter promotes an approach to phonation so that the laryngeal and supraglottic musculature are reflexively relaxed. It is an inborn reflex; vocal training is not a variable. The full, pleasing, and enriched resonance heard in most laughing phonation is a result of optimal coupling between the vocal folds and the supraglottic resonating spaces. Laughing phonation can be a facilitator of efficient vocal physiology in voice therapy. Occasionally, laughter involves inspiratory phonation and is especially contagious during this stage.

Coughing

Following a wide abduction of the folds for a large intake of air, the vocal folds adduct sharply to signal the early phase of a cough reflex. Subglottic pressure rises rapidly, and the entire larynx may go into a sphincteric closure with both the vocal and ventricular folds in tight adduction. The larynx is elevated. Subglottic air pressure violently opens all sphincters, causing the folds to oscillate and wave uncontrollably in the powerful breath stream, much like a flag whipped about by a wind storm. There is considerable abuse to the folds during coughing phonation, for the speed of the breath stream is above hurricane forces at the glottis. Light coughing, controlled coughing, must be used to suppress the vocal abuse in patients who have fold pathology. Careful vocal education by voice clinicians is needed to prevent compounding the poor health of the vocal folds in patients with dysphonias (Fig. 8-8).

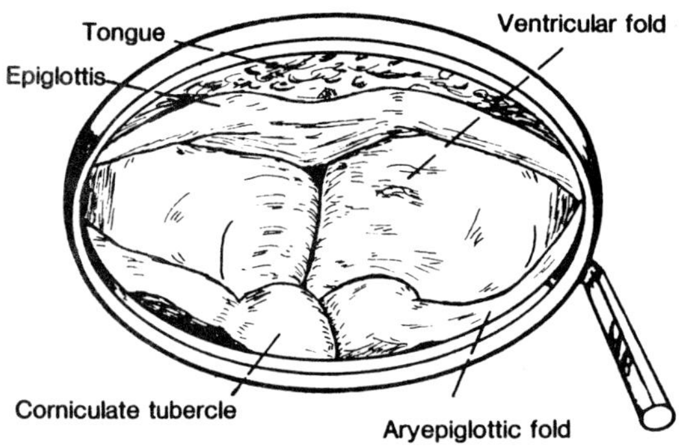

Figure 8-8. Mirror view of the larynx just prior to production of a cough. Notice the appearance of hyperkinesia of the entire larynx.

The grunt is perhaps most closely associated with coughing from a physiological viewpoint. It is a form of vocal hyperfunction. Weight lifting, pole vaulting, tennis playing, football playing, and even jogging, to name but a few, are sports where stressful phonation is heard due to the strenuous mental and physical efforts involved. Many laborers frequently use a grunt in the same fashion as those exerting strain while participating in sports. But chronic pain or physical discomfort may also be expressed as periodic grunts. Physiologically, the grunt is most likely to be a mild, controlled cough. The inhalation may be normal, but the exhalation is achieved by a sound burst as an elevated subglottic pressure opens the folds. There is no strong sphincteric closure of the larynx except in the most stressful grunts. Moaning is perhaps best classified as a mild, prolonged grunt. Patients on vocal rest must avoid these abusive vocalizations.

Throat Clearing

Throat clearing fits into the same abusive category as coughing, since the entire larynx goes into stressful movements. Vocal education may prevent vocal nodules and similar medical problems associated with laryngeal abuse associated with coughing and throat-clearing reflexes, which occur frequently in some patients.

Hiccupping

Hiccupping is more often mentioned under extrarespiratory reflexes due to the role of the diaphragm in producing this phenomenon. A deep inspiration first occurs with widely abducted folds. While the lungs are filling with air, a sudden contraction of the diaphragm causes a sharp closure of the vocal folds — an inspiratory Bernoulli effect. This abrupt termination of the inspiratory cycle produces the hiccup or hiccough. A variety of interesting and even amusing hiccup patterns may be heard. This inspiratory phonation can be heard during speaking as well as during quiet breathing.

Yawning

Everyone is familiar with the relaxing nature of yawning. The technique is inborn and is a natural reflex. After the mouth opens widely, the folds abduct widely to dilate the cavity of the larynx while a considerable volume of air is inhaled. Optional, partial, or soft adduction of the folds results in a relaxed, breathy phonation during the expiratory cycle. In the process, stale air in the lungs is expelled and replaced by fresh air. Additionally, the whole of the vocal tract is relaxed if social inhibitions are laid aside. Social situations often cause the yawn of the adult to be suppressed so that no phonation is heard (the "church" yawn). Therefore, employment of the yawn as a device to facilitate relaxation of the vocal tract musculature may be only limitedly successful for highly social conscious individuals. The soft attack is relatively easy to demonstrate through yawning phonation, since the larynx is relaxed and the folds adduct gently (Fig. 8-9).

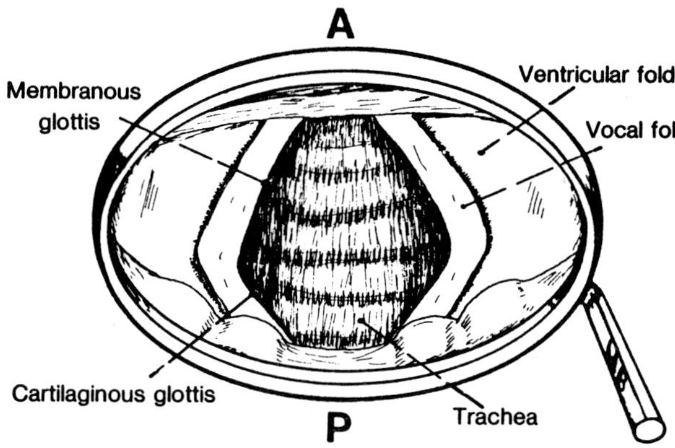

Figure 8-9. Mirror view of the larynx during deep inhalation. The glottal area is widest from anterior (A) to posterior (P).

INTRODUCTION TO VOCAL FOLD VIBRATION

Rapid modulation of the expiratory breath stream at the glottis produces phonation. In a simplistic description, we might infer that a series of surges of air pressure, perhaps like rapid pulsations, are created by the opening and closing of the vocal folds. Going from a closed position to an open glottis while releasing air and back to the closed position constitutes one vibratory cycle. This is modulation of the breath stream into a sound stream. While the glottis may not be completely closed during each vibratory cycle, a train of pulses of air is produced by the adducted vocal folds. The activity at the glottis is not unlike blowing air through loosely approximated lips or stretching the neck of an inflated balloon to cause the escaping air to be modulated into a pulse train of air. This vibratory energy excites air molecules distal from the vibrator so that the sound stream is transmitted from the vocal folds into the upper vocal tract.

Voicing

The sound stream from the glottal area rises into the pharynx and is immediately modified by the vocal tract's resonators and articulators. But it is vocal fold vibration, the act of *phonation*, that has the basic responsibility for attributes of vocal pitch, loudness, and quality in speech and singing. The folds are essential in adding a distinctive feature of speech — voicing or unvoicing, e.g. *z* versus *s*.

To produce voicing for normal conversational speech, the air stream arriving in the larynx from the lungs is constricted and must flow through a smaller passageway because of a valvular structure — the vocal folds. Sometimes the valve tightly constricts the breath stream, which sets the folds into vibration to produce voicing. At other times, the valve is less constrictive so that the folds do not vibrate but cause only turbulence in the escaping air stream; this is how voiceless sounds are produced, e.g. *s, sh, f, p,* (Fig. 8-10).

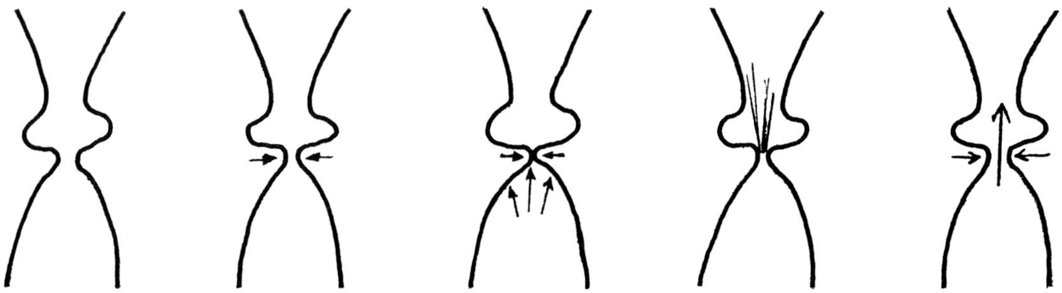

1 2 3 4 5

Figure 8-10. Schematic cross sections of the vocal folds during one cycle of vibration. (1) Normal respiration, folds abducted. (2) Folds begin adduction. (3) Subglottic air pressure is created against the adducted folds. (4) Subglottic pressure blows the folds apart, releasing a pulse of air into the supraglottic area. (5) The outrushing air past the folds sucks (Bernoulli principle) them together, assisted by muscle tension on the folds and tissue elasticity.

THE VOCAL FOLD ADDUCTOR AND TENSOR MECHANISM

Adduction of the folds toward a median position is the initial step in phonation. The point should be made that the intrinsic laryngeal muscles do not open and close the glottis with each cycle of vibration. Opening and closing the folds by the intrinsic muscles would not begin to approach the speed required to produce a glottal pulse train heard in normal phonation, i.e. 100-300 pulsations of air per second.

Adductors

Interarytenoid Muscles

The mechanism of glottal closure is the responsibility of the adductors of the folds. Those intrinsic laryngeal muscles responsible for directly setting the vocal folds in readiness to vibrate are the *transverse* and *oblique arytenoids*, which slide the arytenoid cartilages medially at the posterior commissure and cause the folds to approximate, and the *lateral cricoarytenoideus*. Actually, the arytenoids are said to be pulled a bit over the medial boundary of the cricoid articular facet, so that the last part of the adductor-slide involves displacement (subluxation) at the cricoarytenoid articulation and not a true sliding motion (Fig. 8-11).

Lateral Cricoarytenoid Muscle

To aid closure of the folds, the lateral

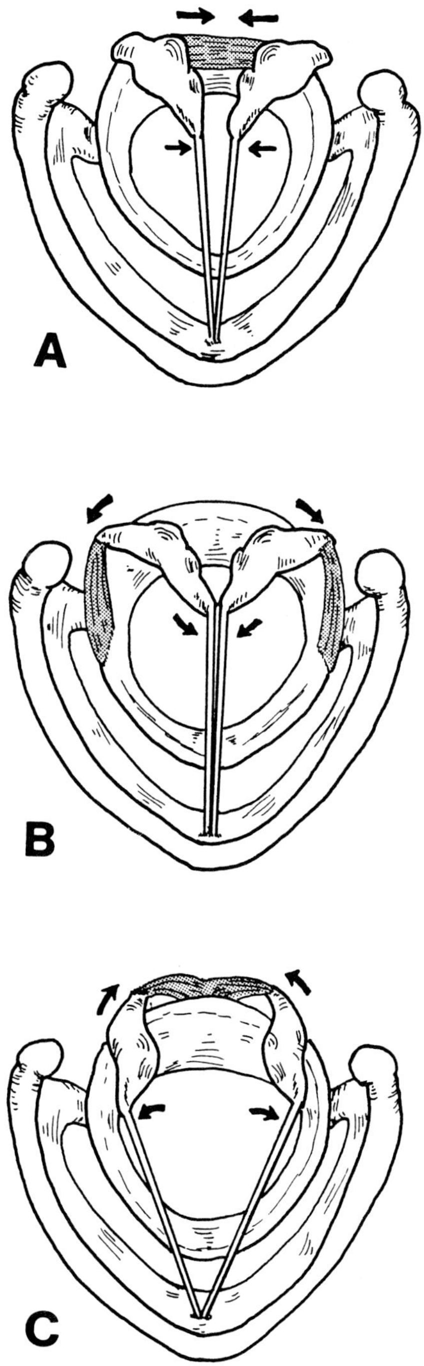

Figure 8-11. Intrinsic muscles affecting the shape of the glottal opening. (A) Transverse arytenoid muscle. (B) Lateral cricoarytenoid muscle. (C) Posterior cricoarytenoid muscle.

cricoarytenoids pull the muscular process of each arytenoid forward (toward the folds) and downward, which aids in moving the vocal processes of the arytenoids together more tightly. Thus, although the lateral cricoarytenoideus assists in approximation of the folds at the midline, only the interarytenoid muscles may be needed for much of the voicing in conversational speech.

Vocal Fold Tensors

Control of the tension of the folds will affect glottal closure, but the role of the vocal fold tensors is concerned principally with the characteristics of the sound produced. This group of muscles includes the vocalis and thyroarytenoideus and the cricothyroideus. These muscles were separately described and illustrated in the previous chapter; only functional considerations are included here.

Vocalis Muscle

The vocalis is often regarded as the medial portion of a larger muscle, the thyroarytenoideus. The controversy as to whether the vocalis is a separate muscle was discussed in the previous chapter. Because of its strategic position on the medial edge in the vocal folds, the vocalis is sometimes termed an *internal vocal fold tensor*. It appears to be able to achieve quick, fine, and varying isometric tensions in the folds that stiffens the folds and contributes to a pitch elevation of the voice.

But the vocalis may also contribute to lowering of the pitch through isotonic activity. If the vocalis contracts to bunch up the mucosa covering the folds and its underlying membrane, conus elasticus, the slackening of the tension in the folds would be an example of the vocalis acting to oppose the muscles that lengthen the folds — the cricothyroideus muscles (Fig. 8-12).

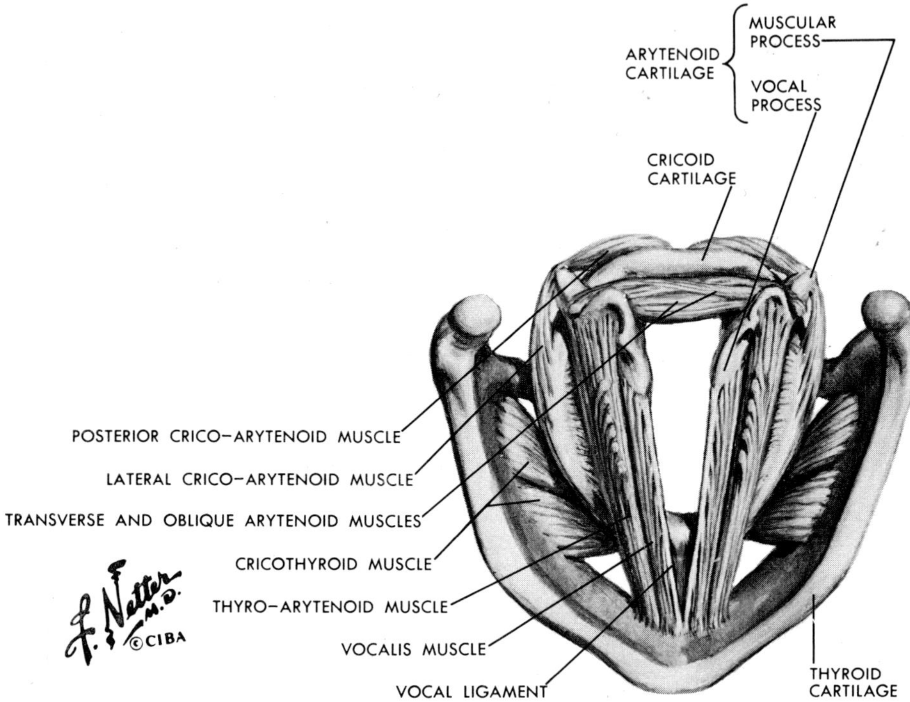

Figure 8-12. Intrinsic muscles associated with the glottal area as viewed from above. Muscle fibers surround the triangular glottal space and interconnect with the four cartilages (2 arytenoids, thyroid, and cricoid). (© Copyright 1964, CIBA Pharmaceutical Company, Division of CIBA-GEIGY Corporation. Reprinted with permission from CLINICAL SYMPOSIA illustrated by Frank H. Netter, M.D. All rights reserved.)

Thyroarytenoid Muscle

Sometimes the thyroarytenoideus is named the *thyroarytenoideus externus*, since the vocalis has origins and insertions overlapping on the same cartilages, namely, the thyroid and arytenoid cartilages. The comparable fibers, those directly forming the mass of the vocal folds, of each muscle group course in a parallel pattern between the thyroid and arytenoid cartilages. The thyroarytenoideus fibers are more numerous than those of the vocalis. Upon contraction, the thyroarytenoid muscle's external fibers would act antagonistically to the action of the cricothyroid, the prominent external tensor of the folds. By pulling forward on the arytenoids, which are held on the posterior rim of the cricoid cartilage, the anterior arch of the cricoid cartilage is rotated downward to increase the "visor angle" between the cricoid and thyroid cartilages. This action shortens the folds and lengthens the cricothyroid muscle (Fig. 8-13).

Cricothyroid Muscle

There are two divisions of the cricothyroid muscle, a vertical part and an oblique part. Upon contraction, the vertical part of the cricothyroid pulls the cricoid arch toward the thyroid (or sometimes vice versa if the cricoid cartilage is fixed by its muscular bindings) to close the visor angle (see Fig. 8-13). In the process, the vocal folds are stretched. It may

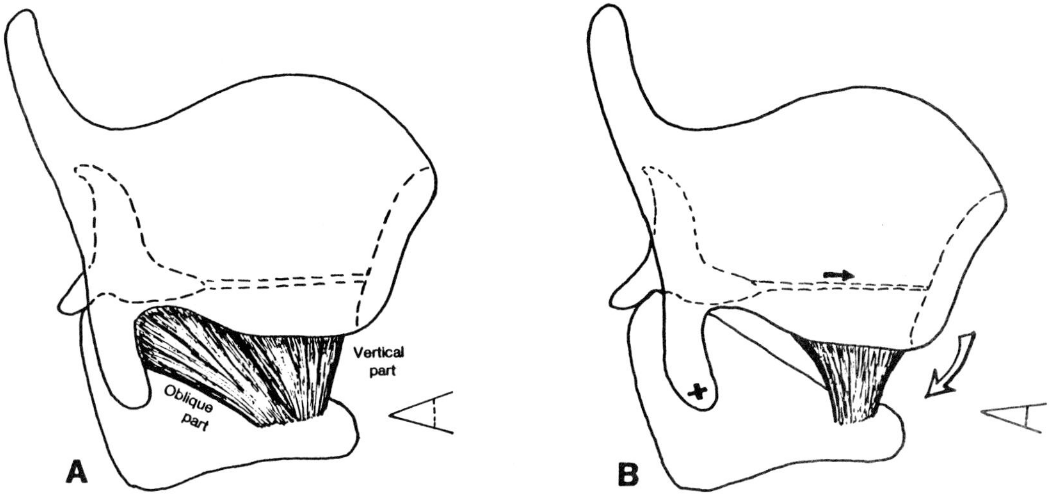

Figure 8-13. The visor angle. The vertical part of the cricothyroid muscle is responsible for a rotational movement between the cricoid and thyroid cartilage. In B, the (+) on the inferior cornu indicates the pivoting point to cause the visor to drop closer to the arch of the cricoid cartilage, which tenses the vocal folds.

be descriptively termed the *external tensor of the folds*.

This muscle has principal responsibility for adjusting the joint between the inferior cornua of the thyroid cartilage and the articular facets of the cricoid cartilage, which pivots for the rocking action that changes the visor angle anteriorly.

As the anterior arch of the cricoid cartilage is tilted back during contraction of the cricothyroideus, the distance between the angle of the thyroid cartilage and the arytenoids is increased, which lengthens the vocal folds. Such a stretching and elongating of the vocal fold tissue will cause them to act more responsively when the folds are separated by a surging pulse of air passing through the glottis. Thinner folds would be more agile and would be able to vibrate at higher rates resulting in elevated pitch.

The oblique division displaces the inferior horn of the thyroid posteriorly, stretching the vocal folds in yet another manner for tensing the folds. The anterior slide of the thyroid cartilage is discussed later in this chapter (see Fig. 8-19).

Vocal Fold Stretching and Tension

It should be stressed that the action of elongating the folds does not mean that there is a corresponding increase in vocal fold tension. The previous discussion of the vocalis and thyroarytenoideus described how these muscles may relax and counter attempts by the cricothyroideus to tense the folds as they are elongated. Therefore, the vocalis and thyroarytenoideus together act antagonistically and synergistically with the cricothyroideus to achieve proper tension of the folds during phonation. Elongation is not a requirement, then, for tense folds, but the external tensor in opposing the internal tensors enables the folds to be tensed with no significant increase in length.

MECHANICS OF VOCAL FOLD VIBRATION

To produce sound, the vocal folds are adducted and the subglottic air pressure parts the folds because the air pressure above the folds is less.

First, the subglottic air pressure blows the folds apart beginning at

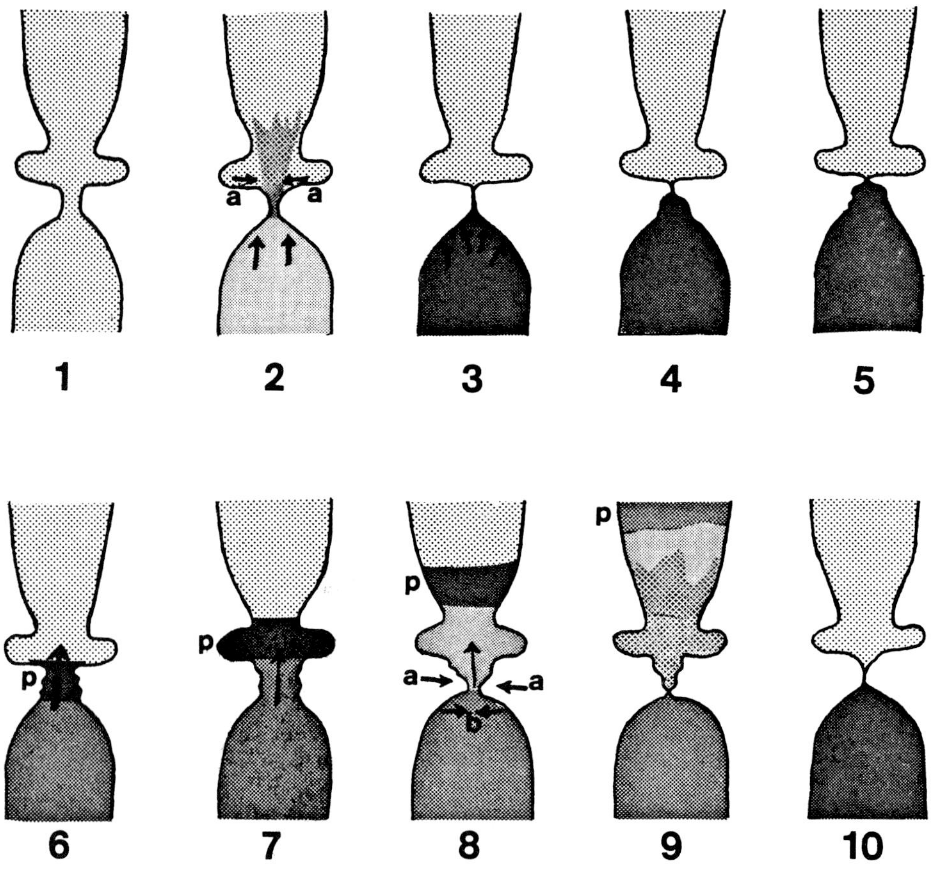

Key:

1. Position of the vocal folds during quiet exhalation.
2. Initial movement of the folds toward adduction accompanied by constriction of elevated subglottic air pressure causing the folds to be sucked medially (a).
3. Full adduction of the folds; increasing subglottic air pressure against the folds.
4. The mucous membrane of the folds begins to separate at the bottom.
5. A rippling effect continues to separate the folds.
6. A pulse of air (p) is released into the supraglottic space.
7. The top edges of the folds separate maximally as the pulse (p) leaves the glottis.
8. Due to the Bernoulli principle (b) and elastic recoil of the folds (a), the folds close.
9. The folds close along the bottom edge to the top as the pulse of energy (p) rises higher in the supraglottic space.
10. The cycle is ready to begin anew.

Figure 8-14. Schematic representation of vocal fold vibration.

the *bottom edge* of the folds.

Second, the mucosal edges of the folds next proceed to open toward the top edge in somewhat of a rippling pattern until air escapes through the glottis.

Third, as air rushes through the slit between the edges of the folds, a special principle of physics comes into play. This is known as the *Bernoulli effect*. The constriction in the airway created by the slit in the glottis causes the velocity of the outgoing air to increase. According to the Bernoulli principle, such an increase results in a drop in the pressure *perpendicular* to the air flow, which would mean in the direction of the glottal margins in this case. The result is a sucking in (medially) on the walls of the glottal chink, beginning with the lower edges of the folds and moving to the top edge. The elasticity of the folds permits them to be sucked toward each other by the sudden drop in pressure that first forced the folds apart.

Fourth, a quantity of air escapes into the supraglottal area, directed toward the upper aspects of the resonatory-articulatory tract.

Fifth, the mucosal edges of the folds meet in the closure phase at the bottom edge first and then upwards to the top edge for complete closure.

Sixth, the subglottic air pressure immediately builds again and the whole cycle is repeated. That is, the subglottic air pressure blows the folds apart; then the folds close under influence of the Bernoulli principle and the inherent elasticity of the vocal folds. The folds are now vibrating to produce phonation, or voicing, for speaking or singing (Fig. 8-14).

Myoelastic-aerodynamic Theory of Phonation

The widely accepted theory of voice production, which was described previously, is named the myoelastic-aerodynamic theory of phonation. This theory proposes that muscle (myo-) tension and elasticity coupled with air flow/pressures set up dynamic alternating forces that open and close the glottis. Both the *vocal folds* and the *air stream* propelled from the lungs through the trachea by expiratory muscles set up an active interplay. According to the aspects of phonation above and to the features of the myoelastic-aerodynamic theory of voice production, phonation occurs as a result of the interaction of many meticulously coordinated events (Fig. 8-15).

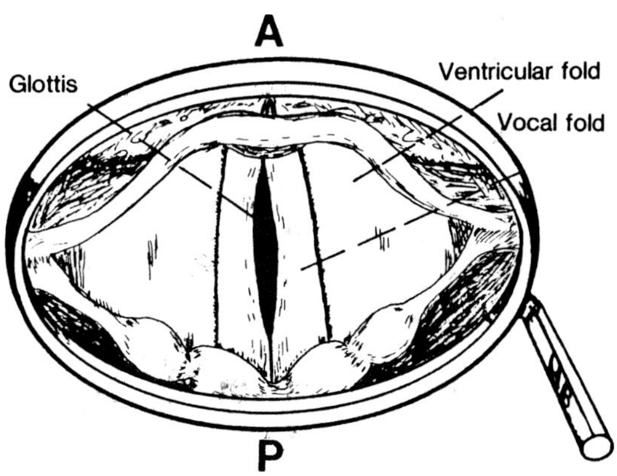

Figure 8-15. Mirror view of the glottis during the opening phase of phonation.

The rapid series of minute surges of air generated by the oscillating folds in the breath stream represents a chain of acoustic energy that excites the resonating cavities of the pharynx, mouth, and nose. The articulators modulate this sound stream for production of voiced speech sounds.

Air Stream Modulation Without Complete Vocal Fold Closure

Individuals who cannot achieve full closure of the glottis will usually have a characteristically *breathy vocal quality*. Openings at the cartilaginous glottis or along the membranous glottis may result in excessive wastage of air in the breath stream. The consequence is a breathy voice that lacks power and prevents the speaker from being able to project when speaking to groups or in noisy surroundings. These individuals are said to have hypofunctional voices due to inefficient modulation of the breath stream.

The Onset of Phonation

The vocal folds need not be *tightly* approximated at the midline to initiate phonation. Even before the folds meet, the edges begin to vibrate, changing the breath stream into a series of pulses of air or sound energy. The first movement of the folds toward closure for vibration is ideally gentle and not strongly reflexive. By contrast, the folds close quickly in the protective mode when foreign matter threatens to enter the respiratory tract below the pharynx.

As the folds are adducted close to each other in a stream of increasing subglottal air pressure, there will be a rise in the velocity of the air flow escaping through the progressively narrowing glottis. The Bernoulli principle will begin at this point to pull the folds toward the midline. But *elastic recoil* of the folds increases and the air flow velocity through the glottis drops. The recoil of the folds is lateralward, away from the midline. Once again, air flow increases and again initiates the Bernoulli effect to draw the folds closer together. Consequently, more of a *lateral* rather than a *vertical* pattern of vibration is set up in this description of the onset of phonation. But the folds move closer to one another with each successive lateral vibratory cycle until a full cycle of opening and closing results in conventional closure of the glottis between the cycles (Fig. 8-16).

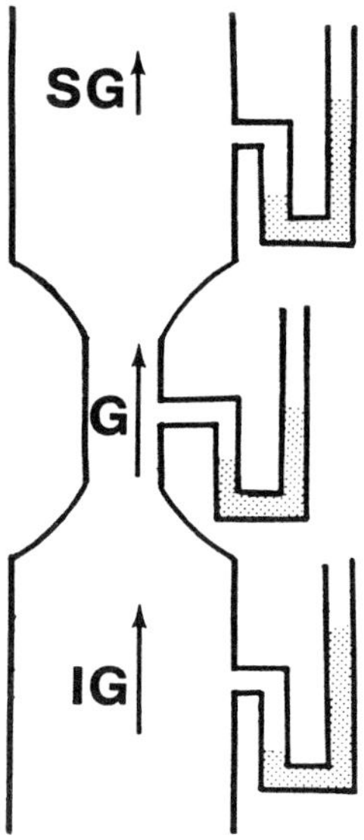

Figure 8-16. Schematic representation of variations in pressure through the larynx. Air rising from the lungs into the infraglottic space (IG) has the highest relative pressure. As the airstream is constricted by the glottis (G), the velocity is greatest, but the lateral pressure is lowest. In the supraglottic space (SG), where the space is unconstricted, the pressure is low.

The onset of phonation is but a few cycles of trembling of the folds. Therefore, only a few breathy glottal pulses emerge before the folds settle into modulation of the breath stream into a distinctively voiced sound stream with vertical fold activity rather than lateral vibrations.

Indeed, the pressure variations within the larynx are quite complex during vocal fold vibration. The Bernoulli effect created by the narrowing of the laryngeal cavity at the level of the folds is a concept of pressure differences. That is, the air around the edges of the folds moves at different velocities than the center of air flow. Thus, pressure differentials from the fold edges to the midline of the glottis are present. These eddies, multiple currents, etc., participate in creating the strong suction to pull the fold margins medially for closure of the folds.

Two characteristic types of phonation onset are associated with soft and hard attacks of the vocal folds. That is, the onset of phonation may lack effective coordination between the adduction of the folds and the initiation of air flow through the larynx. These two types of vocal fold attacks are termed *soft vocal attack* and *hard vocal attack*.

Soft or Breathy Attack

If the initiation of air flow through the larynx precedes the adduction of the folds too early, a quantity of air will be wasted. This pronounced hiss is heard as some stutterers struggle to initiate speech. Also, some cerebral palsied victims lack proper phonatory-breath coordination to prevent an outrush of lung air before the vocal folds are adducted. It is simple to demonstrate normal use of controlled glottal aspiration by saying *hot*. The light hissing of the *h* sound is produced without vocal fold closure, but the folds rush into firm closure to produce voicing for the vowel. The *o* will usually be produced with a soft attack even by those who appear to have a rather hard vocal attack in their usual speech pattern.

Hard or Glottal Attack

The hard attack in the onset of phonation is also related to the less-than-optimal coordination of the adduction of the folds with the rise in subglottic air pressure. If the vocal folds are tightly closed *before* a rise in subglottic air pressure is underway, more effort is likely to be required to blow the folds apart and there will be no trembling or lateral vibratory movements of the folds as a forerunner to the first normal vibratory cycle. This vocal quality is usually termed a *harsh attack* and is associated with hyperfunctional voices. Excessive use of the harsh attack makes the voice a good candidate to become a clinical voice from excessive abuse of the vocal fold margins.

Whispering

That whispered speech is audible without vocal fold vibration means that there must be some other source to generate the audible whisper. It should be pointed out that there is constriction at the level of the glottis during whispered speech production. This point is evident in the preceding discussion. A strong, funneled air stream passing through the glottis with stiffened folds causes a turbulence to be created just above the glottis. The turbulence constitutes a random noise generator just as a punctured tire hisses as the air leaks out. This type of sound, a brief voluntary whisper, is heard breaking up much of our speech as we produce *voiceless fricatives* such as the /s/ in voicing, but also *plosives* and *stops*, respectively, as each /t/ in *tot*.

The vocal folds enable us to produce voluntary whispering of varying degrees. During normal phonation or voicing, there is a pulsing of air flow through the vocal fold generator. No phonation is present during whispering, so all sounds are *unvoiced*. However, there may be vocal fold activity, as in a stage whisper, even though ordinary vibratory patterns may not be expected. There is no movement of the folds in some individuals. Production of forced whisper is apt to cause the folds to vibrate in a manner similar to breathiness. The folds seem to spend more time executing the open phase with a relatively short, even fleeting, closed phase (Fig. 8-17).

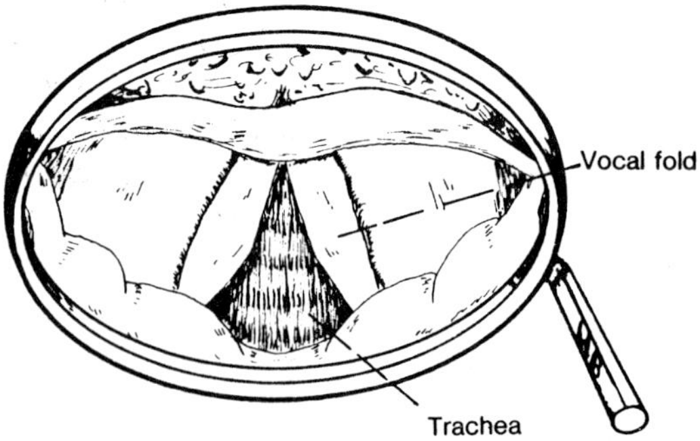

Figure 8-17. Mirror view of the glottis during one method of producing a whisper. The folds are abducted along the entire margin.

The folds may be in a slightly abducted position all along their length during production of voluntary whispered speech (Fig. 8-17) or they may be in contact anteriorly beginning at the anterior commissure but separated between the arytenoid cartilages at the posterior commissure (Fig. 8-18). The posterior cricoarytenoideus and also perhaps the thyroarytenoideus act to set the arytenoids apart in a triangular opening at the posterior commissure. Each of these two positions can also yield phonated sound; therefore, it is postulated that the absence of vibration during whispering is not due to the location of the folds in relation to the breath stream. Rather, the difference is presumed to be the result of the internal tensors, which create a stiffness sufficient to inhibit vibration.

Just as internal fixating stiffness may be involved in preventing phonation, longitudinal tension of the folds can inhibit vocal fold vibration. In fact, the anterior contact between the folds while there is an aperture at the posterior commissure may aid the longitudinal tension in stiffening the folds and preventing voicing.

Aphonia

Some patients experience severely inflammed vocal folds as one of the symptoms of laryngitis. The swelling stiffens and greatly increases the mass of the folds to a point that they are unable to vibrate even with a strong air stream. The effect is somewhat like stopping (damping) a vibrating guitar string by placing a finger lightly on the string. A consequence of edematous (swelled) vocal folds yields another symptom of laryngitis, involuntary whispered speech. The patient's voice is termed *aphonic*.

In addition to weighted, irregular folds due to edema, paralysis of the vocal folds resulting from an interruption of the

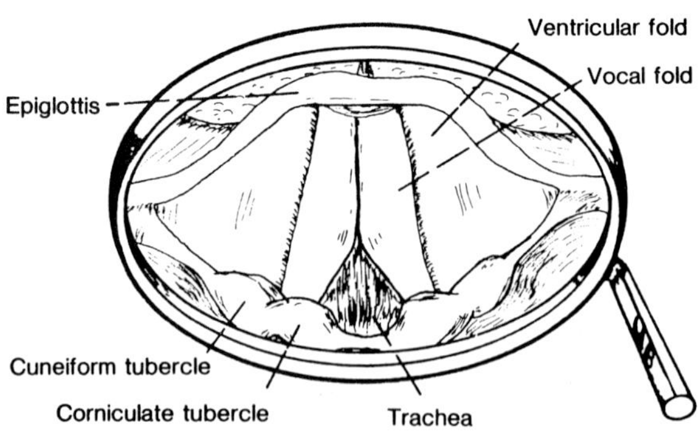

Figure 8-18. Mirror view of the larynx during voluntary whisper with the membranous glottis uninvolved.

nerve supply to the larynx may prevent adduction of the folds and contracting of the internal and external tensors of the fold musculature. There are a number of other possible causes of aphonia, including growths on the glottal margins or in the posterior commissure.

Involuntary whispering that cannot be related to an organic problem is possible. Hysterical aphonia is the label used when the patient exhibits involuntary whispered speech in the absence of an organic cause.

Both vocal folds must be incapacitated to produce aphonia. If only one fold is unable to participate in phonation, then vibration is still possible and voicing, although poor in quality, will be produced.

Production of Unvoiced Sounds

There are two kinds of vocal fold adjustments that may be used to achieve unvoiced sounds: (1) changes in the elasticity of the folds from antagonistic muscles that stiffen the folds and (2) the positioning of the vocal folds out of the force of the breath stream to open the glottis and to reduce the effects of the breath stream on the folds. At the same time, the abducted folds are prevented from vibrating by longitudinal tension along the folds.

PITCH PRODUCTION

By comparison with the usual sounds heard in the environment, the pitch of the voice is rather low. However, the sound produced by the vocal folds contains many overtones that make for a complex vocal tone.

Rapid opening and closing of the vocal folds releases a barrage of interconnected air bursts into the upper vocal tract, which are audible pressure waves. The frequency of those bursts constitutes the perceived pitch of the voice. As we will see later, the quality of voice is dependent on the fundamental frequency of vocal fold vibration and the number and distribution of overtones (or harmonics) of the vocal fold sound generator.

Fundamental Frequency and Pitch

When we speak or sing, the pitch of the voice changes continuously. Asking questions and expressing feeling for emphasis are everyday uses of the changing fundamental frequency of the voice. *Pitch* is the perceived highness or lowness of sound and is related to fundamental frequency, which is the physical attribute of sound. This aspect of sound pertains to the number of vibrations per second.

Thus, we can describe the voice in psychological terms using pitch to designate the highness or lowness of vocal fold vibrations. We can also use a term from physics, *fundamental frequency*, to label the voice by specifying the measured number of vibrations per second that are being produced by rapid opening and closing of the folds. Because each opening of the folds during vibration releases a surge of air escaping from the glottis, it is possible to use electronic equipment such as microphones, amplifiers, and analyzers to detect the number of these pulses per second and thereby find the fundamental frequency.

Role of Vocal Fold Tension in Pitch

While the thickness of the vocal folds decreases with pitch elevation, there is also a tendency for subglottic air pressure to increase with pitch deviation. Stretching the folds for a rise in pitch not only increases tension and elastic recoil but decreases the vibratory mass of the folds. Consequently, stretching causes the folds to stiffen, which increases resistance to their being blown apart by subglottic air pressure. Extra effort is then needed to increase the subglottic air pressure to produce higher pitches.

There is an intricate interplay of subglottic pressures and vocal fold tension during phonation. That there is a complex interaction of the folds and the expiratory musculature for varying pitch production is a point that must be remembered by speech clinicians who work with patients having vocal problems.

Role of Cricothyroid in Pitch Regulation

ROTATION OF THYROID CARTILAGE. The cricothyroid muscle is responsible for narrowing the cricothyroid visor angle during pitch elevation (see Fig. 8-13). As the folds are stretched, there is also a prominent rotation of the thyroid cartilage at the cricothyroid articulation with contraction of the vertical portion (pars recta) of the cricothyroid muscle. The oblique portion (pars oblique) pulls forward on the thyroid cartilage to increase tension in a different manner. The cricothyroid, a tensor of the vocal folds, is thus able to regulate the tension of the vocal ligaments by two distinct mechanisms: rotating and forward sliding of the thyroid cartilage.

ANTERIOR SLIDE OF THYROID CARTILAGE. Trained singers purportedly seem to utilize the translatory (sliding forward) movement more often than untrained singers. Since the arytenoids are bound to the upper posterior border of the cricoid cartilage by ligaments, they are able to anchor the posterior attachment of the fold. The anterior attachment is on the thyroid cartilage, which is being moved forward away from the arytenoids in a translatory movement as the oblique portion of the cricothyroideus contracts (Fig. 8-19).

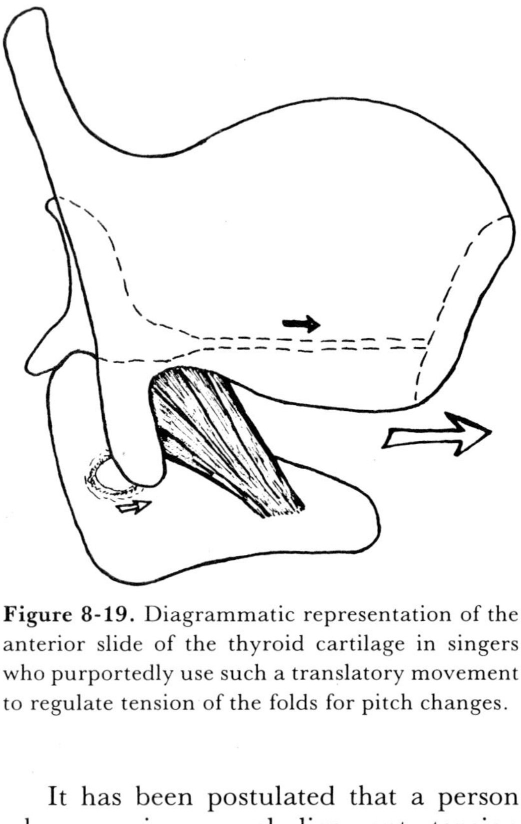

Figure 8-19. Diagrammatic representation of the anterior slide of the thyroid cartilage in singers who purportedly use such a translatory movement to regulate tension of the folds for pitch changes.

It has been postulated that a person who exercises vocal ligament tension through translatory movement of the thyroid cartilage rather than a rotation to change the visor angle has better developed vocal and cricothyroid muscles because of added practice. In the same way, bicyclists develop stouter leg muscles. Well-exercised vocal fold musculature may (1) reduce both the size and duration of the open phase in the vibratory cycle and therefore utilize air flow more effectively and efficiently, (2) result in an increase in resistance to the subglottic air, which enables an increase in acoustic power output, (3) enable the folds to acquire an acoustically more efficient contour, perhaps streamlining the airway above the glottis to facilitate the egress of the vocal tone from the vibrating folds, and (4) elicit additional proprioceptive feedback information that could increase the sensitivity and accuracy of adjustments of the folds for a given intonation.

Characteristics of the Folds During High Pitch

To produce higher pitches, the frequency of the glottal air pulsations must be increased. The major determinant of the frequency of the opening and closing of the glottis during phonation is vocal fold tension.

At higher pitches, the vocalis tends to become disassociated from the vocal ligament and mucosa. Only the ligament and mucosa appear to vibrate. The tense folds are relatively stiff at the higher pitches and the glottal openings are small in each vibratory cycle, which explains the "decoupling" of the vocalis from the ligament and mucosa for even less mass at the vibrating fold edge. As discussed previously, increasing the tension of the folds enables them to vibrate more rapidly since (1) the folds are thinner at the vibrating edges and there is less mass to slow down the movements of the folds, and (2) the folds are more elastic and are therefore able to recoil or snap back to the closed position more quickly.

It was earlier pointed out that adduction of the arytenoid by the interarytenoid muscles squeezes the folds together. This medial compression of the folds against each other during higher pitches can raise the pitch of the voice by shortening the portion of the fold that participates in the vibration. No changes in the A-P tension of the folds are needed when the mass is decreased, since only a small portion of the folds is opened and closed (Fig. 8-20).

Characteristics of the Folds During Low Pitches

The vibrating edges of the vocal folds tend to be shortest in length, least in stiffness, and thickest when producing the lowest pitches. In fact, the folds are *shorter* in the low pitches than at rest, which suggests active muscle contraction to shorten the folds for production of low pitches. The mechanical requirements for pitch lowering of the vocal fold generator would specify thicker (more massive), relaxed, and shorter folds to execute slower cycles of openings and closings. Energy pulses of air released through the glottis, then,

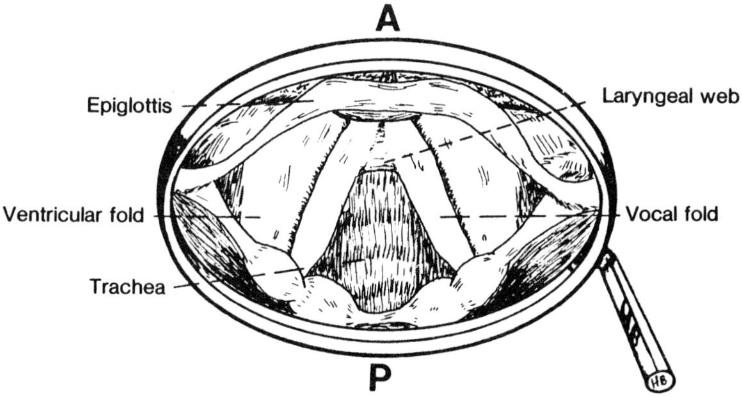

Figure 8-20. Mirror view of the larynx showing a restricted anterior-posterior vibrating zone of the vocal folds due to a web in the anterior commissure.

would be farther apart than when high pitches are produced. These requirements appear to be met with the bunching up of the fold as the vocalis and thyroarytenoid muscles contract to shorten the folds. Providing there is no antagonistic stretching of the folds by the cricothyroid muscle, the folds are thus set to produce the lowest pitches.

Role of the Extrinsic Muscles in Pitch

The extrinsic muscles move the larynx up and down during voicing; however, this movement may be more important in the production of vowels during speech than in the regulation of fundamental frequency. It is usually considered a sign of incorrect voice production that could damage the vocal mechanism if the larynx rises during production of higher pitches, but there is controversy over this view. Whether the extrinsic muscles contribute to pitch changes or simply accompany them is not certain. Based on the extensive treatment that most speech anatomy texts devote to the extrinsic musculature, a deceptive implication of the role of these muscles in phonation seems to be indicated. More information appears warranted about the role of the extrinsic muscles in the pitch-regulating mechanism. Their importance in protecting the lower airway by raising the larynx to effect closure of the larynx's entrance during swallowing is well-known.

Average Fundamental Frequencies of Male and Female Voices

Adult male vocal folds produce an average fundamental frequency of around 130 vibrations per second, which corresponds to the note of C below middle C on the piano. Female vocal folds are shorter and have an average fundamental frequency around 260 vibrations per second, or an octave (eight notes or double the frequency) higher than males, which corresponds to middle C. Hence, longer, more massive folds of the male open and close to produce around 130 air bursts per second, while short female folds produce around 260 per second. Children are apt to have voices with a fundamental frequency of over 300 Hz. (These frequencies, 130 and 260, are actually *cycles* of vibration, and the fundamental frequency may be described as a sinusoidal wave. More commonly, instead of writing out vibrations or cycles per second or number of openings and closings of the glottis, we say 130 Hz, the abbreviation attributed to and being derived from Hertz, the German physicist.)

Pitch Range

The pitch range of the voice extends from below 60 Hz (slightly more than two C's below middle C) for basses to around

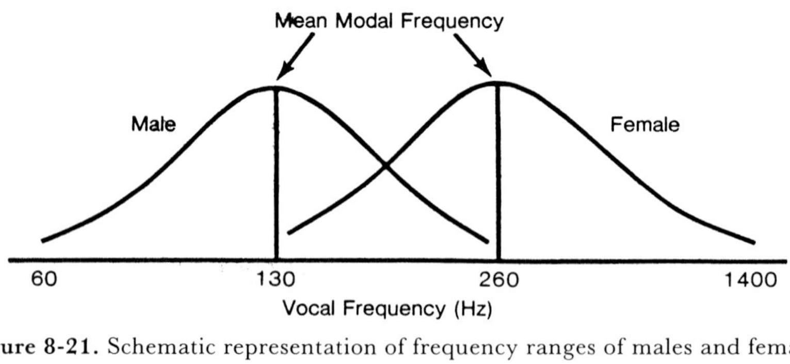

Figure 8-21. Schematic representation of frequency ranges of males and females.

1400 Hz (slightly more than two C's above middle C) for sopranos. If the whistle register of the soprano is included, the highest note will be three C's above middle C, or about 2050 Hz, altogether a compass of five octaves for the human voice. Of course, individual voices have more restricted ranges. Children before puberty may exhibit a range of only a few notes as compared with two octaves for individual adults (Fig. 8-21).

The male voice, combining basses with tenors, would encompass a range from about 60 Hz to 520 Hz (the C above middle C). Altos would set the lower range for females to begin around 100 Hz to the very high note of 1400 Hz for sopranos (Fig. 8-22).

REGISTERS OF THE VOICE

Often the term *register* appears in association with pitch ranges. Although used by musicians more frequently than any other group, registers of the voice have captured the attention of speech scientists, laryngologists, and voice clinicians. Yet, there lacks a good understanding of the term and of the phenomenon of registers in the human voice.

Definition of Voice Register

By way of introduction, then, the definition often cited would describe a voice register as a series of succeeding notes of *equal quality* on a scale from low to high, produced by the application of the *same mechanical principle*, so that the acoustic nature of this group of tones differs basically from another series of succeeding sounds of equal quality produced by another mechanical principle.

Chest, Middle, Head, Whistle Registers

Conventionally, the first series of tones is termed the *chest register*. Next, there is a *middle* or *mixed register*. Finally, there is a *head register*. Male voices extend the head register to the falsetto voice. Some females also may have an extension of the high head register that is known as the *whistle register*, but, unlike the falsetto, the whistle register is produced without vibration of the folds. The escape of high-pressure subglottic air through a narrow glottis under much tension produces a turbulence that characterizes the whistle register.

Pulse, Modal, Loft Registers

Departing from the traditional musical descriptions of the voice registers, Hollien (1977) has given us three registers with a basis other than singing. The first is the low-frequency or *pulse register*, followed by a mid-frequency or *modal register*, and concluding with a high-frequency or *loft register*. The loft range is analogous to the falsetto voice. During pulse phonation, the folds are short, thick, and compressed. Air escapes in a bubbling fashion without active vibration of the folds as seen in the chest register. Loose, floppy fold margins permit subglottic air to blow by in low-frequency puffs (3-50 Hz) in an irregular pattern. The modulation of the air stream occurring in the pulse register may also be described as glottal fry. Vibration of the folds in the modal register produces "optimal" phonation. The pitch and quality are efficiently generated and are pleasing to listeners.

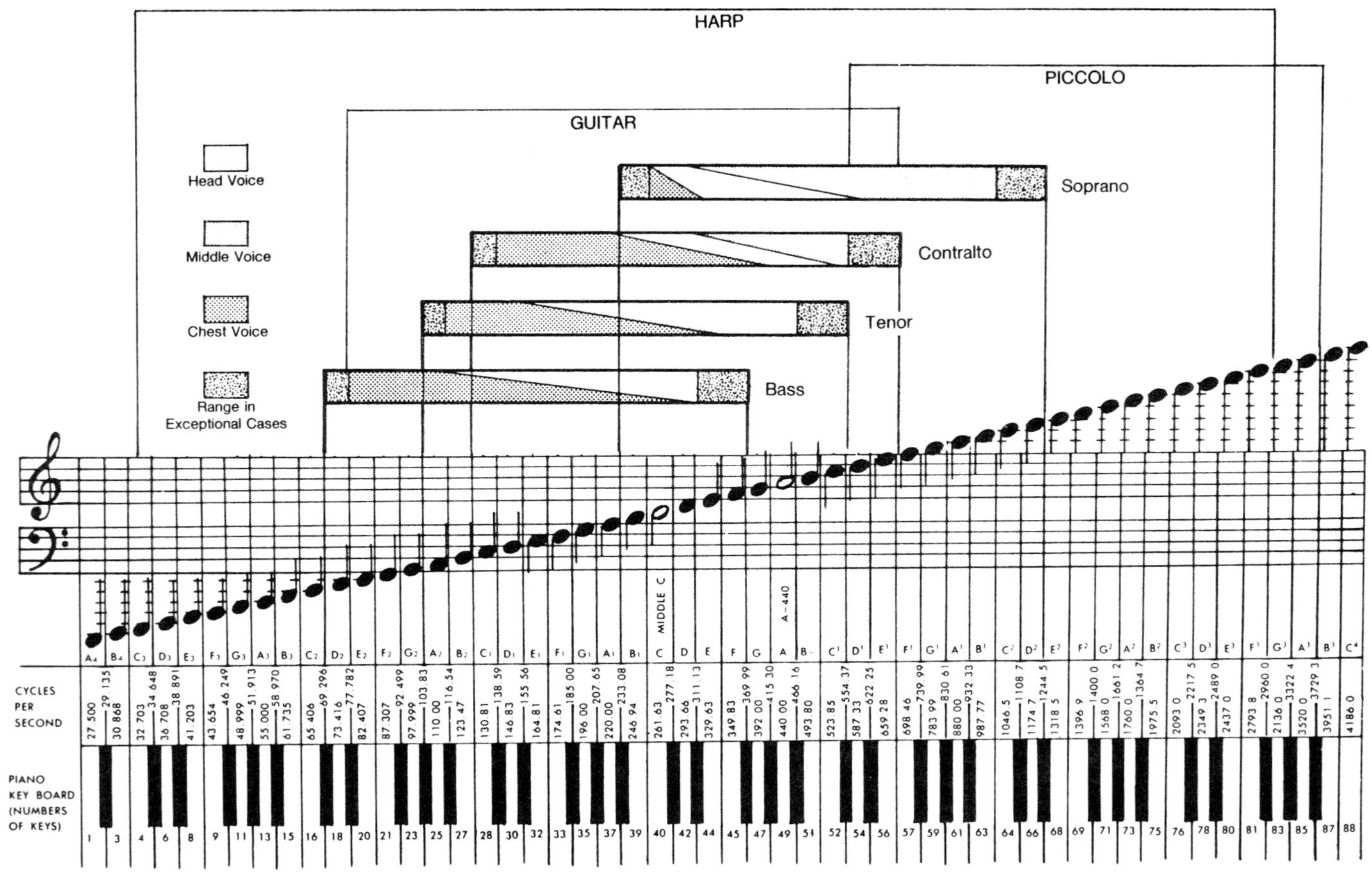

Figure 8-22. Range of voices and some musical instruments. The key on the right of the vocal ranges indicates the registers for the respective voices. Frequencies of each note are shown adjacent to the full piano keyboard. (From *Psychology of Music* by C.E. Seashore. Copyright © 1983. Used with the permission of McGraw-Hill Book Company.)

Transition between Registers

The concept of vocal registers is not accepted by all. Among those subscribing to the existence of the register phenomenon, there is disagreement on the number of registers. Part of the difficulty is that the subjective experiences about registers may be entirely too dominant. Data from laboratory instrumentation is sorely lacking that would corroborate many of the views expressed by singing teachers concerning vocal registering. Still, a break between the boundaries of the registers may be demonstrated by a subject holding a long tube tightly at the lips while singing up the musical scale. The voice will exhibit audible cracks at the points of register transition. Singing teachers train students to blend the registers into a smooth unit to camouflage the transition from one register to the other so that even trained listeners cannot detect the points of transition.

Principles of Voice Production in the Registers

Why should the speech clinician or laryngologist have interest in the mechanism of the register? There are implications of practical importance. During production of the chest register, the folds vibrate along the entire longitudinal dimensions and exhibit extensive movements. The edges of the folds are rather blunt and rounded. By contrast, production of tones in the head register requires increased tension to stretch the folds so that they may be thinned to reduce mass. The edges of the folds become sharp and only the mucosa and vocal ligament may vibrate because increased stiffness of the folds prevents extensive vibration. Therefore, it is evident that the mechanical principle for the chest register differs in a fundamental way from the head register.

Yet another mechanical principle may be illustrated. If a male voice produces pitches above the head register, up to two-thirds of the length of the vocal folds may be restricted from vibration because of damping due to medial compression of the folds. Only the anterior third, a portion of the membranous glottis, vibrates. The arytenoids are held together tightly, and the tensor mechanism stretches the folds taut so that the glottal margins of the folds part with a slit that extends for only a portion of the total glottal length. Such a mechanism is capable of a high rate of openings and closings so that the breath stream is modulated or set into rapid pulsations. This mode of voice production depicts the falsetto voice. There is a definite overlap of the falsetto and head registers, since the lowest note of the falsetto is well below the highest note of the head register (Fig. 8-23).

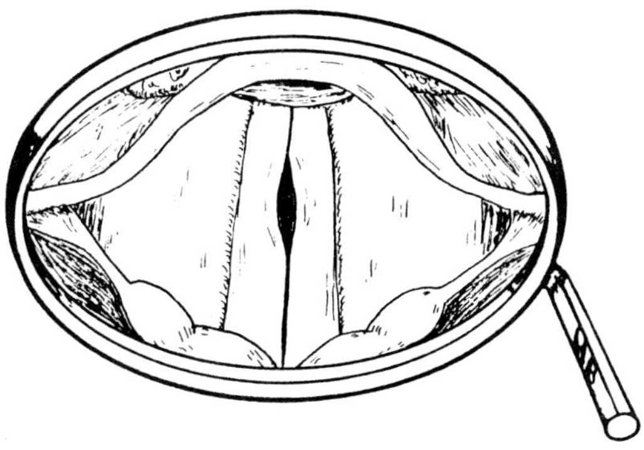

Figure 8-23. Mirror view of the larynx during production of a falsetto voice. Only a limited portion of the membranous glottis is active.

Using various registers may be analogous to our bicyclist who selects a low gear for hill climbing so that he pedals with less effort, but the speed of travel is reduced. On level surfaces, he shifts to a higher gear, pedals with about the same effort but travels faster. Downhill he shifts to the highest gear and further increases the speed of travel without a correlated increase of effort. Different gears are used for different terrain just as the voice registers, or mechanical principles for phonation, are used for slower to faster speeds of vibration of the folds.

Comparison of the Folds of Low and High Voices

From an anatomic and physiologic view, it can be stated that individuals with low voices have long and relatively narrow vocal folds. Those with high voices have short and relatively broad folds. Laryngologists may further indicate that healthy high voices are characterized by more whitish folds; healthy, deeper voices are characterized by more pinkish folds. A pinkish tinge in a soprano, then, may be a sign of mild laryngitis to the laryngologist but a normal color in an alto. Following a concert, the folds will usually present a pinkish coloring with accompanying congestion. They will ordinarily respond to rest and after a few hours return to normal (Fig. 8-24).

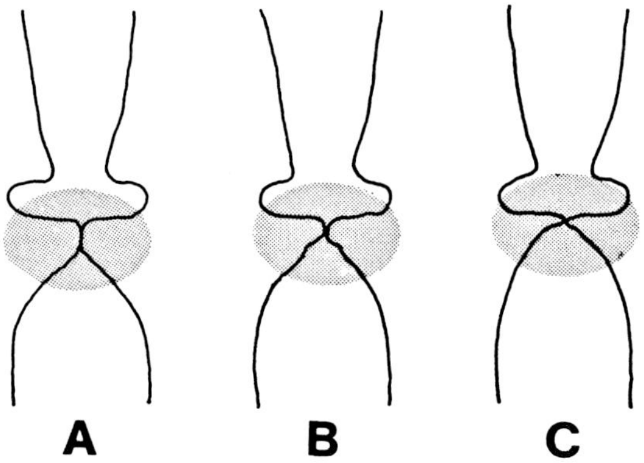

Figure 8-24. Schematic representation of cross sections of the larynx showing variations in vocal fold thickness: (A) low pitch, (B) middle pitch range, and (C) high pitch.

THE INTENSITY MECHANISM OF THE LARYNX

The intensity of phonation refers to the energy generated by the vocal folds as they modulate the breath stream. Pitch and intensity are so inextricably related that it is not practical to think of one without considering the other. In fact, it is probably impossible to isolate one from the other. Of course it is evident that loud phonation is easier when using certain pitches. When we scream from sudden fear, the pitch is not among the lowest we use, nor is it the habitual or optimal pitch.

Factors of Loudness

To produce louder sounds, the folds must open wider during each vibratory cycle. Three factors appear to bear upon the operation of the intensity-changing mechanism: (1) increased subglottic pressure, (2) increased glottic resistance, and (3) increased air flow through the glottis.

Increasing the subglottic pressure usually results in an increase in loudness, assuming other factors contributing to increased energy output of the vocal generator are present. Such an additional factor would include constrictions of the vocal tract, which would adversely impede the vocal folds or part of the supraglottic airway. Added subglottic pressure increases the energy of the subglottic air bursting from the folds (Fig. 8-25).

By doubling the subglottic pressure, only around 9-12 dB will have been added to the intensity of the voice. (Conversational speech at a distance of three feet is usually around 65 dB sound pressure level; and with another 12 dB added, we are up to the level of loud speech as often heard from teachers in noisy classrooms.) Increases in the subglottic pressure tend also to raise the fundamental frequency as noted previously.

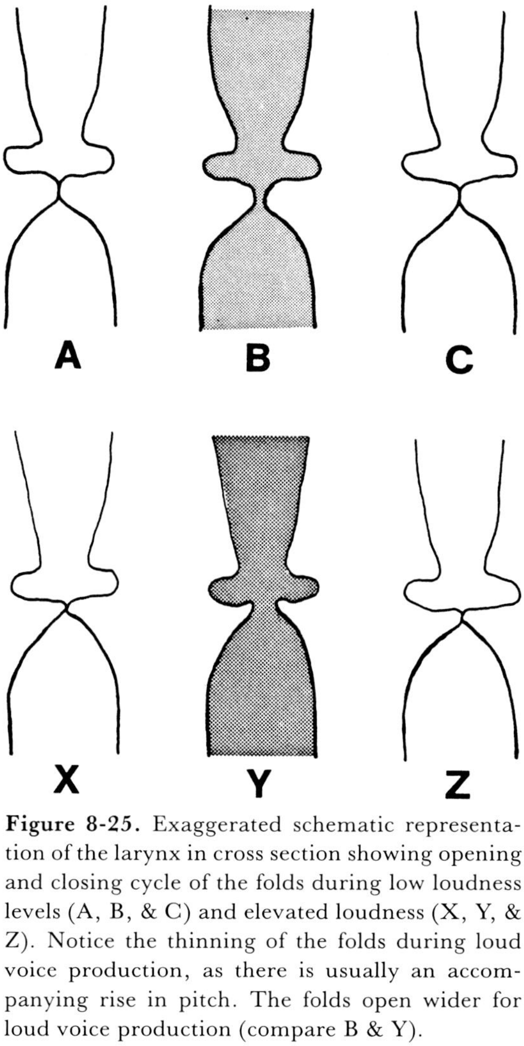

Figure 8-25. Exaggerated schematic representation of the larynx in cross section showing opening and closing cycle of the folds during low loudness levels (A, B, & C) and elevated loudness (X, Y, & Z). Notice the thinning of the folds during loud voice production, as there is usually an accompanying rise in pitch. The folds open wider for loud voice production (compare B & Y).

The vocal folds must make greater excursions to produce louder sounds. Inherent in the vocal fold adjustors is a resistance to the added amplitude the folds travel during each vibratory cycle. The vocal fold tensors account for added resistance, which causes not only an increase in intensity but also a rise in pitch. There is, as a result of the greater tension on the folds, increased medial compression of the folds.

Air flow is said to be an imporant feature in the production of increased loudness. Vocal intensity may be increased by accelerating the air flow while maintaining a constant vocal fold resistance. There is some uncertainty in the literature in the conclusions about the role of air flow versus laryngeal resistance when the resistance focuses on the increased tension of the folds. For example, trained voices are said to utilize increased subglottic pressure and increased air flow through the larynx to raise the intensity, leaving the vocal fold tension relatively unchanged. Untrained singers, on the other hand, may decrease the air flow rate because they increase vocal fold tension, which increases the resistance to air flow through the glottis.

The folds are in the closed phase of the vibration cycle for longer periods as intensity increases. The implication is that the folds meet each other with increased force during closure as the intensity increases, which reduces air flow but also encourages the onset of vocal fold pathology when there is a continuous strain for loud phonation. The site of maximum, abusive contact between the two folds is almost always at the junction of the anterior third and posterior two-thirds along the A-P dimensions of the folds (Fig. 8-26).

Role of Extrinsic Muscles in Loudness

During phonation at loud levels, the laryngeal depressors pull the larynx inferiorly. The purpose of lowering the larynx seems to relate to resonance. A more favorable resonating cavity in the vocal tract may facilitate vibratory activity of the folds through an improved coupling of the vocal fold generator with the upper resonating cavities. Even adjustments in the resonating cavities well above the folds have an influence on the action of the air flow through the glottis. *Voiced stops*, such as the /b/ and /d/, shut down vibration of the folds. Also, other sounds which are less constrictive in the articulatory mechanism may increase resistance at the level of the vocal folds. To overcome this resistance, further increases in subglottic pressure and tension of the folds may be applied up to a point.

Clinical Implications of Loud Phonation

Every voice clinician is familiar with the hyperfunctional voice that has incurred *nodules* or irregular contours of the fold margins due to regular stormy use of the voice. Children develop screamer's nodules. Soloists, cheerleaders, preachers, teachers, and salesmen, as well as numerous other individuals whose personalities are projected with a loud voice, are seen frequently in offices of laryngologists and voice clinicians for amelioration of their vocal handicap. Often there is a combina-

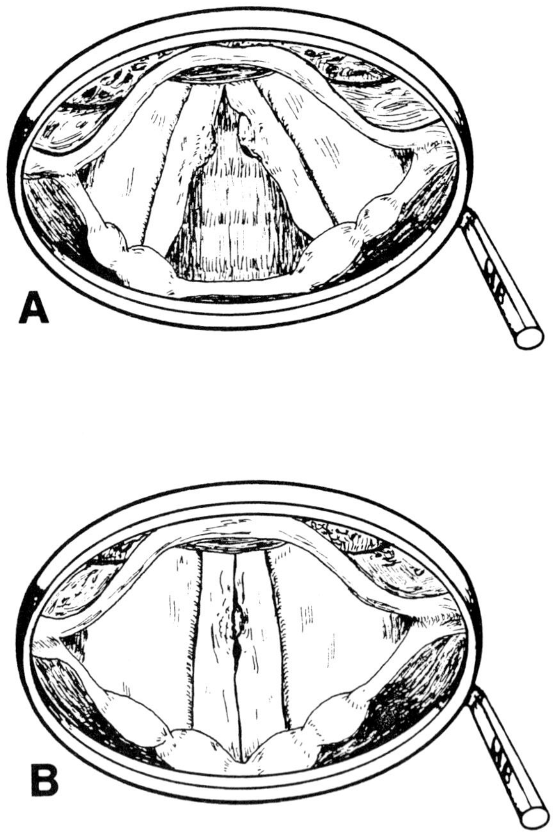

Figure 8-26. Mirror view of a larynx showing bilateral vocal nodules at the junction of the anterior third (posterior two-thirds) of the vocal folds. The masses disturb the evenness of the fold margins (A) and prevent effective adduction of the folds during phonation (B).

tion of using a pitch level that is inappropriately high as well as strain to produce a loud voice. Fortunately, many of these can be rehabilitated if they are willing to learn and practice use of efficient phonatory-breath controls, appropriate pitch levels, sensitive auditory feedback monitoring, and general vocal hygiene (Fig. 8-26).

Loud voices produced without strain are heard as part of many professions without adverse problems arising in the vocal generator. Acquisition of a hygienic vocal skill is possible for those who do get into trouble if the clinician is well aware of and teaches well the physiology of phonation.

MUTATION OF THE VOICE

Table 8-I

PRE AND POST PUBERTY CHANGES IN LARYNGEAL MEASUREMENTS

	Girls		Boys	
	Pre	Post	Pre	Post
Horizontal dimension	30 mm	35 mm	30 mm	40 mm
Vertical dimension	30 mm	40 mm	30 mm	70 mm
Length of folds	15 mm	17 mm	15 mm	23 mm

A significant change in pitch occurs during puberty. This change of voice, known as mutation of the voice, is due to the accelerated growth of the larynx.

Children's voices differ little between the sexes prior to puberty. The range in early childhood extends over only several notes. Just prior to the age of puberty, the range may have reached one-and-one-half octaves. Many 9-14-year-olds have ranges of 2-3 octaves.

In the male, the folds will lengthen by about 10 mm (almost half an inch) but only around 3-4 mm in the female. Since this spurt of growth is more pronounced in the male, the pitch drops noticeably, about an octave. Females maintain a more even growth pattern from prepuberty to postpuberty, which coincides with the general maturing process; hence, there is a drop of only 2-3 notes during puberty (Table 8-I).

Growth of the Larynx During Mutation

Until puberty, the larynx of the child grows slowly. When the sex hormones begin to influence the larynx, growth increases dramatically, causing the cartilaginous framework and vocal folds to grow.

Adjustments During Voice Maturation

While most youngsters adjust to the change without having to contend with frequent pitch breaks as they adjust to the new mechanism, some are not so fortunate. For the less fortunate ones, they seem to have a period of cracking of the voice due to inability of the vocal fold musculature to exert fine control as the larynx grows rapidly. It is as though the proprioceptive senses are giving too little feedback to the brain and the brain is confused about the proper pattern of nerve impulses to send to the various intrinsic musculature of the vocal folds to maintain precise control for even phonation during the constantly changing pitches needed for ordinary conversation.

These involuntary shifts are sudden and involve both pitch and quality shifts. The voice may abruptly jump up or down a full octave, to reach either the falsetto voice or the chest register of a bass. But even these unsettling, difficult mutational periods in the adolescent need not be considered clinical in nature. It is thought that a shift to tenor is less likely to be as difficult or prolonged as a shift to a bass voice.

There is often an emotional impact associated with conspicuous pitch breaks in the voice of adolescents, but the breaks may be reflecting a general state of excessive self-consciousness during this period. It is important when the voice clinician is taking the vocal history of adults that the characteristics of mutation be carefully noted. This may be more significant for patients who have encountered voice problems as a result of practice while enrolled in singing lessons.

The authorities are usually quick to stress the dangers of singing for both boys and girls during the mutational period. Most will not sanction serious voice training to begin until 17 years for girls and 18-19 years for boys. Practice in extending the singing range of the voice during adolescence is strictly prohibited if damage to the vocal instrument is to be avoided. The mutational period of the singing voice lasts longer than the mutation of the speaking voice, a vital point for all interested in the development and protection of the voice.

Around age 25 the cartilages of the larynx in many individuals begin to calcify and lose their elasticity. However, the calcification may be inconsistent and include only a portion of the thyroid and be absent in others. Interestingly, Zemlin (1981) reports no evidence of ossification in arytenoid cartilages among the larynges taken from a geriatric population. Those who preserve their voice into old age are presumed not to have much, if any, calcification.

It should be noted here that *senescence* describes the normal aging process in man while *senility* implicates the pathological condition. Atrophy of the laryngeal muscles and stiffening of the laryngeal joints, if they occur in old age, result in a narrow voice range, a shrinking of phonation time, and a tremulous voice on occasion. The ventricular folds may narrow to make the vocal folds more visible. A hoarse and deteriorated voice ages significantly an individual in the ears of listeners and perhaps to the possessor of such a voice. It is generally thought that pitch rises with age, but this impression is not consistently supported in a number of studies.

VOCAL QUALITIES

One may consider the registers of the voice after traditional musical terminology: chest, mixed, and head. Also, Hollien's classification may be contemplated: pulse, modal, and loft. In either case, the terminology point to *qualities* of the voice.

Vocal quality relates to the *pattern* of vibration of the vocal folds. However, *resonance* in the upper vocal tract also yields "qualities" that are clearly distinguishable. Nasal qualities, for example, are a product of channeling the sound stream into the nasal cavities. Here, our focus is on those qualities attributable to types of phonation.

The terms *breathy, harsh,* and *hoarse* have been used for some time, but decreasingly, to refer to voice qualities. There are literally dozens of other such subjective descriptive terms used also. Further, there is a problem in finding agreement among listeners as to how severe a given quality must be in deviation from "normal" before it is characterized as a clinical voice. Age, sex, and social environment, among other factors, of a given individual weigh heavily on the judgment of normal versus abnormal vocal qualities. Clinicians, then, often have a dilemma if they follow only community standards for classification of voice qualities. The clinician's role in examining persons with vocal complaints must include physiological efficiency and adequacy for use in various communication needs.

Breathiness

When the folds fail to close adequately during each vibratory cycle, the stream of pulses being released at the glottis will contain accompanying unmodulated air, which is heard as noise turbulence. This combination of unmodulated air mixed with glottal pulsations may be classified as breathiness. It is characteristic of hypofunctional laryngeal valving. It was pointed out earlier that if the folds fail to adduct sufficiently or are prevented from vibration due to mechanical restraints (swelling, growths, paralysis, etc.), the whispered quality heard is termed *aphonia*. Therefore, many gradations between aphonia and a normal glottal tone are possible. Somewhere in between these extremes is the fine line delineating an acceptable from an unacceptable amount of air wastage. Speakers with an excessively breathy voice quality are difficult to hear in large classrooms and in noisy environments. Most listeners are more tolerant of breathy voices in females than in males (Fig. 8-27).

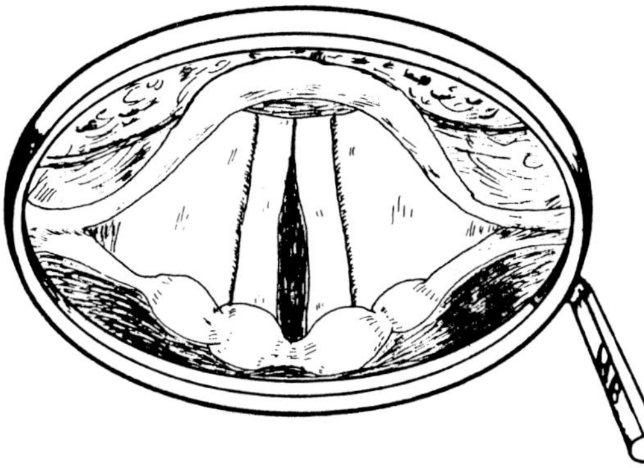

Figure 8-27. Mirror view of the larynx during attempted phonation showing paralysis of the folds. Air wastage and inadequate loudness are major features in the vocal quality associated with bilateral paralysis.

Harshness

On the other extreme of the breathy voice quality is the harsh voice characterized by excessive closure of the folds during phonation. There is elevated tension associated with this hyperfunctional voice. An extreme condition of overapproximation is found in *spastic dysphonia*. (Dysphonia refers to any disorder of phonation). Laryngeal tension appears so great that the breath stream seems to strain to part the glottis for phonation. This hyperfunctional voice sounds strained. The patient may appear to be struggling to breathe because so much effort is focused on the larynx. Interestingly, this quality may vanish if the patient sings, laughs, or talks to himself with no one else present.

Harshness, like breathiness, has many gradations, from spastic dysphonia to normal. The harsh quality most frequently heard is described by many adjectives such as glottal fry, unpleasantness, grating, strident, or metallic. These speakers may often present highly restrictive lip and mandibular movement during speech, revealing that the hypertonicity at the level of the larynx is not the only site of hyperfunction. The pulse register described earlier is an example of the glottal fry quality in the harsh voice.

Hoarseness

Hoarse voice qualities are highly variable. Perceptual standards will greatly influence whether this vocal quality is out of bounds when the vocal quality is not severely deviant. If we construct a continuum with the qualities of aphonia at one extreme and spastic dysphonia at the other extreme, the voice quality in the middle will be a modal quality, or normal. But if air wastage is mixed with overapproximation of the folds, most listeners will perceive the quality to be "hoarseness."

This is the quality heard in association with cancer of the vocal folds. The folds are unevenly contoured due to progressive aggregate masses so that they cannot close adequately during vibration. Unmodulated air escapes as noise turbulence. Additionally, swelling adds mass to the folds, which lowers the pitch and adds resistance to vibration. Extra effort is needed to achieve phonation, so the patient employs increased subglottic pressure in combination with increased glottal tension. The patient often must strain to initiate and maintain phonation, creating the harsh, rough component of the voice. Thus, both components of breathiness and harshness are present in this hoarse voice production.

Of course, the hoarse voice may be mild or severe. While the voice influenced by carcinoma may sound characteristically rough and breathy, only a roughness may be apparent for vocal fold polyps or nodules. Hoarseness resulting from laryngitis may be acoustically rough and low pitched or just low pitched. However, with well-developed nodules, edema, or polyps, adequate closure of the folds cannot be attained (see Fig. 8-26). Thus, the resulting voice quality is breathiness. After cheerleading camps, students may have a hoarse voice that lingers for some time, even after vocal rest. This vocal quality is considered a medical management problem.

The Objective of the Voice Clinician

It is tempting for voice clinicians to latch on to labels for voices rather than attempting to describe the physiology of a patient's voice and concomitant features. There is a need to include such physiologic data in the interest of uniformity due to variations in perceptual judgments. In essence, the clinician cannot expect to find the problem in the quality of the voice but must focus on the structure and function of the apparatus producing the quality.

Physical Properties Related to Voice Quality

There are several interrelated physical properties of the vocal folds that influence the quality of the sound produced. These may be listed as (1) compliance, (2) mass, (3) elasticity, and (4) stiffness. To produce variables in voice qualities, the folds vibrate in various modes, which is the aggregrate result of the four properties listed.

Compliance refers to the manner in

which the folds react to forces acting upon them. It is unlikely that all regions along one vocal fold will be uniform in elasticity, density of the soft tissues, stiffness, or precise contour of the edge. Neither is it likely that the two folds would mirror one another in these characteristics. As a result, the two folds will react subtly different from each other, while portions of each fold react differently from portion to portion. Regions of the folds that offer less resistance to displacement are said to be more compliant and will yield to pressures of the expired breath stream or muscular forces, stretching the folds sooner than areas having less compliance. The folds are twin vibrators with complex and varied vibrating patterns possible, since compliance and other properties must be considered with two vibrators in mind.

Mass is the density of a body. More massive folds are slower to initiate a vibratory pattern and, once started, will open and close at a slower rate than less massive folds. Short, thick folds are used for production of low pitches.

Elastic properties of the vocal folds enable the folds to restore themselves to their original position when displaced. Stretching of the vocal folds increases tension and thus the elastic recoil force. Tense, long, and thin folds will snap back closed quickly during phonation, enabling a higher rate of air pulses to be released from the glottis. Thus, the speed of recovery, the elasticity, of the folds is proportional to the tension of the folds, other properties unchanged. Consequently, pitch usually rises with increases in elasticity.

Stiffness of the folds is related to their resistance to movement or displacement. Increased subglottic air pressure is required to open folds in which the resistance to their opening is great. The folds are least stiff during low pitches, but they are also shortest in length and thickness.

Unfortunately, the properties above cannot be investigated as single entities. If they could, their separate influences on vibration of the vocal folds could be assessed and the effect upon phonation postulated. Instead, we are left with but a general description of some of the mechanical and dynamic properties that are present in vocal fold action in the production of voice quality and the other attributes of speech. In summary, we can surmise that there are many factors influencing the behavior of the folds during vibration. Therefore, the complexity of the vocal qualities emanating from the larynx are the product of vocal folds' subtle and greatly variable vibratory patterns. These patterns are highly individualistic. They may reveal much about our reactions to our internal and external environments.

VIBRATO PHYSIOLOGY

Little is actually known about the phenomenon of vocal vibrato, although it is included as an integral activity in singing pedagogy and is usually heard clearly in the voices of trained singers. Vibrato contributes to the enrichment of sung tones. It is, therefore, a desirable artistic coloring of the vocal tone. The speaking voice is also enhanced by the use of some vibrato.

Vibrato is perceived as a combination of pitch and loudness variations. The average intensity fluctuation is about 2 dB. The amount of pitch fluctuation is around one-half musical tone. There is a rhythmic variation of pitch occurring with an average frequency of around five times per second and a variation in intensity of

1-5 per second. Listeners usually describe voices void of vibrato as dull, hard, flat, or metallic. Both singing and speaking voices may be expected to exhibit vibrato, but the pitch and loudness variations are especially emphasized for the singing voice.

Quivering voices that seem sickly are referred to as tremulous or palsied voices. The vibrato increases in frequency and appears irregular and uncontrolled. Highly restricted pitch ranges and weak intensity may be associated with greater tremulo frequency in the voice. The number of variations increases from 5 to 12 per second. It has been postulated that persons over sixty who develop a tremulous voice are likely experiencing stiffening of the laryngeal joints and a weakening of laryngeal muscles. For singing, male voices seem to tolerate age better than females before becoming excessively unsteady.

Vibrato physiology of the voice is characterized by both pitch and loudness fluctuations that are rapid, continuous, and discrete. The speed of these pitch and loudness variations suggests a highly agile vocal generator. How subtle frequency and intensity perturbations are produced by the vocal folds is seldom addressed in speech and hearing literature.

Unfortunately, most information about vibrato physiology is subjective, derived from singing teachers and singers. Their subjective observations often are actually introspective speculations about the technique employed by the larynx to produce vibrato.

Several points may be listed to describe the physiology from the investigations of a few researchers: (a) Vibrato takes place at the level of the larynx, although it is not confined to the larynx; (b) It is the consequence of reciprocal contractions of the cricothyroid muscle; (c) Activity of the mylohyoid muscle, and perhaps other suprahyoid muscles, contributes to the vibrato; (d) The vocal folds appear to function identically during vibrato phonation versus nonvibrato phonation, based on analysis of high-speed photography of the folds; (e) There is speculation that the mechanism (cricothyroideus) for producing vibrato at low pitches is more of a laryngeal phenomenon, at higher pitches is more of a vocal tract phenomenon (suprahyoid muscles and the tongue), and at mid-range pitches is a combination of the two phenomena; and (f) The expiratory musculature and laryngeal depressor strap muscles do not appear to be importantly linked to vibrato production.

Vocal Trill

The principal difference between vocal vibrato and trill is that a greater degree of pitch modulation is present in the trill. Instrumental music, such as the piano scores of Bach and Beethoven, often employs the trill; they are ±1 half tone and have exact, prescribed time or note value. It is considered a triumph of vocal training by many who master this vocal ornament. Especially are sopranos expected to learn to trill. The vocal trill is a musically prescribed variant calling for regulated changing of vocal pitch back and forth between the written note and the adjacent note above. While the number of changes in pitch is around five variations per second, the same as in vibrato, the amount of pitch modulation is greater than in vibrato. At least some accomplished singers employ intensity fluctuations in addition to pitch modulation. High-speed photography of the larynx during trill production indicates rhythmic contractions and relaxations of the laryngeal structures, including the epiglottis, for trained singers.

Yodeling

Viewed by many as an ethnologic peculiarity, yodeling is an intriguing vocal physiology. Like the trill, yodeling is a musical entity. It is characterized by fluctuations between the chest and head registers while singing vowels. The larynx appears at a considerably lower position for yodeling than during singing the same tones with a normal voice. The resonant cavities appear enlarged during yodeling, especially the supraglottic area, which is wider and longer. There is usually more respiratory movement during yodeling than in singing with abdominal movements dominating over chest activity. Yodeling seems to require more breath volume per second than does regular singing, but the chest movements show a smoother and more ordinary course of contraction than in normal singing.

A good yodeling style cannot be viewed as vocal abuse any moreso than a good singing style. Therefore, it appears to be an appropriate use of the voice and may be aligned with a physiologically natural style of artistic singing. This view may be derived from the following list of physiological characteristics of phonation during yodeling: (1) low position of the larynx, (2) absence of laryngeal elevation while leaping from low notes to high notes, (3) widening (dilation) of the supraglottic resonators, (4) dominance of abdominal expiratory breath support, (5) use of the soft vocal attack in the initiation of phonation. Yodeling, then, is the prototype of efficient vocal physiology.

That yodeling conforms to the rules of vocal hygiene and a natural singing style might suggest this vocal technique as a possible therapeutic approach for certain patients with dysphonic voices. Or, yodeling might be incorporated into singing lessons for those students who need special assistance to develop correct singing physiology in a natural manner.

BIBLIOGRAPHY

Balfour, H.H., and Ayoub, E.M.: Hoarseness as the presenting symptom of mitral insufficiency. *JAMA, 204*:1190-1193, 1968.

Brodnitz, F.S.: *Vocal Rehabilitation*. Rochester, Whiting Pr, 1959.

Dickson, D.R., and Dickson, W.M.: Function of the Human Larynx. Presented at annual meeting of Pennsylvania Academy of Ophthalmology and Otolaryngology, Bedford, Pennsylvania, May, 1970.

Fink, B.R., and Demarest, R.J.: *Laryngeal Biomechanics*. Cambridge, MA: Harvard U Pr, 1978.

Greene, M.C.L.: *The Voice and Its Disorders*, 4th ed. New York, Pitman, 1979.

Hartman, D.E., and Aronson, A.E.: Clinical investigators of intermittent breathy dysphonia. *J Speech Hear Disord, 46*:428-432, 1981.

Hollien, H.: The registers and ranges of the voice. In Cooper, M., and Cooper, M.H. (Eds.): *Approaches to Vocal Rehabilitation*. Springfield, IL: Thomas, 1977.

Kaplan, H.M.: *Anatomy and Physiology of Speech*, 2nd ed. New York, McGraw, 1971.

Landman, G.H.M.: *Laryngography and Cinelaryngography*. Baltimore, Williams and Wilkins, 1970.

Luchsinger, R., and Arnold, G.: *Voice-Speech-Language*. Belmont, CA: Wadsworth Pub, 1965.

Pressman, J.J., and Kelemen, G.: Physiology of the larynx. *Physiol Rev, 35*:506-554, 1955.

Zemlin, W.R.: *Speech and Hearing Science*, 2nd ed. Englewood Cliffs, NJ: P-H, 1981.

PART THREE: THE VOCAL TRACT – UPPER ASPECTS

Chapter 9

THE INTACT SKULL

INTRODUCTION

INTELLIGENT study of the soft tissues of the head, which are responsible for speech reception, processing, and production, is impossible without some knowledge of the skull. Liberal use of a complete skull with a removable top should therefore be undertaken before the disarticulated skull is considered. The emphasis here should be to understand the general features of the skull. After the bones and major landmarks of the intact skull have been mastered, the disarticulated bones may be considered. However, a detailed study of *all* the separate bones of the skull is not necessary for the medical student, nor for a speech pathologist or audiologist.

Although there are several ways of tallying the bones of the skull, a total of 29 will be considered here as members of the skull. These will be examined as members of the *cranial* and *facial* aspects of the skull.

The posterior-superior portion of the skull is the cranial or neural part, and the anterior portion is the facial part. No distinct external demarcations separate the two parts of the skull. Except for the mandible and the auditory bones, the bones of the skull unite in specialized joints known as *sutures*.

Cranial Part

The cranium forms a brain case for containing the brain and housing the auditory mechanism. Including the six auditory ossicles, there are 14 bones in this portion of the skull. Functions pertaining to communication include housing and protecting the brain (the speech-coding mechanism), the sense of hearing, and the master gland of the body, the hypophysis (pituitary). Attachments for some important musculature of the articulatory mechanism are provided by certain bones of the cranial skeleton (Fig. 9-1).

Facial Part

The facial skeleton contains the framework for housing the eyeballs and the upper parts of the respiratory and digestive tracts and includes a total of 15 bones, including the hyoid bone. Sometimes the facial skeleton is divided into the upper facial skeleton, which is related to the orbital and nasal cavities; the bony palate, including the upper teeth; and the lower fa-

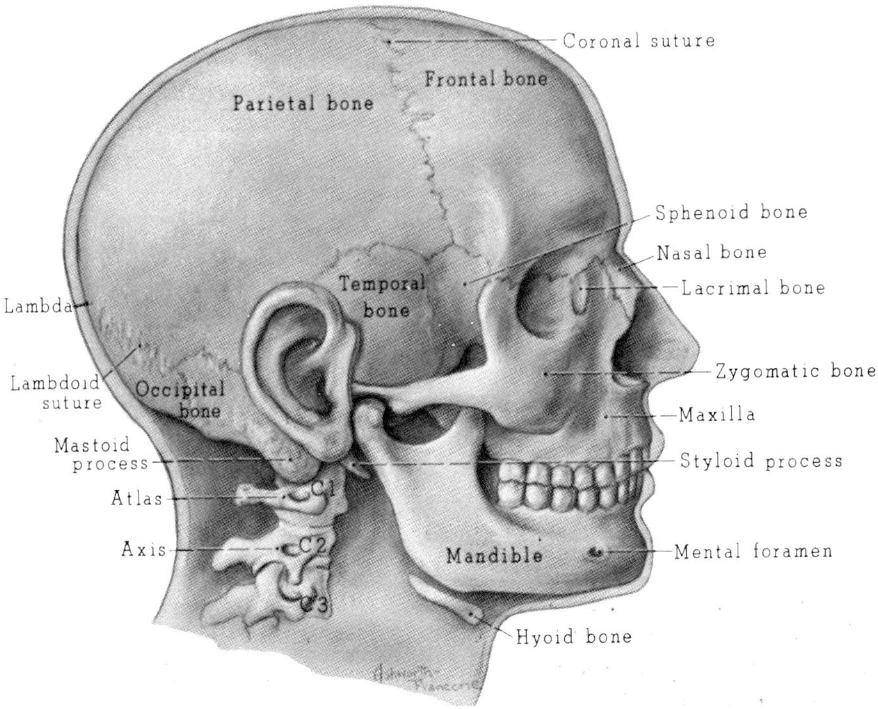

Figure 9-1. Outline of the head, face, and neck in relation to the skull and cervical vertebrae. The hyoid bone is attached only by soft tissue. (From Jacob, S.W., and Francone, C.A.: *Elements of Anatomy and Physiology*, 1976. Courtesy of W.B. Saunders Co., Philadelphia.)

cial skeleton, which includes the mandible plus the lower teeth (Table 9-I).

The basic framework for producing speech sounds rests with the facial skeleton. The upper and lower jaws and the teeth serve the articulatory functions, while the nasal cavity serves as a resonant chamber. In addition, attachments of speech-related musculature pertaining to the soft palate, pharynx, and tongue are associated with the facial skeleton.

THE EXTERIOR OF THE SKULL

Exteriorly, the skull may be studied from the front, from the side, from the back, from the top, and from beneath. An examination of the interior is possible after removal of the cap of the skull (Fig. 9-2).

Frontal View

From this view, the facial part of the skull appears to predominate. Both upper and lower aspects of the facial skeleton are clearly present. The general outline of the frontal aspect of the skull includes the *frontal bone* superiorly, the *zygomatic bones* laterally, and the *mandible* completing the lower part. The borders of the orbital cavities are formed by the *supraorbital margin* of the frontal bone, the *frontal process* of the max-

Table 9-I

SUMMARY OF THE BONES OF THE SKULL

Division	*Bone*	*Number*	*Function*
Facial Skeleton		15	
	Nasal	2	Form bridge of nose
	Maxilla	2	Form upper jaw; palatine processes contribute to oral, nasal cavities; frontal processes contribute to orbital cavities
	Zygomatic	2	Form cheek bones
	Mandible	1	Forms lower jaw
	Vomer	1	Forms lower part of nasal septum
	Palatine	2	Form part of lateral wall of nasal cavities and posterior part of hard palate
	Inferior concha	2	Form portion of lateral wall of nasal cavities
	Lacrimal	2	Form portion of medial wall of orbital cavities
	Hyoid	1	Forms muscle attachments of tongue and larynx
Cranial Skeleton		14	
	Frontal	1	Forms forehead
	Parietal	2	Form midportion of skull and skull cap
	Occipital	1	Forms back and part of underside of skull; provides articulation with vertebral column
	Temporal	2	Form side of skull and houses sense organs for hearing and balance; the middle ear cavities contain the ossicles: Malleus (2), Incus (2), Stapes (2)
	Ethmoid	1	Forms small portion of anterior floor of cranial cavity; contributes to nasal septum with perpendicular plate, middle and superior nasal cavity with cribriform plate
	Sphenoid	1	Forms bulk of floor of midportion of cranial cavity

illa, and the zygomatic bone. The width of the nasal bridge formed chiefly by the paired *nasal bones* between the orbital cavities may show considerable variation from subject to subject.

The bony margins of the nasal cavities often vary significantly in size and form. They are formed by the two *nasal bones* and two *maxillae*. Some correlation has been purported to exist between the width of the external nasal openings and climatic conditions. The basic framework of the nose is completed by cartilages protruding from the bony margins. Within the nasal cavity are the *inferior nasal conchae*, extending from the lateral walls. Dividing the two nasal cavities is a vertical bony partition called the *bony nasal septum* (see Figs. 9-3 & 2-22). Usually, there is some left or right deviation of the septum in most skulls. The bony septum does not approach the front entrance of the nasal cavities. In the living subject, a cartilaginous septum completes the partition anteriorly; it, too, is seen to deviate to one side slightly in the majority of individuals.

On the front of the skull, three pairs of *foramina* (openings for the blood vessels and nerves) plus several *fissures* (clefts) are evident. The *supraorbital and infraorbital foramina* are located in the frontal and maxillary bones, respectively. The *mental*

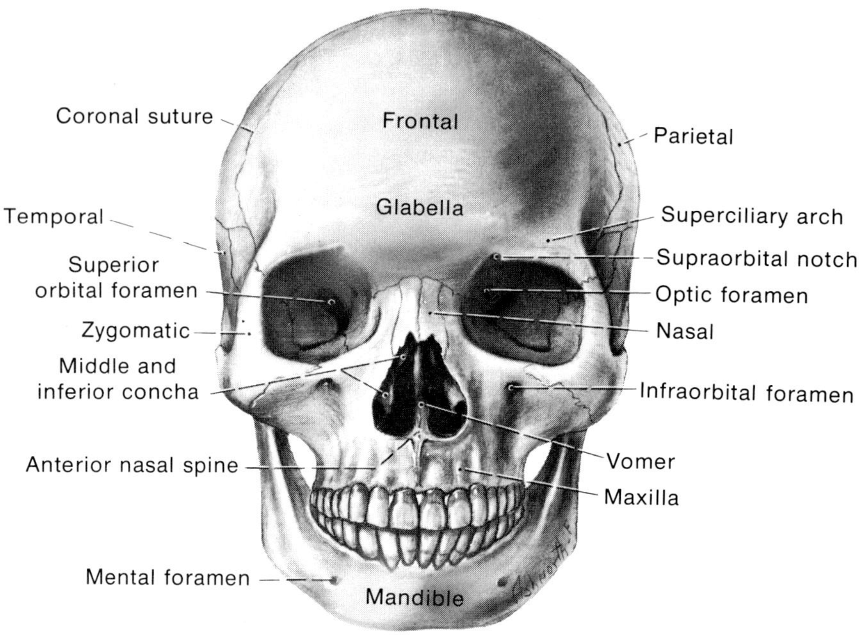

Figure 9-2. Frontal aspect of the skull. The facial skeleton dominates in this view. (From Jacob, S.W., and Francone, C.A.: *Structure and Function in Man*, 4th ed., 1978. Courtesy of W.B. Saunders Co., Philadelphia.)

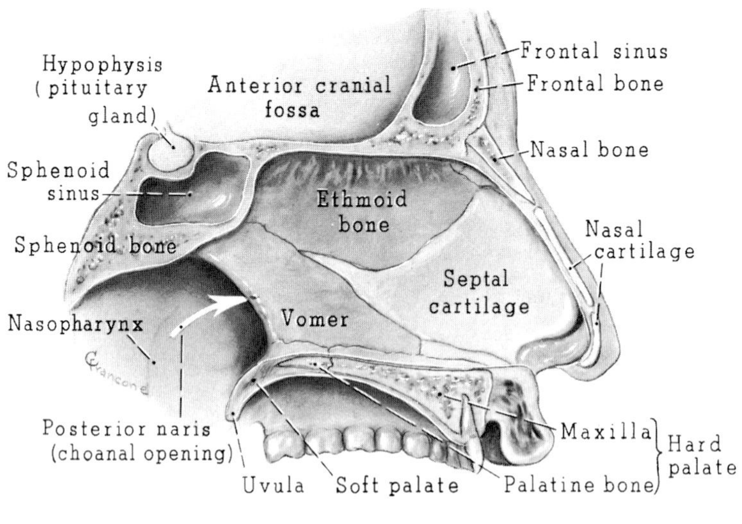

Figure 9-3. Sagittal section of the nose showing the components of the nasal septum, the hard palate, and the entrance into the posterior choana (funnel-shaped cavity). (From Jacob, S.W., and Francone, C.A.: *Elements of Anatomy and Physiology*, 1976. Courtesy of W.B. Saunders Co., Philadelphia.)

foramen is found on the mandible.

Within the orbital cavities are the *inferior* and *superior orbital fissures*. The nerves innervating muscles associated with eye movement pass through the superior fissures along with the ophthalmic veins. Other nerves and blood vessels pass through the inferior fissure.

The orbital fissures are formed as converging processes of surrounding bones fail to unite completely. These include the *ethmoid, sphenoid,* and *lacrimal* bones. Those bones comprising the margin of the orbital cavity named above (frontal, zygomatic, maxilla) also aid in forming the cavity walls (Fig. 9-4).

The upper jaw is formed by the two maxillary bones. Each is braced by arching supports, extending to the bony forehead, so that the teeth transmit biting and chewing forces over a wide area (Fig. 9-39). The mandible is a very strong bone of one-piece construction. Biting and chewing forces are generated by action of this bone. By contrast, the crocodile's upper jaw is hinged.

Lateral View

Both the facial and neural portions of the skull are present in the lateral aspect. Of interest are the *sutures* (Latin for seam), which indicate the limits of a number of cranial bones. (Sutures or suturae are a type of immovable fibrous articulation or joint, occurring only between the bones of the skull.) The squamosal suture arches across the flat portion of the lateral part of the brain case. It lies primarily between the *temporal* and *parietal* bones. Posteriorly, it meets the lambdoidal suture, which travels around the posterior of the skull to intersect the squamosal suture on the opposite side. The lambdoidal suture is between the occipital and parietal bones. Coursing inferiorly from the junction of the squamosal and lambdoidal sutures is the *occipitomastoid suture* (Fig. 9-5).

The frontal bone and parietal bone are joined with the *coronal suture*. At the apex of the skull, the *coronal suture* meets the *sagittal suture*, which travels posteriorly, separating the parietal bones. The sagittal

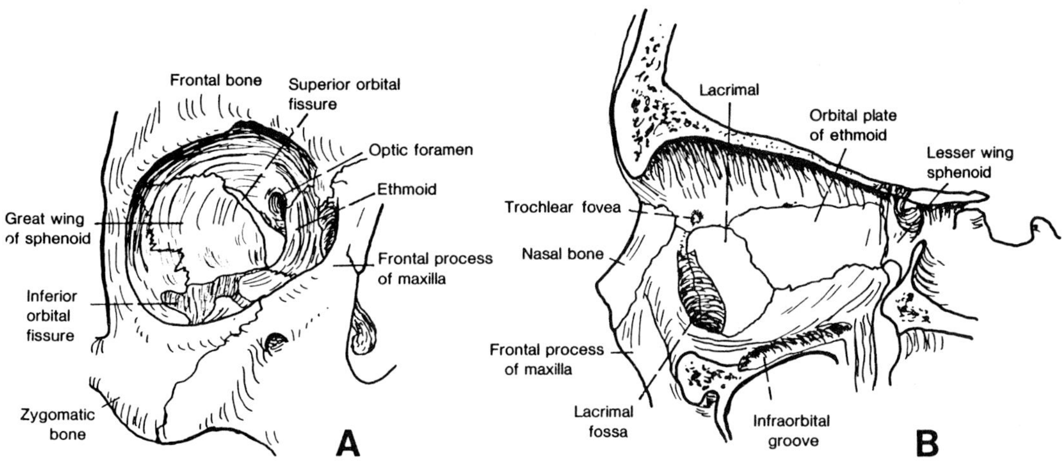

Figure 9-4. Bones of the orbital cavity. (A) Anterior view of the right orbit. (B) Medial wall of the left orbit.

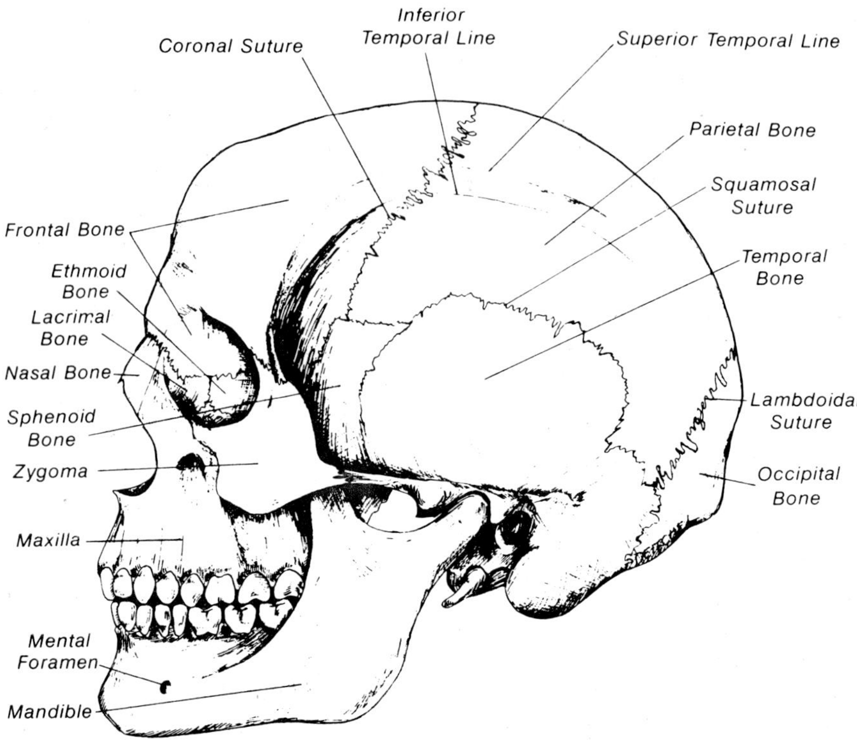

Figure 9-5. Lateral aspect of the skull. The cranial skeleton dominates in this view. (From Fried, L.A.: *Anatomy of the Head, Neck, Face, and Jaws*, 2nd ed., 1980. Courtesy of Lea & Febiger, publishers.)

suture ends as it intersects the lambdoidal suture. The reader will recall that the coronal and sagittal planes of the body were described in reference to these sutures (see Chap. 2).

Occasionally, small islands of bone arise as the normal suture lines divide and form irregular, isolated bones termed *wormian bones* (Fig. 9-6). These bones occur most frequently along the lambdoidal suture. Over one hundred have been found in the skull of an adult hydrocephalic subject.

The microscopic anatomy of the cranial sutures is interesting. The sutures in the adult are comprised of minute, elongated particles of bone known as *sutural bones*. Their random arrangement in direction provides excellent stability for support of adjacent bones. An interesting analogue to this design is the surgical procedure of bone grafting into a maxillary arch cleft for purposes of stabilizing the maxillary arch. The graft is taken either from the iliac crest (hip bone) or from a floating rib (11th and 12th ribs). The graft is prepared for the graft site by whittling the graft into bone chips. The chips are implanted in an irregular arrangement as a strategy for supporting the cleft portion of the maxilla from multidirectional stress. It would appear that the anatomy of the cranial sutures and the anatomy of the surgically created maxillary arch form are similar in the sense of functional design.

Several important landmarks of the temporal bone are shown in the lateral view. The *external auditory meatus* or external auditory canal is an elliptical orifice

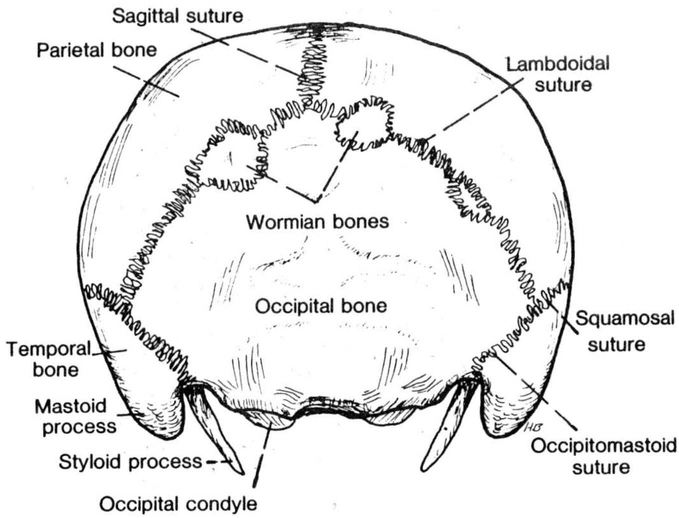

Figure 9-6. Posterior aspect of the skull with the mandible removed showing the sutures. The lambdoidal suture contains three wormian bones. The skull articulates with the vertebral column through the occipital condyles. Several muscles associated with tongue and pharyngeal movement take their origins from the styloid processes.

bounded by the *tympanic part* of the temporal bone. An extension of cartilage in the living subject fans out to form the basic framework of the external ear. The *mastoid process*, a nipple-shaped protuberance, is situated posteriorly and inferiorly to the *mandibular* or *glenoid fossa* (*fossa* = hollow or depressed area). It is this fossa that provides for articulation between the mandible and the temporal bone.

Arising anteriorly to the mandibular fossa is the *zygomatic arch*, which serves as one of the support pillars of the maxillary skeleton. The *styloid process* is a slender, tooth-like structure pointing inferiorly. It serves as an important origin for muscles that influence the position of the tongue during articulation and swallowing.

The *temporal fossa* is an area above the zygomatic arch and behind the lateral margin of the orbital cavity. It contains the origin of the *temporalis muscle*, which is inserted onto a process of the mandible. The area comprising the temporal fossa includes a narrow strip of the parietal bone, most of the *squamous portion* of the temporal bone, the temporal surface of the frontal bone, and the temporal surface of the *greater wing* of the sphenoid bone. The superior limits of the temporal fossa are two parallel lines, the *temporal lines*, arching over the lateral surface of the cranial skeleton. The field of the temporal fossa is better studied with the zygomatic process and the lateral aspect of the zygomatic bone removed (Fig. 9-7).

Just inside the zygomatic process of the temporal bone and protruding inferiorly, the *lateral pterygoid plate* of the sphenoid bone is situated. Here, internal to the position that is held by the zygomatic arch is a relatively deep fossa, the *infratemporal fossa* (Fig. 9-8). This fossa contains a rich network of blood vessels and nerves being sent to or from the brain case. A surgeon doing a maxillary osteotomy (see Chap. 20) will attempt diligently to stay away from this area.

On the undisturbed skull, it is possible

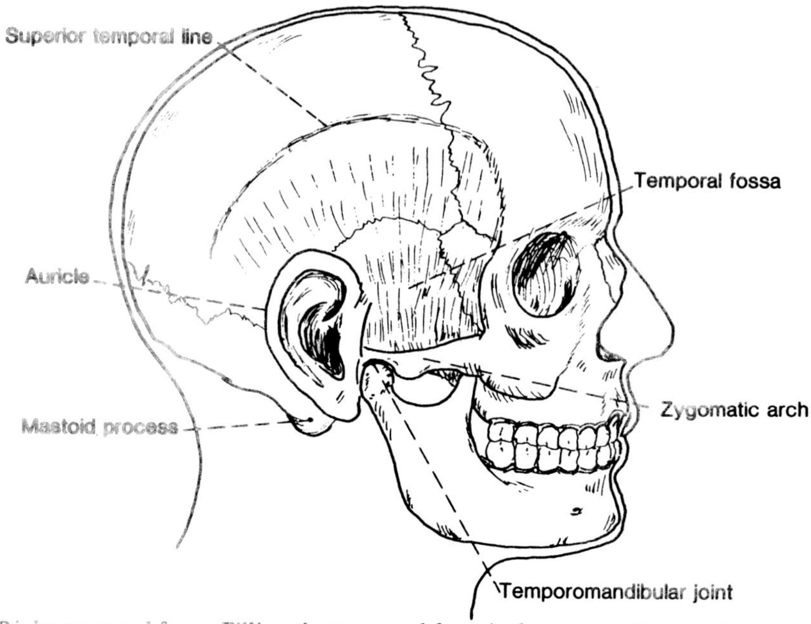

Figure 9-7. Right temporal fossa. Filling the temporal fossa is the temporalis muscle, a prominent muscle of mastication that takes its origin along the temporal line of the skull, courses down the side of the skull, and inserts onto the coronoid process of the mandible.

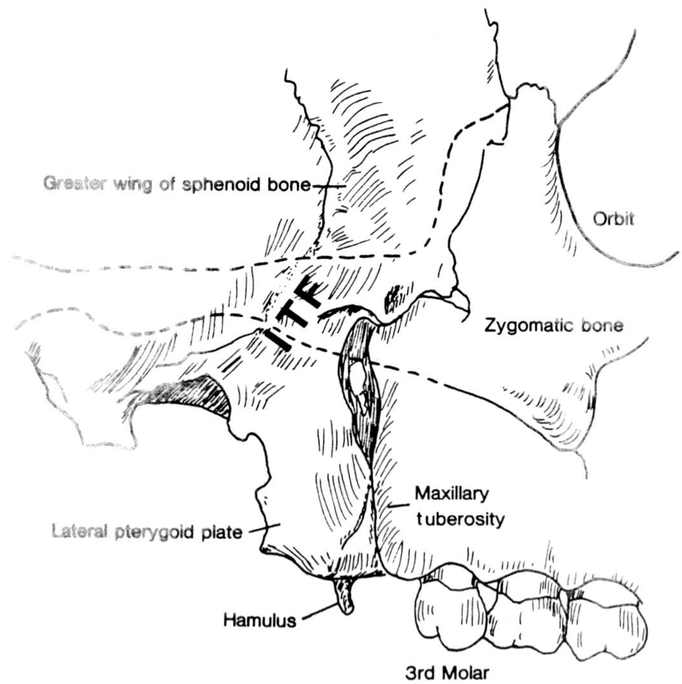

Figure 9-8. The medial wall of the right infratemporal fossa. The mandible has been removed. The zygomatic arch (outlined) passes over this region (ITF). (From Gardner, E., Gray, D.J., and O'Rahilly, R.: *Anatomy*, 4th ed., 1975. Courtesy of W.B. Saunders Co., Philadelphia.)

to see portions of the pterygoid plate through the *mandibular notch*, which lies between the *coronoid* and *condyloid processes* of the mandible.

Inferior View

When viewed from beneath, the structures forming the skull bone are more readily observed with the mandible removed.

The basilar view of the skull demonstrates well that the human skull consists of a primary part (neurocranium), with the facial skeleton or secondary part attached directly and with lateral buttresses.

Since early anatomists were restricted to a gross anatomical approach to studying the skull, one can appreciate some of the influences on their lives that such study might have motivated. As an example, one should hold the skull such that a basilar view is obtained, with the facial skeleton facing the floor. One should then make an imaginary midsagittal cut of the skull into two parts and rotate these parts so that the segments of the facial skeleton are positioned laterally, with the two bases of the occipital bones juxtaposed. In this position, an architectural form is ob-

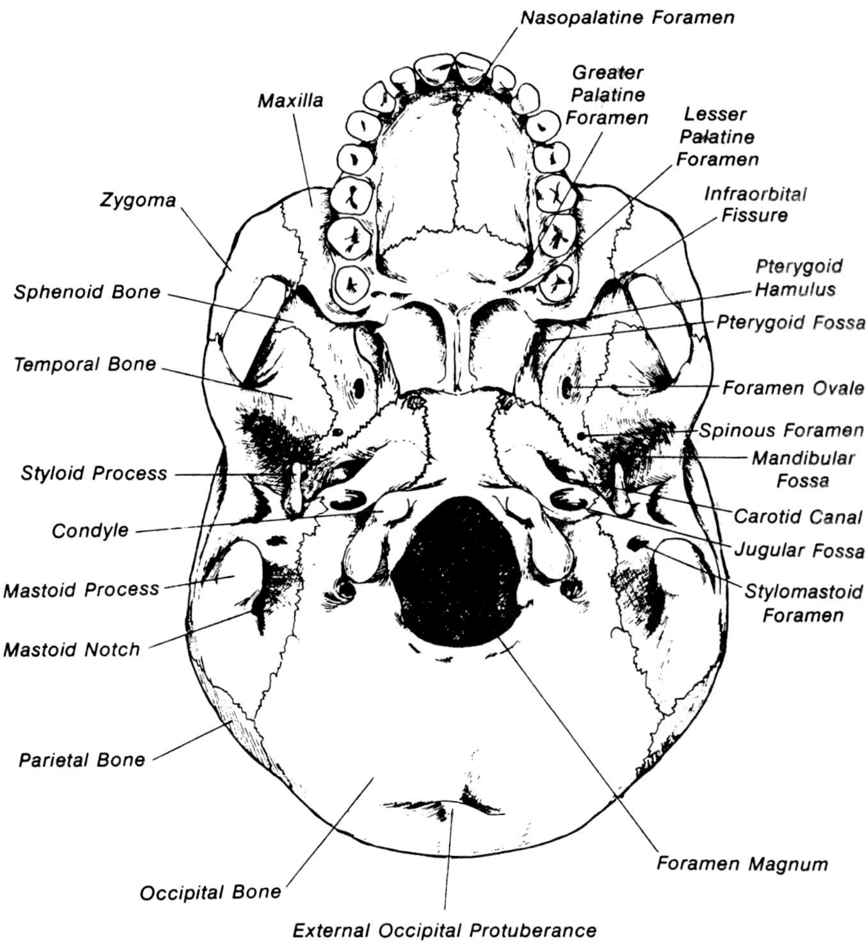

Figure 9-9. Inferior aspect of the skull with the mandible removed. (From Fried, L.A.: *Anatomy of the Head, Neck, Face, and Jaws*, 2nd ed., 1980. Courtesy of Lea & Febiger, publisher.)

served that consists of two lateral, secondary structures that are stabilized to the primary structure by flying buttresses. When one considers that some of the world's famous cathedrals were designed by anatomists such as Michelangelo, it is interesting to ponder the possible influence of anatomic observations on their architectural contributions. At any rate, this design is a time-tested arrangement that has served well in the skull and various cathedrals throughout the world (Fig. 9-9).

The predominant feature of the facial skeleton is the *bony palate* (Fig. 9-10), which is made up by the *palatal process* of four bones: two *maxillae* and two *palatine bones*. The midline palatine suture is not fixed until about age sixteen, an important fact to the orthodontist and plastic surgeon.

An opening in the bony palate near the incisors called the *incisive foramen* or *canal* permits the passage of blood and nerve vessels onto the palatal surface of the oral cavity. Medial to the third molar is another opening, the *palatine foramen*.

Each half of the upper (or lower) adult dental arch contains eight teeth: the central and lateral incisors, the cuspid, the first and second bicuspids, and the first, second, and third molars. Alternatively, they may be listed as: two incisors, a canine, two premolars, and three molars. A complete treatment of the dentition is found in Chapter 10.

As the two palatine bones meet medially, the *posterior nasal spine* is formed at the posterior limit. Below the spine in the inverted skull is the *vomer*, a part of the bony nasal septum forming a pillar between the *body* of the *sphenoid* and the palatine bones. On either side of the vomer are the two posterior entrances into the nasal cavities. They are known as the *choanae* (funnel-shaped opening) and connect the nasopharyngeal cavities (Fig. 9-11).

The *zygomatic processes* of the maxillae, leading to the zygomatic arches, are situated on the frontolateral portions of the skull. The zygomatic process of the temporal bone completes the arch and intro-

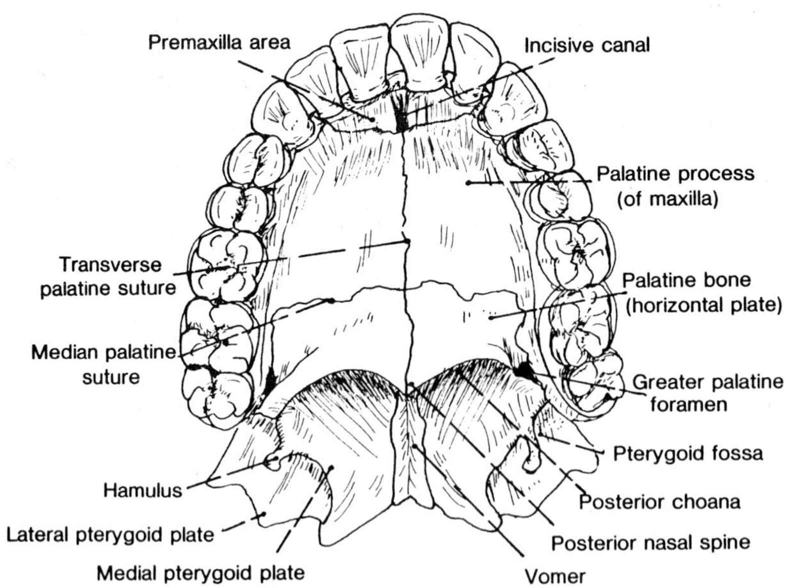

Figure 9-10. The hard palate, maxillary alveolar process, vomer, and pterygoid plates. Four bones contribute processes to make up the palate.

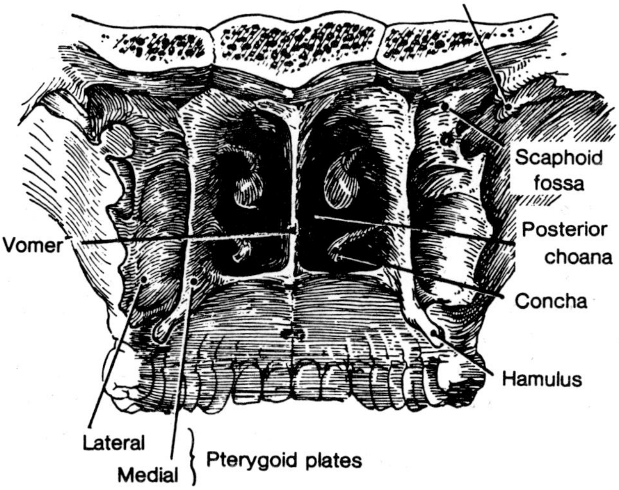

Figure 9-11. The posterior entrance into the nasal cavities viewed from the nasopharynx. (From Nagle, Raymond J., and Sears, Victor H.: *Denture Prosthetics*, ed. 2, St. Louis, 1962, The C.V. Mosby Co.)

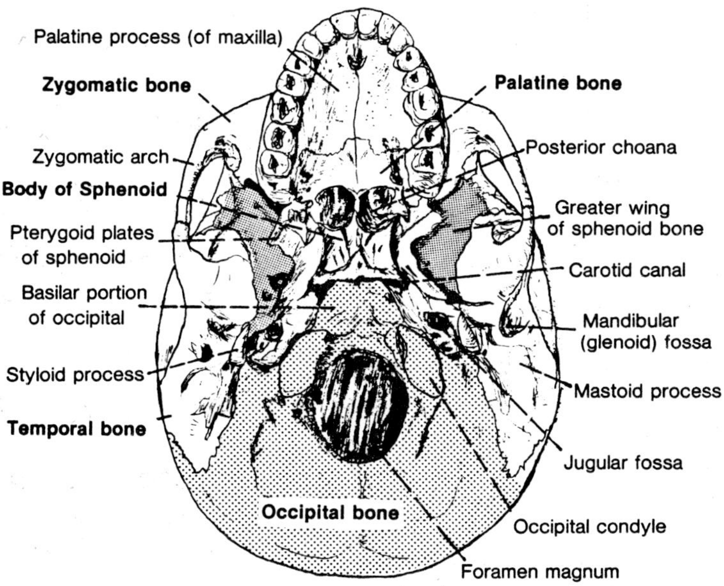

Figure 9-12. Inferior aspect of the skull emphasizing the separate bones (bold type) and some of their parts (light type). The greater wing of the sphenoid bone is shaded as is the whole of the occipital bone.

duces the inferior parts of the temporal bone. Medially, on the inferior surface of the temporal bone, is the styloid process and the *carotid canal* for entrance of the *carotid artery*. The *jugular foramen*, located posteriorly to the carotid canal and formed from juxtaposed notches on the temporal and occipital bones, permits the exit of blood vessels and nerves.

Under the zygomatic arch is a portion

of the sphenoid bone, which participates in the formation of the brain case. Spaces on the lateral walls in the base of the cranium are sealed by the greater wings of the sphenoid bone. The two greater wings emanate from a medial area on the skull base. The area contains other parts of the sphenoid bone, the *body* of the sphenoid bone and the two pairs of *pterygoid plates*, which project inferiorly from the body of the sphenoid bone. The *hamulus*, a small, delicate process of the *medial pterygoid plate*, is visible. The tendon of the tensor veli palatini muscle passes around this process (Fig. 9-11).

The *occipital bone* comprises the major area of the posterior section of the base of the cranium. The *foramen magnum* pierces the occipital bone. Through this opening, the spinal cord is continuous with the brain. Two knuckles arise on the anterolateral margins of the foramen magnum. These are the *occipital condyles*, where the skull is attached to the vertebral column. If these condyles are excessively flattened, flexion and extension of the head on the vertebral axis is restricted. Note that the condyles are positioned closer to the front than the back of the foramen magnum. This is because the articular facets of the first *cervical vertebra* (atlas) are positioned over the anterolateral part of the atlas. Also, the foramen magnum and the adjacent condyles are not on center with respect to the anterior-posterior (A-P) dimensions of the skull. This rearward placement of the cervical vertebral column permits a central position of the pharynx. Extending anteriorly from the foramen magnum is the *basilar portion* of the occipital bone, which articulates with the body or corpus of the sphenoid bone. The lateral margins of the occipital bone meet the temporal bones (Fig. 9-12).

Superior View

The skull cap or calvarium has three sutures traversing it. The coronal suture follows a rounding path of the anterior aspect, dividing the frontal bone from the two parietal bones. The nearly transverse suture appearing on the posterior area is the lambdoidal, which serves as the anterosuperior margin of the occipital bone. Actually, this suture has the appearance of an inverted V when viewed posteriorly (see Fig. 9-6). The sagittal suture courses across the top of the skull from front to back, interconnecting the coronal and lambdoidal sutures in the median plane of the body. These three sutures form the configuration of the letter H (Fig. 9-13).

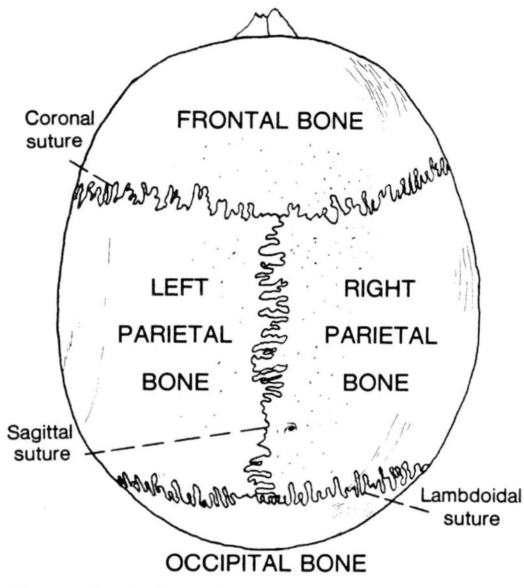

Figure 9-13. Superior aspect of the skull. Note the H-shaped pattern of the sutures in this view.

The point of junction between the coronal and sagittal sutures is the site of the *anterior fontanelle*, or soft spot in the infant skull. A *posterior fontanelle* exists at the junction of the sagittal and lambdoidal sutures. Ossification does not eliminate these fontanelles until about the end of the second year (Fig. 9-14).

Since the opposing edges of the cranial

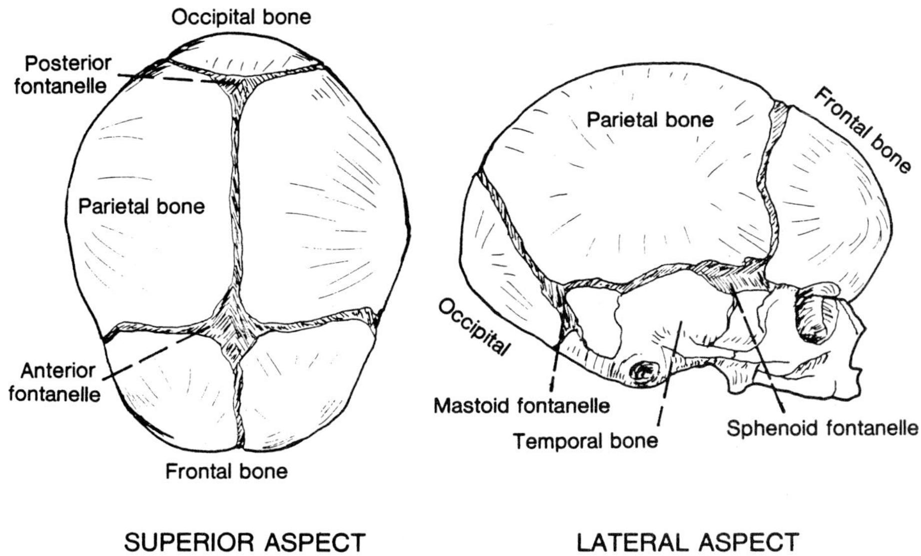

Figure 9-14. Skull of newborn infant showing fontanelles. It is apparent that the infant's cranial skull is made up of groups of separate bones in contrast to the solid cranial cavity in the adult.

bones are not ossified at term (culmination of pregnancy), the bones are permitted to override one another during birth. This accommodation effectively reduces the size of the head and aids its passage through the birth canal.

The membranous nature of the contiguous edges of the flat bones of the calvarium may sometimes close prematurely, limiting further growth of the skull. Consequently, normal development of the brain is disturbed. The individual has an abnormally small head in a condition known as *microcephaly*. This condition typically results in severe mental retardation.

A contracondition of microcephaly is hydrocephaly, in which there is marked enlargement of the skull. The cause of this anomaly is an excessive amount of cerebrospinal fluid in the cavities within the brain. The result is a significant elevation of the intracranial pressure. While the sutural joints of the calvarium are still partially movable, as is the case in young children, excessive intracranial pressures may cause the skull to enlarge. These children require either that the surplus fluid be drawn off periodically or that a shunt is surgically created to reduce pressure on the brain thereby preventing brain damage (Fig. 9-15).

The rounded topography of the normal skull offers a great deal of protection to the brain from trauma. Since the angle of incidence of a traumatic blow to the skull equals the angle of reflection, forces are dissipated greatly from oblique blows contacting the skull. Further, the presence of cerebrospinal fluid in tissue spaces interposed between the skull case and the brain surface provide additional displacement of traumatic forces away from the brain. This design should be appreciated by the student of anatomy for the logic and utility of the configuration of the skull case.

Posterior View

Most prominent from a posterior view is the *occipital bone*, which forms the posterior wall of the skull. The upper borders

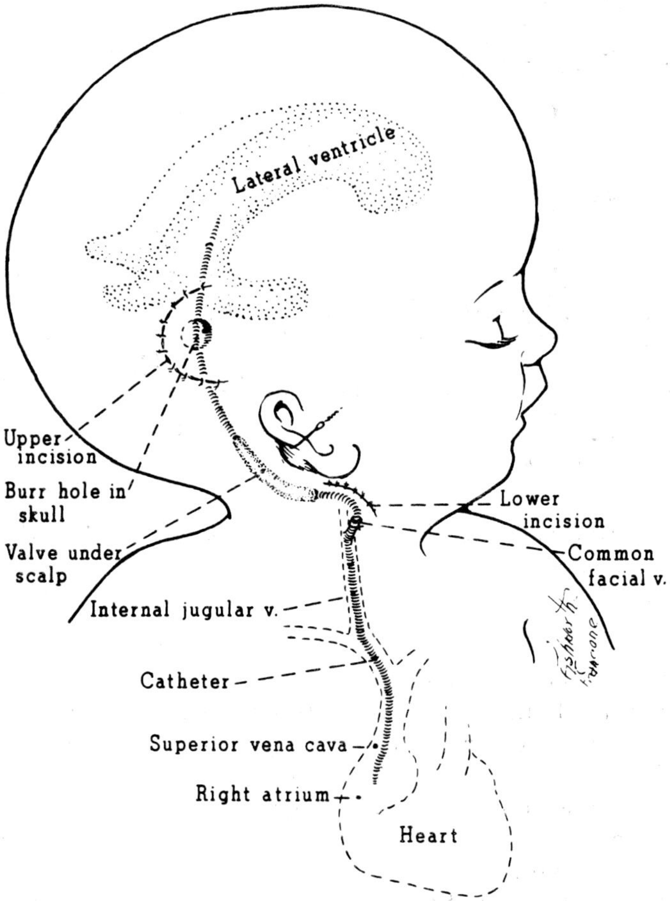

Figure 9-15. Corrective procedure for hydrocephalus to prevent high intracranial pressure. A catheter drains the ventricular system into the right atrium of the heart. (From Jacob, S.W., and Francone, C.A.: *Elements of Anatomy and Physiology*, 1976. Courtesy of W.B. Saunders Co., Philadelphia.)

of the occipital are laced to the posterior border of the parietal bones by the lambdoidal suture. The lambdoidal suture is the posterior boundary of the parietal bone on either side. At the superior midline, the posterior limit of the sagittal suture can be seen (Fig. 9-16).

The mastoid processes of the temporal bones can be recognized along the lower portion of the skull. The styloid process of the temporal bone is also visible in a complete skull.

THE INTERIOR OF THE SKULL

The floor of the cranial cavity is formed by the superior surface of the base of the skull. With the cap removed, the floor of the brain case is revealed. It may be divided, for disarticulated purposes, into three paired areas termed the *cranial fossae*. These fossae are arranged in steps; the *posterior cranial fossa* is the lowest and

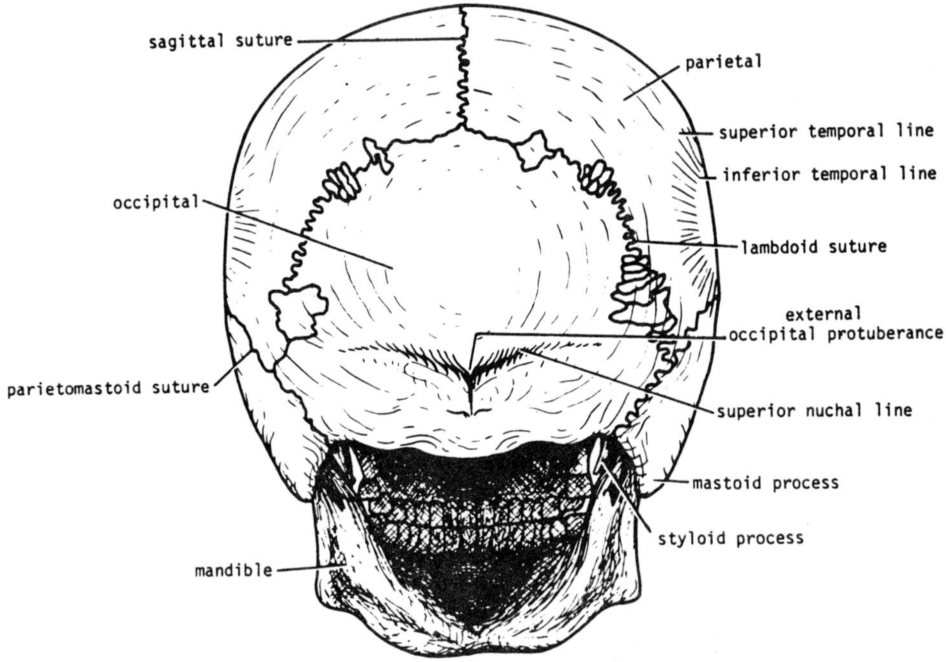

Figure 9-16. Posterior aspect of the skull showing the prominence of the occipital bone and the mastoid and styloid processes of temporal bone. (From Snell, R.S.: *Clinical Anatomy for Medical Students*, 2nd ed., 1981. Courtesy of Little, Brown and Co., Boston.)

the *anterior cranial fossa* is the highest, with the *middle cranial fossa* situated between. The anterior and middle cranial areas house a portion of the cerebrum or main part of the brain, while the posterior cranial fossa houses the cerebellum (Fig. 9-17).

Anterior Cranial Fossa

The *frontal* and *ethmoid* bones contribute to the formation of this fossa. The posterior border of this fossa is the *lesser wing* of the *sphenoid bone*. At the midline, the *crista galli* (thick triangular process) of the ethmoid bone divides the paired anterior cranial fossae. Lateral to the crista galli, the *cribriform plate* (plate with small openings like a sieve) of the ethmoid forms a small portion of the floor of the cranial cavity and the anterior fossa. The *orbital plates* of the frontal bone are the principal contributors to the formation of the anterior cranial fossa.

Middle Cranial Fossa

This fossa (there is one on either side of the *sella turcica* or *hypophyseal fossa* of the sphenoid bone) is formed by the *lesser* and *greater wing* of the sphenoid, by the *squamosal* and *petrous portions* of the *temporal bones*, and by the inferior portions of the parietal bones. The sella turcica is the notable landmark associated with this fossa since it houses the hypophysis or pituitary, the master gland of the body.

There are several important communications of the middle cranial fossa with adjacent cavities and spaces of the skull. The *optic nerve* and ophthalmic artery pass through the *optic foramen*, which is found at the root of the lesser wing of the sphenoid. The *superior orbital fissure*, a perforation between the wings of the sphenoid, allows the passage into the orbit of three nerves for the muscles of the eyeball, plus the ophthalmic nerve and the first division of the *trigeminal nerve*. The rotundum (round) and ovale (oval) *foramina* emit divisions of the trigeminal nerve. The *carotid artery* gains entrance into the cranial cavity by means of the orifice of the carotid canal at the top of the *petrous portion* of the temporal bone (see Fig. 9-32) and not through the *foramen lacerum*, which is closed by a fibrocartilaginous plate in the living subject. However, the carotid artery, an important supplier of blood to the brain, does bend and continue along on the cranial surface of the fibrocartilage of the foramen lacerum.

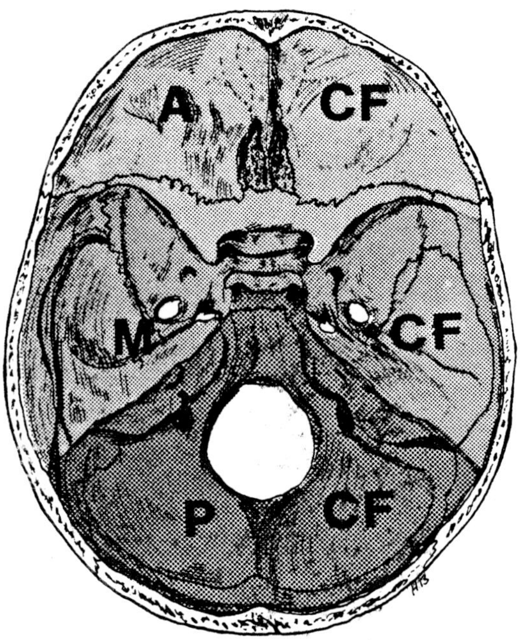

Figure 9-17. Floor of the skull viewed from above to show the cranial fossae: (ACF) anterior cranial fossa; (MCF) middle cranial fossa; and (PCF) posterior cranial fossa.

On the anterior surface of the petrous portion of the temporal bone is a slight eminence, arcuate eminence, caused by the *superior semicircular canal*. The roof of the tympanic cavity lies in front and slightly lateral to the projection caused by the semicircular canal (Fig. 9-18).

Posterior Cranial Fossa

Perhaps this pair of fossae represent the least complex of the three. Nonetheless, the posterior fossa is the largest and deepest. This fossa is bordered by the *dorsum sellae* of the sphenoid anteriorly, the *petrous* and *mastoid parts* of the temporal bone anterolaterally, and the occipital bone posteriorly, along with the mastoid angles of the parietal bone.

The two opposing notches between the edges of the petrous portion of the temporal and occipital bones form the *jugular foramen*. In addition to the passage of the internal jugular vein, several cranial nerves pass through the sections of this opening (Fig. 9-19).

The *internal acoustic meatus* is situated anteriorly to the jugular foramen. Through this meatus pass the *facial* and *acoustic nerves*, along with the *internal artery*.

The *foramen magnum* is central in the posterior cranial fossa. The spinal cord is continuous through this foramen with the *medulla oblongata*, or lowermost stem of the brain. Vertebral arteries and spinal accessory nerves pass through this the largest of the foramina into the cranial cavity, along with the spinal cord. On either side of the foramen magnum is the *hypoglossal canal*, which transmits the *hypoglossal nerve*.

Fractures of the Cranial Fossae

Accidents that damage the floor of the skull may directly impair speech and hearing. The most common wound to the bones of the cranial floor is a fracture.

Fractures involving the anterior cranial fossa may injure the cribriform plate

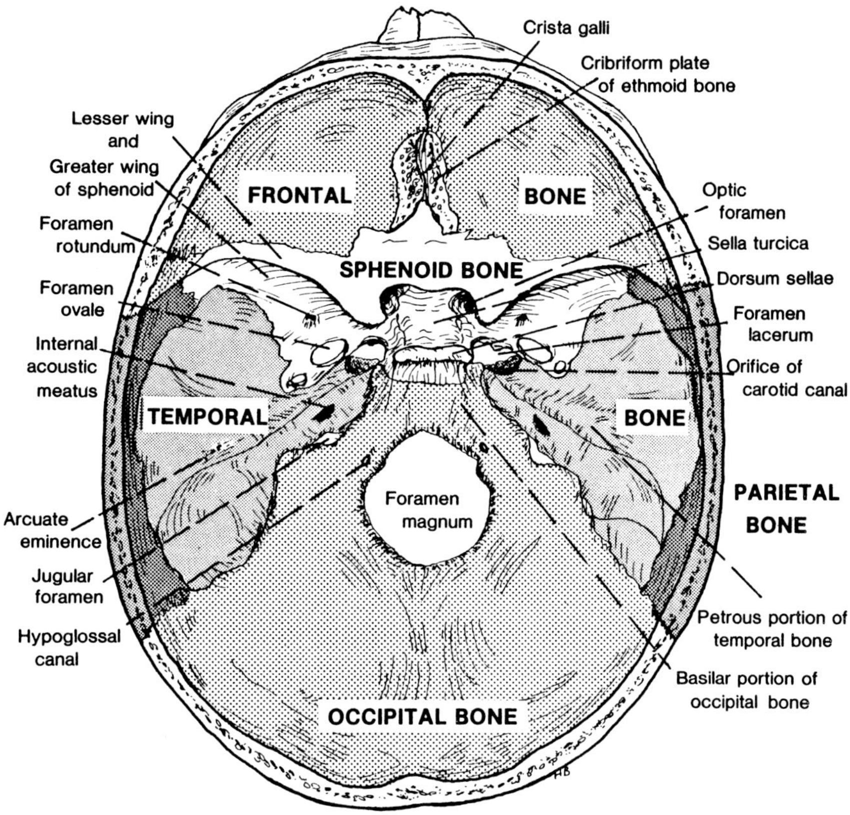

Figure 9-18. Floor of the skull with the bones of the cranial fossae identified.

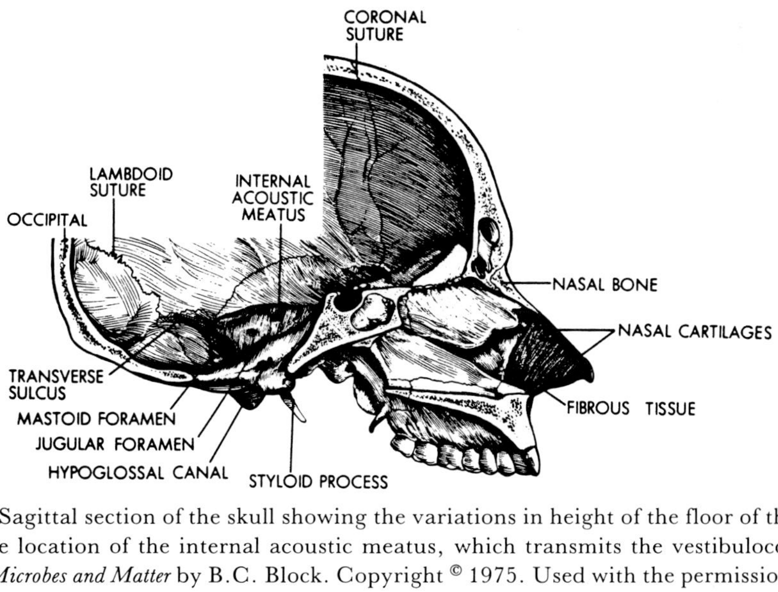

Figure 9-19. Sagittal section of the skull showing the variations in height of the floor of the cranial cavity. Notice the location of the internal acoustic meatus, which transmits the vestibulocochlear nerve. (From *Man, Microbes and Matter* by B.C. Block. Copyright © 1975. Used with the permission of McGraw-Hill Book Company.)

of the ethmoid bone. Since the coverings of the brain and mucoperiosteum are likely to be damaged, symptoms may include nose bleeding and leaking of cerebrospinal fluid into the nose. When the orbital plates of the frontal bone are fractured, hemorrhage into the orbital cavity causes the eye to bulge.

The most fragile of the cranial fossae is the middle fossa. It is the weakest part of the base of the skull due to the numerous perforations (foramina and canals) in this region. Both the cavities of the middle ear and sphenoidal sinuses are susceptible to fractures. Cerebrospinal fluid and blood may leak from the external auditory meatus in response to fractures. Damage to important nerves in this region, particularly the facial, is possible, as well as grave hearing loss and equilibrium disturbances.

In fractures of the posterior cranial fossa, leakage of blood invades the neck between the muscles. If the mucous membrane of the roof of the nasopharynx is broken, blood may escape there. Sometimes the fracture will involve the jugular foramen, with damage to cranial nerves in the vicinity (the glossopharyngeal, vagus, and spinal accessory). Articulation and phonation may be seriously impaired from such wounds.

Roof of the Cranial Fossae

The interior surface of the skull cap, or calvaria, presents an irregular surface resembling knuckle prints and fingerprints left in dough after kneading. These digitated impressions, or *digital markings*, are caused by growth of the cerebral convolutions of the brain, which results in growth of the bony capsule. Grooves for arteries and veins of the *dura mater* — that covering of the brain next to the bony case — are seen coursing about the cap of the neurocranium. Other features include the sagittal, coronal, and lambdoidal suture lines (Fig. 9-20).

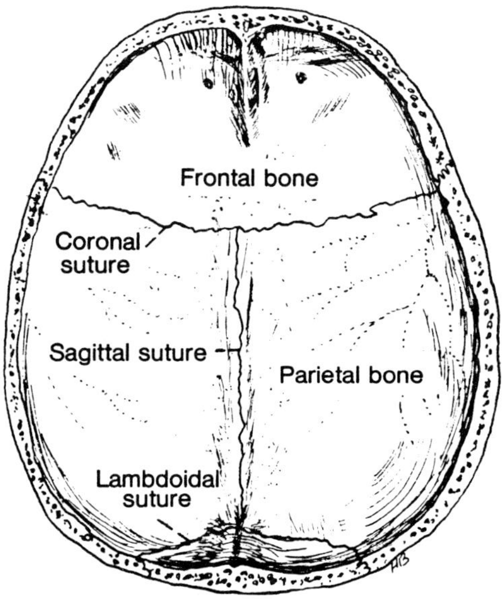

Figure 9-20. The skull cap as viewed from beneath.

BONES PERTAINING TO THE BRAIN AND EAR

The cranial portion of the skull consists of a total of fourteen separate bones, if the auditory ossicles of the temporal bones are included. Most of these are characteristically flat bones with a limited marrow cavity. In addition to housing the brain, hearing mechanism, and hypophysis or pituitary gland, the cranium also provides articulation for the skull with the rest of the skeleton through the vertebral column. Modification of the anterior portion of the cranial bones assists with accommodation of the eye through partial formation of the orbital cavity (Table 9-II).

Here, as in a subsequent part of the facial skeleton and the structures pertaining to articulation, a detailed consideration of the individual bones of the cranial skeleton is presented as an elaboration of structures housing the brain and the organ of hearing. Some of the bones of the cranial skeleton serve as an adjunct to the articulatory and resonant structures of the facial skeleton. The reader should refer to the figures of the intact skull bones shown earlier in this chapter for relative placement of the individual bones described here (Fig. 9-21).

The bones of the cranial vault are made up of three layers. Two layers of compact bone, the outer and inner surfaces, are divided by the *diploe*, or middle spongy layer. This layering enhances the skull's resiliency and is protective to the brain. The adult skull splinters if the blow to the vault is severe. The outer layer will break into linear fractures, with the inner

Table 9-II

BONES OF THE CRANIAL SKELETON

Name	*Number*	*Number and Name of Articulations*	
Frontal	1	12:	both parietals, both nasals, both maxillae, both lacrimals, both zygomatics, sphenoidal, ethmoidal
Parietal	2	5:	occipital, frontal, temporal, sphenoidal, other parietal
Occipital	1	6:	both parietals, both temporals, sphenoidal, atlas of vertebral column
Temporal	2	5:	parietal, sphenoidal, occipital, zygomatic, mandible
Ethmoid	1	13:	both nasals, both maxillae, both lacrimals, both palatines, both inferior nasal conchae, frontal, sphenoidal, vomer
Sphenoid	1	12:	both parietals, both temporals, both zygomatics, both palatines, vomer, ethmoidal, frontal, occipital

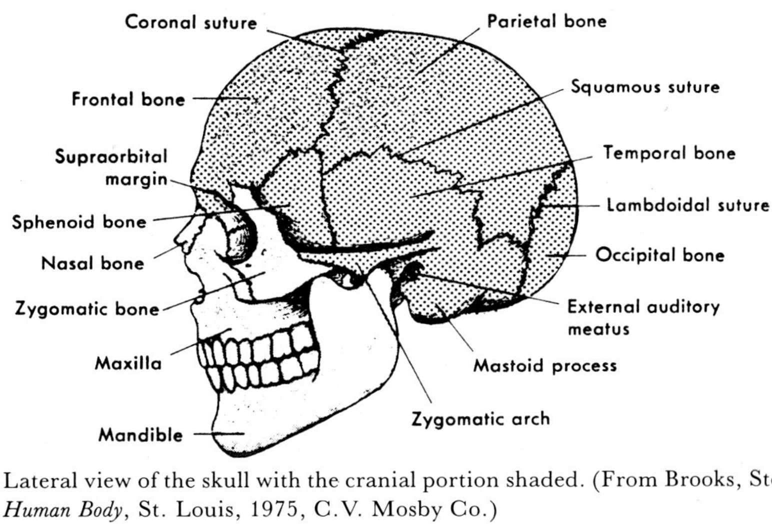

Figure 9-21. Lateral view of the skull with the cranial portion shaded. (From Brooks, Stewart M.: *Basic Science and the Human Body*, St. Louis, 1975, C.V. Mosby Co.)

layer's fragments penetrating the brain tissue. However, the vault of a young child is much more resilient, and the result is a depression of the traumatized area, termed a *pond fracture*.

The Occipital Bone

The back of the skull is formed by the *occipital bone*. It continues underneath the cranial cavity, with the *basilar portion* medially positioned and projecting anteriorly. The basilar part is a thick, wedge-shaped structure located in front of the *foramen magnum*. Its anterior limit is squared to join a like surface on the sphenoid bone to form the *spheno-occipital synchondrosis*. This important cartilaginous junction permits cranial growth, culminating in ossification between the sixteenth and twentieth years (Fig. 9-22).

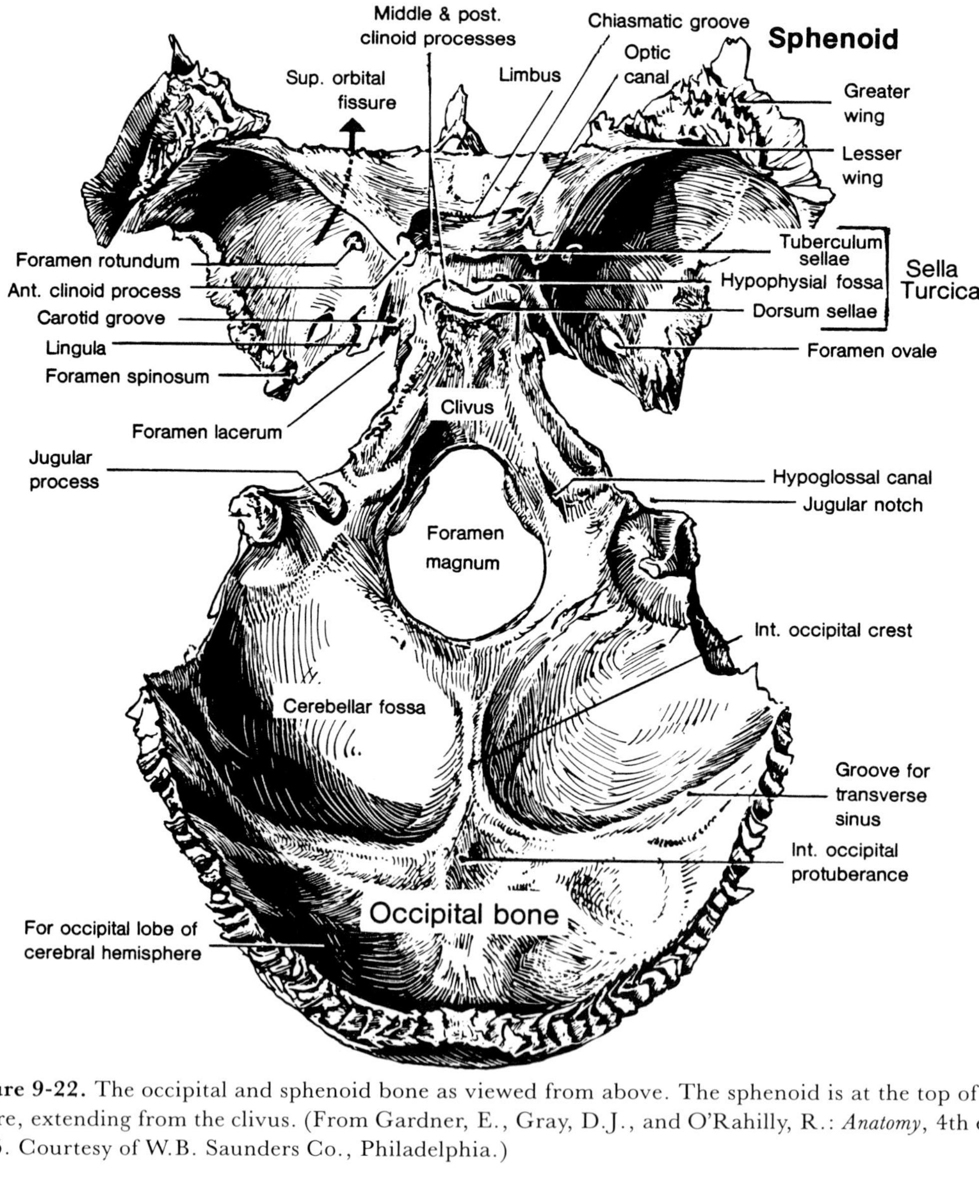

Figure 9-22. The occipital and sphenoid bone as viewed from above. The sphenoid is at the top of the figure, extending from the clivus. (From Gardner, E., Gray, D.J., and O'Rahilly, R.: *Anatomy*, 4th ed., 1975. Courtesy of W.B. Saunders Co., Philadelphia.)

The principal feature of the occipital bone is the large elliptical opening, the *foramen magnum*, which allows communication between the cranial cavity and the vertebral canal. The connection between the brain and spinal cord is established via the foramen magnum. Vertebral arteries and spinal accessory nerves also pass through this opening.

On the lower anterior rim of the foramen magnum is a small tubercle, the *pharyngeal tubercle*, which provides for an attachment for the pharyngeal raphe. This raphe is associated with the pharyngeal constrictor muscles of the pharynx. Thus, the pharynx is suspended, in part, from the occipital bone. The pharyngeal tubercle is a highly variable finding that might not be readily identified on all skull specimens.

A pair of knuckle-like or jelly-bean-like prominences known as the *occipital condyles* are located on the inferior surface at the frontolateral edges of the foramen magnum. These elliptical articulating facets fit into the superior articular facets of the first cervical vertebra, the *atlas* (see Chap. 6).

Articulations with the temporal and parietal bones are made with the basilar part, while the posterior vertical part joins the parietal bones across the lambdoidal suture and the mastoid plate of the temporal bones on the lower lateral aspect.

The outer surface of the posterior portion of the occipital bone is divided into a large upper and smaller lower field by a rough line termed the *superior nuchal line*, a landmark made prominent by neck muscles attached along this line. The origin of the superior nuchal line is the *external occipital protuberance*, a bony prominence at the midline. Location is usually easily made by palpation. The surface above this line is generally smooth (Fig. 9-23).

An *inferior nuchal line* runs parallel to the superior line. Lying on a roughened portion of the *exterior* surface of the occipital bone, the inferior nuchal line and other similar external features are dependent upon the development of the neck musculature. Such markings are, therefore, usually more distinct in the skulls of males than in females.

An *internal* occipital protuberance is found at the center of four shallow fossae on the inner surface of the occipital bone. The two lower fossae are smooth and contain the cerebellar hemispheres of the brain. The upper fossae contain the occipital lobes of the cerebrum.

The Sphenoid Bone

The *sphenoid bone* is a butterfly-shaped structure forming the bulk of the floor of the midportion of the cranial cavity. A most complex bone, it is made up of a body and three paired processes. The posterior border of the body is squared and continues as the basilar part of the occipital bone. As stated previously, the joint between the sphenoid and occipital bones is known as the spheno-occipital synchondrosis, with the connecting cartilage turning to bone at the end of adolescence.

On the upper or *cranial surface* of the sphenoid body is a deep fossa, variously termed the *hypophyseal fossa, pituitary fossa*, or the *sella turcica* (Turkish saddle). In olden days, when the Turks travelled long distances for warring and plundering, their saddles were constructed with high backs for support. The hypophyseal fossa reminded anatomists of these old saddles. The name *sella turcica* is an example of anatomy as a descriptive science. It is in this fossa that the master gland of the body, the hypophysis or pituitary, is housed on the floor of the cranial cavity. This tiny gland, about 1.5 cm in diameter, is connected by a stalk to the hypothalamus of the brain. The anterior part of the saddle is termed the *tuberculum sellae*,

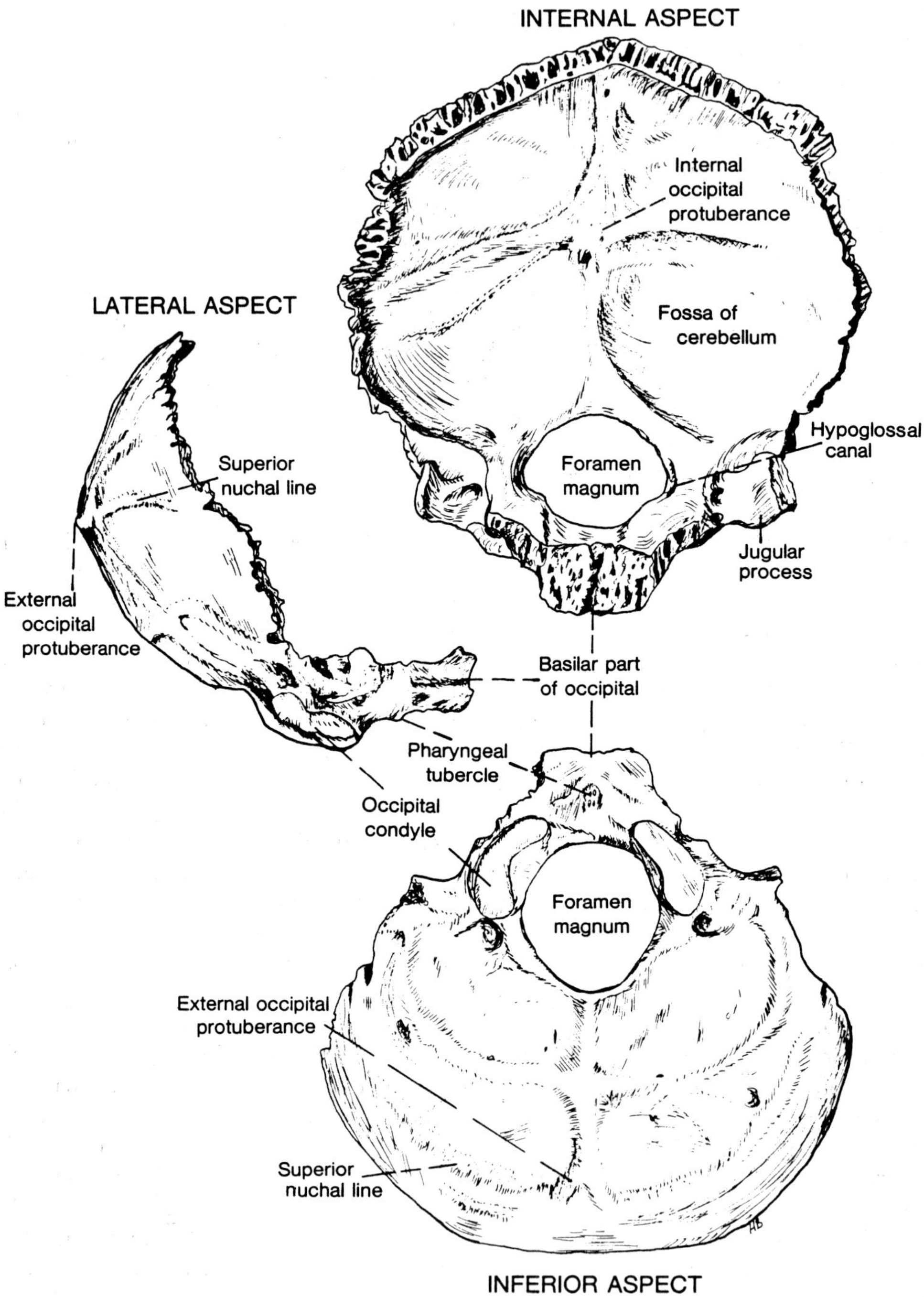

Figure 9-23. Three views of the occipital bone. (From Spalteholz, W., and Spanner, R.: *Atlas of Human Anatomy*, 16th ed., 1961. Courtesy of Bohn, Scheltema and Holkema, Amsterdam.)

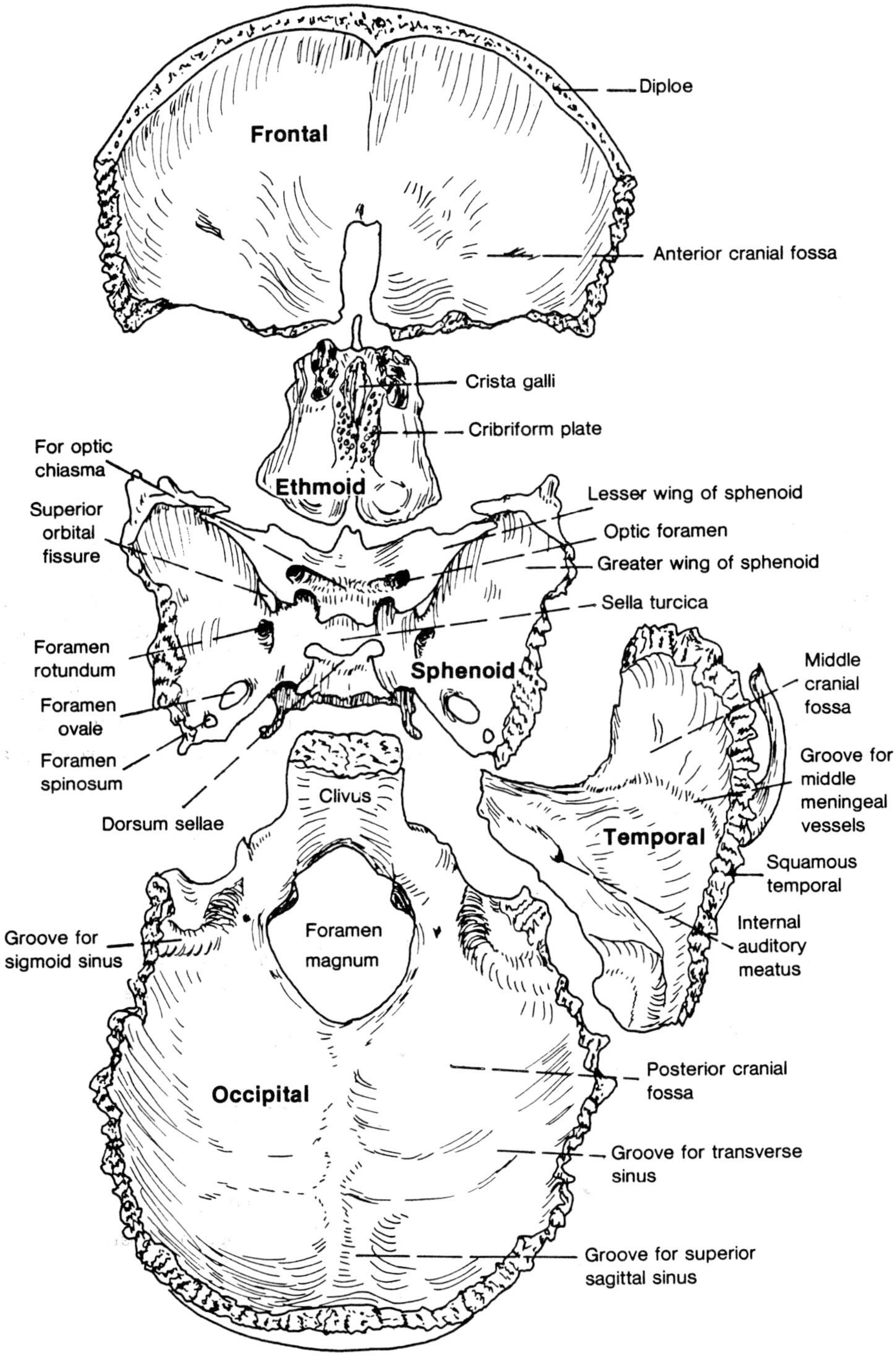

Figure 9-24. Bones of the floor of the skull separated and viewed from above.

while the posterior part of the saddle is the *dorsum sellae.* There are paired, wing-like projections off the superolateral border of the dorsum sellae; these are the *posterior clinoid processes.* The posterior border of the dorsum sellae is continuous with the posterior body of the sphenoid. This border continues past the spheno-occipital synchondrosis to the anterior border of the foramen magnum by way of the basilar portion of the occipital bone. The *clivus* is the term to describe the line from dorsum sellae to foramen magnum. The anterior portion of the brain stem rests against the clivus (slope) (see Figs. 9-22 & 9-24).

Underneath the hypophyseal fossa, within the body of the sphenoid, are the two *sphenoid sinuses.* These two air-filled cavities communicate with the nasal cavity and are divided into two compartments by a thin, asymmetrical sphenoidal septum.

The anterior border of the hypophyseal fossa is a blunt, transverse ridge trailing off medially from the posterior part of the lesser wings. It bends around on either side to form small knobs at the ends, the *anterior clinoid processes.* A groove, the groove of the *optic chiasma,* lies beneath the posterior limit of a ridge joining the anterior clinoid processes. The optic chiasma, an X-shaped junction of the two optic tracts, rests just behind this groove and not in it as the name implies. Leading from this groove on each side are the *optic canals,* which pass the optic nerves and ophthalmic arteries into the orbital cavities (Fig. 9-25).

The *lesser wings* of the sphenoid bone arise from the sphenoid body by two roots on the upper anterior surface (see Figs, 9-22 & 9-24). These are triangular plates, which join the frontal bone at their anterior border. The posterior border of each wing is smooth and ends in the knobby posterior projection already labeled the anterior clinoid process. The lesser wings serve as roofs for the posterior part of the orbital cavities. They also provide a per-

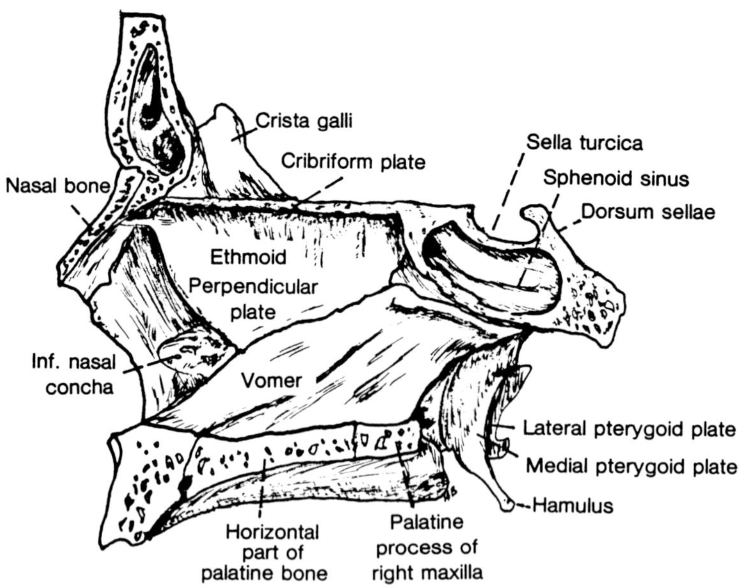

Figure 9-25. The sphenoid air sinus and right pterygoid plates, sagittal section. (From Spalteholz, W., and Spanner, R.: *Atlas of Human Anatomy,* 16th ed., 1961. Courtesy of Bohn, Scheltema and Holkema, Amsterdam.

foration through which the optic nerve passes.

The greater wings arise from the lower lateral aspects of the posterior part of the sphenoid body. Each greater wing joins other neighboring bones, the frontal and zygomatic bones, to seal a portion of the back of the orbital cavity. The smooth medial area of the greater wing is separated from the lesser wing on the posterior wall of the orbit by the *superior orbital fissure*. Several important nerves associated with muscles responsible for moving the eyeball, along with pertinent veins, pass through this fissure into the socket for the eye. Other significant foramina (foramen rotundum and foramen ovale — also the round foramen and oval foramen, respectively) pass branches of the trigeminal nerve (the maxillary and mandibular nerves, respectively). Altogether, the sphenoid bone is related to six of the twelve cranial nerves; however, not all of these six nerves pass through this bone; some pass over its surfaces.

The superior margin of the greater wing connects with the parietal, frontal, and temporal bones. On this area of the surface of the skull is the *temporal fossa*, a shallow lateral depression.

Arising at the junction between each greater wing and the sphenoid body is a rather complex, inferiorly projected process, the *pterygoid process*. A pair of plates, the *medial* and *lateral pterygoid plates*, is present on either side, although they are fused at their anterior edges in the upper portions. Inferiorly, the plates are separated by the *pterygoid notch*, which, in the living subject, is filled by a process of the palatine bone (see Figs. 9-11 & 9-43).

A deep fossa known as the *pterygoid fossa* is found on either side between the medial and lateral plates. It provides, in part, the origin for a muscle of mastication, the *medial pterygoid muscle*. The *tensor veli palatini muscle* takes its origin in a shallow fossa located at the root of the pterygoid process, referred to as the *scaphoid fossa* see Fig. 9-11).

At the inferior end of the medial plate is a thin, curved process, the *pterygoid hamulus*. A deep notch separates the hamulus from the main part of the medial plate and gives freedom for the tendon of the tensor veli palatini muscle to pass around the hamulus on its way to the soft palate.

The position of the sphenoid bone in relation to the internal aspects of the facial skeleton, then, is that the pterygoid plates descend on either side of the nasopharyngeal cavity at the anterior portion. This forms an inverted U, with the inferior surface of the sphenoid body at the top. In this area also, the sphenoid sinus empties into the ethmoid sinus and into the superior nasal recess. Infections of sphenoid and ethmoid sinuses are identified by posterior rhinoscopy, not by discharge from the nares.

The inferior surface of the body carries an elevation in the midline, termed the *sphenoid rostrum* — so named because its oblique direction points toward the front of the head. It is linked to the vomer, the plowshare bone of the bony nasal septum, and the perpendicular plate of the ethmoid bone. As previously stated, the perpendicular plate along with the vomer forms the bony portion of the nasal septum.

The Frontal Bone

The greater part, the squama (platelike), of the *frontal bone* serves as the anterior portion of the cranial vault. The frontal bone also contributes the greater part of the roof of the orbital cavities that house the eyeballs (Fig. 9-26). Although the frontal bone is regarded as an unpaired bone, it develops as a paired bone, with the two halves joined by a frontal suture at birth. The suture typically ossifies

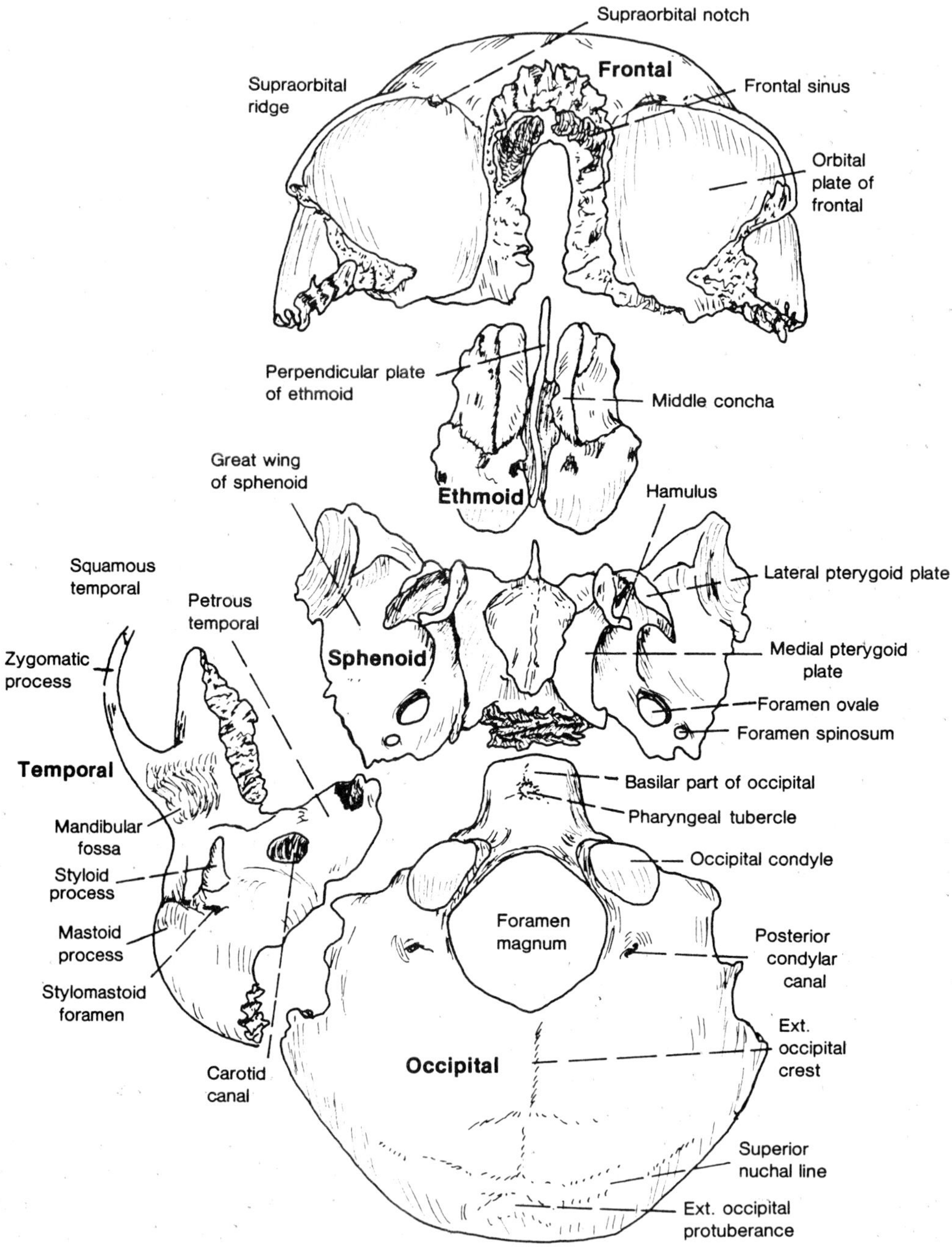

Figure 9-26. Bones of the base of the skull separated and viewed from below.

by age two but may persist in rare cases as the metopic (forehead) suture. It is traditionally considered as an unpaired bone, however.

Each orbital part of the frontal bone is separated from the other by the ethmoid notch, which is a wide space parallel to the hard palate and filled by the cribriform plate of the ethmoid in the articulated skull. The anterior portion of the ethmoid notch presents a very rough area, with a nasal spine projecting forward and downward. This spine unites with the perpendicular plate of the ethmoid bone, and together they form the upper portion of the bony nasal septum. The sharp nasal spine extends from the nasal part of the frontal bone. From the nasal part of the frontal bone, the nasal bones arise, and their medial borders join the nasal spine to form the bridge of the nose.

A smooth *supraorbital rim* serves as a boundary between the squama and orbital portions of the frontal bone. The rim ends laterally by a downward projection to form the zygomatic process, which joins the cheekbone. Toward the medial end of the rim is a notch, the *supraorbital notch*. The notch may be transformed into a *supraorbital foramen* if the ligament that normally closes the supraorbital notch ossifies. The supraorbital rim affords protection of the eye by its overhang. Through the supraorbital foramen nerves pass onto the surface brow region.

The smooth elevation usually present just above the nasal part and medial to the supraorbital rims is termed the *glabella* (smooth). Lateral to the glabella on either side and related to the eyebrows are the *superciliary arches*, or ridges, slight ridges that may be expected to be better developed in males than females. The degree of prominence of the glabella and superciliary arches may be dictated by the size of the frontal air sinuses, although there is not always a high correlation between these observations.

As the supraorbital margins end laterally in the zygomatic process, a strong arch marks the beginning of the *temporal line*. Extending upward and backward, the temporal line divides into upper and lower lines as it approaches the posterior border of the frontal bone. In the intact skull, the lines are continuous with the corresponding lines on the parietal bone and outline the area of the temporal fossa (see Fig. 9-5).

The lower surface or *orbital plate* of the orbital part faces inferiorly. The *lacrimal fossa* is situated in a concavity located in the lateral area of the orbital part near the root of the zygomatic process. The *lacrimal gland*, or tear gland, is housed in this fossa. Another depression is found on the orbital plate; it is near the nasal part and is named the *trochlear fovea*. It may contain a small, sharp bony process (Fig. 9-4). Here, a cartilaginous pulley for the superior oblique muscle of the eyeball is fastened (see Fig. 13-8).

At the posterior border, the frontal bone meets the parietal bones across the coronal suture. This border is prominently serrated. Continuing downward, the junction with the greater wing of the sphenoid bone is found. In front of this area is the zygomatic process of the frontal bone, mentioned previously.

Around the *ethmoid notch*, the wide articulating border presents several incomplete *air cells*, which are completed by similar, corresponding ethmoid air cells when the ethmoid bone is in place. The notch is thereby filled by the ethmoid bone's cribriform plate, which seals the anterior floor of the brain case. Forward of the ethmoid notch are the *frontal air sinuses*. These sinuses represent a separation between the inner and outer plates of the frontal bone. An irregular bony septum divides the frontal sinuses, which are usually larger in males than in females. A passage, the frontonasal duct, permits the sinuses to communicate with the nasal cavities. The growth of the frontal sinuses is highly variable. Usually they do not begin to develop until well after age five. Frontal headaches in young children are,

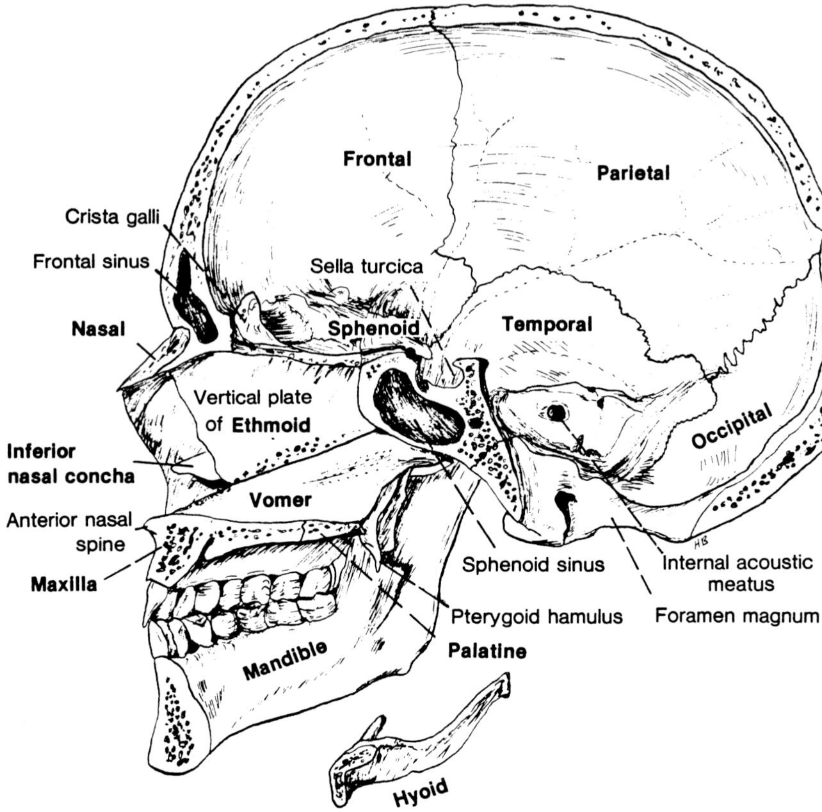

Figure 9-27. Sagittal aspect of the skull.

therefore, rare.

The *cranial surface* of the frontal bone is cup-shaped. A number of branching furrows are found on the cranial surface for the branching subdivisions of a principal cerebral artery. Other depressions on the inner surface are for the convolutions of the two frontal lobes of the brain (Fig. 9-27).

The Ethmoid Bone

The ethmoid bone is the smallest bone contributing to the formation of the cranium. Its general configuration from a frontal view reminds one of a stack of forks with most of the shaft of the handles removed. While it is classified as a bone of the cranial skeleton, it also contributes to the facial skeleton by forming most of the walls of the upper part of the nasal cavity. Its role in sealing the anterior part of the base of the cranium medially to the two orbital cavities allows it to be included in the cranial bones (see Fig. 9-26).

Altogether the ethmoid bone articulates with thirteen bones of the skull. The ethmoid bone is a very light, irregular bone. It presents four parts, the cribriform plate, a perpendicular plate, and two lateral masses of labyrinths.

The *horizontal* or *cribriform plate* fills the ethmoid notch of the frontal bone to form part of the anterior base of the cranium, hence the description indicating likeness to a crib. The inferior surface of the cribriform plate forms part of the roof of the nasal cavity.

When viewed from the end, a *vertical plate* above and below the cribriform plate is most apparent. That part of the vertical plate reaching into the cranial cavity is named the *crista galli*, or cock's comb (see Fig. 9-25). At the base of the crista galli, on either side, the cribriform plate is deeply grooved and narrow to accommodate the *olfactory bulbs*, which are located on the underside of the frontal lobes of the brain. The *olfactory foramina*, perforations in the cribriform plate, pass the olfactory nerves into the cranial cavity from the nasal cavity.

The large vertical plate seen extending inferiorly from the cribriform plate in the midline is the *perpendicular plate*. It is part of the bony nasal septum. Anteriorly, the vertical plate articulates with the nasal spine of the frontal bone and the crest of the nasal bones. The lower half joins the vomer, and the anterior border provides for the medial attachment of the septal cartilage, which completes the nasal septum (Fig. 9-28).

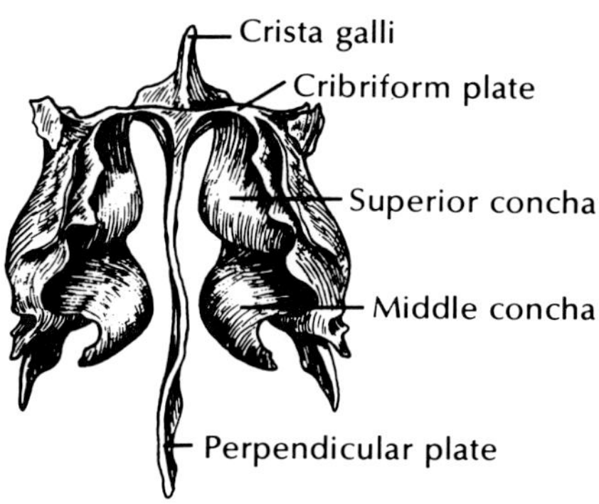

Figure 9-28. The ethmoid bone as viewed from behind. The perpendicular plate is the upper part of bony nasal septum; it is seldom smooth and straight. The cribriform plate forms a small part of the floor of the anterior cranial fossa and transmits olfactory fibers into the nasal cavity. (From Francis, C.C., and Martin, A.H.: *Introduction to Human Anatomy*, ed. 7, St. Louis, 1975, The C.V. Mosby Co.)

Two relatively large lateral masses, the *labyrinths*, are situated on either side of the perpendicular plate. The medial surface of the labyrinth contributes additional parts of the nasal cavities by supplying the *middle* and *superior nasal conchae*. The *superior* and *middle meatus* are narrow spaces underneath the respective conchae, directing air through the nasal cavities.

The front, central location of the ethmoid enables it to articulate with the largest number of adjacent bones: both nasals, maxillae, lacrimals, palatines, and inferior nasal conchae; also, the frontal, sphenoidal, and vomer bones.

The Temporal Bone

The *temporal bone* is paired and forms part of the sides and base of the cranium. Lying in the midportion of the floor of the brain cavity, each temporal bone articulates with five neighboring bones: the occipital posteriorly, the parietal superiorly, the sphenoid's greater wing anteriorly, the zygomatic through the zygomatic arch, and the other temporal medially. Within the cavities of this bone are the *hearing* and *equilibrium* sense receptors.

The basic role provided by the internally housed structures of the temporal bone is to decode incoming sound waves into patterns of neural impulses, the language of the brain. An important fossa receives the condyloid process of the mandible to form the temporomandibular joint, the only movable joint between bones of the skull's framework.

Each of the temporal bones has four parts: (1) a *squamous portion*, a flat, platelike section at the side of the skull which includes the *zygomatic process*, extending anteriorly to meet the temporal process of the zygomatic bone and the *mandibular fossa*, also known as the glenoid fossa, on its inferior surface; (2) a *tympanic portion*, a

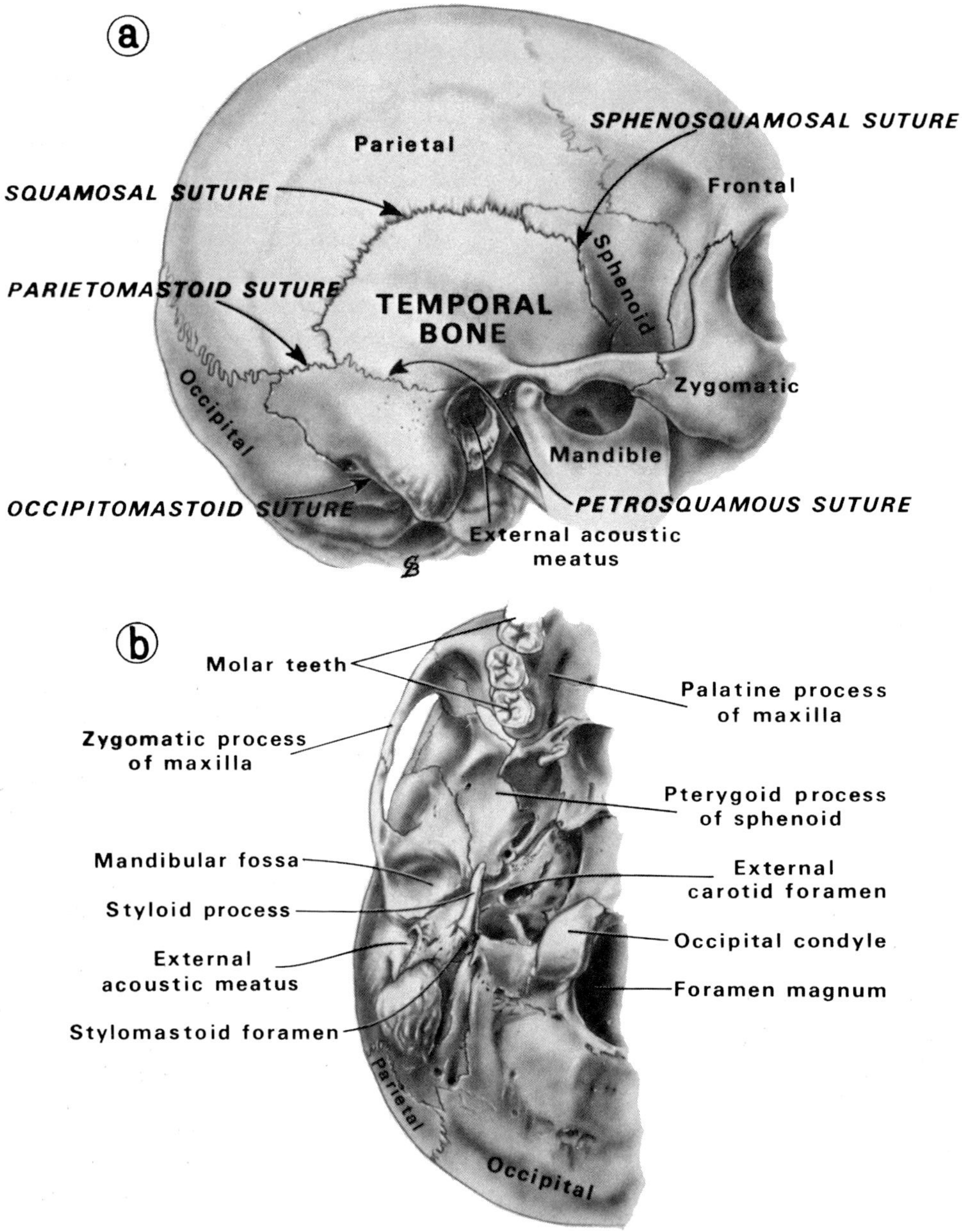

Figure 9-29. Right side of the cranium. Lateral (a) and inferior (b) views of the cranium emphasizing the relationship of the temporal bone to other skeletal components. (From Anson, B.J., and Donaldson, J.A.: *Surgical Anatomy of the Temporal Bone and Ear*, 3rd ed., 1981. Courtesy of W.B. Saunders Co., Philadelphia.)

curved plate of bone surrounding the external auditory meatus; (3) a *mastoid portion*, an area lying posterior-inferior to the external auditory meatus and ending in a prominent nipple-shaped process, the mastoid process; and (4) a *petrous portion*, projecting medially to form a prominent ridge in the floor of the cranial cavity, housing the middle and inner cavities (Fig. 9-29).

Squamous Portion

The *squamous part* of the temporal bone is a scale-like, thin plate of bone forming the lower side wall of the cranial portion of the skull. It presents a strong beveled edge at its superior border and meets the corresponding beveled margin of the parietal bone across the *squamosal suture*. Anteriorly, it articulates with the greater wing of the sphenoid bone. The zygomatic process projects from the lateral surface of the squamous portion to meet the temporal process of the zygomatic bone. The arch thus formed is the *zygomatic arch*.

The fibers of the *masseter muscle*, a strong muscle of mastication, take their origin on the zygomatic arch. Behind the arch passes the *temporalis muscle*, which takes its origin along the temporal line, converges inferiorly across the squamous plate behind the arch, and inserts on the upper part of the ramus of the mandible at the coronoid process. The infratemporal fossa is the relatively deep depression located behind the arch (Fig.9-8).

Beneath the zygomatic arch is the mandibular fossa, which is formed by the squamous and the tympanic parts of the temporal bone and provides, with the condyloid process of the mandible, a sy-

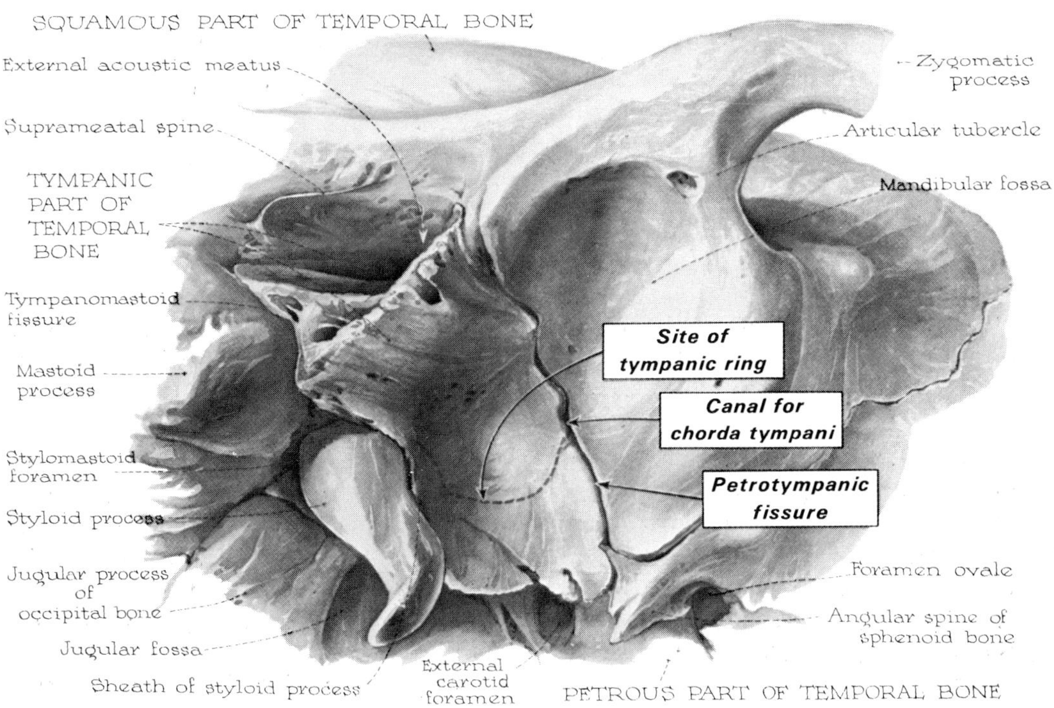

Figure 9-30. Close up of the inferior surface of the right temporal bone emphasizing the tympanic portion. The location of the eardrum within the external auditory meatus is shown by a dotted line. (From Anson, B.J., and Donaldson, J.A.: *Surgical Anatomy of the Temporal Bone and Ear*, 3rd ed., 1981. Courtesy of W.B. Saunders Co., Philadelphia.)

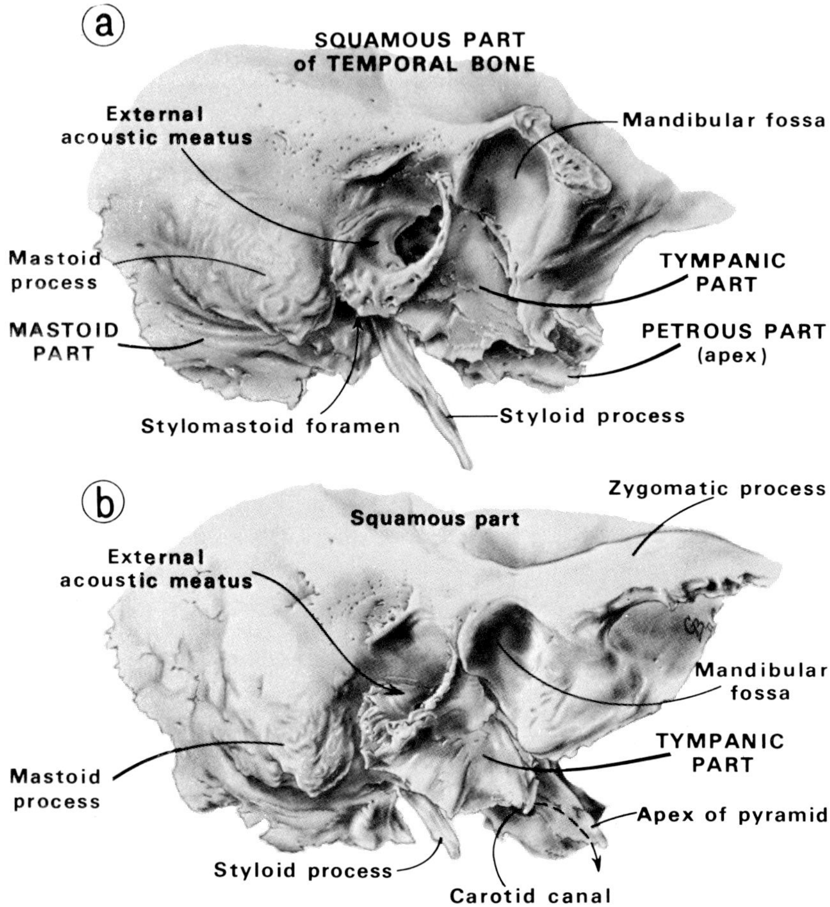

Figure 9-31. Two specimens showing variations in the temporal bone, inferolateral view. Notice particularly the large styloid process (a) and the less prominent styloid process (b). (From Anson, B.J., and Donaldson, J.A.: *Surgical Anatomy of the Temporal Bone and Ear*, 3rd ed., 1981. Courtesy of W.B. Saunders Co., Philadelphia.)

novial joint capable of gliding and hinge actions. The *tympanic part* of the temporal bone separates this fossa from the external auditory meatus.

The inside surface of the squamous portion is marked by depressions corresponding to the convolutions of the brain's temporal lobe.

Tympanic Portion

The *tympanic portion*, a plate-like part, forms the anterior wall and floor of the external auditory meatus. The *external acoustic meatus* is about 2.5 cm in length, ending medially at the tympanic membrane, or eardrum. It is directed slightly anteriorly and is seen forming a slight curve in the adult but is relatively straight in the very young.

On its anteroinferior surface, the tympanic part contributes the posterior portion of the mandibular fossa, although this is the nonarticulating aspect of the fossa. The rough, free lateral border provides attachment of the cartilaginous portion of the external auditory meatus.

Mastoid Process

The *mastoid portion* is characterized by a projected nipple-shaped process, the *mastoid process*, which is situated behind and below the *external acoustic meatus orifice*. Attached to the mastoid process is the sternocleidomastoid muscle as well as other muscles of the posterior neck region. Within the mastoid process, hollowed-out spaces communicate with the middle-ear cavities. From a coronal section of the temporal bone, it is possible to see that the *mastoid air cells* are larger at the superior part where they communicate with a large irregular cavity termed the *tympanic antrum*, which in turn communicates with the *tympanic cavity* or middle-ear cavity by opening anteriorly into a portion of the tympanic cavity known as the *attic* or *epitympanic recess* (see Fig. 17-4). The tegmen tympani is a partition of thin bone that separates the tympanic antrum from the cranial cavity (Fig. 9-31).

The *mastoid notch*, a deep notch, separates the mastoid process from the inferior medial surface. Here in the mastoid notch, the posterior portion of the digastric muscle takes its origin.

Petrous Portion

The *petrous portion* extends inward and forward on the base of the skull, comprising a wedge of bone between the greater wing of the sphenoid bone on its outer side and the basilar part of the occipital and body of the sphenoid on its inner side. If one holds a skull up to the light, it is noted that the petrous portion of the temporal bone does not transilluminate. It is rock-hard (petrous) because of the important role it plays in protecting the inner-ear mechanism. The term *petrous* is an apt description of this process of the temporal bone and is another example of anatomy as a descriptive science. This essential part of the temporal bone houses the inner ear containing organs of *hearing* and *balance*. It articulates with the occipital bone posteriorly and medially and with the greater wing of the sphenoid bone anteriorly.

The undersurface of the petrous portion is a rough area, pierced by openings and having irregular edges. Just medial to the styloid process, that slender tooth-like process pointing downward and forward, are two openings. The *carotid canal* passes the major artery by the same name into the brain case and is situated anteriorly to the jugular fossa. In the undisturbed skull, the occipital bone encloses the posterior aspect of this fossa, forming the *jugular foramen* (see Fig. 9-18). Major vessels exit the cranium here. There may often be bony spines dividing a small portion of the jugular foramen from the larger part that transmits the jugular vein. The smaller area, the neural jugular foramen, passes the *glossopharyngeal* and *vagus nerves*, which are essential to functions of the tongue, pharynx, and larynx.

At the outer side of the inferior surface is a groove separating the temporal and sphenoid bones. With the addition of a cartilaginous floor, the groove is converted into a canal, forming the terminal cartilaginous portion of the *internal auditory tube*, also termed the *pharyngotympanic* or *eustachian tube*; it opens into the nasopharynx. Between the styloid and mastoid processes is found the *stylomastoid foramen*, which transmits the facial nerve out of the skull.

The cranial surface of the petrous portion forms part of the floor of the *middle cranial fossa* (see Fig. 9-17). It is continuous with the inner surface of the squamous part. Near the center is the *arcuate eminence*, which denotes the location of the *superior semicircular canal* within housed. It is identified by a slight prominence on the superior surface of the petrous portion. Located anterolaterally to this eminence is a slight depression, which is the *tegmen tympani*, the roof of the middle ear.

At the medialmost portion of the petrous part is the termination of the carotid canal, allowing the internal artery to exit the temporal bone at the *foramen lacerum*, which is formed with the junction of the sphenoid bone in most subjects (see Figs. 9-18 & 9-32).

On the posterior surface inside the cranial cavity, the petrous portion contributes to the posterior cranial fossa. A large opening here, the *internal acoustic meatus*, provides passage for the facial and vestibulocochlear nerves plus an arterial branch. The facial nerve passes through the temporal bone to emerge at the *stylomastoid foramen*.

The middle-ear cavity, along with the ossicles and the inner ear, both located within the cavities of the petrous portion, are considered under the division pertaining to the ear.

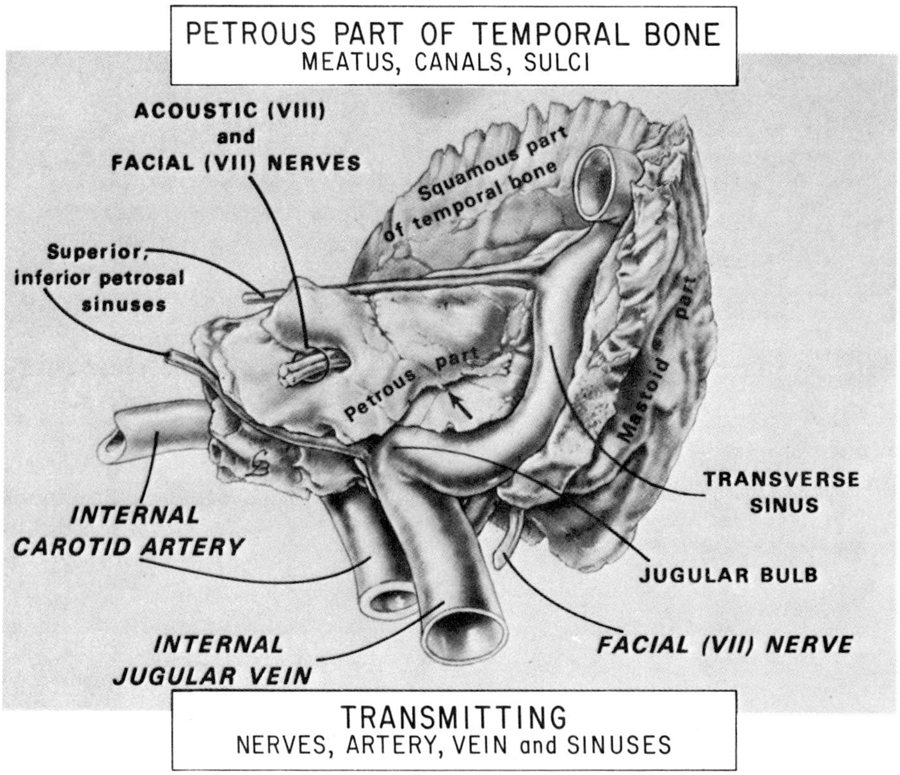

Figure 9-32. Posteromedial view of the temporal bone, petrous portion, and its role in transmitting nerves and vessels. (From Anson, B.J., and Donaldson, J.A.: *Surgical Anatomy of the Temporal Bone and Ear*, 3rd ed., 1981. Courtesy of W.B. Saunders Co., Philadelphia.)

BONES PERTAINING TO ARTICULATION

When viewing the complete skull, at once it is evident that the upper and lower jaws make up the greater part of the facial skeleton. They contain the dental arches, and the lower jaw provides the only movable part of the skull. The upper jaw, the *maxilla* or *maxillary bone*, is a paired bone, while the lower jaw, the *mandible*, is unpaired. Together, the maxillae and the mandible provide the bony framework for the oral cavity (Table 9-III).

Some of the disarticulated bones of the facial framework are considered here as a follow-up to views of the intact skull in a previous section. Features not ordinarily observable in the articulated skull are best

Table 9-III

BONES OF THE FACIAL SKELETON

Name	*Number*	*Number and Name of Articulations*	
Mandible	1	1:	both temporal bones
Maxilla	2	9:	frontal, ethmoidal, nasal, zygomatic, lacrimal, inferior nasal concha, palatine, vomer, other maxilla
Palatine	2	6:	maxilla, ethomoidal, sphenoidal, inferior nasal concha, vomer, other palatine
Inferior nasal concha	2	4:	maxilla, ethmoidal, lacrimal, palatine
Vomer	1	6:	both maxillae, both palatine bones, sphenoidal, ethmoidal
Nasal	2	4:	frontal, ethmoidal, maxilla, other nasal
Zygomatic	2	4:	maxilla, frontal, temporal, sphenoidal
Lacrimal	2	4:	frontal ethmoidal, maxilla, inferior nasal concha
Hyoid	1	none	

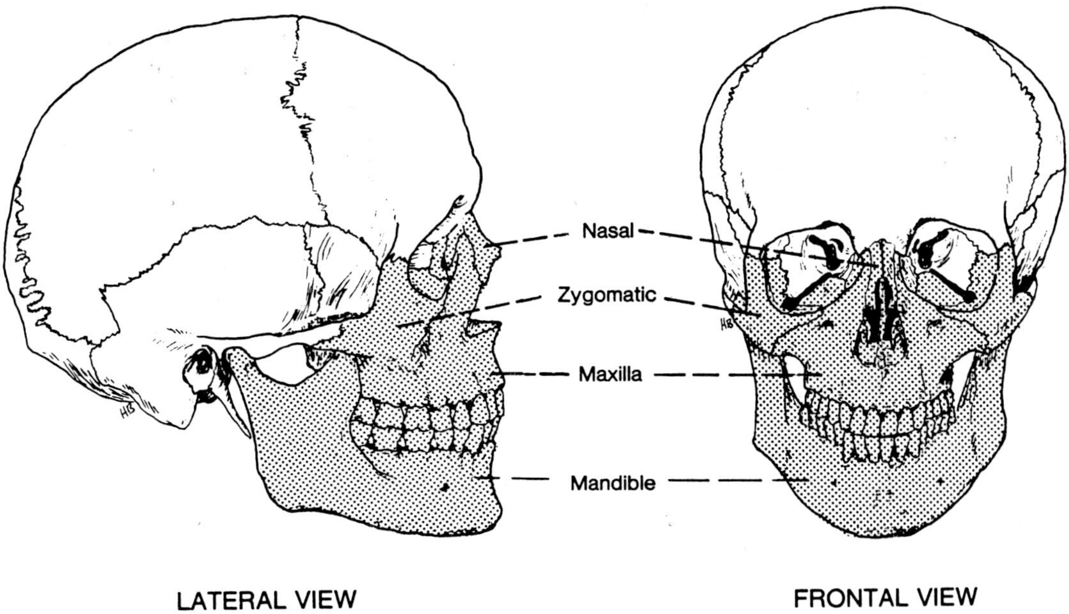

Figure 9-33. The facial skeleton (shaded) from two views.

examined in the disarticulated bones, or at least in close examination of the bone *in situ* (Fig. 9-33).

The Mandible

The mandible plus the lower teeth constitute the lower facial skeleton. In the fetus, this bone is separated into right and left halves. During the first year, the two halves unite at the midline, and the *mental symphysis* is formed. The two major parts of the mandible are the body and the ramus. Important processes include the alveolar process on the body and the coronoid and condyloid processes of the ramus.

Body of the Mandible

The body, or corpus, of the mandible is a U-shaped configuration of significant strength. The outer surface is designated as the facial surface and gives origin to a number of facial muscles. A prominent landmark is the *menial foramen*. This opening is located in an area just below the first bicuspid (or sometimes the second bicuspid) and allows for passage of the mental (chin) vessels and nerve.

At the upper portion of the corpus of the mandible is the *alveolar process*. Found on the top portion of the alveolar process or alveolar border are sockets, the *alveoli*, for the teeth. There are eight alveoli on either half of the lower dental arch in the adult (Fig. 9-34).

The anterior portion of the corpus of the mandible is marked by a triangular area. At the point of the chin and top of the triangle is the *mental protuberance*. The lateral points of the triangle are formed by two prominences called *mental tubercles*. Just above each tubercle is a depression referred to as the *mental fossa*, which provides for muscular attachments in the area. The several muscle attachments to the facial aspect of the mandible are considered in the section on muscles of facial expression.

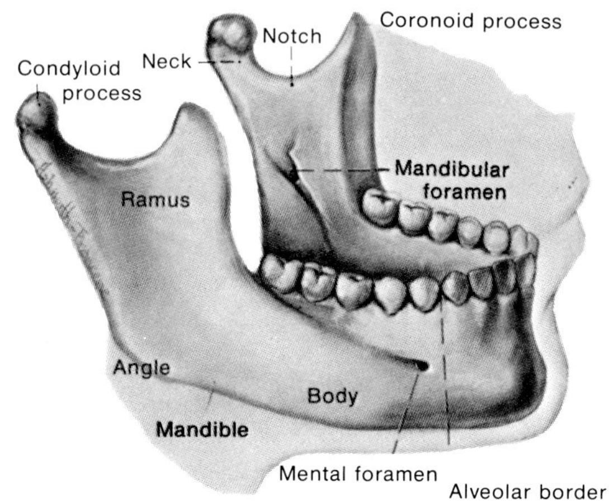

Figure 9-34. The mandible in relation to the chin and mouth. (From Jacob S.W., and Francone, C.A.: *Structure and Function in Man*, 4th ed., 1978. Courtesy of W.B. Saunders, Philadelphia.)

On the inner surface, the *lingual surface*, the margin of the floor of the oral cavity is formed. Here the lingual surface relates to the tongue, two salivary glands, and a number of muscles. On the anterior lingual surface are variably defined spines, the *mental spines* or *genial tubercles*. This area serves as the origin of the *geniohyoid muscle* and the *genioglossus muscle*.

The *digastric fossa* is a shallow, roughened depression at the lower border and close to the midline of the mandibular body. There is one such oval depression on either side of the symphysis for the attachment of the *digastric muscle*.

In the region of the third molar on the inner surface of the mandibular body is a diagonally placed crest known as the *mylohyoid line* or sometimes as the mylohyoid ridge. If well developed, the anterior aspect of this line is continuous at the midline with that of the opposite side as the two lines converge at the symphysis, passing between the digastric fossae and mental spines (Figs. 9-35 & 9-36).

From the mylohyoid line, the *mylohyoid muscle* takes its origin and forms the floor of the oral cavity. The bone below this line forms an attachment site for a portion of the *submandibular salivary gland.* At the superior end of the mylohyoid line, near the alveolar border, a small part of the *superior pharyngeal constrictor* is attached that relates the mandible to the pharynx.

Ramus of the Mandible

Projected in a perpendicular direction from the body of the mandible is the ramus. Each ramus carries two processes and has an inner and outer surface and an anterior and posterior border. The *mandibular angle* is easily palpated on the lower border as an irregularly rough area of the ramus. Alternate depressions with knob-like elevations at the border result from the attachments of the two muscles, the

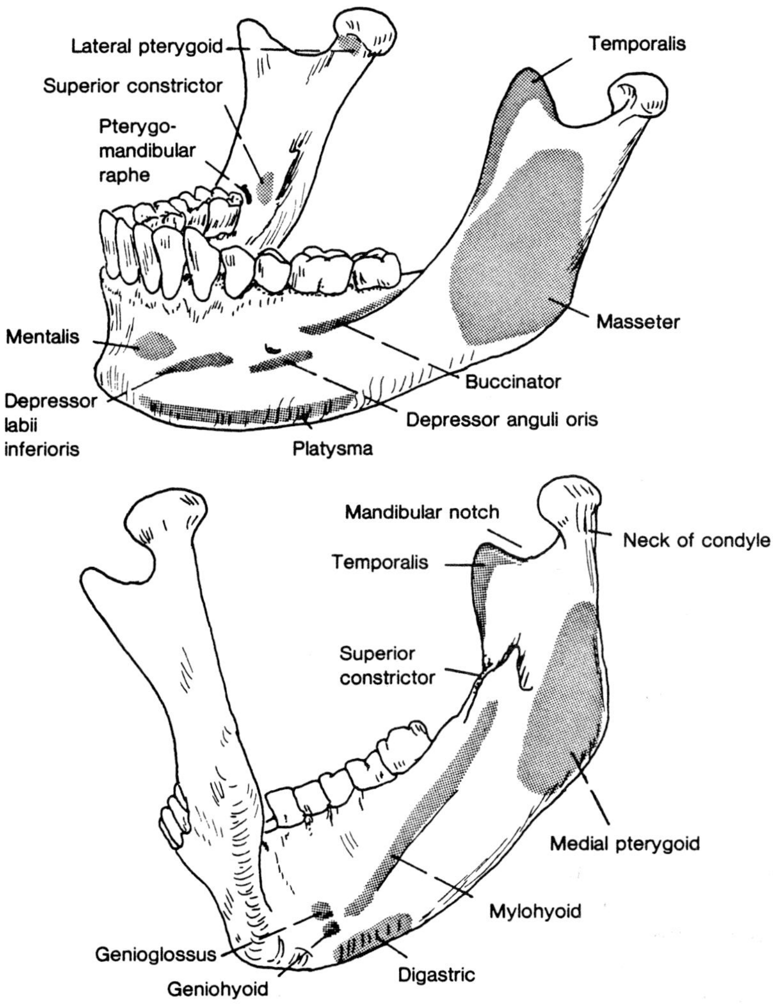

Figure 9-35. The mandible showing its muscle attachments.

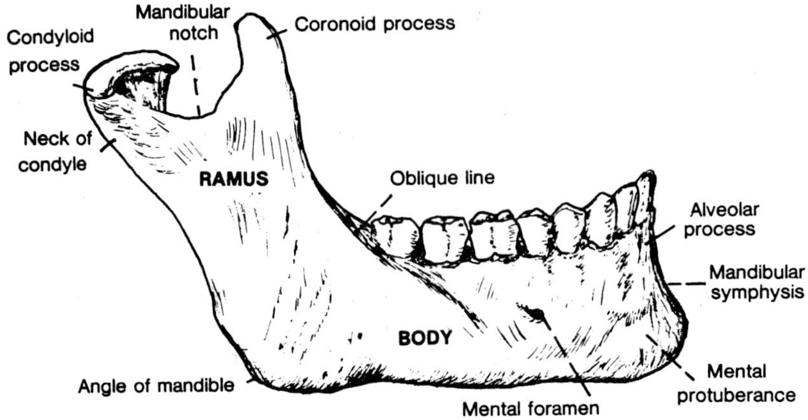

LATERAL ASPECT

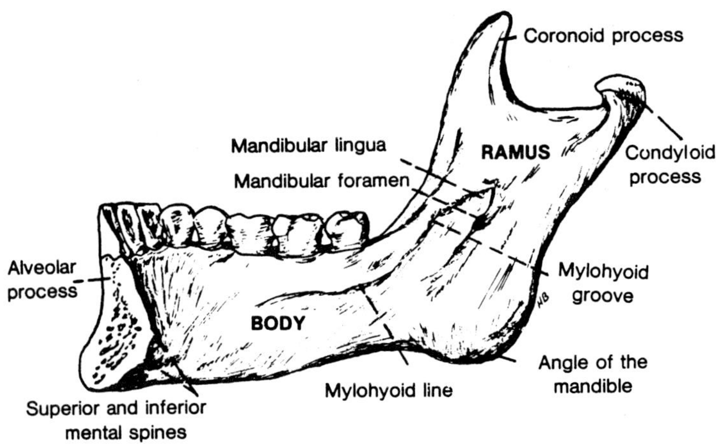

MEDIAL ASPECT

Figure 9-36. Outer and inner surfaces of the mandible. (From Spalteholz, W., and Spanner, R.: *Atlas of Human Anatomy*, 16th ed., 1961. Courtesy of Bohn, Scheltema and Holkema, Amsterdam.)

masseter muscle on the outer side and the *medial pterygoid muscle* on the inner side. However, each of these muscles extend their area of attachment over a considerable portion of the lower part of the ramus and are not necessarily confined to the mandibular angle.

On the center of the medial surface of the ramus is the location of the *mandibular foramen*. A variable lip of bone, the lingula (small tongue), is found to cover partially the anterior circumference of the foramen. A ligament, the sphenomandibular ligament, is attached on its lower end in the area of the mandibular foramen and on its superior end to the spine of the sphenoid bone. Another ligament, the stylomandibular ligament, courses from the styloid process to the angle and inner border of the mandible. The *mylohyoid groove* is a sharply demarcated groove, running in a line downward and forward from the mandibular foramen and ending under the posterior part of the mylohyoid line. Beginning at the mandibular foramen is the *mandibular canal*, which allows passage of the alveolar nerve and vessels. The mandibular canal curves downward and forward and follows a horizontal course below the roots of the molars until it splits into two canals in the region of the bicuspids. One canal opens at the mental foramen, and the other splits into smaller canals and serves the incisors (Fig. 9-36).

The superior end of the ramus is divided into two processes separated by the *mandibular* or *semilunar notch*. The anterior process is the *coronoid* (crow's beak) *process*. From the tip of the coronoid process commences a crest, the *temporal crest*, which runs downward on the medial surface of the ramus. It ends on the alveolar border with the formation of the *retromolar triangle*. The designation, *temporal crest*, is made because of the attachment of the deep tendon of the *temporal muscle* on the vertical part of this crest.

The *condylar process* is usually shorter than the coronoid process; however, it is thicker and consists of two parts, the *mandibular condyle*, or knuckle, and the supporting stalk, the *neck*. The rounded articular portion is adapted to form, with the temporal bone, the *temporomandibular joint*. An articular disk between the mandibular head and the *mandibular* or *glenoid fossa* — a depression in the temporal bone — cushions and frees the action of the joint. The articular disk is a tough, oval fibrocartilage plate.

The *mandibular crest* is a ridge following a trajectory that begins on the medial surface of the upper neck of the condylar process and extends in a forward and downward course to the region just above the mandibular foramen. This crest is the main buttress of the mandibular ramus, transmitting chewing forces from the base of the alveolar process to the base of the skull. Just anterior to this crest is the insertion of the fibers of the *lateral pterygoid muscle*.

Clinical Aspects of the Mandible

It is possible for the mandible to be decreased in thickness (height) by about one-half if the alveolar process is absorbed after loss of the permanent teeth. This usually is found in *edentulous* (toothless) persons who do not acquire dental plates. The result is that the chin appears much more prominent.

Infants and children have short rami. This results in a high-riding tongue. For this reason, it is difficult to see the palatine tonsils and posterior pharyngeal wall in a young child's mouth. Also, the tongue is on a more accelerated, differentiated growth pattern than the mandible in youth.

Related to deviations of the mandible are the *prognathous* and *retruded mandible* (see Chap. 10). These are skeletal variations or discrepancies characterized by protrusion of the mandible or retrocession of the mandible relative to the upper facial structures. In the case of the protruded mandible, the lower face is prominent with a jaw that appears aggressive. In the condition of a retruded mandible, the upper lip appears insufficient and is drawn up over the protruded upper teeth (see Fig. 2-24).

The Maxilla

The whole of the upper jaw is formed by the union of two bones, the maxillae. Except for the mandible, they are the largest bones of the facial skeleton. Each maxilla consists of a body containing a large cavity called the *maxillary air sinus*. Four processes extend from the maxillary body: alveolar, palatine, frontal, and zygomatic. The maxilla contributes to the formation of the nasal, oral, and orbital cavities. Each bone has a *facial, posterior, orbital*, and *nasal surface*.

Facial Surface

The facial surface of the maxillary bone reveals an *alveolar border* inferiorly, the nasal aperture medially, the lower orbital margin superiorly, and the zygomatic process laterally. Certain muscles of facial expression are attached to this surface. The opening onto the facial surface below the midregion of the orbital margin is the *infraorbital foramen*, which passes the branches of the infraorbital nerve and blood vessels onto the surface of the face. On the anterior part, at the lower margin of the nasal opening, is a pointed process that, with its match from the opposite side, forms the *anterior nasal spine*.

Two shallow depressions or fossae appear on the facial surface. The *incisive fossa* is located above the sockets of the incisors and medially to the *canine eminence*, which is formed by the prominent outer wall of the socket of the canine tooth. The *canine fossa* is a larger depression lateral to and behind the incisive fossa. It lies between the alveolar process below, the infraorbital foramen above, and in front of the zygomatic process. There may be considerable variation in the development of these fossae in different individuals (Fig. 9-37).

Posterior Surface

The posterior surface is separated from the anterior surface by the zygomatic process of the maxilla. This aspect of the maxillary body is also known as the *infratemporal surface* because it is part of the anterior wall of the infratemporal fossa — an important space below the cranial base and behind the maxilla. A rounded eminence on the posterior surface, the *maxillary tuberosity*, is the posterior limit of the bone and becomes especially prominent

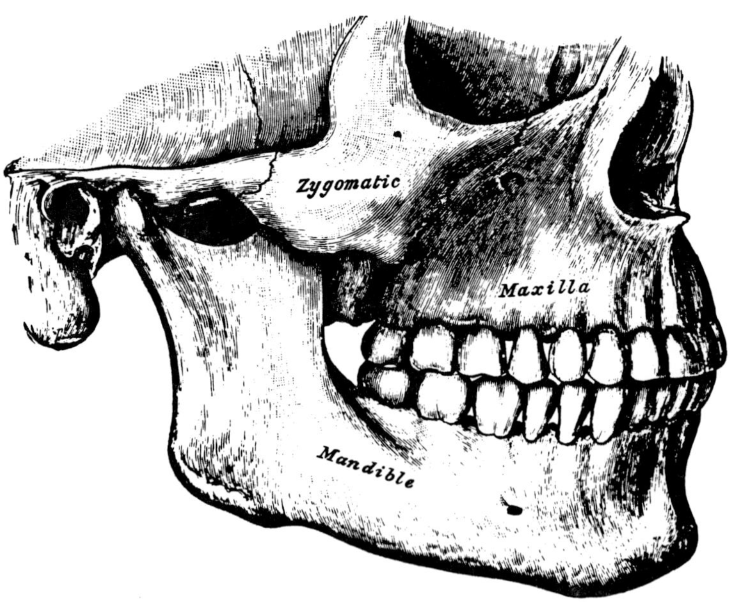

Figure 9-37. The jaws. (From Goss, C.M. (Ed.): *Gray's Anatomy of the Human Body*, 28th ed., 1973. Courtesy of Lea & Febiger, publisher.)

after growth of the third molar. A number of small foramina pierce the maxillary tuberosity, transmitting alveolar vessels and nerves. The medial surface is rough where articulation with the palatine bone is made.

Orbital Surface

The orbital surface presents a smooth surface, which forms the greater part of the floor of the orbital cavity. It is grooved by the *infraorbital groove*, allowing the infraorbital vessels and nerve to be passed onto the facial surface through the infraorbital foramen. The groove courses anteriorly from the middle of the posterior part of the orbital surface carrying its vessels (see Fig. 9-40).

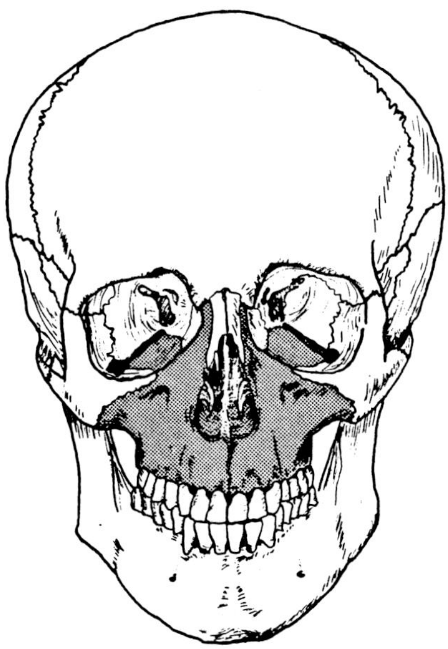

Figure 9-38. Frontal view of the skull emphasizing the maxillae (shaded).

Nasal Surface

The nasal surface of the disarticulated maxilla shows the opening into the *maxillary sinus* or *antrum of Highmore*. This air sinus communicates with the nasal cavity by means of a large irregular *maxillary hiatus* (opening) leading into the sinus on the nasal surface of the maxillary body. The opening into the maxillary sinus is greatly reduced in the articulated skull by parts of other bones that contribute to the lateral wall of the nasal cavity, namely, the ethmoid, inferior nasal concha, and palatine bones (see Fig. 9-40).

Behind the maxillary hiatus is a rough area of the nasal surface where the palatine bone articulates with the maxilla. The *lacrimal groove*, positioned anteriorly to the maxillary hiatus, becomes the *nasolacrimal canal* with the addition of the lacrimal bone and inferior nasal concha. This canal transmits the *nasolacrimal duct* into the nasal cavity (see Fig. 2-25).

Processes of the Maxilla

The processes of the maxilla are extensions from the main body of the basic

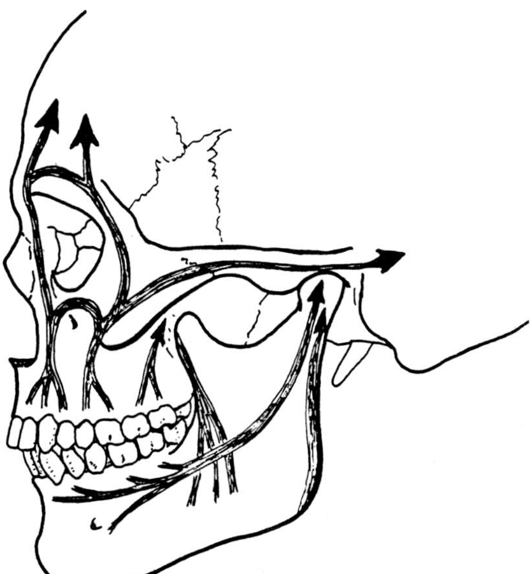

Figure 9-39. Diagrammatic representation of the facial buttresses. The arrows show the direction of thrust from the teeth to the cranial base.

bone. These are the zygomatic, frontal,

alveolar, and palatine processes, which are described next.

The *zygomatic process* of the maxillary bone articulates with the zygomatic bone at the lateral edge of the facial skeleton. A slight flaring of the maxilla to meet the zygomatic bone is evident.

The *frontal process* articulates with the nasal process of the frontal bone. It forms part of the lateral boundary of the nose as it passes between the nasal and lacrimal bones. Through this process, chewing forces are transmitted to the frontal bone of the skull (Fig. 9-39).

The *alveolar process* is that region on the lower border of the maxilla. Each maxillary bone constitutes one-half of the upper alveolar arch and contains eight teeth of the permanent dental arch. Protruding from the alveolar process are two incisors, a cuspid, two bicuspids, and three molars. As previously indicated, the deep sockets for the reception of the teeth along the alveolar border are known as *alveoli* (Fig. 9-40).

The *palatine process* is a horizontal projection from the medial surface of the maxillary bone (Fig. 9-10). Together with the palatine process from the opposite bone, it forms three-fourths of the hard palate (up to the incisive foramen) and a considerable portion of the roof of the oral cavity and the floor of the nasal cavity. On each palatine process is a canal, commencing near the first molar and running in the nasal floor to the midline. When the two maxillary bones are articulated, the canals open anteriorly into a central, funnel-shaped opening, the *incisive canal.* The canal transmits terminal branches of nerves and an artery to the surface of the hard palate. The *midline palatine suture* marks the union of the two palatine processes.

As the midline suture reaches the region of the incisive foramen in a young person, it gives rise to two very fine *incisive sutures*, which run outward on the bony palate to the space between the lateral incisor and cuspid teeth on either side. These incisive sutures do not appear on the facial surface and disappear with maturity. In animals other than man, the incisive sutures form the *premaxilla*, which is a separate bone before and after birth.

Most of the muscles of facial expression take their origin on the upper and lower jaws and upper facial areas. The maxillary bones in particular play a prominent role in providing attachments for muscles pertaining to the upper lip and corners of the mouth. It must be remembered that facial expression is a powerful medium for communication of emotion in combination with speech or silence. Additionally, the maxillae contribute to the formation of the oral and nasal speech resonant chambers.

The nine articulations, in summary, are with the frontal, ethmoidal, nasal, zygomatic, lacrimal, inferior nasal concha, palatine, vomer, and the other maxilla.

The Palatine Bone

The two palatine bones, in conjunction with the palatine processes of the maxillary bones, form the bony palate. In addition to its role in forming the posterior portion of the bony palate, the palatine bone contributes to the side wall of the nasal cavity and the floor of the orbital cavity. It consists of a *horizontal* and *vertical plate* and three notable processes, i.e. the *sphenoidal, orbital*, and *pyramidal processes*.

The basic configuration of each palatine bone is an L-shape when viewed from behind. The *horizontal plate* of the palatine bone joins the palatine process of the maxillary bone at the *transverse palatine suture*. The palatine surface comprises, with the opposite bone, the back part of the floor of the nasal cavity and bony roof of the oral cavity. The posterior border is

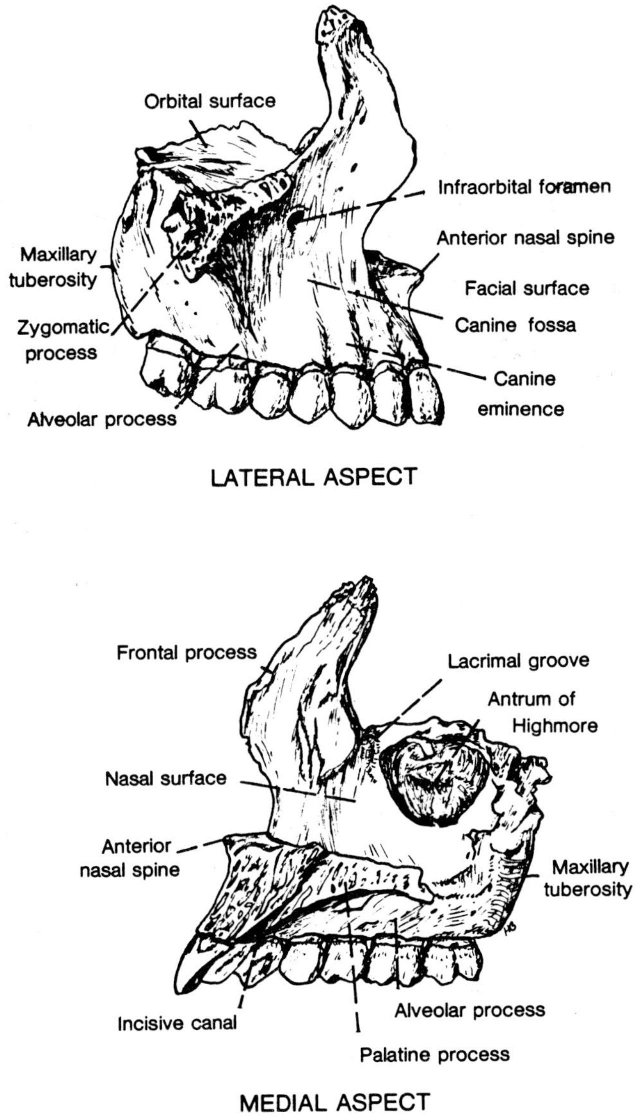

Figure 9-40. Outer and inner surfaces of the maxilla. (From Sicher, H., and Tandler, J.: *Anatomie fur Zahnarzte*, copyright 1928. Springer-Verlag, Heidelberg.)

scalloped from lateral to mesial, and it rises sharply at its medial limit and forms, with its contralateral bone, the *posterior nasal spine*. As the medial borders of the two horizontal plates connect at the midline, each superior edge is elevated to a sharp crest, the *palatine nasal crest*, which articulates with the vomer above to form a portion of the bony nasal septum (Fig. 9-41).

The *vertical plate* is the upright part of the L-shape to the palatine bone. The vertical plate forms part of the side wall of the nasal cavity and aids in narrowing the entrance into the maxillary sinus.

The pyramidal process is opposite the horizontal plate but on the vertical part.

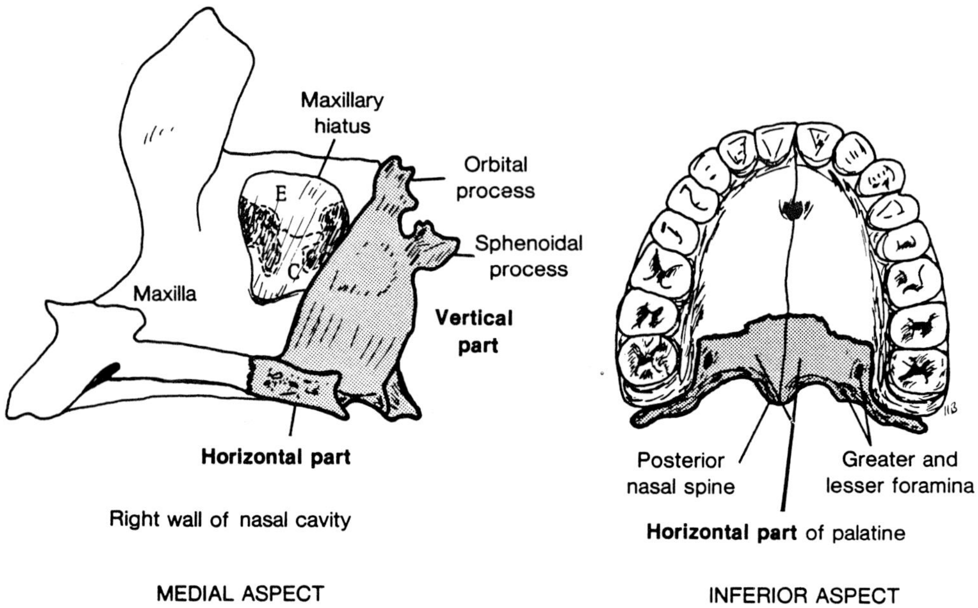

Figure 9-41. Two views of the palatine bone (shaded).

Some fibers of the *medial pterygoid muscle* attach here. On the base of this process are the lesser palatine foramina, which transmit the lesser palatine nerves and vessels supplying the soft palate.

The Vomer

The vomer forms the inferior and posterior part of the bony nasal septum. While it is placed in the median plane, it is frequently deviated to one side. The shape of the vomer is that of a plowshare.

The vomer articulates laterally with the medial pterygoid plates and the sphenoidal process of the palatine. The vomer is also attached to the lower surface of the sphenoid body (Fig. 9-25). The anterosuperior border meets the descending *perpendicular plate* of the ethmoid bone, which contributes the anterosuperior portion of the bony nasal septum. Anteriorly, the septum is completed by the cartilage of the nasal septum (Fig. 9-42).

The posterior border is free and slightly concave. It is covered with nasal mucous membrane and separates the posterior nasal apertures or the *choanae*. The inferior border articulates with the maxillary crest anteriorly and the palatine crest posteriorly.

The Inferior Nasal Concha

Also known as the *inferior turbinate bone*, the *inferior nasal concha* consists of an oval plate, having a blunt anterior end and a sharp posterior end. Unlike the superior and middle nasal conchae, which are part of the ethmoid, the inferior nasal concha is thought of as a separate bone. It is attached to the lateral wall of the nasal cavity and projects medially. Anteriorly, the blunt plate joins the *conchal* or *horizontal crest* of the maxilla. The posterior part articulates with the palatine bone. The middle part gives rise to the maxillary process, which articulates with the maxilla and fills in the lower part of the opening to the maxillary sinus. It is interesting

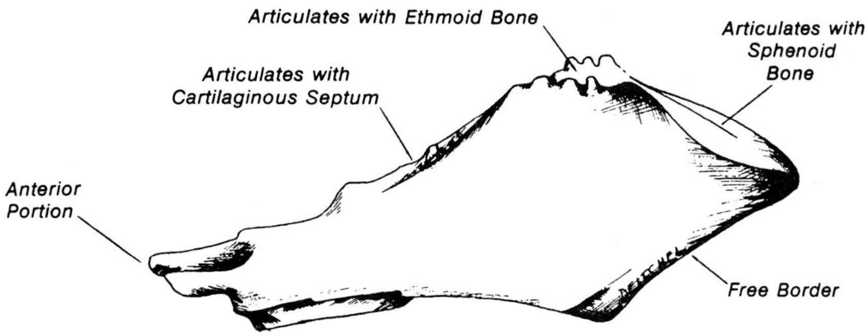

Figure 9-42. The vomer, lateral aspect. (From Fried, L.A.: *Anatomy of the Head, Neck, Face, and Jaws*, 2nd ed., 1980. Courtesy of Lea & Febiger, publisher.)

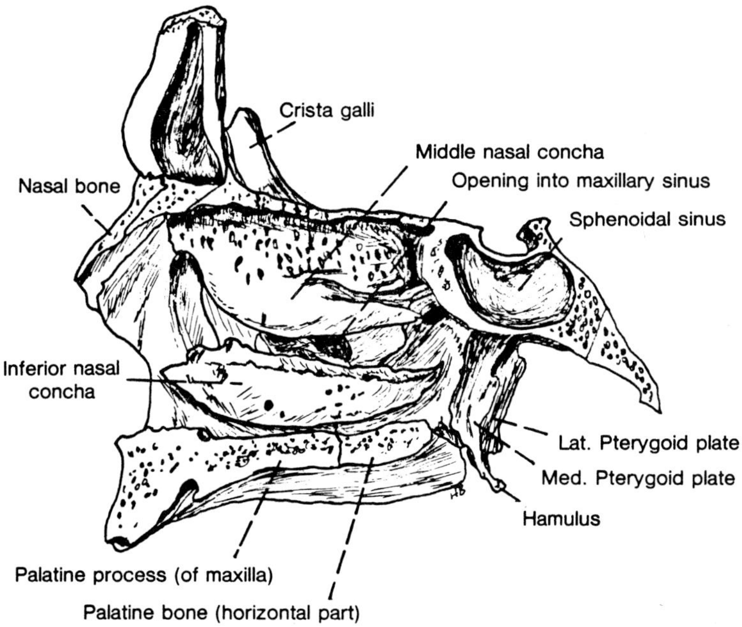

Figure 9-43. The right inferior nasal concha and related structures. (From Spalteholz, W., and Spanner, R.: *Atlas of Human Anatomy*, 16th ed., 1961. Courtesy of Bohn, Scheltema and Holkema, Amsterdam.)

that the concha is tilted inferiorly as one considers its anteroposterior dimension. This tilting aids in the flow of drainage toward the pharynx (Fig. 9-43).

The *lacrimal process* is an anterior vertical process of the inferior concha rising to join the descending process of the lacrimal bone. In so doing, it assists in forming the canal for the *nasolacrimal duct*, which opens below the inferior nasal concha into the *inferior nasal meatus*. The nasolacrimal duct transmits tears into the nasal cavity. During crying or eye irritation, the excessive tear secretions draining into the nose prompt the need to blow the fluid from the nasal cavity.

Another vertical process, the *ethmoid process*, meets the descending *uncinate* (hooked) *process* of the ethmoid, crossing the maxillary hiatus to narrow further this opening (Fig. 9-43). The inferior nasal concha articulates with the maxilla, ethmoid, lacrimal, and palatine as it seals part of the nasal wall defect entering the maxillary sinus.

The Zygomatic Bone

Each zygomatic bone is readily observed from a frontal view of the skull. Its name is derived from its shape (*zygo* = yoke). One of the pair is also visible in the lateral view of the skull. This bone forms the lateral rim of the orbit and the bony prominence of the cheek. Along with the frontal process of the maxillary bone, chewing forces are transmitted to the frontal bone and the top of the skull (see Fig. 9-39).

The zygomatic bone has three surfaces. The rounded surface, the *outer surface*, forms the prominence of the cheek (sometimes called malar); the *orbital surface* forms a part of the lateral wall of the orbit and cannot be seen from the side; the *temporal surface* or the inner surface exhibits a rough triangular area for articulation with the maxilla and a lateral, smooth, concave surface, which forms the anterior boundary of the temporal fossa and a forward portion of the infratemporal fossa.

There are three processes extending from the diamond-shaped body of the zygomatic bone. The *frontal process* articulates with the zygomatic process of the frontal bone close to the roof of the orbit along the outer margin. The posterior margin of the frontal process is serrated for articulation with the greater wing of the sphenoid bone. A *temporal process* projects freely posteriorly in almost a horizontal plane to articulate with the zygomatic process of the temporal bone on the right side of the skull. Union of these two processes forms the zygomatic arch. The *maxillary process* presents a rough edge, which meets the maxillary bone.

Articulations by the zygomatic bone with adjacent bones include the maxillary bone medially, the frontal bone at the superolateral rim of the orbit, the sphenoid bone below the frontal articulation, and the temporal bone through the zygomatic arch.

Several muscles of facial expression originate on the outer surface of the zygomatic bone. Fibers of the masseter, an important muscle of the mandible, arise from the lower inner surface at the masseteric border (Fig. 9-44).

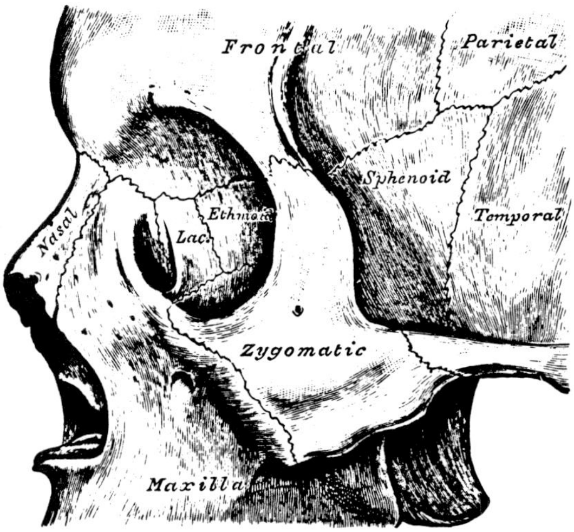

Figure 9-44. The left zygomatic bone. (From Goss, C.M. (Ed.): *Gray's Anatomy of the Human Body*, 28th ed., 1973. Courtesy of Lea & Febiger, publisher.)

The Nasal Bone

Each nasal bone fits between its contralateral fellow and the frontal process of the maxillary bone. It meets the frontal bone superiorly. Together, the nasal bones form the upper part of the bridge of the nose and the anterior part of the roof of the nasal cavity.

This small bone may vary widely with respect to curvature, length, width, and symmetry between the two bones. One basis for structural variation is racial.

The Lacrimal Bone

The lacrimal bone, a paired bone, is said to be the most fragile bone of the face; it is the smallest. It is visible in a lateral view of the skull as it forms a portion of the anterior part of the medial wall of the orbit. Often, on skulls used as study specimens the lacrimal bones are absent, disappearing during the acid-cleaning process when the soft tissues are removed. The lacrimal bone is an irregularly rectangular thin plate of bone, having an orbital and a nasal surface. This bone contributes the *lacrimal sulcus*, housing the lacrimal sac where tears collect after circulating across the cornea. (The lacrimal gland is located in the upper outer corner of the orbit, just above the bones forming the frontozygomatic junction.)

The *lacrimal hamulus* on the orbital surface articulates with the maxilla and completes the orifice of the *lacrimal canal*. Through the lacrimal canal, tears descend to emerge into the inferior meatus of the nasal cavity. A fold of tissue, the *lacrimal fold*, protects the opening by not allowing air to be forced into the lacrimal sac from the lacrimal canal during nose blowing.

The lacrimal bone articulates with four bones. Two, the frontal and ethmoidal bones, are part of the cranium, while the other two, the maxilla and the inferior nasal concha, are part of the face.

The Hyoid Bone

The hyoid bone is included here under the discussion of the facial skeleton since it can be designated as the skeleton of the tongue. This unpaired bone is suspended from the tips of the styloid processes of the temporal bones by stylohyoid ligaments. There are no bony articulations. In addition, a number of muscles support the hyoid, permitting great variability of movement. Altogether about thirty muscles join the hyoid bone directly to other structures. The implications for affecting speech production are quite significant. Also, since the tongue is so difficult to measure in living subjects, changes in position of the hyoid bone as seen by lateral cephalometric x-ray films are often used in research studies as criteria to explain various phenomena associated with the tongue (see Figs. 9-27 & 9-45).

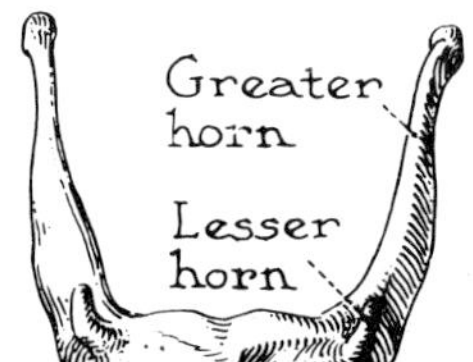

Figure 9-45. The hyoid bone as viewed from the front. (From King, B.G., and Showers, M.J.; *Human Anatomy and Physiology*, 1969. Courtesy of W.B. Saunders Co., Philadelphia.

This U-shaped bone possesses a *body* or *corpus* that is approximately rectangular and has a thick upper and sharp lower border. It is crossed by well-marked transverse ridges, which serve as crests, marking the fields of attachment of several muscles, all of which relate to speech functions. Two pairs of horns, the *greater* and *lesser cornua*, project posteriorly from the body of the hyoid bone.

LOOKING AHEAD

In this chapter the skeletal framework — the supporting structures — for articulation and audition and the brain were detailed. The reader should refer to this chapter frequently to relate new material pertaining to the speech and hearing structures to the bony framework.

Subsequently, chapters on dentition; the oral, nasal, and pharyngeal cavities; musculature of the tongue, soft palate, and pharynx; musculature of the face and mandible; the brain and cranial nerves; and audition will relate and build from the present chapter. Additionally, a number of examples of applied anatomy of the head in the final division of this text, Part Five, will rely on an understanding of the skull.

BIBLIOGRAPHY

Ackerman, J.L., and Profitt, W.R.: The characteristics of malocclusion: a modern approach to classification and diagnosis. *Am J Orthod, 56*:443-454, 1969.

Aitken, J.T., Causey, G., Joseph, J., and Young, J.Z.: *A Manual of Human Anatomy: Head and Neck*, 2nd ed. London, Churchill, 1976, vol. II.

Basmajian, J.: *Primary Anatomy*, 8th ed. Baltimore, Williams and Wilkins, 1982.

Crafts, R.C.: *A Textbook of Human Anatomy*, 2nd ed. New York, Wiley, 1979.

Crouch, J.E.: *Functional Human Anatomy* 3rd ed. Philadelphia, Lea and Febiger, 1978.

DuBrul, E.L.: *Sicher's Oral Anatomy*, 7th ed. St. Louis, Mosby, 1980.

Enlow, D.H.: *Handbook of Facial Growth*, 2nd ed. Philadelphia, Saunders, 1982.

Fried, L.A.: *Anatomy of the Head, Neck, Face, and Jaws*, 2nd ed. Philadelphia, Lea and Febiger, 1980.

Gardner, E., Gray, D.J., and O'Rahilly, R.: *Anatomy: A Regional Study of the Human Structure*, 4th ed. Philadelphia, Saunders, 1975.

Gardner, W.D., and Osburn, W.A.: *Structure of the Human Body*. Philadelphia, Saunders, 1967.

Grant, J.C.B.: *An Atlas of Anatomy*, 6th ed. Baltimore, Williams and Wilkins, 1972.

Gray, H.: *The Anatomy of the Human Body* (Charles M. Goss, Ed.), 28th ed. Philadelphia, Lea and Febiger, 1973.

Lore, J.M., Jr.: *An Atlas of Head and Neck Surgery*, 2nd ed. Philadelphia, Saunders, 1973, vol. I.

Mason, R.: Orofacial anomalies and speech. In Wingo, J., and Holloway, G. (Eds.): *An Appraisal of Speech Pathology and Audiology: A Symposium*. Springfield, IL: Thomas, 1973.

Massler, M., and Schour, I.: *Atlas of the Mouth*, 2nd ed. Chicago, Am Dental, 1958.

O'Rahilly, R.: *Basic Human Anatomy*. Philadelphia, Saunders, 1983.

Paff, G.H.: *Anatomy of the Head and Neck*. Philadelphia, Saunders, 1973.

Romanes, G.J. (Ed.): *Cunningham's Textbook of Anatomy*, 12th ed. New York, Oxford U Pr, 1981.

Rankow, R.M.: *An Atlas of Surgery of the Face, Mouth and Neck*. Philadelphia, Saunders, 1968.

Scott, J.H., and Dixon, A.D.: *Anatomy for Students of Dentistry*, 4th ed. London, Churchill, 1978.

Snell, R.S.: *Clinical Anatomy for Medical Students*, 2nd ed. Boston, Little, 1981.

Steen, E.B., and Montagu, A.: *Anatomy and Physiology*. New York, Har-Row, 1959, vol. I.

Woodburne, R.T.: *Essentials of Human Anatomy*, 6th ed. New York, Oxford U Pr, 1978.

Chapter 10

THE DENTITION

THE teeth may be viewed as a part of the articulatory mechanism. They serve as important boundaries within the oral cavity, where the breath stream is varied to form numerous speech sounds.

Each tooth develops in the alveolar process of the maxillary bones or mandible. The *alveolar process*, a ridge of bone on the two jaws, contains sockets or *alveoli* that firmly anchor the teeth within the jaws. It is important for the speech clinician to appreciate that *all* teeth are housed within the alveolar ridges. Thus, in communicating clinical information to a dentist, use of the term *alveolar ridge* should be qualified to identify the *portion* of the ridge being referred to. Usually, speech clinicians are referring to tongue contacts with the anterior maxillary alveolus, or the area of the incisive papilla.

TOOTH STRUCTURE AND SUPPORT

The principal parts of a tooth are the enamel, dentin, pulp, and cementum. The *enamel* covers the *crown* of the tooth (that part which is visible above the gums). Enamel is the hardest substance in the body; this unique tissue is made up almost entirely of inorganic salts. The enamel is especially suited to its role of mastication, which requires a tough tissue capable of resisting abrasive wear (Fig. 10-1).

Enamel hardness is uniform throughout. The formation of enamel is greater in areas of occlusal or incisal wear, providing a thickness that usually can supply a lifetime of service. Once destroyed, enamel is incapable of repair or regeneration. This is because enamel becomes completely cut off from any cellular elements when it reaches the end of its functional growth. Enamel does *not* continue to form throughout life. No more enamel is deposited when the thickness proper for a tooth has been reached. This occurs before the tooth erupts.

Enamel undergoes a posteruptive aging process called *maturation*. During this process, the permeability of enamel decreases with the gradual incorporation of various agents (ions) from the oral environment such as fluorides, iron, zinc, and others. As a product of this maturation, the enamel becomes more yellow with advancing age, topical fluorides pen-

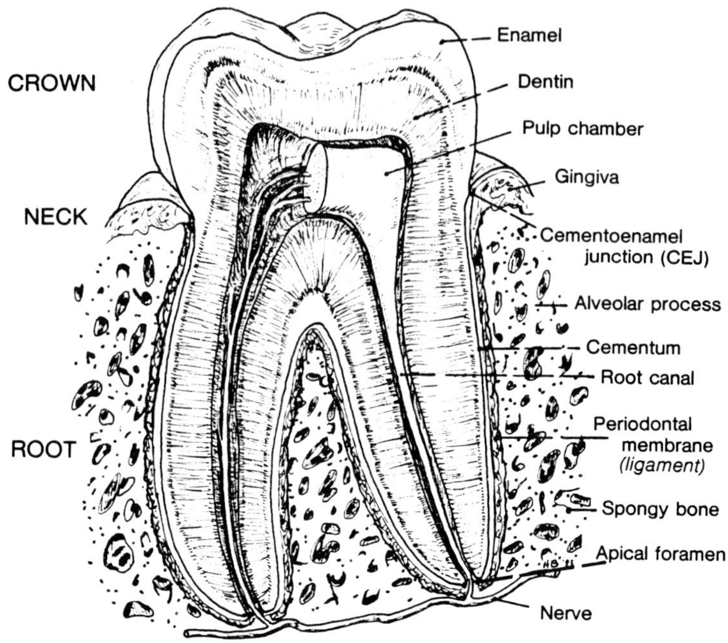

Figure 10-1. Cross-sectional view of a molar and related structures. (Redrawn from Jacob, S.W., and Francone, C.A.: *Laboratory Manual of Structure and Function in Man*, 5th ed., 1982. Courtesy of W.B. Saunders Co., Philadelphia.)

etrate less deeply as the enamel matures.

To withstand the stresses of mastication, enamel requires a supportive base of *dentin*. Dentin forms the major bulk of the tooth. It is not as hard as enamel, but is harder than cementum or bone, having a greater proportion (70-75%) of inorganic material incorporated in it. The tensile strength of dentin is greater than enamel, providing an important supportive base. Formation of dentin can continue throughout life, thus increasing its bulk. This is seen as a progressive narrowing of the size of the pulp chamber with advancing age.

Many *tubules* penetrate the dentin. These are protoplasmic processes, extending from odontoblast cells in the pulp of the tooth. Through the tubules, irritations (pain) may be transmitted to the nerves of the pulp. Such pain irritations arise from bacteria, chemical, and thermal changes, or the dentist's drill.

Although dentin can be formed throughout life, it is incapable of restoring any loss of its own substance. When a *carious lesion*, or cavity, penetrates the enamel and reaches the dentinoenamel junction, the dentin already there can initiate a slow alteration of the dentinal tubules to seal off the pulp from the cavity. A transformation of dentin already produced is called *sclerotic* or *eburnated* dentin. It protects the pulp from future irritations.

The soft tissue, including the nerve of the tooth, is the *pulp*. The pulp cavity is located within the tooth and generally corresponds to the basic shape of the tooth. The dental pulp cavity is composed of nerves, veins, arteries, connective tissue cells, lymph passageways, odontoblasts, and intercellular substance. The odontoblasts (cells that give rise to dentin) lining the pulp cavity are capable of depositing secondary or new dentin into the dentinal tubules to act as a protective barrier against caries. This is a mechanism by which the pulp chamber decreases in size over time as the pulp responds to advancing caries and other stimulating processes by sealing itself off (Fig. 10-2).

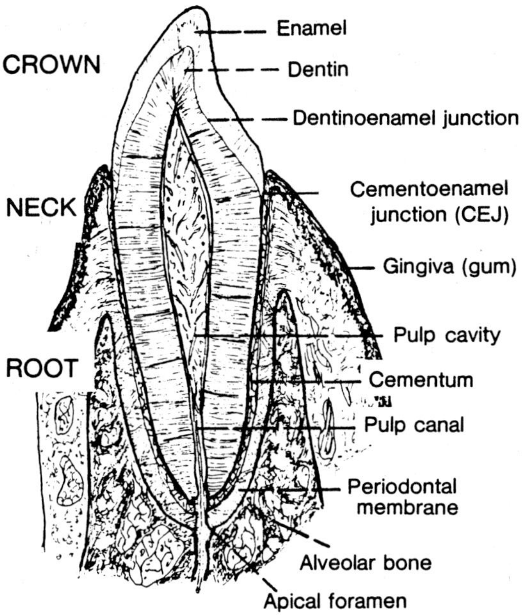

Figure 10-2. Cross-sectional view of a canine (cuspid) and related structures.

That portion of the tooth embedded in the tooth socket of the alveolar process of the jaw is the *root*. In the root, the *apical foramen* or apex permits the pulp cavity to communicate with the surrounding tissues at the bottom of the alveolus. The *root canal* is that passage from the apical foramen through the root of the tooth. Eventually, the canal widens upon reaching the pulp chamber in the crown. *Cementum* is a thin bone-like covering of the tooth's root. Cementum is composed of 65 to 70 percent inorganic material. It is slightly softer than dentin but has the ability to repair any loss of its substance from resorption by the deposition of new cementum. This capability, like that of dentin, is a formation process that can continue throughout life.

The cementum meets the enamel at the *cementoenamel junction*, commonly referred to as the *CEJ* or *cervical line*. Fibers that attach the tooth to supporting bone, the *periodontal ligament*, are imbedded in cementum. For descriptive purposes, the periodontal ligament apparatus has been widely referred to in dentistry as the *periodontal membrane*. This may be a misleading term, since the connective tissue between the root and bone can only be accurately described as ligamentous. Nonetheless, the membrane designation will be allowed in this text because of the usefulness of the image this terminology provides.

The periodontal membrane is comprised of a series of fiber bundles that are attached between the cementum of the tooth and the wall of the socket. These fibers are nonelastic and arranged in a variety of directions as a means of resisting and cushioning the stresses placed upon the teeth. The pressures of mastication lead to a stretching of some or all of the fiber bundles, resulting in traction that acts on cementum and bone. This transformation of forces is essential for the maintenance of the normal life of a tooth.

The tissues that surround and invest the tooth, including the gingiva (gums), periodontal membrane, cementum, and alveolar bone, make up a structural and functional unit referred to as the *periodontium*. *Periodontal disease* refers to a destructive process involving the attachment apparatus of the teeth. Gingival disease (or *gingivitis*), by contrast, involves only the gum tissues. Gingivitis can progress into periodontitis if left untreated.

Clinical Notes

The epithelial attachment of gum tissue to tooth and alveolar bone is a vulnerable site in the periodontium. Inflammatory changes in the epithelium, as with gingivitis, can cause the epithelium to migrate along the cementum toward the apical area of the tooth. This is at the expense of the periodontal membrane fibers and alveolar bone, which can be destroyed along the way. Over time, there is loosening of teeth in their sockets. In the adult, more teeth are removed due to periodontal disease than are lost due to pulpal abscess and caries. Unnecessary loss of teeth due to disease conditions should be prevented at all ages.

The older dental patient is more susceptible to caries as the immunologic system decreases in efficiency. Cavities become most prevalent at the surface adjacent to the gums (cervical area of crown). Surgery on the gums, or gingiva, is common in the aged patient.

The elderly are more likely to be taking medications that could affect dental treatment. There are about 400 drugs that affect saliva secretions, most often lowering saliva output. Aging itself lowers saliva output. This can impair lubrication of the oral cavity, resulting in cracked or decayed teeth.

In the older patient, front teeth are most important in preserving facial appearance and some speech articulation. Consequently, proper care of teeth should be practiced throughout life.

It is pertinent to mention one last clinical perspective relative to the structure of teeth. The commercial advertisement of toothpastes is usually accompanied by comments about fluoride protection and, often, the whitening capabilities of the dentifrice. It is true that the surface of enamel is somewhat permeable to topical application of fluoride, whether in a sodium or tin (stannous) base. As pointed out previously, however, the permeability of the teeth to fluoride decreases with advancing age. This does not imply that anyone should give up the regular use of fluoride toothpaste, nor ignore the benefits to dental health from systemically acting fluoride protection such as from fluoridated water.

Tooth Color

As to the color of teeth, most individuals are naturally interested in having the whitest teeth possible. A common problem in denture construction is disagreement between dentist and patient about the recommended coloration of the denture teeth. Most individuals select a shade that is much whiter than their natural teeth, thus calling attention to the artificial nature of the denture.

The color of natural teeth is dependent upon the crown of the tooth, and, specifically, the enamel. Normal enamel is a light yellowish color and is usually translucent enough to permit the darker yellow of the dentin to aid in determining the shade of the tooth. This is especially the case where the enamel is thin. That is, the thinner the enamel, the darker the shade of the tooth. Thus, the thinner portions of enamel close to the gums (the cervical portion of the crowns) tend to be darker than the incisal or occlusal portions.

The teeth of a child are generally whiter than adult teeth. Adult teeth that are truly white often signal a problem of translucency, such as from imperfect calcification of enamel. The bottom line regarding whiteness of teeth is that once stains are removed, there is little that can be done to whiten teeth short of veneer crowns or some other more extensive dental treatment. Bleaching or topical resin applications may improve the situation for a relatively short period of time. Such procedures should obviously not be attempted by anyone other than a dentist, i.e. don't attempt to self-correct this problem. Also, the consumer should be advised to be wary of toothpaste advertisements that claim great success with teeth whitening. Occasional use of a toothpaste or dentifrice with a high abrasive content is acceptable, but regular use of any abrasive compound to remove stains is not recommended.

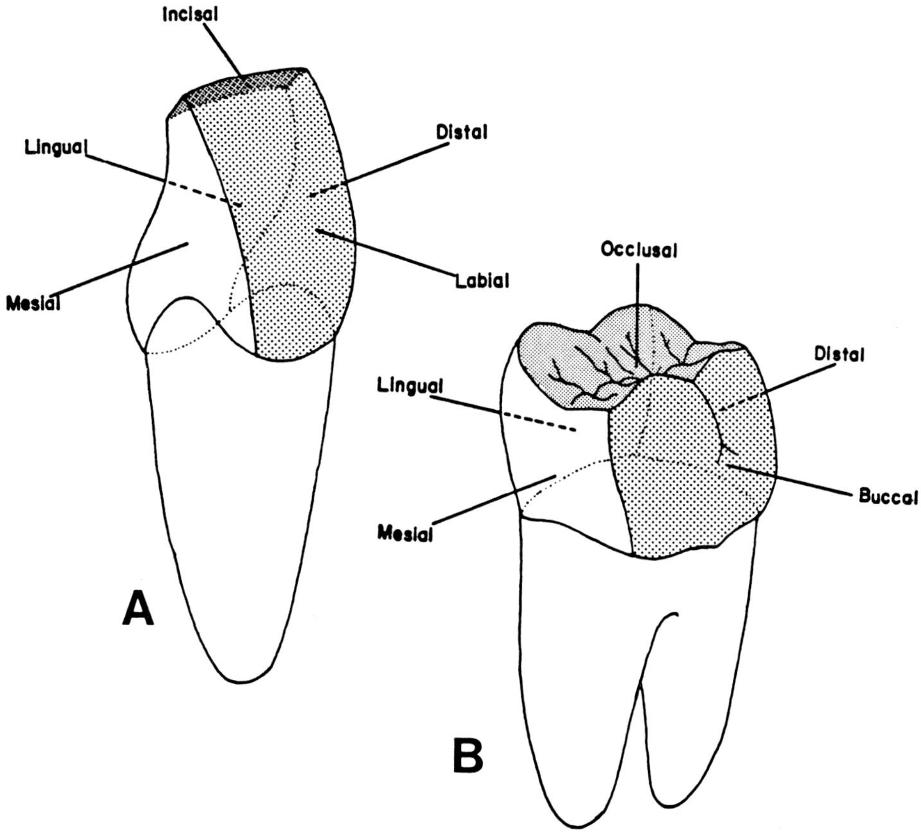

Figure 10-3. (A) Surfaces of an incisor tooth. (B) Surfaces of a molar tooth. (Reproduced with permission from Füller, J.L., and Denehy, G.E.; *Concise Dental Anatomy and Morphology*. Copyright © 1977 by Year Book Medical Publishers, Inc., Chicago.)

CHARACTERISTICS OF TEETH

The crown of each tooth has five surfaces: lingual, facial, occlusal, mesial, and distal. (1) The *lingual surface* faces the tongue or oral cavity, (2) the *buccal surface* is on the outside of the molars and premolars, and the *labial surface* describes the outside area of the incisors and canines. (Most orthodontists use the term *facial* to describe either the buccal or labial surface of a tooth.) (3) For posterior teeth (premolars and molars) that surface in contact with the teeth of the opposite dental arch is the *occlusal surface*. Anterior teeth (incisors and canines), however, do not have a true occlusal surface. These teeth are said to have an *incisal surface* (Fig. 10-3).

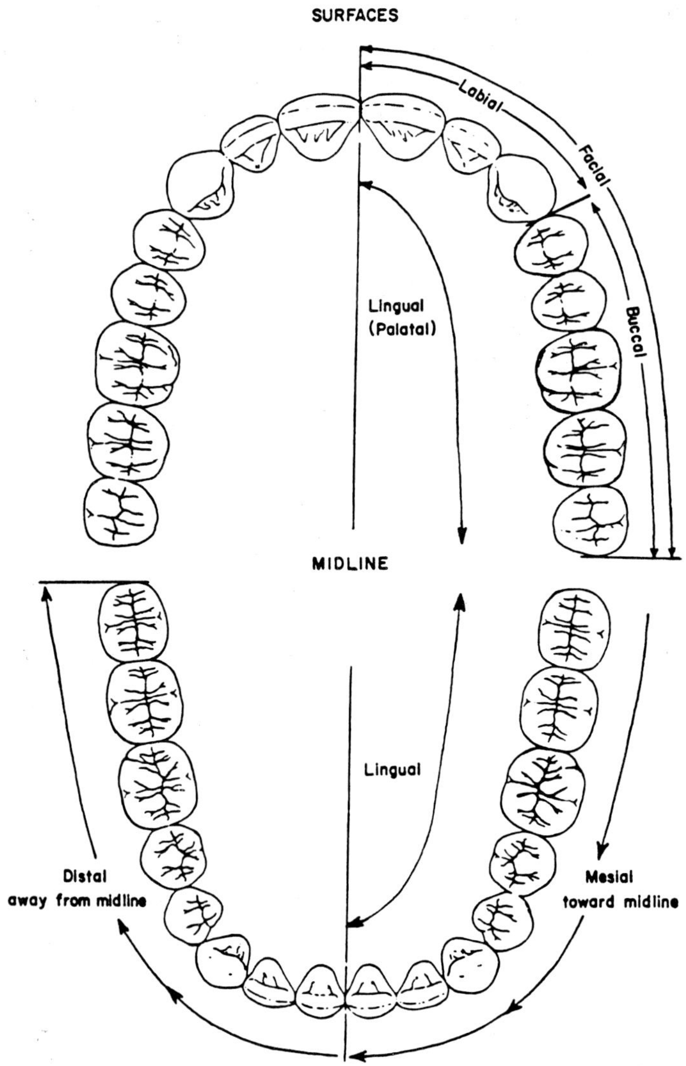

Figure 10-4. Terminology related to tooth surfaces and position for the dental arches. (Reproduced with permission from Fuller, J.L., and Denehy, G.E.: *Concise Dental Anatomy and Morphology*. Copyright 1977 by Year Book Medical Publishers, Inc., Chicago.)

The pointed process on the occlusal surface of back teeth is the *cusp*. The other two surfaces face and contact adjacent teeth and are termed (4) mesial or (5) distal (interproximal) surfaces. The *proximal* or *mesial* surface would be closest to the midline, and the *distal* surface is farthest from the midline. In this regard, it is convenient to picture the dental arch as a straight line, with molars at each end and incisors at the middle. The use of the terms *mesial* and *distal* is readily appreciated from this perspective (Fig. 10-4).

Directions or relations to different areas on the crown are expressed by use of the terms *occlusally* (toward the incisal edge or biting surface) or *cervically* (toward the gum line). For roots, apically means

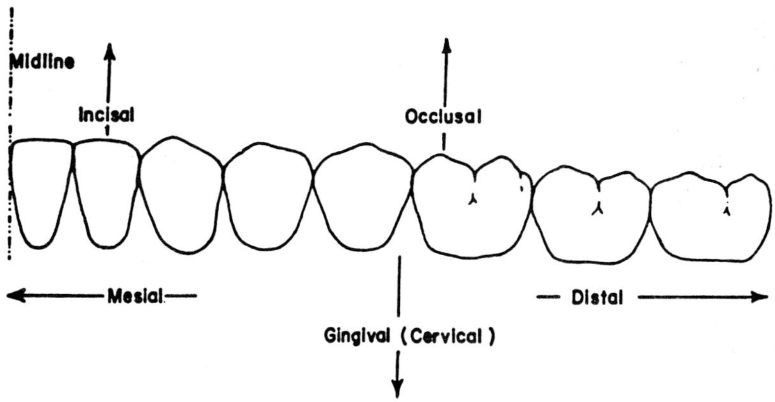

Figure 10-5. Terminology related to tooth crowns (lateral view). (Reproduced with permission from Fuller, J.L., and Denehy, G.E.: *Concise Dental Anatomy and Morphology*. Copyright 1977 by Year Book Medical Publishers, Inc., Chicago.)

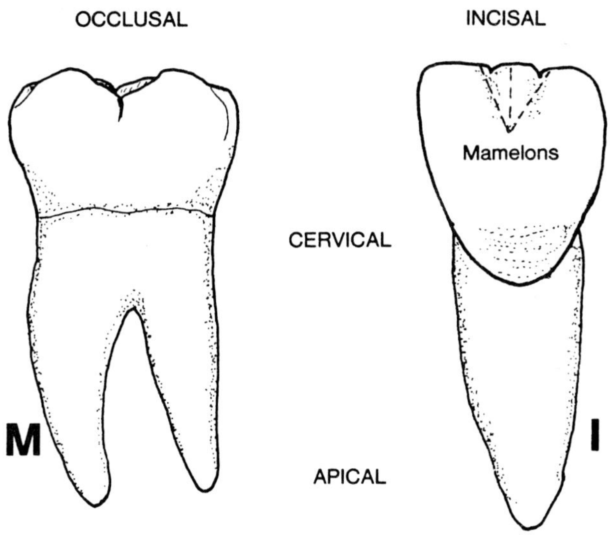

Figure 10-6. Characteristic shape of a lower molar tooth (M) and an incisor (I).

toward the apex, while cervically means toward the CEJ or gum line (Fig. 10-5).

Permanent *incisors* are characterized by a broad crown, ending in a sharp incisal edge. When the incisors erupt, they exhibit the presence of three rounded bumps or *mamelons* on the incisal edges. These mamelons wear away with time, unless the individual has an openbite malocclusion that precludes the normal functional relationships between upper and lower incisors (Fig. 10-6).

On the lingual surface of the upper and lower incisors is a concavity that abruptly turns into a convexity toward the cervical portion of the tooth. This convexity is known as the *cingulum* of the tooth. The cingulum on upper incisors is much more extensive than on lower incisors, extending more incisally also. In normal occlusion, the incisal edge of the lower incisors rests at the concavity occlusal to the upper cingulum (Fig. 10-7).

The upper *lateral incisors* are the most variable teeth in the dentition. Small, pegged or cone-shaped lateral incisors are a frequent variation seen in the adult dentition (see Fig. 10-17). In the patient with a cleft through the maxillary alveolus, the lateral incisor may be absent, present and distorted in size and shape, and located on either side of the cleft site.

The *canine* teeth, also referred to as *cuspids*, have pointed cusps instead of an occlusal edge. Because of their long roots and their strategic location at the point of greatest bend of the dental arch, they are very important teeth to retain in the dentition. Dentists are extremely reluctant to recommend extraction of canine teeth. If a single canine is removed or absent, the dental midline will almost always shift to the side of the missing tooth. When a canine is unerupted or otherwise blocked out of the dental arch (such as facial and infraerupted), an orthodontic plan to manage this situation may involve expansion of the arch or extraction of the first premolar to make room for the canine to be brought into the arch orthodontically (Fig. 10-8).

The canines also serve an important role in most individuals in guiding the opening and closing of the bite in chewing. The lateral excursions (side-to-side movements) of the lower jaw involved in normal mastication are often guided solely by the canines. That is, when the bite opens in a side-to-side movement by contacts of an upper and lower canine on one

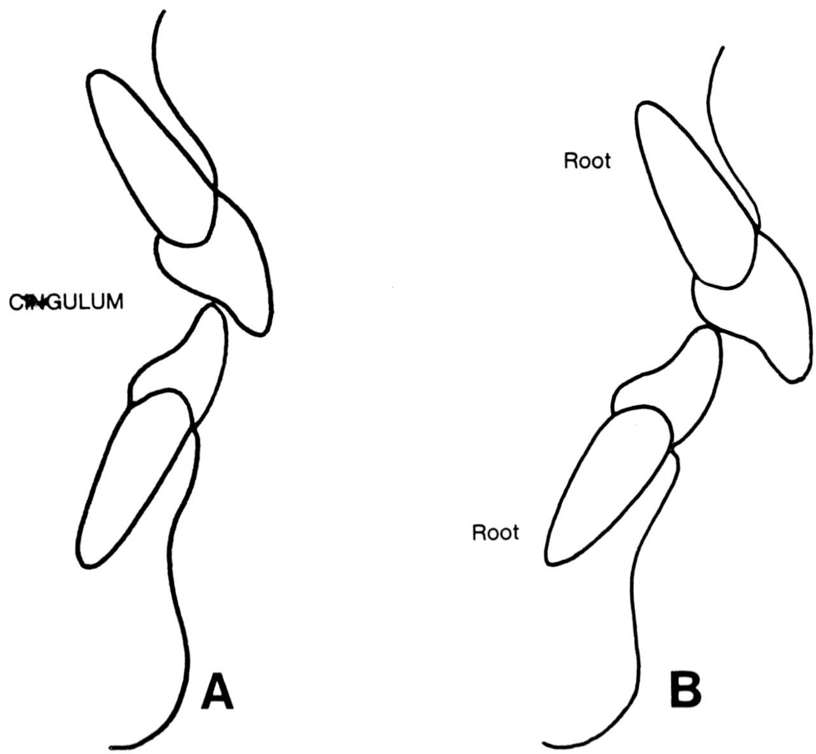

Figure 10-7. Lateral view of normal (A) and abnormal (B) occlusal contact of incisor teeth. Normally, the lower incisal edge fits in the concavity incisal to the upper incisor cingulum.

side, this is referred to as *canine rise*. When all posterior teeth participate together in the opening of the jaw during lateral excursions (that is, a contact of upper and lower teeth on one side during opening), this is referred to as *group function*.

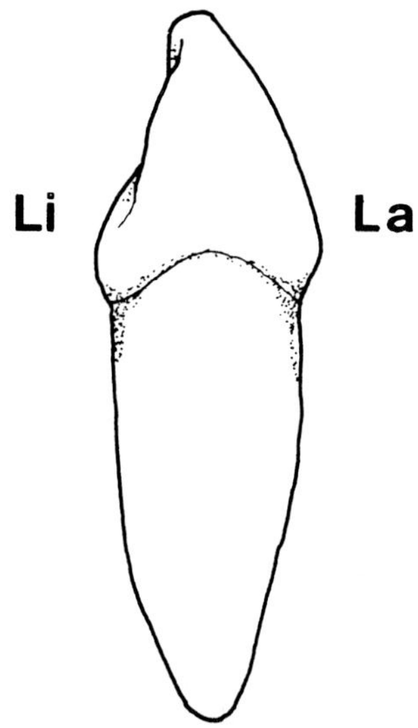

Figure 10-8. Lateral view of a lower incisor tooth showing the lingual (Li) and labial (La) surfaces.

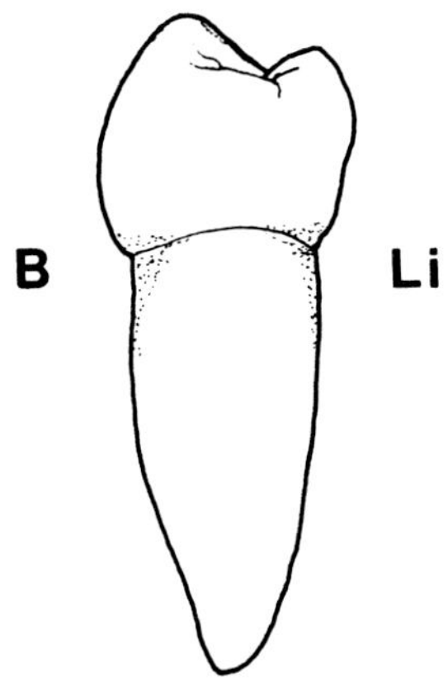

Figure 10-9. Lateral view of a lower premolar tooth showing the buccal (B) and lingual (Li) surfaces.

The *premolar* teeth are also known as *bicuspids*. These teeth in the permanent dentition replace the primary molars in the primary dentition. Since the premolars are smaller than the primary teeth they replace, this is one area of the dental arch where some space can be gained in the transition from primary to permanent teeth.

Premolar teeth have two cusps: a buccal (facial) and a lingual cusp. The lingual cusp of the lower premolars is not as well developed as the uppers. By virtue of their anatomy, the premolars are able to aid the molars in grinding food in preparation for swallowing. Their location behind the canines and in front of the molars also makes them somewhat suitable to aid in the tearing of food that is accomplished by the incisors and canines. Actually, the canines in man have lost some of their special function and often work with the incisors, while at other time they aid the premolars.

Molar teeth are primarily responsible for grinding and crushing of food. It is said that the forces of mastication at the molars can exceed 300 pounds per square inch. An appreciation for this force can be gained by considering an individual standing on a few small ice cubes without breaking them. A similar ice cube placed in the mouth can be crushed by the masticatory forces applied to the molars. The crushing

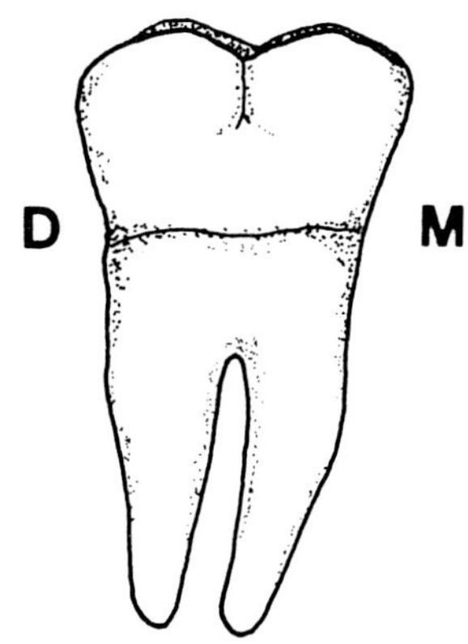

Figure 10-10. Facial (buccal) view of a lower molar tooth showing the distal (D) and mesial (M) surfaces.

of ice cubes or smaller pieces of ice is a dangerous habit and one performed at the expense of the health of the dentition. Fracture lines on the teeth, broken cusps, or damaged dental restorations are the typical sequelae of biting into ice.

Molar teeth are characterized by a wide, convoluted occlusal surface (see Fig. 10-3). Cusps constitute the hills, while pits and fissures constitute the valleys on the molar teeth. The wide occlusal surface area of molars facilitates the masticatory processes necessary for preparation of foods for swallowing (Fig. 10-10).

ERUPTION OF TEETH

Teeth have the capacity to erupt throughout life. Despite extensive research in dentistry, the cause of eruption of teeth remains unknown. Somehow, an eruptive force is generated that moves a tooth through bone and continues to move it after it has broken through into the oral cavity. Eruptive forces remain somewhat active even after a tooth is in full occlusion. The processes of eruption continue actively, along with the vertical growth of the face. A maxillary first molar may erupt more than a centimeter between ages six and sixteen years, as the jaws grow during this time and dental eruption keeps apace with vertical jaw growth. Even after years of being in a stable position, a tooth may erupt again if its antagonist in the opposite arch is removed (Proffit, 1978).

The theories to account for tooth eruption can be divided into three groups: (1) theories based on cell proliferation at the apex of the tooth root; (2) theories based on blood pressure/blood flow differentials in the periodontal membrane; and (3) theories based on metabolic changes in the periodontal membrane. Current thinking in dental research appears to favor the theory that eruptive forces are generated in the periodontal membrane, but the mechanism by which this is accomplished remains unclear (Proffit, 1978).

It is well known in dentistry that the teeth erupt in a predictable sequence and manner, except where there are environmental or other impeding factors involved. Maxillary teeth typically erupt downward and forward, while mandibular teeth erupt upward and forward.

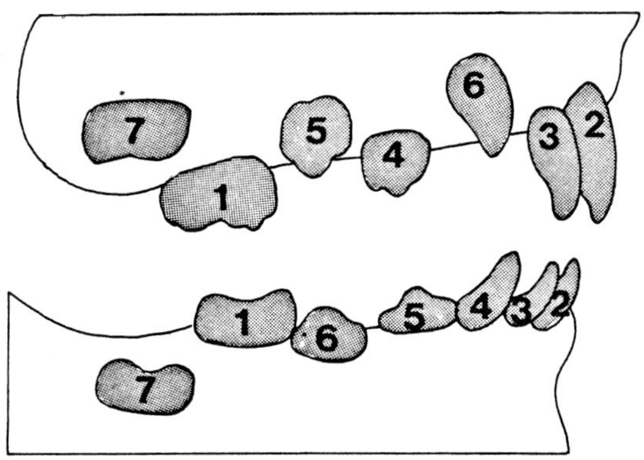

Figure 10-11. A favorable eruption sequence of permanent teeth shown in an eleven-year-old. (Modified from Barrett, Richard H., and Hanson, Marvin L.: *Oral Myofunctional Disorders*, ed. 2, St. Louis, 1978, The C.V. Mosby Co.)

Determinants of Tooth Position

Information about the eruption of teeth has a great deal of clinical significance. For example, the individual who has a mouth-open, lips-apart resting posture, with the lateral surfaces of the tongue resting over the occlusal surfaces

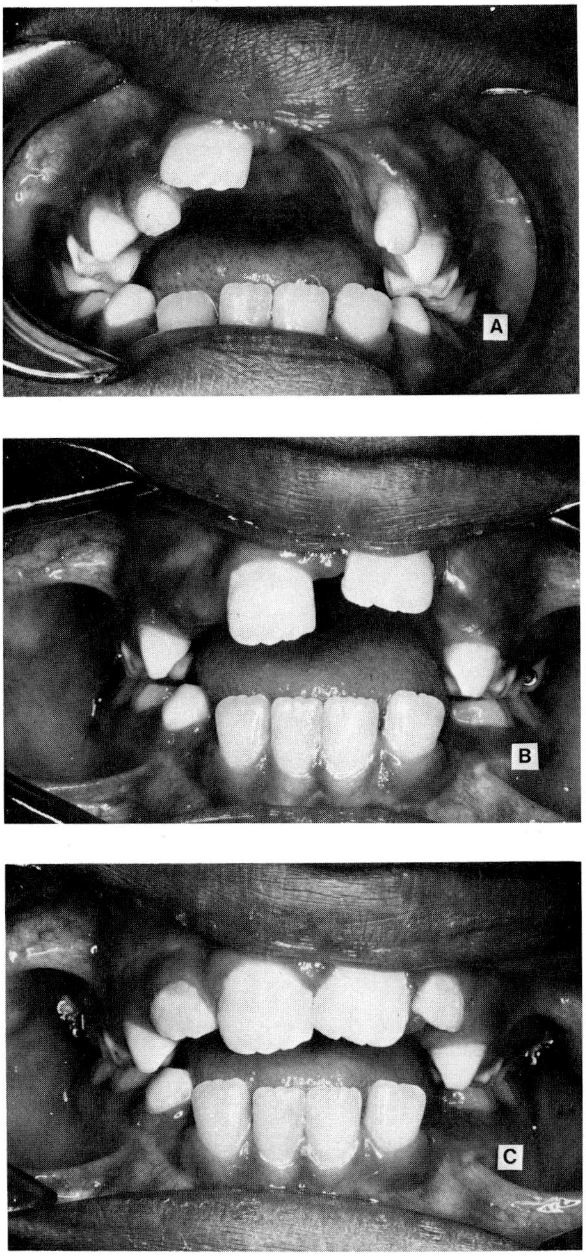

Figure 10-12. Anterior openbite and failure of eruption of incisors in an eleven-year-old thumb sucker with an anterior tongue posture: (A) before appliance therapy, (B) during appliance therapy, and (C) near completion of therapy. A removable orthodontic appliance with a smooth palatal bar was used to discourage the continuation of thumb sucking and anterior tongue posture.

of the lower posterior teeth, is likely to develop a malocclusion whereby the upper teeth are positioned farther forward than the mandibular teeth. This is related to the added, unimpeded stimulation of the upper teeth to supraerupt in a downward and forward direction. Hence, a Class II dental *malocclusion*, which describes a disparity in tooth and jaw position in the *horizontal* plane, can be caused by dental eruptive variations in the *vertical* plane of space (see Fig. 10-25).

Openbite

An anterior openbite malocclusion can also be the product of eruptive variations. The individual with a mouth-open, lips-apart resting posture whose tongue is also seen to rest forward between the anterior teeth is likely to develop an anterior openbite. This is related to supraeruption of posterior teeth and infraeruption of anterior teeth and is most often seen in the *mixed-dentition* stage where the anterior permanent teeth are just beginning to erupt. It is estimated, based on transducer research, that only 5 grams of force are needed to impede the eruption of anterior teeth (Fig. 10-12).

Tongue Thrust and Posture

The reasons for a differential eruption sequence of the teeth and the development of various malocclusions have been a source of much speculation in dentistry. Such speculations have resulted in some active controversies among clinicians who attempt either to prevent or remediate situations where environmental factors are thought to contribute to tooth position.

One area of controversy has been the possible role of the tongue in creating or maintaining a malocclusion. Current thinking is that resting tongue posture has much more of a link to tooth position than functional activities of the tongue, as in speaking or swallowing. While there is no impressive evidence that teeth are pushed into a new position by tongue functions, the long-acting forces of a resting tongue can influence tooth position in the dental arch. It is quite possible that most individuals classified as having a *tongue thrust* (a forward gesture pattern of the tongue in speaking or during the initiation phase of a swallow) *also* have an anterior resting tongue posture. This probable association of anterior resting posture and tongue thrusting has not been clearly demonstrated.

Thumb Sucking

There is little doubt in dentistry that persistent finger or thumb sucking past the age of six years can have a deleterious effect on the eruption of teeth or on the configuration of the anterior maxillary dental arch. The presence of a prominent thumb or finger in the oral cavity can have the same effect as an anterior resting tongue in impeding eruption of incisors. It can also distort the maxillary anterior alveolus in the vertical plane of space. If a child is a tongue thruster and sucks his thumb, the tongue thrusting is not expected to diminish as long as the thumb sucking persists. Even so, tongue thrusting is not thought to be a single cause of openbite (Fig. 10-13).

Orofacial Muscle "Balance"

Some clinicians contend that the teeth are positioned in the middle of a dynamic muscle force field, with the musculature of the tongue on one side, and the opposing and balancing muscles of facial expression and mastication on the other. This muscular relationship around the facial and lingual sides of the teeth is said, by some, to represent an equilibrium. Where a malocclusion is seen, an equilibrium is thought to be absent due to an

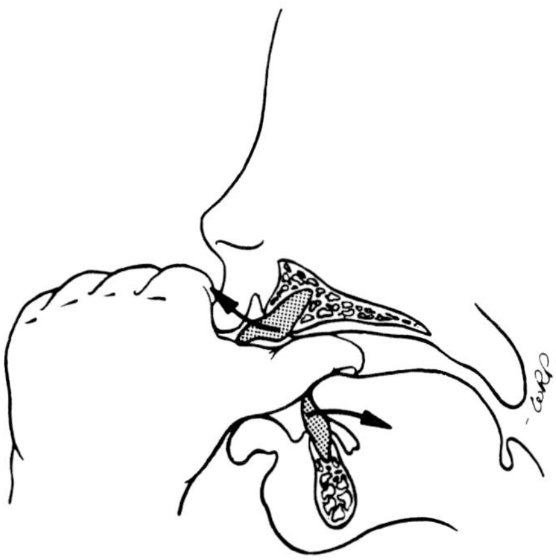

Figure 10-13. Effects on incisor teeth from a persistent thumb-sucking habit. (From Enlow, D.H.: *Handbook of Facial Growth*, 2nd ed., 1982. Courtesy of W.B. Saunders Co., Philadelphia.)

orofacial muscle imbalance.

Current research data in dentistry suggests that the concept of an equilibrium of muscles determining tooth position is *not* an adequate explanation for either dental stability or the development of a malocclusion. A host of scientific experiments have been reported that demonstrate *no* equilibrium to be present in the functional or resting muscle forces on each side of the dental arches. Consequently, terms such as "*orofacial muscle imbalance*" should be avoided in clinical use, since balance or equilibrium cannot be demonstrated in individuals with normal occlusion.

One of the missing links in the development of an equilibrium equation for teeth may be, in part, the role of the periodontal membrane apparatus. The responses of the periodontal membrane, especially in the vertical plane of space, contribute to the basis for tooth position stability. At present, it appears that determinants of tooth position involve forces that exert long-standing (not short-acting) influences on the dentition in the vertical, horizontal, and transverse planes of space (Proffit, 1978).

THE DENTAL ARCHES

The Primary Dentition

The first teeth become well-developed prior to birth, beginning development *in utero* in the fourth month. But specific eruption schedules are difficult to formulate unless a *range* of ages is given for eruption of the various teeth. Otherwise, any times outside the specified range could then be considered abnormal (Table 10-I).

There is a total of twenty "baby" teeth, which should be referred to as primary or deciduous teeth in the clinical reports of speech pathologists, but most dentists prefer the use of *primary*. All are present by the child's second birthday or shortly thereafter and are fully erupted before age three. Shedding begins around age six and is completed by age eleven or twelve as the permanent teeth erupt simultaneously in the mixed-dentition period. One particular feature is that the deciduous teeth are *small*, often resulting in gaps between the teeth as the child grows. These spaces are normal, positive findings in the deciduous dentition.

While the permanent anterior teeth are larger than the primary teeth they may replace, this is not the case in the posterior dentition. The primary or deciduous molars are located over the eruption paths of the permanent (and smaller) premolars. At age six, the first permanent molar erupts distal to the second primary molar. The long line of molar teeth seen at this age (two primary molars plus one

Table 10-I

SCHEDULE FOR ERUPTION AND SHEDDING OF THE PRIMARY TEETH

Tooth		*Eruption*		*Shedding*	
		LOWER	UPPER	LOWER	UPPER
		Age (months)		Age (years)	
Central Incisors		6	7½	6	7½
Lateral Incisors		7	9	7	8
Canines (*Cuspids*)		16	18	9½	11½
First Molars		12	14	10	10½
Second Molars		20	24	11	10½
	Incisors	Range ± 2 months		Range ± 6 months	
	Molars	Range ± 4 months			

SCHEDULE FOR ERUPTION OF THE PERMANENT TEETH

Tooth	LOWER	UPPER
	Age (years)	Age (years)
Central Incisors	6 – 7	7 – 8
Lateral Incisors	7 – 8	8 – 9
Canines	9 – 10	11 – 12
First Premolars (*Bicuspids*)	10 – 12	10 – 11
Second Premolars	11 – 12	10 – 12
First Molars	6 – 7	6 – 7
Second Molars	11 – 13	12 – 13
Third Molars	17 – 21	17 – 21

permanent molar) in each quadrant of the dental arch presents an impressive array of enamel (Fig. 10-14).

The transition from primary to permanent teeth involves a possible space discrepancy in the dental arches. The anterior permanent teeth need more room. Nature handles this situation in a variety of ways: (1) spaces are seen in the primary dentition, especially between anterior teeth; (2) permanent incisors flare facially (labially) upon eruption, thus providing more space mesiodistally; and (3) the mesiodistal widths of the permanent premolars are less than the primary molars they replace.

The critical considerations in dentistry regarding tooth size and the circumference of the dental arch result often in the placement of a space-maintenance appliance when a deciduous tooth is shed. The decision for placement of such an appliance is usually based on the dentist's measurement of x-ray films of unerupted permanent teeth in comparison to arch circumference. An *edentulous space* can, in the absence of a space maintainer, be filled in by mesial drift of teeth posterior to the space. Stainless steel crowns, distal shoes, and other dental appliances are devices used to maintain space either in restoring a tooth (i.e. stainless steel crown) or in maintaining the dimensions of an edentulous area.

The Mixed Dentition

The mixed-dentition stage is characterized by the presence of some primary teeth and some permanent teeth. It typi-

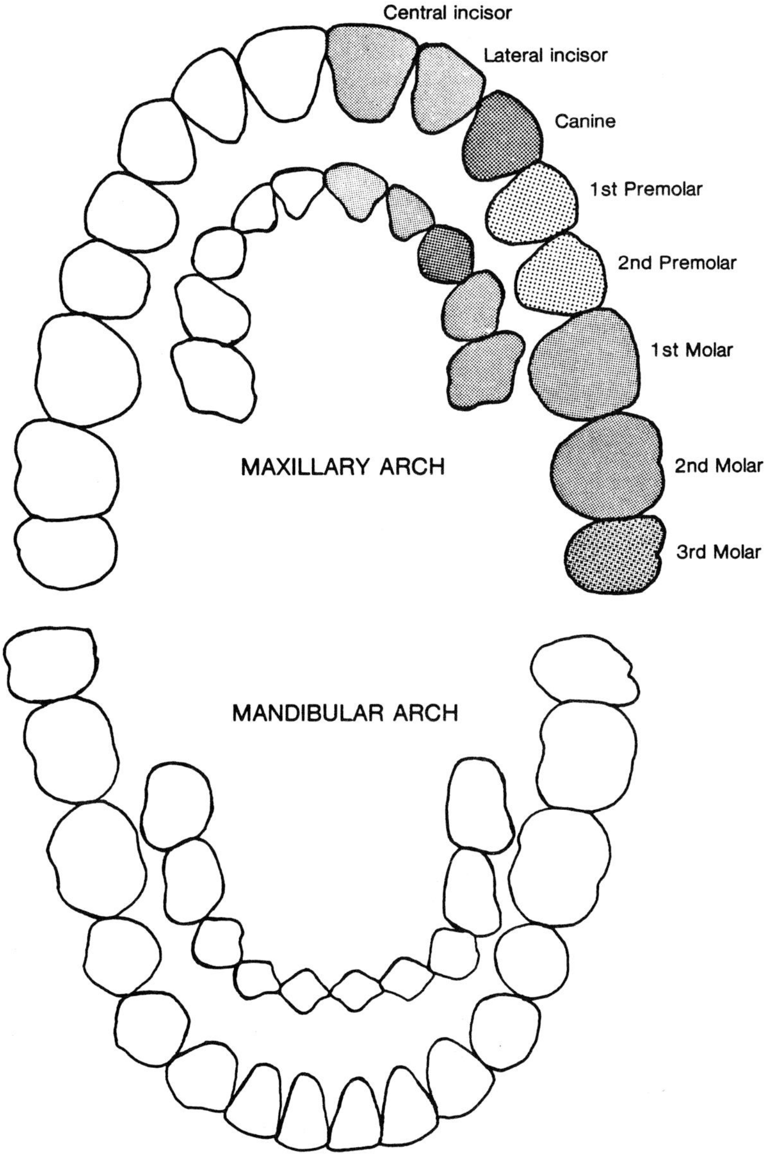

Figure 10-14. Comparison of primary and permanent dental arches.

cally covers six or seven years, beginning with the eruption of the first permanent molar and ending with exfoliation of the second primary molars. During the early part of the mixed dentition, almost every child will *not* have the vertical overlap between upper and lower incisors, since these teeth are in the initial processes of eruption. Accordingly, all children have what appears to be an *openbite* malocclusion at some point, although this is not an appropriate label to associate with the incisor eruption stage. Knowledge of eruption stages aids the clinician in avoiding mislabelling some of nature's normal processes.

It is quite important for the speech clinician to be familiar with the eruption schedule of the permanent teeth in order to understand the possible influence on articulation skills during the period of mixed dentition when several teeth may be missing at one time. For example, the speech pathologist should be able to inspect the anterior aspect of the dental arches and discover during the speech evaluation if the exfoliation and eruption schedule will interfere with articulation therapy. Production of the *th* (voiced and voiceless), *f*, and *v* are quite difficult, if not impossible, for some children to produce without distortion when the front teeth are absent. Since the dates from Table 10-I showing the average time for eruption of the permanent teeth indicate averages of large populations, each child must be studied regarding the timing of eruption of any given tooth, with timing between eruption of the individual teeth considered.

A general perspective, however, is that children should be capable of producing normal articulation, in spite of tooth loss, in the mixed dentition. The variations in a child's ability to do so seems related to speech adaptability. Such differences represent the challenge of speech therapy for this age range. The point is that structural variation does not automatically dictate a given speech status.

There are numerous examples of adults who have lost teeth and exhibit distorted articulation for many sounds, particularly the sibilants. Ill-fitting dentures or partial plates may have the same effect on articulation. Even the acoustic quality of the vowels may be affected in the case of inadequate or missing dental structures. Perhaps the speech adaptability of the adult is not as efficient as the child's accommodation. This is often seen in the speech adaptability difficulties experienced by adults fitted with an orthodontic retainer (Fig. 10-15).

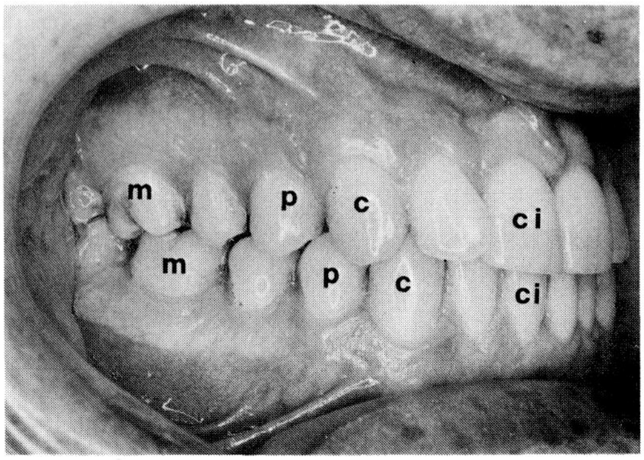

Figure 10-15. Lateral view of an adult dental arch highlighting molar teeth (M), premolars (P), canines (C), and central incisors (Ci).

The Permanent Dentition

The exfoliation of the last primary tooth signals the end of the period of mixed dentition. Females usually attain a full complement of permanent or adult teeth sooner than males. A total of *thirty-two* teeth comprise the complete permanent (or succedaneous) dentition, although few adults have this many functional teeth. The third molars, commonly referred to as *wisdom teeth*, frequently fail to erupt and become impacted due to lack of space. Even in a fully erupted state, their natural history usually includes extraction at some point in time, as they are prone to caries due to the difficulties in cleaning adequately in the distant recess of the oral cavity.

The appearance of the permanent dentition is quite different than the primary dental arches. As pointed out earlier, the anterior permanent teeth are larger than the teeth they replace; consequently, the permanent teeth erupt with the incisal edges flared facially. In contrast, the primary incisors are upright.

The canines and premolars also have positional differences from their primary counterparts. This is characterized by a slight tipping of the crowns mesially, or the roots distally, if seen radiographically. The prominence of the canines is quite apparent in the adult dentition.

The prominence of all of the anterior teeth in the normal adult dentition provides additional lip support. That is, the teeth determine to a large extent the esthetic acceptability of the lips with the other facial parts. This is a useful feature of the adult dentition, especially as the nose and chin continue to grow past the time the adult teeth have erupted. Thus, facial cosmesis is a continuously evolving and slightly changing phenomenon that continues up to full maturation, with the teeth playing an important role in the process.

Orthodontists are confronted with the dilemma of rearranging the dental arches in orthodontic treatment without sacrificing facial esthetics. Sometimes the imprudent extraction of permanent teeth for treatment of a crowded dentition has a deleterious effect on facial proportions. In such cases, lip support is reduced from the orthodontic retraction of the anterior dentition, and the nose or chin areas are made to look more prominent. Orthodontic treatment planning, like any good treatment considerations, should involve a thorough assessment of the structures in question as they relate to surrounding anatomy and functions.

Some pertinent clinical guidelines for evaluating the teeth in relation to facial form will be presented in Chapter 21.

Anomalies of the Teeth

A number of abnormal conditions of the teeth may exist relative to (1) developmental variations, (2) faulty positioning, or (3) faulty usage. Some representative conditions will be described:

Developmental Variations

This category includes anomalies in shape and size, structure, and number.

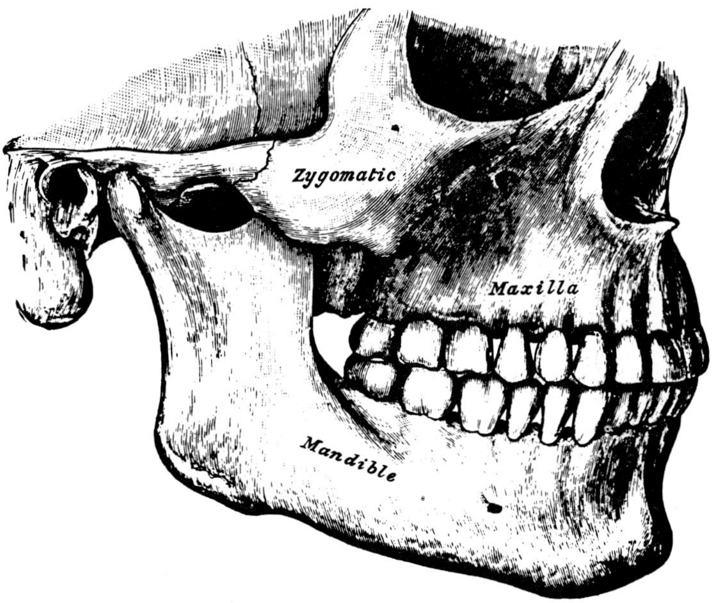

Figure 10-16. Lateral view of the dental arch showing acceptable intercuspation. (From Goss, C.M. (Ed.): *Gray's Anatomy of the Human Body*, 28th ed., 1973. Courtesy of Lea & Febiger, publisher.)

Anomalies in Shape and Size

Microdontia. A condition in which the teeth are small in relation to the dental arch. This may affect all teeth or be limited to a single tooth, such as a peg-shaped lateral incisor (Fig. 10-17).

Macrodontia. A condition of unusually large teeth in relationship to jaw size. The usual result is dental crowding. Macrodontia, or *megadontia*, may be associated with an early fetal disorder or a genetic factor.

Gemination. A developmental disturbance of a single tooth root developing two incompletely formed crowns. The crowns, separated by a median cleft, converge into a single root and canal.

Fusion. The union of two separate tooth germs. The fusion of crowns is called *complete fusion*, whereas the fusion of roots is *partial fusion*. It is difficult to differentiate diagnostically (clinically) between gemination and fusion.

Anomalies in Structure

Amelogenesis Imperfecta. A hereditary disturbance of enamel formation where the enamel is very thin or missing. The enamel appears very chalky or displays a yellow-brown appearance due to the dentin showing through.

Mottled Enamel. A brown staining and pitting of enamel caused by the ingestion of water containing excessive amounts of fluoride during the time that the enamel is developing.

Dentinogenesis Imperfecta. A hereditary disturbance where the dentin is abnormal and the enamel and cementum are normal. The pulp chamber is very small.

Tetracycline Stains. Discoloration of teeth ranging from light yellow to brown associated with demineralization from ingestion of the antibiotic tetracycline in the early formative stages of tooth development.

Anomalies in Number

Supernumerary Teeth. Extra teeth, usually appearing much smaller than the other teeth. They occur most frequently in the maxillary central incisor area and in the premolar region.

Faulty Positioning

Diastema. Abnormally wide spacing, especially between the upper central

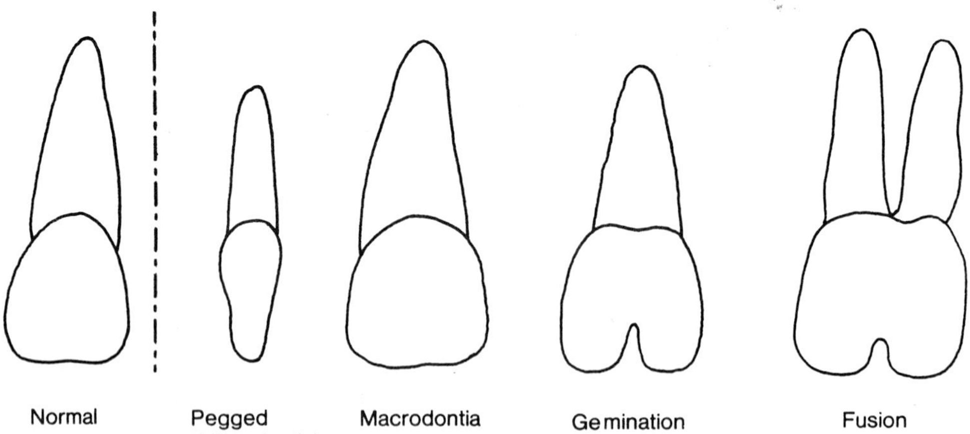

Figure 10-17. Some developmental variations in tooth shape and size for incisors.

incisors. In some instances, the superior labial frenulum may be attached to the alveolar process between the teeth (Fig. 10-18).

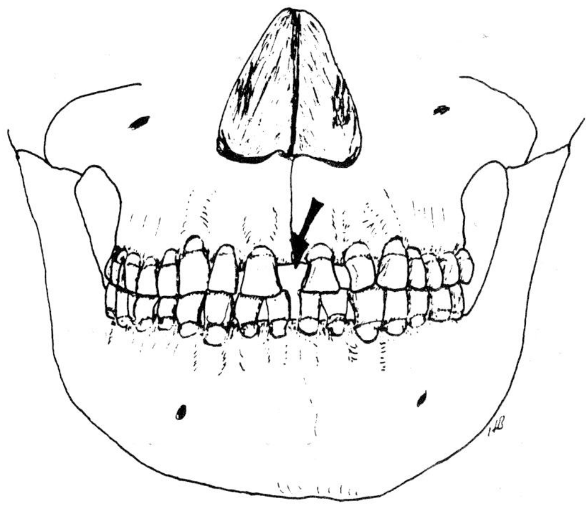

Figure 10-18. Maxillary central diastema in a skull specimen.

Labioversion. The abnormal slanting of a tooth (incisors or canines) toward the lips. Most orthodontists would prefer to describe this as "facially flared" or "blocked out of the arch facially"—if a space problem is causing the tooth to be out of position (ectopic).
Linguaversion. The slanting of a tooth toward the tongue. An orthodontist would prefer "lingually inclined" or "blocked out of the arch lingually."
Supraversion. Overeruption of a tooth. The tooth extends past the normal line of occlusion.
Infraversion. Failure of a tooth to erupt to the line of occlusion. This can occur if the tooth is blocked out or if it is ankylosed (fixated).
Axiversion. A tipped tooth, or the wrong axial inclination. An orthodontist would simply say "tipped."
Torsiversion. A tooth rotated on its long axis. Most orthodontists would simply say "rotated."

Faulty Usage

Attrition. A normal process of wear that can become abnormal by masticatory activity. Attrition can be caused by chewing abrasive foods, bruxism (grinding habit), chewing tobacco, or other factors.
Erosion. Whereas attrition is related to mechanical wear, erosion is associated with chemical wear or dissolution of tooth structure. This is seen where the patient ingests excessive amounts of acidic acids, as from sucking lemons, has excessive vomiting, or ingests medications that chemically erode the teeth.
Abrasion. A mechanical wearing away of the teeth,usually the facial or lingual surfaces, and usually associated with toothbrushing improperly. The use of an abrasive toothpaste or too vigorous brushing with a hard-bristled toothbrush have been implicated. Abrasion is most often seen in the cervical third of the crown, quite often on the facial surface of the upper premolars.

OCCLUSION OF THE TEETH

With the closing of the jaws, contact between the upper and lower teeth occurs. This is termed *occlusion.* To accomplish this closing, a number of factors have to work in harmony for an even contact to occur between all upper and lower teeth. The opening and closing pattern of the mandible in relationship to the *glenoid fos-*

sae of the temporal bone is an important component of occlusion, as is the angle of mandibular excursions in and out of the fossae. The angle of the maxillary occlusal plane in relationship to other planes of the head is also significant in some patients. Altogether, occlusion is not a static relationship between maxillary and mandibular teeth but a dynamic state that depends upon adaptive functioning in daily life (Fig. 10-19).

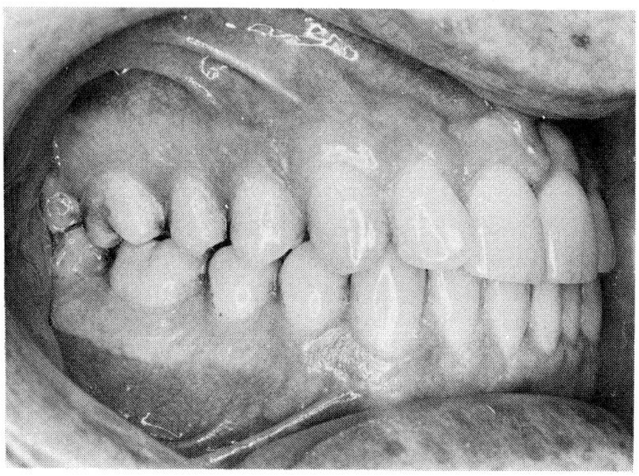

Figure 10-19. Lateral view of occlusal relationships of the teeth showing acceptable interdigitation and overlap of the dental arches.

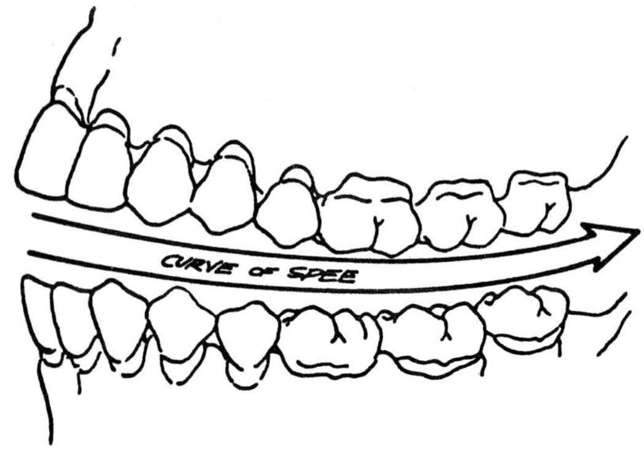

Figure 10-20. Lateral view of the gradual slope of the dental arches referred to as the Curve of Spee. (Reproduced with permission from Fuller, J.L., and Denehy, G.E.: *Concise Dental Anatomy and Morphology*. Copyright © 1977 by Year Book Medical Publishers, Inc., Chicago.)

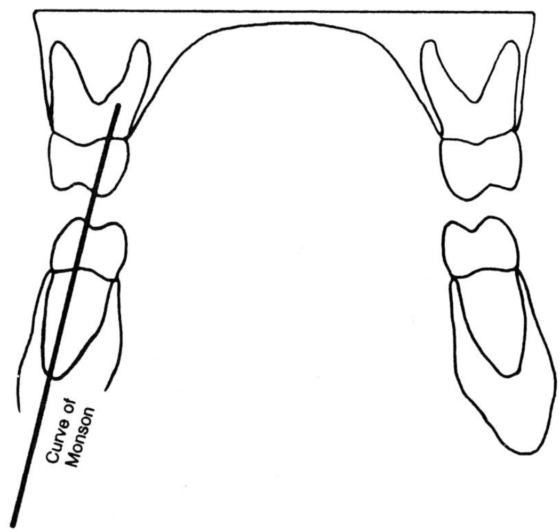

Figure 10-21. Coronal view of the tipping of the crowns of posterior teeth referred to as the curve of Monson. Lower posterior crowns normally tip lingually, while upper posterior crowns show a slight buccal inclination.

The arrangement and alignment of the upper teeth can be thought of as a target for articulation of the lower teeth during closing of the jaws. In the normal situation, the upper dental arch is longer and wider than the lower dental arch. Consequently, achieving a contact of all upper and lower teeth finds the lower arch positioned about one-half tooth ahead of the upper arch as seen at the molars. This basic requirement has led to a classification of malocclusions based on molar relationships that will be reviewed presently.

The biting surfaces of the teeth do not, in either arch, usually conform to a flat plane. Most often the maxillary occlusal plane, as seen in profile, inclines slightly downward from the canine to first molar and then rises again in the molar region. The resulting curve, which is slightly convex for the maxillary teeth and concave in the lower arch, is called the *curve of Spee*. In the coronal plane, the posterior teeth are arranged so that the upper molars are tipped somewhat buccally and the lowers somewhat lingually. This is known as the *curve of Monson* (Figs. 10-20 & 10-21).

Examination of the articulated dental arches reveals that in occlusion, the upper incisors cover the incisal one-third of the lower incisors. This normal relationship is referred to as an *overbite* (vertical overlap). The upper incisors are also seen to project farther facially than the lower incisors. The contact of the lower incisal edges is on the lingual side of the upper incisors at a point just below the convexity of the cingulum (see Fig. 10-7A). The 2-3 millimeter forward projection of the upper incisors over the lowers is referred to as an *overjet* (horizontal overlap).

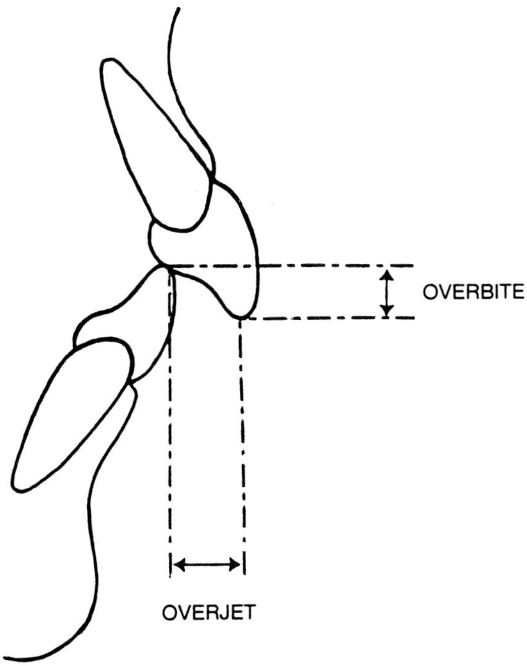

Figure 10-22. Excessive horizontal overlap (overjet) and vertical overlap (overbite) for incisor teeth.

Distally along the facial surface of the dental arch, the overbite and overjet relationship decreases but is normally maintained throughout the dental arch. That is, there is normally an overjet and overbite involving all upper and lower teeth, but the dimensions of overlap are greatest at the front of the dental arch. Since overbite (vertical overlap) and overjet (horizontal overlap) are normal findings, any variation in these dimensions needs to be qualified by an adjective, such as *excessive overbite*. Where the upper incisors completely obscure a view of the lower incisors with closing of the jaws, the term *closed bite* can be used as a synonym for excessive overbite. An *openbite*, by contrast, is a term applied when there is absence of occlusion in one area when occlusion is evident in other parts of the dental arch. Like excessive overbite or overjet, an openbite can occur anywhere along the dental arches. Like most other dental variations, an openbite may be associated with a variety of causative factors, as was pointed out earlier in this chapter.

Crossbite: Clinical Considerations

Another tooth position variation that speech clinicians should be knowledgeable about is the condition of *crossbite*. Crossbite occurs when a maxillary tooth loses its normal horizontal overlap (overjet) relationship with its mandibular counterpart. Hence, either a maxillary tooth is positioned lingual to a mandibular tooth or a mandibular tooth is facial to a maxillary tooth (Fig. 10-23).

Crossbite can occur at a single tooth location in the dental arch or may involve an entire dental quadrant. When crossbite involves multiple teeth in a quadrant of the dental arch, the condition is often a *skeletal crossbite*. Such a patient would eventually be referred for dental consultation and treatment. One procedure in dentistry for correcting a crossbite is expansion of the maxillary arch with an appliance. In individuals under age sixteen, this is a relatively simple procedure, since the midpalatal suture yields to the expansion appliance, separates, and then develops new bone at the midline to fill in the expansion area. After age sixteen, the midpalatal suture is rigid and, when indicated, has to be surgically opened to accomplish expansion of the maxillary shelves. Accordingly, the speech clinician can play an important role in screening for those children who may experience problems in dental alignment and proper contact of opposing teeth. Crossbite conditions may occasionally pose such problems.

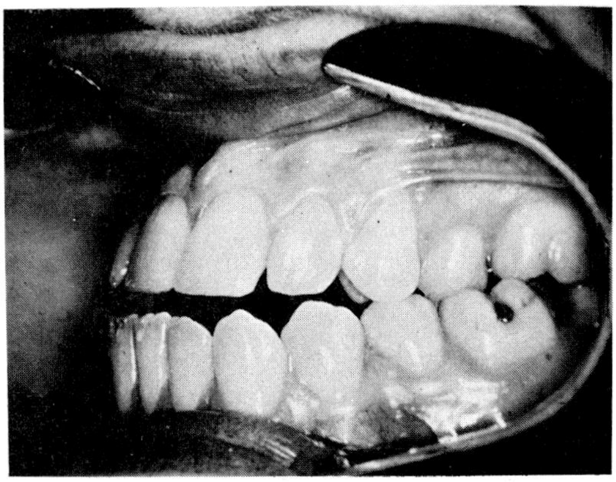

Figure 10-23. An individual with a posterior crossbite, anterior openbite, and a lack of anterior overjet (negative overjet). Notice that the upper canine is absent.

There is a growing divergence of opinion in dentistry as to the optimal time to correct a crossbite. Some dentists and orthodontists feel that crossbites should be corrected as soon as they are observed, while others are reluctant to treat a crossbite unless it creates some functional problem. Recent evidence indicates that children and adults tolerate crossbites surprisingly well and that the persistence of a crossbite into the adult dentition will not automatically lead to temporomandibular joint pain or early loss of teeth. Those practitioners who do not advocate early treatment of crossbites would treat a child whose crossbite creates a functional shift. This denotes an obligatory shifting of the mandible to one side or the other during closing of the jaws due to the interference of an upper and lower tooth or teeth. Most children with a crossbite do not possess such a functional problem, however.

As stated above, the concept of occlusion is based primarily on contactual relationships of the first molar teeth. In the adult arch, these teeth are the first permanent molars. Designation of occlusal class or biting relationships of upper and lower teeth is determined by examining the facial (buccal) surfaces of the first molar teeth in the adult dentition. In a child, the second primary molar is usually used to evaluate occlusal status. In the event that the primary molar or permanent molar is absent in the child or adult, the canines can be used as substitutes for determining occlusal relationships.

Normal Occlusion

Normal occlusion is characterized by a relationship in which the mesiobuccal cusp of the first permanent maxillary molar occludes over the buccal groove on the first permanent mandibular molar (Fig. 10-24). Additionally, there is no rotation or crowding of other teeth in the dental arches. Such ideal occlusion is termed *Class I normal*. Most individuals fall into a situation of having normal molar relationships bilaterally, but they have some variation in the anterior portion of the dental arch. Hence, most people have *Class I malocclusions*.

The normal occlusal position described above represents maximum efficiency in mastication. Such a structural state enhances a pleasing appearance of

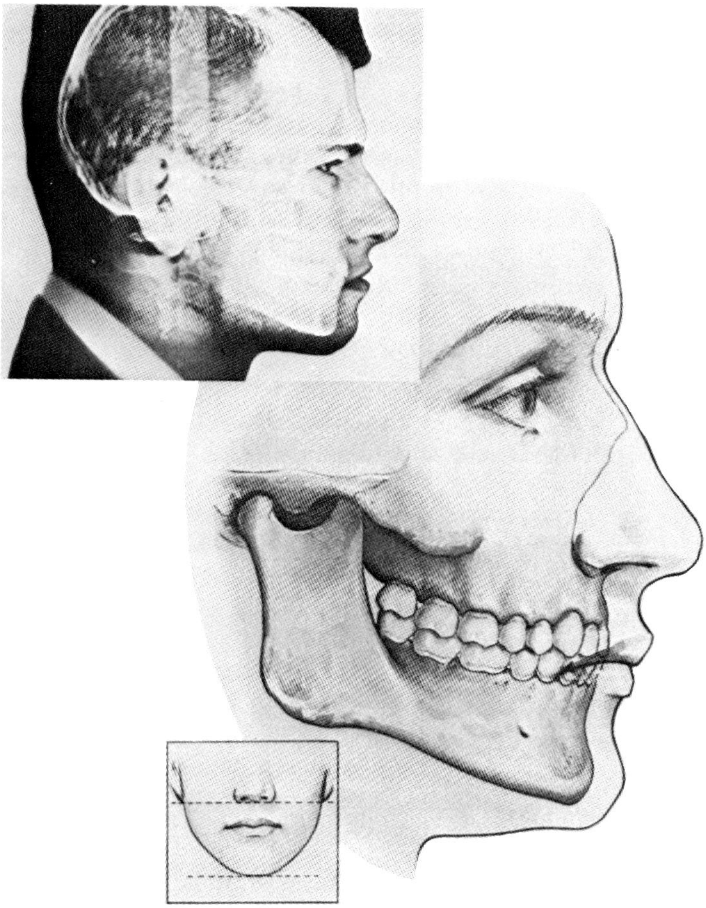

Figure 10-24. Normal occlusion. The skeletal and dental profile is shown along with soft tissue and bony relationships. The lower inset shows lower face height, with the lips positioned at the upper third of the lower face. (From Massler, M., and Schour, I.: *Atlas of the Mouth*, 1958. Copyright by the American Dental Association. Reprinted by permission.)

the entire face. This is the position of the jaw assumed in a jaw-jerk reflex (the response of the open jaw to being tapped on top of the prominence of the chin during a neurological examination). Yet the mandible may move forward in a protruding or backward in a retruding movement of the jaw. Thus, the normal occlusal position of the teeth indicates a median or rest position.

However, the teeth are not in actual contact with each other very much. In the period of one full day, the molars touch less than *six* minutes. In the case of a mouth-open posture, they are in contact much less.

Abnormal Occlusion

Any abnormal positioning of the teeth, arising either from faulty tooth positions or from abnormality in growth of the jaw, results in *malocclusion*. Stated another way, abnormal occlusion is any problem with the contacting teeth of the jaws.

Classification of Malocclusion

Edward H. Angle, an orthodontist, provided the first classification of dental malocclusion in 1889. Angle's classification has also been used to evaluate the adequacy of the facial skeleton. Hence, an individual may have a deficient dental profile, or skeletal profile, or both. Often, the designation of skeletal profile type is the same as the dental profile type. The three classes he proposed for dental malocclusion are as follows:

Class I. Normal position of the molar teeth of the individual arches, the jaw relationship usually being normal. The condition is usually characterized by a crowding and rotation of some teeth, since this class of malocclusion usually refers to a problem of the teeth and the alveolar processes without involvement of the jaws.

Class II. A distal position of the first mandibular molar in relation to the first maxillary molar. Often, the chin gives an underdeveloped cosmetic effect to the mouth area. However, the mandible may actually be normal. The situation is further

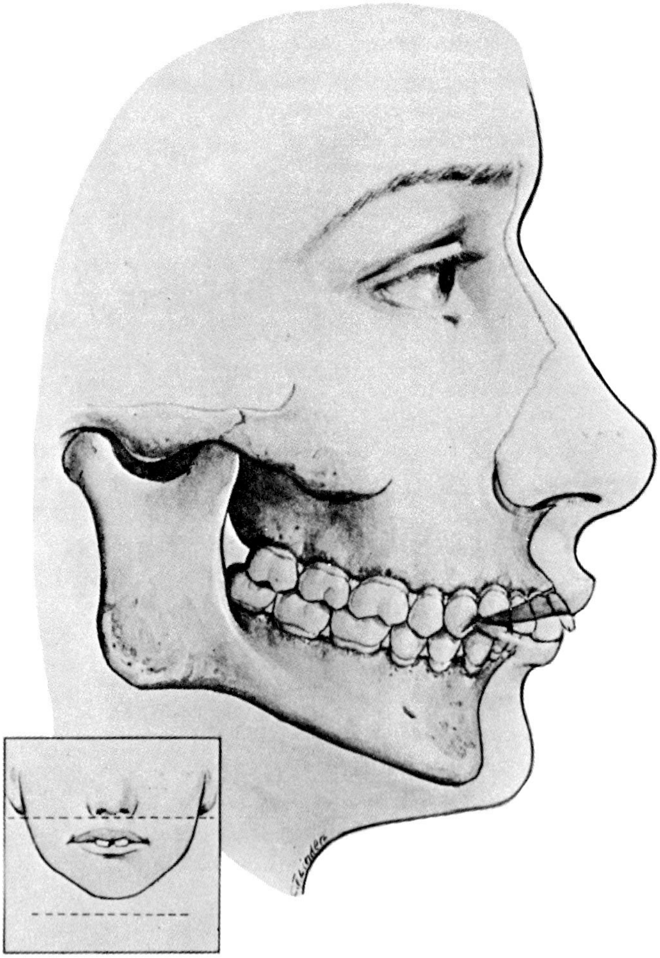

Figure 10-25. Abnormal skeletal and dental occlusion characterized by retrusion of the mandible and protrusion of the maxilla. Notice the decreased lower face height in the inset view. An Angle Class II malocclusion. (From Massler, M., and Schour, I.: *Atlas of the Mouth*, 1958. Copyright by the American Dental Association. Reprinted by permission.)

complicated if the upper incisors are protruded. Characteristically, the upper lip appears to be inadequate to cover the upper teeth (Fig. 10-25).

Class III. A mesial position of the first mandibular molar in relation to the first maxillary molar. The aggressive appearance of the mandible is usually prominent (skeleton Class III).

A Class I occlusion has also been described as a *neutroclusion*, a Class II malocclusion as a *distoclusion*, and a Class III as a *mesioclusion*. While most dentists are aware of these terms, they are not part of current usage and should probably be avoided (Fig. 10-26).

An appreciation of the differences between a *skeletal* and *dental* malocclusion can be gained by examining diagrammatic representations of the facial parts. The skeletal malocclusions illustrate that the jaws can deviate or rotate in a variety of directions, and that the attached dental bases may adapt to these changes or follow along rather passively. While some

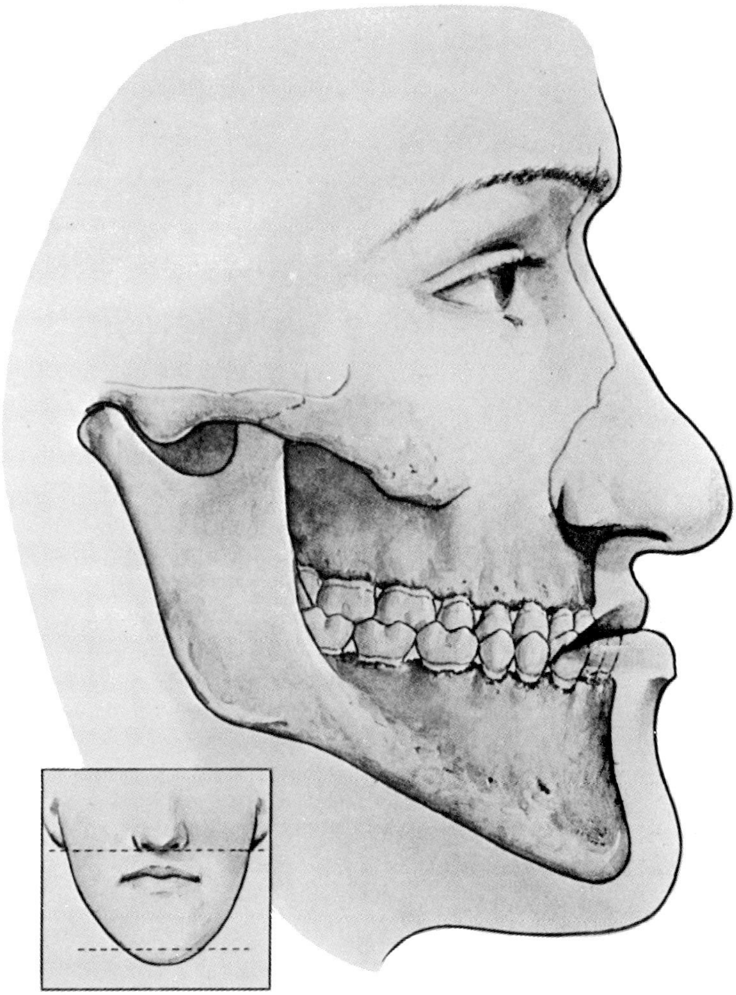

Figure 10-26. Abnormal skeletal and dental occlusion characterized by mandibular prognathism. Notice the increased lower face height and crossbite condition in this individual with an Angle Class III malocclusion. (From Massler, M., and Schour, I.: *Atlas of the Mouth*, 1958. Copyright by the American Dental Association. Reprinted by permission.)

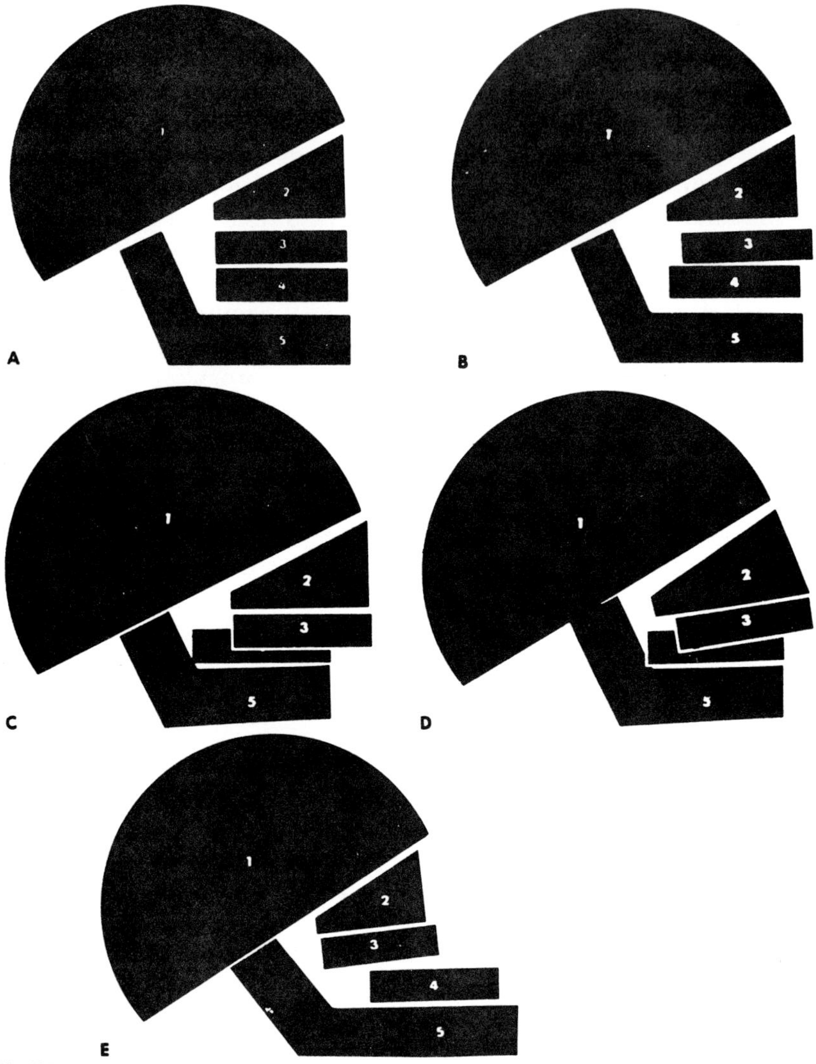

Figure 10-27. Diagrammatic representations of dentofacial relationships. (1) Upper face and cranium, (2) nasomaxillary complex, (3) maxillary dentition, (4) mandibular dentition, (5) mandible. (A) Normal (orthognathic) skeletal and dental relationships. (B) Malocclusion due solely to maxillary dental protrusion. (C) Mandibular skeletal retrusion plus deep bite with mandibular overclosure. (D) Maxillary skeletal protrusion and skeletal openbite tendency due to upwardly inclined palatal plane. This type of skeletal sagittal-vertical interaction occurs frequently in Class II openbite problems. (E) Combination of maxillary skeletal retrusion and mandibular skeletal protrusion. The maxillary teeth are shifted forward on their base, whereas the mandibular teeth are shifted backward. Thus, there is a partial dental compensation for the skeletal malocclusion. This dentofacial pattern is typical of some adult repaired cleft lip and palate patients who may require facial osteotomies to correct the skeletal dysplasias of the jaws. (From Bell, W.H., Proffit, W.R., and White, R.P.: *Surgical Correction of Dentofacial Deformities*, Vol. I, 1980, Courtesy of W.B. Saunders Co., Philadelphia.)

skeletal malocclusions can be identified by facial and intraoral inspection, radiographic assessment is the procedure of choice in identifying the characteristics of the dentition and skeletal framework with measurable diagnostic accuracy (Fig. 10-27).

It should be made clear that while there is a high correlation between dental class and skeletal class, they are not always synonymous. Angle's classification system deals with teeth relationships in the anteroposterior plane only. More current methods of analysis are available, however. Clinically, the investigation should begin with counting the teeth to eliminate many potential errors. Occlusion should also be looked at on both sides of the arch; it is not unusual, for example, to be Class I on the right and Class II on the left. This would be described as a Class II subdivision left (indicating automatically that the right side is Class I).

The abnormal occlusal positions (namely, those of Classes II and III) may occasionally interfere with normal tongue movements required for articulation. Crowding the tongue too far forward or backward may hamper efficient tongue mobility. Masticatory efficiency may also be poor. To some, the cosmetic impairment is of chief concern.

Almost all types of malocclusion and associated orofacial deformities can be improved by either orthodontic treatment alone or combined with surgical procedures. It is worth noting that, while some orofacial variations are not esthetically pleasing by cosmetic concepts, the condition may not actually be a dental anomaly, since the jaws may be in occlusal harmony while the whole face exhibits prognathism or some other unfavorable appearance.

BIBLIOGRAPHY

DuBrul, E.: *Sicher's Oral Anatomy*, 7th ed. St. Louis, C.V. Mosby, 1980.

Fuller, J.L., and Denehy, G.E.: *Concise Dental Anatomy and Morphology*. Chicago, Year Book Medical Pub, 1977.

Mason, R., and Proffit, W.: The tongue thrust controversy: background and recommendations. *J Speech Hear Disord, 39*:115-132, 1974.

Mason, R.: *Tongue Thrust. Oral Motor Behavior: Impact on Oral Conditions and Dental Treatment*. NIH Publication No. 79-1845, 32-48, 1979.

Moyers, R.: *Handbook of Orthodontics*, 3rd ed. Chicago, Year Book Medical Pub, 1972.

Ness, A.: Movements and forces in tooth eruption. *Advances in Oral Biology, 1*:33-75, 1964.

Proffit, W.: Equilibrium theory revisited: factors influencing position of the teeth. *Angle Orthod, 48*:175-186, 1978.

Rankow, R.: *An Atlas of Surgery of the Face, Mouth and Neck*. Philadelphia, Saunders, 1968.

Scott, J., and Dixon, A.: *Anatomy for Students of Dentistry*, 4th ed. London, Churchill, 1978.

Scott, H., and Symons, N.: *Introduction to Dental Anatomy*. Baltimore, Williams and Wilkins, 1971.

Sturdevant, C., Barton, R., and Brauer, J.: *The Art and Science of Operative Dentistry*. New York, McGraw, 1968.

Chapter 11

THE ORAL, NASAL, AND PHARYNGEAL CAVITIES

THE ORAL CAVITIES

THE role of the oral cavity in speech resonance is important in forming a number of sounds, particularly the vowels. By utilizing the articulators to change the size and shape of the oral cavity, certain harmonics in the laryngeal tone are augmented. The tone quality produced by these harmonic variations is described as a vowel sound. The principal articulator for the vowels is the tongue. Thus, the oral cavity plays a critical role in the vocal tract in regulating various interactions of resonance and articulation. Furthermore, the mouth is biologically important in digestive functions; it may also be involved in breathing, particularly during speaking (Fig. 11-1).

The oral cavity proper is bounded above by the hard palate, which separates it from the nasal cavity. The raised dental ridges of the jaws, the alveolar processes, and teeth serve as the peripheral boundaries. The floor is comprised of muscular tissue. A horseshoe-shaped strip of the floor surrounding the root of the tongue may be identified by visual or finger inspection. This space is known as the *sublingual groove*. At the rear of the mouth, the oral cavity presents a narrow opening, the isthmus of fauces, into the oropharyngeal space. The opening is surrounded by the vertical part of the soft palate above, the palatal or faucial pillars laterally, and the base of the tongue beneath.

The Oral Vestibule

From one point of view, the oral cavity exhibits a major cavity inside the dental arches (the mouth) and a lesser cavity outside of the dental arches, bounded by the cheeks and the lips, the *oral vestibule* (Fig. 11-2). The internal limits of the vestibule are the gums and teeth, while the lateral portion of the vestibule is the buccal cavity, since it is bounded by the cheeks or *buccae*. With the principal cheek muscle, the buccinator, the tongue is able to keep food between the teeth during chewing (Fig. 11-3).

The parotid salivary gland, the largest of the salivary glands, sends its secretions into the vestibule via a duct. Most of the parotid gland is located on the outside of the ramus, or vertical portion of the mandible. Its duct, sometimes referred to as

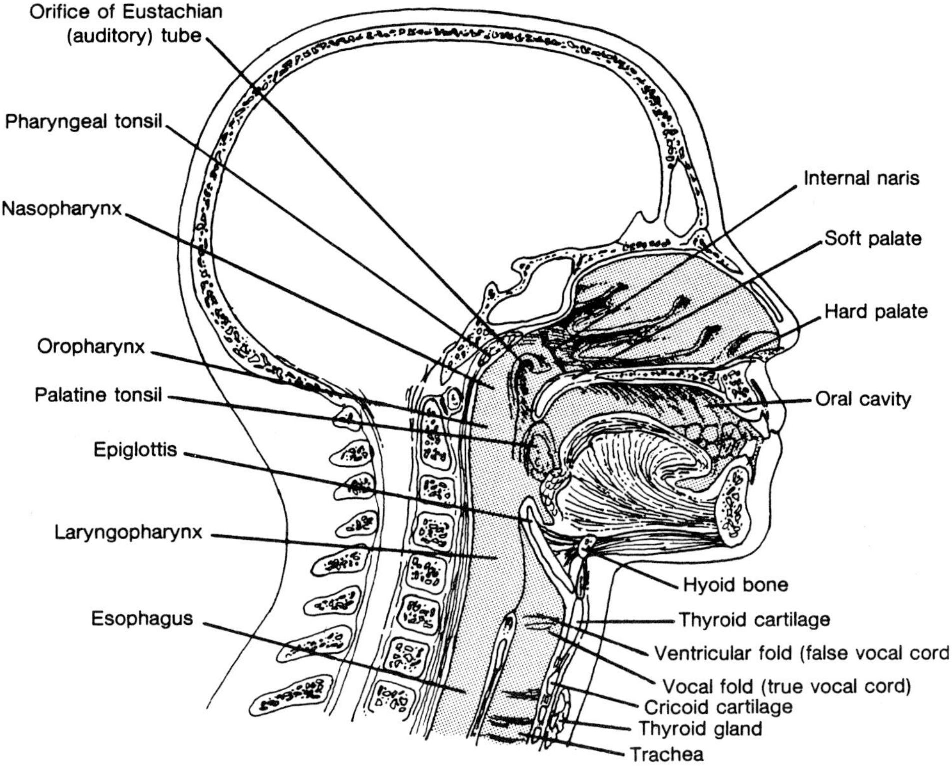

Figure 11-1. Sagittal view of the head showing the oral, nasal, and pharyngeal cavities. (From Tortora, G.J.: *Principles of Human Anatomy*, 1977. Used by permission of Harper and Row Publishers, Inc., New York.)

Stenson's duct, pierces the cheek musculature and sets its orifice in a variably developed fold of mucosa, the parotid papilla, opposite the second upper molar tooth.

Although the major salivary glands (there are three pairs emptying into the oral cavity) serve the digestive tract as accessory organs of digestion, they also provide lubrication for swallowing. Of course, the articulators must also be provided with adequate lubrication for efficient mobility. Most people have experienced the handicap of attempting to speak and swallow with a dry or "cotton" mouth.

The gum or *gingiva* is the mucous membrane covering the necks of the teeth, crowns of unerupted teeth, and a portion of underlying alveolar bone. This membrane is reflected from the internal lining of the cheeks and lips onto the alveolar processes.

Medially, a thin membrane connects the mucosa of the vestibule to the inside of the lip. This median fold of mucous membrane is termed the *labial frenulum* or *frenum*. The upper frenulum is stronger. Sometimes a space exists between the two upper central incisors (termed a *median diastema*), and the frenulum will be found extensively developed and attached down onto the gingiva between the separated teeth (see *lingual* frenum, Fig. 11-4).

Other sickle-shaped folds of varying height may be found in the buccal portion of the vestibule. These folds may some-

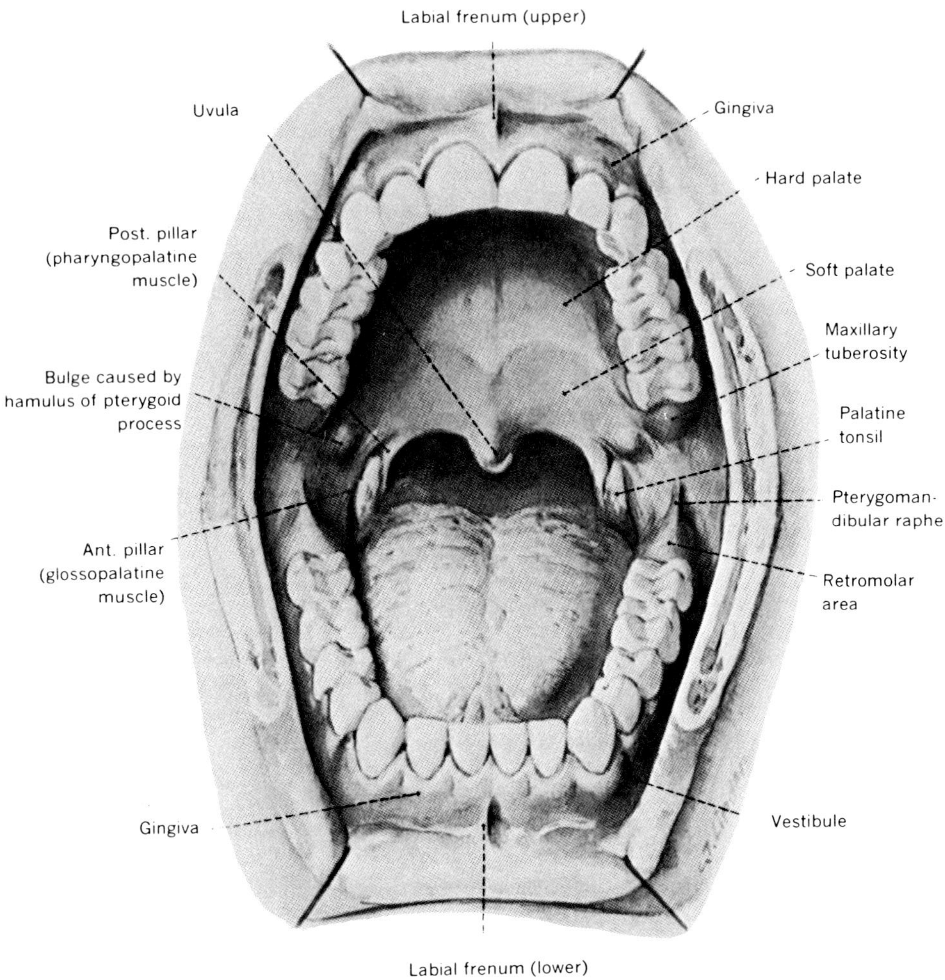

Figure 11-2. The oral cavity. (From Massler, M., and Schour, I.: *Atlas of the Mouth*, 2nd ed., 1958. Copyright by the American Dental Association. Reprinted by permission.)

times be more numerous in persons with a defective soft palate, such as an overt cleft or a submucous cleft. For this reason, extra connective tissue in the lateral portion of the vestibule signals a need for careful inspection of the palates of those patients exhibiting speech articulatory-resonance disturbances (Fig. 11-5).

The Hard Palate

A total of four bones contributes to the makeup of the hard palate. These consist of the palatine processes of the maxillae anteriorly and the horizontal portions of the palatine bones posteriorly (see Fig. 9-10). The hard palate is covered by a tightly adhered periosteal membrane that contains mucous glands. This mucoperiosteum is especially prominent on the slope of the anterior maxillary alveolar ridge, where a series of variable ridges called *rugae* are located. The roughened

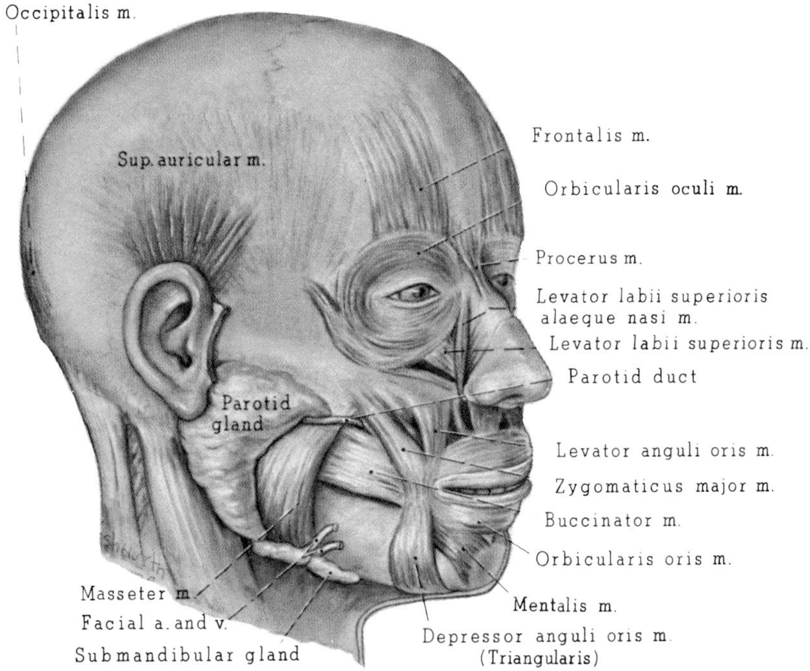

Figure 11-3. Muscles of the face, deep layer. (From Jacob, S.W., and Francone, C.A.: *Elements of Anatomy and Physiology*, 1976. Courtesy of W.B. Saunders Co., Philadelphia.)

surface of the palatal rugae may aid in the production of some lingua-alveolar sounds.

On the anterior rim of the palate on the alveolar arch can be found a small oval-shaped prominence, the palatine or *incisive papilla*, the forward point of which is close to the central incisors. The palatal end of the incisive papilla covers the oral opening of the nasopalatine canals and is known as the *incisive foramen*. A rather distinct ridge takes its origin from the incisive papilla, extending in a sagittal line posteriorly over the entire length of the hard palate. This narrow, low ridge is the *palatine raphe*. Midline coloration of the hard palate is a primary guideline in ascertaining if defects are present. Normal midline colors are pink and white. A blue tint on the midline may signal an abnormal blood supply network due to submucous cleft palate deformities.

Failure of the palatine processes of the maxillary bones to unite at the midline during embryological development may create a cleft of the hard and soft palates (see Chap. 3). This condition creates a direct communication between the oral and nasal cavities, severely handicapping, in most cases, normal speech resonance and normal speech sound articulation. Surgical repair and/or fitting the individual with an appliance to seal the palatal defect are alternatives on the road to creating a functional oral cavity for speech purposes.

Some children exhibit an abnormally high palatal vault, with associated difficulty in producing certain speech sounds. To produce the *sh* sound accurately, for example, compensatory movements may be necessary. Characteristically, individ-

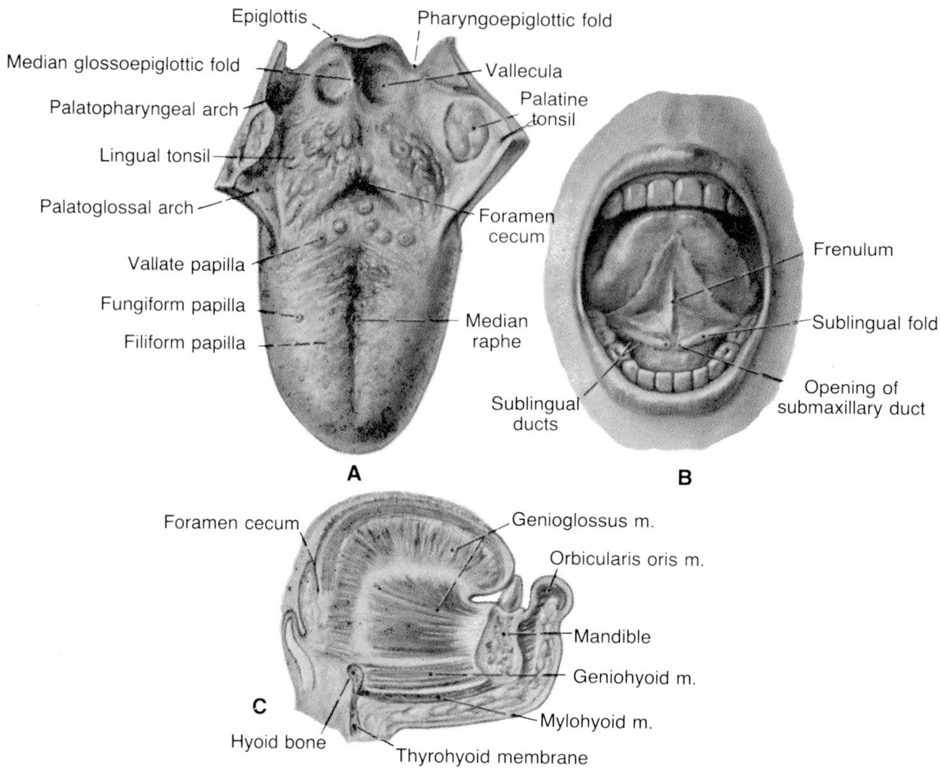

Figure 11-4. (A) Dorsal view of the tongue. (B) Anterior view of the oral cavity with tongue raised. (C) Midsagittal section through the tongue. (From Jacob, S.W., and Francone, C.A.: *Structure and Function in Man*, 4th ed., 1978. Courtesy of W.B. Saunders Co., Philadelphia.)

uals with a wide maxillary dental arch tend to have a low or flat palatal vault. Those with a narrow maxillary arch, conversely, exhibit a high palatal contour or vault.

About one in five to seven persons may exhibit a protuberance or rounded swelling, termed a *torus palatinus*, on the hard palate. It is usually at the junction of the palatine and maxillary bones of the palate, but it should not affect normal articulation in any way.

The Soft Palate

Guarding the superior aspect of the posterior entrance into the oral cavity is the soft palate. The posterior margin of the soft palate or velum is free and marked by a medially located central extension, the *uvula*. Covered with a thick fold of mucous membrane, the soft palate contains an intricately arranged musculature that extends into the nasopharyngeal walls and permits the palate to be raised and stretched in a sphincteric action to effect palatopharyngeal or velopharyngeal closure. The palate is also capable of descending to meet the back of the tongue, sealing the rear entrance into the oral cavity (see Figs. 11-4C & 11-1).

The coloring of the soft palate is

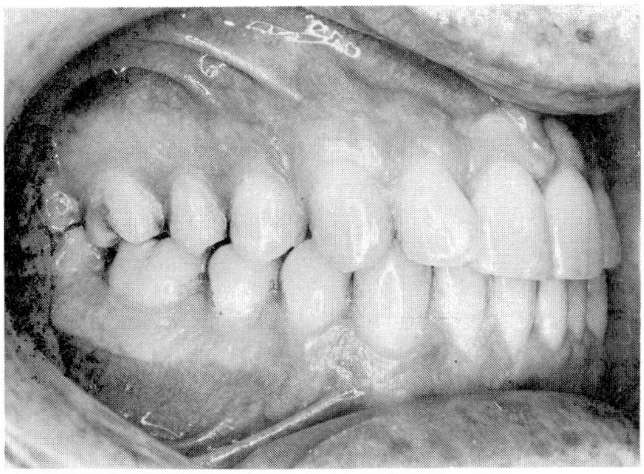

Figure 11-5. The dentition and gingiva of the oral cavity showing lateral gingival buccal frenulae (or folds) in the oral vestibule.

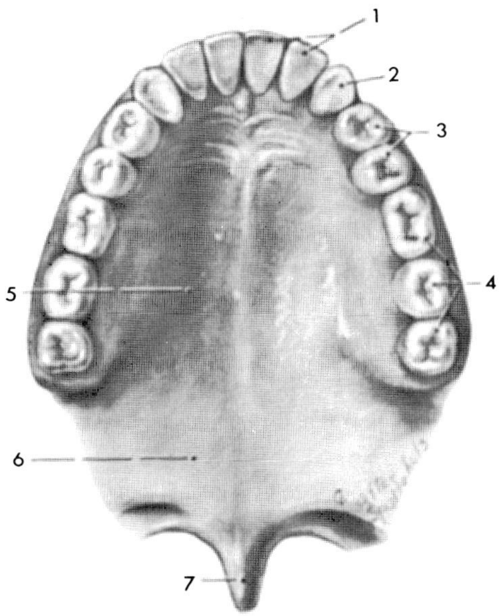

Figure 11-6. Roof of the mouth with adult teeth: (1) incisors, (2) canine, (3) premolars, (4) molars, (5) hard palate, (6) soft palate, (7) uvula. (From Jacob, S.W., and Francone, C.A.: *Laboratory Manual of Structure and Function in Man*, 5th ed., 1982. Courtesy of W.B. Saunders Co., Philadelphia.)

characteristically different from the hard palate and is related to differences in texture of the mucosa and its underlying bed. On the soft palate, the mucous membrane is far thinner than that of the hard palate. This thinner and more loosely textured mucosa lends the character of translucency. With a rich vascularization and massive, densely packed glands (minor salivary glands) under the mucosa, the soft palate appears darker red with a yellowish tint. By contrast, the hard palate appears pale pink or white and may also exhibit a general bluish-gray hue because of its mucoperiosteal (mucous lining over bone covering) bed.

Endowed with its own internal and external sets of muscles, the soft palate is an effective organ of articulation. To close the nasopharyngeal cavity from the oropharyngeal cavity, the soft palate musculature, working in concert with the pharyngeal musculature at the palatal level, seals the space through a sphincteric action. The primary muscle of the palatal group working in this closure activity is the *levator veli palatini*. The role of the other palatal muscles, particularly the *tensor veli palatini*, is not established as being functionally significant in palatopharyngeal closure for speech. Damage to the nerve supplying the tensor muscle does not appear to result in a speech deficit. Nonetheless, the tensor palatini can tighten the soft palate somewhat, which may be of limited value in certain instances (see Fig. 11-17).

A pair of tiny salivary orifices on the oral surface may sometimes be visible just behind the junction of the hard and soft palates. These are the ducts of the minor salivary glands of the palate, the *fovea palatinae*. They look like two pinholes in the palate on either side of the midline raphe.

The Fauces

The opening between the oropharyngeal cavity and the oral cavity is called the *fauces* (also isthmus of fauces or faucial isthmus). This aperture is bordered by the soft palate superiorly, the tongue infe-

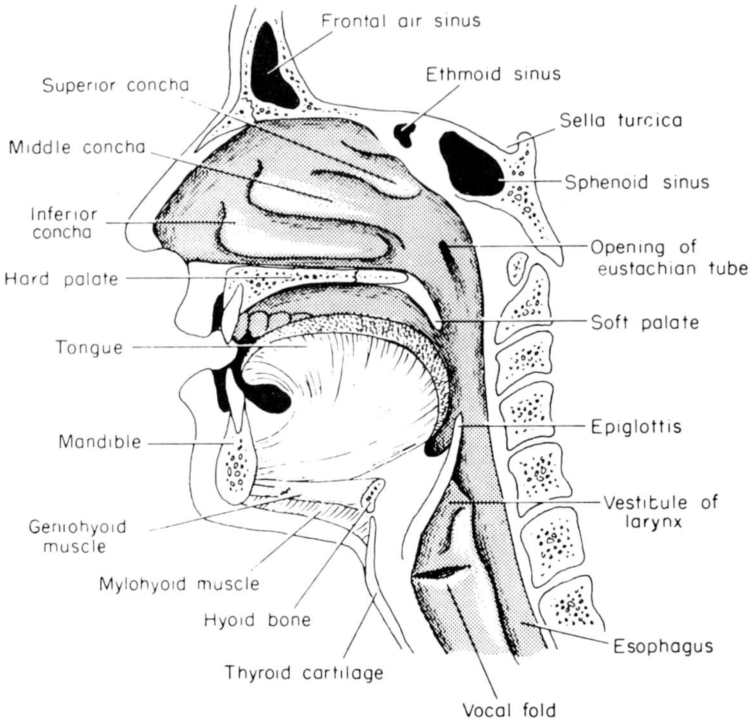

Figure 11-7. Midsaggital view of the head showing the palates in relation to surrounding structures. The septum has been removed showing the right lateral wall of the nasal cavity. (From Brooks, Stewart M.: *Basic Science and the Human Body*, 1975. Courtesy of The C.V. Mosby Co, St. Louis; modified from Francis, C.C.: *Introduction to Human Anatomy*, ed. 6, St. Louis, 1973, The C.V. Mosby Co.)

riorly, and the palatoglossal arch laterally (see also Fig. 11-2).

Arching down from the soft palate on each side are two folds. The anterior fold or arch is formed by the palatoglossus muscle, and the mucosa folded around it is known as the *palatoglossus fold*. This fold extends from the anterolateral portion of the soft palate to the lateral dorsum of the tongue. This communication with the tongue should be remembered in a clinical examination of soft palate function. If the tongue is protruded during velar examination, the palatoglossus muscle is pulled forward and the range of soft palate movement may be restricted. Hence, the clinician should remember to examine soft palate function with the tongue tip behind the lower incisors. Posterior to the palatoglossal fold is the posterior fold, the *palatopharyngeal fold*, which covers the palatopharyngeus muscle. At the base of each arch, the *glossoepiglottic fold* continues the posteroinferior margin (fauces) of the oral cavity, coursing medially to blend with the mucosa of the epiglottis. Often, the two arches are referred to, respectively, as the *posterior* and *anterior faucial pillars*, particularly from a clinical standpoint. Tongue movement and the sphincteric action of the pharynx influence the posterior opening of the oral cavity (Fig. 11-8).

Lying in the bed found between the palatopharyngeal arch and the palatoglossal arch is the *palatine* or *faucial tonsil*. The tonsils are accumulations of lymphoid tis-

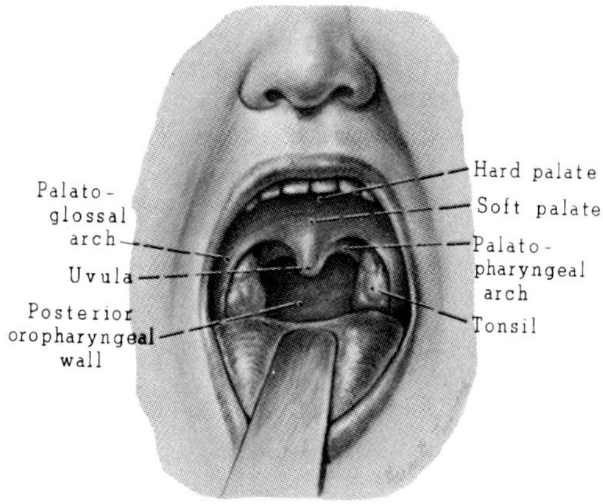

Figure 11-8. Clinical view of the oral cavity highlighting the faucial isthmus. (From Jacob, S.W., and Francone, C.A.: *Elements of Anatomy and Physiology*, 1976. Courtesy of W.B. Saunders Co., Philadelphia.)

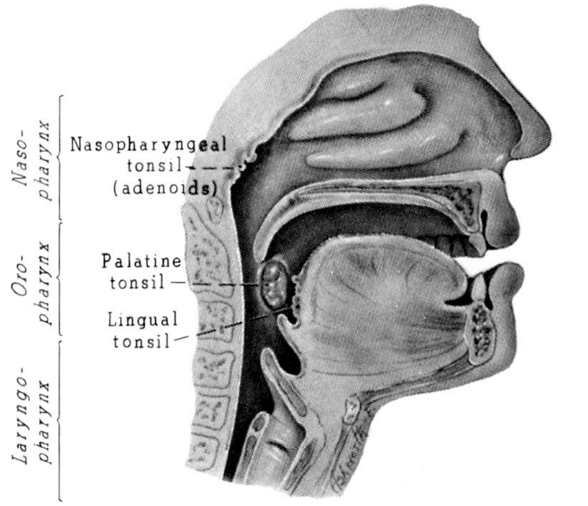

Figure 11-9. The nasopharyngeal tonsils extend from the roof of the nasal pharynx to the free edge of the soft palate; the palatine tonsils are attached to the side walls of the back of the mouth between the anterior and posterior pillars; the lingual tonsils are located on the dorsum of the tongue from the vallate papillae of the tongue to the epiglottis. (From Jacob, S.W., and Francone, C.A.: *Elements of Anatomy and Physiology*, 1976. Courtesy of W.B. Saunders Co., Philadelphia.)

sue, positioned under the epithelial lining of the head and throat cavities. Like the adenoid, it is thought that they release agents into the blood for the purpose of fighting bacteria. The palatine tonsil may become quite enlarged and may highly restrict the posterior opening into the oral cavity. While the tonsils can significantly obstruct the oropharyngeal opening, tonsillar size rarely has an adverse effect on speech articulation or voice quality. Ordinarily, the tonsils are relatively large before puberty and diminish in size thereafter. Thus the tonsils — pharyngeal, palatine, and lingual — may be expected to grow rapidly until about age twelve and then their growth curve falls off sharply as the natural involution process begins. It should be noted that the growth cycles of tonsils and adenoids may differ in the same individual; hence, the speech and hearing clinician should be forewarned not to look at faucial tonsil size and presume about the size of the adenoids (Fig. 11-9).

That portion of the tongue arising

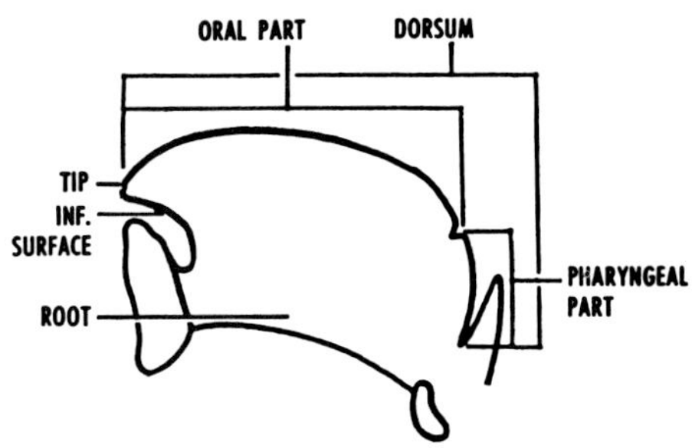

Figure 11-10. Surfaces of the tongue. (From Gardner, E., Gray, D.J., and O'Rahilly, R.: *Anatomy*, 4th ed., 1975. Courtesy of W.B. Saunders Co., Philadelphia.)

from the inferior limits of the posterior opening into the oral cavity is known as the *pharyngeal surface* of the tongue. Numerous small aggregations of lymphoid tissue are found on this surface of the tongue. Collectively, these rough prominences cover the posterior third of the tongue and are termed the *lingual tonsil* (Fig. 11-9).

The Tongue

The tongue is a muscular organ located primarily in the oral cavity but is also partially located in the pharynx. It is attached to the floor of the mouth, with its root anchored to the hyoid bone (sometimes referred to as the bone of the tongue). Other attachments of extrinsic musculature serve to position the tongue in innumerable locations in the mouth and partially out of the mouth cavity proper. By contrast, the front of a frog's tongue is attached to the front of the oral cavity and turns over as it is flicked out of the mouth (see Fig. 11-4).

Functional Considerations

Functions of the tongue are mainly digestive, which involves taste or gustation (complimented by smell), mastication, and swallowing. During mastication, food is moved between the teeth for grinding and mixing. Once broken up, the tongue forms a ball or bolus of food, which is propelled into the oropharynx and into the entrance of the esophagus.

Swallowing or deglutition is a major function of the tongue. Older dental literature reported the presumption that an individual swallows about two thousand times per day; however, more recent studies indicate that this count is exaggerated, that one swallows only about six to eight hundred times per day.

Due to the interest in tongue thrusting and dental occlusal problems by dentists and speech pathologists, swallowing is a pertinent topic in speech anatomy and physiology. The incidence of the tongue thrusting forward against the anterior dental structures during swallowing is 100 percent in infancy. By age six, the incidence is about 50 percent, and it further declines to about 25 percent by age twelve. In adulthood, there is about a 15 percent incidence.

A combination of morphologic factors probably dictates the swallow action used by a given subject. Among the main factors are (1) size of the oral isthmus, (2) length of the mandibular ramus and resulting tongue posture, and (3) area of the pharyngeal airspace as determined particularly by the adenoid mass. There is no one rule by which all are to swallow. There are perhaps as many as ten transitional stages of swallowing through which most humans pass from infancy to adulthood.

How one decides to teach prescriptive swallowing appears to be the pivotal question, for any results claimed in the child may have been forthcoming due to structural and functional maturation.

Altogether, it can be summarized that a swallow is an asymmetrical and changeable phenomenon in children. After a child goes through puberty and the oropharyngeal dimensions are more stable, the swallowing pattern becomes more reproducible but remains asymmetrical. Additional information about tongue thrust is found in Chapter 10, "The Dentition." The mechanics of a swallow are presented in the next chapter.

Another cardinal tongue function is speech. The tongue is the principal organ of speech articulation and is important also in resonance due to its ability to alter the resonant properties of the oral cavity. Consideration of the impressive level of motor and sensory innervation of the tongue provides an appreciation for why man is able to speak. It would appear that, from a standpoint of neurologic design, speech was anticipated in the morphogenesis of the tongue!

Although the tongue is attached to the lower jaw, which imposes limitations on its movements, tongue position remains relatively mobile and versatile, at least in the older child and adult. Until age seven, the child may lack independent movement of the mandible and tongue. During this period there are different growth curves for the tongue and oral neurological maturation, which may indicate a neuromotor development lag for speech. The child that is particularly slow to develop appropriate articulation skills may be suspected of experiencing such a lag.

The tongue boasts an impressive network of intrinsic musculature of its own, enabling it to be humped, curled, grooved, pointed, flattened, and so on to an almost unending variety of such possibilities. This variety of mobility is made possible by intrinsic muscles running lengthwise, crosswise, and vertically and is complimented by multiple extrinsic muscular attachments.

Dorsum of the Tongue

Beginning at the tip, a median furrow, the *median sulcus*, courses straight back down the superior surface or the back of the tongue's body. About two-thirds of the anterior portion of the tongue turns its surface upward, constituting the *dorsum* of the tongue, while the posterior third or base faces posteriorly. A distinct boundary in the form of a shallow, V-shaped groove, pointing backward and called the *terminal sulcus*, separates the body and base of the tongue. Interestingly, the terminal sulcus marks fairly well the zones of the tongue innervated by the trigeminal and glossopharyngeal cranial nerves. At the point of the V-shaped groove is a deep blind pit, termed the *foramen cecum* (Fig. 11-11).

Generally, the upper surfaces of the tongue are responsible for gustation or

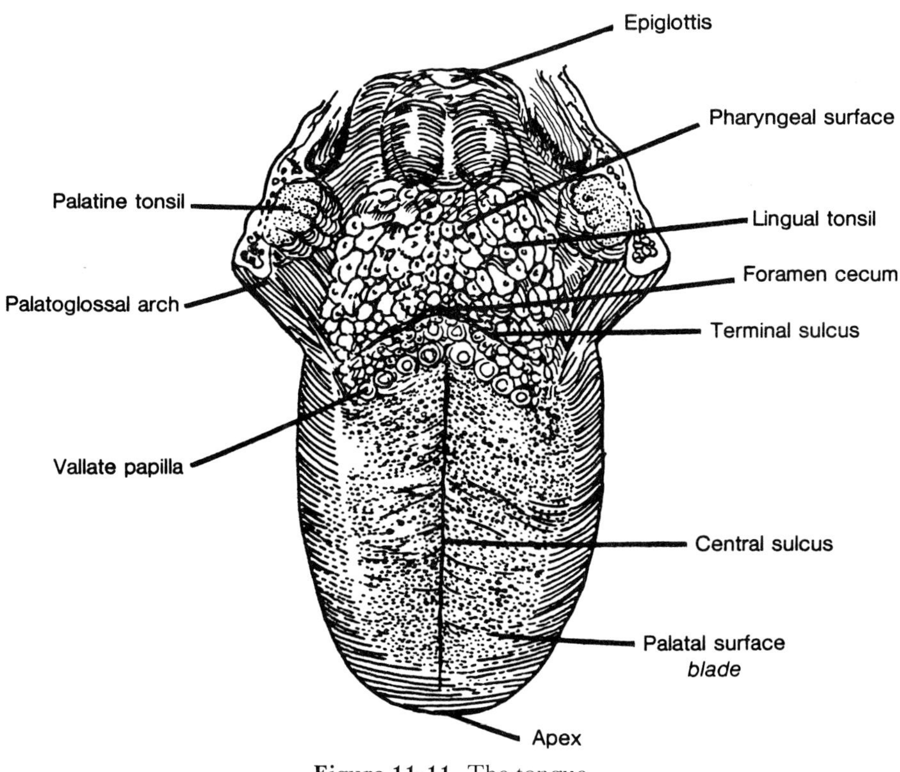

Figure 11-11. The tongue.

taste. As an organ of taste, the tongue has embedded in its surfaces numerous taste buds. The primary taste areas include sweet and salt about the tip, sour on the lateral back part of the body, and bitter on the upper surface along the terminal sulcus. Taste buds deteriorate with age, but sweet sensations appear to persist superiorly. Taste is, of course, related to digestive functions of the tongue. Like the lips, the tongue may also participate in judging the fitness of food that may be too hot or cold.

The dorsum of the tongue is most significant during speech since it is over this surface that the expired breath flows and is altered during articulatory movements. (The dorsum of the tongue is analogous in position to the back of a standing dog.) There is an anterior and a posterior vertical part when viewed in a sagittal plane. The anterior portion is in contact with the palate, and the posterior portion faces the pharynx. Thus the first part may be designated the *palatine surface* and the second as the *pharyngeal surface* (see Fig. 11-10).

The terminal sulcus is the posterior boundary of the palatine surface of the tongue. *Lingual papillae* are abundant on the palatine surface. These are minute projections of various shapes — nipple-like, knob-like, thread-like, or mushroom-like. The papillae give the tongue a roughened appearance. They also tend to collect food particles and particulate material from sinus drainage, necessitating brushing of the tongue when the teeth are brushed.

Strange conditions sometimes arise from the papillae, particularly those papillae that are thread-like. Black hairy tongue is a benign condition characterized by an unusual elongation of these papillae. They may extend 1 to 2 cm, but the malady is generally confined to a triangular area anterior to the terminal sulcus. Cause of hairy tongue is due to excessive blood in the mouth for long periods, associated with extended use of antibiotics, particularly antibiotic lozenges. Coating of the tongue is of little significance from a health standpoint unless the tongue is also dry. Dryness is evidence of failure of salivary glands and also indicates the state of hydration of the body. It may be an early symptom of a number of systemic disturbances, e.g. infectious diseases and prolonged but not acute fevers.

Inferior Aspects of the Tongue

The underside of the tongue lies in contact with the floor of the mouth at rest (see Fig. 11-4). The *lingual frenulum* or *frenum* arises from the floor of the mouth to connect the underside of the tongue at the midline. The lingual frenulum is sickle-shaped and may extend to the tip of the tongue in atypical cases. When such an extensive frenulum is present, two problems may be introduced. One is the common term "*tongue-tied speech*," which is a lay term for most defective speech sounds. This is an abuse of terms, naturally. Usually a tight frenulum, although clearly present, is not a factor in articulation problems. Although it may be a contributing factor in some cases, it can be blamed as a definite cause for only certain consonants.

A second problem pertains to proper growth of the child's mandible. Excessively restrictive frenulums are said by some but denied by others to influence proper growth, since the membrane of the frenulum is fastened directly to the inside surface of the lower jaw. If this is true, then lack of development would result in the mandible not matching the upper jaw, causing a malocclusion between the upper and lower dental arches. A well-informed dentist should be consulted for responses to such problems.

The Floor of the Mouth

The floor of the mouth is covered with a thin mucous membrane, which makes the underlying structures relatively apparent. The floor is a protected location, so the thin covering does not jeopardize this area. Soluble drugs placed under the tongue are rapidly absorbed into the richly vascularized floor of the mouth and underside of the tongue. This is a principal site for injecting epinephrine into the vascular system during heart failure episodes. Also, a patient with angina pectoris (chest pain from heart) places a nitroglycerine tablet under the tongue during an anginal attack.

Two pairs of salivary glands are located on the floor of the mouth. Separated by the attachment of the lingual frenulum on the floor of the mouth are two openings of the *submandibular salivary glands*, which are known as Wharton's ducts or the submandibular ducts. They transport saliva from their respective glands, located on the internal surface of the mandible. The submandibular duct crosses the floor of the mouth to open at the medial end of a transverse fold at the base of the tongue, the *sublingual fold* (see Fig. 11-4).

The other pair of salivary glands, the *sublinguals*, is located in the floor of the mouth, more closely positioned to the tongue. Five to fifteen short ducts open at intervals along each side of the sublingual fold in addition to a principal duct, Bartholin's duct, which opens at or in the vicinity of the submandibular duct orifice. Finally, the floor of the mouth is sealed with a sheet of muscle connecting the hyoid bone and the arch inside the mandible, the mylohyoid muscle.

THE NASAL CAVITIES

The nares, or *nostrils*, provide the anterior opening into the nasal cavities. Two posterior nasal openings, the *choanae*, lead into the nasal cavities from the nasopharynx. Both nasal cavities, then, have orifices located anteriorly as well as posteriorly (Fig. 11-12).

Each nasal cavity is separated from the other by a vertical partition of bone and cartilage known as the *nasal septum*. The roof of each cavity is the cribriform plate of the ethmoid bone. On the superior side of the ethmoid bone is part of the cranial cavity housing the anterior brain. At the top of the nasal cavity are the olfactory (smell) receptors. As the airstream is forcibly inhaled and passes through the upper regions of the nasal cavity, the sense of smell is activated if appropriate chemical particles are contained in the air. The floor of the nasal cavity is also the roof of the mouth; it is composed of bone and divides the nasal and oral cavities (Fig. 11-13).

Each of the lateral walls of the nasal cavities is made very irregular by the presence of three bony protrusions, the *nasal conchae* (shell-like). These conchae, or turbinate bones, project over three channels along the lateral walls. The lowest turbinate is termed the *inferior nasal concha* and represents the largest of the three. The *middle* and *superior conchae* lie above, respectively. Collectively, the three nasal conchae not only make the lateral nasal cavity wall irregular but also rather narrow, particularly near the rear portions of the cavities.

The epithelial lining of the nasal cavities is a mucous membrane. It is richly supplied with blood vessels, which serve the function of a radiator, warming in-

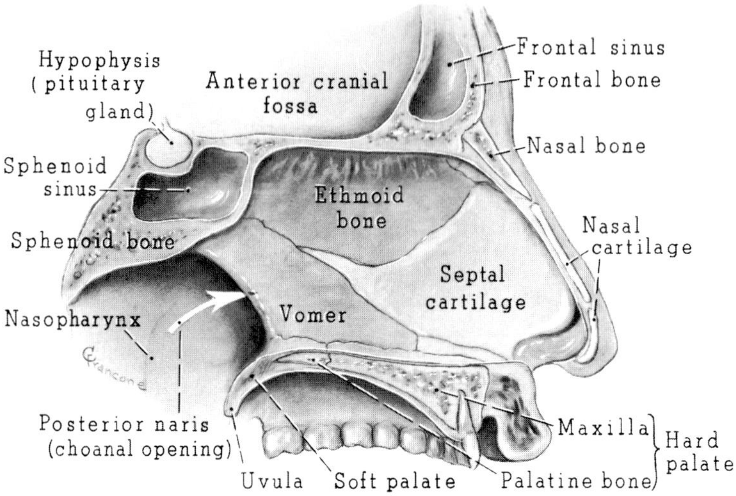

Figure 11-12. Sagittal section through the nose showing the components of the nasal septum. (From Jacob, S.W., and Francone, C.A.: *Elements of Anatomy and Physiology*, 1976. Courtesy W.B. Saunders Co., Philadelphia.)

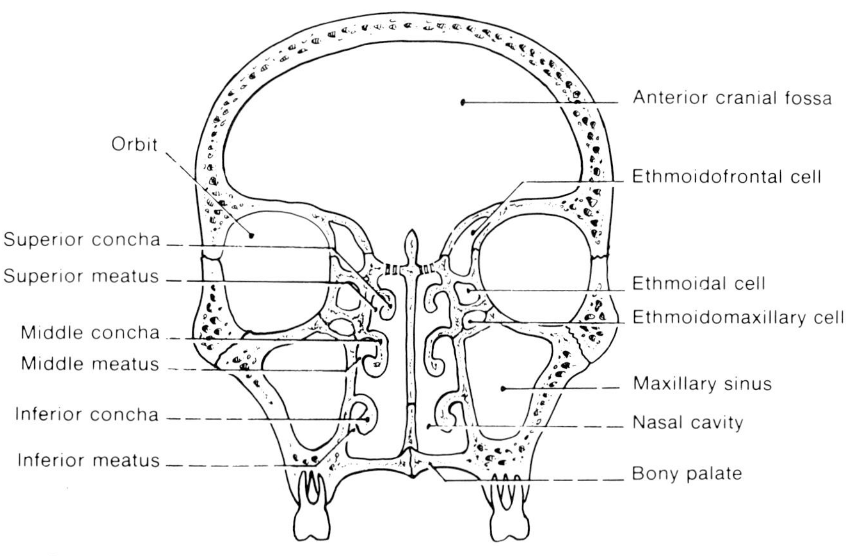

Figure 11-13. Coronal section of the head showing the nasal conchae, or turbinates. (From Bossy, Jean: *Atlas of Neuroanatomy and Special Sense Organs*, 1970. Courtesy of W.B. Saunders Co., Philadelphia.)

spired air passing through each nasal cavity. The conchae greatly increase the surface area available for radiant activity. Anyone who has walked rapidly or ridden a bicycle in a cold environment appreciates the difference between breathing through the mouth and inhaling through the nose under such conditions. The stinging sensation from cold air reaching the lungs via the mouth is a memorable

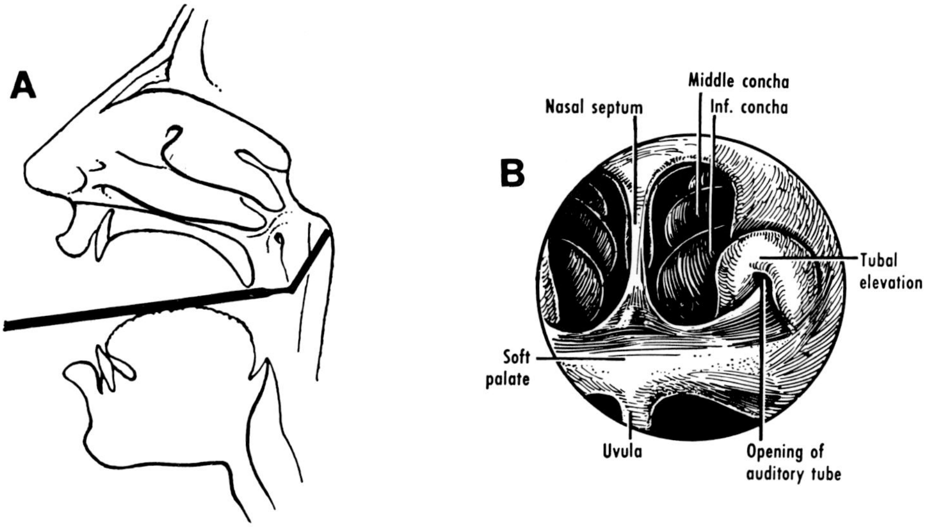

Figure 11-14. Posterior indirect rhinoscopy. Mirror view of the posterior entrance into the nose (the posterior choanae) from the nasopharynx. Notice the nasal conchae and auditory tube orifice. (From Gardner, E., Gray, D.J., and O'Rahilly, R.: *Anatomy*, 4th ed., 1975. Courtesy of W.B. Saunders Co., Philadelphia.)

event. Additionally, mucus is secreted by the lining which moistens inspiratory air. Humidified air is essential for normal laryngeal functions. The upper respiratory airways do not tolerate dry membranes very well. Movements required in swallowing, for example, depend on lubricated surfaces to enable the structures to move against one another without friction. Also, the house-cleaning operation of the mucous membrane is nullified by the drying of the membrane. The moistness of the lining may also slightly influence resonance. Moist membranes tend to be thicker. Dryness causes the membranes to shrink over the turbinates, enlarging the volume of the cavity (Fig. 11-15).

The nasal cavities provide for a filtering activity for the respiratory tract. The inferior position of the nostrils, together with some hair near the entrance, aids as a first-line screening of inspired air. Behind the anterior third of the nasal cavity, the epithelial lining has cilia beating rhythmically about 250 times per minute in a posterior direction. The mucous coat is located above the cilia and acts as a vehicle to collect foreign particles, moving the material along about 10 mm per minute. Collected foreign matter is then eliminated through movements of the underlying cilia, either by the chemical agents contained within the mucus or through being carried into the oropharynx, where it is swallowed and eliminated through the intestinal tract.

In speech production, three sounds, the *m, n*, and *ng*, require participation by the nasal resonant chambers. If other sounds which are not normally nasalized are directed into the nasal cavity, the intelligibility of the speech may markedly decrease. This condition, in pathologic proportion, is known as *hypernasality*.

Several pathologic conditions of the nasal cavities have the potential of affecting speech by interfering with the balance

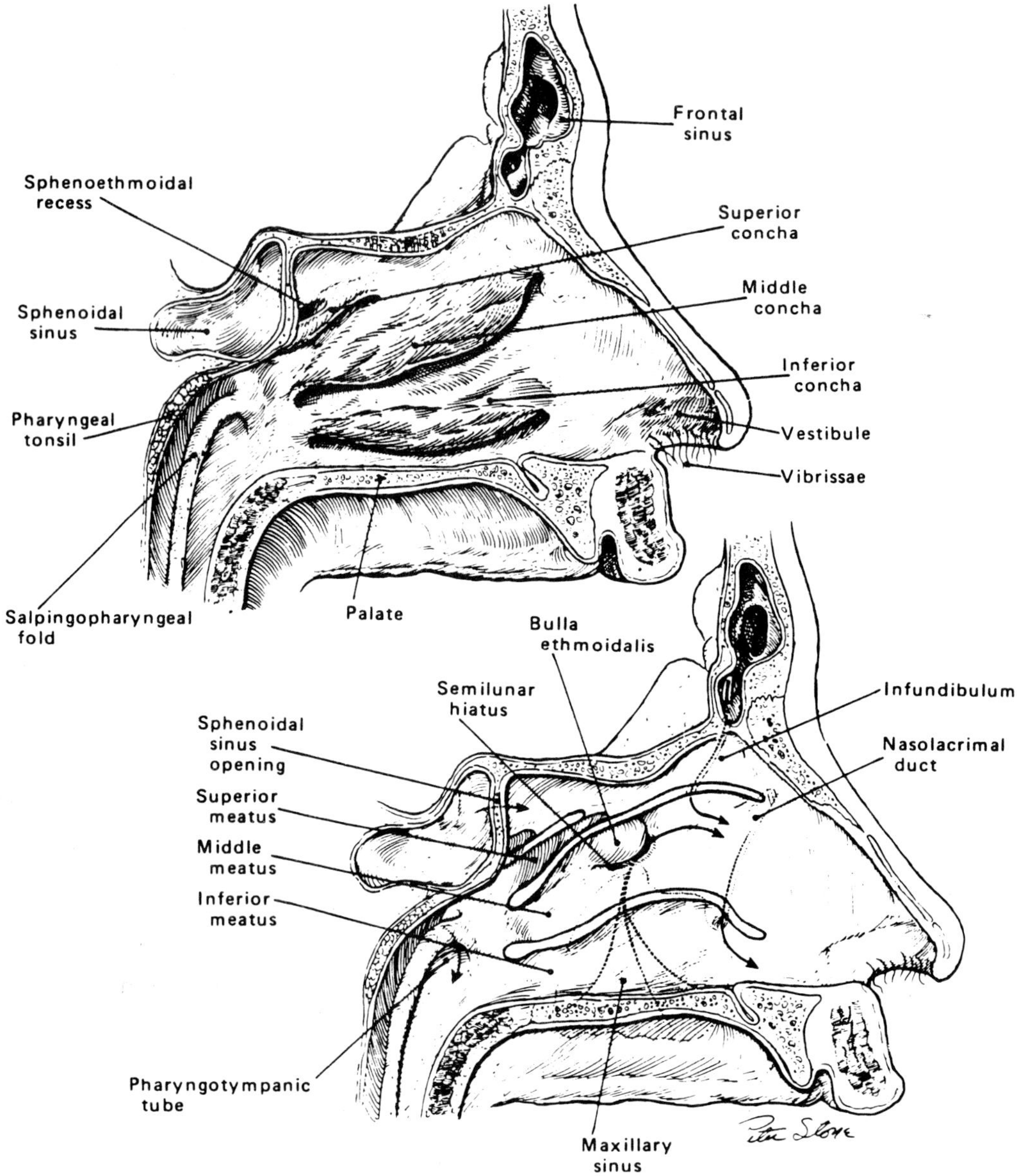

Figure 11-15. Midsagittal section through the head showing the left lateral nasal cavity (top) and lateral wall of nasal cavity after removal of nasal conchae (bottom). (From *Dynamic Anatomy and Physiology*, 3rd ed. by L.L. Langley, I.R. Telford, and J.B. Christensen. Copyright © 1974. Used with the permission of McGraw-Hill Book Company.)

of resonance between the cavities. Some conditions result in excessive nasal resonance. Clefts of the palates may preclude sufficient control of the sound stream and balance of resonance. Grave resonant disturbances to the speech sound stream may be expected if the anatomic features of the floor of the nasal cavity are grossly inadequate. In addition to clefts, paralyses and congenitally short soft palates are other conditions causing severe disturbances in resonance. Insufficient resonance by the nasal cavity may disturb articulation. About 85 percent of the normal adult population has a *partial* deviation of the septum. However, a significantly deviated or deflected septum may block or impede air flow through the nasal cavities. The hollow sound associated with occluded nasal cavities may also be associated with deviated septum conditions. A person with completely occluded nasal cavities would produce *bad* for *man, dabig* for *naming*. These are hyponasal conditions. It should be mentioned that an occlusion of the nasal cavities at the posterior choanae results in a different type of nasal sound from that produced when the occlusion is near the nostrils.

A number of diseases affecting the nasal mucosa will cause swelling and an increased secretion from the mucous glands, perhaps as much as a liter a day. Inflammation associated with a head cold (rhinitis) frequently closes the nasal passages and diminishes their participation as a resonant system. Changes in the tissue at the posterior nasal aperture, especially the pharyngeal tonsil, have the potential to block the nasal cavities from the oropharyngeal cavity.

THE PHARYNGEAL CAVITIES

The irregularly shaped cavities of the head and neck compose the resonant system in the vocal tract. Its cavities are coupled together in a variable fashion. The basic biological function of the resonators — the nasal, oral, and pharyngeal cavities — is to serve as passageways through which air is carried from outside the body into the larynx and lower respiratory airways. In additon, the oral cavity and pharynx have another primary function — the intake of food, mastication, and swallowing. The resonators thereby participate in the vocal tract in an additional role to their responsibilities in the respiratory and digestive tracts (Fig. 11-16).

The pharynx is a tube-like resonator standing vertically on top of the esophagus (the passage leading to the stomach). The pharynx is a funnel-shaped space with muscosa-covered muscular side and rear walls, capable of some movement in a sphincteric action. This movement may, in a few instances produce a bulging effect (Passavant's bar) on the belly of the posterior pharyngeal wall in the area where the soft palate rises to close the nasal cavity from the oral cavity.

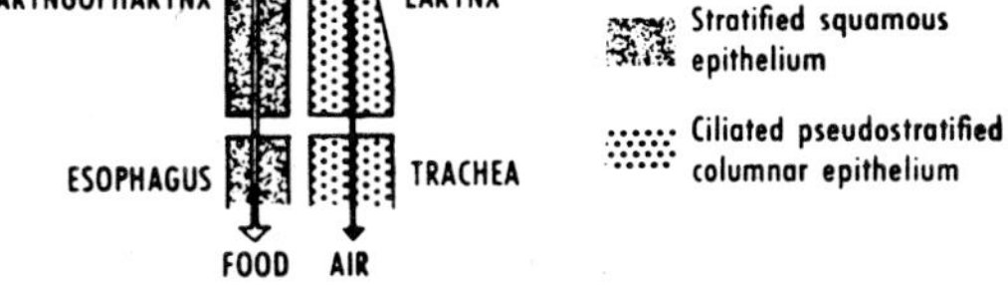

Figure 11-16. The functional cross-over of the respiratory and digestive pathways. (From Gardner, E., Gray, D.J., and O'Rahilly, R.: *Anatomy*, 4th ed., 1975. Courtesy of W.B. Saunders Co., Philadelphia.)

The vertebral bodies of the cervical section of the spinal column lie immediately behind the posterior wall. It is interesting to examine the pharynx in a lateral x-ray; the posterior wall of the pharynx is normally straight with the exception (when present) of a mass of adenoid tissue bulging forward at the superior aspect. This tissue may be important as a target site for contacts of the soft palate to close off the nasal cavities.

The anterior wall is broken by the opening into the posterior entrances of the oral and nasal cavities, and the cavity of the larynx opens into the lower anterior part of the pharynx. Since the pharynx communicates with three cavities at various levels, it may be arbitrarily divided into nasal, oral, and laryngeal parts. These are termed the *nasopharyngeal cavity*, the *oropharyngeal cavity*, and the *laryngopharyngeal cavity*. Structurally, the anterior wall is irregular, made up of sound-absorptive soft tissue and varying shapes of musculature.

Nasopharyngeal Cavity

The uppermost part of the pharyngeal tube extends from the base of the skull. On the lateral walls of the nasopharynx are found the orifices of the pharyngotympanic tubes, also referred to as the internal auditory tube, eustachian tube, or just the *auditory tube* (Fig. 11-18).

The pharynx is widest in the upper part of the nasopharynx. This is true because of the *pharyngeal* or *Rosenmuller's recess*, a deep and narrow recess located behind the prominence of the cartilaginous auditory tube (Fig. 11-18). The lateral limit of the pharyngeal recess is the bony auditory tube at the base of the skull. This remote site is sometimes the location of pharyngeal tumors or nasopharyngeal cancer. A posterior indirect rhinoscopy with a dental or laryngeal mirror is used to reflect this region for inspection whenever changes in vocal resonance signal a need for such an examination (see Fig. 11-14).

The eustachian tubes provide a passage that communicates between the nasopharynx and the middle-ear cavities. The nasopharyngeal end of the auditory

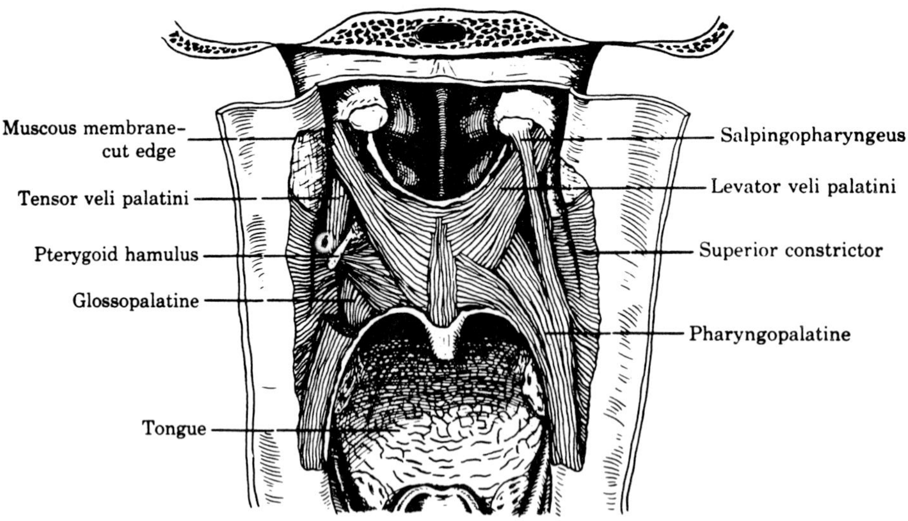

Figure 11-17. Posterior aspect of the nasal and pharyngeal cavities with the posterior pharyngeal wall reflected away. (From *Basic Human Anatomy* by C.E. Tobin. Copyright © 1973. Used with the permission of McGraw-Hill Book Company.)

tube is cartilaginous, and the mound of mucous membrane at the opening is termed the *torus tubarius*. At the lower end of the torus tubarius, the mound of mucous membrane appears as a prominent and sharp fold, extending downward on the lateral wall of the pharynx. This fold is termed the *salpingopharyngeal fold* after the muscle which it covers.

The pharyngeal tonsil, or adenoid mass, is a diffuse aggregation of lymphatic tissue that is found on the posterior and lateral walls of the nasopharynx. An enlarged adenoid may occlude the eustachian tube's opening, resulting in a lowering of the air pressure inside the middle-ear cavity as the oxygen contained in the trapped air is absorbed by the membranous lining. Such a situation often calls for surgical removal of adenoid tissue (adenoidectomy). The same type of discomfort (acute pain in some cases) is experienced by persons with head colds while flying in pressurized or unpressurized aircraft cabins and continuing on even after landing.

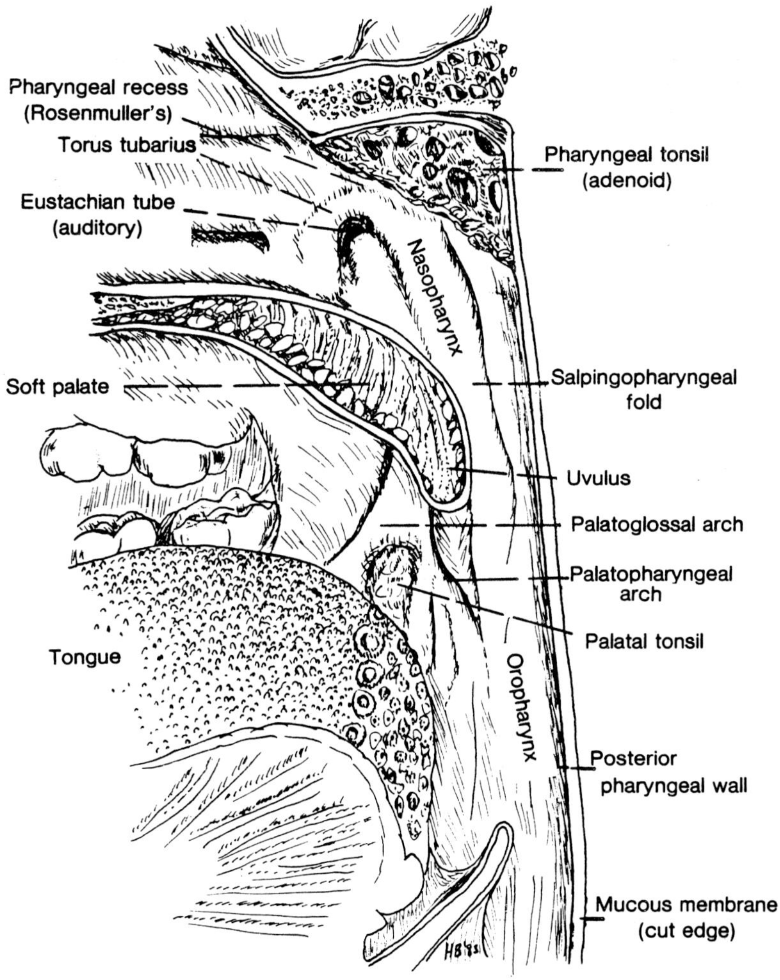

Figure 11-18. Sagittal view of the oro-naso-pharyngeal cavities.

Laryngopharyngeal Cavity

The laryngopharynx is the lowest portion of the pharyngeal cavity (see Fig. 11-1). The larynx opens into the lower anterior portion of the pharynx, with the aditus (entrance) formed principally by the epiglottis and aryepiglottic folds. The recess located on either side of the laryngeal entrance is known as the *piriform recess* or sinus. This recess is formed due to the protrusion of the aditus into the pharynx, resulting in a deep groove on each side between the aryepiglottic fold and the lateral pharyngeal wall. On occasion, during the act of swallowing a pill, one may have the feeling that it "caught in the throat." This may be due to the pill being temporarily lodged in this recess.

Continuing with the laryngeal part of the pharynx, the pear-shaped aditus exhibits an upper rounded border. This border is formed by the epiglottis. Along the lateral borders of the entrance are the aryepiglottic folds, which point downward, causing the apex of the opening to be located between the tips of the arytenoid cartilages. The epiglottis, a lid-like structure to the aditus of the larynx, could, perhaps, meter the freeness of the vocal tone to rise into the pharynx and determine the coupling of the laryngeal cavity and the pharynx. A crude demonstration of this concept can be performed easily by the reader. While prolonging an *ah* sound with the mouth widely opened, quickly protrude the tongue maximally to pull the epiglottis forward and open the larynx. There will be a noticeable modification in the tonal quality. The use of a higher pitch produces a more distinguishable change in the sound. Whether the change is due to the epiglottis opening the laryngeal entrance more fully or to the alteration of the oral cavity's dimensions by moving the tongue is not clear. It is probably a combination of both.

Oropharyngeal Cavity

Between the nasopharynx and laryngopharynx is the oropharynx (Fig. 11-18). It is an area with a variable opening and participates with the soft palate in coupling the oral cavity with the pharynx in resonance. The major feature of the oropharynx is its anterior valve, the faucial isthmus, that was discussed earlier in the section on the "Oral Cavity."

BIBLIOGRAPHY

Calnan, J.: Palatopharyngeal incompetence in speech. In Pruzansky, S. (Ed.): *Congenital Anomalies of the Face and Associated Structures.* Springfield, IL: Thomas, 1961.

Crouch, J.E.: *Functional Human Anatomy*, 3rd ed. Philadelphia, Lea and Febiger, 1978.

Dickson, D.R.: Anatomy of the normal velopharyngeal mechanism. *Clin Plast Surg, 2*:235-248, 1975.

Dickson, D.R.: Normal and cleft palate anatomy. *Cleft Palate J, 9*:280-293, 1972.

Dickson, D.R., and Dickson, W.M.: Velopharyngeal anatomy. *J Speech Hear Res, 15*: 373-381, 1972.

Gardner, E., Gray, D.J., and O'Rahilly, R.: *Anatomy: A Regional Study of the Human Structure*, 4th ed. Philadelphia, Saunders, 1975.

Hagerty, R. Hill, M., Pettit, H., and Kane, J.: Posterior pharyngeal movements in normals. *J Speech Hear Res, 1*:203-210, 1958.

Kaplan, H.M.: *Anatomy and Physiology of Speech*, 3rd ed. New York, McGraw, 1981.

Mason, R., and Grandstaff, H.L.: Evaluating velopharyngeal mechanisms in hypernasal speakers. *Language, Speech and Hearing Services in Schools, 4*:53, 1971.

Mason, R., and Proffit, W.: The tongue thrust controversy: background and recommenda-

tions. *J Speech Hear Disord, 39*:115-132, 1974.
Paff, G.H.: *Anatomy of the Head and Neck*. Philadelphia, Saunders, 1973.
Zemlin, W.R.: *Speech and Hearing Science*, 2nd ed. Englewood Cliffs, NJ: P-H, 1981.

Chapter 12

MUSCLES OF THE SOFT PALATE, PHARYNX, AND TONGUE

IT is appropriate to consider the musculature of the soft palate, pharynx, and tongue together, since these organs are structurally and functionally related. Functionally, the pharyngeal, palatal, and lingual muscles play an important and cooperative role in speech, mastication, and deglutition (swallowing).

Like the tongue, the soft palate and pharynx are muscular organs. Additionally, all have distinct extrinsic muscles associated with them, which are partly responsible for positioning these organs.

THE SOFT PALATE

The soft palate is anchored to the posterior margin of the horizontal plates of the palatine bones, which serve as the posterior limit of the hard palate. Laterally, the soft palate is continuous with the fauces. The posterior border is free, with its characteristic conical projection in an inferomedian position, the *uvula* (Fig. 12-1).

The *palatine aponeurosis* is known as the fibrous skeleton of the soft palate. Derived from a tendinous expansion of the tensor palatini muscle, the aponeurosis spreads throughout the soft palate from the hard palate to the free border. This plate of fibrous connective tissue is attached to the posterior limit of the hard palate. The tendons of both tensor palatini muscles meet at the midline, effectively covering the entire soft palate. The aponeurosis is, upon contraction of the palate's tensor and levator muscles, transformed into a firm, horizontal structure to participate in closing the nasopharynx from the oropharynx. Essentially, the aponeurosis acts as a firm plate to which the palatini muscles may attach and pull against it.

During swallowing and sucking, the soft palate may execute maximal closure of the pharyngeal isthmus — the connecting channel between the nasal and oral parts of the pharynx. A similar degree of closure is required for high pressure intraoral speech sounds such as the plosives, *p, d, t,* etc., especially during loud speech. Relatively tight closure is also used for the tense, high back vowel, *oo* as in *tool*, contrasted with the lax, low back vowel, *a* as in *ah*. The terms tense and low, and lax and high refer to relative states and positions of the tongue, respectively.

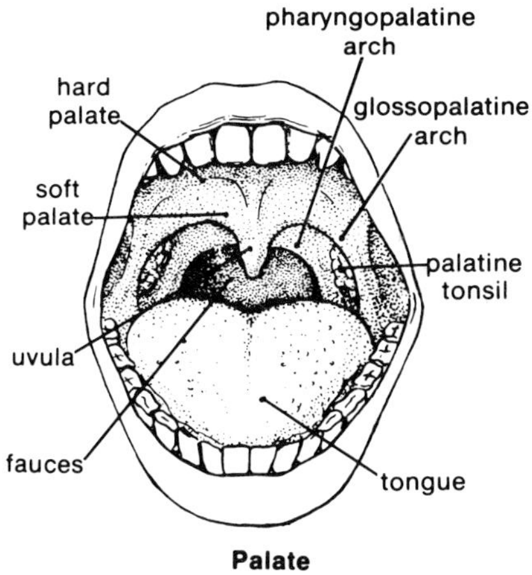

Figure 12-1. The oral cavity. (From Nicolosi, Lucille: *Terminology of Communication Disorders.* Copyright 1978, The Williams and Wilkins Co. Reproduced by permission.)

Relaxed, the soft palate hangs in a position continuous with the roof of the mouth. This is the position for normal breathing or in forming the nasal consonants. In closure, the soft palate is raised by action of the levator palatini on either side. These pull on the palatal aponeurosis in synchrony with the upper fibers of the superior pharyngeal constrictor through a sphincteric action, pulling the posterior pharyngeal wall slightly forward. The palatopharyngeal muscles narrow the faucial isthmus like draw drapes being closed. Collectively, these movements effectively separate the pharyngeal isthmus as the soft palate or velum buckles about 80 percent toward the back and contacts the posterior pharyngeal wall. Or, in some cases, the pharyngeal surface of the palate may contact the adenoid tissue. However, it should be noted that some individuals do not present evidence of any posterior pharyngeal wall movement during palatopharyngeal or velopharyngeal closure.

Muscles of the Soft Palate

Altogether there are five muscles. Three are considered *primary* palatal muscles, since they do not originate or insert into surrounding musculature of other muscular organs. These primary muscles are the tensor palatini, levator palatini, and uvular muscle. Usually, the term *veli* is included as part of the name of the first two palatal muscles. This appears needlessly redundant and is not employed here. All are considered paired.

The other palatal muscles, palatoglossus and palatopharyngeus, relate the soft palate to the tongue and pharynx, respectively. Their action is highly dependent upon the response of other muscles attached to the tongue and pharynx. Sometimes these two muscles are variously described as belonging, respectively, to the tongue and pharynx (Table 12-I).

Tensor Palatini

The tensor palatini muscle has multiple origins and a variable incidence, as found in dissections. Fibers arise from the scaphoid fossa at the base of the pterygoid process of the sphenoid bone, from the wall of the cartilaginous auditory tube (pharyngotympanic or eustachian tube), and from the sphenoidal spine, which is at the base of the greater wing of the sphenoid adjacent to the petrous portion of the temporal bone (see Fig. 9-11). This flat muscle courses downward between the medial pterygoid plate and the medial pterygoid muscle. On its descent, the muscle narrows and ends in a tendon, which loops around the pterygoid hamu-

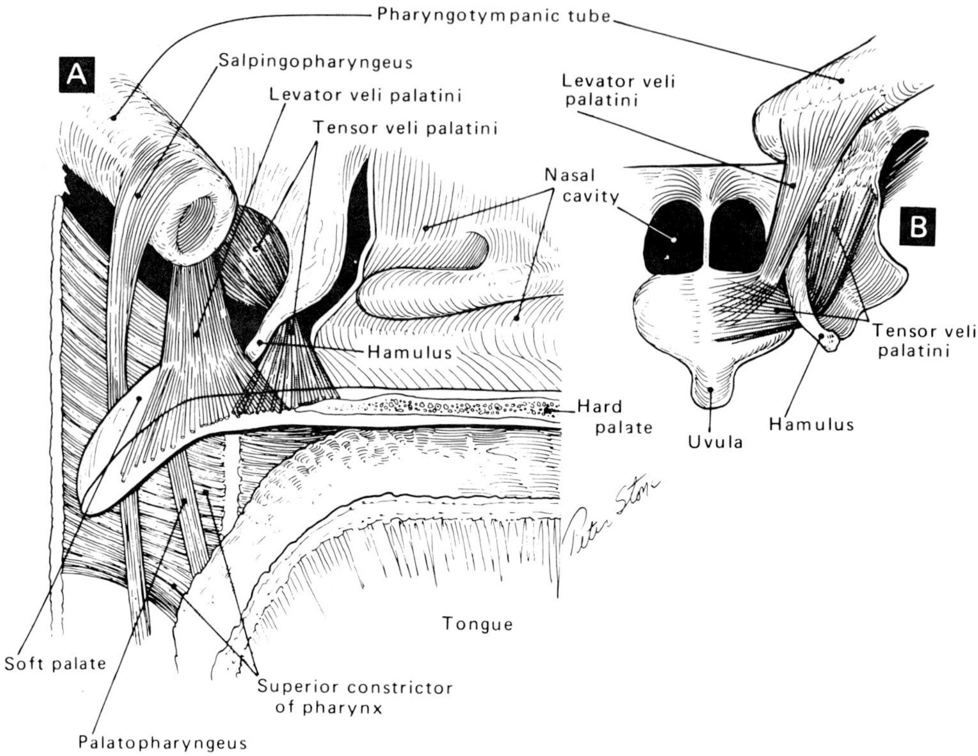

Figure 12-2. Muscular slings associated with the soft palate. (From *Dynamic Anatomy and Physiology* by L.L. Langley, I.R. Telford, and J.B. Christensen. Copyright © 1969. Used with the permission of McGraw-Hill Book Company.)

Table 12-I

MUSCLES OF THE SOFT PALATE

Muscle	*Origin*	*Insertion*	*Action*
Tensor palatini	scaphoid fossa, auditory tube cartilage	palatine aponeurosis	tenses soft palate
Levator palatini	petrous portion of temporal bone, auditory tube cartilage	palatine aponeurosis	raises soft palate
Uvulus	posterior nasal spine and palatine aponeurosis	mucosa of uvula	shortens uvula
Palatoglossus	undersurface of soft palate in palatine aponeurosis	dorsum sides of tongue	elevates tongue; narrows palatine arches
Palatopharyngeus	palatine aponeurosis	walls of pharynx and thyroid cartilage	constricts palatopharyngeal folds

lus, turns sharply, and continues into the soft palate in a horizontal plane. As the horizontal portion of the tendon reaches the lateral aspect of the soft palate, it fans out in the formation of the palatine aponeurosis, with its mate from the opposite side contributing the other half of the aponeurosis. This important aponeurosis not only gives structural basis to the soft palate but also serves as a point of attachment of other palatal muscles (Fig. 12-2).

The tensor palatini tends to tighten the palatal aponeurosis during contraction, but this is not thought to be a necessary or contributory component of velopharyngeal closure. A second function is also present. This involves participation in opening the nasopharyngeal orifice of the auditory tube by virtue of those fibers of the tensor palatini which take their origin in the anterolateral wall of the auditory tube. Sometimes these fibers are termed the *dilator tubae muscle*, which is responsible for opening the auditory tube orifice. Hence, upon contraction, the tensor palatini may help to refill the tympanic cavity with air and equalize air pressure on the inside and outside. There are other pharyngeal muscles, however, which are more effective in this operation of relieving the dull tension caused by altitude changes and other causes of internal/external pressure differences. The nerve supply is from a branch of the trigeminal (V) cranial nerve.

Levator Palatini

The levator palatini muscle operates in conjunction with the superior constrictor to achieve velopharyngeal closure. The levator muscle elevates the posterior part of the soft palate, the velum, and pulls it slightly backward, which effectively closes the nasal from the oral pharynx. This muscle is highly active in speech, swallowing, and blowing activities (Fig. 12-3).

The levator palatini arises from the under surface of the apex of the petrous portion of the temporal bone. A few fibers arise from the medial wall of the cartilaginous auditory tube. The muscle courses obliquely downward and forward inside the upper border of the pharyngeal superior constrictor muscle and ends in the palatal aponeurosis. Neural innervation (supply nerve fibers to) is derived from a branch of the vagus (X) cranial nerve by way of the pharyngeal plexus.

Uvulus

The uvula — the posterior midline appendage of the soft palate — contains part of the uvulus muscle, as well as mucous glands and areolar tissue covered with mucous membrane.

The uvular muscle is a variably paired band of muscle fibers on either side of the midline of the nasal surface of the soft palate. The fibers begin at the posterior nasal spine of the hard palate and in the palatine aponeurosis. They course through the soft palate to the uvula and insert into the uvula's mucous membrane. Thus, the fibers of the uvulus muscle run throughout the soft palate, longitudinally at the midline and superiorly in the palate, and end in the uvula itself.

Contraction of the musculus uvulae draws the uvula up to shrink it. The uvulus is also thought to be the main contributor to the midline longitudinal velar bulge seen often by nasendoscopic examination of the nasal (superior) surface of the velum during function. Since the uvula extends from the posterior free margin of the velum, it appears that it could, in some cases, play a potential minor role in the following: (1) swallowing, by assisting in the sealing operation during palatopharyngeal closure and di-

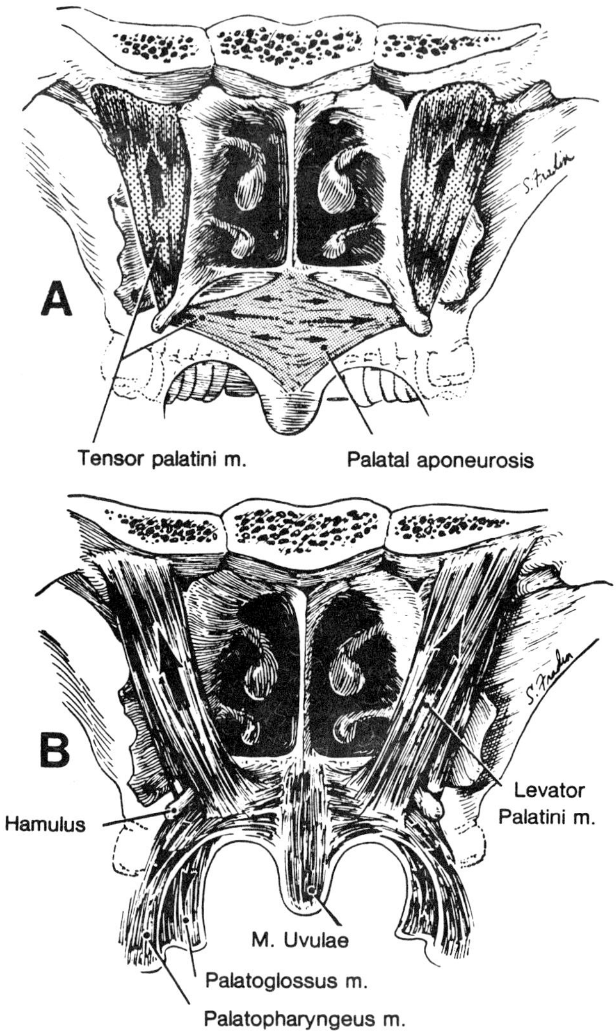

Figure 12-3. Muscular connections and interactions on the nasal surface of the soft palate from a posterior view. (From Nagle, Raymond J., and Sears, Victor H.: *Denture Prosthetics*, ed. 2, St. Louis, 1962, The C.V. Mosby Co.)

recting food downward; (2) directing drainage from the nasal cavities to the base of the tongue and away from the posterior pharyngeal wall and the auditory tube orifice; (3) resonance, by directing the sound stream from the larynx into the oral cavity to prevent an excessive nasal tone; and (4) articulation, for production of the uvular *r* sound (a trilled sound not a part of the American speech sounds).

Seldom must the uvula be excised. Malignancies or significant mechanical interference are the basic reasons for excision. Yet some physicians have removed the uvula during tonsilloadenoidectomy. The surgical procedure is termed a *uvulectomy* or *staphylectomy*. Many speech clinicians and vocal music teachers become alarmed over excessively long uvulas that are seen clinging to the sides of palatal arches or the back of the tongue. Perhaps such sticking, caused by the wetness of the structures, prevents the uvula from reacting rapidly enough for speech or singing resonance control, but this is questionable. Normally, short uvulas appear not to be significant in speech interference. However, following a uvulectomy, speech and swallowing interference may possibly result in rare instances.

The uvular muscle is innervated by the same nerve supply as the levator palatini.

Palatoglossus

By means of the palatoglossus muscle, located on the anterior oral surface in the soft palate, the palate is related to the tongue. The origin is the anterior undersurface of the soft palate, where it is tied to the palatal aponeurosis. Fibers from both palatoglossus muscles meet at the midline of the palate. From here, the muscle fibers descend inside the palatoglossal arch (the anterior faucial pillar) in front of the palatine tonsil. The insertion is into the sides of the tongue, where the fibers interlace with those of the transverse fibers of the tongue.

The action of the palatoglossus pulls the root of the tongue up and back. If the tongue is fixed, the soft palate is pulled downward. The action of the two palatoglossus muscles may be termed a constrictor of the anterior faucial pillars, the palatoglossal arches, due to their action in pulling them medially to narrow the port between the oral and pharyngeal cavities. Fibers from the pharyngeal plexus innervate this muscle with the vagus (X) dominating among the nerves supplying the plexus (Fig. 12-4).

Palatopharyngeus

Although sometimes designated as a muscle of the pharynx, the palatopharyngeus muscle is also classified as a palatal muscle. It is described as a longitudinal muscle in relation to the pharynx. It arises from the midline of the palatine aponeu-

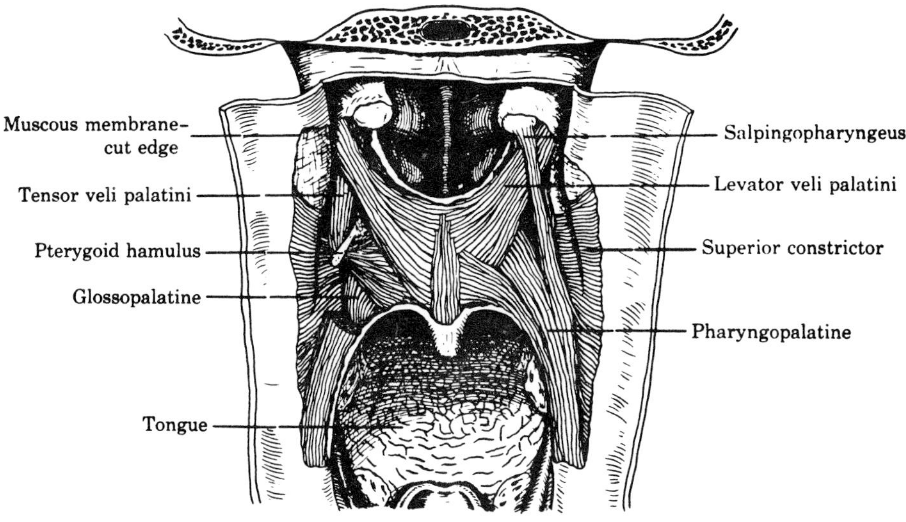

Figure 12-4. The naso-oro-pharyngeal areas viewed from behind. The pharyngopalatine muscle has been cut on the left to expose the glossopalatine muscle and the course of the tensor palatini around the hamulus on its way to insert into the palatine aponeurosis. (From *Basic Human Anatomy* by C.E. Tobin. Copyright © 1973. Used with the permission of McGraw-Hill Book Company.)

rosis, where it is continuous with its mate from the opposite side and forms the posterior border of the hard palate. Leaving the soft palate, the palatopharyngeus narrows as it descends inside the palatopharyngeal arch just behind the palatine tonsil. Upon reaching the base of the arch, the muscle again widens, mixing with other longitudinal muscles of the pharynx, namely, the stylopharyngeus and salpingopharyngeus. The fibers of these muscles intermingle with each other in the pharyngeal wall. In fact, some anatomists feel that the palatopharyngeus and salpingopharyngeus are branches of the same muscle. Suffice it to say that there is great variation in the distribution of these muscles in the pharynx, especially the salpingopharyngeus. Insertion is gained by the palatopharyngeus on the posterior edge of the thyroid cartilage of the larynx and the mucous membrane on the side of the pharynx; a few fibers pass across the posterior midline of the pharynx to mix with those of its contralateral mate.

The palatopharyngeus pulls the wall of the pharynx upward. Acting together, both muscles constrict, or approximate the posterior faucial pillars by moving them toward the midline. This action lowers the soft palate, further narrowing the posterior oral orifice. This muscle may also have an influence on the larynx by helping to raise it if the palate is fixed. The muscle is supplied from the pharyngeal plexus by means of the accessory (XI) cranial nerve, although this innervation is not clear.

THE PHARYNX

The long, tapered pharyngeal cavity is wrapped in muscle fibers and lined with a mucous membrane. Already several references to muscle fibers from surrounding structures have been made in light of their contribution to the pharyngeal musculature. There are three primary muscles of the pharynx that are designated the constrictors: superior, middle, and inferior. Other pharyngeal muscles, the stylopharyngeus and salpingopharyngeus, are undisputed members of the pharyngeal musculature. However, palatoglossus and palatopharyngeus are variously designated as muscles of the pharynx or of the other organ to which they are attached. In this treatment, these latter two muscles are not classified as members of the pharyngeal muscles; each of these is found under muscles of the tongue and soft palate, respectively. All muscles associated with the pharynx are paired.

The pharynx communicates with the left and right nasal cavities through the *choanae*, the term referring to the posterior openings of the nasal cavities; with the tympanic cavity by means of the auditory tube orifice; with the oral cavity; with the laryngeal cavity; and with the esophagus, for a total of seven interruptions of the boundaries of the pharynx. Roughly about 12.5 cm long, the pharynx is greater in its transverse dimension than in its anterior-posterior dimension (sometimes termed the *A-P position*). Overall, the pharynx is largest at the base of the skull and narrows as it terminates in the esophagus.

The posterior wall is continuous at the midline from its superior aspect to the inferior limit. It is marked by a vertical median raphe, extending from the pharyngeal tubercle on the basilar portion of the occipital bone. The cervical vertebrae lie just behind. To the fibrous raphe are attached the pairs of each of the three pha-

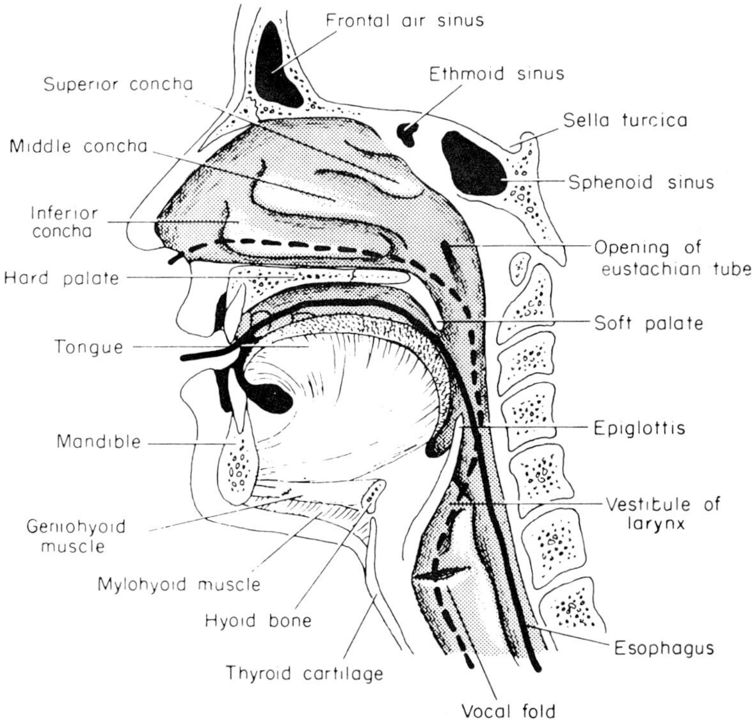

Figure 12-5. Functional cross-over of respiratory and digestive pathways. (From Brooks, Stewart M.: *Basic Science and the Human Body*, St. Louis, 1975, The C.V. Mosby Co.; modified from Francis, C.C.: *Introduction to Human Anatomy*, ed. 6, St. Louis, 1973, The C.V. Mosby Co.)

ryngeal constrictor muscles. They usually arrive at this point of insertion by sending their bands of fibers toward it in a more or less circular course from various origins in the anterolateral plane. The other pharyngeal muscles are characteristically longitudinal muscles and are unrelated to the sphincteric actions of the pharynx, as are the constrictors.

While the pharynx acts as a variable tube resonator, it is also responsible for passing the sound stream from the larynx into the oral and nasal cavities as required for speech sound articulation and sound transmission. Persons without a larynx (laryngectomees) use the pharynx to direct air into the esophagus, where it is trapped and sent back through the pseudoglottis of the inferior portion of the pharynx, transforming the sound into a sound stream for the speech articulators to act upon. Lay persons term this "speaking on a burp": speech pathologists and other professionals refer to this as esophageal speech. In normal anatomy, the pharynx does not send air into the esophagus during breathing; rather, the air is directed into the entrance of the larynx for passage into the lungs.

Deglutition

Deglutition is a fundamental activity of the pharynx in addition to passing air to and from the lungs. The first stage of swallowing (oral stage) is purely voluntary. Food is ground and rolled into a bolus soaked with saliva in the oral cavity.

The bolus is then forced into the pharynx, ending the *voluntary* or *oral phase* of swallowing. But the manner with which this phase of swallowing occurs is highly unstable across ages. Swallowing in children is normally extremely variable, but in adults it is stable and individualistic. Hereafter, purely reflexive mechanisms control the remaining stages initiated by the voluntary act. Automatically, the soft palate is raised to prevent the bolus from entering the nasopharynx in the *pharyngeal stage* of swallowing. The larynx also rises to protect its entrance, while the vocal folds are adducted to guard the trachea and lungs. Wave-like contractions of the pharynx, as required by the fingers during milking, propel the bolus into the esophagus. After the food passes into the esophagus, the final (*esophageal*) stage of swallowing is entered. Peristaltic waves (clasping and compressing movements of the muscles of the esophagus) move the bolus from one part of the tube to the next until the bolus reaches the stomach. This segmental contracting and relaxing enables one to swallow even while upside down (Figs. 12-6 & 12-7).

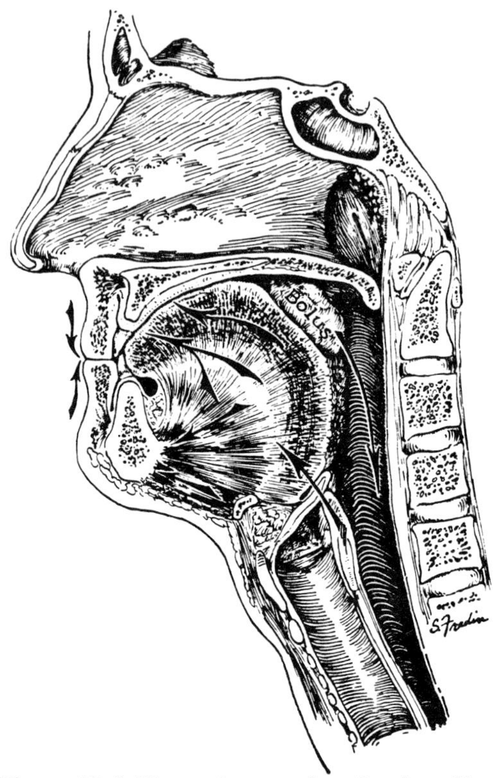

Figure 12-6. The oral stage of swallowing. (From Nagle, Raymond J., and Sears, Victor H.: *Denture Prosthetics*, ed. 2, St. Louis, 1962, The C.V. Mosby Co.)

Muscles of the Pharynx

Superior Pharyngeal Constrictor

The origin of the superior pharyngeal constrictor muscle is the most complex of the pharyngeal muscles. Five points of origin may be considered for this complex muscle. The irregular design of the pharyngeal space requires the superior constrictor to form the broken cavity into a smooth fibrous tube for respiratory and swallowing efficiency (Table 12-II).

The uppermost fibers arise from the posterior border of the median pterygoid plate and from the tip of its hamulus. This group of fibers may be termed the *pterygopharyngeal muscle* portion of the superior constrictor. It is said to be responsible for a bulging of the posterior pharyngeal wall, which has been called *Passavant's pad* the better part of a century. This is a variably present ridge that forms at the approximate level of the velopharyngeal valve and occasionally may assist in closure. It is most noticeable in persons with a cleft of the soft palate and has been seen to appear variously in some normal adults. The sphincteric action of the superior constrictor, in combination with palatal elevation, displays this prominence on the posterior pharyngeal wall during forceful contraction. Occasionally, the palatopharyngeal muscle is designated as being responsible for giving rise to Passa-

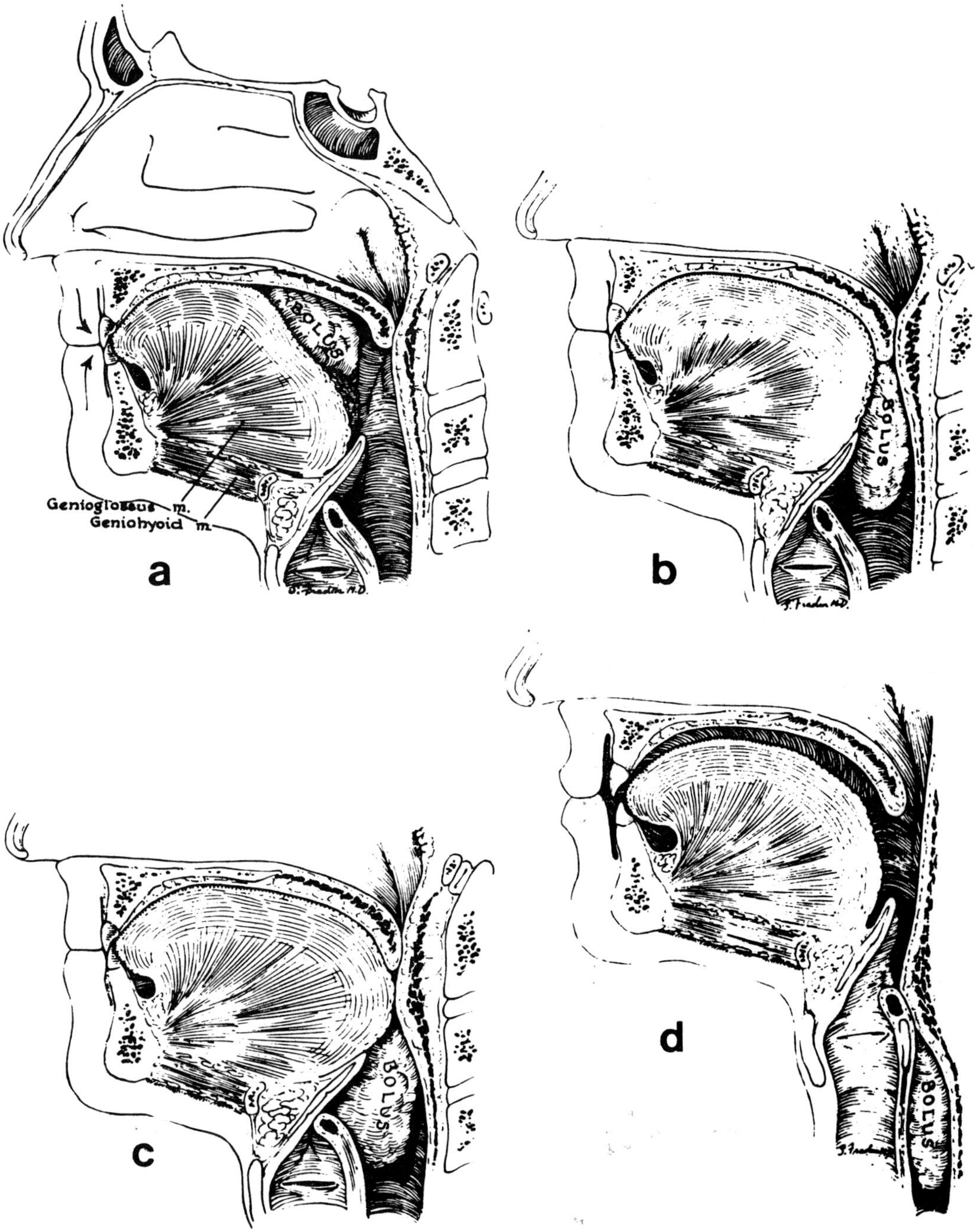

Figure 12-7. The stages of a swallow: (a) the oral stage; (b) the pharyngeal stage; (c) transition between pharyngeal and esophageal stages; (d) esophageal stage. (From Silverman, S.I.: *Oral Physiology*, 1961. Reproduced here with the permission of Doctor Silverman. Originally published by The C.V. Mosby Co., St. Louis.)

Table 12-II

MUSCLES OF THE PHARYNX

Muscle	*Origin*	*Insertion*	*Action*
Superior pharyngeal constrictor	pterygomandibular raphe, medial pterygoid plate, mylohyoid ridge of mandible, mucous membrane of floor of mouth, side of tongue, pharyngeal tubercle of basilar portion of occipital bone	median raphe	constricts pharynx, aids in velopharyngeal closure
Middle pharyngeal constrictor	greater and lesser cornu of hyoid bone, lower end of stylohyoid ligament	median raphe of posterior wall of pharynx	constricts pharynx
Inferior pharyngeal constrictor	oblique line of thyroid cartilage, side of cricoid cartilage	median raphe of posterior wall of pharynx	constricts pharynx
Stylopharyngeus	styloid process of temporal bone	wall of pharynx, thyroid cartilage	raises and opens pharynx
Salpingopharyngeus	cartilage portion of auditory tube	posterior part of palatopharyngeus	opens auditory tube; raises oropharynx and nasopharynx

vant's pad. A controversy on this point seems unwarranted until questions as to its function are better resolved (Fig. 12-8).

A second group of fibers arise from the posterior part of the pterygomandibular raphe, which is composed of a narrow tendinous band passing from the pterygoid hamulus to the retromolar triangle of the mandible. The fibers of the buccinator arise from the front of this raphe and, posteriorly, the superior pharyngeal constrictor fibers arise. With the orbicularis oris anteriorly, the two buccinators on either side, and the two superior constrictors at the back, the mouth is encircled in a horizontal plane with a broad band of muscle fibers.

A third point of origin for the superior constrictor is found on the posterior aspect of the mylohyoid ridge of the mandible, although it may be argued that the fibers actually arise from the membranous floor of the oral cavity and are only indirectly attached to the lower jaw.

A fourth group of fibers, though small in number, arises from the sides of the tongue. They may be continuous with the intrinsic lingual musculature, particularly the longitudinal muscle. Some writers do not list this fourth point of origin of the superior constrictor.

On the basilar portion of the occipital, just in front of the foramen magnum, is a fifth point of attachment, the pharyngeal tubercle, although this is not always considered a point of origin. This tubercle serves as the superior point of attachment of the pharyngeal aponeurosis and median raphe of the three pharyngeal constrictors. At this level, which is immediately below the base of the skull, there are no muscle fibers. This area is sometimes referred to as the sinus of Morgagni, a nonmuscular space lined by the pharyngeal aponeurosis. Instead, the superior fibers arch downward between the pterygoid plate and the pharyngeal tuber-

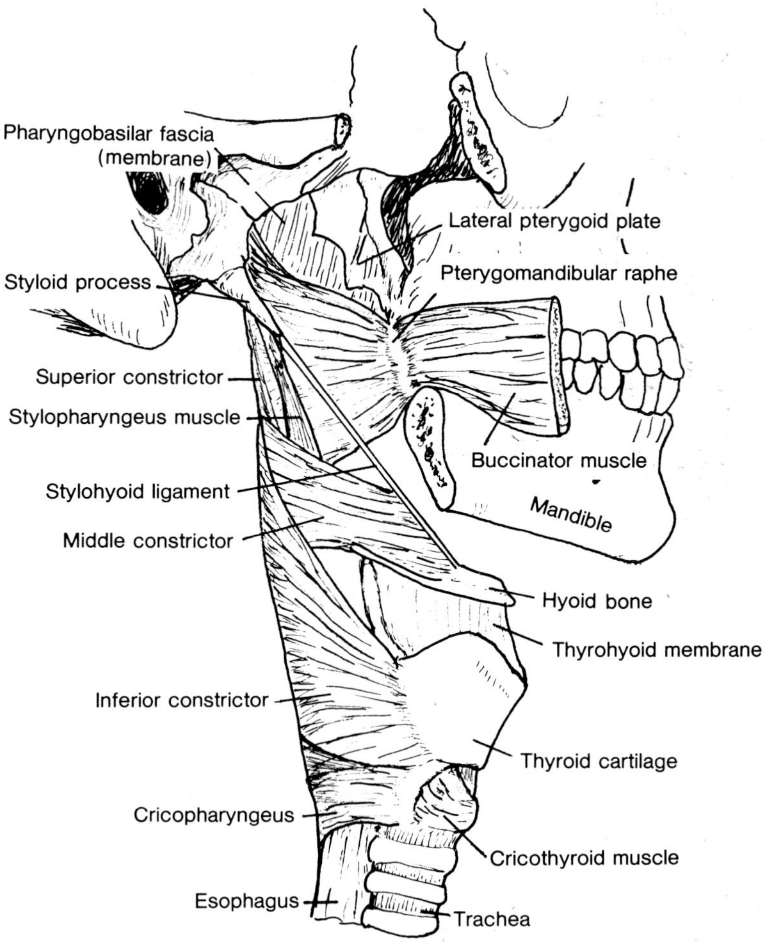

Figure 12-8. Constrictor muscles of the pharynx and their anterior attachments.

cle to permit the levator palatini and the auditory tube to pass into the nasopharyngeal cavity. Hence, the levator palatini and the cartilaginous auditory tube enter the upper part of the pharynx by passing between the skull and the uppermost fibers of the superior constrictor.

The walls of the pharynx are sealed in its upper part by a membrane of connective tissue and not muscle fibers. It may be termed the *craniopharyngeal membrane* or *pharyngeal aponeurosis*. The mucous membrane lining the pharynx is continuous with the aponeurosis. The pharyngeal aponeurosis serves as the uppermost part of the nasopharyngeal cavity, which is fixed in size due to the lack of muscle fibers and due to the spaced points of origin from which the pharynx is suspended (Fig. 12-8).

As mentioned earlier, all the pharyngeal constrictors have a common insertion, the median raphe. The distinguishing feature of the insertion points is that of height on the median raphe, although this is not reliable, since there is considerable overlap of the various constrictors. They telescope into each other like stacked pa-

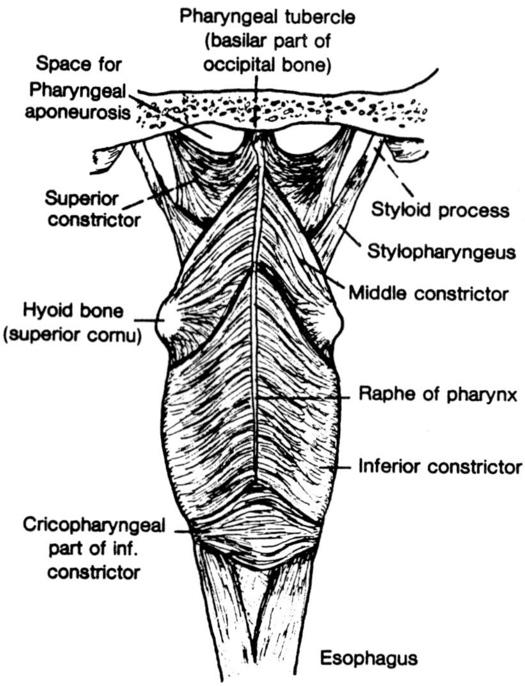

Figure 12-9. Muscles of the pharynx as viewed from a posterior view.

per cups. The superior constrictor inserts at the superior section of the median raphe.

The action of the superior pharyngeal constrictor is circular constriction, pulling forward the posterior pharyngeal wall. This movement, present during swallowing, becomes involuntary after the bolus of food has entered the pharynx. The pharyngeal constrictors react by progressive sphincteric action, known as peristalsis, to force the bolus into the esophagus. This sphincteric action is probably not considered significant for acts of speech by normal speakers but is present during gagging and regurgitation or similar causes of forcibly closing the palatopharyngeal valve. This is to say there may be activity during speech, but how important this movement is for articulation and resonance aspects of speech production is not clear. For insufficient palates, short or cleft, the movement is typically found and is deemed important. Innervation is by the pharyngeal plexus, with the vagus (X) cranial nerve supplying most of the fibers.

Middle Pharyngeal Constrictor

Typically, the fibers of the middle pharyngeal constrictor take their origin from the greater and lesser horns of the hyoid bone. Additionally, some fibers arise from the inferior part of the stylohyoid ligament. The fibers course in a transverse direction, the upper fibers curving steeply upward and overlapping those of the superior constrictor. Those below are overlapped by the inferior constrictor. Insertion is then gained at the median raphe.

When viewed posteriorly, the middle constrictor has the appearance of a large diamond-shaped field. Between the superior and middle constrictors is an opening in the lateral wall to permit passage of the stylopharyngeal muscle into the pharynx.

The action of the middle constrictor is parallel to that of the superior constrictor. Its function in deglutition has been outlined previously. The vagus (X) and glossopharyngeal (IX) cranial nerves supply this muscle through the pharyngeal plexus.

Inferior Pharyngeal Constrictor

The lowest section of the pharyngeal constrictors, the inferior pharyngeal constrictor, relates the pharynx to the larynx. There are two parts. The upper division arises from the oblique line on the lamina of the thyroid cartilage, with some fibers arising from the lesser horn of the hyoid bone. A lower division arises from the lateral circumference of the cricoid cartilage of the larynx. These two divisions may be termed, respectively, the *thyropharyngeal* and *cricopharyngeal muscles* or portions.

This is the thickest of the constrictor muscles. Its upper fibers run obliquely upward and cover the lower part of the middle constrictor. Fibers of the cricopharyngeus are horizontal, and the lowest fibers are continuous with those of the esophagus. All the fibers arrive at the posterior median line of the raphe for insertion.

Upon contraction, the inferior constrictor narrows the lower part of the pharynx from above downward to the esophagus during swallowing. During a laryngectomy, removal of the larynx, the surgeon must cut the cricopharyngeus and thyropharyngeus away from either side to release the larynx. Particular attention is paid to the suturing together of the two free ends of the cricopharyngeus at the midline, for it is believed that a good approximation of the muscle bundles will aid the patient in obtaining a good pseudoglottis for esophageal speech.

Innervation of the inferior constrictor is provided by the accessory (XI) cranial nerve through the pharyngeal plexus.

Stylopharyngeus

The stylopharyngeal muscle begins on the medial side at the base of the styloid process of the temporal bone. Coursing downward as a longitudinal muscle of the pharynx, the stylopharyngeus enters the pharynx in an opening between the superior and medial constrictor muscles. Inside the pharynx, it spreads its fibers to cover a wide area on the surface of the middle constrictor and the superior and posterior border of the thyroid cartilage of the larynx.

This long, slender muscle has the task of widening and elevating the pharynx. The dilating action is caused by the angle formed by the wide point of origin converging to the point of insertion. The glossopharyngeal (IX) cranial nerve is said to supply the stylopharyngeus. However, no one has demonstrated paralysis of this muscle or any functional deficit from damage to the motor branch of cranial nerve IX.

Salpingopharyngeus

The salpingopharyngeal muscle is frequently discussed as a part of the palatopharyngeal muscle. Regardless of the classification, the fibers, when present, are seen as a group arising from the lower portion of the cartilaginous pharyngotympanic or auditory tube near its orifice in the nasopharynx. They descend in the salpingopharyngeal fold formed by the mucosal lining of the pharynx covering this muscle. Next, they join those fibers of the palatopharyngeal muscle and spread in the lateral pharyngeal wall.

This variable muscle is thought to aid in elevation of the pharynx, but the fibers are not strong. It is also associated with distortion of the *torus tubarius*, the mound of mucous membrane at the orifice of the auditory tube, during swallowing and yawning. This action aids in permitting the air pressure in the middle ear to equalize with the outside air. The vagus (X) cranial nerve supplies the salpingopharyngeus through the pharyngeal plexus.

THE TONGUE

The tongue is associated with the oral cavity in vital speech functions, among other responsibilities. It is a muscular organ. With its anterior and lateral margins free, the tongue is capable of a wide variety of shapes and positions. The hyoid bone is related to the tongue by the hyoglossus muscle, an extrinsic muscle of the tongue.

Gross movements by the tongue are affected by the position of the hyoid bone. A number of suprahyoid and infrahyoid muscles hold the hyoid bone in suspension above the larynx and below the mandible. These muscles, however, are not directly associated with the tongue, precluding their classification as extrinsic tongue muscles. On the other hand, their influence on the underside of the tongue justifies considering muscles of the hyoid bone along with the tongue.

Earlier it was emphasized that much variability may be expected in the form and presence of the facial musculature. This is also true of the tongue and hyoid musculature. For example, the hyoglossus, which interconnects the hyoid bone and tongue, may sometimes include fibers from the sternum and larynx that bypass the hyoid bone and travel directly to the tongue, possibly affecting tongue position.

Muscles of the Tongue

There are two types of muscles associated with the tongue: the *extrinsic* and *intrinsic muscles*. All are paired except for the superior longitudinal muscle. The purpose of the extrinsic muscles is to position the tongue in the oral cavity; the intrinsic muscles shape the organ itself by grooving, flattening, rounding, etc. Perhaps a comparison may be drawn between these two sets of lingual muscles and a hand puppet. The arm holding up the puppet and the wrist serve as an extrinsic control to move the puppet's head around. Intrinsic control is provided by the fingers to shape the puppet's mouth and nose area. This is an oversimplification but serves as a starting point for understanding the

Table 12-III

MUSCLES OF THE TONGUE

Muscle	*Origin*	*Insertion*	*Action*
		EXTRINSIC MUSCLES	
Styloglossus	styloid process	side of tongue	retracts and elevates tongue
Hyoglossus	hyoid bone	posterior half of tongue along side	depresses and retracts tongue
Genioglossus	upper genial tubercle of mandible	tip to base of tongue; hyoid bone	protrudes the tongue; raises and protracts hyoid bone
Palatoglossus	soft palate	back and side of tongue	elevates tongue and narrows fauces
		INTRINSIC MUSCLES	
Superior longitudinal	root of tongue	tip of tongue	shortens and curls tip of tongue upward
Inferior longitudinal	root of tongue	tip of tongue	shortens and curls tip of tongue downward
Transverse	lingual septum	dorsum and lateral margins of tongue	elongation; narrows and raises sides of tongue
Vertical	dorsum of tongue	lower surface of tongue	flattens and broadens tongue

functions of the extrinsic and intrinsic muscles of the tongue (Table 12-III).

Extrinsic Muscles of the Tongue

Typically, three muscles — styloglossus, hyoglussus, and genioglossus — are classified as extrinsic muscles of the tongue. A fourth, palatoglossus, is sometimes included, but it is more aptly placed as a muscle of the soft palate. Therefore, the palatoglossus is not discussed in this presentation as an extrinsic muscle of the tongue; rather, it is classified and discussed as an extrinsic muscle of the soft palate (Fig. 12-10).

All the extrinsic muscles of the tongue are paired and are innervated by the hypoglossal (XII) cranial nerve.

The various positions of the tongue in the oral cavity are vital to speech sound resonance, i.e. certain consonants and vowels depend on tongue placement to effect a specific resonance as that sound's distinctive feature. Further modification of the sound stream is achieved by the intrinsic muscles of the tongue as they shape the organ itself.

HYOGLOSSUS. The hyoglossus muscle is a quadrilateral-shaped muscle arising from below the tongue on the hyoid bone. The area of attachment extends from the upper border of the hyoid body to the greater cornu of the bone. A group of fibers arising from the lesser cornu and body of the hyoid and having the same insertion as the hyoglossus is sometimes designated as the *chondroglossus muscle*. The chondroglossus is not viewed here as a separate muscle from the hyoglossus. From their origin, the fibers course

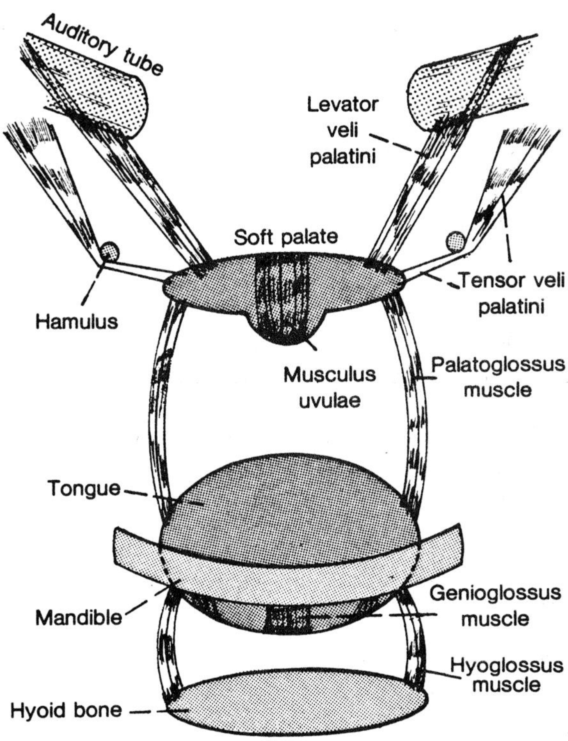

Figure 12-10. A schematic view of the extrinsic muscles of the tongue and their connections to movable structures.

directly upward to insert along the sides of the posterior half of the tongue. Fibers from the styloglossus and hyoglossus interlace at the side of the tongue.

The hyoglossus depresses the tongue. Using a tongue blade to keep the tongue flat on the floor of the oral cavity during inspection of the oropharyngeal port duplicates the effect of the hyoglossus on the tongue.

GENIOGLOSSUS. Through the genioglossus, the tongue is related to the mandible. This muscle takes its origin from the superior genial tubercle behind the mental symphysis of the mandible. From here its fibers course in a broadening fan shape back toward the tongue. The superior fibers reach into the tip of the tongue, the middle fibers travel almost horizontally to insert into the base of the tongue, and the inferior fibers attach both to the base of the tongue and to the hyoid bone. The genioglossus is responsible for forming the major bulk of the substance of the tongue.

Of the three extrinsic muscles of the tongue, the genioglossus is the strongest. It is principally a protractor of the tongue, with its horizontal fibers responsible for this movement. If only one of the pair of this muscle contracts, the tongue is drawn forward, protruding the tip to the side *opposite* the active muscle. Moving the tongue along the lips in a licking action involves this type unilateral contraction of one of the two genioglossus muscles. Contracting in unison, the pair protrudes the tongue in the midline. The anterior superior fibers, because of their vertical course, depress the tongue. The lowest group of fibers protract the hyoid bone and raise it (Fig. 12-11).

PALATOGLOSSUS. The palatoglossus is also classified as a muscle of the soft palate and is described in detail there. Nevertheless, it is listed here to emphasize its potential role for influencing tongue position in certain instances.

STYLOGLOSSUS. Of the three muscles taking their origins from the styloid process of the temporal bone, the styloglossus is the shortest (the other muscles originating from the styloid process are not associated with the tongue directly). Arising from the tip of the styloid process, the styloglossus fibers course downward and forward on the lateral surface of the superior constrictor muscle of the pharynx. The fibers enter the tongue at the base of the palatoglossal arch. From this point, the muscle turns in an almost horizontal direction, with some of the fibers continuing to the tip of the tongue.

The action of the styloglossus retracts and elevates the tongue, particularly during swallowing. During speech sound production, especially isolated sounds, this muscle assists in positioning the tongue for several sounds, including *k* and *g*.

Intrinsic Muscles of the Tongue

For convenience in identifying the intrinsic muscles of the tongue, they have been named after the course of the muscle fibers: vertical, longitudinal, and transversal. As intrinsic muscles, none are attached to a bone; all are confined to the lingual organ, contributing to its mass. The hypoglossal (XII) cranial nerve provides neural innervation for all the intrinsic muscles.

Altogether there are four intrinsic muscles of the tongue. There are two longitudinal muscles (superior and inferior), a transverse muscle, and a vertical muscle. Working as synergists with the extrinsic muscles of the tongue, they alter the shape of the tongue by shortening, narrowing, and curving it in various directions.

LONGITUDINALIS SUPERIOR. The superior longitudinal muscle is the only unpaired muscle of the tongue. It is situated

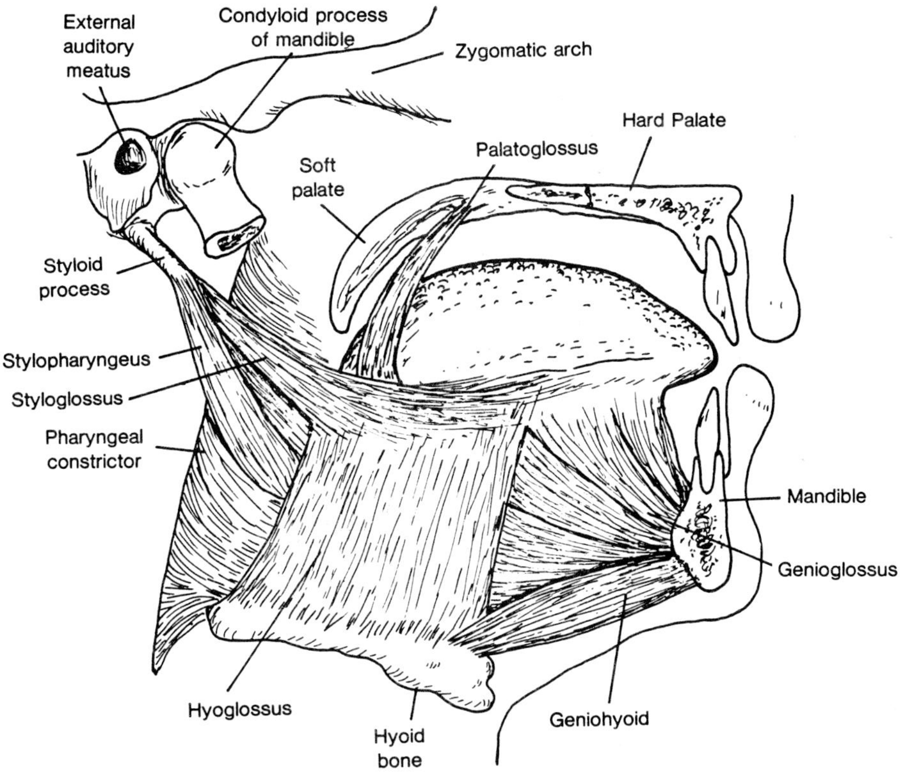

Figure 12-11. Extrinsic muscles of the tongue and their connections.

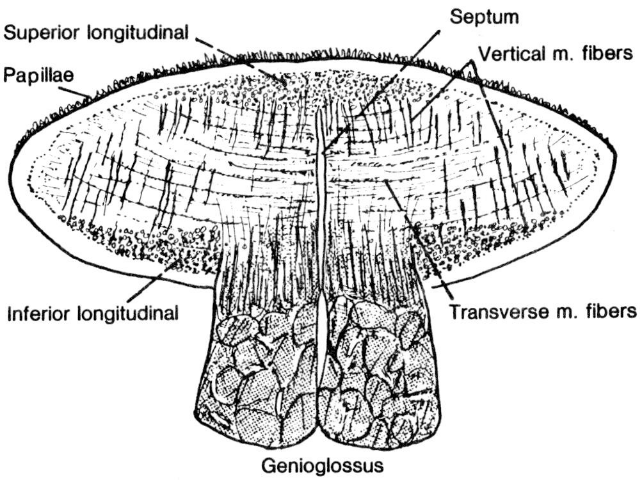

Figure 12-12. Coronal section of the tongue showing the intrinsic muscles.

just under the mucous membrane on the dorsum of the tongue. From a cross section of the tongue, it is evident that the lingual septum, a sickle-shaped plate of connective tissue in the middle of the tongue, does not extend to the dorsal surface. Hence, it does not separate the fibers of the superior longitudinal muscle into separate muscles. The fibers of this muscle arise from the submucous layer posteriorly near the epiglottis and from the median fibrous septum of the tongue. Coursing anteriorly, the fibers run anterior to the edges of the tongue and its apex. However, these attachments are indefinite.

Acting with its lower mate, longitudinalis inferior, these upper fibers shorten the tongue. The superior fibers alone curl the tip and sides upward making the dorsum of the tongue concave (Fig. 12-12).

LONGITUDINALIS INFERIOR. Like the superior longitudinal muscle, the inferior longitudinal muscle travels in a posterior-anterior direction. It is located on the inferior surface of the tongue between the genioglossus and hyoglossus muscles of the tongue. It arises from the mucous membrane of the root of the tongue, with a few of the fibers coming from the hyoid bone. The insertion is the apex of the tongue in the mucosa on the lower and dorsal surface. These two inferior longitudinal muscles, one on either side of the genioglossus, are rather widely separated.

The function of the inferior longitudinal muscle is to shorten the tongue, especially in combination with the superior longitudinal muscle. Acting alone, the longitudinalis inferior pulls the tip downward, making the dorsum of the tongue convex. Hence, to move the tongue tip up and down between the open lips, the two longitudinal muscles would alternately contract and relax.

TRANSVERSUS. The transverse muscle of the tongue arises from the lingual septum, which is composed of a thin plate of connective tissue buried in the midline of the tongue. Since the septum does not reach the dorsal surface nor the tip, the superior longitudinal fibers are undivided as they course from the back to the apex of the tongue. Underneath the superior lon-

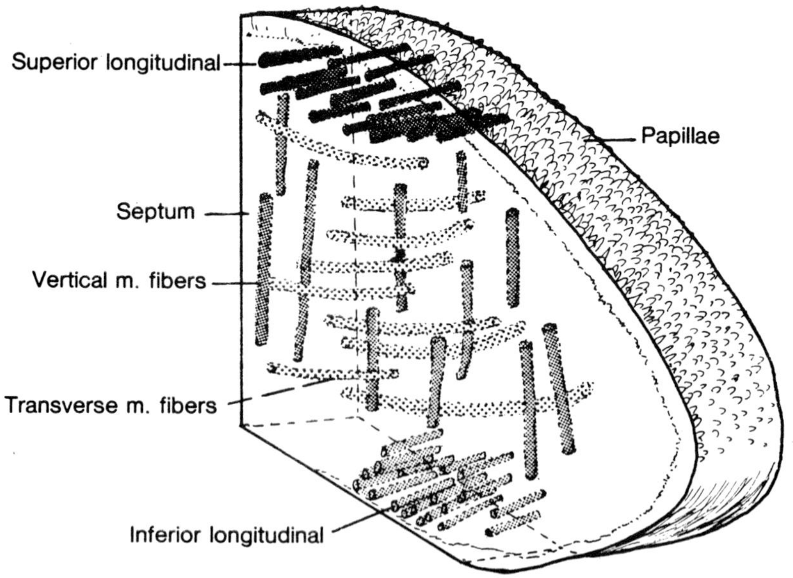

Figure 12-13. Schematic of the intrinsic muscles of the tongue.

gitudinal fibers, the transverse fibers travel from the septum in a lateral, horizontal direction to insert into the dorsal and lateral margins of the tongue, where they interlace with fibers of other muscles. The transverse muscle contributes a large portion to the fleshy part of the tongue (Fig. 12-13).

Principal functions of the transverse muscle include thickening (narrowing) of the tongue to elongate and to bulge the dorsal portion upward and lifting the lateral margins to help create a central groove.

VERTICALIS. A paired muscle, the *vertical muscle*, exists as vertical fibers at the borders of the tongue. These fibers extend from the upper to the lower surface of the tongue. Upon contraction, this muscle flattens and broadens the tongue.

BIBLIOGRAPHY

Calnan, J.: Palatopharyngeal incompetence in speech. In Pruzansky, S. (Ed.): *Congenital Anomalies of the Face and Associated Structures* Springfield, IL: Thomas, 1961.

Dickson, D.R.: Anatomy of the normal velopharyngeal mechanism. *Clin Plast Surg, 2*:235-248, 1975.

Dickson, D.R.: Normal and cleft palate anatomy. *Cleft Palate J, 9*:280-293, 1972.

Dickson, D.R., and Dickson, W.M.: Velopharyngeal anatomy. *J Speech Hear Res, 15*:373-381, 1972.

Dubrul, E.L.: *Sicher's Oral Anatomy*, 7th ed. St. Louis, C.V. Mosby, 1980.

Dworkin, J.P., and Culatta, R.: Tongue strength: its relationship to tongue thrusting, openbite, and articulatory proficiency. *J Speech Hear Disord, 45*:277-282, 1980.

Gardner, R., Gray, D.J., and O'Rahilly, R.: *Anatomy: A Regional Study of the Human Structure*, 4th ed. Philadelphia, Saunders, 1975.

Iglesias, A., Kuehn, D.P., and Morris, H.L.: Simultaneous assessment of pharyngeal wall and velar displacement for selected speech sounds. *J Speech Hear Res, 23*:420-447, 1980.

Langley, L.L., Telford, I.R., and Christensen, J.B.: *Dynamic Anatomy and Physiology*, 4th ed. New York, McGraw, 1974.

Massler, M., and Schour, I.: *Atlas of the Mouth*, 2nd ed. Chicago, Am Dental, 1958.

Paff, G.H.: *Anatomy of the Head and Neck*. Philadelphia, Saunders, 1973.

Porter, R.J., Jr., and Lubker, J.F.: The "Linguameter": a device for investigating tongue-muscle control. *J Speech Hear Res, 23*:490-494, 1980.

Scott, J.H., and Dixon, A.D.: *Anatomy for Students of Dentistry*, 4th ed. London, Churchill, 1978.

Silverman, S.I.: *Oral Physiology*. St. Louis, C.V. Mosby, 1961.

Snell, R.S.: *Clinical Anatomy for Medical Students*, 2nd ed. Boston, Little, 1981.

Chapter 13

MUSCLES OF THE FACE AND MANDIBLE

PRIMARILY, the muscles of the face surround the facial orifices — eye, ear, nose, and mouth. They have the ability to express a variety of emotions and consequently play a prominent role in communication of feeling in speech carried out in face-to-face situations.

The muscles. of mastication are also referred to as the muscles of the mandible. They permit the mandible to move and guide its functions in mastication and speaking.

MUSCLES OF THE FACE AND MOUTH

Most of the facial muscles are associated with the lips. They arrive at the circumference of the mouth from all directions, enabling numerous movements of the lips for facial expression and articulation of speech sounds.

The insertion of these muscles, especially those of the mouth, is usually to the deeper layers of the skin. The muscles often appear in scattered bundles so that the action of a single muscle is modified by other members of the group. Those muscles of the lower facial area contribute to a number of speech sound articulations.

Superficial Muscles of the Mouth Area

The superficial layer of muscles around the mouth may be termed *radial muscles* of the lips since they act upon the skin around the lips from various directions (see Fig. 13-2). All are paired, except for the sphincteric muscle of the lips, the *orbicularis oris*. Branches of the facial (VII) cranial nerve innervate (supply nerve fibers to) these expressive muscles (Fig. 13-1).

Platysma

By observing the neck region in a mirror while clenching the jaws and pretending to relieve pressure of a tight collar, one can see the sheet of muscle covering each side of the neck. This is the *platysma*. Platysma means broad, which describes the thin, sheet-like shape of this muscle (Table 13-I).

This muscle is extensive, extending from a wide area of the upper chest's su-

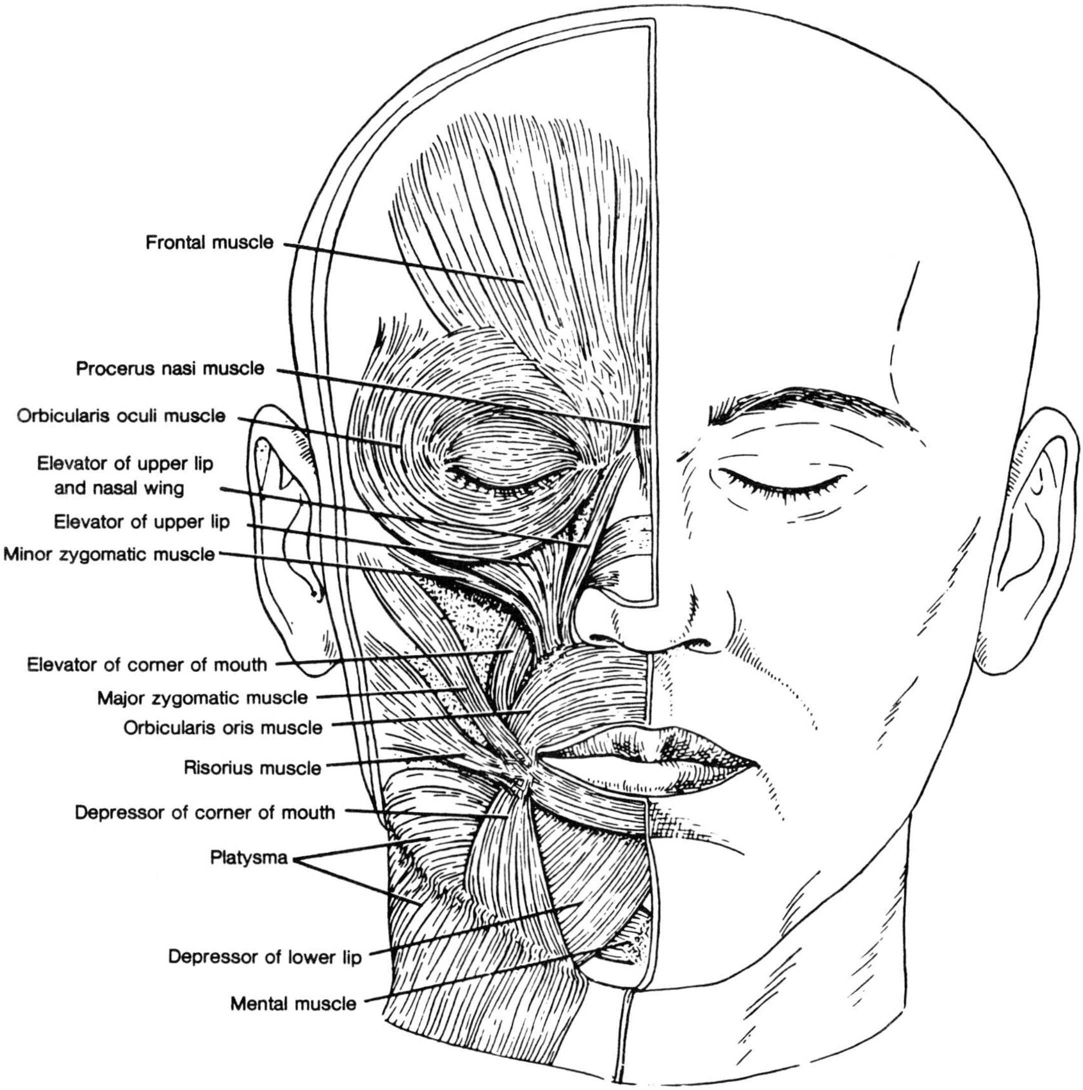

Figure 13-1. The muscles of facial expression. The muscles are diagrammatically illustrated and are subject to considerable variation, including unification of neighboring muscles and sometimes absence of some of the muscles. (From Sicher, H., and Tandler, J.: *Anatomie fur Zahnärzte*, copyright 1928. Courtesy of Springer-Verlag, Heidelberg.)

perficial fascia and overlapping the facial surface of the mandible. Its action in facial expression affects the corners of the mouth, the lower lip, and the mandible by pulling down on these. It may also alter the appearance of the neck by smoothing the skin under moderate contraction. Strong vertical ridges are revealed during forcible contraction.

Depressor Labii Inferioris

This muscle is a dilator of the lips, since its action separates the lips. Its fibers

Table 13-I

SUPERFICIAL MUSCLES OF THE MOUTH AREA

Muscle	*Origin*	*Insertion*	*Action*
Platysma	upper chest fascia, over clavicle	chin, mandible, skin around mouth	opens mouth, tightens skin of neck, may pull corner of mouth downward and lateralward
Depressor labii inferioris	chin of mandible	lower lip along from the midline	depresses lower lip
Depressor anguli oris	mandible	corner of mouth, lower lip	draws corners of mouth downward and possibly medialward
Levator labii superioris	three heads: two from maxilla and one from zygomatic bone	upper lip	raises upper lip
Zygomaticus	zygomatic bone	corner of mouth	raises corner of mouth upward and lateralward
Risorius	fascia of ramus over masseter muscle	corner of mouth	retracts corner of mouth lateralward
Mentalis	chin of mandible	skin of chin	raises and protrudes lower lip
Orbicularis oris	muscles adjacent to mouth send fibers encircling the mouth	skin around lips	closes and protrudes lips, compresses lips against teeth

radiate from the lateral aspect of the lower lip near the corner of the mouth. This broad, short muscle is sometimes referred to as the *square muscle of the lower lip*. Upon contraction, the depressor labii inferioris might assist in pulling the lower lip down away from the lower teeth for examining the gums of the lower jaw or for articulation of labial sounds when the teeth are kept tightly occluded.

Depressor Anguli Oris

The depressor anguli oris has a characteristic triangular fiber pattern and is sometimes called the *triangularis muscle*. This muscle originates from the outer surface and above the lower border of the mandible, interdigitating with many fibers of the platysma at this site. The fibers converge on the corner of the mouth, where they enter the tendinous node at the angle of the mouth. Due to its location near the corner of the mouth, it is included as a radial muscle of the lips. Its action on the corner of the mouth also places it as a dilator of the lips.

Facial expressions denoting sadness or the expression seen on a baby's face, with the corners of the mouth turned down just before he begins to cry after being insulted, may involve this muscle.

Levator Labii Superioris

The principal elevator of the upper lip is the levator labii superioris. This muscle is similar in configuration to the depressor labii inferioris; hence it carries an alternate label as the *square muscle of the upper lip*. It is also known as *quadratus labii superioris*.

At the insertion, the fibers appear as a unitary band interlacing with fibers of the orbicularis oris, but by following the common band of fibers up toward the origins, one finds the band splitting into three separate divisions of smaller bands, radiating under the orbit. In a few persons, the muscle fibers do not split distinctly.

The three points of origin are termed *heads*. Each head denotes a separate muscle, for the individual bundles of fibers may be traced all the way into the insertion regardless of the fusing taking place as the three bands converge into the upper lip as a single muscle plate. The levator labii superioris will be treated here as being composed of three separate muscle bundles: *zygomaticus minor, levator labii superioris*, and *levator labii superioris alaeque nasi*.

The zygomaticus minor is near the major zygomatic muscle on the lateral portion of the face. Its origin is on the prominence of the cheekbone. Due to its smaller size and lesser strength, it is also termed the *lesser zygomatic muscle*. It is also typically the weakest of the three composite muscles of the levator labii superioris and may not be developed in about 20 percent of the population.

The levator labii superioris is the center band of fibers of the three bands. It is somewhat confusing that the three muscles collectively are referred to by the same term as this middle group of muscle fibers. The origin of this middle muscle is just below the orbit; it is sometimes designated as the *infraorbital head*. Usually, this middle muscle of the three is the most well developed.

Medially is the *levator labii superioris alaeque nasi*. Its action elevates the upper lip and the wing (ala) of the nostril. This group of muscle fibers originates on the frontal process of the maxilla. The origin may also be termed the *angular head* of the composite levator labii superioris. The fibers descend steeply, perhaps as two groups of fibers, with one ending in the skin of the upper lip medial to the levator labii superioris or central muscle and the other splinter fibers inserting into the lateral circumference of the nostril, reaching as far as the posterior border. Some fibers of this medialmost part may be traced into the upper lip, ending in the area of the philtrum where they interweave with the orbicularis oris.

This composite muscle is one of the radial muscles of the mouth with the capability of dilating the lips upon contraction. It assists in facial expressions, such as sneering, and aids pipe smokers who hold the pipe clenched between the teeth at the corner of the mouth by raising the lip to enable them to puff smoke from the oral cavity through the teeth around the pipe stem. No doubt this extensive muscle may be involved in a host of facial expressions. The action of this group of muscles causes a fold to appear in the skin running laterally and inferiorly from the nasal wing. This fold is termed the *nasolabial groove*.

Zygomaticus

Zygomaticus major, or *greater zygomatic muscle*, overshadows zygomaticus minor in degree of development. It is not a division of a larger complex as is the minor muscle. It is often the most well-developed muscle of the midfacial zone. The zygomaticus major arises some distance behind the origin of its kin, zygomaticus minor. The insertion is at the corner of the mouth, after coursing from the lateral surface of the face where the zygomaticus takes its origin on the temporal process of the cheekbone.

Zygomaticus major is one of the radial muscles of the mouth, pulling the corner of the mouth toward the cheek (Fig. 13-2). Through its action, it is involved in ex-

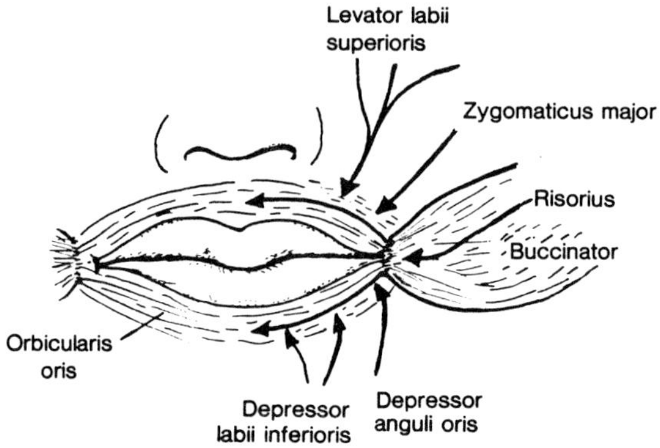

Figure 13-2. Radial muscles of the mouth (compare with Fig. 13-1).

pressions of smiling and possibly as a fixator for activity by the orbicularis oris during speech.

Risorius

This is a highly variable radial muscle of the mouth and is known as the smiling or grinning muscle (*risus* = laughter). Sometimes it is absent or is made up of only a few bundles of unconnected muscle fibers. Its origin is the fascia covering the masseter muscle of the ramus of the mandible. As the muscle converges toward the corner of the mouth, the muscle takes almost a horizontal course and becomes thickened. A few bundles enter the lower lip, while the majority end in the upper lip's mucous membrane and skin as well as at the corner of the mouth.

With this muscle, smiling is the principal facial expression; however, retraction of the corner of the mouth in other facial expressions may involve the risorius. It may also have the function of thinning the lips in expression. Careful articulation of the vowel *e* will result in some retraction of the corners of the mouth from risorius involvement.

Mentalis

The mental muscle or chin muscle is more commonly termed the *mentalis*. Its name is derived from its location (*mentum* = chin). The origin is the incisive fossa of the mandible. Some fibers course toward the midline as others arch toward the lower margin of the chin, but all eventually end in the skin of the chin area.

Upon contraction, the mentalis elevates the skin of the chin and causes the lower lip to turn outward. Since its origin is about the same level as the anterior oral vestibule, any reflexive contraction may significantly hamper dental work on the labial surface of the lower front teeth. This is especially true of dental cleaning procedures and orthodontic work.

Orbicularis Oris

The orbicularis oris is the most complex of the superficial muscles of facial expression. Its purpose is to regulate the opening and closing of the lips. Sucking, chewing, swallowing, expression, and speech are essential functions of this muscle of the mouth. It is important particularly in closing the lips for the labial plosives, i.e. *p*, *b*, and a number of other sounds.

The ability to suck liquids through a straw is a simple task due to the sealing effect of this muscle acting in a drawstring fashion around the straw. But unrepaired clefts of the lip preclude the normal sphincteric action of this muscle. The surgeon endeavors to reconstruct the integrity of the sphincter to effect the closing action served by this muscle in speech and eating functions.

There are no direct attachments to any bone by the fibers of the orbicularis oris. Instead, its fibers are derived from various muscle sources, including the radial muscles entering the lips from the several directions surrounding the lips and the buccinator muscle (cheek muscle) (see Fig. 13-3). Most of the fibers are placed on either side of the upper and lower lips, with heavy interlacing at the midline. Therefore, there is the appearance of a circular or sphincteric pattern coursing around the mouth opening. This is the only unpaired muscle of the expressive muscles.

Deep Muscles of the Mouth Area

The deep layer of muscles of the mouth, like the superficial layer, is involved in facial expression. There are fewer muscles in the deep layer, and only one, the buccinator, is considered as a significant muscle of the group (Table 13-II).

It may be of interest to the speech clinician to know that descriptions of muscles as presented here come from cadaveric dissection as presented by various writers. The facial muscles vary widely in form and presence in various specimens. The smaller facial muscles, in particular the levator anguli oris and the incisive muscle, are especially challenging to dissect out.

Levator Anguli Oris

Arising from the canine fossa below the infraorbital foramen, the levator anguli oris courses downward to insert in the corner of the mouth. Here its fibers interlace with the radial muscles of the mouth and the orbicularis oris.

Sometimes termed the *canine muscle* because of its origin in the canine fossa, this small muscle, as a rule, is well developed. Its action elevates the corner of the mouth while pulling it slightly medially.

Incisive Muscles

The *superior incisive* and *inferior incisive muscles* are closely related to the orbicularis oris. Their courses, after arising from the alveolar process at the canine eminence on the lower jaw and a similar location on

Table 13-II

DEEP MUSCLES OF THE MOUTH

Muscle	*Origin*	*Insertion*	*Action*
Levator anguli oris	maxilla near canine fossa	corner of mouth at upper lip	moves corner of mouth up and medially
Incisive muscles (upper and lower)	Upper: alveolar process at canine eminence of maxilla Lower: alveolar process at canine eminence of mandible	corner of mouth, fibers intermingling with orbicularis oris	presses on the respective vestibules, upper and lower, making them shallow
Buccinator	alveolar processes of maxilla and mandible, pterygomandibular raphe	mucous membrane of the cheek, corner of mouth	retracts corner of mouth, keeps cheeks taut against teeth

the upper jaw, closely follow the marginal bundles of orbicularis oris fibers. The insertion is the region of the corner of the mouth.

Their action, like the mental muscle, may press upon the upper and lower anterior oral vestibules, making them shallow and potentially interfering with dental work.

Buccinator

The buccinator is also referred to as the *cheek muscle* and, infrequently, the trumpeter's muscle (*buccinator* = trumpeter). This important muscle of the cheek is the basic structure of the lateral cheek wall. It is composed of a flat, thin, but strong set of fibers in contact with the mucous membrane of the lateral oral vestibule, also called the *buccal cavity*.

It arises from three locations, a horseshoe-shaped line along the base of the alveolar processes of the maxilla, the mandible, and the pterygomandibular raphe. This latter attachment is composed of a tendinous strip attached to the pterygoid hamulus and extending inferiorly and laterally to a point on the mandible behind the third molar at the base of the temporal crest (see Fig. 12-8).

The fiber bundles of the buccinator tend to converge as they course toward the corner of the mouth. Some of the fibers interlace around the corner of the mouth with the muscles ending there. Others terminate in the skin and mucous membrane of the upper and lower lips and the cheek.

The parotid duct is carried among the muscle bundles of the buccinator. The orifice is opposite the second upper molar and is usually readily observed with the

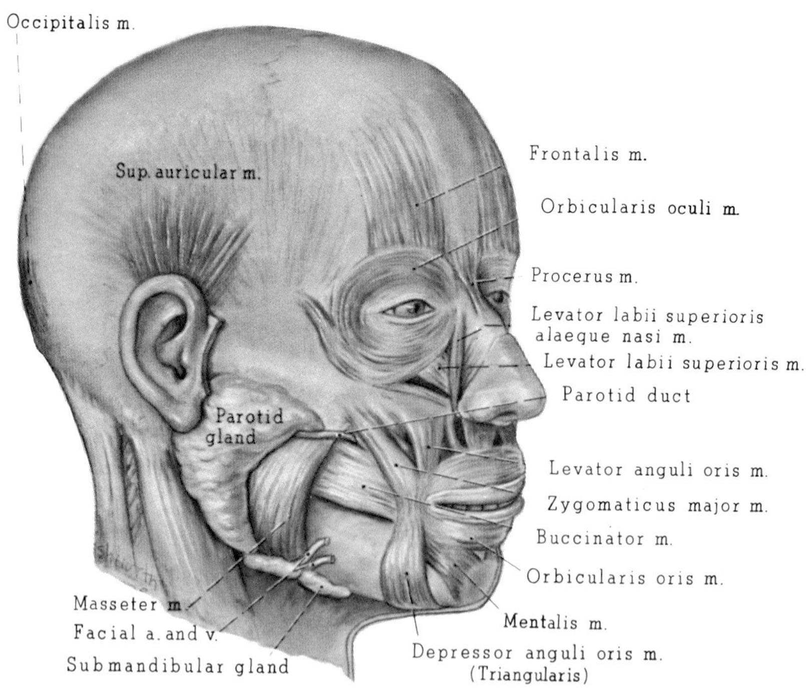

Figure 13-3. Deep and superficial muscles of the face. (From Jacob, S.W., and Francone, C.A.: *Elements of Anatomy and Physiology*, 1976. Courtesy of W.B. Saunders Co., Philadelphia.)

aid of a tongue blade to move the cheek away from the teeth.

Additionally, there is the *buccal pad of fat* (Bichat) lying in contact with the buccinator muscle. This rounded capsule of fat protrudes in front of the masseter muscle over the mandible and extends forward to the buccinator, filling in the space between the two muscles. It appears to have some basic duties related to mastication, since during progressive emaciation, this pad exists for a period considerably longer than do the subcutaneous deposits of fat.

The function of the buccinator is principally mastication. It moves food across the grinding teeth by tightening the lateral cheek wall to push the food onto the molars. In facial expression, it affects the corner of the mouth by pulling it laterally and posteriorly.

The buccinator muscle fibers at the pterygomandibular raphe are seen to surround the ligament and continue as fibers of the superior constrictor in impressive numbers. A case could be made for a claim that the buccinator and superior constrictor are the same muscle. However, for descriptive purposes, it remains convenient to designate the two as separate, especially since their innervation differs (Fig. 13-4).

The speech clinician should appreciate the interconnections of the buccinator with the superior constrictor and orbicularis oris because of the functional significance. The oral orifice valve bounded by orbicularis oris enjoys an intimate anatomical and physiological relationship with the velopharyngeal valve. Lateral cinefluorographic x-ray studies of individuals speaking indicate that velopharyngeal valving activity in some persons is influenced by anterior oral activity. The therapeutic technique of increasing the range of movement of anterior oral structures as a means of facilitating more active and consistent velopharyngeal contacts is logical from an anatomic and functional standpoint. The correlative activities between the front of the mouth and the back of the mouth are fruitful areas for study in clinical research.

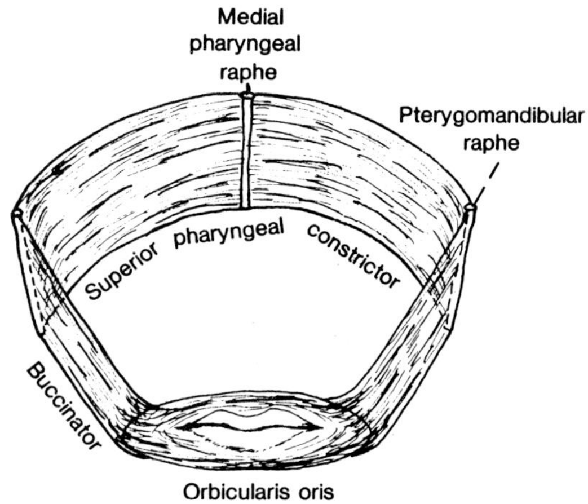

Figure 13-4. Muscular interconnections between the mouth and pharynx, a schematic concept.

MUSCLES OF THE NOSE

The nasal muscles do not participate in a prominent role in facial expression, but the subtleness revealed by their action is notable in communication. Sometimes speech clinicians see patients who exhibit a marked constriction of the nostrils and related facial movements as the person attempts to prevent excessive nasal air emission due to a cleft of the palate or due to an incompetent velopharyngeal closure. Alternately dilating and compressing the nostrils is related to respiratory functions

Table 13-III

MUSCLES OF THE NOSE

Muscle	*Origin*	*Insertion*	*Action*
Dilator naris	lateral incisive fossa and canine eminence of maxilla	posterior of nostril	dilates nostril
Compressor naris	lateral incisive fossa and canine eminence of maxilla	bridge of nose	constricts nostrils
Procerus	lower part of nasal bone	skin between eyebrows	wrinkles skin over bridge of nose

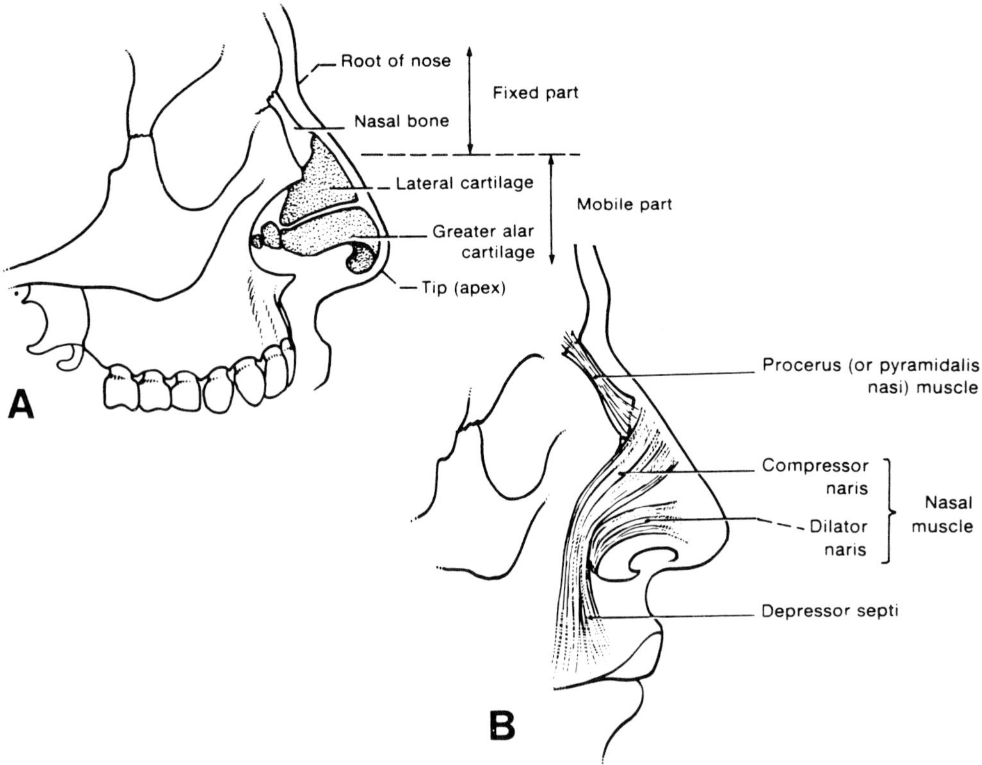

Figure 13-5. The cartilaginous framework for the nose (A) and overlying musculature (B). (From Bossy, Jean: *Atlas of Neuroanatomy and Special Sense Organs*, 1970. Courtesy of W.B. Saunders Co., Philadelphia.)

and to reflexive reactions as when in dusty areas.

All the nasal muscles are paired and are innervated by branches of the facial (VII) cranial nerve. The primary muscles of the nose are described separately below (Table 13-III).

Dilator Naris

As the name implies, this little muscle of the nose acts to enlarge or flare the nostril. It is aided in its function by the levator labii superioris alaeque nasi, previously presented as a muscle of the mouth (levator labii superioris), which inserts in the skin of the lip, and the alar (*ala* = wing) cartilage of the nose. Sometimes it is referred to as the *alar division* of the nose muscles due to its fibers inserting in the skin of the wing of the nose. It arises from the alveolar eminence of the lateral incisor (sometimes the incisive fossa) and canine of the maxilla (Fig. 13-5).

Compressor Naris

This small muscle acts to narrow the nostril by compressing the lateral margins of the nostril, drawing the mobile nasal cartilages medially. This muscle is also referred to simply as the *nasalis* or *depressor septi*. Rarely, it is termed the transverse nasal muscle, although its course does follow this descriptive pattern. Together with its contralateral mate, it forms a sling-like band across the anterior part of the nasal bridge.

The compressor naris is the muscle responsible for the clinical observation of a *nasal grimace* seen in some patients with velopharyngeal incompetence.

Procerus

The procerus is described as the slender muscle of the nose. It courses between the nasal bone and the skin between the eyebrows (Fig. 13-6). A mean expression can be achieved with the action of this muscle by pulling down the medial limits of the eyebrow.

MUSCLES OF THE EYELID

Altogether, there are three principal muscles which may be grouped under muscles of expression pertaining to the eyelid. The largest of these is the orbicularis oculi, the closer of the eyelids. The other muscles are the corrugator supercilii and the levator palpebrae superioris. On occasion, the procerus is listed with an expanded heading to include muscles of the eyebrow.

The muscles of the eye included here must not be confused with the intrinsic and extrinsic eye muscles which directly involve attachments to the eyeball and not the eyelid. However, the levator palpebrae superioris is traditionally classified as an extrinsic muscle of the eye and not part of the facial expressive musculature around the eye. The departure here is to emphasize the role of this muscle in facial expression (Table 13-IV).

Orbicularis Oculi

The orbicularis oculi is the sphincteric muscle of the eyelid. It is composed of two principal aspects, the *orbital* and *palpebral parts*.

The orbital part is located on the surface of the orbital margin of the forehead and cheek. It arises from the medial palpebral ligament and from the medial corners of the orbital cavity of the skull. From this source, the fibers set off on a

Table 13-IV
MUSCLES OF THE EYELIDS

Muscle	*Origin*	*Insertion*	*Action*
Orbicularis oculi	maxillary, frontal, and lacrimal bones; ligament of lid	skin encircling the eye	closes the eyelid and tightens skin of forehead; dilates lacrimal sac
Levator palpebrae superioris	rear of orbital cavity	tarsal plate of upper lid	raises upper lid
Corrugator supercilii	brow ridge of frontal bone	eyebrow and skin of forehead	pulls eyebrow medially

wide, arching course, forming a series of concentric loops around the top of the lid and continuing uninterrupted around the lateral portion of the orbit. The orbital fibers of the orbicularis oculi draw the eyebrows forcibly downward on contraction, producing a series of heavy wrinkles over the orbital margins.

The palpebral part begins at the medial palpebral ligament and the medial corner of the orbital cavity, as does the orbital part. Its fibers pass across both the upper and lower lids in front of the superior and inferior *tarsal plates* and end in the lateral corner of the eye socket. Some of the palpebrae muscle fibers form the fascia covering the lacrimal sac in the medial portion of the orbital cavity. This bundle is sometimes known as *Horner's muscle*, which dilates the lacrimal sac. The palpebral fibers of the orbicularis oculi act to gently close the eyelids as in sleep. Wrinkling of the skin around the circumference of the eye is accomplished by the orbital part of the orbicularis oculi as well as tightening the skin of the forehead (Fig. 13-6).

Corrugator Supercilii

The corrugator supercilii is a small band of muscle fibers lying deep to the orbicularis oris. It takes its origin from the superciliary arch. After traveling laterally on a horizontal course, the fibers end in the skin of the eyebrow.

The action of the corrugator wrinkles the eyebrow by pulling the brow medially, which causes vertical wrinkles of the forehead between the brows and the upper bridge part of the nose. Expressions of frowning rely on this muscle.

Levator Palpebrae Superioris

This muscle, as stated earlier, is usually classified as an extrinsic muscle of the eye but is being included here as one of the muscles of facial expression due to its role in raising the upper eyelid. Hence, the classification in this instance is based on the function of the muscle being linked to those around the eyelids and not the eyeball. The levator palpebrae superioris is not responsible for movement of the eyeball as are the extrinsic eye muscles, which further justifies its inclusion here. It is true that it is a muscle of the orbital cavity but not of the eyeball (Fig. 13-7).

The levator palpebrae superioris arises at the rear of the orbital cavity in the same area as the extrinsic muscles of the eyeball. From this posterior point, it extends forward along the roof of the orbital cavity and above the other muscles in the cavity. Upon reaching the eyelid, it diverges into the upper tarsal plate and skin of the upper eyelid near its margin.

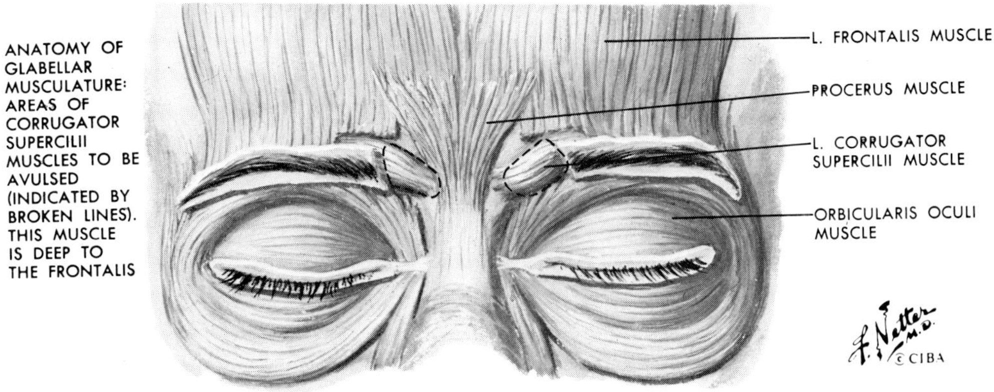

Figure 13-6. The procerus muscle and surrounding musculature of the eyelid. (© Copyright 1967, CIBA Pharmaceutical Company, Division of CIBA-GEIGY Corporation. Reprinted with permission from CLINICAL SYMPOSIA illustrated by Frank H. Netter, M.D. All rights reserved.)

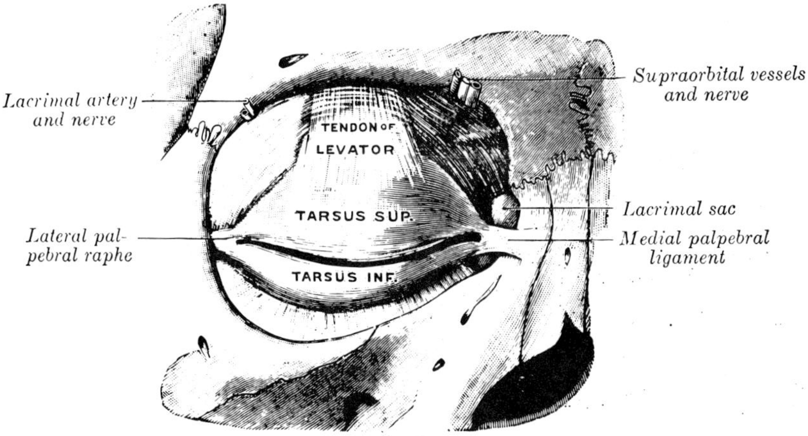

Figure 13-7. The levator palpebrae superioris muscle and connections. (From Goss, Charles M. (Ed.): *Gray's Anatomy of the Human Body*, 28th ed., 1973. Courtesy of Lea & Febiger, publisher.)

Contraction of this muscle raises the eyelid. *Check ligaments* — the medial and lateral palpebrae ligaments — are anchored to the bony orbital walls to prevent the drawing of the eyelid completely into the orbital cavity. Occasionally, the lateral tension on the lids is tight enough to cause difficulty to some wearers of contact lens. Putting in the lens is extra troublesome. Often, these individual's eyeballs protrude more than usual from the orbital cavities, which may relate to the extra lateral tautness of the eyelids. A number of nuances are reflexively conveyed in visual communication by the action of the eyelids, as in the "saucer-eyed" appearance upon being presented with shocking news, e.g. contest winners on television. Unlike the innervation of all other facial muscles, the nerve supply is derived from the oculomotor (III) cranial nerve.

MUSCLES OF THE EYE

Table 13-V
EXTRINSIC MUSCLES OF THE EYE

Muscle	*Origin*	*Insertion*	*Action*
Rectus muscles Superior Inferior Medial Lateral	all arise from around the optic foramen at the rear of the orbital cavity	all insert into the eyeball respective to each name	each turns the eyeball toward the respective rectus: upward, downward, inward, outward
Superior oblique	rear of orbit near optic foramen	eyeball deep to superior rectus after passing through a trochlea	aids in abduction, depression, and medial rotation of the eyeball
Inferior oblique	floor of orbit	lower and lateral surfaces of the eyeball	aids in abduction, elevation, and rotation of the eyeball
Levator palpebrae superioris	roof of orbital cavity at rear	tarsal plate of upper lid	raises upper lid

The eye may be significant in the role of facial expression, at least in certain instances. Muscles pertaining to eyeball movements are treated under the discussion of "Extrinsic Muscles of the Eye." All are considered extrinsic eye muscles due to the fact that one end of each muscle is not attached to the eyeball (Table 13-V).

Extrinsic Muscles of the Eye

The eyeball is a spherical structure with an anterior swelling or bulge called the *cornea*. It is able to turn in all directions in its socket-like housing, the orbital cavity. There are seven extrinsic muscles of the eye associated with each orbital cavity. Sometimes they are termed the *ocular muscles*. Their function is to cause the eyeball to rotate in several planes in order to position rapidly the pupil for varied fields of viewing without moving the head. Their contribution to communication is apparent if the listener looks away or if the speaker or listener rolls his eyes upward to punctuate meaning. The latter behavior might represent absurdity in the communication.

One of the ocular muscles is not attached to the eye but lies within the orbital cavity. This is the elevator of the upper eyelid. It was previously presented as a muscle of facial expression due to its role in registering and communicating visually emotional reaction to others in face-to-face communication. All other extrinsic eye muscles insert in the sclera, the white (outer coat) covering of the eyeball.

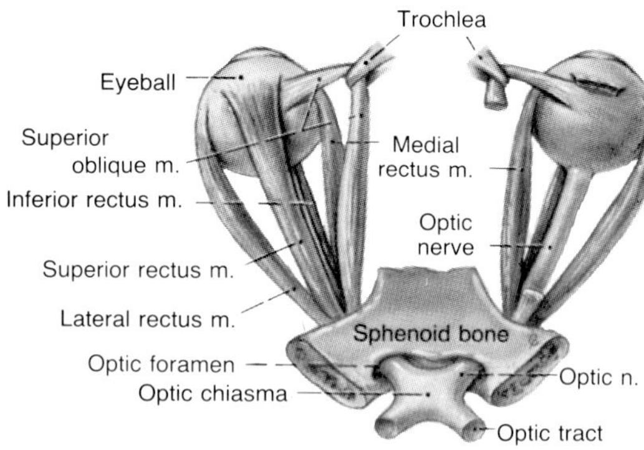

Figure 13-8. Extrinsic muscles of the eye (inferior oblique not shown). (From Jacob, S.W., and Francone, C.A.: *Structure and Function in Man*, 4th ed., 1978. Courtesy of W.B. Saunders Co., Philadelphia.)

Rectus Muscles

There are four rectus muscles, *inferior rectus, superior rectus, lateral rectus*, and *medial rectus*. Each is named according to its insertion on the eyeball. *Rectus* means straight and denotes the direct course the rectus muscles take from their point of origin to their insertions. All arise from a common location at the rear of the orbital cavity from a ring or cuff of fibrous connective tissue, the *annular tendon*, which surrounds the optic foramen and is part of the superior orbital fissure. These short muscles course directly from their origins to their place of insertion, in contrast to the indirect courses of other extrinsic eye muscles (Fig. 13-8).

Each rectus may turn the eyeball in the direction of the muscle's insertion, (Fig. 13-9). Typically, they function in combination with other extrinsic eye muscles. All except the lateral rectus are supplied by the oculomotor (III) cranial nerve. The lateral rectus is, instead, innervated by the abducens (VI) cranial nerve.

Superior Oblique

Complicated movements of the eyeball are possible with each of the two oblique muscles, due to their unusual courses. The delicate adjustments required for true stereoscopic vision are a function of the recti acting with the oblique muscles. Upon contraction, the superior oblique pulls the eyeball to aid in downward and downward-outward movements.

Of the two oblique muscles of the eye, the superior oblique presents the most unusual course. It arises at the optic foramen from the sphenoid bone along with the rectus muscles, just above the annular ring. It extends initially straightforward along with the rectus muscles. As it reaches the anterior medial orbit, it becomes tendinous and passes through the *trochlea*, a small, fibrous pulley attached to the frontal bone on the roof of the orbital cavity. From here, the superior oblique reverses its direction to insert into the upper lateral aspect of the eyeball. It is served by the trochlear (IV) cranial nerve.

Inferior Oblique

The inferior oblique takes its origin from the medial-front part of the orbital cavity. It courses obliquely and laterally below the eyeball. Continuing, it passes beneath the eyeball to insert into the lower lateral surface of the eyeball near the attachment of the superior oblique.

Like the superior oblique, the inferior oblique is responsible for aiding in delicate, complicated movements of the eyeball. Upward and lateral positions of the cornea are achieved by the action of this muscle, although it works to assist the recti. Innervation is via the oculomotor (III) cranial nerve.

Levator Palpebrae Superioris

This muscle of the orbital cavity is not attached to the eyeball. Rather, it passes over the eyeball on its course from the rear of the orbit and inserts in the upper eyelid or, specifically, the superior tarsal plate. Its point of origin is in the same region as the rectus muscles of the orbital cavity at the rear of the orbit. Its forward end widens into an aponeurotic sheet, which reaches into the upper eyelid to raise the lid.

The medial and lateral palpebrae ligaments serve as limiting ligaments to prevent the *levator palpebrae superioris* from drawing the upper eyelid into the orbital cavity upon contraction. It is served by the oculomotor nerve.

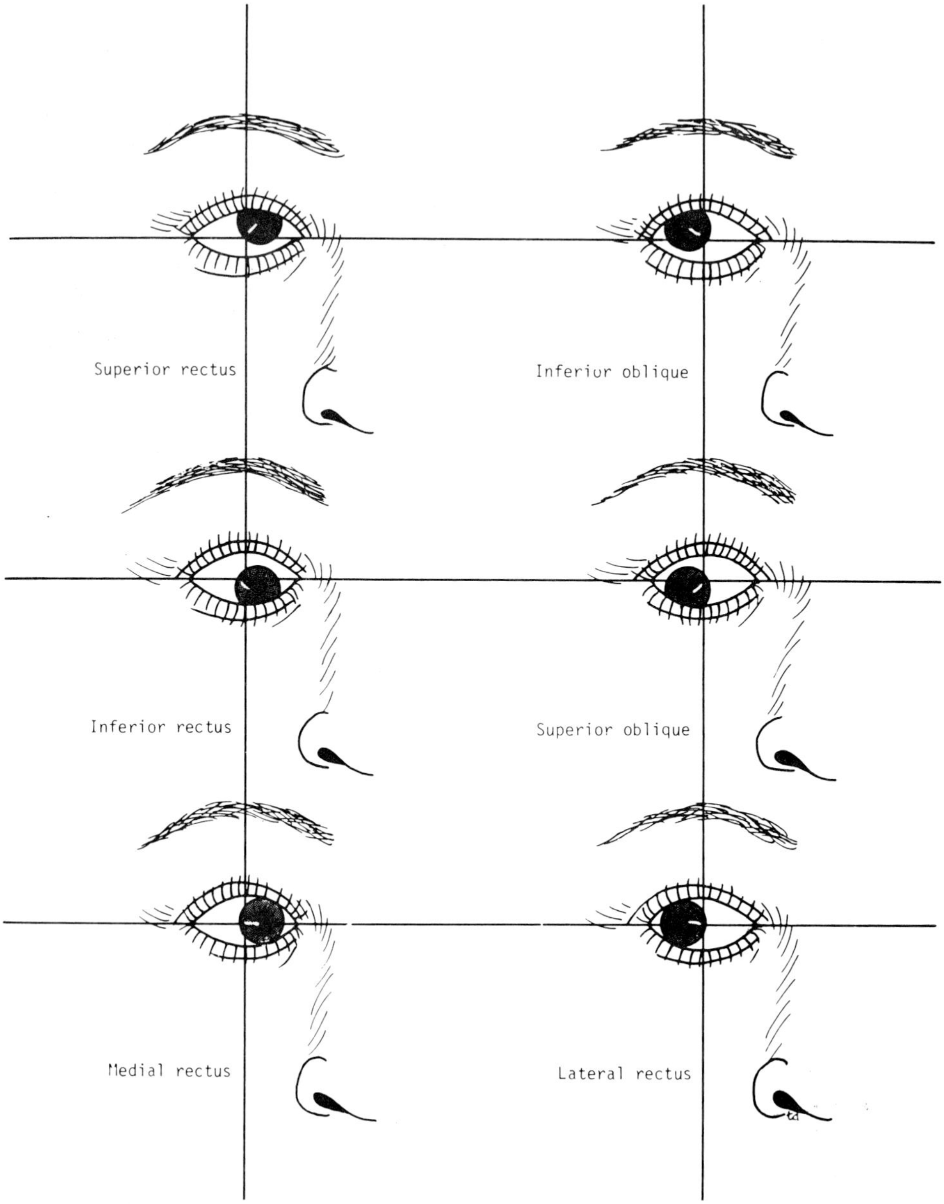

Figure 13-9. Cooperative interactions of ocular muscles in moving the eyeball. (From Snell, R.S.: *Clinical Anatomy for Medical Students*, 2nd ed., 1981. Courtesy of Little, Brown and Co., Boston.)

Clinical Aspects of the Ocular Muscles

The tonus of the ocular muscles maintains the primary position of rest for the eye. This is the same position as when the person gazes straight ahead into the distance without focus on any specific point in space. In the rest position, the visual axes of the eyes are parallel. When focused on a particular point, the eyes converge so that the axes meet at the definite object. The degree of convergence is dependent on the nearness of the object. The interplay of the ocular muscles enables rapid, reciprocally controlled movement of the eyeballs.

In paralysis of the nerves innervating the ocular muscles, the deviation of the visual axes are termed *tropias*, while the general term for such anomalies is *strabismus*. Diplopia is double vision resulting from asynchronous visual axes of the eyes. Drooping of the upper eyelid (ptosis) is due to paralysis of the *levator palpebrae superioris*. Nystagmus is an involuntary rapid movement of the eyeball; it may appear as oscillation of the eyeball in response to disturbances of the vestibular apparatus.

If the lens of the eye does not focus the image precisely on the retina, visual clarity is disturbed. It should be noted that this class of clinical conditions is unassociated with the ocular muscles (Fig. 13-10).

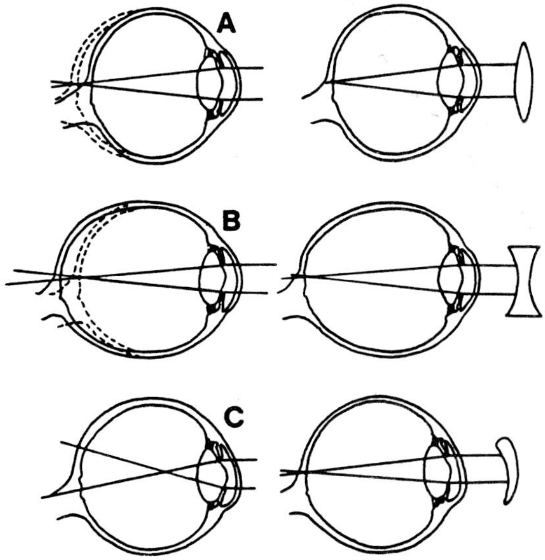

Figure 13-10. (A)Hyperopia. The focus of the image extends beyond the retina until corrective lens are added (right). (B) Myopia. The focus of the image is in front of the retina with corrective lens added (right). (C) Astigmatism. Uneven focusing of the image resulting from distortion of the curvature of the lens with corrective lens added (right). (From Snell, R.S.: *Clinical Anatomy for Medical Students*, 2nd ed., 1981. Courtesy of Little, Brown and Co., Boston.)

MUSCLES OF THE SCALP

Except as they affect the skin of the forehead, the scalp muscles may not be thought of as important muscles of facial expression. Included are two pairs of muscles at the front and back of the scalp. These move the scalp, usually to a limited degree, depending on the individual. The other muscles associated with the scalp belong to the external ear and are known as the *auricular muscles*. These muscles are often highly functional in animals other than man. Their purpose is to turn the animal's auricle in the direction of attention-commanding sounds to enhance sound-gathering from the air. However, this is not possible in man. The muscles are rudimentary and typically nonfunctional in human ears. All muscles of the scalp area are paired (Table 13-VI).

Occipitofrontalis

Two pairs of muscles comprise the occipitofrontalis muscle. One pair begins at the back of the skull, courses up, and disappears into the *epicranial aponeurosis*, a thin, dense sheet of connective tissue connecting the two muscle bellies. The other pair emanates from the epicranial aponeurosis at the front of the scalp and courses to the eyebrow line. The anterior pair of muscles may be termed the *frontalis*

Table 13-VI
MUSCLES ASSOCIATED WITH THE SCALP

Muscle	*Origin*	*Insertion*	*Action*
Occipitofrontalis			
Occipital Division	occipital bone at superior nuchal line	epicranial aponeurosis	moves scalp and works with frontalis by fixating epicranial aponeurosis
Frontal Division	anterior margin of the epicranial aponeurosis	skin of forehead	raises eyebrows; wrinkles forehead
Auricularis			
Anterior Aspect	temporal fascia	front of helix	feeble movements of the auricle
Superior Aspect	epicranial aponeurosis	triangular fossa	feeble movements of the auricle
Posterior Aspect	mastoid process	medial surface of concha	feeble movements of the auricle

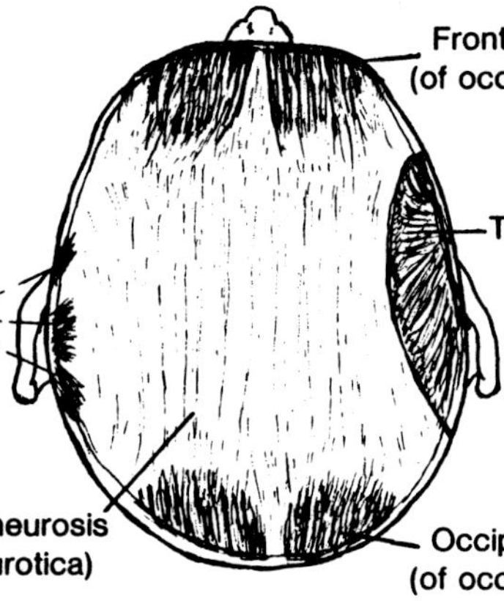

Figure 13-11. Muscles of the scalp.

muscle, or the frontal belly of the occipitofrontalis muscle. The posterior belly is known as the *occipital muscle*, or occipital belly of the occipitofrontalis. As noted above, the frontal belly takes its origin from the brow line; the occipital belly arises from the superior nuchal line of the occipital bone (Fig. 13-11).

Special consideration should be given the epicranial aponeurosis interconnecting the separate parts of the occipitofrontalis. Known also as the *galea aponeurotica*, the aponeurotic cranial cap serves as a common tendon between the frontalis and occipitalis. The skin is tightly bound to this aponeurosis and moves with it whenever the muscles at either end contract (Fig. 13-12).

The technique of "scalping" in America's early history was performed by cutting the skin of the forehead and pulling back the epicranial aponeurosis with its upper layer of skin and hair to expose the periosteum of the cranial bones. Lacerations of the scalp, even small ones, may result in severe loss of blood due to the rich vascular supply to the hair follicles of the scalp. Because of the tightly stretched skin of the scalp, blows from blunt objects, fists or baseball bats, may result in a split in the scalp similar to a cut. The pull of the *occipitofrontalis* is the source of tension, which will, in turn, pull such wounds open, necessitating sutures. These wounds heal readily due to the profuse blood supply. The action of the frontal belly pulls up the skin of the forehead from the brow region, causing transverse wrinkles, which denote surprise and similar emotional expressions. Acting together, the two parts of the occipitofrontalis move the scalp in varying amounts from one individual to another. The facial (VII) cranial nerve innervates this muscle.

Auricular Muscles

Three muscles, located in front, above, and behind the ear, comprise the muscles of the external ear. They are classified as scalp muscles due to their origins in the epicranial aponeurosis. Furthermore, they lie on the lateral surface of the scalp (see also Fig. 2-27). Each one ends at the cartilaginous root of the auricle. This group of muscles is highly variable in man, owing to their vestigial classification, that is, these muscles are a remnant of a previously functional structure. The loss of some former function is suggested by their presence; perhaps man could once duplicate movements of the auricle after the fashion of cats and dogs.

Based on their location, the individual members of this muscle group are named the *anterior, superior*, and *posterior auricular muscles*. The anterior auricular might be termed the "protractor of the ear," and the posterior auricular acts as the "retractor." In terms of strength, the posterior auricular is strongest and the superior auricular is typically the largest. Voluntary move-

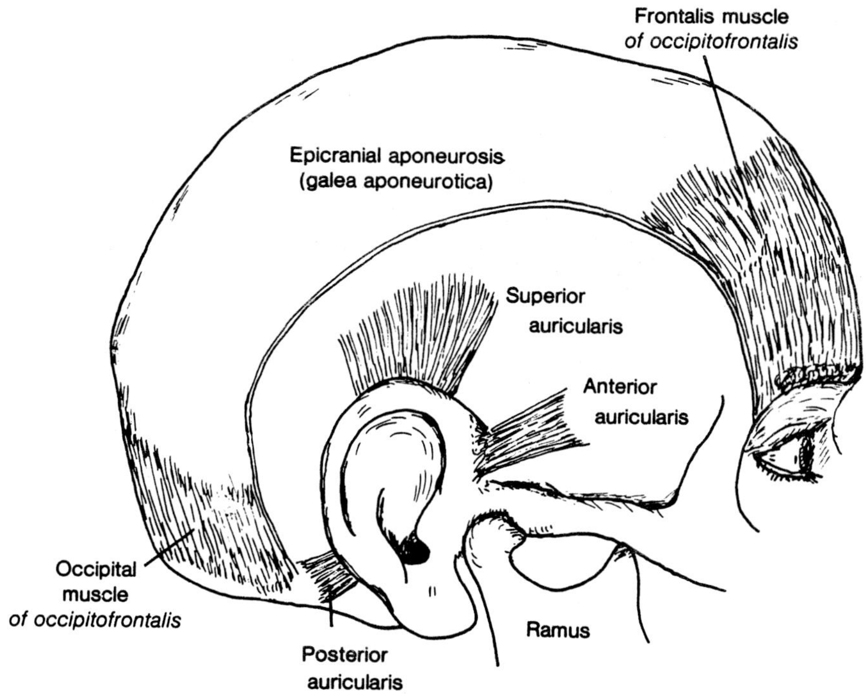

Figure 13-12. Muscles of the scalp and auricle.

ments are usually feeble, except in a few gifted and practiced individuals. The muscles may frequently contract involuntarily as other muscles of facial expression act. Cranial nerve VII (facial) innervates the auricularis muscles.

MUSCLES OF MASTICATION

Table 13-VII
MUSCLES OF MASTICATION

Muscle	*Origin*	*Insertion*	*Action*
Masseter	zygomatic arch	ramus of mandible	raises and may retract mandible
Temporalis	temporal fossa	coronoid process of mandible	raises and retracts mandible
Lateral pterygoid	lateral pterygoid plate and greater wing of sphenoid	neck of mandible and temporomandibular joint capsule	lowers and protrudes mandible; lateral movements with alternate contraction
Medial pterygoid	lateral pterygoid plate, pterygoid fossa, and maxillary tuberosity	ramus of mandible	raises and protrudes mandible; lateral movements with unilateral contraction

To enable mandibular movement, there are four pairs of primary muscles of the mandible designated as muscles of mastication. Each is innervated by branches of the trigeminal (V) cranial nerve, whereas the muscles of the face receive innervation from the facial nerve (cranial nerve VII). This difference in innervation is often useful for the patient with damage to one of these nerves. For example, when the trigeminal nerve is damaged, some ability to open and close the mouth is taken over by the muscles of facial expression. The partial reciprocity between cranial nerves V and VII permits the maintenance of some functions that would otherwise be completely impaired if the entire facial area were innervated by a single nerve. This is important to remember in generating a treatment strategy for a neurologically impaired patient (Table 13-VII).

Besides the muscles of mastication, other muscles exert influence on mandibular movement, such as the platysma and digastric. They are not, however, properly included with the primary masticatory muscles.

Masseter

Masseter means "chewer." It is one of the two typically powerful muscles closing the jaw, as in biting, and is the most superficial of all the muscles of mastication. As it elevates the mandible, it draws it slightly backward; thereby, a combination movement of elevation and retrusion is present (Fig. 13-13).

Upon contraction, as in forcibly clenching the teeth, the masseter is easily palpable over the ramus. It arises primarily from the lower border of the zygomatic arch. Afterwards, the heavy concentrations of muscle fibers travel down over the lateral surface of the ramus. The insertion includes the outer surface of the coronoid process to the angle of the mandible.

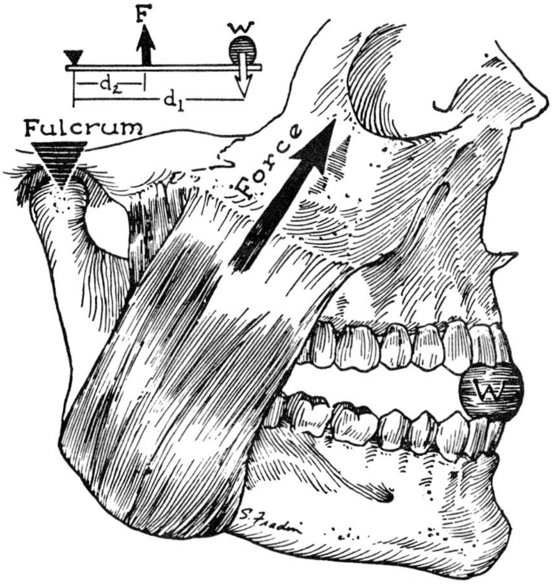

Figure 13-13. The masseter muscle exhibiting a Class III lever system. (From Nagle, Raymond J., and Sears, Victor, H.: *Denture Prosthetics*, ed. 2, St. Louis, 1962, The C.V. Mosby Co.)

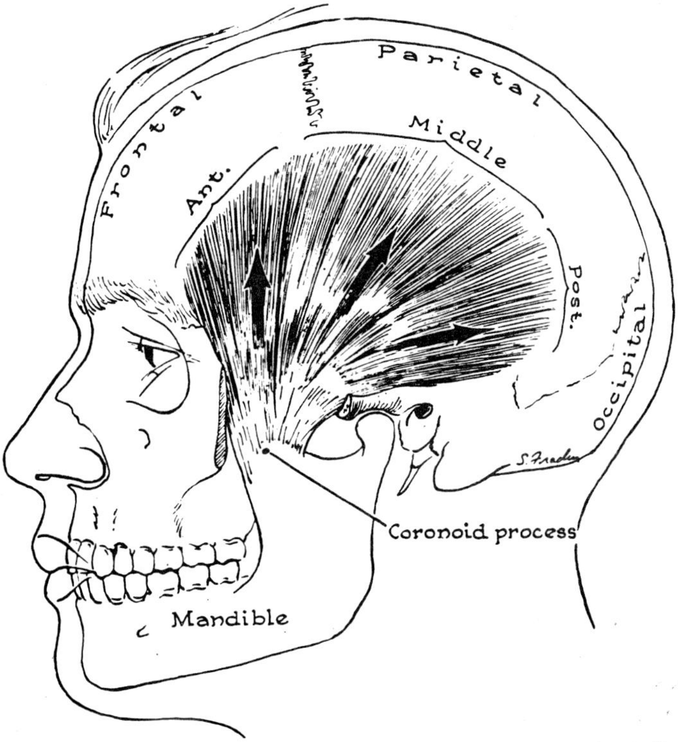

Figure 13-14. The three divisions of the temporalis muscle. (From Nagle, Raymond J., and Sears, Victor H.: *Denture Prosthetics*, ed. 2, St. Louis, 1962, The C.V. Mosby Co.)

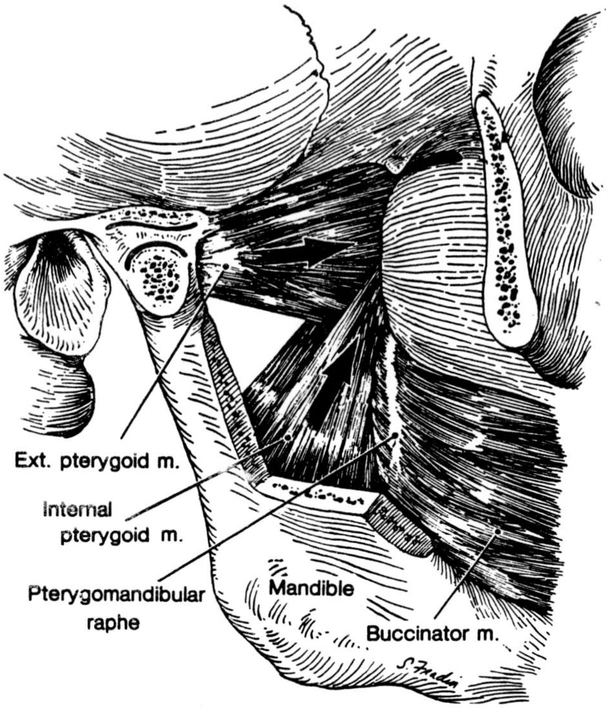

Figure 13-15. The pterygoid muscles. (From Nagle, Raymond J., and Sears, Victor H.: *Denture Prosthetics*, ed. 2, St. Louis, 1962, The C.V. Mosby Co.)

Temporalis

The temporal muscle, like the masseter, may be palpated upon clenching the jaw strongly. It is located in the temporal fossa. Hence, the origin is a very wide field, limited by the inferior temporal line of the skull's lateral aspect. The fibers converge, passing behind the zygomatic arch to insert on the coronoid process of the mandible, which is located inside the zygomatic arch. The field of insertion extends from the tip of the coronoid process down onto the anterior part of the ramus (Figs. 13-14 & 13-17).

Somewhat in contrast to the masseter, the *temporalis* is designed for movement rather than power. It elevates the mandible and pulls it back. This action is understandable when one examines this fan-shaped muscle. The anterior and superior fibers have a vertical or near-vertical pull on the coronoid process, and the posterior fibers have an oblique pull, which retracts the mandible.

Lateral Pterygoid

The lateral pterygoid or external pterygoid arises from two heads attached to the outer surface of the lateral pterygoid plate and the infratemporal surface of the greater wing of the sphenoid bone. Its course is horizontal and outward, closely following the cranial bone. The fibers fuse in front of the temporomandibular joint and insert into the front of the neck of the condyloid head on the mandible and capsule, including the articular disk of this joint. Retracting fibers are attached directly to the mandible (Figs. 13-15 & 13-17).

Upon contraction, the lateral pterygoid protrudes and depresses the mandible. It may also move the mandible from side to side in combination with action of the ipsilateral (same side) medial pterygoid muscle.

Medial Pterygoid

The medial pterygoid is also termed the *internal pterygoid* and, sometimes, the *internal masseter*. It too is located on the inside of the ramus and functions as a weak counterpart to the masseter (Fig. 13-16).

The medial pterygoid arises primarily from the pterygoid fossa on the inner surfaces of the lateral pterygoid plate but also from the maxillary tuberosity of the maxilla and pyramidal process of the palatine bone. After coursing downward, backward, and outward, the fibers insert into

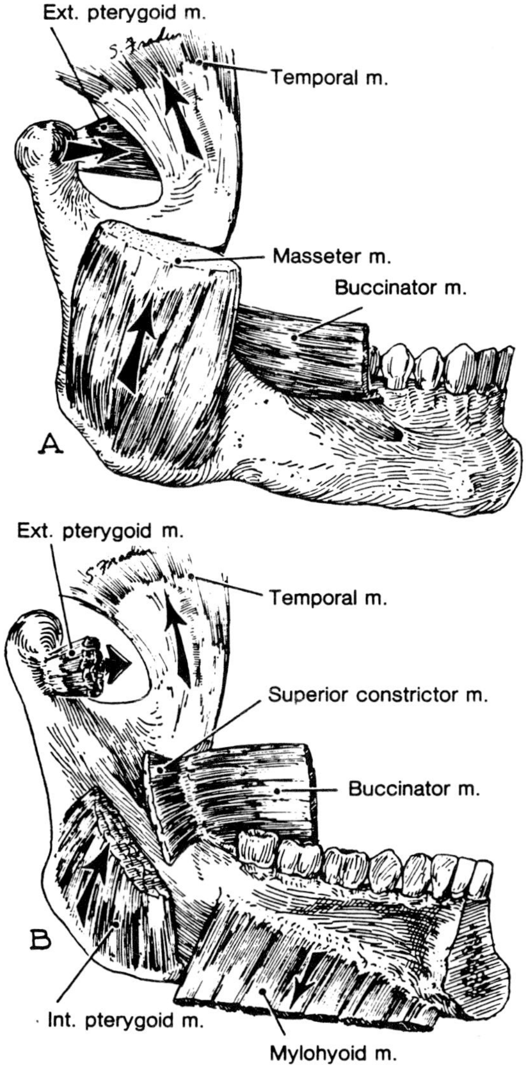

Figure 13-16. The pterygoid sling as seen from facial (A) and lingual (B) views. (From Nagle, Raymond J., and Sears, Victor H.: *Denture Prosthetics*, ed. 2, St. Louis, 1962, The C.V. Mosby Co.)

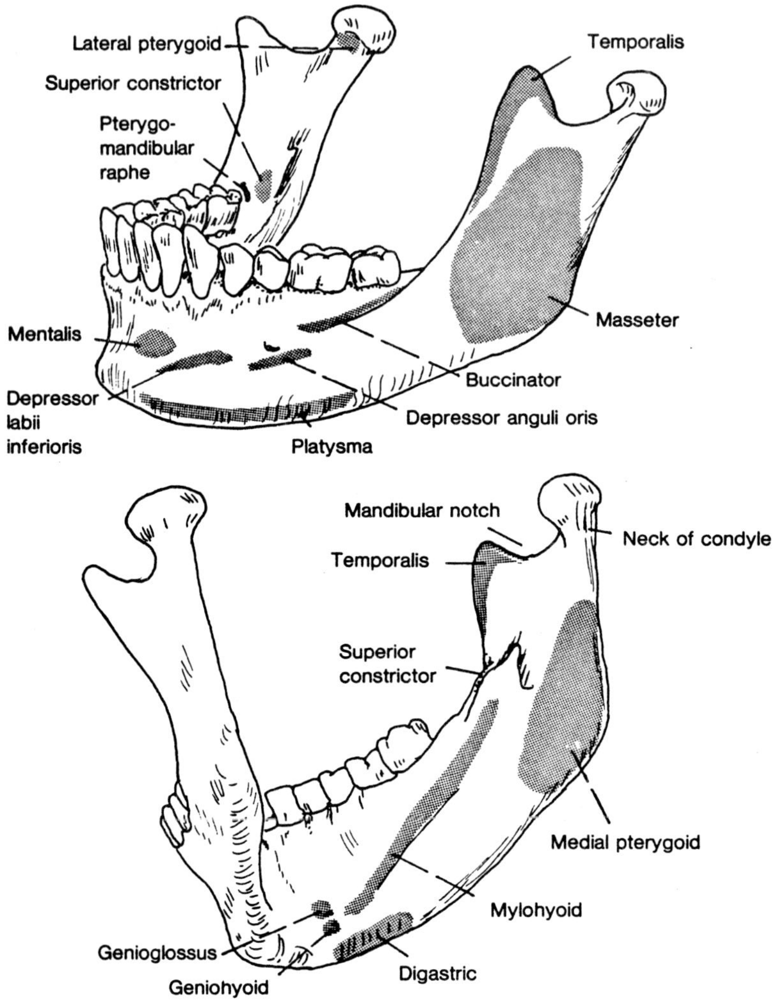

Figure 13-17. Pterygoid and other muscle attachments on the mandible.

the medial surface of the angle of the mandible in opposition to the masseteric fibers. The field of insertion includes an irregular area on the lower medial surface of the ramus (Fig. 13-16).

The action of this muscle is first as an elevator of the mandible and then as a protractor. In so functioning, it acts as a synergist of the masseter muscle. The principal direction of the pull of the fibers is upward. If only one of the two medial pterygoids contracts, the mandible is shifted to one side.

BIBLIOGRAPHY

Bassett, D.L.: *A Stereoscopic Atlas of Human Anatomy*, Section II. Plates 35.1-36.1. Baltimore, Williams and Wilkins, 1954.

Dawson, H.L.: *Basic Human Anatomy*, 2nd ed. New York, Appleton, 1974.

DuBrul, E.L.: *Sicher's Oral Anatomy*, 7th ed. St. Louis, C.V. Mosby, 1980.

Folkins, J.W.: Muscle activity for jaw closing during speech. *J Speech Hear Res, 24*:601-615, 1981.

Kennedy, J.G., and Folkins, J.W.: Standardization of lip muscle nomenclature. *J Speech Hear Res, 21*:603-605, 1978.

Gardner, E., Gray, D.J., and O'Rahilly, R.: *Anatomy: A Regional Study of the Human Structure*, 4th ed. Philadelphia, Saunders, 1975.

Langley, L.L., Telford, I.R., and Christensen, J.B.: *Dynamic Anatomy and Physiology*, 4th ed. New York, McGraw, 1974.

Massler, M., and Schour, I.: *Atlas of the Mouth*, 2nd ed. Chicago, Am Dental, 1958.

Pickrell, K.L.: *Reconstruction Plastic Surgery of the Face*. Summit, CIBA, 1967.

Rankow, R.M.: *An Atlas of Surgery of the Face, Mouth and Neck*. Philadelphia, Saunders, 1968.

Scott, J.H., and Dixon, A.D.: *Anatomy for Students of Dentistry*, 4th ed. London, Churchill, 1978.

Silverman, S.I.: *Oral Physiology*. St. Louis, C.V. Mosby, 1961.

Snell, R.S.: *Clinical Anatomy for Medical Students*, 2nd ed. Boston, Little, 1981.

Chapter 14

THE TEMPOROMANDIBULAR JOINT

TMJ CHARACTERISTICS

ONE of the interesting features of the mandible is that while it is a single bone, it has two joints that require intricate coordination by the pairs of muscles responsible for its movements. The twin articulations, the temporomandibular joints (TMJ), must be coordinated as a single joint in all movements. Hence, in function, the right and left temporomandibular joints should be thought of as but a *single* articulation even though it is anatomically bilateral (Fig. 14-1).

The temporal bone provides the mandibular fossa, also called the *glenoid fossa*, for this joint. An eminence, the articular eminence, forms the anterior portion of the fossa. The articular surface of the mandible is the mandibular head or condyle. Between the two articulating bones (the temporal and mandible) is a dense, oval fibrous disk that is commonly referred to as the *articular disk* or meniscus (cresent-shaped) (Fig. 14-2).

Around the joint are ligaments that form a capsule to enclose the joint. These capsular ligaments prevent dislocation of the condyle. However, the capsule is relatively weak in the front, and dislocations are most likely to occur in this direction. Lining the joint cavity is a synovial membrane, which secretes a lubricant.

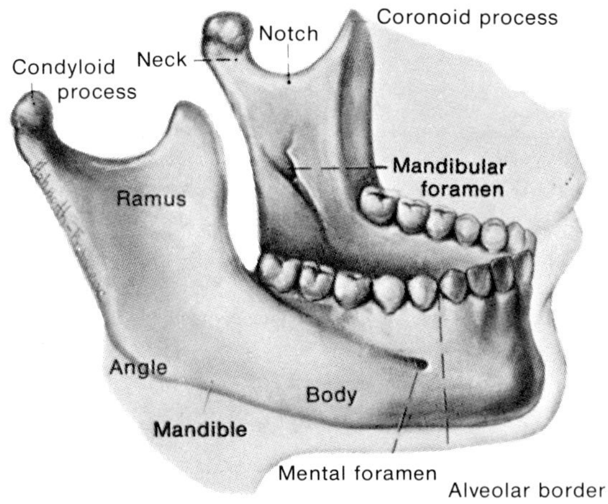

Figure 14-1. The mandible. (From Jacob, S.W., and Francone, C.A.: *Structure and Function in Man*, 4th ed., 1978. Courtesy of W.B. Saunders Co., Philadelphia.)

Each joint is actually a compound articulation; that is, there are two joints in a single capsule. The upper part of the joint enables a sliding movement of the condyle and the articular disk, moving together against the articular eminence on the

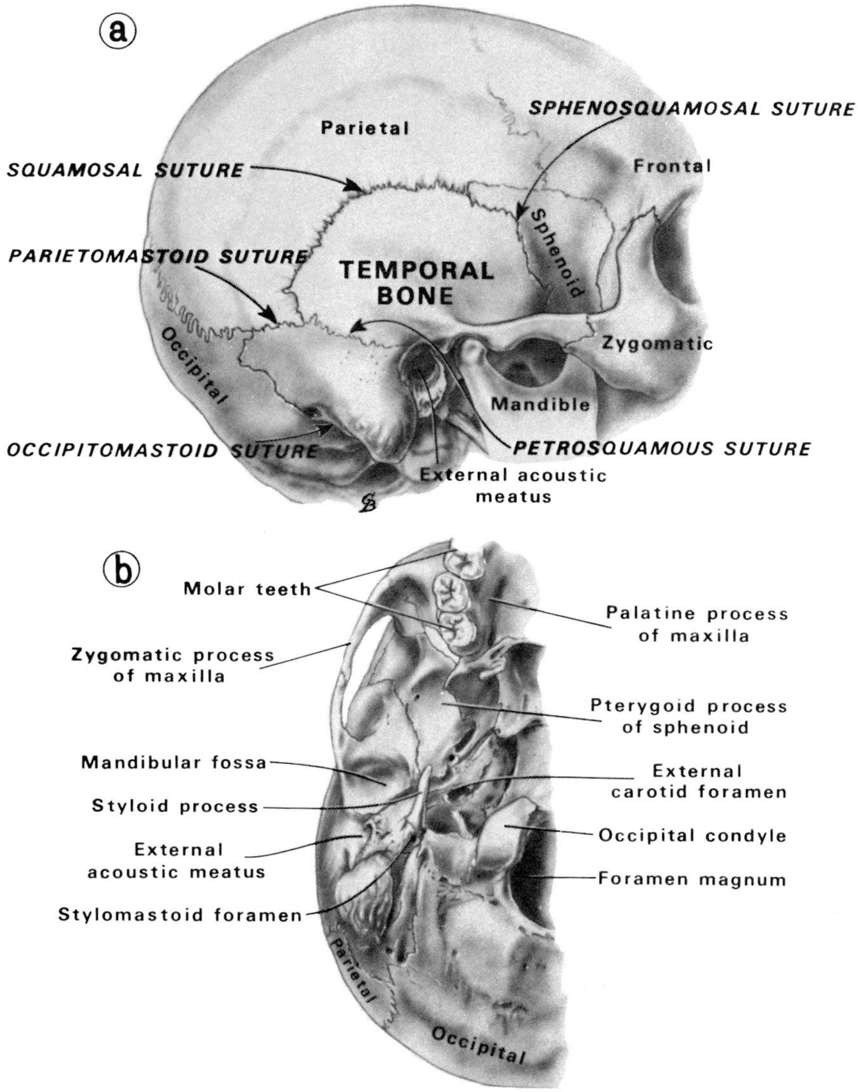

Figure 14-2. The cranium showing the squamous, mastoid, and tympanic parts of the temporal bone (a) and emphasizing the relationship of the temporal bone to other skeletal elements (b). (a) seen from the right. (b) viewed from below (right half of the cranium). (From Anson, B.J., and Donaldson, J.A.: *Surgical Anatomy of the Temporal Bone and Ear*, 3rd ed., 1981. Courtesy of W.B. Saunders Co., Philadelphia.)

forward part of the mandibular fossa. In the lower part of the joint, the head of the condyle rotates beneath the undersurface of the articular disk in a hinge action between the disk and the condyle. The temporomandibular joint is classified as a diarthrodial ginglymus; it is also classified as a synovial joint (Fig. 14-3).

The compound feature of the temporomandibular joint permits a variety of movements. The first phase of opening the mouth requires no more than a simple

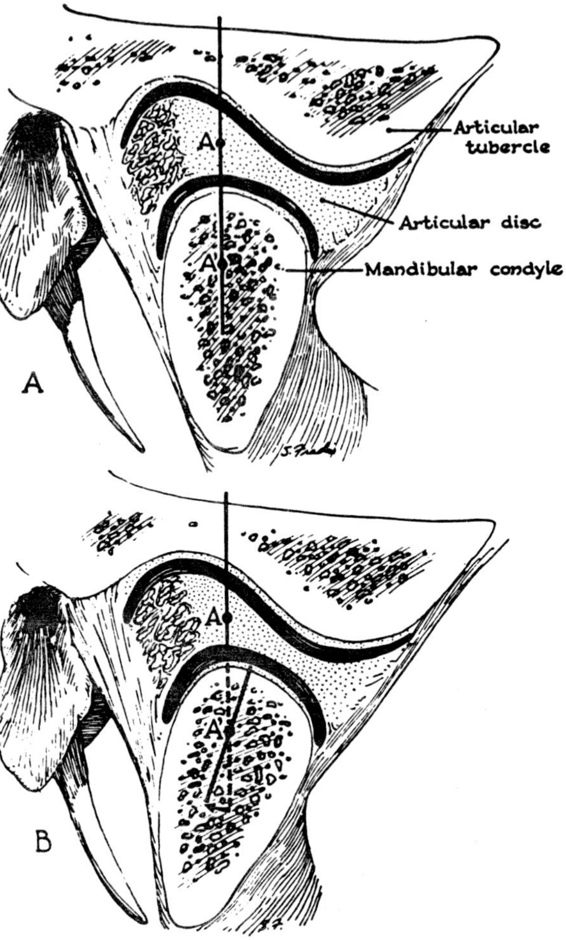

Figure 14-3. (**A**) Sagittal section through the mandibular condyle shown at rest in the glenoid fossa. (**B**) An opening movement of the mandible showing the hinge action of the condyle. (From Nagle, Raymond J., and Sears, Victor, H.: *Denture Prosthetics*, ed. 2, St. Louis, 1962, The C.V. Mosby Co.)

hinge action. The lower half of the joint is used for this movement. During jaw protrusion and wide openings of the mouth, both the upper and lower parts of the joint are involved. Since the hinge action may be halted as the mandible compresses the soft tissue between the ramus and the mastoid process, the mandible must be brought forward in a protracted state to allow the hinge movement to continue (Fig. 14-4).

The mandible is the only movable bone of the skull. Mandibular movement is basically related to chewing. Most of the muscles moving the mandible are classified as muscles of mastication. These muscles are located at the back of the mandible, specifically on the various surfaces and processes of the ramus. They provide considerable power for biting and chewing and can shift the mandible laterally, as well as protrude and retract the mandible. While the amount of travel is typically very slight, the mandible moves

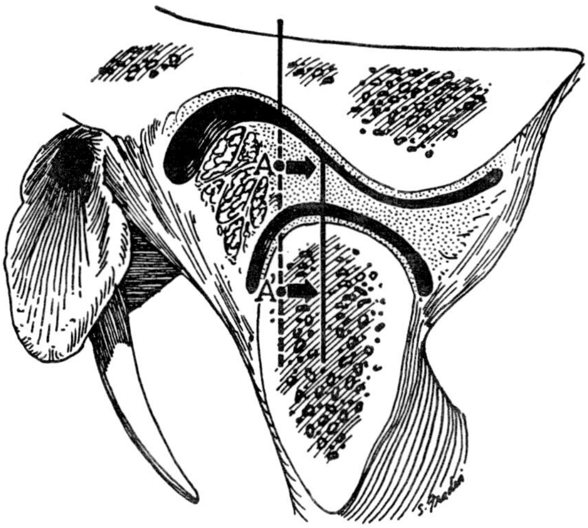

Figure 14-4. Protrusive movement of the mandible. (From Nagle, Raymond J., and Sears, Victor H.: *Denture Prosthetics*, ed. 2, St. Louis, 1962, The C.V. Mosby Co.)

up and down rapidly during speech. Singers often open the mouth quite wide while performing; speakers usually do not.

Mandibular motion has been studied by x-ray motion pictures (cinefluorography). Tongue movement as well as palatopharyngeal valving may also be examined in relation to movement of the mandible by cinefluorography, but this technique is no longer used for clinical study of mandibular motion. By lowering the mandible, the tongue is lowered in the oral cavity, permitting additional freedom of movement for articulation. In addition, resonance features of the oral cavity are altered by lowering the mandible.

From the standpoint of facial expression, it may be pointed out that the various positions of the mandible have significance in visual communication of feelings. The open mouth, resulting from a lowered mandible, may suggest shock or surprise. A protruded mandible may indicate pride or bullishness, etc. The teeth clenched in an even bite unmistakably denote an intensely angry mood. Nonetheless, all mandibular movement in facial expression is seen as ancillary to the principal muscles of expression surrounding the facial orifices.

CLASSIFICATION OF MANDIBULAR MOVEMENT

The movements of the mandible may be classified under five categories:

1. *Protrusion*. During protrusion or protraction the mandibular condyles slide forward on the articular eminences and the teeth remain in gliding contact.

2. *Retrusion*. Retrusion or retraction occurs as both condyles move rearward to settle in the mandibular fossa, the teeth remaining in gliding contact. The most retruded biting position of the mandible is called *centric relation* (CR).

3. *Opening*. The mandible rotates downward in a hinge action as the condyles move forward.

4. *Closing*. Closure may be achieved with the teeth in occlusion while the condyles are in extreme protrusion (prognathic position) or while the condylar head is seated at the extreme posterior aspect of the mandibular fossa (retruded position). The biting position that produces maximum interdigitation or intercuspations of teeth is called *centric occlusion* (CO). In many individuals, the bite in centric relation is the same as centric occlusion (CR = CO). Typically, the bite produced in maximum intercuspation (CO) is with the mandi-

ble slightly ahead of its most retruded position (CR) in the glenoid fossa (Fig. 14-5).

5. *Lateral movements*. A side-to-side movement of the mandible is accomplished by moving the chin to one side of the midline. The key factor is that one of the mandibular condyles remains relatively fixed by tonic contraction of the muscles on that side while the other condyle is drawn forward on the articular eminence by the contraction of the lateral and medial pterygoids of that side with relaxation of the posterior fibers of the temporalis. The chin is then drawn back to the midline by contraction of the posterior fibers of the temporalis and relaxation of the pterygoids so that the foward moving condyle is retracted. Moving the chin to the opposite side involves fixating the previously moving condyle at the back of the mandibular (glenoid) fossa while the previously fixed condyle moves forward.

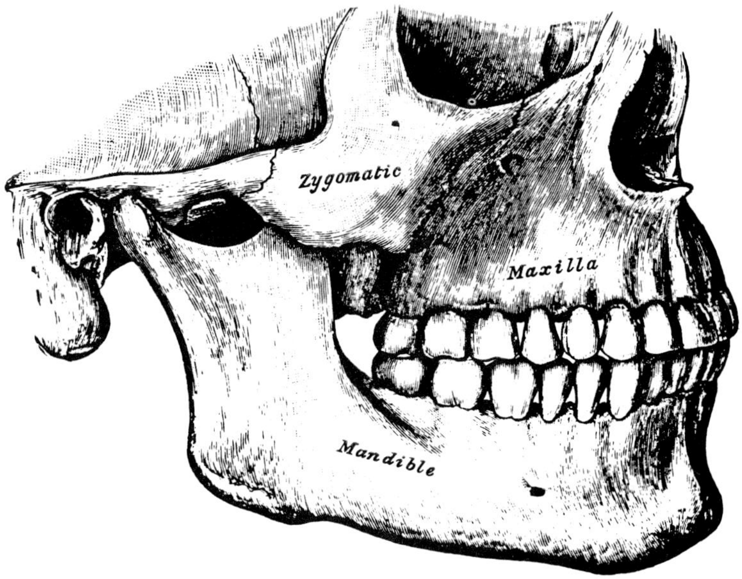

Figure 14-5. The teeth in centric occlusion (maximum intercuspation). (From Goss (Ed.): *Gray's Anatomy of the Human Body*, 28th ed., 1973. Courtesy of Lea & Febiger, publisher.)

PATHOLOGY OF THE TEMPOROMANDIBULAR JOINT

Dislocation of the mandible occurs from a number of causes. These include excessive opening of the mouth during yawning, during prolonged surgical operations in the oral cavity (especially under general anesthesia), or from attempting to bite an abnormally large piece of food. The capsular ligaments are stretched as the condyle is forced over the articular eminence forward into the zygomatic fossa. The raised articular eminence prevents the condyle from returning to the mandibular fossa. Frequently, the muscles of mastication will freeze (trismus) in re-

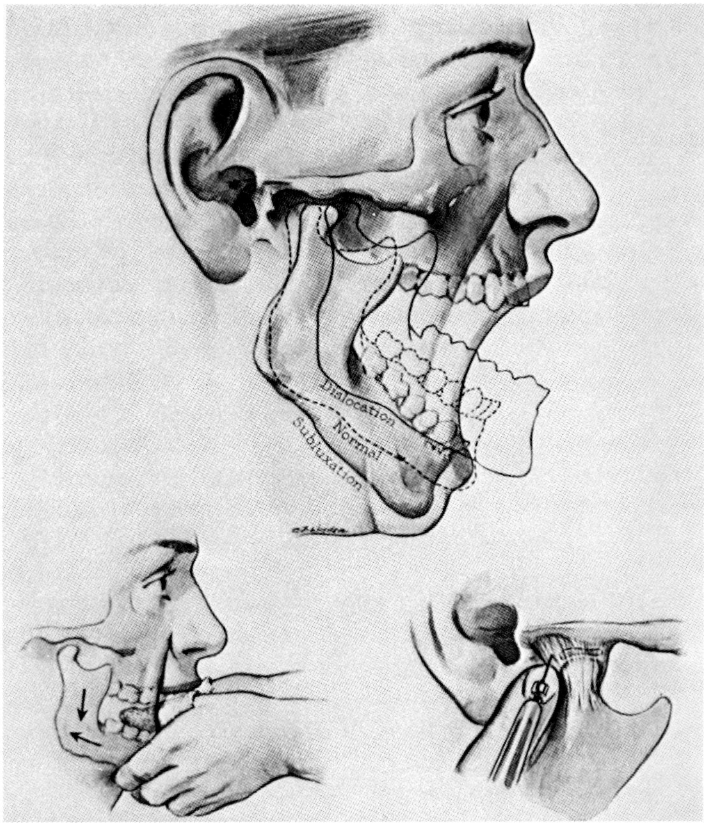

Figure 14-6. Dislocation of the mandible and its treatment. (Upper middle) Positions of opening. (Lower left) Reduction of dislocation (subluxation). (Lower right) Treatment by injection. (From Massler, M., and Schour, I.: *Atlas of the Mouth*, 2nd ed., 1958. Copyright by the American Dental Association. Reprinted by permission.)

sponse to the painful dislocation, but after a lapse of time the muscles relax and permit a hinge action in the abnormally protruded position. Assistance may be required to slide the mandible back to its normal position. During the dislocation, speech production is very difficult due to restricted tongue mobility (Fig. 14-6).

A noise is sometimes heard while certain individuals chew. This may be the result of a displaced or worn articular disk.

Ankylosis, an abnormal immobility of a joint, may affect one or both of the temporomandibular articulations. If only one side is affected as the mouth is opened, there will be a deviation of the midline, since the gliding motion of the joint is lost. The chin will, as in cases of muscle paralysis, deviate toward the affected side as the individual opens the mouth wide. In the same category is facial asymmetry resulting from traumas, for example, caused by obstetrical forceps being placed over the joint area during birth or falls in which the chin strikes the floor. The growing cartilage of the mandibular condyle is injured in such cases, which disturbs the synchrony of growth between the mandible and maxillae. This asynchrony comes about from the influence of the pressure forces caused by function. Failure of the affected side of the mandible to elongate results in a deviation in the midline, which is most evident during wide opening of the mouth. Foreign objects or tumors in the mandibular fossa will cause a similar deviation; however, this deviation will be noticeable while the mandible is closed. The mandible returns to the midline during wide opening of the mouth as the condyle slides away from the occupied mandibular fossa.

Classifying Temporomandibular Joint Disorders

The temporomandibular joint has been implicated as the etiologic factor for a variety of aches, pains, and noises reported by patients in the dental office. These complaints have many differing etiologies that can affect the surronding functional anatomy of the temporomandibular joint or the joint itself. Conversely, in many patients, problems that begin extrinsic to the TMJ apparatus end up developing pathology and malfunction of the joint.

Dentists and physicians have become intensely interested in the diagnosis and treatment of the variety of problems associated with pain in the head, neck, and facial areas. The search for careful differential diagnostic measures to elucidate the nature of the problem has led to several classification systems for TMJ disorders. Some of the possible contributing factors on any classification system would include: congenital pathology, such as aplasia (lack of development) of the condyle; developmental pathology, such as hyperplasia (abnormal overgrowth) of the condyle; trauma or injuries; myofascial pain dysfunction syndrome (to be explained later); arthritis, such as in degenerative joint disease (osteoarthritis) or inflammatory (rheumatoid) arthritis; neoplasms; or a host of other diseases (Akin, 1980).

One of the primary purposes of the history and examination for TMJ-associated disorders is to distinguish between *complaints of pain* and *signs of dysfunction*. The examination process would include a dental examination for malocclusions that may contribute to the disorder; soft tissue examination, especially of the muscles of mastication; neurologic evaluation, especially of the cranial nerves; palpation and auscultation (listening for sounds) at the condylar heads; and radiographic studies. In addition, anesthetic blocks to identify specific areas of pain by manipulating or blocking these areas may be used. Diagnostic laboratory tests may also be indicated, such as the erythrocyte sedimentation rate test, which can be used to differentiate between rheumatoid arthritis and osteoarthritis. Radiographic examination may reveal evidence of arthritic changes, tumors, fractures, hyperplasia or hypoplasia of the condyles, as well as other variations in the anatomy of the TMJ area. Any changes in joint contours, spaces, or translatory movements during condylar displacement may be revealed with radiography (Akin, 1980).

Conventional radiographic assessments do not always demonstrate evidence of disease. In recent years, *arthrography* has provided an additional diagnostic method for studying the functional and internal disorders of the TMJ. Arthrography consists of injecting radiopaque contrast media into a synovial space and studying the injected area radiographically. This technique provides a means of studying soft tissue problems of the TMJ indirectly (Akin, 1980).

Whether the etiology of the TMJ disorder is *extrinsic* to the joint (*extracapsular*) or within the joint area (*intracapsular*), the manifestations to the patient are not highly predictable. The reason for this is that the complaints are similar in both disorders. Patients complain of muscle pain, headache, clicking, popping and grating noises in the joint, earaches, dizziness and locking of the jaw — all of which interfere with mastication.

Myofascial Pain Dysfunction Syndrome (MPD Syndrome)

Dental practitioners have realized that not all pains linked to the TMJ area have their primary etiology in this area. Sinus pathology, tumors, stressful habits, and other factors as mentioned above can lead to the production of pain in the ear or tenderness of the masticatory muscles that can radiate to many other areas. The theory purported by many is that the pain is radiated along the fascial sheaths between muscles. Fascial sheaths are sheets of fibrous tissue investing (surrounding) individual muscles. These areas are rich in blood vessels and nerve fibers that can potentially transmit pain signals from one area to another through these fasciae. Many clinicians feel that there is insufficient evidence that this is the mechanism of radiated pain; hence, the term *myofascial* has recently been replaced with *myofacial* among some clinicians. The patient who does not have an intrinsic or intracapsular problem with the TMJ, but has pain and dysfunction in the facial, head, or neck areas, is said to possess a myofascial pain dysfunction syndrome (MPD).

The MPD syndrome usually begins with a dull, unilateral pain in front of the ear, which seems to radiate, and is accompanied by TMJ noises and limitation in jaw movement. Headaches are common findings in these patients. The pain is relatively constant but increases during the day and may worsen at mealtime (Bell, 1969; Akin, 1980).

The MPD patient usually starts out with no radiographic signs of organic pathology or changes in the TMJ. Dental-bite adjustments do not seem to help, nor do dental biteplates or appliances achieve lasting relief.

Typically, the MPD patient is female (in 80-90% of patients studied), under 40 years of age, and reports pain over the TM joints that is elicited by jaw movements. Extensive studies of such patients reveal generally that the majority of TMJ problems experienced by the MPD patient are disturbances of function rather than structure. However, problems that initially are functional may develop into organic change eventually. The cycle of changes and developments, with stress, muscle fatigue, myospasm and myofascial pain, can lead to occlusal disharmony and altered chewing patterns (Laskin, 1969).

Many avenues of therapy are recommended variously for the patient with a diagnosed problem of MPD. Most clinicians feel that a soft diet, the application of heat, and various medications are important components of treatment. Ultrasound is an effective method of applying deep heat. Myotherapy and psychotherapy may also be used as a means of increasing a patient's adaptive capacity. If anxiety is considered a major component of the problem, a variety of drugs may be indicated, such as Valium® (a psychotherapeutic agent), analgesic agents, muscle relaxants, or anti-inflammatory agents, such as steroids. Biofeedback is often used as a part of relaxation training in combination with other forms of therapy (Zide, 1980). If there is a dental component to the problem such as from a premature contact of opposing tooth cusps, it is possible that an occlusal adjustment or equilibration can resolve some of the pain and restore the joint area to normal function. An equilibration is accomplished by reshaping the occlusal surfaces of the teeth, especially the cusps and inclined planes leading from the cusps, by use of dental burs and a high-speed handpiece. The selection of spots for such shaping is determined by having the patient bite on dye paper that registers the nature and area of opposing tooth contacts for the occlusal surfaces.

In other patients with complaints of myofascial pain dysfunction, there are occlusal problems such as loss of the vertical dimension of occlusion. That is, the bite is overclosed as a function of mutilated or worn cusps, destructive habit patterns such as bruxism, or other factors that may change the position of the mandibular condyle in the mandibular (glenoid) fossa. In such patients, the use of some appliance therapy, such as a biteplate, nightguard or other dental splint that changes the vertical level of the bite, can often alleviate the functional complaints. In selected patients, the rebuilding of the entire dental arch is indicated (Fig. 14-7).

Surgical treatment for MPD has been recommended. For example, condylotomy, with shortening of the lateral pterygoid muscle, has reduced the functional complaints in jaw movement in some patients (Fig. 14-8). In others, orthognathic surgery followed by reconstructive dental crown and bridge work may be recommended. Altogether, a large number of surgical procedures is available to the oral surgeon for any associated TMJ disorder, such as capsular plication, wedge meniscoplasty, synovial scarification, capsular rearrangement, eminectomy, and condylar shave (Kent, 1980).

As might be expected, the exact nature of the myofascial pain dysfunction syndrome problems remains relatively obscure. A team approach is recommended for diagnosing and treating it. The team may include a neurologist, otolaryngologist, psychologist or psychiatrist, oral surgon, radiologist, orthodontist, general dentist, and others.

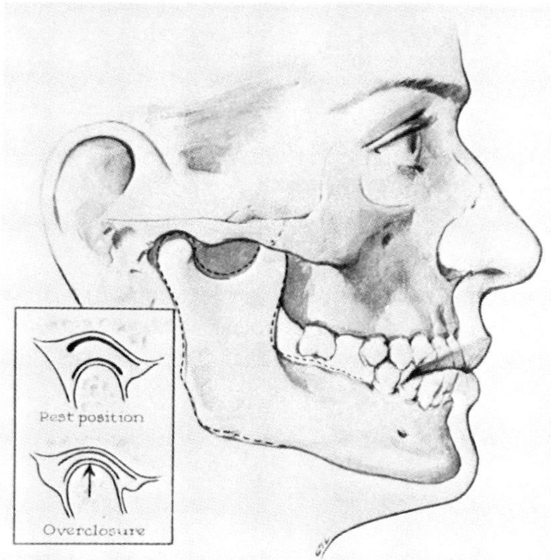

Figure 14-7. Overclosure of the mandible from loss of the posterior vertical component of occlusion. (From Massler, M., and Schour, I.: *Atlas of the Mouth*, 2nd ed., 1958. Copyright by the American Dental Association. Reprinted by permission.)

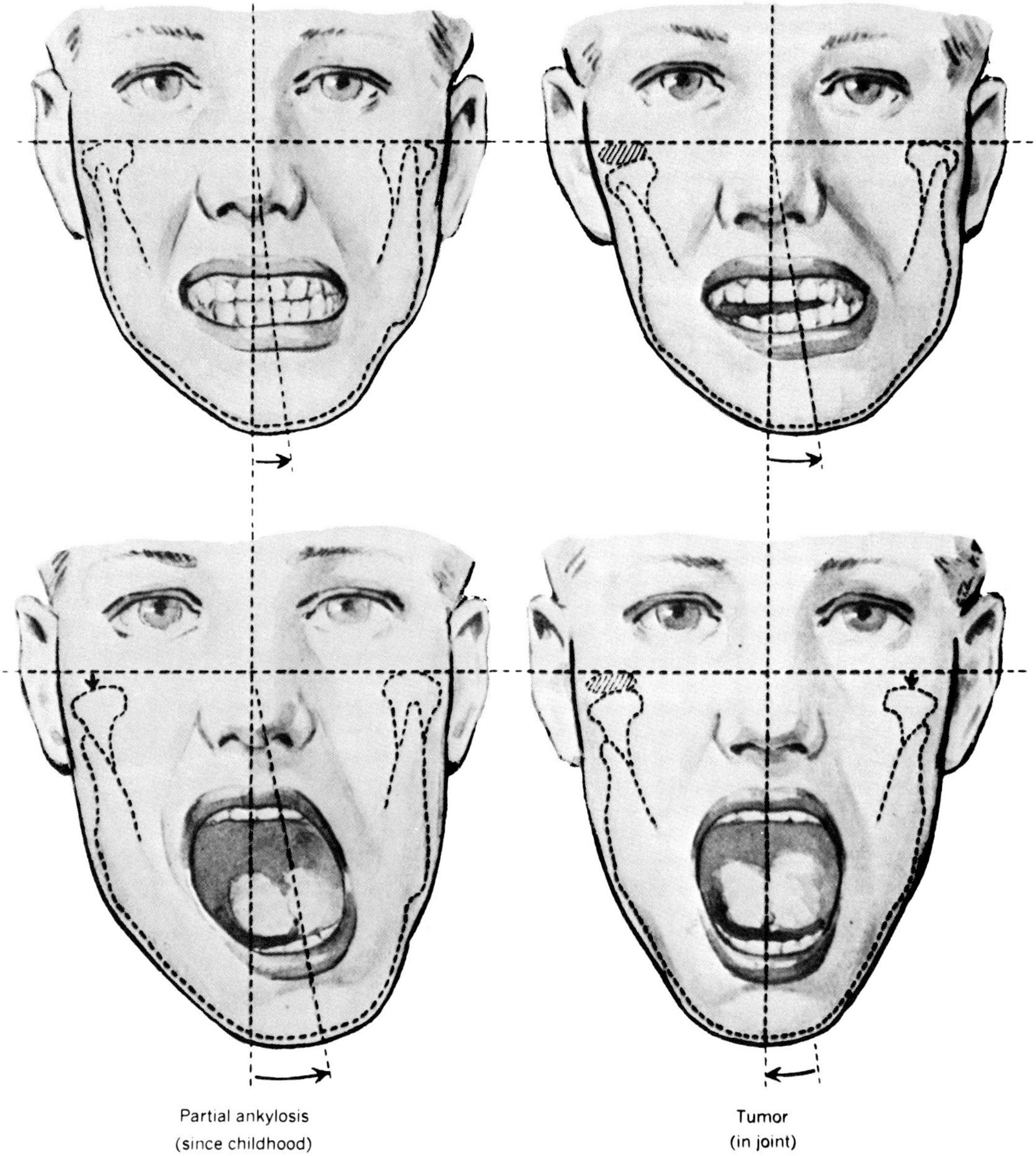

Figure 14-8. Abnormal jaw opening. MPD may relate to asymmetrical jaw openings caused by factors changing the position of the mandibular condyle in the mandibular (glenoid) fossa of the temporal bone. (From Massler, M., and Schour, I.: *Atlas of the Mouth*, 2nd ed., 1958. Copyright by the American Dental Association. Reprinted by permission.)

BIBLIOGRAPHY

Akin, R.K.: Temporomandibular joint surgery: classification of disorders. In Syllabus for Review Course in Oral and Maxillofacial Surgery, Louisiana State University Medical Center, November, 15-21, 1980.

Bell, W.E.: Clinical diagnosis of the pain dysfunction syndrome. *J Am Dent Assoc, 79*:154-160, 1969.

DuBrul, E.L.: *Sicher's Oral Anatomy*, 7th ed. St. Louis, C.V. Mosby, 1980.

Irby, W.B. (Ed.): *Current Advances in Oral Surgery*. St. Louis, C.V. Mosby, 1980, vol. III.

Kent, J.N.: Temporomandibular joint surgery: TMJ reconstruction. In Syllabus for Review Course in Oral and Maxillofacial Surgery, Louisiana State University Medical Center, November 15-21, 1980.

Laskin, D.M.: Etiology of the pain-dysfunction syndrome. *J Am Dent Assoc, 79*:147-153, 1969.

Zide, M.F.: Temporomandibular joint surgery: myofacial pain dysfunction syndrome. In Syllabus for Review Course in Oral and Maxillofacial Surgery, Louisiana State University Medical Center, November 15-21, 1980.

PART FOUR: NEURAL MECHANISMS OF SPEECH AND HEARING

Chapter 15

INTRODUCTORY NEUROANATOMY AND NEUROPHYSIOLOGY

GENERAL NEUROLOGY

A STUDY of the structure of the nervous system is known as *neuroanatomy*, while the study of neural functions is termed *neurophysiology*. All speech, language, and hearing functions are based in neuroanatomical and neurophysiological considerations. Of most interest to speech pathologists and audiologists is that part of the nervous system familiar to all, the brain.

No one can pretend that a study of neuroanatomy and neurology is easy. Indeed, the central nervous system (CNS) is an elusive system, harboring its mysterious functions better than other organs and systems. Our larynxes, for example, differ only very little from those of much more lowly beasts, but our brains are our crowning glory and set us apart from and above the rest of the whole animal kingdom. In examining our own brains, we might try and begin to comprehend what it is that makes us "only a little lower than the angels." This challenge will lead us into a study of neuroanatomy and neurophysiology of the CNS that will serve as a foundation for understanding the neural aspects of speech and hearing.

A steady and constant stream of information arrives at the brain from scores of input transmission lines during the waking hours. These are strung between all the sensory receptors and the brain, including fibers carrying unconscious information on muscle contractions, which makes it possible to monitor movement without having to watch the activity. The brain must somehow channel these input data into appropriate regions where incoming information may be utilized if recognized as meaningful. The recognition of *meaningfulness* is a major brain function.

In addition to acting as an *interpreter*, the brain functions as a *dispatcher*. There are scores of outgoing transmission lines carrying signals that will cause muscles to act and produce patterns of general and specific movement.

In a very general way, the nervous system has been outlined. The brain (which includes the cerebral hemispheres, brain stem, and cerebellum), along with its extension down the vertebral column, the spinal cord, comprises the *central nervous system*. Those tracts of neural fibers carry-

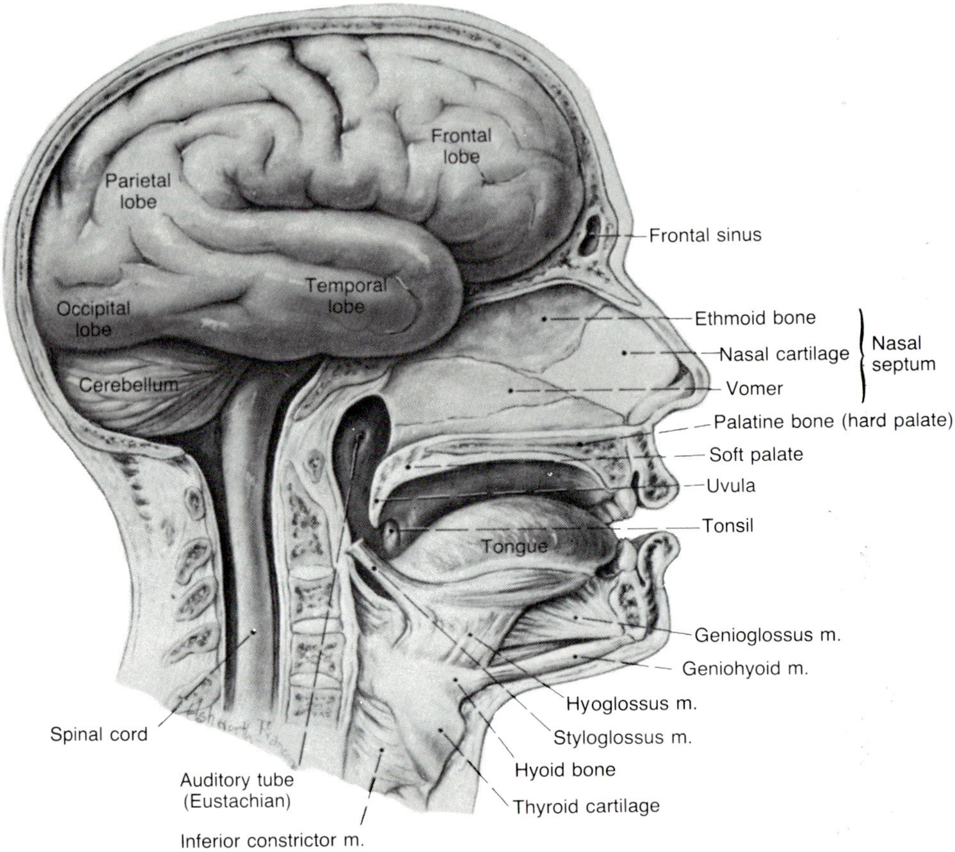

Figure 15-1. Relation of the brain and spinal cord to the skull and neck. The brain is intact; the skull and neck are sectioned sagittally. (From Jacob, S.W., and Francone, C.A.: *Structure and Function in Man*, 4th ed., 1974. Courtesy of W.B. Saunders Co., Philadelphia.)

ing information between the CNS and muscles, glands, the skin, or other organs make up the *peripheral nervous system* (Fig. 15-1).

As a unit, the nervous system enables one to regulate the internal environment for maintaining balance among the numerous functions of the body. In another similar mode, it relates man to the external environment through sensory awareness and preparation of the internal activities for responding to the outside world. The nervous system is able to accomplish these assignments because it is extensively spread throughout the body. Additionally, the network of neural tracts contains a great number of intersections permitting exchange of neural impulses so that activity in one region of the body may affect the events of another.

TERMINOLOGY

The vocabulary of neuroanatomy and neurophysiology can be awesome to the beginning student of anatomy. A good memory and perceptive set would certainly be helpful in piecing together terms, locations, and functions. The reader should refer to Chapter 3 for a brief review of the embryonic development of the nervous system, since there is free use of embryonic terminology in anatomic descriptions.

Selected vocabulary from neurology are listed here to facilitate the reading of this unit. Before proceeding with this and subsequent chapters, it would be worthwhile to become familiar with these general terms.

Afferent. Toward the brain, spinal cord, or a center. Such fibers are also known as *centripetal*. Sensory nerves are afferent nerves (Fig. 15-2).

Ascending. Running in a cranial or rostral direction.

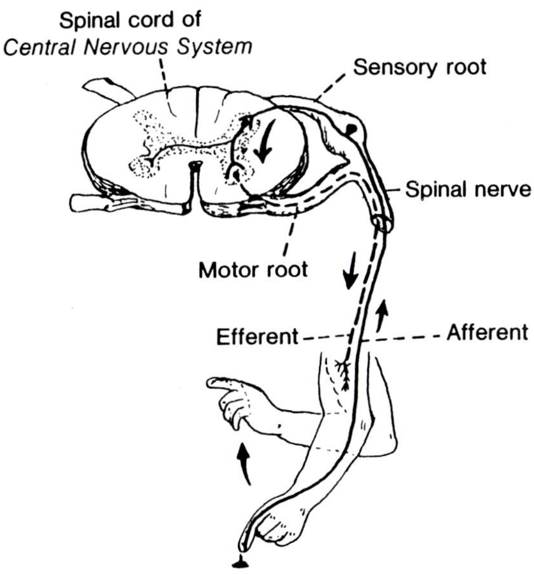

Figure 15-2. Afferent and efferent tracts. Afferent nerve fibers carry sensory information toward the CNS; efferent nerve fibers enable the CNS to control muscles and glands. (Modified from Anderson, P.D.: *Clinical Anatomy and Physiology for Allied Health Sciences*, 1976. Courtesy of W.B. Saunders Co., Philadelphia.)

Autonomic nervous system. Centers and pathways in both the central and peripheral nervous system associated with visceral controls.

Centers. A collection of nerve cell bodies.

Central nervous system. The brain and spinal cord.

Cerebellum. Little brain. That structure situated in the posterior cranial fossa.

Cerebrum. Head. The main part of the brain occupying the upper part of the cranium and forming the largest part of the central nervous system (Fig. 15-3).

Choroid. Resembling a delicate membrane. The choroid plexus of the brain produces cerebrospinal fluid.

Colliculus. Little hill. Rounded eminence of midbrain. The superior and inferior colliculi are associated with vision and audition, respectively.

Decussation. A crossing over, as in a nerve tract crossing the midline, such as the pyramid decussation of the medulla.

Descending. Running away from the brain or in a caudal direction.

Diencephalon. That part of the brain consisting of the thalamus, epithalamus, subthalamus, and hypothalamus; the posterior of the two brain vesicles formed embryonically from the prosencephalon (Fig. 15-4).

Effector. A nerve end organ that receives nerve impulses and produces an external or internal dynamic response.

Efferent. Away from the brain, spinal cord, or a center. Such fibers are also known as *centrifugal*. Motor nerves are efferent nerves.

Encephalon. Refers to the brain.

Ganglion. Swelling or enlargement. An enlargement or mass of nerve tissue containing the cell bodies of nerves, e.g. the spiral ganglion of the cochlea.

Geniculate. Little knee or bend. As in the medial geniculate body; important in

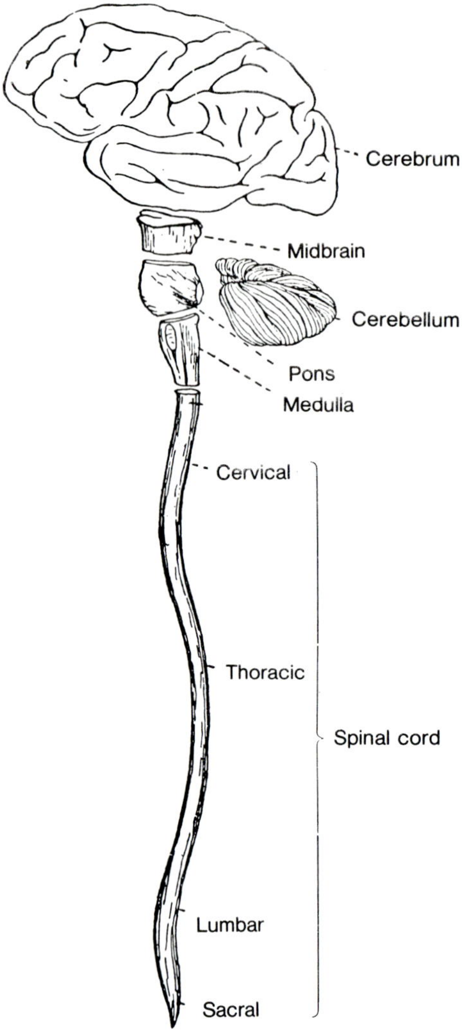

Figure 15-3. Major parts of the CNS. (From King, B.G., and Showers, M.J.: *Human Anatomy and Physiology*, 1969. Courtesy of W.B. Saunders Co., Philadelphia.)

the central auditory pathway.

Gyrus. Circular. The raised or convoluted ridge between grooves on the surface of the brain.

Kinesthesia. Conscious sensation of joint movement.

Lemniscus. Hanging ribbon or band. A bundle of nerve fibers in the CNS.

Limbic. Hem or border. The limbic lobe forming a border around the upper end of the brain stem.

Mesencephalon. Middlebrain or midbrain.

Metencephalon. Afterbrain or part of the hindbrain.

Myelencephalon. Marrowbrain or the medulla oblongata.

Myelin. Marrow. Fat-like substance forming a sheath around certain nerve fibers.

Nerve. A cord-like bundle of nerve fibers interconnecting the central nervous system and other parts of the body.

Nervous system. A composition of nerve cells and associated cells organized into the brain, brain stem, spinal cord, nerves, ganglia, and parts of receptor organs that correlate and regulate the internal reactions of the body and control its adjustment to the environment.

Neurilemma. Husk. The thin, spiraling sheath surrounding a peripheral nerve fiber.

Neuroglia. Nerve glue. The cells that support or hold nervous tissue together and which provide nutrition for the nerve cells.

Nucleus. Nut. As used in this unit: A general term designating a group of nerve cells located within the central nervous system, e.g. the caudate nucleus.

Peduncle. Little foot. A stem attaching another part, as the cerebellar peduncles.

Peripheral nervous system. The nerves and ganglia outside the brain and spinal cord.

Plexus. Braid. The interweaving of nerves in a specific area.

Pons. Bridge. That part of the brain stem between the midbrain and medulla oblongata.

Proprioception. Unconscious nervous impulses set up by lengthening or shorten-

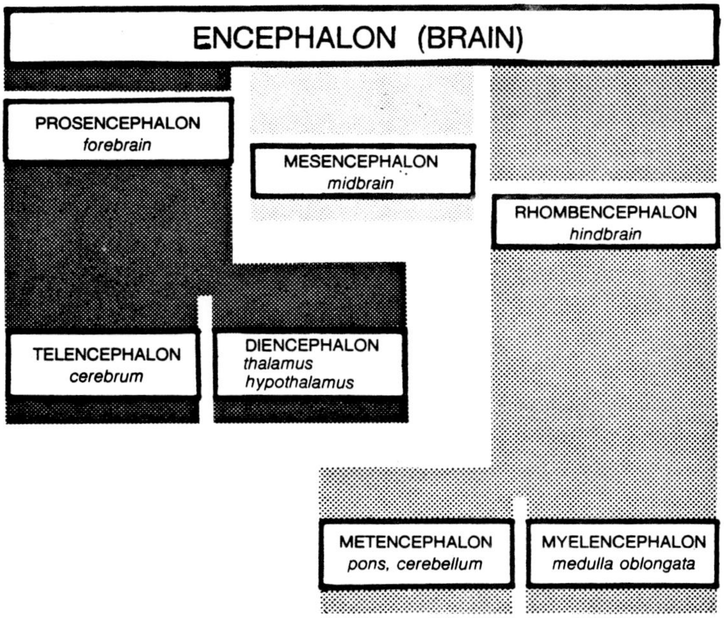

Figure 15-4. Basic divisions of the brain. Those terms ending in *on* are more closely associated with the vocabulary in embryology, but these terms are freely used by clinicians and anatomists. (From Brooks, Stewart M., and Paynton-Brooks, Natalie: *The Human Body*, ed. 2, St. Louis, 1980. The C.V. Mosby Co.)

ing of muscles.

Receptor. A specialized organ that includes nerve termination and functions to change a stimulus into a nerve impulse.

Rhombencephalon. The hindbrain or posterior primary brain vesicle. Includes the pons, cerebellum, and medulla.

Sheaths. Coverings that protect nerves.

Sulcus. A groove or furrow between the gyri on the surface of the brain.

Telencephalon. The endbrain.

Thalamus. Inner chamber. A major sensory integrative center in the diencephalon.

Ventricle. Belly. A cavity. The brain contains four ventricles holding cerebrospinal fluid.

NERVE TISSUE

In the brain, there are over 15 billion neurons, with numerous *processes* matted about like the roots of several plants in one pot. But this puzzling matrix allows multitudinous functional connections that enable impulses to travel varied pathways along these processes as an enlivened network of consciousness. Neurons and their supporting tissue, *neuroglia*, comprise most of the brain.

The component part of nerve tissue is the nerve cell, which, with all its processes, is called a *neuron* or *neurone*. These specialized cells possess the basic properties of irritability and conductivity to perform the specialized functions of receiving stimuli and conducting impulses throughout the body. The long protoplasmic processes of the neurons make possible functional connections, termed *synapses*, both close and remote to the nerve cell.

Nerve tissue, along with sense organs, has the ability to utilize and transmit the electrical impulses that regulate the activity of the body, provide the body with awareness of the environment, and permit adaptation to environmental changes. While nerve cells are specialized for the conduction of impulses, they sacrifice other biological functions since they cannot reproduce themselves and their metabolism is so simple and immediate that they cannot live for more than a few minutes without oxygen. They are the most delicate cells in the body, requiring a chemical environment even more constant than that of the plasma that bathes other cells (Fig. 15-5).

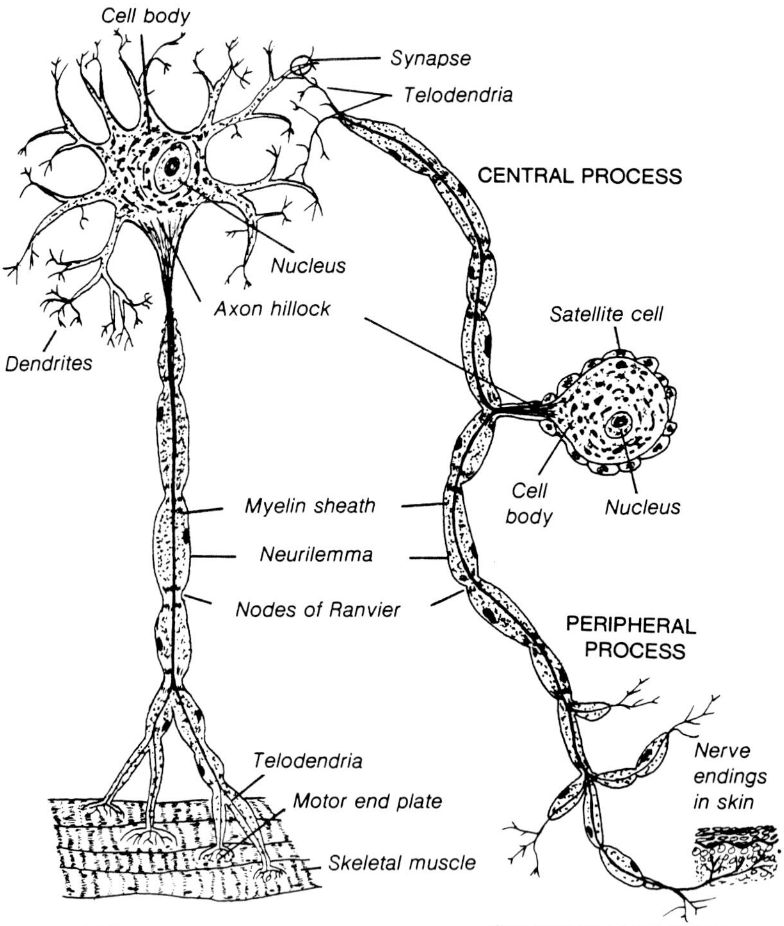

Figure 15-5. Schematic representation of two neurons. The sensory neuron is bipolar. The motor neuron is multipolar since it has many dendritic processes plus an axon. Multipolar neurons abound in the CNS; bipolar neurons are typically found in ganglia outside the CNS. The spiral and vestibular ganglia of the ear contain bipolar neurons.

The elongated processes of nerve cells are frequently bundled together for carrying impulses pertaining to a given function or region of the body. These bundles, lying outside the central nervous system, are termed *peripheral nerves*. There are two types of nerves, *efferent* and *afferent*, with respect to the central nervous system. They are often designated as *motor* and *sensory* nerves, respectively.

Structurally, a neuron differs from the cells of other tissues. Its principal parts are listed below, but not all neurons exhibit the same number and pattern of processes.

CELL BODY. The cell bodies of nerve cells contain, in addition to the chief constituents of the cytoplasm (the physico-

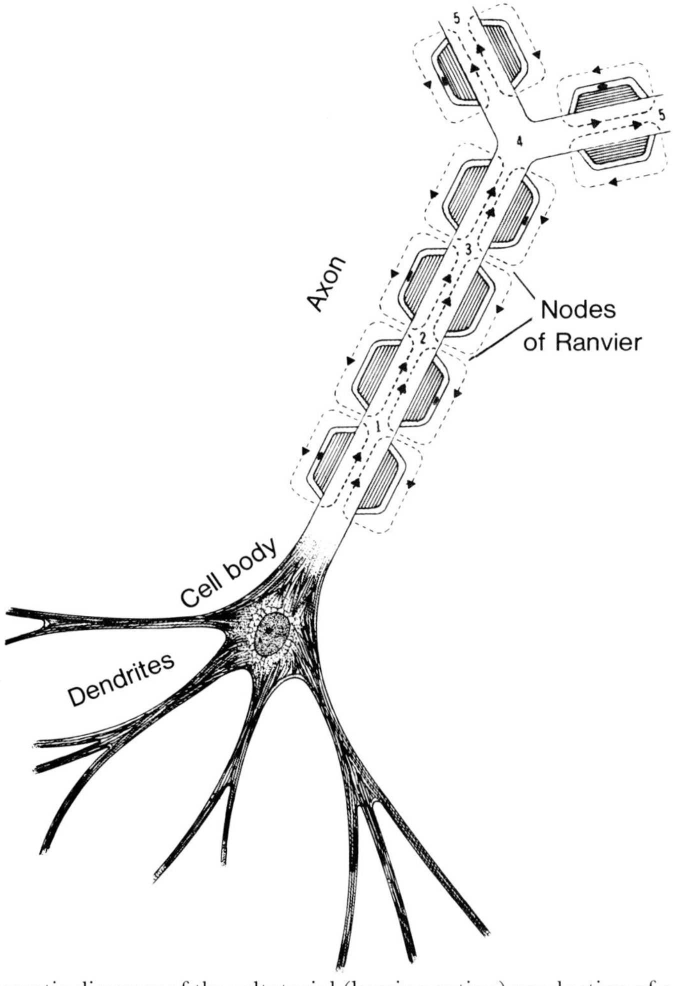

Figure 15-6. Schematic diagram of the saltatorial (leaping action) conduction of a nerve impulse along a myelinated axon. First, the ionic electric current flows from one node of Ranvier to the next as shown by the arrows and dotted lines. Next, each region of the covered axon is successively depolarized until the end is reached. The numbers in the nodes of Ranvier (on the axon) indicate the sequence of depolarization. (From *Man, Microbes and Matter* by B.C. Block. Copyright © 1975. Used with the permission of McGraw-Hill Book Company.)

chemical basis of living matter exclusive of the nucleus of the cell), masses of tigroid material called *Nissl substance* (after Franz Nissl), a chromatin-like material.

NERVE PROCESSES. The extension of the cell is made possible through the neuronal processes it contains.

DENDRITES usually consist of numerous short, branched processes for the purpose of conducting impulses *toward* the cell body. These are the afferent fibers transmitting sensory and other types of impulses to other cells.

AXONS are typically long, delicate processes for the purpose of conducting impulses away from the cell body. There is but a single axon per cell. Of the two processes of a neuron, the axon may be considerably longer, sometimes extending over four feet in man. Axons act as efferent fibers to send impulses to other neurons, muscles, and glands. Typically, axons synapse with dendrites in the CNS, but they may synapse with cell bodies of other neurons.

Two accessory structures, a *myelin sheath* and *neurilemma*, may be found encasing the neuron's processes. The myelin sheath, a fat-like substance, is responsible for the white matter of the CNS but is not part of the neuron although it is applied to its processes. Cutting into a fresh brain during dissection reveals a glistening white appearance within; this is due to the myelin sheaths. The gray matter of the central nervous system consists of cell bodies and unmyelineated fibers. Only axons in the CNS contain myelin sheaths. Myelin is a poor conductor of electric current and is continuous over the fiber it encases except for small breaks that occur every few millimeters, the *nodes of Ranvier* (Fig. 15-6.).

Neurilemma is another kind of cellular sheath, which is important to the formation of myelin. Its presence is necessary for the regeneration of injured peripheral nervous system (PNS) fibers. However, neurilemma is not found on central nervous system fibers. Known also as *cells of Schwann*, neurilemma encases both axons and dendrites and sometimes cell bodies in the PNS (Fig. 15-7).

Nonnervous Cells

Nervous tissue in the central nervous system is composed of neurons and nonnervous cells. In fact, a major portion of the central and peripheral nervous system is composed of nonnervous cells. About half of the brain's bulk, for example, is made up of neuroglial cells (nerve glue), sometimes shortened to *glia*. They do not participate in the transmission of impulses nor do they exhibit synapses. Neuroglia outnumber neurons in the ratio of about ten to one.

NEUROGLIA contain the essential components for metabolically active cells; they can divide throughout life, in contrast to nerve cells, which are unable to reproduce. It is this reproductive feature that may become a liability, since neuroglial cells are the primary source of tumors in the central nervous system. Additionally, neuroglial cells can divide and become phagocytes, which consume or ingest dead tissue and cells.

There are two categories of neuroglia, *macroglia* and *microglia*. Macroglia consists of astrocytes and oligodendrocytes. Astrocytes are star-shaped cells, with processes ending primarily on blood vessels or arranged like satellite cells around a neuron. Astrocytes occupy injured areas cleaned by phagocytes, sometimes producing the glial scar after healing.

Oligodendrocytes are typically found arranged in rows along nerve fibers and around neuronal cell bodies. These are the myelin-forming cells of the CNS. The myelin sheaths of axons within the CNS consist of oligodendroglial processes wrapped around the axons in a helical fashion, not unlike a Swiss roll. Nodes of Ranvier represent the interruptions, or boundaries, between the processes of separate oligodendroglial elements (Fig. 15-8).

The CNS is richly supplied with blood vessels and demands about one-fifth of the heart's total output. Those wandering macrophages, the phagocytes (*phago* = to eat), normally migrate in and out of capillaries to perform their scavenging func-

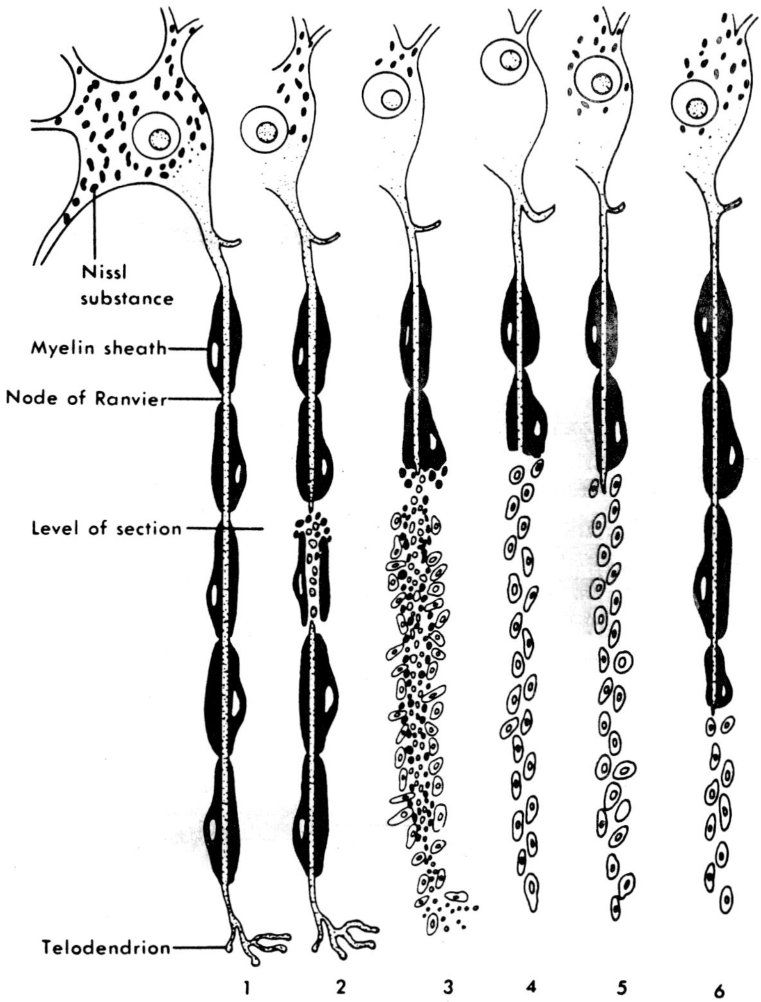

Figure 15-7. Schematic representation of degeneration and regeneration of a peripheral (motor) axon. (1) Axon is cut (note level of section). (2) Axon and myelin begin disintegration at section. (3) Nissl substance in the cell body shifts away from axon. Neuroglial and adjacent neurilemma cells become phagocytes to perform cleaning of the area. (4, 5, & 6) The axon begins to grow distally through the cord of phagocytic cells, and neurilemma cells begin to form a myelin sheath. Nissl substance reappears. A similar degenerative process occurs in the CNS but without regeneration to any significant degree. The process is known as wallerian degeneration. (From Gardner, E.: *Fundamentals of Neurology*, 6th ed., 1975. Courtesy of W.B. Saunders Co., Philadelphia.)

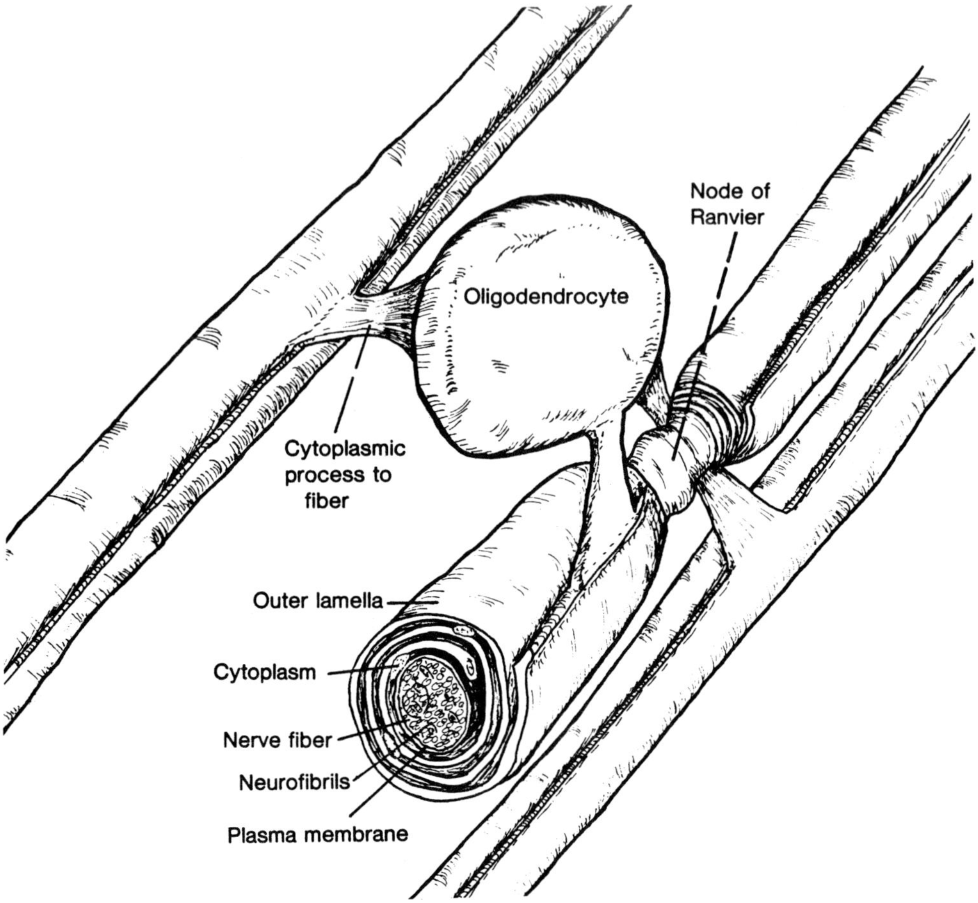

Figure 15-8. Myelineated fibers of the CNS showing a neuroglial cell (oligodendrocyte) supplying cytoplasmic processes that form the myelin sheath around nerve fibers. (Modified from Gardner, E.: *Fundamentals of Neurology*, 6th ed., 1975. Courtesy of W.B. Saunders Co., Philadelphia.)

tions. Termed microglia, the macrophages take on a peculiar morphology in the CNS.

Electrical Transmission and the Synapse

An electrical impulse is propagated along myelineated fibers by leaping in spurts from one node to the other, where sodium and potassium ions are in motion. This means that the large intervening space between the nodes is left untouched by the process of impulse transmission. The concentration of sodium and potassium on the two sides of the neuronal membrane, including the processes, may shift rapidly to cause the cell and its processes to become electrically active (Fig. 15-9).

As the processes of neurons approach each other, a *synapse* or point of functional contact is formed. Across such junctions, electrical impulses may be conducted. These interconnections are exceedingly numerous in certain areas (perhaps over a thousand for a single neuron), permitting neurons to communicate with each other

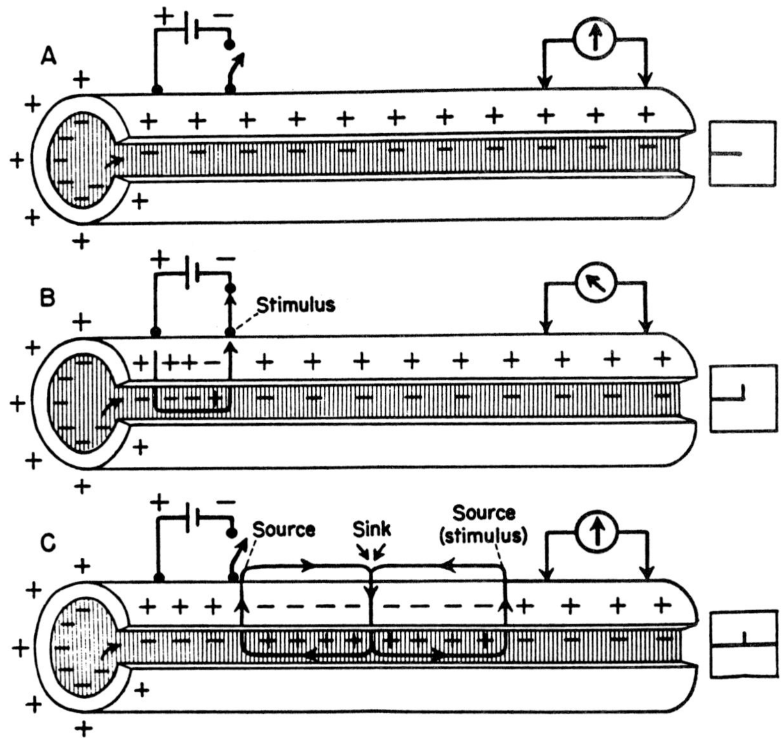

Figure 15-9. Diagram illustrating the initial changes following application of a stimulus to a nerve fiber. (A) The fiber is represented as a tube with a strip cut away so as to expose the interior. The stimulating electrodes, indicated by (+) and (−), are at one end and the recording electrodes at the other. The electrical charges internal and external to the membrane are also indicated. (B) When the switch is closed, a battery current flows from the positive electrode into the fiber and out through the membrane back to the negative electrode (cathode), bringing about the change in membrane conductance. Some of the stimulating current leaks down the nerve, instantaneously for all practical purposes, and is recorded as a small deflection (inset at right). This is the shock artifact and indicates the moment of stimulation. (C) The effect of current flow out through the membrane is to initiate the conduction of a nerve impulse. The impulse is represented as two dipoles, each consisting of current flow (not the stimulating current from the battery) within a region indicated by the reversal of electrical charges. The current flow out through the membrane in the leading dipole acts as a stimulus, just as did the artificial current from the battery. According to conventional methods, the recording instrument will not show a deflection until the region of negativity passes under the first electrode. (From Gardner, E.: *Fundamentals of Neurology*, 6th ed., 1975. Courtesy of W.B. Saunders Co., Philadelphia.)

in very complex patterns. The synapse is a specialized interconnection between two processes in which the fibers never touch but are separated by a *synaptic cleft*. The *presynaptic fiber*, or axon, contains a transmitter substance that the *postsynaptic* fiber, or dendrite, does not. For this reason, the impulse passes only in one direction: from the presynaptic fiber to a postsynaptic fiber (Fig. 15-10).

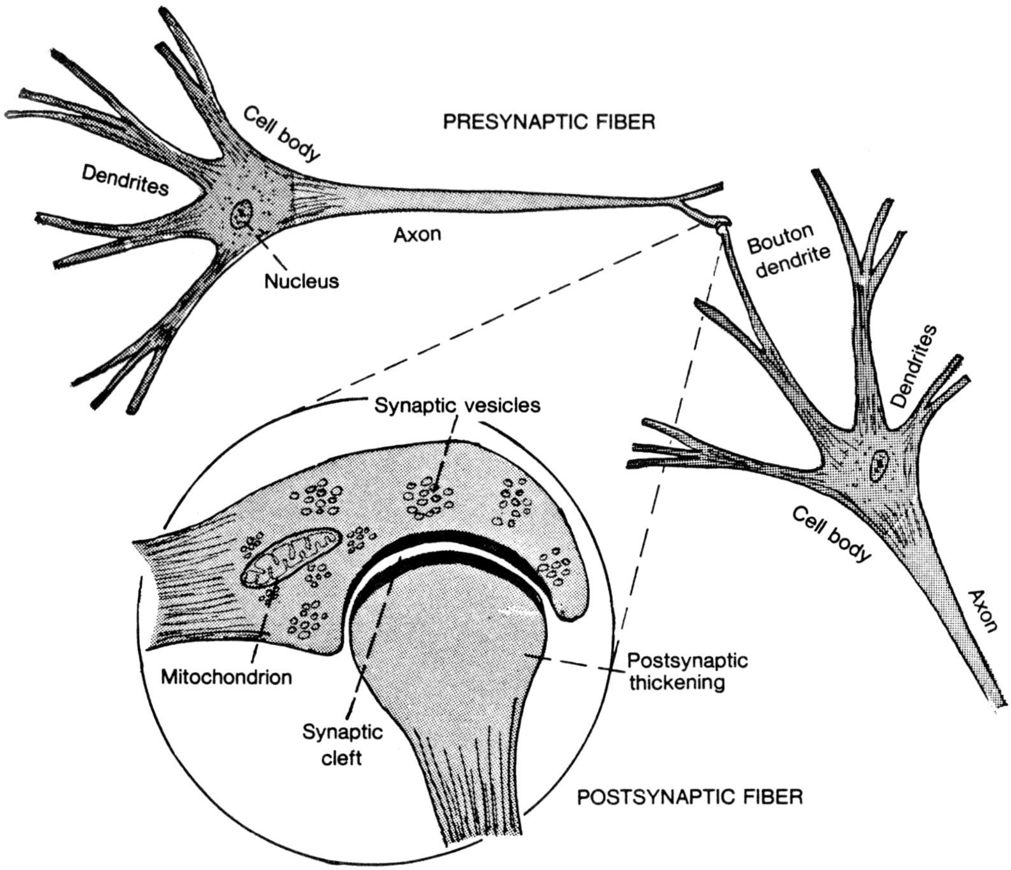

Figure 15-10. The synapse, a point of functional contact between two neurons.

CHARACTERISTICS OF THE BRAIN

General Structural Characteristics

The brain's control over the body is principally contralateral, that is, the left hemisphere of the brain controls the right side of the body and vice versa. Contralaterality of motor control is a function of crossing over (decussation) of motor tracts from one side of the body midline to the other. The same general plan of contralaterality holds for sensory control. In the instance of auditory and visual feedback, most sensory fibers decussate but many do not (Fig. 15-11).

Another characteristic of brain function is *cerebral dominance*. The perception of language, speech symbols, and writing, and also the production of speech and writing, are located in the left hemisphere for about 95 percent of right-handed and most left-handed individuals. The right hemisphere is said to function as the primary reception area for spatial events and nonspeech auditory signals, such as environmental noises and music. However, there is mounting evidence that the right hemisphere may have more potential for language than was previously thought.

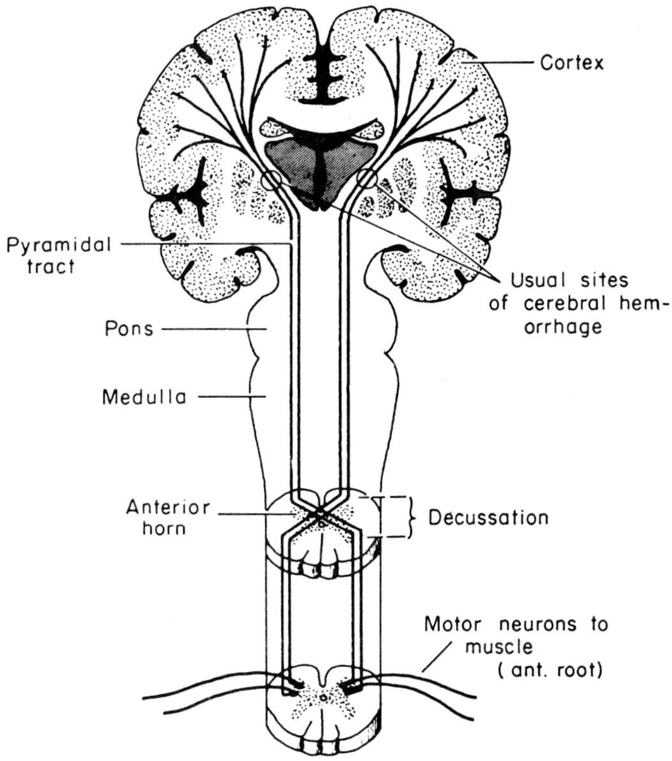

Figure 15-11. Schematic of the corticospinal tracts showing decussation in the lower medulla. Note the site of potential cerebral hemorrhage, which results in cerebral apoplexy (stroke) and which would impair contralateral functions because of nerve tract decussation. (From Brooks, Stewart M.: *Basic Science and the Human Body*, St. Louis, 1975, The C.V. Mosby Co.)

Nonspeech motor signals to the muscles of the articulatory apparatus are said to emanate from the right hemisphere.

The implications of cerebral dominance to clinical speech pathology are considerable. For example, there is little rationale for working on nonspeech movements of the articulators as a means of developing facility in speech activities. Even though the muscles involved are the same, the difference in cerebral control of speech and nonspeech activities should discourage assumptions of learning reciprocity. This is substantiated in a number of experimental studies.

The brain weighs about 1350 gr in man and about 4500 gr in an elephant. The weight of the brain as compared with the weight of the spinal cord is about 50:1 in man; it is about 15:1 in the gorilla; and the brain weighs less than the spinal cord in the frog. At birth, man's brain weighs about 350 gr; at age one, it weighs 800 gr (Fig. 15-12).

Functional Characteristics

The brain's functions may be categorized into three general activities. First, it acts as a *reception center* for the input of sensory impulses from the major and minor sense organs. Normal language and

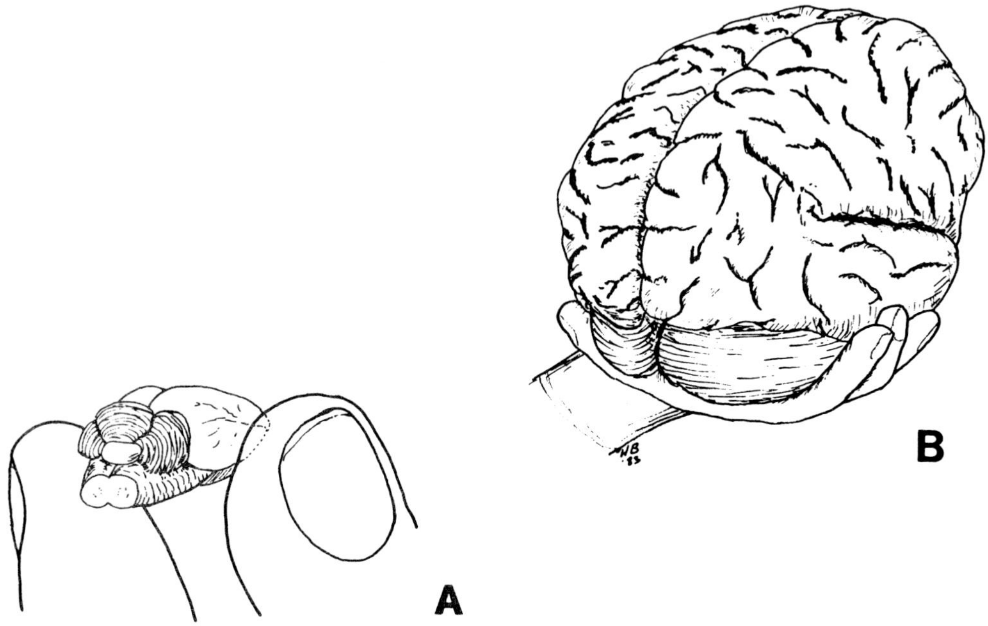

Figure 15-12. Brains vary in size, weight, and configuration among classes. The size of a rat's brain (A) is dwarfed by the fingers. The human brain (B) (not to scale) is considerably larger and more complex.

speech acquisitions are dependent primarily on the proper operation of those sense modalities sending appropriate auditory and visual input information to the brain and its ability to receive these data.

Processing these incoming impulses is largely an *associative function* of the brain. In essence, the brain is communicating with itself in this second general activity. Incoming impulses are interpreted according to information stored from previous experiences. It is through these association pathways within the brain that integrating networks of nerve fibers communicate with areas that have as their specific purpose associative activity. Attaching meaning to sensory impulses, daydreaming, or deep, theoretical cogitation is the result of association. The number and variety of possibilities for associative integration is a feature of man's cortical behavior that permits language to occur uniquely in his/her species. Through the *corpus callosum* — a mass of some 200 million fibers interconnecting the two cerebral hemispheres — each half of the brain complements the functions of the other in a complex associative function (Fig. 15-13).

The third general activity assigned the brain is *muscular control*. This output function utilizes two distinct systems arising from the brain and extending into the spinal cord. These bundles of fibers are known as the *pyramidal* and *extrapyramidal systems*.

The pyramidal system stimulates muscle contraction directly from impulses originating in the cortex, or outer covering of the brain. The pyramidal system involves nerve tracts that primarily act to excite or facilitate motion. The extrapyramidal system is less well-defined, but its function is essentially the same; there are, however, fewer extrapyramidal tracts. Some are excitatory and others inhibit a muscle from responding. Such in-

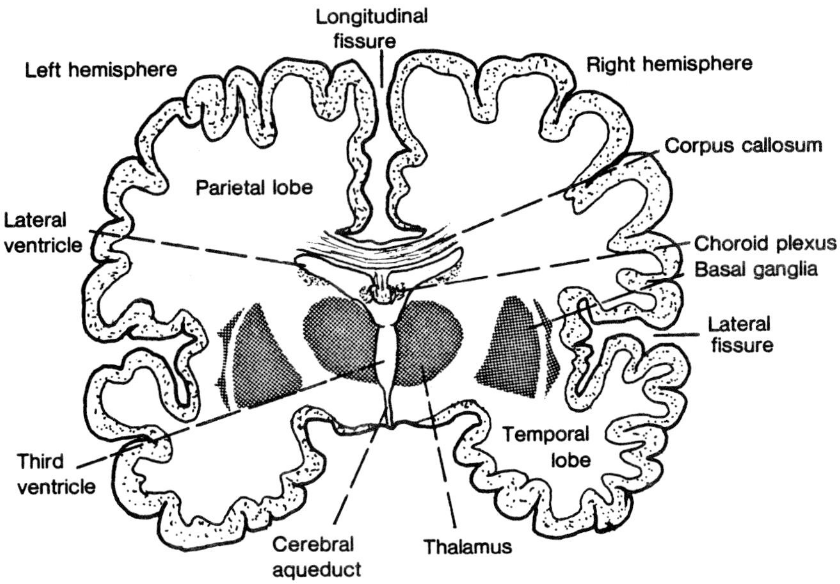

Figure 15-13. Diagram of a coronal section of the brain showing certain features of the internal and external parts. Notice that the corpus callosum interconnects the two hemispheres.

hibition is important for synergistic activity of opposing muscles. Extrapyramidal tracts originate from a diversity of structures in the brain, including the cortex. Both send their outgoing fibers, termed *motor* fibers, to the spinal cord, which transmits the impulse to the designated muscle through its spinal nerves.

Muscular activity is smooth and free from jerkiness because there are structures that coordinate the flow of impulses within the brain. Such structures are associated with the extrapyramidal system and the reticular formation of the brain stem. The motor nerves deliver an uncoordinated train of impulses when the coordination centers are disturbed. Individuals with some types of cerebral palsy are afflicted with this type of impairment. In many instances, the problem is damage to extrapyramidal tracts, which results in a lack of inhibition of opposing muscle groups.

With every category of communicative disorders, there is a *possibility* that the cause of the problem is with the nervous system. The phonatory apparatus may have an interrupted nerve supply, thus causing a voice disorder due to paralysis of the vocal folds. An articulation impairment characterized by slow, slurred speech may result from a partial paralysis of the tongue and soft palate. Nerve damage in the cochlea is the etiology of various problems related to audition, which range from a mild impairment of sibilant sounds to profound levels of hearing loss. Faulty programming in the brain can disturb perception or language learning in general, giving rise to a host of problems such as dysphasia and specific learning disabilities.

There is one point that must be mentioned before leaving this introductory discussion of the brain and areas of disorders. This concerns the ability of the brain to bounce back after severe assaults from traumas or disease. Functions lost due to the insult sometimes return in the face of remarkable odds. Head wounds, for example, may permanently destroy brain tissue, but important functions initially lost gradually reappear, although the previous preciseness of the function may never be fully regained. This phenomenon illustrates the *plasticity* of the nervous system. Such conformability to an insult is not the result of regenerated nerve cells in the injured area but due to the brain as a whole adapting to the deficit.

PROTECTIVE FEATURES

Encased in bone, the central nervous system is provided a location of maximum protection. The brain is protected by the bony brain case of the skull (the calvarium). The spinal cord runs through a spacious canal within the vertebral column.

There is also a protective mechanism against a deficient food supply. In conditions of starvation, the central nervous system does not lose weight and become emaciated even though the organism may die from lack of nourishment. However, this protective mechanism apparently does not operate in cases of malnutrition of infants when the brain is growing rapidly and the fibers are attempting to form profuse synapses.

The Meninges

A protective covering of three nonneural membranes surrounds the brain and spinal cord. Together, with the cerebrospinal fluid contained between two of the layers, the *meninges* act as a protective agent of the central nervous system. Listed from outer to inner layer, the membranes are the *dura mater, arachnoid mater*, and *pia mater*. These layers are graded in thickness and consistency to bridge the gap between the bone and the soft exterior of the CNS (Fig. 15-14).

Dura Mater

The outer layer is truly a durable covering. Dura mater is from Latin and means *hard mother*. It lines the cranial bones, serving as the periosteum of the inner surface, and extends inside the vertebral foramina to also protect the spinal cord. This layer is a tough, strong, leathery membrane and folds into the major fissures of the brain (Fig. 15-15).

The *falx cerebri* (*falx* = sickle) is a sickle-shaped fold of the dura mater, projecting into the longitudinal fissure of the brain with a free border arching over the corpus callosum. The posterior limit of the falx cerebri meets at right angles the *tentorium cerebelli* (*tentorium* = tent), which passes transversely between the cerebrum and cerebellum. Another fold, the *falx cerebelli*, extends into the fissure between the cerebellar hemispheres. These partitions give support to the brain parts by preventing them from shifting about and so avoiding injury.

There are two layers of the dura that sometimes separate to form *venous channels* or *sinuses* for returning blood from the brain to the heart. These are found in association with the folds described above. An abundant supply of sensory endings exists in the dura. Perhaps many headaches result from irritation to these sensory fibers (Fig. 15-16).

Arachnoid Mater

The arachnoid mater is a weblike tissue layer found immediately beneath the dura. Arachnoid is from Greek and means resembling a spider's web. Fine trabeculae (beams) extend to the inner meningeal layer, the pia mater, creating

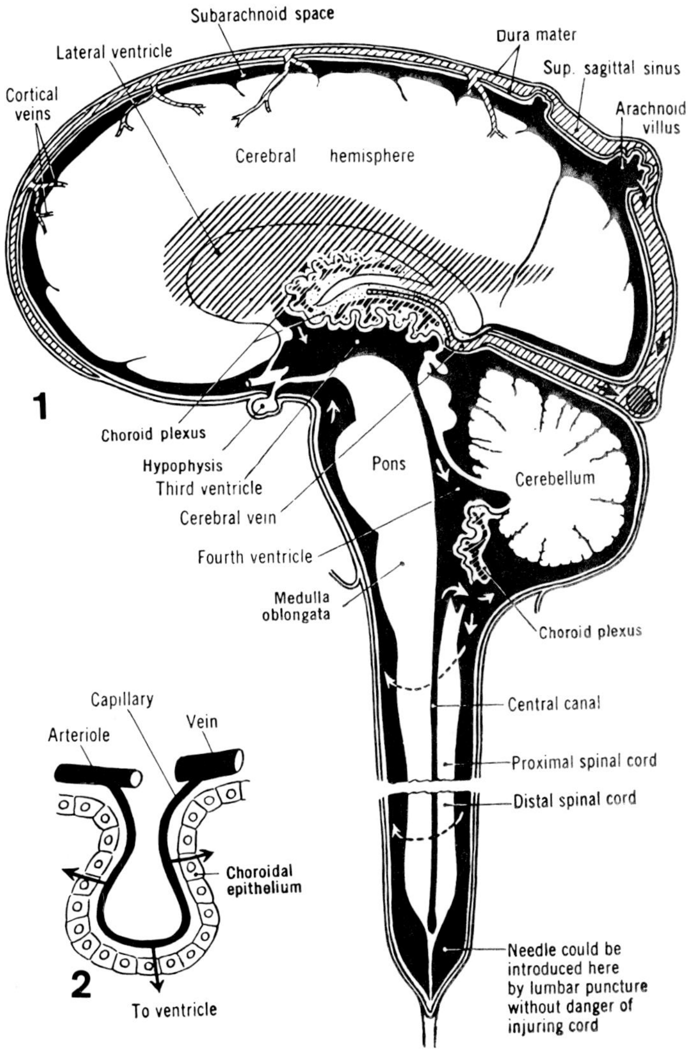

Figure 15-14. (1) Circulation of cerebrospinal fluid (in black); arrows indicate the direction of circulation. The drawing represents a median section of the nervous system and therefore shows the third ventricle, cerebral aqueduct, fourth ventricle, and central canal, with the approximate size and location of one of the lateral ventricles indicated by oblique lines. Notice the aperture in the fourth ventricle by which fluid reaches the subarachnoid space. Notice also one of the arachnoid villi through which fluid enters venous blood in a dural sinus. (2) Fundamental plan of choroid plexuses. Cerebrospinal fluid is formed from blood plasma and passes through choroidal epithelium into the ventricular space. (From Gardner, E.: *Fundamentals of Neurology*, 6th ed., 1975. Courtesy of W.B. Saunders Co., Philadelphia.)

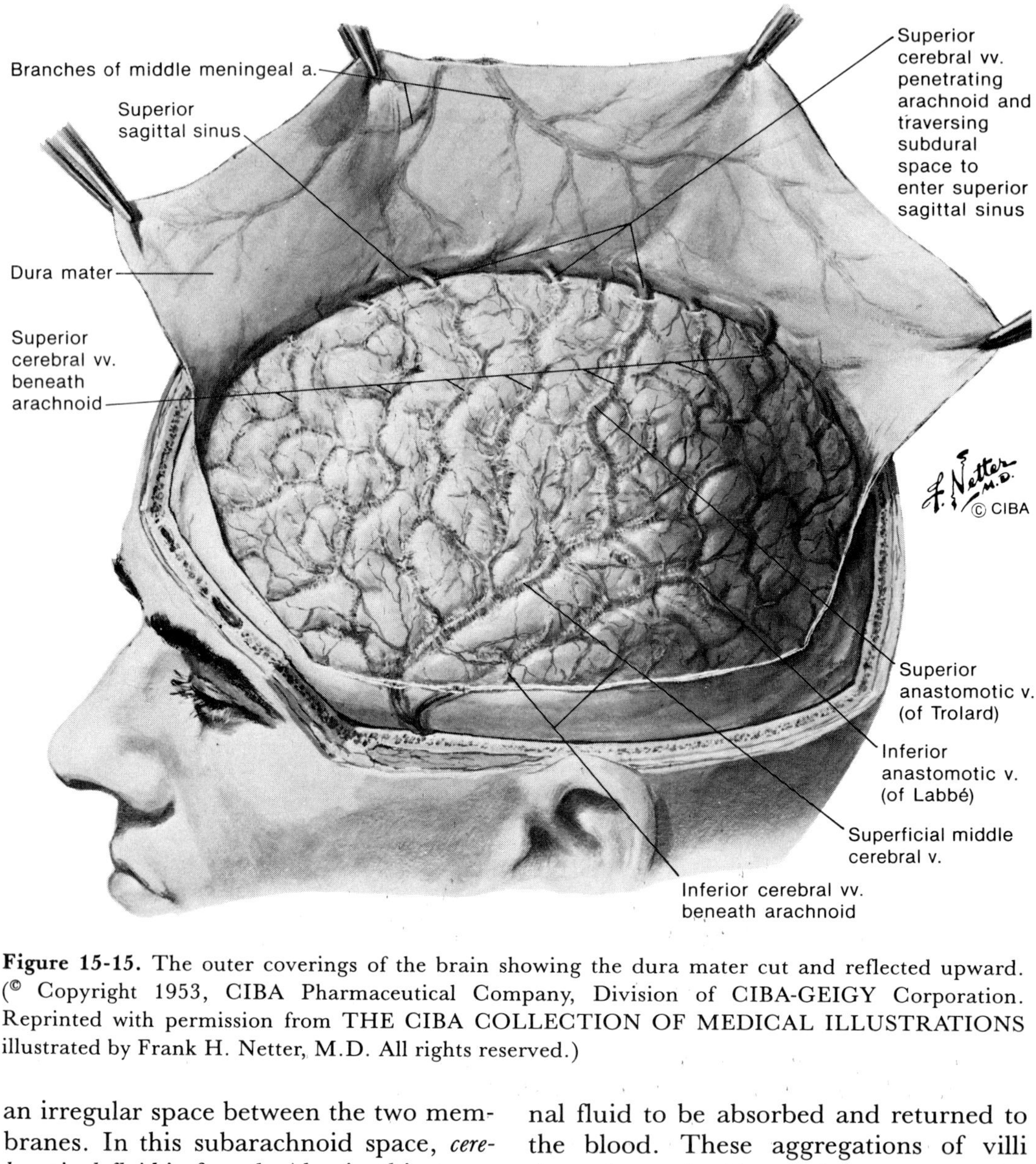

Figure 15-15. The outer coverings of the brain showing the dura mater cut and reflected upward.

an irregular space between the two membranes. In this subarachnoid space, *cerebrospinal fluid* is found. Also in this space are the principal arteries and veins of the brain. Unlike the double-layered dura, the arachnoid membrane does not fold into the fissures of the brain (Fig. 15-17).

Occasionally, the arachnoid mater will send finger-like projections into the *venus sinuses* of the dura mater. Such projections, termed *villi*, permit the cerebrospinal fluid to be absorbed and returned to the blood. These aggregations of villi through *arachnoid granulations* empty the fluid into the superior sagittal sinus (Fig. 15-18).

Pia Mater

The brain and spinal cord are invested with the *pia mater*, which in Latin means *tender mother*. It is a thin but richly vascular membrane. So closely does it adhere to

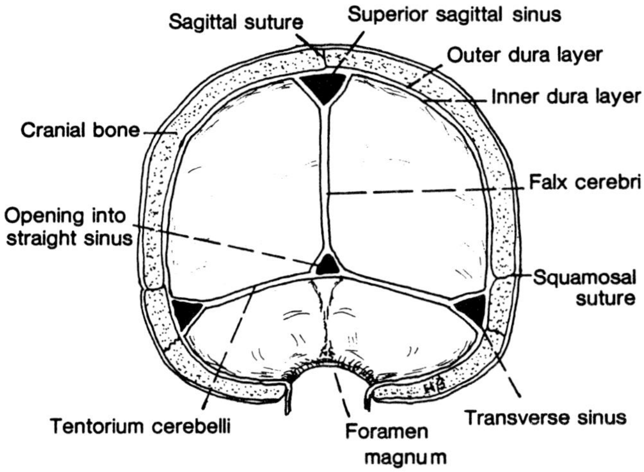

Figure 15-16. Diagram of the venous channels formed by the layers of dura mater. Membranes, acting as partitions, interconnect the sinuses and provide stability for the brain.

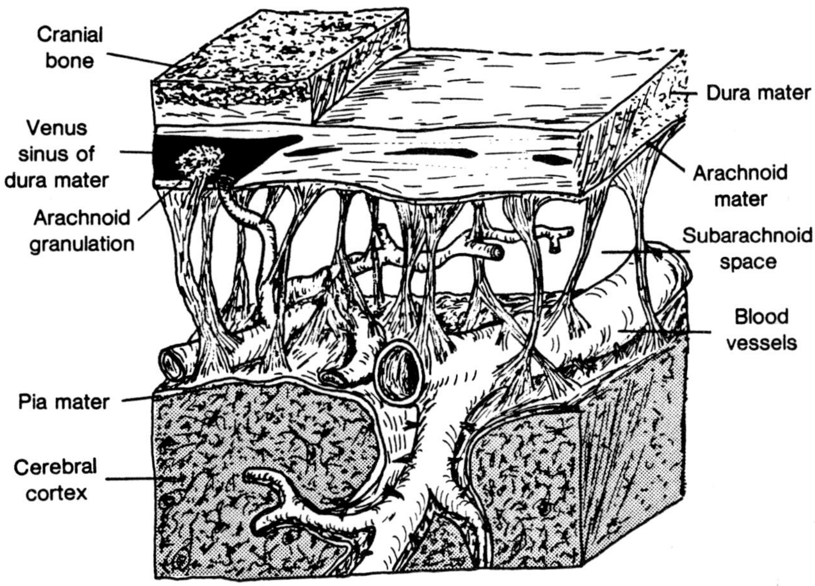

Figure 15-17. Schematic representation of a section of the meninges showing the arachnoid space. (From Spalteholz, W., and Spanner, R.: *Atlas of Human Anatomy*, 16th ed., 1961. Courtesy of Bohn, Scheltema and Holkema, Amsterdam.

Meninges and Superficial Cerebral Veins

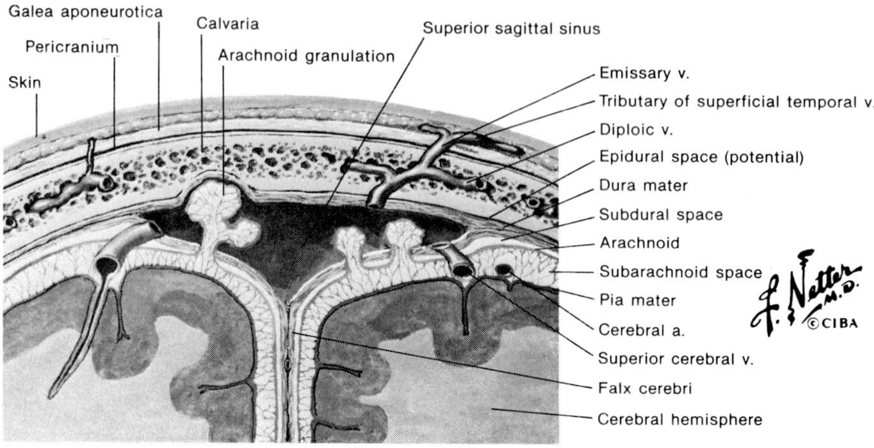

Figure 15-18. The superior sagittal sinus in cross section and related structures, including arachnoid granulations. (© Copyright 1953, CIBA Pharmaceutical Company, Division of CIBA-GEIGY Corporation. Reprinted with permission from THE CIBA COLLECTION OF MEDICAL ILLUSTRATIONS illustrated by Frank H. Netter, M.D. All rights reserved.)

the convolutions of the cerebral cortex that it cannot be dissected away without nerve damage. The pia mater is, then, literally the membranous skin of the brain and spinal cord.

Cerebrospinal Fluid

Cerebrospinal fluid is produced at tufts of modified capillaries within the *ventricles* (see Figs. 15-13 & 15-14). Derived from blood, this fluid has been described as being like strained blood. It is an important liquid cushion or shock absorber for the brain and spinal cord. All four ventricles in the interior of the brain contain *choroid plexuses*: the site of production of cerebrospinal fluid. Together, they produce about 500 ml per day, which is absorbed into the blood stream through the arachnoid villi. The amount of the cerebrospinal fluid that can be held in cavities of the system is about 124 ml.

To trace the circulation of cerebrospinal fluid, one would start with the two lateral ventricles of the cerebral hemispheres (Fig. 15-19). Fluid leaves each of the *lateral ventricles* to enter the *third ventricle* by way of the *interventricular foramen*. The *cerebral aqueduct* passes fluid into the *fourth ventricle*, which contains three foramina for releasing fluid from inside the brain. There are two lateral openings: the *foramen of Luschka* and a *central foramen of Magendie*. These foramina release fluid from the fourth ventricle into the subarachnoid space. The fourth ventricle tapers sharply into a canal in the caudal aspect of the medulla to continue into the spinal cord as the *central canal* (Fig. 15-20).

The cerebrospinal fluid, in addition to floating the brain to protect it from trauma, serves as a vehicle for removing waste products associated with neuronal activity. Withdrawing fluid through a lumbar puncture needle between the third and fourth lumbar vertebrae poses no danger of injury to the spinal cord, since the cord does not extend this far down. The meninges continue on below this point, forming a pocket that is filled with fluid. By attaching a manometer to the

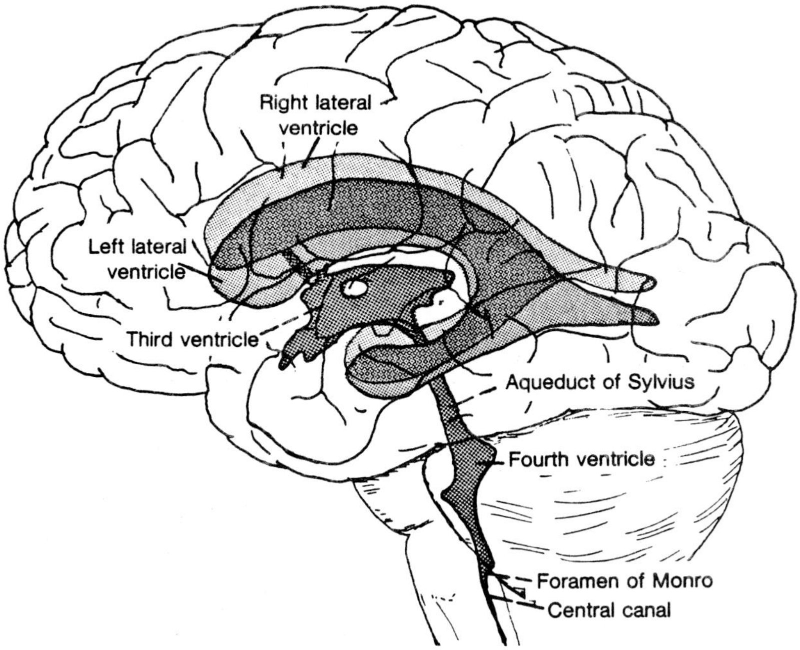

Figure 15-19. Schematic phantom representations of the ventricles of the brain.

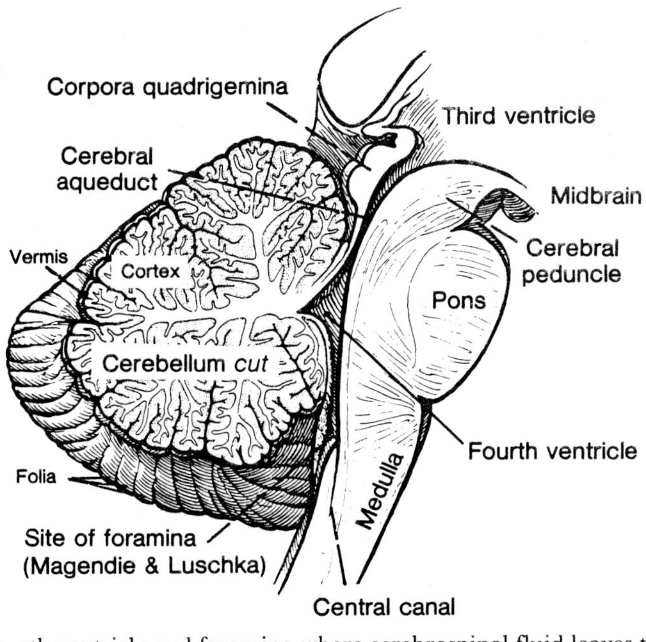

Figure 15-20. The fourth ventricle and foramina where cerebrospinal fluid leaves the ventricle. (From *Man, Microbes and Matter* by B.C. Block. Copyright © 1975. Used with the permission of McGraw-Hill Book Company.

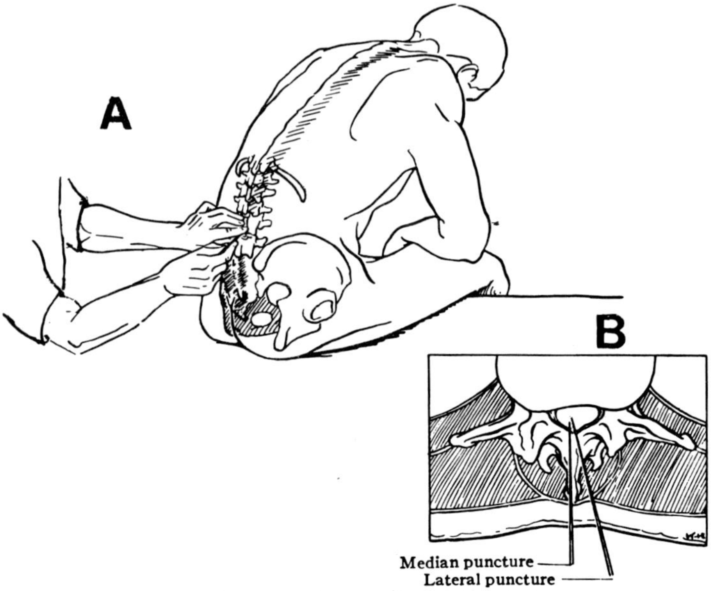

Figure 15-21. Lumbar puncture or spinal tap. (From Callander, C.L.: *Surgical Anatomy*, 2nd ed., 1966. Courtesy of W.B. Saunders Co., Philadelphia.)

needle, the cerebrospinal fluid pressure may be measured (Fig. 15-21).

Hydrocephalus (water in the cranium) is a disorder relating to excess cerebrospinal fluid, which raises intracranial pressure. Usually in such cases there is interference in the circulation or absorption of the fluid. If the condition is not corrected, the brain may be damaged due to the compressional effects of the fluid within the ventricles (see Fig. 15-19). In the infant, the skull will also increase in size, since the cranial skull sutures are not yet rigid from ossification.

To prevent brain damage and possible death, excess cerebrospinal fluid must be removed. Up to 500 ml of fluid may be produced daily. A syringe is sometimes used to draw off fluid from the fourth ventricle. Adults and also some children may require the surgical installation of a shunt to drain the fluid from the brain into a lymph vessel, the internal jugular vein, or the esophagus (Fig. 15-22). Speech clinicians may see in medical reports the notation of a trephine procedure. To trephine the skull case implies drilling a hole through the skull to obtain drainage for release of cerebrospinal fluid.

DIVISIONS OF THE BRAIN

There are three main divisions of the brain: the forebrain, midbrain, and hindbrain. Each member of these divisions represents a group of structures comprising the brain. Each division will be discussed from a structural and functional standpoint.

Forebrain

The forebrain is the largest part of the brain, its mass being considerable in com-

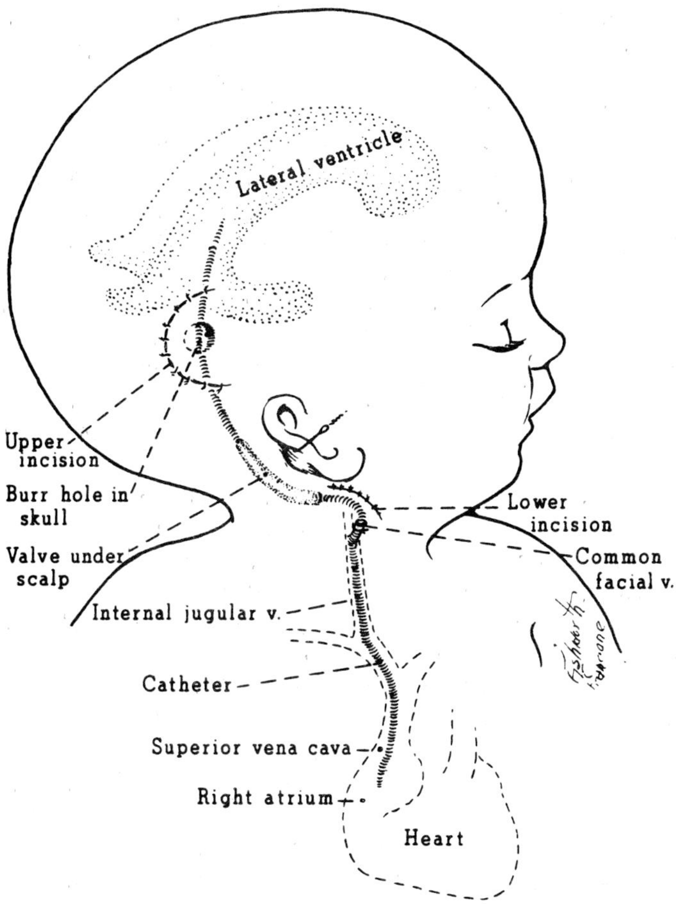

Figure 15-22. Operative procedure used to drain excess cerebrospinal fluid to prevent brain damage associated with hydrocephalus. (From Jacob, S.W., and Francone, C.A.: *Elements of Anatomy and Physiology*, 1976. Courtesy of W.B. Saunders Co., Philadelphia.)

Table 15-I
DIVISIONS OF THE BRAIN

	Major Parts of the Brain	*Cavities of the Brain*
FOREBRAIN	Cerebrum Diencephalon	Right and left lateral hemisphere Third ventricle
MIDBRAIN		Cerebral aqueduct
HINDBRAIN	Pons Medulla oblongata Cerebellum	Fourth ventricle and central canal

parison with the midbrain or hindbrain. Due to its role in many complex behaviors, including speech and hearing, it is designated as the highest division of the brain. From a developmental standpoint, the forebrain is the last section of the brain to develop. It is chief of the brain parts and is able to dominate the other, older divisions. The forebrain forms the telencephalon, which gives rise to the cerebrum and the diencephalon, or interbrain (Fig. 15-23).

Upon viewing the brain, one cannot avoid observing the extent to which the cerebrum structurally dominates the brain. This wrinkle-covered structure is divided into left and right cerebral hemispheres by a deep *longitudinal fissure* (Fig. 15-13). The two hemispheres are joined by prominent bands of commissural nerve fibers (bands of myelinated fibers that cross the midline between the hemispheres) known as the *corpus callosum*.

Cerebral Cortex

The outside covering of the cerebral hemispheres is known as the cerebral cortex, or "bark of the brain." It is about 50-100 cells thick, averaging about 2.5 mm thickness, and contains around 13 billion cells. It is made up principally of nerve cell bodies, which give it its gray coloring. Six functional layers are generally recognized (Fig. 15-24).

The cortex is marked by numerous folds or convolutions with grooves in between them. The convolutions are known as *gyri*, and the depressions in between the gyri are termed *sulci* (singular forms are *gyrus* and *sulcus*). Such an arrangement is an old biological trick that permits a large surface area, which is estimated to be about 200,000 to 250,000 square centimeters total area. Only about one-fourth of its surface is visible; the remainder is buried in the convolutions. It is housed in the cramped cranial cavity. In essence, then, a greater surface area with increased functional potential is obtained without demanding greater brain cavity volume. Nonetheless, the gyri and sulci are laid out in a recognizable pattern even though minor variations are usually present from brain to brain. There are several particularly large sulci or fissures that are used in marking the surface of each cerebral hemisphere into lobes. Such markings provide a convenient means of communication

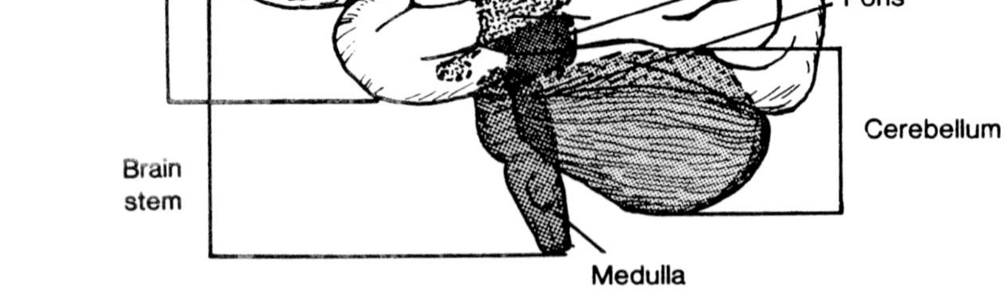

Figure 15-23. Parts of the brain.

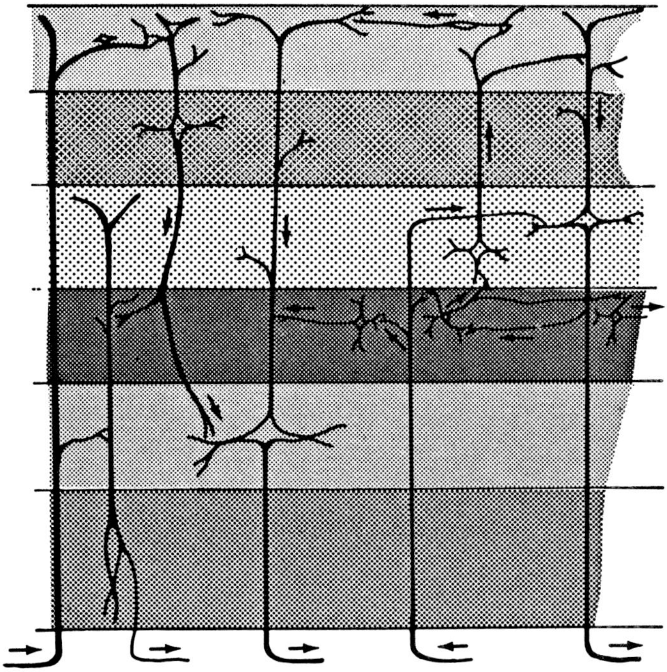

Figure 15-24. Layers of the cerebral cortex. (From Minckler, Jeff: Thalamus and cerebrum. (In Minckler, Jeff (Ed.): *Introduction to Neuroscience*, St. Louis, 1972. The C.V. Mosby Co.; after Lorente de Nó, R.: Cerebral cortex. In Fulton, J. (Ed.): *Physiology of the Nervous System*, ed. 3, London, 1949, Oxford University Press.)

about the various areas of the cerebrum (Fig. 15-25).

LOBES. Generally, the lobes of the cerebral cortex derive their names from the bones of the cranium just above them. Altogether, five lobes may be identified on the lateral view of the brain. All the lobes are paired, since the longitudinal fissure separates the brain into halves, or hemispheres.

Underneath the frontal bone are the two **frontal lobes**. They comprise about 41 percent of the surface area of the cerebral cortex and extend back toward the parietal bone until the *central sulcus* is encountered. This posterior boundary of the frontal lobe is also known as the *fissure of Rolando*. A corresponding arch across the top of the skull is the coronal suture, which is roughly parallel to and positioned well ahead of the central fissure. The longitudinal fissure and sagittal suture share a similar relationship (Fig. 15-26).

The lobe of the cortex located behind the central sulcus is the **parietal lobe** which, together with its mate, comprises about 21 percent of the surface area of the cerebral cortex. It is bounded medially by the longitudinal fissure, and the anterolateral limit is formed by the *lateral sulcus*, also known as the *fissure of Sylvius*, which extends alongside the posterior aspect of the frontal lobe. Below the lateral fissure is the **temporal lobe**. The surface area covering the temporal lobes constitutes

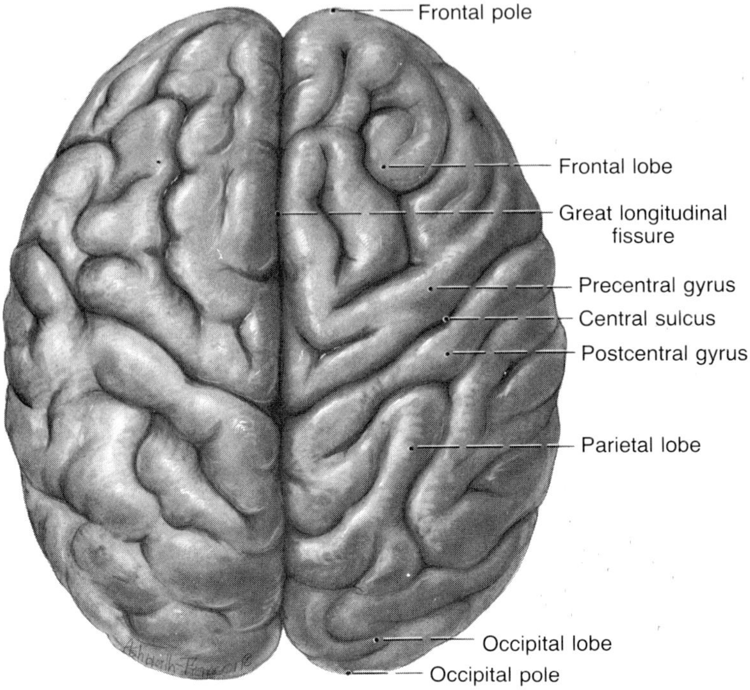

Figure 15-25. Superior aspect of the brain. The longitudinal fissure divides the cerebral hemispheres. (From Jacob, S.W., and Francone, C.A.: *Structure and Function in Man*, 4th ed., 1978. Courtesy of W.B. Saunders Co., Philadelphia.)

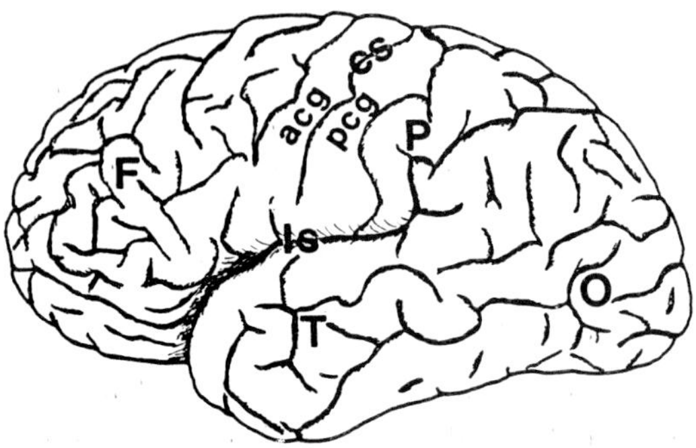

Figure 15-26. Diagram of the left side of the brain showing the frontal, parietal, occipital, and temporal lobes. Additionally, the central sulcus (cs) and lateral sulcus (ls) are shown. The anterior central gyrus (acg) and posterior central gyrus (pcg) are identified on either side of the central sulcus.

around 21 percent of the total cerebral cortex.

If the lateral fissure is widened by retraction of its walls, one can inspect the normally hidden **insular lobe** or **island of Reil** at the bottom of the groove (Fig. 15-27). However, some anatomists do not consider this a lobe of the brain. In a transverse section of the brain in this area, a T-shaped fissure lying on its side is observed. The stalk of the T is from the lateral fissure, and the crossbar marks the surface of the insular lobe (see Fig. 15-13).

At the posterior aspect of the brain is the **occipital lobe**. About 17 percent of the surface area of the cerebral cortex is devoted to the occipital lobes. It lies posterior to the parietal and temporal lobes and is without clear landmarks for definite boundaries. But a longitudinal fissure makes a clear division between the two occipital lobes. It is interesting to note that all lobes along this fissure throw their folds into the fissure to extend the potential functional area of the cortex.

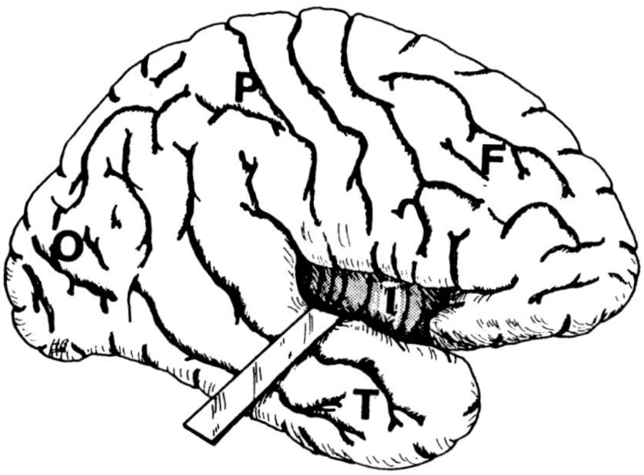

Figure 15-27. The right lobes of the brain. Notice that the temporal lobe (T) has been retracted to expose the insular (I) lobe at the base of the lateral fissure. Just above the insular lobe are the frontal (F) and parietal lobes (P). The occipital lobe (O) lies posteriorly to the temporal lobe.

The cortical surface comprising the lateral wall of the longitudinal fissure has a gyrus running along the lower aspect of the fissure known as the *cingulate gyrus*. It is just dorsal to the corpus callosum and is a part of the limbic system, a complex and important center in emotion and motivated behavior (Fig. 15-28). (For a further discussion of the limbic system, see Chapter 22, "Applied Clinical Anatomy and Physiology.")

It appears by implication that the two cerebral hemispheres and their "bark," the cortex, are mirror images of each other, though such is not the case. If a coronal-transverse section is cut through a human brain in the plane of the Sylvian fissure to expose and compare the two temporal lobes at their upper surface regions, the *planum* (flat) *temporale*, the left planum temporale would be larger in about 65

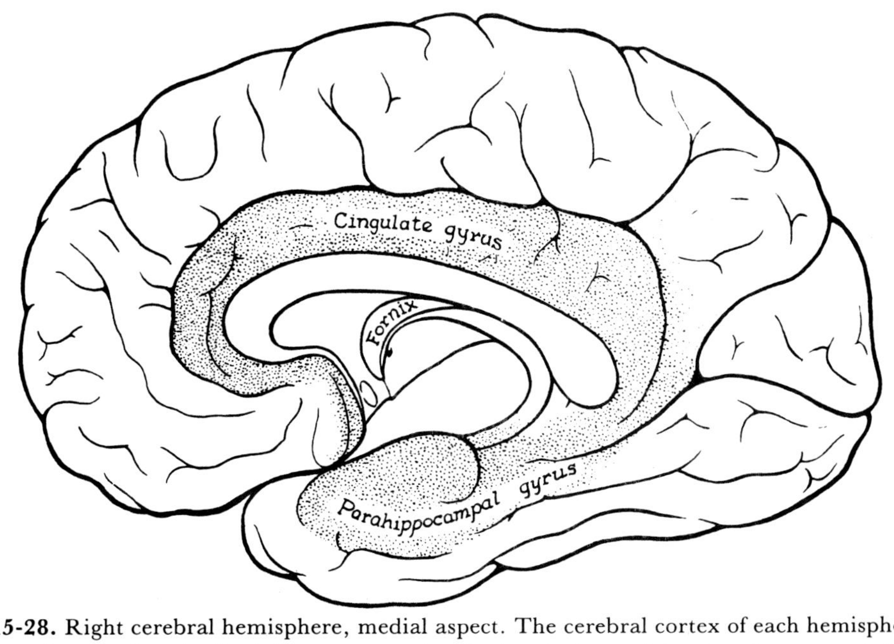

Figure 15-28. Right cerebral hemisphere, medial aspect. The cerebral cortex of each hemisphere continues down into the longitudinal fissure separating the hemispheres. The cingulate gyrus is found on the lateral wall of the longitudinal fissure. (From Truex, R.C., and Carpenter, M.B.: *Human Neuroanatomy*, 6th ed. Copyright 1969, The Williams and Wilkins Co. Reproduced by permission.)

percent of a lot of cadaveric brains; the right planum temporale would be larger in about 11 percent of the brains; and the two would be about the same size in 24 percent of the brains (Geschwind and Levitsky, 1968). Additionally, the gyri and sulci over the cortex show considerable individual variation is size, shape, and position. Campain and Minckler (1976) have also reported on the disparity between the left and right temporal lobes in the area of the auditory cortex. The halves of the same brain are, therefore, structurally unlike each other, and they in turn are structurally unlike from every other brain (Fig. 15-29).

FUNCTIONAL AREAS OF THE CORTEX: GENERAL. Many areas on the cortex have been mapped in detail by assigning numbers to them. Over one hundred such numbered areas are recognized. It is possible to number brain areas due to the rather stable pattern of the gyri and sulci from one brain to another and consistency of general functions for a given area. It should be pointed out that there is pronounced opposition by some to the assignment of localized areas for a given function. Surgical evidence appears to support a view of localization, even though the phrenologists first introduced the concept from a bogus basis. Although some functions can be localized on the brain, other processes such as language are extremely diffuse with regard to brain function and defy absolute description and identification by localization approaches.

As presented in Chapter 22, three basic functional parts of the brain have been proposed. These focus on structures in the brain stem (Part 1), the posterior aspects of the cerebral hemispheres (Part 2), and

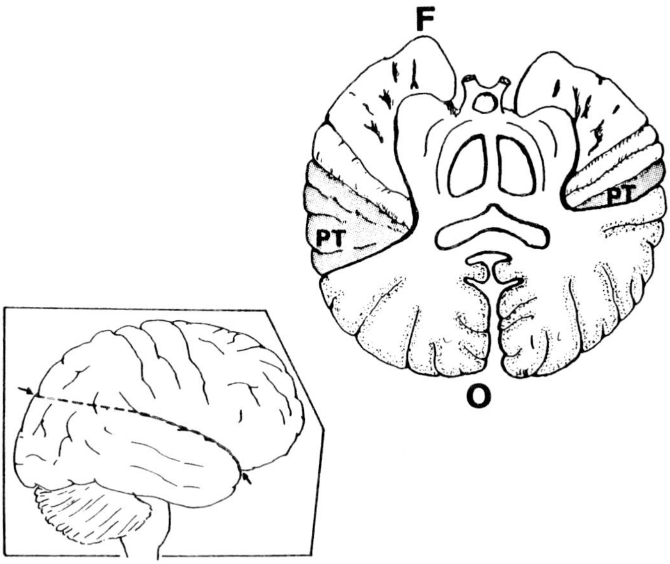

Figure 15-29. Asymmetry of the superior temporal lobe. The upper surfaces of the temporal lobes have been exposed by a cut on each side, following the broken line shown in the inset. Typical left-right differences are shown by the area of shading, with the left planum temporale (PT) clearly dominant in size. The occipital lobe (O) and frontal lobe (F) are identified. (From Geschwind, N., and Levitsky, W.: Human brain: left-right asymmetries in temporal speech region, *Science, 161*:186-187, 1968. Used by permission.)

the anterior parts of the cerebral hemispheres (Part 3). With each functional part, there is a more sophisticated role to play in language and cognitive behavior in general. Each of these proposed functional parts of language processing involves use of the previously outlined categories of brain activity: reception center, associative function, and muscular control. Comparing levels of brain function to levels of sleep is a general analogy but not an entirely appropriate one.

The postcentral gyrus lies immediately posterior to the central sulcus and marks the anterior border of the parietal lobe. It is known as the *sensory strip*, owing to its role in serving the *somatesthetic* (bodily consciousness) sensory areas. The nerve cells in this strip of the cortex receive and interpret sensations from the opposite side of the body (Fig. 15-30).

Three areas along the postcentral gyrus are recognized. Those portions of this gyrus that have been numbered are areas 1, 2, and 3, although there are no clear functional differences between the numbers. These areas are concerned with sensations of pain, temperature, touch, and pressure (Fig. 15-31). In a similar pattern to the motor strip, the various parts of the body are represented by points and sections along the sensory strip. The face is represented laterally, and the leg is represented medially (Fig. 15-32).

It is interesting to note that the brain tissue itself is insensitive to sensation but gives rise to consciousness as sensory stimuli arrive at the postcentral gyrus. For this reason, it is possible to perform brain surgery under local anesthesia in or-

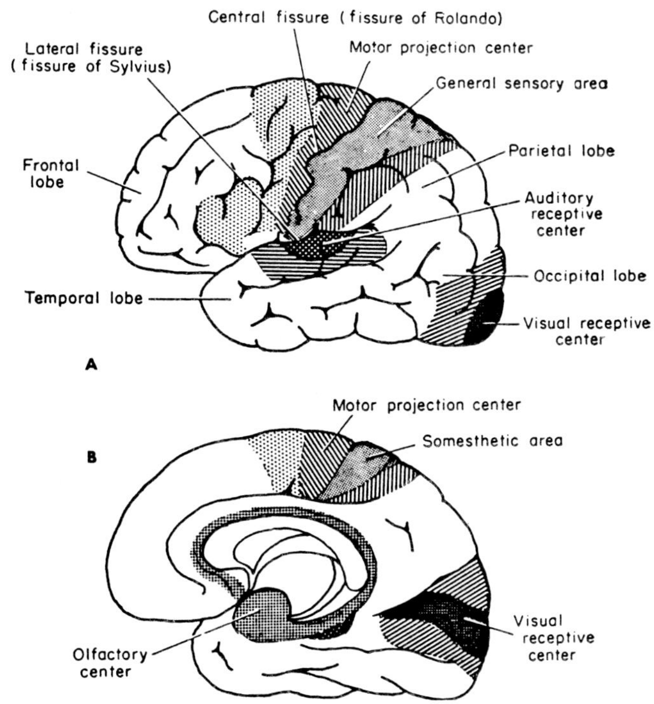

Figure 15-30. Diagram of brain showing various functional areas and centers: (A) lateral view, (B) medial view. (From Brooks, Stewart M., and Paynton-Brooks, Natalie: *The Human Body*, ed. 2, St. Louis, 1980, The C.V. Mosby Co.)

der that the patient may be able to respond to stimulation of the brain as the surgeon might determine functional features of the diseased site in elective surgery.

The *precentral gyrus* is the rearmost gyrus of the frontal lobe and is anterior to the central sulcus. It is a motor area controlling voluntary muscles to effect movement of the opposite side of the body. Large motor nerve cells in this strip sometimes called the "*motor strip of the cortex,*"

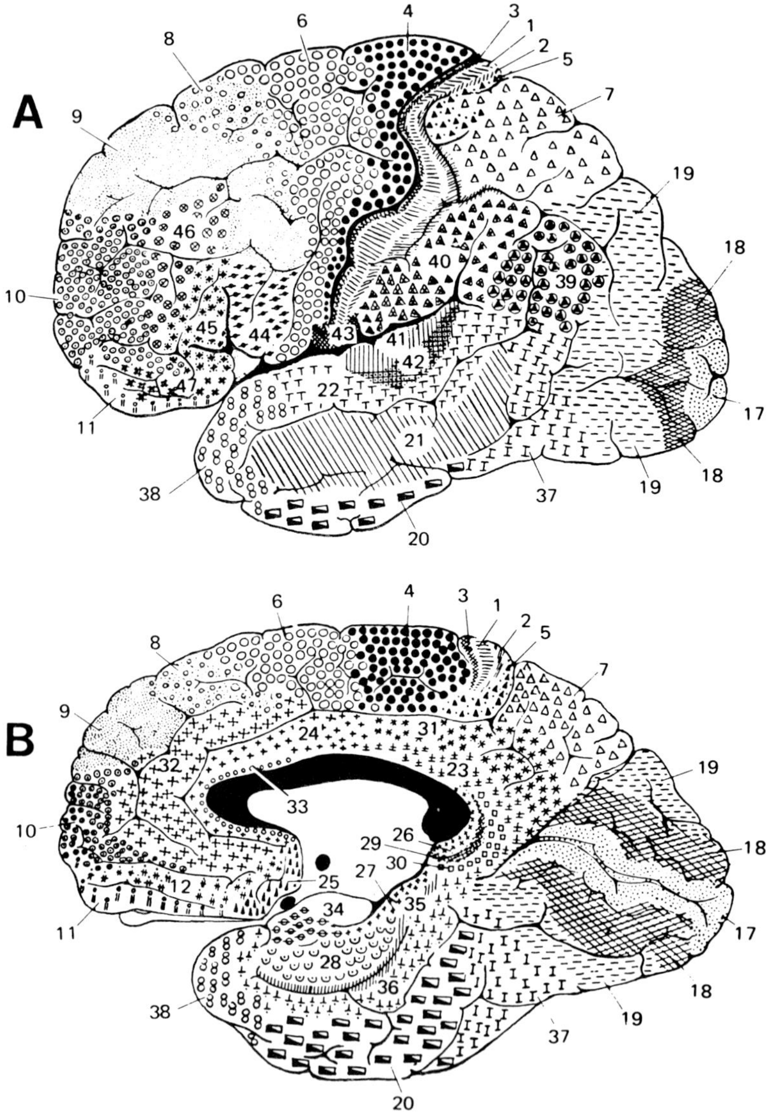

Figure 15-31. Brodmann's numbering of functional areas of the cortex: (A) lateral view, (B) medial view. (Reprinted by permission of the publisher from Martin, John H.: Somatic sensory system II: Anatomical substrates for somatic sensation. In Kandel and Schwartz, *Principles of Neural Science*, page 180. Copyright 1981 by Elsevier Science Publishing Co., Inc.)

send their fibers down through the brain to the medulla oblongata of the hindbrain, where bundles of fibers cross over to the other side. From this point, they descend to the spinal cord on the side contralateral to their origin; this crossing over is known as *decussation*.

The precentral gyrus is designated as area 4. If an electrode is touched to this motor strip while the subject is conscious but under local anesthesia for the surgical wound in the scalp and skull, a corresponding group of muscles will contract. For example, stimulation by an electrical current to the lower surface of area 4 will result in contraction of the facial muscles. Higher up toward the midline, stimulation will cause the hand muscles to respond upon application of the stimulus. Tongue control is represented near the lowest section of area 4, close to the lateral fissure. Lesions anywhere along the precentral gyrus will yield a corresponding deficit in movement of the muscle group represented by the section (Fig. 15-32).

FUNCTIONAL AREAS: SPEECH AND LANGUAGE. The gyrus in front of the lower portion of the precentral gyrus is the *motor speech area* or *Broca's area* (see Fig. 15-33). It is recognized as areas 44 and 45 and is dominant in the left hemisphere for most persons. Broca's area, sometimes termed the *anterior speech cortex*, serves to control the movements required in speech production. Fibers from areas 44 and 45 project by way of the corpus callosum to the right hemisphere for bilateral control of the articulatory musculature. Conversely, the companion area on the right hemisphere has bilateral projections but does not duplicate Broca's area when the left hemisphere is dominant. In essence, Broca's area is responsible for programming the articulators for speech. Without this area, the speech musculature cannot be coordinated sufficiently to produce speech, even though the speech muscles may still be consciously contracted. (See the section "Speech Disorders: Dysarthria, Apraxia, and Aphasia" in Chap-

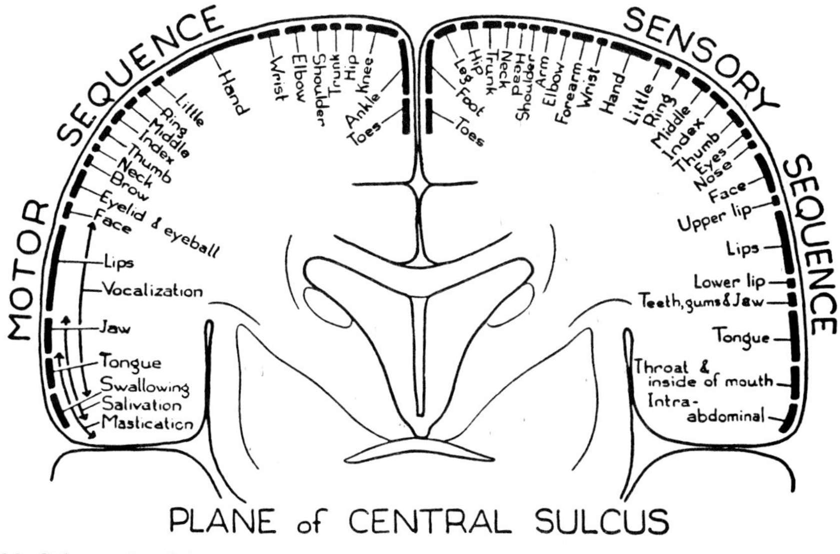

Figure 15-32. Schematic of the motor and sensory strips of the precentral gyrus (motor sequence) and the postcentral gyrus (sensory sequence). (From Rasmussen and Penfield: Further Studies of the Sensory and Motor Cerebral Cortex of Man. In *Federation Proceedings*, 1947, *6*:452-460.)

ter 22.)

A supplementary motor area for speech is situated at the peak of the frontal lobe and folds down medially into the longitudinal fissure. This area, the uppermost aspect of area 6, termed the *superior speech cortex*, is the most easily dispensable (see Fig. 15-30). If it is removed, loss of speech functions are only temporary in most individuals.

The *superior temporal gyrus* has been designated as area 41 (see Fig. 15-31). It lies adjacent to the lateral sulcus on the temporal lobe. Area 41 is known as the auditory area for its role in sound reception via the auditory pathway fibers and for its role in interpretation of this sensory data. The representation of some music stimuli and sound discrimination, for example, are found on this and nearby regions of the temporal lobe.

The posterior speech cortex is not confined to one lobe (see Fig. 15-33). This wide zone, the so-called language formulation area and known as *Wernicke's area*, encompasses an area around the upper limit of the lateral fissure on the parietal lobe (*angular gyrus*) and spreads down onto the temporal lobe. Areas 22, 37, 39, and 40 include most of the postspeech cortex. These areas lie in the parietotemporal field. Destruction of this region in the left hemisphere for most persons, regardless of handedness, can produce the gravest effect on language formulation. The term *aphasia* (loss or disturbance of language) is used to designate such a condition.

Through an arching bundle of association fibers, the *arcuate fasciculus* (see also Chap. 22), the posterior speech cortex communicates with the anterior speech cortex (Broca's area). Wernicke's area is responsible for understanding oral language and language formulation for exte-

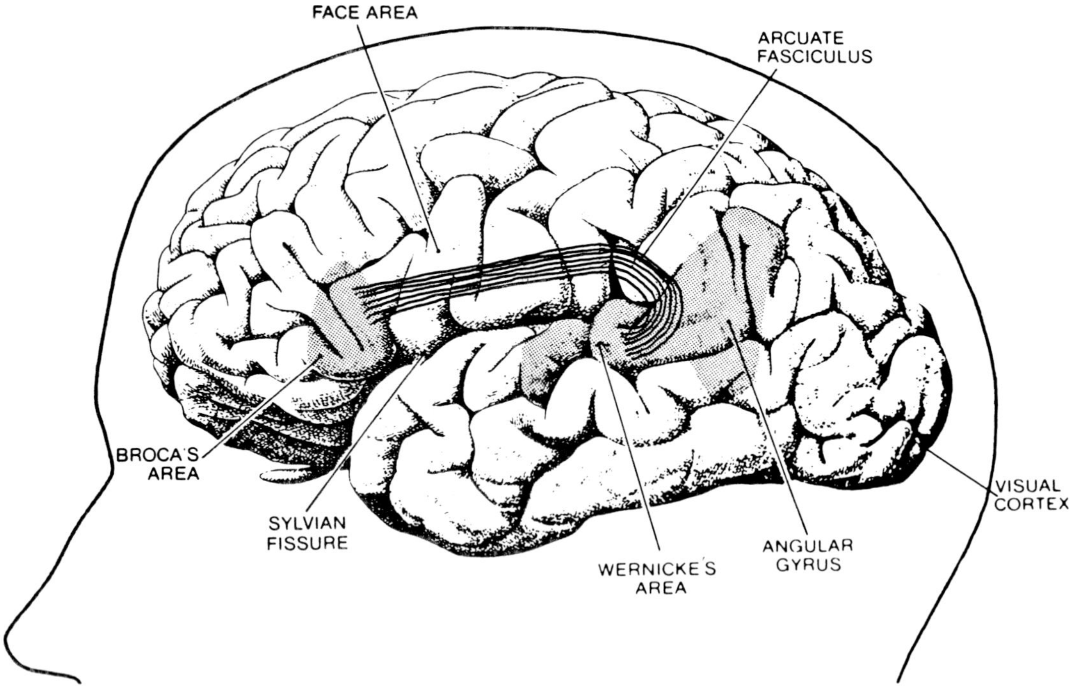

Figure 15-33. Left aspect of the brain, showing the language and speech centers. (From Language and the brain by Norman Geschwind. Copyright © 1972 by *Scientific American*, Inc. All rights reserved.)

riorization. When these signals are projected to the motor speech programmer in Broca's area, articulate speech is possible (Fig. 15-33).

Even though various areas of the brain have been associated with language formulation, the student of speech and language should be reminded of the complexity and diffuseness of central nervous system functioning in the language integration process. Many concomitant abilities develop along with language and contribute to the language learning process as correlates of language. Such an ability includes the adjustment to sensory impressions received from the environment, that is, a set-to-attend. As a consequence of the diffuse, integrative nature of language formulation in the central nervous system, it is only possible at this time to delineate *areas* of the brain pertinent to language. Nonetheless, neurosurgeons persist in attempting to localize specific language functions because of the necessity of surgical penetration of possible language sites during brain surgery. Perhaps it would be wise for the student to concentrate with greater focus on *levels of organization of language* or mental processes in the brain rather than look for their localization.

A *visual area* is represented by area 17 on the posterior aspect. This area is located primarily on the medial surface of the occipital lobe along the *calcarine fissure* (inside the longitudinal fissure). The re-

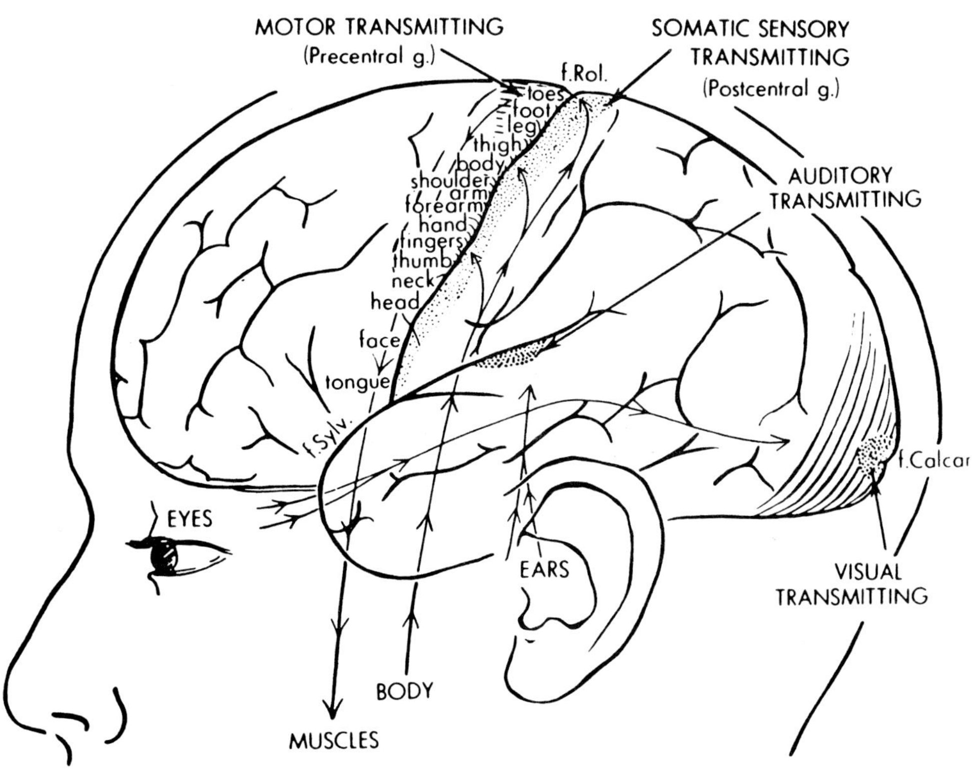

Figure 15-34. Diagram of the left hemisphere of the brain. The motor strip (precentral gyrus), auditory, and visual transmitting centers are illustrated. (From *The Understanding of the Brain* by J.C. Eccles. Copyright © 1975. Used with the permission of McGraw-Hill Book Company.)

tina of the eye sends impulses to this area by way of the dorsal thalamic complex. If destroyed, blindness results, although light and dark may continue to be distinguished. Area 17, then, may be called the *primary visual sensory area.*

Two other primary visual areas, 18 and 19, lie immediately adjacent to area 17. While area 17 appears on the posterior pole of the occipital lobe, area 18 encompasses a broad, lateral strip surrounding 17, and area 19 is forward to 18. These comprise the visual association cortex. Here the retina is clearly represented on the cortex for recognition of patterns and for integration of the stimuli into more complex phenomena and with other senses (Fig. 15-34).

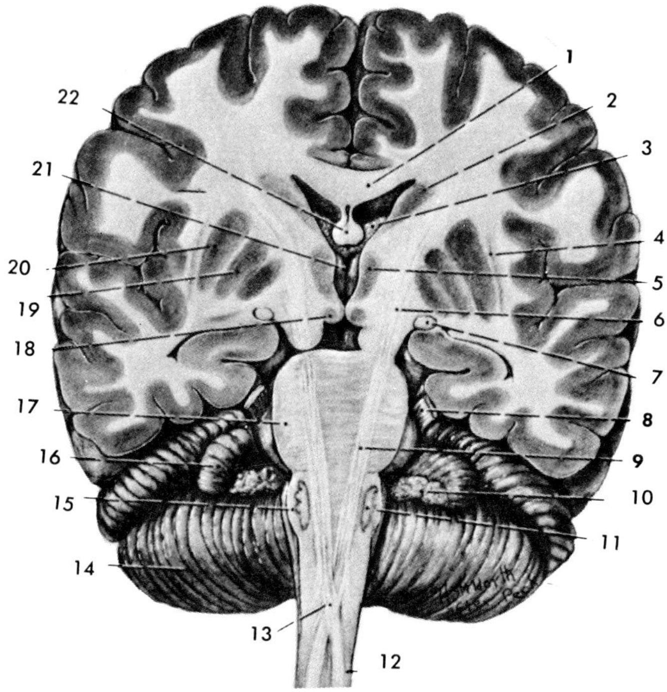

Key:

1. Corpus callosum
2. Caudate nucleus
3. Choroid plexus of lateral ventricle
4. Claustrum
5. Thalamus
6. Internal capsule
7. Optic tract
8. Trigeminal n.
9. Pyramidal tract
10. Choroid plexus
11. Nucleus of olivary body
12. Spinal cord
13. Decussation of pyramids
14. Cerebellum
15. Olive
16. Flocculus
17. Pons
18. Mammillary body
19. Globus pallidus } Lentiform nucleus
20. Putamen }
21. Third ventricle
22. Anterior pillar of fornix

Figure 15-35. Frontal section through the brain. (From Jacob, S.W., and Francone, C.A.: *Laboratory Manual of Structure and Function in Man*, 1974. Courtesy of W.B. Saunders, Co., Philadelphia.)

White Matter of the Cerebrum

From the above description of a few of the functional areas of the brain, one might conclude erroneously that each area carries out a specific duty without influence from any other area. No single area functions independently. *Association* and *commissural* fibers interconnect the many cortical areas in order that they may function in concert. The epitome of such interactions in humans may very likely be language processes. The automatic integration of auditory and visual stimuli, especially in response to a startling situation, requires that the specific areas of the brain exchange information in order to prepare the individual for a response. This task is carried out by the association and commissural fibers of the cerebrum, which integrate the various activities of the areas involved. These fibers are amassed below the cortex and have a characteristic white appearance as contrasted with the gray color of the cortex. As mentioned earlier, myelin sheaths covering axons are responsible for the white portions of the brain (Fig. 15-35).

The most prominent band of commissural fibers interconnecting the two hemispheres is the *corpus callosum* (see Fig. 15-39), which was introduced previously. In a sagittal view, this structure has the appearance of the outline of an auricle or C shape. The corpus callosum enables communication between most parts of the cortex from one hemisphere to the other. Coordination of right- and left-sided movements and integration of sensory information is permitted by these reciprocal connections of the hemispheres (Fig. 15-36).

The commissural paths, therefore, relate the activity of one side of the brain to the other. Still, there are numerous paths within a given hemisphere relating adjacent and remote convolutions with one another. These are association pathways. Fibers interconnecting contiguous gyri are called *short association fibers* and form a layer of white matter just underneath the cortex. A deeper group of fibers appears as bundles traveling from one area of a hemisphere to a more distant area of the same hemisphere. These are the *long association fibers* (Fig. 15-37).

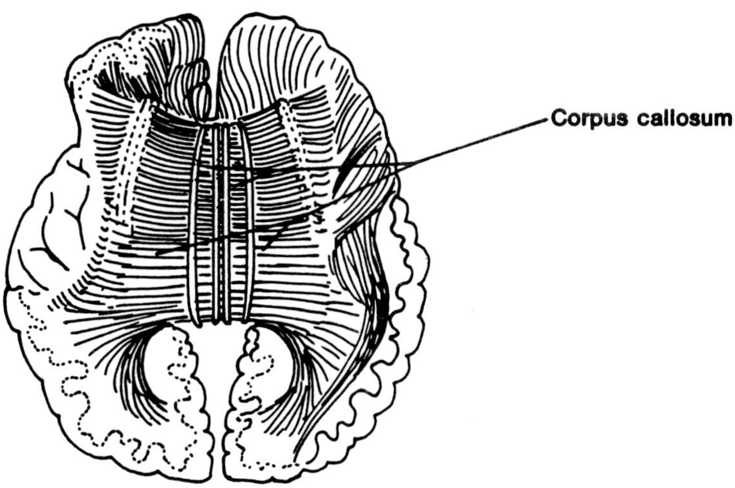

Figure 15-36. Corpus callosum. (From *It's Your Body* by L.M. Elson. Copyright © 1975. Used with the permission of McGraw-Hill Book Company.)

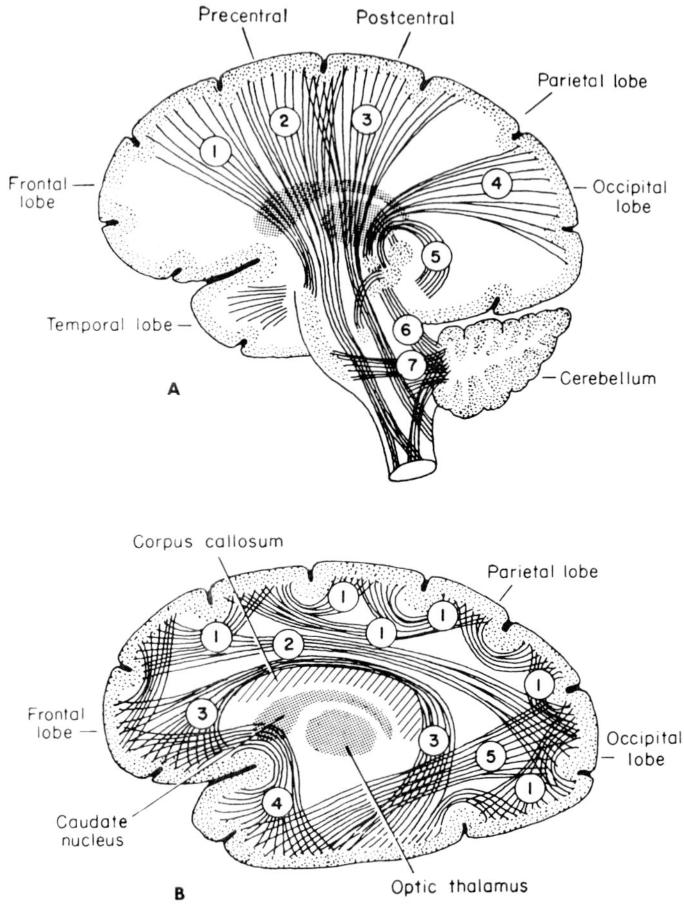

Figure 15-37. (**A**) Projection tracts of the brain: (1) tract connecting frontal convolutions with cells lying in pons; (2) tract descending into cord after decussating in medulla; (3) sensory tract carrying impulses to postcentral gyri; (4) visual tract; (5) auditory tract; (6) superior cerebellar peduncle; and (7) middle cerebellar peduncle. (**B**) Association tracts of brain: (1) tracts between adjacent convolutions; (2) tract between frontal and occipital areas; (3 & 4) tracts between frontal and temporal areas; (5) tract between occipital and temporal areas. Notice the corpus callosum through which the commissural tracts pass between hemispheres. (From Brooks, Stewart M., and Paynton-Brooks, Natalie: *The Human Body*, ed. 2, St. Louis, 1980, The C.V. Mosby Co.)

Joining other parts of the central nervous system to the cerebrum are special groups of pathways termed *projection tracts*. Fibers emanate from practically all parts of the cortex to merge into these tight tracts for passing through the narrow, crowded region in the base of the cerebral hemispheres. Due to the interesting pattern of these fibers, they are called the *corona radiata* (radiating crown). As the fibers pass through this compacted area, they form the *internal capsule*, a part of which encases the *lentiform nucleus*, a mass of gray matter. The term *internal capsule* is appropriate for this encased portion of the projection tracts (Fig. 15-38).

Efferent tracts, delivering muscle impulses from the cortex, are sometimes in-

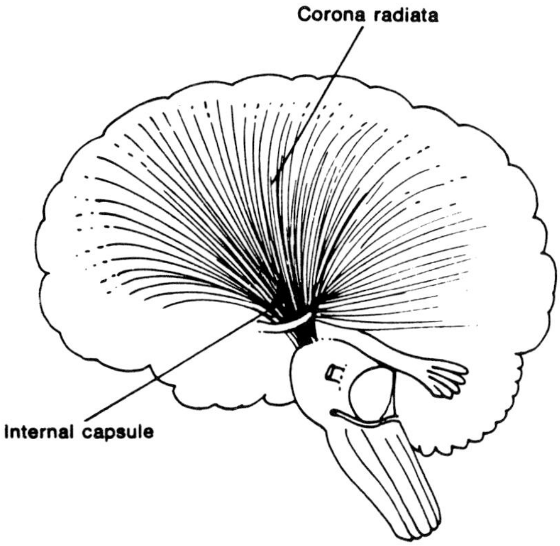

Figure 15-38. Schematic of the corona radiata. (From *It's Your Body* by L.M. Elson. Copyright © 1975. Used with the permission of McGraw-Hill Book Company.)

terrupted by rupture of a small blood vessel or blockage due to a clot in the capsule region. The result is hemiplegia or paralysis on one side of the body. This cerebral vascular accident (CVA) is known as a *stroke*, or *apoplexy* (see Fig. 15-11). Since the fibers for the arms, legs, and trunk are carried in separate bundles through the capsule, the site of the CVA may be suggested by the corresponding disability.

The Lateral Ventricles

Arching within each cerebral hemisphere in the white matter is a cavity, the lateral ventricle of the cerebrospinal fluid circulatory system. Each contains three projections known as the anterior, posterior, and inferior horns (see Fig. 15-19). Most of the lateral ventricle is located in the parietal lobe, but the cavity extends

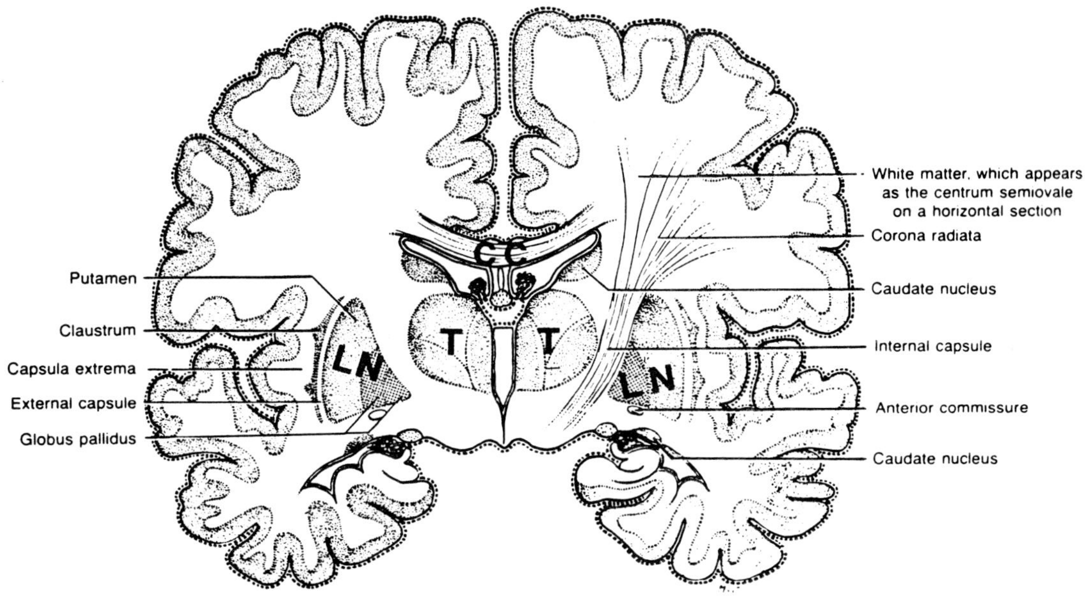

Figure 15-39. Diagram of a coronal section of the brain showing the internal capsule and related structures: (cc) corpus callosum, (T) thalamus, (LN) lenticular nucleus (putamen and globus pallidus). (From Bossy, Jean: *Atlas of Neuroanatomy and Special Sense Organs*, 1970. Courtesy of W.B. Saunders Co., Philadelphia.)

into all the major divisions of the hemisphere by means of the anterior and posterior horns of the ventricle.

Basal Ganglia

Located within the white matter of each cerebral hemisphere are masses of gray matter composed of cell bodies. Collectively, these masses are known as the *basal ganglia* or *basal nuclei*. In a coronal sectional cut of the brain, the gray structures are clearly differentiated from the white matter around and between them. Each mass of gray matter is named. The medialmost mass is the *caudate nucleus* and lateral to it is the *lentiform nucleus* (Fig. 15-39).

An extensive, C-shaped structure, the *caudate nucleus* is a deep gray mass and follows the shape of the cerebral hemisphere. There is a large globular *head* and a narrowing *body* and *tail*. Note in Figure 15-39 how the narrowing body courses upward and alongside the ventricle in its respective hemisphere. The caudate nucleus clamps vertically around the *thalamus* near its external border (Fig. 15-40).

The lentiform nucleus is shaped like a short cone and lies parallel with the lateral surface of the hemisphere. It may be divided into two members, the *globus pallidus* and *putamen* (see Fig. 15-39). The globus pallidus (*globus* = sphere, and *pallidus* = pale) is the smaller, is subdivided into two divisions, and is continuous with structures of the midbrain. Putamen is the Latin word for *shell* and is the relatively dark-colored outer segment of the lentiform nucleus. When considered together (i.e. the caudate and lentiform nuclei along with the internal capsule), the group may be termed the *corpus striatum* due to the alternate gray and white strips characterizing the region in question.

At the outside limit of the basal ganglia is a thin strip of gray matter regarded by some as a member of the basal ganglia; this is the *claustrum*, which lies close to the insular lobe (see Figs. 15-39 & 15-41). It exhibits some undulations corresponding to those of the insular lobe; due to this fact, the claustrum is said to be a detached portion of the insular lobe.

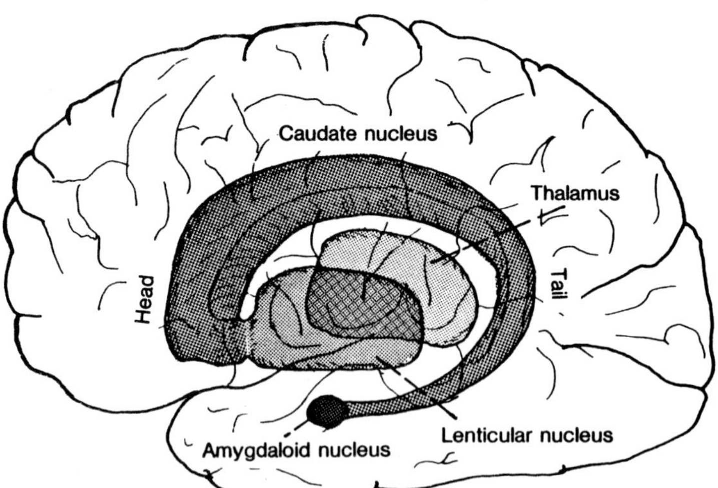

Figure 15-40. Schematic representation of the parts of the caudate nucleus in relation to the thalamus, lenticular nucleus, amygdaloid nucleus, and whole of the brain.

Positioned medially to the basal ganglia is the *thalamus*, a deeper structure, which is considered under a following, separate section. It is mentioned here because it shares with the basal ganglia the role of transporting information between other parts of the nervous system and the cerebral cortex. Also, the basal ganglia are linked with the thalamus and with the cortex.

The globus pallidus is the major motor or efferent center of the basal nuclei. It serves to increase (facilitate) muscle tone. The corpus striatum is concerned with the production and control of automatic association such as rhythmic movements, especially in suppression (inhibition) of random cortical activity. Thus, these subcortical nuclei appear to be valuable coordination centers (Fig. 15-41).

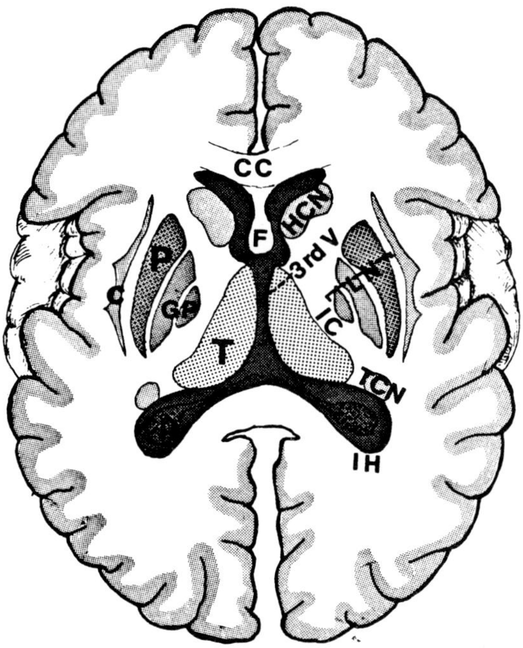

Key:

CC – Corpus callosum	LN – Lenticular nucleus (putamen and globus pallidus)
HCN – Head of caudate nucleus	IC – Internal capsule
F – Fornix (crux of)	T – Thalamus
P – Putamen	TCN – Tail of caudate nucleus
3rdV – Third ventricle	CP – Choroid plexus
C – Claustrum	IH – Inferior horn (of lateral ventricle)
GP – Globus pallidus	

Figure 15-41. Transverse section of the brain.

Although the specific functions of the basal ganglia are poorly understood, lesions in this area of the brain produce several rather well-known abnormalities. It is known that disorders of the basal ganglia result in involuntary, uncontrollable, purposeless movements, which suggests that the activity needs refinement by suppression of random cortical excitation. The cerebellum below the base of the brain is also concerned with coordination of motor impulses but produces different effects in response to lesions. (Several clinical aspects of the basal ganglia are listed in Chapter 22 under "Applied Clinical Anatomy and Physiology.")

Diencephalon — Thalamic Structures

The widely spread cerebral hemispheres clearly dominate in the forebrain. The other part of the forebrain is the diencephalon, which is made up principally of the thalamic portion of the brain. Literally enveloped by the cerebral hemispheres or telencephalon, the diencephalon is centrally located in a position medial to the basal ganglia (see Figs. 15-39 & 15-41). A narrow midline space for the circulation of cerebrospinal fluid (the third ventricle) separates the diencephalon into halves, with a bridge known as the *interthalamic connection* or *massa intermedia* joining the left and right thalamus (see Fig. 15-44). This commissure is clearly present in a sagittal section of the brain.

The thalamus itself is a major relay station. It is regarded as a major sensory center of the brain and is second only to the postcentral gyrus of the cortex. Its function is regarded as a sensory relay station under dominance of the cortex. Sensory impulses are received from a variety of inputs, e.g. smell, visual, auditory, proprioception (muscle sensations), plus touch, pain, temperature, and pressure. The thalamus coordinates the sensations into an integrated pattern and projects the pattern to the somatic sensory strip (postcentral gyrus) of the brain.

There are several thalamic nuclei (collections of cell bodies) within the diencephalon, and each has an assigned function as a member of the major subcortical coordinating center for sensory impulses from over the body. The *medial* and *lateral geniculate bodies* within the dorsal thalamus contain a final synapse before the pathway continues onto the cerebral cortex for auditory and visual fibers, respectively.

Below the thalamus is the *hypothalamus*, which participates in the regulation of a number of basic bodily functions such as eating, blood pressure regulation, emotions, and heat regulation.

Suspended from the hypothalamus is the master gland of the body, the *hypophysis* or *pituitary*. It hangs by a stalk, with its body housed in the hypophyseal fossa of the sphenoid bone. Neural connections are found between the hypothalamus and the posterior lobe of the hypophysis. These neurosecretions with the hypophysis involve oxytoxin (stimulates smooth muscle of the uterus) and vasopressin (vasoconstriction and antidiuretic action). The anterior part is without neural connections (Fig. 15-42).

Midbrain

The midbrain is strikingly small com-

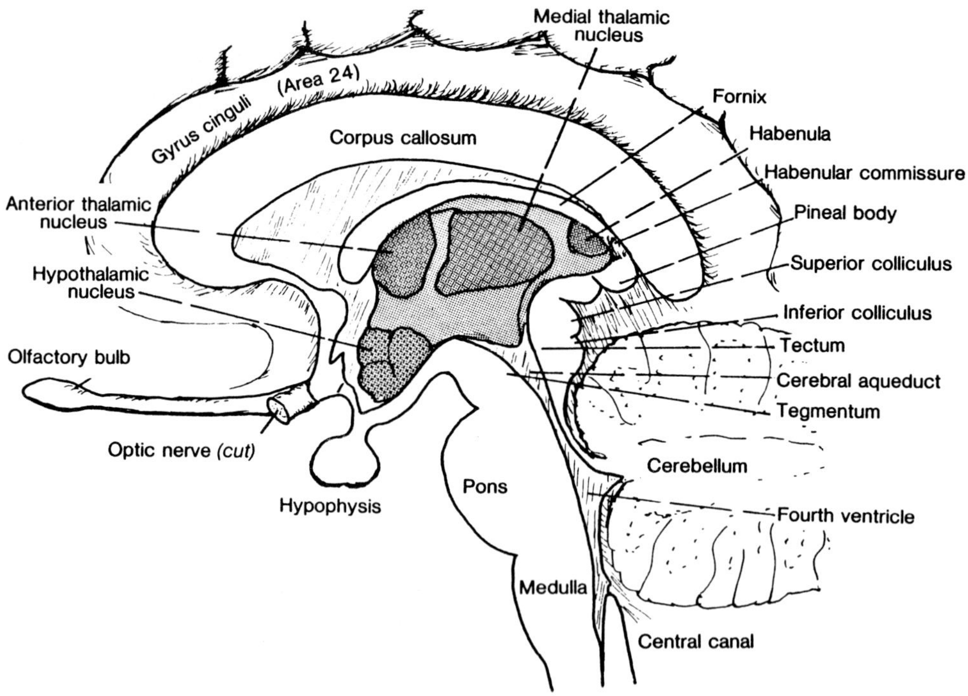

Figure 15-42. Diagram of a section of the upper brain stem, illustrating nuclei of the thalamus and hypothalamus and relation to surrounding structures.

pared with the forebrain. Numerous nerve fibers are forced tightly together to pass through the midbrain. The major structures comprising the midbrain, or mesencephalon, are the *cerebral peduncles* and the *corpora quadrigemina*. Additionally, there are several smaller components, including the *pineal body*, a narrow cavity or canal known as the *cerebral aqueduct*, and the *tegmentum*. The size of the midbrain is small in comparison with the cerebrum. It projects about 1.25 cm above the pons of the hindbrain, which is located inferiorly.

Cerebral Peduncles

The *cerebral peduncles* comprise the lateral halves of the midbrain. They appear as two rounded bands on the ventral aspect of the diencephalon and midbrain and are composed of efferent nerve tracts traveling from the motor areas of the cerebral cortex via the brain stem to the spinal cord. These tracts are known as the *pyramidal tracts*, the name derived from the pyramid-shaped bulge caused by the amassed fiber bundles on the anterior aspect of the medulla oblongata (Fig. 15-43).

Colliculi

The corpora quadrigemina are comprised of two pairs of rounded eminences located on the dorsal surface of the midbrain (see Fig. 15-42). They are termed *colliculi*, which means *little hills*. The upper two prominences are called the *superior colliculi* and serve as a relay for visual impulses from the retina to the lateral geniculate body of the thalamus (the primary relay structure) and on to the occipital lobes. Most writers indicate that the superior colliculi serve as a visual reflex center. In lower animals, it is the principal visual area. The blinking reaction caused by someone suddenly waving his hand near the eyes is a common type response engineered by the superior colliculi.

The lower pair of prominences pertains to audition. They are termed the *inferior colliculi*. Like the superior pair, the inferior pair is relegated as a reflexive center, but for audition. Clapping the hands behind the head of an unsuspecting victim illustrates the role of the inferior colliculi as a reflexive auditory center. The subject may turn or quickly lower his head, or a general startle response may be

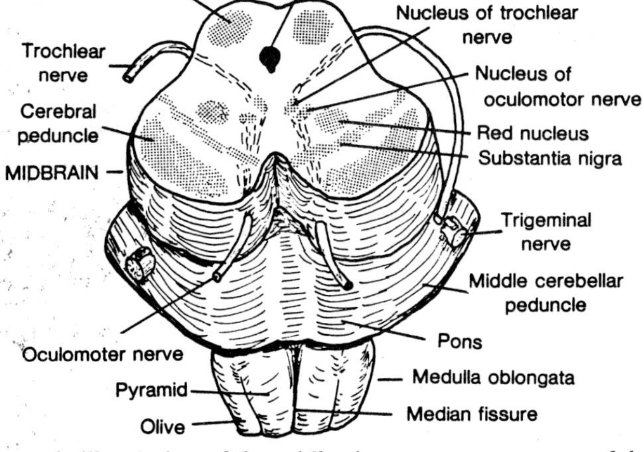

Figure 15-43. Schematic illustration of the midbrain cut to expose some of the internal aspects.

exhibited. The inferior colliculi receive the fibers from the ascending auditory paths from the medulla oblongata and lower midbrain level. From the inferior colliculi, the fibers pass to the medial geniculate body of the thalamus and then to the temporal lobe.

The corpora quadrigemina contain interconnections, a fact that posits a functional relationship between the auditory and visual reflex centers of the midbrain. These four colliculi also make up the rooflike structure or tectum on the posterior aspect of the midbrain (Fig. 15-44).

Pineal Body

A small ovoid body, the *pineal body* (shaped like a pine cone), is situated at the junction of the posterior limit of the diencephalon and midbrain just above the superior colliculi (Fig. 15-44). Frequently, it becomes calcified with age and serves as an important landmark in radiographs of the brain. Its role may relate to the endocrine system, although it ceases to be a functioning structure at puberty. It could possibly act as a supplementary maturational control during prepuberty stages based on evidence from clinical cases of patients who have had a related pathology before puberty.

Cerebral Aqueduct

A channel between the third ventricle of the diencephalon above and the fourth ventricle of the hindbrain below transports cerebrospinal fluid from the upper to the lower level (see Figs. 15-19 & 15-44). The canal, a part of the midbrain, is the cerebral aqueduct; it is also referred to

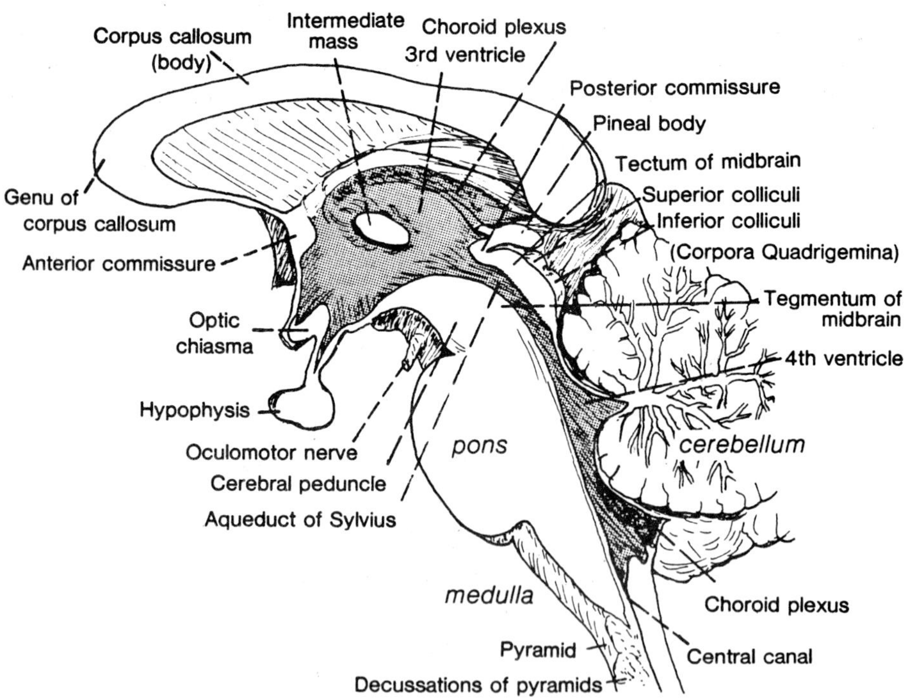

Figure 15-44. Sagittal section through the brain stem and the cerebellum. The cavities associated with the cerebrospinal fluid system are shaded.

as the *aqueduct of Sylvius*. It is surrounded by the cerebral peduncles anteriorly and the corpora quadrigemina posteriorly.

Tegmentum

A mass of gray substance, *substantia nigra* (black substance), with a number of neurons containing a dark (melanin) pigment, constitutes a zone within the midbrain just behind the cerebral peduncles and separates the cerebral peduncles and the tegmentum (Fig. 15-44). Another nucleus, the *red nucleus*, is located posteriorly to the substantia nigra. When the dissected specimen is fresh, the red nucleus appears pinkish. Both structures are considered components of the motor systems and are described in further detail later.

The tegmentum (*tegmen* = cover), located just behind the cerebral peduncles, consists of several substances: (1) nuclei associated with the limbic midbrain area, (2) the red nucleus described above, (3) nuclei of the oculomotor and trochlear cranial nerves, and (4) fibers of the medial longitudinal fasciculi (bundles), which are groups of fibers passing through the posterior aspect of the midbrain. Dorsal to the tegmentum are the cerebral aqueduct and tectum comprising the corpora quadrigemina.

Hindbrain

This section of the brain is often referred to as the most primitive part of the brain, owing to its basic responsibilities pertaining to life functions. It also represents the lowest portion of the brain stem. Three prominent structures make up the hindbrain or rhombencephalon. The pons and cerebellum, known also as the metencephalon portion, are the uppermost parts. The medulla oblongata or myelencephalon is the most caudal part. The cavity of the hindbrain is the fourth ventricle.

Pons

The midbrain is joined onto the hindbrain by the *pons* (bridge). These two anterior prominences appear as salient, band-like segments coursing transversely across the superior part of the brain stem (see Fig. 15-43). This fibrous belt continues around the lateral margins of the brain stem as the *middle cerebellar peduncles* to connect the two halves of the cerebellar hemispheres. Two other cerebellar peduncles, the *superior* and *inferior*, connect the cerebellum to the dorsal hindbrain and medulla oblongata, respectively.

Numerous cells are located in this ventral aspect of the pons; they are known as the *pontine nuclei*. Many of these synapse with fibers from the cerebral cortex. It is the axons of these pontine cells that project laterally to form the middle cerebellar peduncles. Vertically directed fibers, the corticopontine fibers, are found in this region, linking the cerebellum with the midbrain and the medulla oblongata via the superior cerebellar peduncles.

Positioned posteriorly to the pontine nuclei and the corticopontine tracts are scattered groups of neuronal cell bodies comprising a portion of the *reticular formation*, a projection system having a function somewhat parallel to that of the thalamus. This important formation pertains to arousal of the cortex or attentiveness (Fig. 15-45) (See the section in Chapter 22 on the "Reticular Formation.")

Lateral to the reticular formation cells are two small groups of cell bodies. These comprise the motor nuclei of the trigeminal cranial nerve. Other cranial nerves connect at or near the junction of the pons and medulla oblongata. These include the abducent, facial, and vestibulocochlear cranial nerves. The superior complex, a part of the pons, receives some auditory fibers directly and others as part of the auditory pathway passing to the auditory cortex. It is a paired, wrinkled gray area deep within the pons.

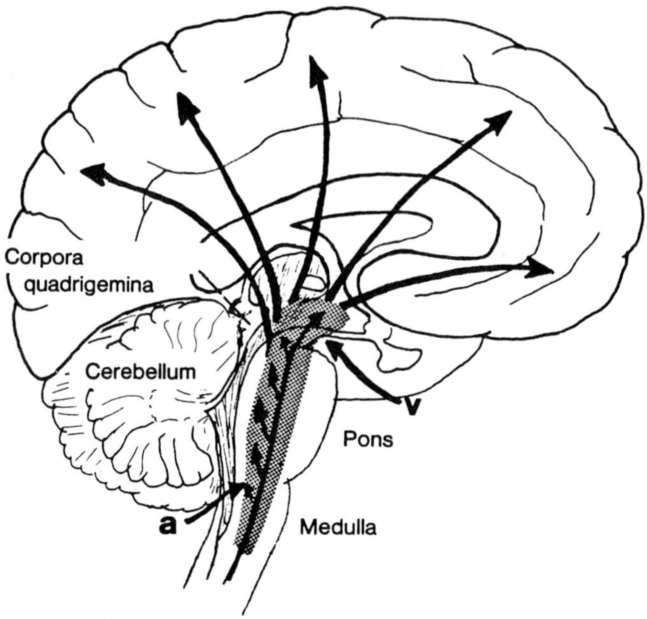

Figure 15-45. Schematic diagram of the reticular formation (shaded). Radiations to the cortex are shown projecting from the reticular formation. Auditory impulses (a) and visual impulses (v) are input to the formation as indicated, along with ascending sensory tracts from the spinal cord area.

At the junction of the pons with the medulla oblongata below, the eighth cranial nerve enters the hindbrain. The ventral and dorsal cochlear nuclei, located on the ventrolateral and dorsal aspect of the inferior cerebellar peduncle, receive auditory fibers. Thus, cochlear fibers have their first synapse with cell bodies in these nuclei before passing superiorly on their way to the cerebral cortex.

Most of the fourth ventricle is located in the pons, with the remaining part extending into the medulla oblongata below (see Fig. 15-44). Its roof is formed by the superior cerebellar peduncles, giving the impression that the superior cerebellar peduncle is an extension of the inferior colliculi of the midbrain. On the ventral side of the fourth ventricle are two groups of vertical fiber bundles known as the medial and lateral longitudinal fasciculi (bundles). These fasciculi arrive at the pons from the cerebral peduncles of the midbrain and split into smaller bundles. Some, the corticospinal fibers, pass through the pons and are collected together to form the pyramid prominence of the medulla oblongata below; others end in the pontine nuclei.

Cerebellum

The cerebellum, a notably remarkable organ, is intimately involved with a number of muscular activities. It contains complete motor and sensory representations of the body, but lesions in the cerebellum produce neither muscle weakness nor disorders of perception. Rather, the cerebellum controls the timing and the pattern of muscles activated during movement. Other functions involve spinal cord and brain stem mechanisms to achieve postural control. Lesions of the cerebellum, therefore, can result in disturbances in the coordination of eye and limb movements as well as disorders of muscle tone and postures.

The cerebellum, which has the appearance of a little brain, is tucked underneath the occipital lobe and is posterior to the brain stem to which it is connected (see Figs. 15-1, 15-23, & 15-46). It is housed in the posterior cranial fossa on the floor of the cranial cavity. The cerebellum exhibits two hemispheres, with a gray surface layer similar to the cerebral cortex except thinner. Each cerebellar hemisphere exhibits ipsilateral control — the right cerebellar hemisphere affects the right side of the body.

It is connected to the brain stem by its three peduncles, the superior, middle, and inferior cerebellar peduncles. Some neurons within the cerebellum send their axons through the superior cerebellar peduncle up to the red nucleus of the midbrain, which relays the impulses to the medulla oblongata and spinal cord. Others go on to the thalamus, which relays these impulses to the cerebral cortex. Therefore, the superior cerebellar peduncles relate the cerebellum to the midbrain and thalamus.

The middle cerebellar peduncles provide a connection to the pons, and the inferior cerebellar peduncles join the cerebellum to the medulla oblongata. A medial structure, the *vermis*, interconnects the two cerebellar hemispheres and is regarded by some as the third cerebellar lobe. The most prominent gray mass inside the cerebellum is the *dentate* (tooth-shaped) *nucleus*, which is embedded within the white matter. The cortex is marked by

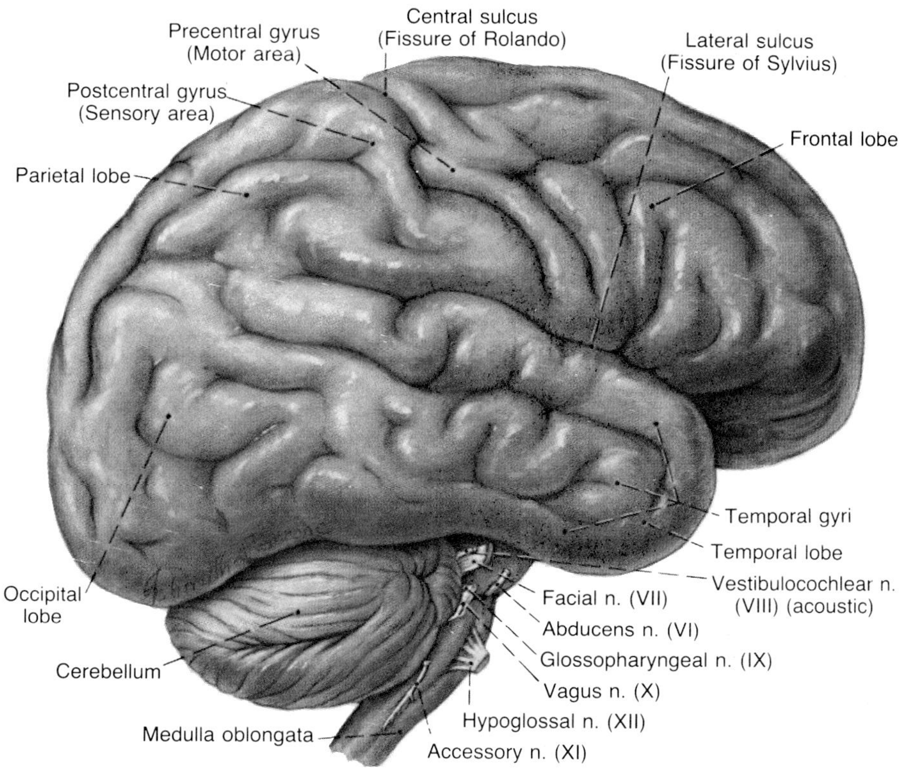

Figure 15-46. Right side of the brain showing the cerebrum, cerebellum, and medulla. Several cranial nerves are evident. (From Jacob, S.W., and Francone, C.A.: *Structure and Function in Man*, 4th ed., 1978. Courtesy of W.B. Saunders Co., Philadelphia.)

narrow transverse fissures that divide the outside layer into numerous folds or *folia*.

Basically, the role of the cerebellum involves movement controls. The functional characteristics of the cerebellum may be listed as follows:

1. It receives proprioceptive impulses from tendons, joints, and muscles and from the semicircular canals for coordinated movements, maintaining posture, balance, and muscle tone.
2. It is dominated by the cerebral cortex.
3. It contains efferent pathways enabling it to produce its motor effects.
4. It allows nerve impulses to be fed back to the same region from which they originated, setting up a feedback circuit.
5. Its functions are below a conscious level.
6. Voluntary movement of skeletal muscles can be carried out without cerebellar involvement, but such movements are clumsy and without organization.
7. The cerebellar hemispheres exhibit ipsilateral control.
8. In essence, it functions as a reflex center for coordination and refinement of muscle activity, but it is not the only center for motor coordination (Fig. 15-47).

Important in speech muscular coordination, the cerebellum makes it possible to carry out breathing and speaking activities for meaningful phrasing and loudness as well as adding facial expressions for corresponding reinforcement of the acoustic product.

Medulla Oblongata

The third major part of the hindbrain

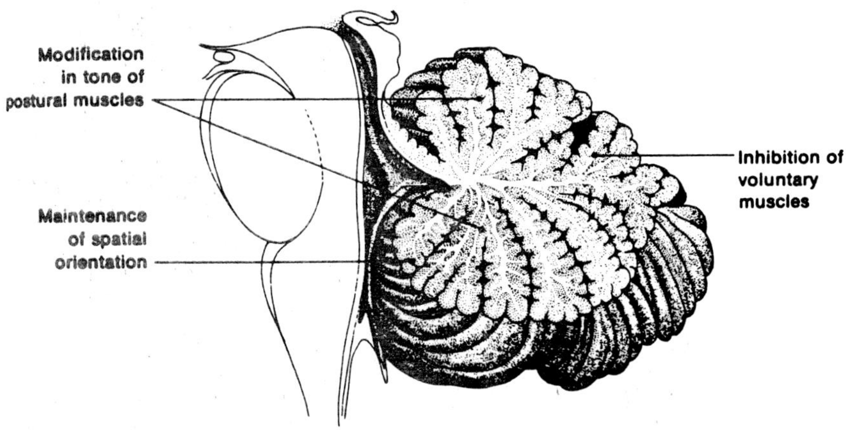

Figure 15-47. Sagittal section through the cerebellum to illustrate several functional areas. (From *Man, Microbes, and Matter* by B.C. Block. Copyright © 1975. Used with the permission of McGraw-Hill Book Company.)

is the *medulla oblongata* or medulla. In the scheme of the central nervous system, it extends the brain to join the spinal cord. From another point of view, the medulla is an upward extension of the spinal cord. There are no external lines of demarcation between the medulla and spinal cord; however, the foramen magnum is used as a line of outward differentiation, since at this level, a sharp change in cross-sectional topography is found. The nuclei become rearranged; a number of nerve tracts are also rerouted and rearranged. As the spinal cord passes into the foramen magnum and becomes the medulla, it enlarges to a bulbar appearance, having the general shape of a cone. A number of tracts and disorders are described after this characteristic appearance, e.g. corticobulbar pathway, bulbar paralysis.

A *median fissure* is evident on the ventral surface, with a rounded prominence on either side, the *pyramids* (see Fig. 15-48). These are composed of bundles of fibers (of the pyramidal pathway) coming from the precentral gyrus of the cerebral cortex. Lower on the medulla, the pyramids taper, with up to about 80 percent of the motor fibers crossing to the opposite side. This is termed the *pyramidal decussation* and denotes the transition of the medulla to the spinal cord. The pyramidal decussation is clearly present on the ventral surface of the medulla, for it obliterates the median fissure. As the fibers continue caudally in the spinal cord, they are known as the lateral corticospinal tracts.

Another pair of prominences on the medulla are the *olives* (Fig. 15-48). They are identified as oval bulges lateral to and above the medullary pyramids, just underneath the pons. Within these elevations are the inferior olivary nuclei, extending deep in the medulla and characterized by numerous folds of gray matter contrasting with the surrounding white matter. Afferent fibers from the neurons of the olivary nuclei pass via the inferior cerebellar peduncle into the cerebellum in the olivocerebellar tract.

The cavity of the medulla is actually a continuation of the fourth ventricle from the pons above. It narrows to continue into the spinal cord as the central canal. At the floor of the ventricle in this section of the brain stem may be found a number of nuclei serving a group of cranial nerves.

Within the medulla are the groups of cell bodies or nuclei of the origin of cranial nerves VIII through XII, or the last five. These nerves, in brief, serve in sensory and motor functions of the ear, tongue, pharynx, larynx, heart, viscera, and some neck muscles.

The presence of the nucleus for cranial nerve VIII in the medulla has been assiduously avoided in most writings. The peripheral connections for the cochlear nucleus enter the brain stem at the junction of pons and medulla, but recent segmental cuts of the brain stem show that the nucleus is located in the medulla. An interesting project for the student of anatomy is a literature survey of statements in neurology texts as to the location of the dorsal nucleus.

The vital centers to control heartbeat, blood pressure, and respiration are contained within the medulla. The hypothalamus also plays a role in controlling heart rate and blood pressure. Vestibular functions or coordination of equilibrium are also carried out by nuclei in the medulla. Connections with cranial nerve nuclei, namely, the vagus, cause the internal organs to be responsive to motion sickness when the vestibular apparatus is overstimulated. This circuit initiates the well-known reactions to motion sickness of nausea and vomiting. The centers for respiration are located in the lower aspect of the medulla, with a separate center for inspiration and expiration. An overdose of barbituates will knock them out, since they are acutely sensitive to this type of drug (Fig. 15-49).

A large number of small collections of cell bodies and white matter located deep within the medulla comprise the reticular formation portion of the medulla. The reticular system is an arousal mechanism relating to consciousness and attention.

Before leaving the medulla, another reference to its fiber tracts is warranted. To avoid overwhelming the new student with entirely too many new names of tracts and other neuroanatomical terminology, the various tracts in the medulla are not listed here. Essentially, these many pathways connect the spinal cord and cerebellum to higher centers of the

EXTERNAL FEATURES

VENTRAL ASPECT

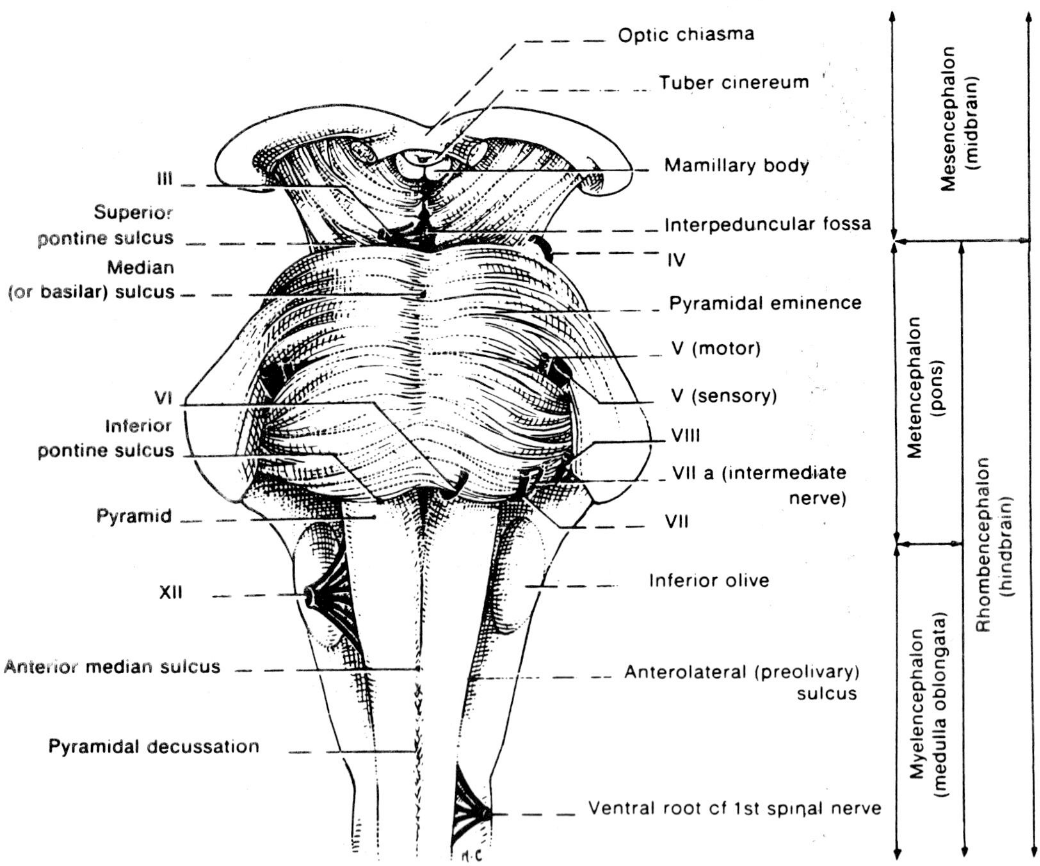

Figure 15-48. The brain stem. (From Bossy, Jean: *Atlas of Neuroanatomy and Special Sense Organs*, 1970. Courtesy of W.B. Saunders Co., Philadelphia.)

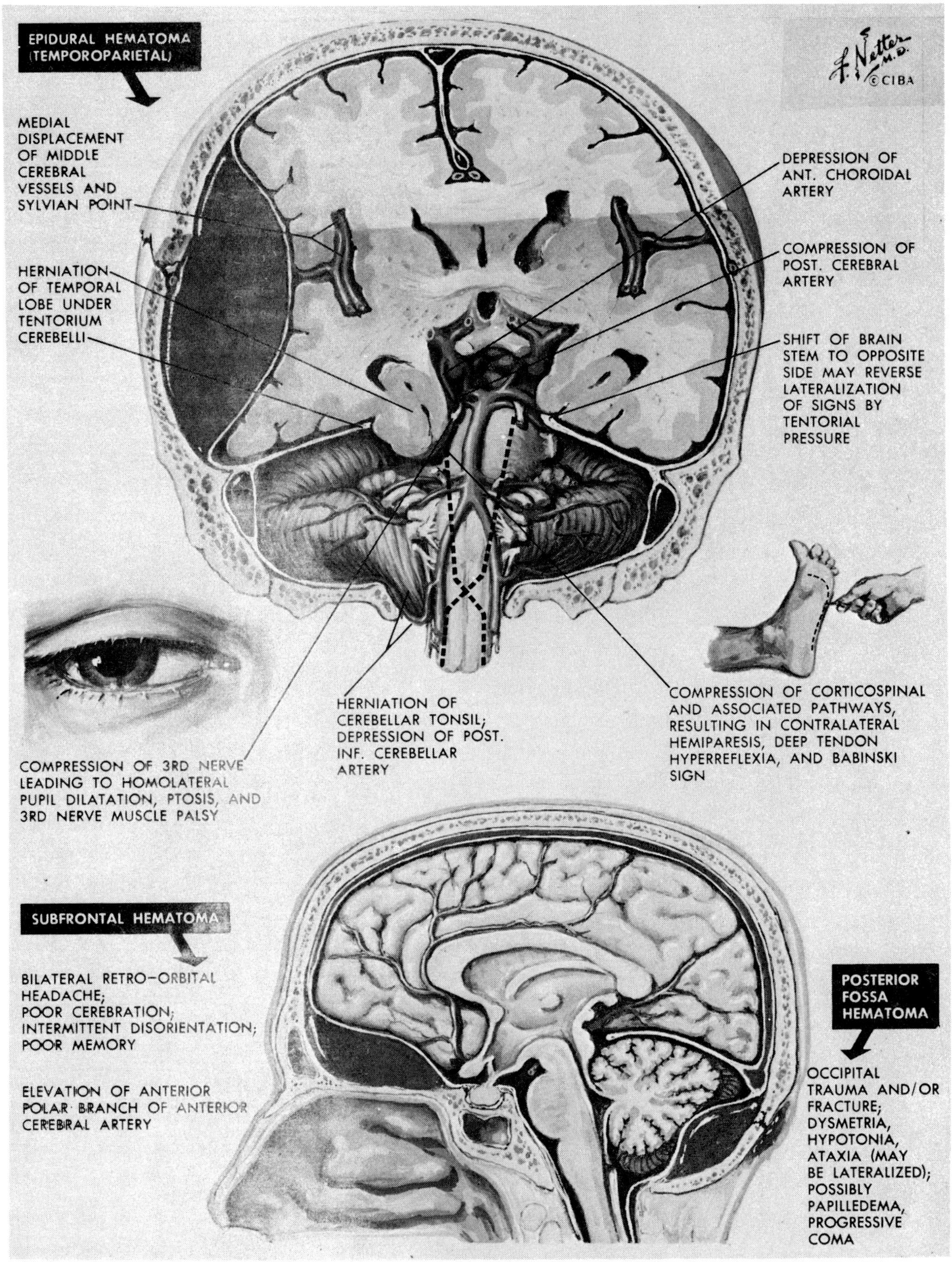

Figure 15-49. Clinical conditions (injuries) displacing or invading areas of the brain resulting in impaired neurological function. Tumors and other lesions have a similar effect. (© Copyright 1966, CIBA Pharmaceutical Company, Division of CIBA-GEIGY Corporation. Reprinted with permission from CLINICAL SYMPOSIA illustrated by Frank H. Netter, M.D. All rights reserved.)

brain, although they are rearranged and sometimes rerouted as they pass through the medulla. It should be understood that both afferent and efferent tracts are involved in these various pathways, i.e. nerve impulses are passing both ways through the medulla. Even the simplest muscular activity brings into use many very complex neural interactions, enabling its performance to be carried out in a smoothly coordinated and purposeful manner. The interconnections via tracks with centers in all divisions of the brain may be employed for even an elementary muscular activity.

BIBLIOGRAPHY

Bossy, J.: *Atlas of Neuroanatomy*. Philadelphia, Saunders, 1970.

Campain, R., and Minckler, J.: A note on the gross configurations of the human auditory cortex. *Brain and Language, 3*:319-323, 1976.

Carpenter, M.B.: *Core Text of Neuroanatomy*, 2nd ed. Baltimore, Williams and Wilkins, 1978.

Eccles, J.C.: *The Understanding of the Brain*. New York, McGraw, 1973.

Gardner, E.: *Fundamentals of Neurology: A Psychophysiological Approach*, 6th ed. Philadelphia, Saunders, 1975.

Gardner, E., Gray, D.J., and O'Rahilly, R.: *Anatomy: A Regional Study of the Human Structure*, 4th ed. Philadelphia, Saunders, 1975.

Geschwind, N., and Levitsky, W.: Human brain: left-right asymmetries in temporal speech region. *Science, 161*:186-187. 1968.

Gray, H.: *The Anatomy of the Human Body* (Charles M. Goss, Ed.), 28th ed. Philadelphia, Lea and Febiger, 1973.

Hausman, Louis: *Illustrations of the Nervous System*, Atlas III. Springfield, IL: Thomas, 1961.

Minckler, J. (Ed.): *Introduction to Neuroscience*. St. Louis, C.V. Mosby, 1972.

Netter, F.H.: *Nervous System*. Summit, NJ: CIBA, 1962, vol. I.

Penfield, W., and Roberts, L.: *Speech and Brain Mechanisms*. Princeton, NJ: Princeton U Pr, 1959.

Schlossberg, L., and Zuidema, G.D.: *The Johns Hopkins Atlas of Human Functional Anatomy*. Baltimore, Johns Hopkins U Pr, 1977.

Scott, J.H., and Dixon, A.D.: *Anatomy for Students of Dentistry*, 4th ed. London, Churchill, 1978.

Silverman, S.I.: *Oral Physiology*. St. Louis, C.V. Mosby, 1961.

Chapter 16

INTEGRATION AND COORDINATION

THE nervous system, together with the endocrine system, controls and coordinates the functions of the component parts of the body. It accomplishes its purpose by means of a vast network of nerves that carry signals into and out of the central nervous system. This network is a diverse and extensive apparatus designed to relate us to our environment, but there are two environments. First, the external receptors, such as the eyes and ears, make us aware of changes in the *external* environment, and the CNS initiates approprite responses through the skeletal muscles to adapt to these changes. Second, internal receptors signal information about the state of our *internal environment* to the vital centers of the brain, which in turn mediate responses to maintain homeostasis, i.e. the tendency toward stability in the normal internal body states.

In this chapter, we will describe the structure and function of the *spinal cord, neural pathways, spinal nerves*, and *cranial nerves* as they extend the involvement of the brain in a number of specific integrating and coordinating activities, especially communication. While the spinal cord is an extension of the brain and therefore a member of the CNS, the spinal and cranial nerves are members of the peripheral nervous system. They extend outside the meninges of the CNS and link the CNS with the internal and external environmental changes leading to physiological and psychological adjustments (Fig. 16-1).

The student will be reminded periodically that there is but *one* nervous system, even though anatomists use the designation *nervous system* to refer to various functional arrangements as nervous systems, e.g. autonomic nervous system, voluntary nervous system, peripheral nervous system, and limbic system.

THE SPINAL CORD

The spinal cord is a continuation of the brain through the foramen magnum of the posterior cranial fossa of the skull into the protective encasement provided by the vertebral foramina of the bony spinal column. Sometimes the spinal cord and brain stem (taken to include the medulla oblongata, pons, midbrain, dien-

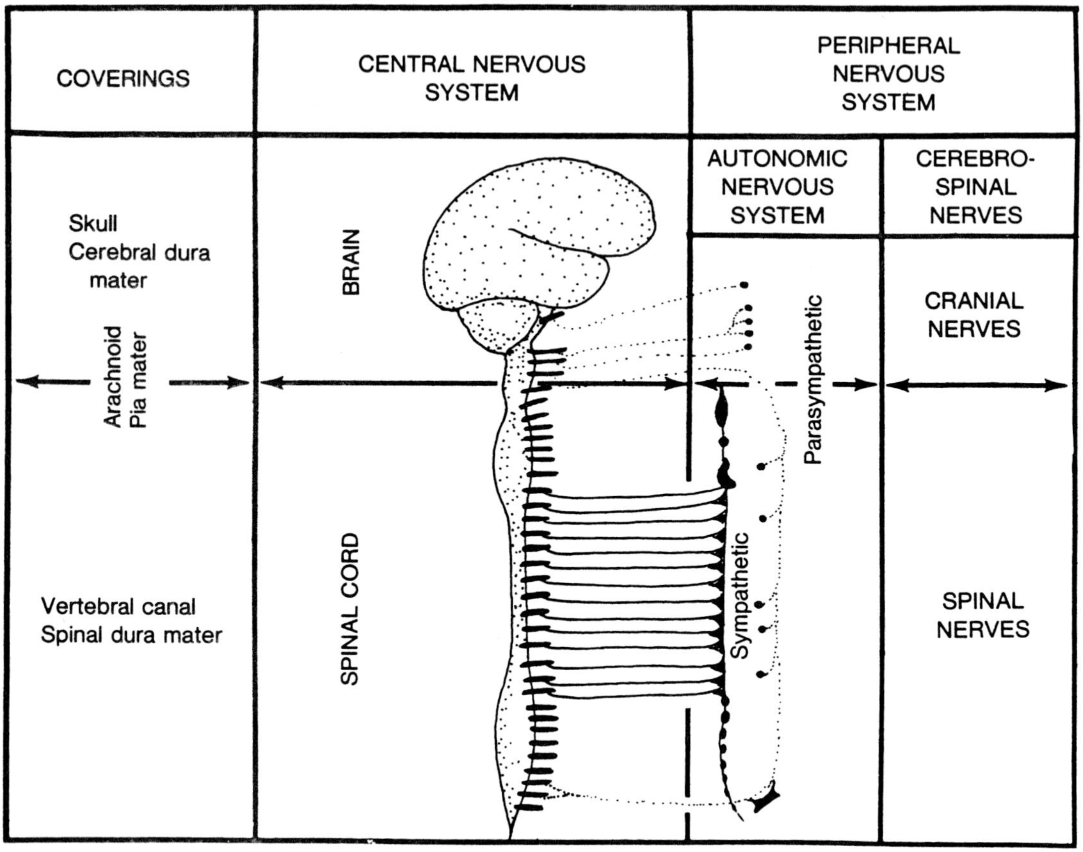

Figure 16-1. Divisions associated with introductory neuroanatomy and neurophysiology (From Bossy, Jean: *Atlas of Neuroanatomy and Special Sense Organs*, 1970. Courtesy of W.B. Saunders Co., Philadelphia.)

cephalon, and basal ganglia) are considered together under neuroanatomic and neurophysiologic approaches. Here, the spinal cord is grouped with its own extensions, the spinal nerves and other nerves, emanating from the central nervous system. While the combination is unorthodox in comparison with most anatomy texts, the purpose of such a grouping here is to emphasize the intricate, functional unity between the CNS and the PNS. The designation of CNS and PNS as divisions is, as noted previously, arbitrary terminology (Fig. 16-2).

General Structure of the Spinal Cord

External Aspects

The spinal cord is tapered near the top (at the level of the foramen magnum) and also at the bottom. It lacks uniformity in diameter or shape throughout its total length (see Fig. 16-2). It is typically the diameter of a pencil, while the average vertebral foramen surrounding the cord, at least in a dried bone, is large enough to insert the index finger.

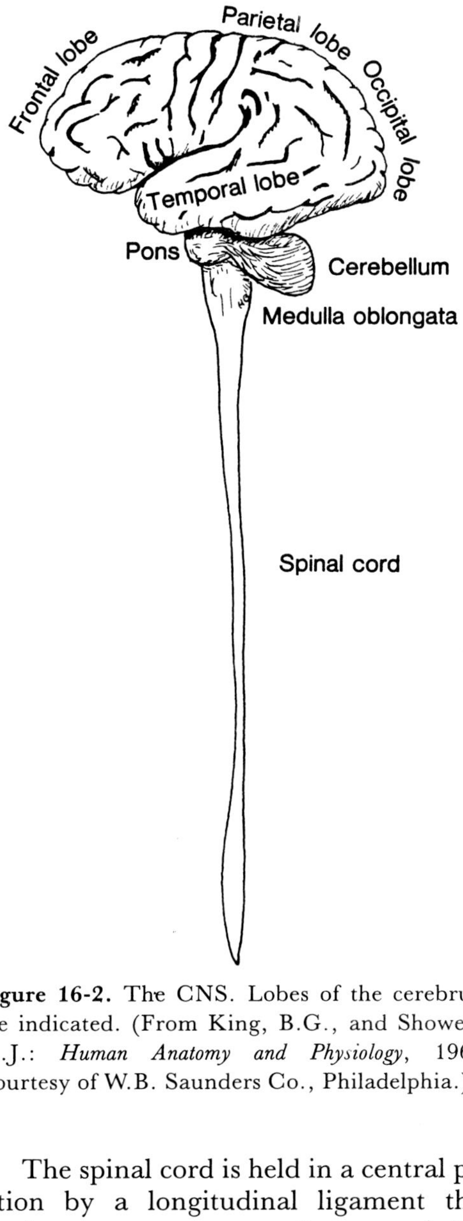

Figure 16-2. The CNS. Lobes of the cerebrum are indicated. (From King, B.G., and Showers, M.J.: *Human Anatomy and Physiology*, 1969. Courtesy of W.B. Saunders Co., Philadelphia.)

The spinal cord is held in a central position by a longitudinal ligament that sends out numerous, small, sawtooth-like ligaments termed *denticulate ligaments*, attaching the cord to the walls of the vertebral canal (Fig. 16-3). Together with the bony encasement and cerebrospinal fluid bathing the cord, the denticulate ligaments serve as a protective system for the spinal cord.

Until the third month of fetal life, the cord extends the length of the vertebral canal. In a child, the canal grows more rapidly and leaves the terminated cord at the level of the first, second, or third lumbar vertebra. In adults, the length from the base of the skull to the second lumbar vertebra is about 40 to 45 cm on the average.

The *meninges* are continuous through the foramen magnum and surround the spinal cord in the vertebral canal in the same fashion as the brain. *Cerebrospinal fluid* bathes all sides of the cord in the subarachnoid space. The pia mater is closely applied to the spinal cord, while the dura and arachnoid layers of the meninges extend a short way around the roots of each spinal nerve as it leaves the vertebral canal through an intervertebral foramen. The pairs of denticulate ligaments extend from the pia to the dural sheath for anchoring the cord; they alternate with the spinal nerves along the length of the cord.

Below the termination of the spinal cord, the vertebral canal continues. This space is occupied by a collection of long nerve roots trailing off the lower aspect of the spinal cord on their way out of the vertebral canal through the intervertebral foramina below the cord. They are known as the *cauda equina* for their resemblence to a horse's tail (Fig. 16-23). The pia is extended below the end of the cord as a single, glistening thread (filum), which is termed the *filum terminale*. This structure is attached inferiorly to the coccyx. The dura and arachnoid membranes continue to line the borders of the vertebral canal, creating the cavity below the spinal cord. This is the site used for lumbar puncture (see Fig. 15-21).

There are thirty-one pairs of spinal nerves leaving the spinal cord at successive levels. Each area of the spinal cord giving rise to a pair of spinal nerves is known as a *spinal cord segment* (Fig. 16-4). According to the level of the vertebral col-

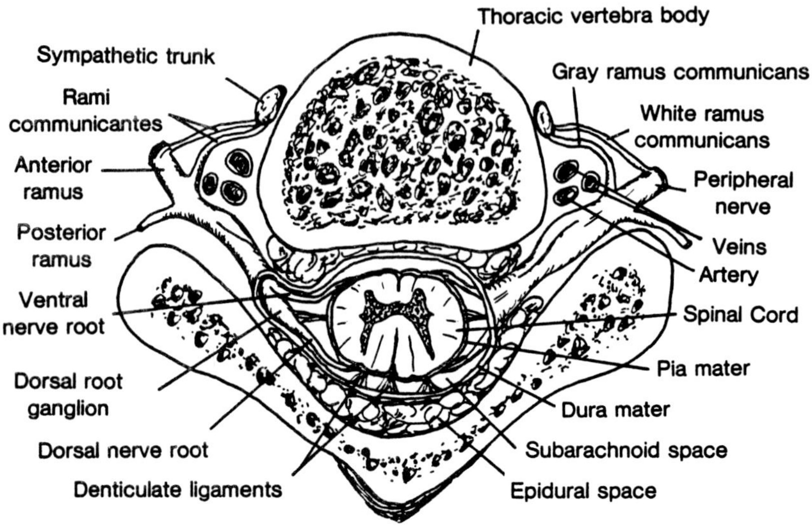

Figure 16-3. Relation of the spinal cord and nerves to a vertebra. Transverse section through the thoracic region.

umn, the spinal nerves may be named as cervical, thoracic, lumbar, sacral, and coccygeal nerves.

Internal Aspects

A typical cross section of the spinal cord reveals a characteristic ovoid shape, bulging at the lateral aspects. A fissure, the *anterior medial fissure*, and the *posterior median septum* divide the cord into left and right halves. Distinct areas of white and gray matter are apparent. A small central canal is surrounded by a large *gray mass* that has the shape of the letter H. The projections of the H represent two *anterior* or *ventral horns* (or columns) and two *posterior* or *dorsal horns* of gray matter. The ventral horns are associated with motor functions, and the dorsal horns are associated with sensory stimuli (Fig. 16-5).

The ventral horns contain the *ventral horn cells*. Axons from these cells comprise the *vental root fibers* of the spinal nerve, which innervate the skeletal muscles.

The dorsal horns consist of cells serving as *interneurons* and *transmission* neurons. These cells receive incoming dorsal root fibers from the spinal nerve bringing in sensory impulses from peripheral receptors. The dorsal cells also receive fibers from higher levels in the central nervous system, which control the sensitivity threshold of these cells. By modulating the receptivity of these dorsal horn cells, they may become more sensitive to specific peripheral receptors. In the thoracic portion of the spinal cord, a *lateral horn* is also present but only as a small, bulging configuration of gray matter (Fig. 16-6).

The *white matter*, by contrast with its location in the brain, lies on the outside of the gray matter in the spinal cord. A large portion of the white matter is devoted to fibers arising and terminating within the spinal cord to *interconnect* various levels and provide for coordinated activity. In other words, they serve as transmission lines. The horns of gray matter divide the white matter into anterior, lateral, and posterior parts of *funiculi* (cord-like structures), also known as fasciculus (small bundle). These funiculi make up the fiber tracts, organized according to ascending

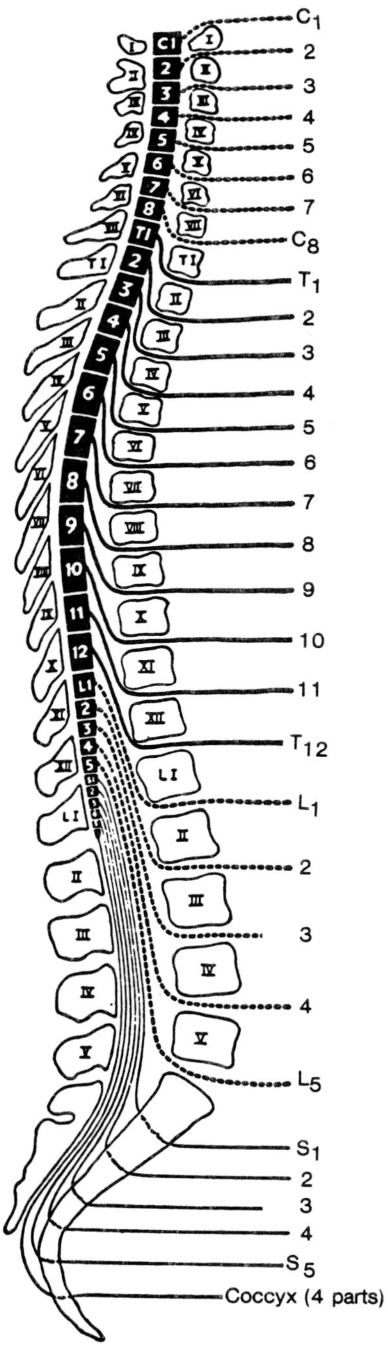

Figure 16-4. Segments of the spinal cord and alignment of the spinal segments with the vertebrae. (From Haymaker, W.: *Peripheral Nerve Injuries*, 1953. Courtesy of W.B. Saunders Co., Philadelphia.)

or descending. Most fibers in the tracts are enclosed in a fat-like covering termed *myelin*, which gives the characteristic white appearance to these parts of the spinal cord.

Tracts ascending transmit sensory or afferent impulses to the brain; tracts descending transmit motor or efferent impulses from the brain to the anterior and lateral horns. Fiber tracts grouped within a functionally related area are sometimes termed *fasciculi*.

One basic point of information might be helpful concerning the names of the tracts. They are often given names after the location of the cells of origin and the point at which the axons synapse with the next neuron in the pathway. Thus, the corticospinal tract originates with cells in the cortex of the brain, and their axons synapse in the spinal cord. Terms such as *lateral, ventral*, and *dorsal* precede the names of some of the tracts as a clue to the location of the pathway in the spinal cord. Several important tracts are illustrated in Figures 16-7 to 16-12.

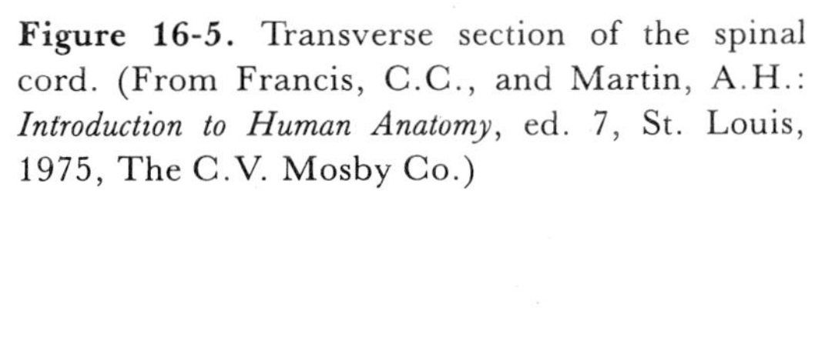

Figure 16-5. Transverse section of the spinal cord. (From Francis, C.C., and Martin, A.H.: *Introduction to Human Anatomy*, ed. 7, St. Louis, 1975, The C.V. Mosby Co.)

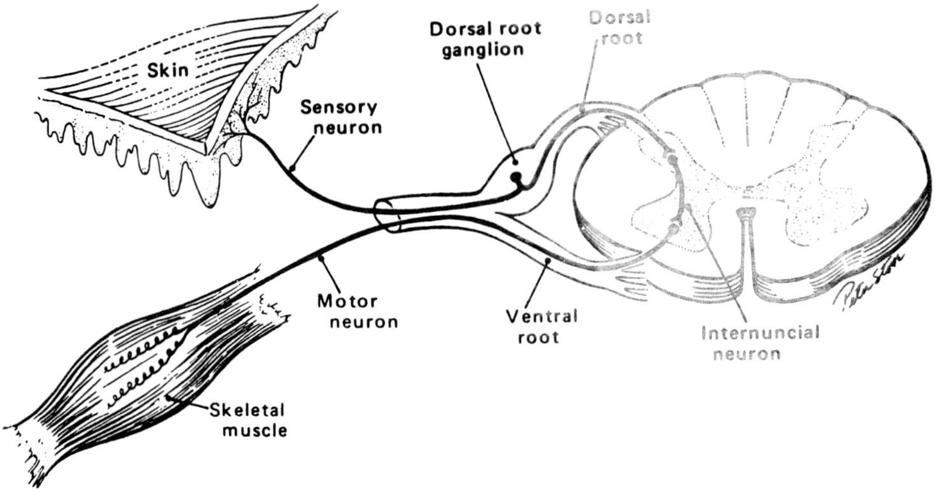

Figure 16-6. Components of a simple reflex. Afferent and efferent path into and out of the spinal cord through an internuncial neuron. (From *Dynamic Anatomy and Physiology*, 3rd ed. by L.L. Langley, I.R. Telford, and J.B. Christensen. Copyright © 1969. Used with the permission of McGraw-Hill Book Company.)

NEURAL PATHWAYS AND MOVEMENT

Elaborate pathways provide for nerve tracts for transmitting neural impulses to and from the spinal cord as well as up and down the spinal cord. The tracts extend into the brain, making pathways to and from the spinal cord continuous with parts of the brain. These pathways, designated as afferent and efferent tracts, are sometimes described as sensory and motor tracts, respectively.

The route from the brain to a remote point on a limb is long, but some nerve cells cover the distance without the assistance from an interneuron.

First, we will examine the afferent pathway and its role in some aspects of perception; then reflexes and the motor tracts will be described.

Afferent Pathways

Receptors — specialized nerve endings that are sensitive to a specific change in environment — provide the basis for all knowledge acquired by the individual. The process is termed *perception*, and several receptors may be used simultaneously to enable the person to gain information. It is the brain that serves as the organ for decoding incoming sensory data and making associations with previous inputs. Also, sensory data is required for carrying out a simple skilled movement that has been well learned and practiced repeatedly, as in playing an organ with the hands and feet simultaneously.

The gap between the brain itself and receptors concerned with the internal or external environment of the person is bridged by the *afferent pathways*. Impulses carried over these pathways transmit pain, temperature, touch, pressure, and proprioception (movement and position awareness). In tracing the general afferent pathway, one begins with a receptor that is stimulated by an adequate stimulus. Next, an afferent neuron transmits the impulse toward the central nervous

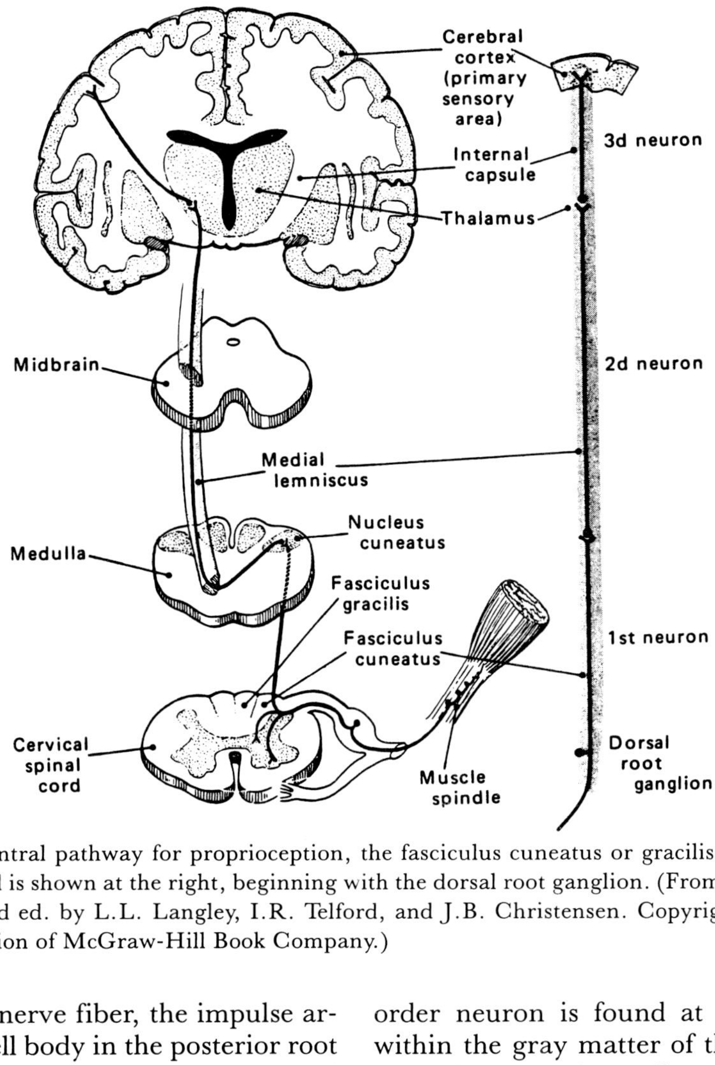

Figure 16-7. Central pathway for proprioception, the fasciculus cuneatus or gracilis. The number of neurons involved is shown at the right, beginning with the dorsal root ganglion. (From *Dynamic Anatomy and Physiology*, 3rd ed. by L.L. Langley, I.R. Telford, and J.B. Christensen. Copyright © 1969. Used with the permission of McGraw-Hill Book Company.)

system over a nerve fiber, the impulse arriving at the cell body in the posterior root spinal ganglion. It is sent on to the spinal cord over the central process of the sensory neuron, which is termed the *first-order neuron* or *primary neuron* in the afferent pathway (Fig. 16-7).

Upon entering the spinal cord through the posterior root, the fiber sets up connections for a spinal reflex arc or ascends to synapse with another neuron. In either case, a *second-order neuron* may be expected to be present in order to convey the impulse within white matter up the cord to the thalamus. The cell body of the second-order neuron is found at various levels within the gray matter of the spinal cord or brain stem, depending on the type of receptor being served. Sometimes the second-order neuron crosses the gray matter to the opposite side of the cord.

A *third-order neuron* transmits the impulses to the cerebral cortex. Its cell body lies in the thalamus. This is the final neuron, since three neurons comprise the sensory chain of the afferent pathway.

In a cross section of the spinal cord, the ascending as well as descending pathways may be identified. These divide the cord into major regions and appear as white matter due to myelinization of the nerve fibers carried in the areas (see Fig. 16-5). As mentioned earlier, some tracts are descriptively named; thus, the ventral spinothalamic tract or ventral spinocerebellar tract originates at spinal levels but terminates in the structure comprising the second part of the tract's name. Pain and temperature are transmitted up the spinal cord via the lateral spinothalamic tract; touch and proprioception are transmitted via fasciculi of the posterior funiculus (Fig. 16-8).

A pathway within the brain stem, the *medial lemniscus*, enables the impulse to be propagated to the thalamus and onto the cerebral cortex; however, this does not represent an added neuron in the sensory chain. Rather, it is a continuation of the second-order neuron, but at the same time, it could be its beginning (Fig. 16-9).

Faulty perception permits an interesting condition. An illusion is a false perception of a stimulus, while a hallucination is perception without a stimulus present. The inebriated fellow who listens to

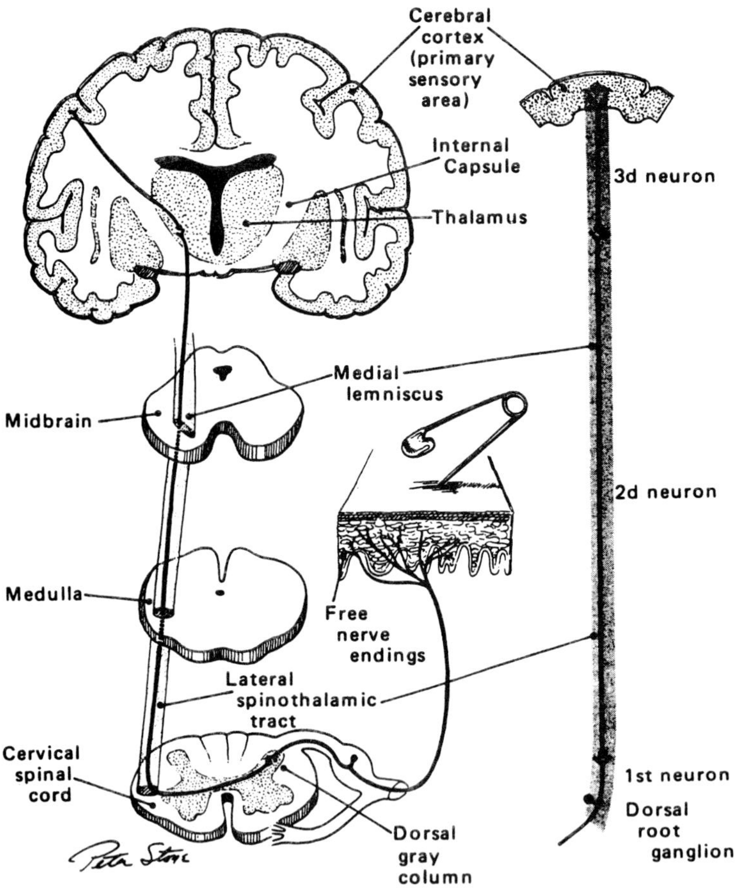

Figure 16-8. The central pathway for impulses perceived as pain, the lateral spinothalamic tract. Fibers cross upon entering the spinal cord. (From *Dynamic Anatomy and Physiology*, 3rd ed., by L.L. Langley, I.R. Telford, and J.B. Christensen. Copyright © 1969. Used with the permission of McGraw-Hill Book Company.)

his wife tell an elaborate story to friends and then requests that she sing another song is suffering an illusion. If he hears singing without a vocal sound of any type present within earshot, he is experiencing a hallucination.

Reflexive Movement

There may be two types of movement mediated by the nervous system, *reflex* and *voluntary*. Involuntary muscular contraction in response to stimulation of a sense organ is termed a reflex. (Touching the tongue to a spoon of very hot soup brings about such a reflexive reaction.) Willed movement is directed by the brain and results in voluntary musculature contraction. In each case, movement is dependent on muscle contraction as a result of neural stimulation. The two pathways traveled by the nerve fibers serving the

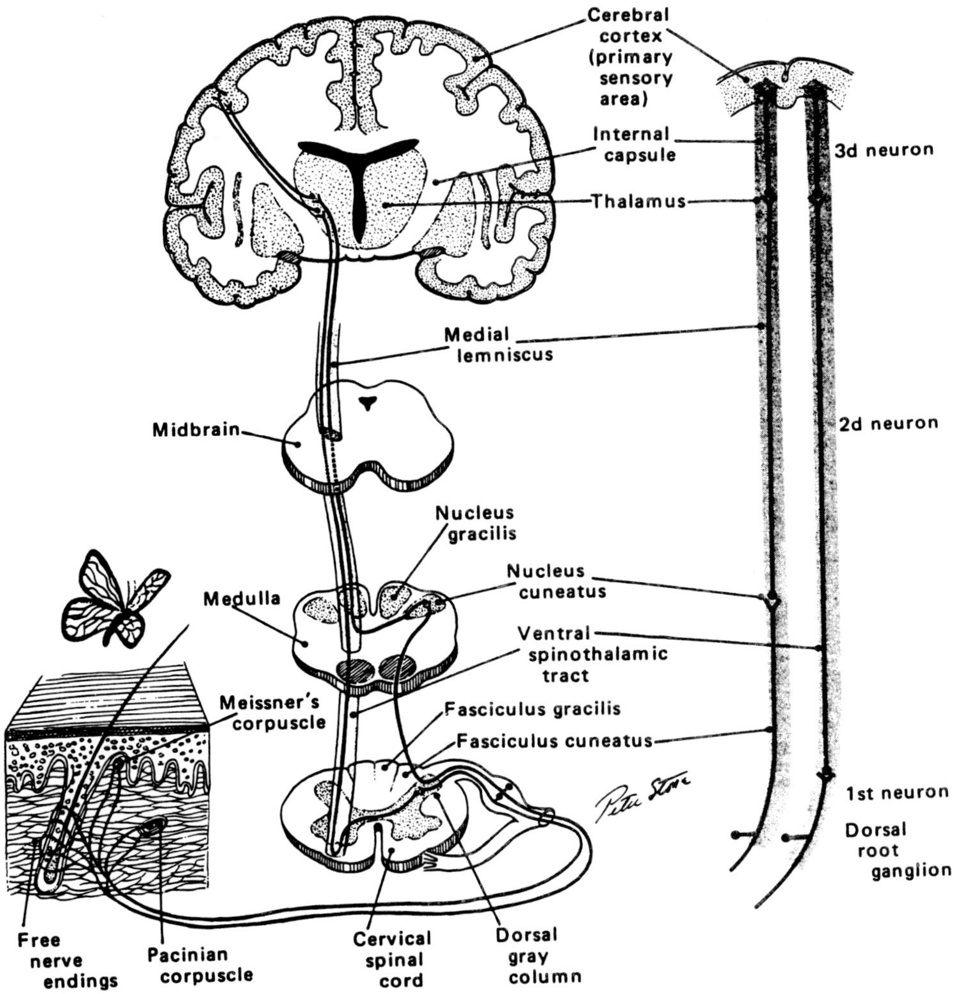

Figure 16-9. Central pathways for touch and light pressure perception, the ventral spinothalamic and the fasciculus cuneatus or gracilis. Notice that some fibers cross upon entering the spinal cord while others cross in the medulla at the nucleus gracilis or cuneatus. (From *Dynamic Anatomy and Physiology*, 3rd ed., by L.L. Langley, I.R. Telford, and J.B. Christensen. Copyright © 1969. Used with the permission of McGraw-Hill Book Company.)

muscles, both voluntary and involuntary, vary significantly.

The peripheral nervous system provides for efficiency in response to certain stimuli through the *reflex arc*. When the patellar tendon is tapped just below the knee, the foot kicks forward. This reflexive movement is *involuntary*. Using the same muscles, one may *voluntarily* kick a ball.

In essence, the reflex is an automatic response to a stimulus. The person may be aware of the movement, and he may be able to inhibit it, but it occurs involuntarily and without direction from *consciousness*. Glandular tissue may also exhibit an involuntary response to a stimulus as illustrated by the fact that when a speck of foreign matter gets into the eye, the lacrimal gland attempts to flush it out with tears.

The automaticity suggested by the speed and precision of the speech mechanism in producing articulate speech is intimately related to the concept of reflexive control systems. Movement controlled through reflexive pathways is the basis for the speech act, overriding the vegetative functions of the autonomic controls normally present. Learning and practicing articulatory placement certainly is a voluntary motor act, but the reflexive implications are quite dissimilar to those found in the completely automatic, involuntary reflex arcs such as the spinal reflex.

The type of spinal reflex arc most frequently described in the literature is made up of a *three-neuron arc*. An interconnecting neuron is found between the afferent and efferent neurons of the arc, hence a three-neuron reflex arc. This pattern of a spinal reflex arc presented in Figure 16-10 is as follows:

1. A receptor (proprioceptive, tactile, pain, etc.).
2. An afferent neuron, its cell body located in the dorsal root ganglion outside the spinal cord.
3. A synapse.
4. An *internuncial neuron* lying within the gray matter of the spinal cord, its cell body located in the posterior gray column of the spinal cord.
5. A synapse.
6. An efferent neuron, its cell body located in the anterior horn of the gray matter of the spinal cord; its axon exiting the anterior root of a spinal nerve.
7. An effector organ — a skeletal muscle.

Thus, reflex indicates that a discharge by an afferent neuron is reflected back from the central nervous system to the periphery of the body.

It should be pointed out that most reflex arcs are more complicated than the one cited here. For instance, there may be many *internuncial neurons* affording connections to different levels in the cord and crossing from one side to the other. The number of internuncial neurons may be indicated in a relative sense by measures of *reflex time*: the time required in the passage of impulses from the receptor to the effector. This stimulus-to-response measurement yields a faster time when there are fewer synapses along the pathway since synaptic conduction is slower than nerve fiber conduction.

In the reflex arc, synaptic transmission can be influenced by higher centers involving cerebral activity. The cerebral impulse descends the brain stem and spinal cord to end at a synapse between the afferent and efferent neurons of the reflex arc in question. In some cases, these corticospinal fibers are employed to transmit volitionally originated impulses. Other pathways are utilized merely to control reflex activity. The higher centers provide controls through facilitation and

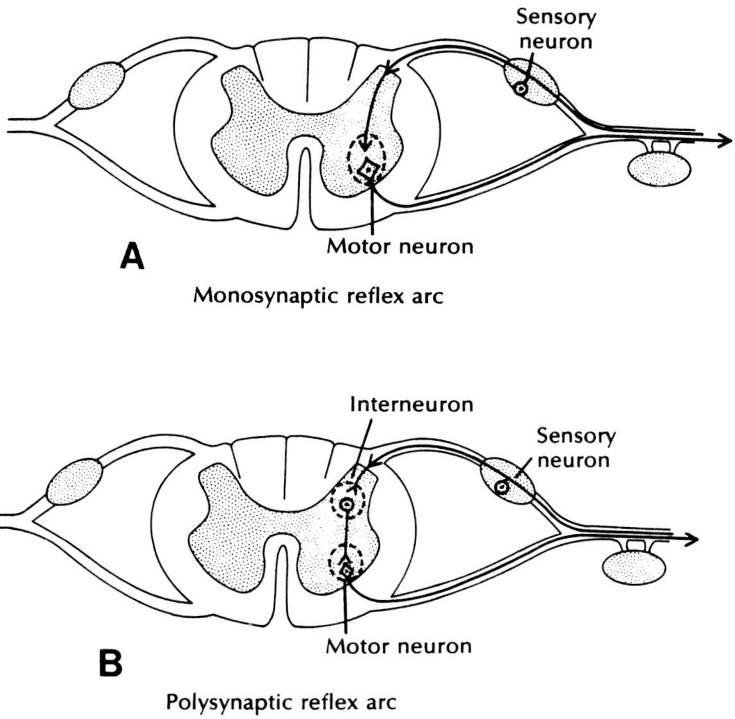

Figure 16-10. Reflex arcs: (A) monosynaptic arc; (B) polysynaptic arc with internuncial neuron. (From Francis, C.C., and Martin, A.H.: *Introduction to Human Anatomy*, ed. 7, St. Louis, 1975. The C.V. Mosby Co.)

inhibition of synaptic transmission. That is, the higher centers make it easier (facilitation) for an impulse to cross a synapse in a reflex arc or impede or prevent (inhibition) the impulse from crossing the synaptic junction in the reflex pathway.

Conditioned Reflexes

Conditioned reflexes play an important role in habit formation and reactions. Their establishment is made possible by the cerebrum. To set up a conditioned response, previous training is required for forming associations. The production of speech is made smooth and rapid with the establishment of conditioned reflexes.

Humans perform numerous other activities through the benefit of conditioned reflexes, which make for economy and efficiency in expended energy. Some examples of these reflexes are shifting gears, turning off a light upon leaving a room, tying shoelaces, playing a musical instrument, and writing one's name. These represent but a miniscule sample of such reflexes.

The reader is encouraged to explore this fascinating topic through readings in basic psychology texts. Of special applicability to speech and hearing are the mechanisms for the learning and practicing of language, emotional, and reasoning patterns as related to the principles of conditioning.

Injury to any segment of a reflex arc is fatal to the integrity of the whole arc. A severed mandibular nerve, for example, may lead to sensory loss, which influences the motor activity of the masticatory muscles. As a result, mastication is not performed adequately and adjacent tissues may be damaged, as in cheek biting. In normal mastication, the tongue and cheek are constantly active in repositioning the food bolus on the molar occlusal table after each bite in the chewing act. This illustrates a complex, interlocked system of arcs.

The so-called *gag reflex* is a test of the pharyngeal reflex and is carried out by touching the pharynx, soft palate, or tongue with the tip of a tongue depressor. A normal response is the contraction of the pharynx and elevation of the soft palate. Such clinical features as speed and accuracy of movement and muscle strength, tone, and steadiness are included in neurologic assessment related to dysarthrias (see also Chaps. 22, 23, & 26).

Voluntary Movement

Volitional actions (those movements under voluntary control) emanate from the motor strip on the cortex. There is a point-to-point relationship along area 4 of the cerebral cortex (the gyrus anterior to the central fissure) and muscular regions of the body with each area of the body represented on the motor cortex (see also Chap. 22). The size of each cortical area is correlated with the amount of mobility and dexterity of the body part, e.g. the more complex the movement potential associated with a given part of the body, the larger the area along the motor cortex devoted to this structure.

An axon innervating a given muscle may follow one of two structural patterns. On the one hand, it may have little spread of its axon feet (terminal arborizations) to stimulate only a few muscle fibers. Upon transmitting an impulse to the structure, only a few muscle fibers will contract. On the other hand, an axon may arborize into numerous terminal branches, supplying many muscle fibers from only the single neuron.

In supplying the posterior muscles of the back, the axons arborize profusely, innervating numerous muscle fibers. Thus, one axon causes many muscle fibers to contract with the firing of the nerve cell. There is little arborization of the axons supplying the extrinsic eye muscles, for example. As a result, it is possible to contract a few small bundles of muscle fibers in an eye muscle as contrasted with the ability to contract only large bundles in the back.

Consequently, one can expect to find far greater cortical representation for the nerve fibers that are able to execute fine movement in a muscle, such as the eye, than for nerve fibers serving a wide area of the body. It appears, then, that the more dexterous a muscle is, the greater the area devoted to this muscle along the cortical motor strip. Appropriately, the tongue and lips are well accounted for on the motor strip.

Major Motor Pathways

There are two mechanisms of *efferent* or *motor pathways* originating in the brain. One system effects movements of a volitional nature, particularly skilled movements involving only a few muscles. The other system, according to traditional viewpoints, contributes synchronization of impulses arriving at the muscles. It should be noted that these are not the only efferent and descending complexes in the brain. Further, these two systems, as well as others that might be listed, are divided into separate systems for descriptive purposes. None can function independently, and many influences are typically focused on a given motor activity.

PYRAMIDAL SYSTEM. This system is composed of pyramidal corticospinal tracts. The latter name describes the ori-

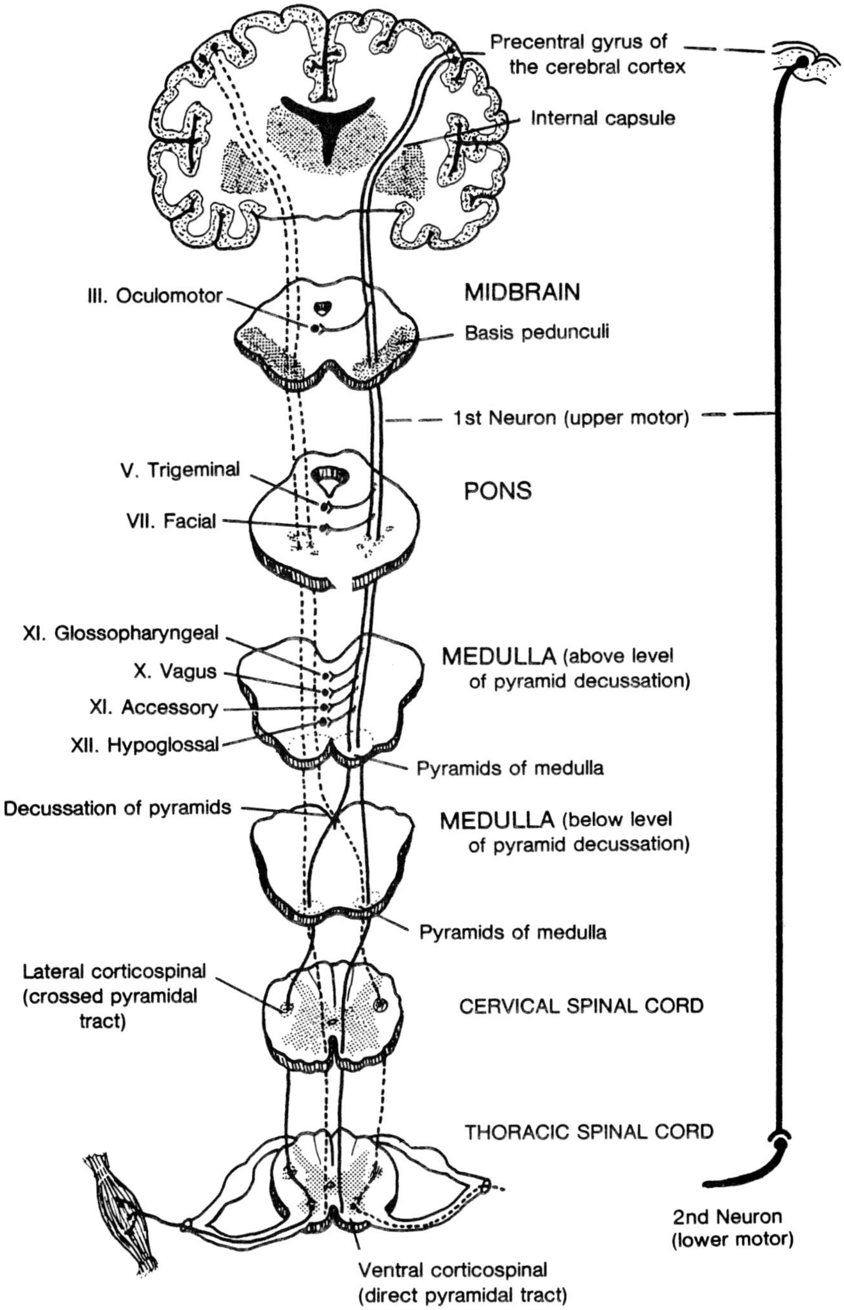

Figure 16-11. Efferent pathway – corticospinal tracts. The pyramidal system. Tracts originate in pyramidal cells of the cortex. Fibers that cross at the medulla form the lateral corticospinal tracts, while the remaining fibers form the ventral corticospinal tracts. The basis pedunculi are part of the cerebral peduncles. Notice that the group of cortical fibers terminating in the medulla are descriptively termed the *corticobulbar tract.* (From Crouch, J.E.: *Functional Human Anatomy*, 3rd ed., 1978. Courtesy of Lea & Febiger, publisher.)

gin (cortex of the cerebrum) and its destination (spinal cord). The cells of this efferent tract are found in the precentral gyrus, the motor strip of the cerebral cortex. Large pyramidal cells (Betz cells) in the fifth layer of the cortex provide the axons comprising the corticospinal tract (Fig. 16-11).

From the *cortical motor strip* (area 4), the axons leave in a pattern corresponding to a topographical arrangement of the body (see Chap. 15). They twist on their course down through the white matter within the cerebral hemispheres. Next, they pass through the *internal capsule* in a compacted fashion, yet retaining a topographical arrangement as in the cortex.

Leaving the hemisphere, the fibers descend into the *cerebral peduncles* of the midbrain. After passing through the midbrain, the fibers continue on through the *pons* and emerge on the anterior surface of the *medulla oblongata* in an eminence appearing as a vertically rounded column termed the *pyramid*. About a million or so fibers are embedded in the medullary pyramid. It is for this special pyramid-shaped landmark on the lower portion of the brain stem that this efferent system is named.

It is noteworthy that the corticospinal tracts convey impulses to the cerebellum as they descend through the brain stem. This is done via collaterals to pontine nuclei, whose cells send fibers across the midline to the opposite hemisphere. The posterior lobe of the cerebellar hemisphere relays information back to the cerebrum from the dentate nucleus by way of the thalamus. Some dentate fibers synapse in the red nucleus on the way. The collaterals sent by the pyramidal tracts to the cerebellum permit some conditioning and control influences of the cerebellum on both upper and lower neurons.

Those fibers emanating from the cortex and ending in the medulla are termed corticobulbar fibers. (Due to the bulb shape of the medulla oblongata in general, tracts related to this structure may be designated as bulbar tracts.) To innervate muscles of the larynx, face, and neck areas, the corticobulbar fibers travel to the motor nuclei of the brain stem of the cranial nerves that supply these structures. Sometimes these fibers are described as the corticobulbar pathway, as if they were a subdivision of the pyramidal system. Regardless, the fact that these muscles are associated with speech production and control *places these fibers of the pyramidal tract in a significant role for the speech and hearing clinician*. The specific classes of muscles can be seen in Figure 15-32, the motor sequence.

Those fibers committed to the corticospinal pathway do not get sidetracked in the medulla oblongata. They continue down the medulla in the pyramids until they reach the lower end of the medulla. Here, up to 80 percent cross over to the opposite side. These are the *decussating fibers*. The anterior medial fissure of the medulla is interrupted by this crossing, which is clearly visible on the surface. The uncrossed fibers continue to descend toward the spinal cord.

Below the level of the pyramidal decussation, the fibers continue in the lateral funiculus of the spinal cord. For these decussating fibers, the name *lateral corticospinal tract* is given. Uncrossed fibers continue downward in the anterior funiculus as the *ventral corticospinal tract*. All of these fibers eventually terminate in a spinal cord segment. The lateral corticospinal tract is related to the internuncial neurons, which indirectly influence the motor neurons, especially those serving the hands and the feet. The ventral corticospinal tract terminates in the anterior horn of the spinal cord, synapsing with the motor neurons innervating muscles of the trunk.

All corticospinal and corticobulbar fibers are known as *upper motor neurons*, since they have their cell bodies lodged in the cerebral cortex. The neuron that synapses with the cortically based neuron and serves a muscle is termed a *lower motor neuron*. Hence, all motor neurons of the cranial and spinal nerves are lower motor neurons. The upper motor neurons initiate voluntary movement through facilitating or inhibiting the transmission of an impulse at a synapse. Lower motor neurons serve only to cause contraction of a specific muscle.

The peripheral structures of one side of the body are represented in the opposite side of the brain along the precentral gyrus. Abnormalities of movement from lesions in the pyramidal system are discussed in Chapter 22 under "Applied Clinical Anatomy and Physiology."

EXTRAPYRAMIDAL SYSTEM. The pyramidal system was described as a voluntary motor pathway with a well-defined origin (area 4 of the motor cortex). The *extrapyramidal system*, by contrast, arises from a wide area of the *cerebral cortex*. The paths of this system appear to originate frequently from the motor areas of the *frontal lobe*, although a number of nuclear masses belong to the extrapyramidal system. These include some of the *basal ganglia, brain stem nuclei* (in particular, the subthalamic nucleus), *red nucleus*, and *nuclei* of the *reticular formation*.

The functions of the extrapyramidal system are closely tied to *associated automatic activity* such as extra movements occurring during general complex movements (e.g. swinging the arms while walking). It is a motor regulatory system, and it appears to operate independently of the direct motor system — the pyramidal system. The extrapyramidal system influences motor acts through inhibition and facilitation to assure smooth, coordinated, and purposeful movements. It serves as a type of complicated servomechanism for the final common pathway to the spinal effector mechanism (pyramidal tracts) (Fig. 16-12).

Movements that are essentially reflex in character (that is, postural adjustments and gross movement) involve the extrapyramidal system. In essence, then, impulses traveling the fibers of the extrapyramidal system may modify the pyramidal pathways. An example of the interplay of muscles that must be controlled for successful, coordinated execution of movement concerns the need to relax progressively the flexor muscles as extensor muscles are contracted, as when extending the leg from a drawn-up position. Otherwise, the action would be jerky and exaggerated. The two sets of muscles seldom contract simultaneously. Rather, movement is produced by keeping the two from working against one another. In this way, only one reflex at a time is dominant.

The relaxation of antagonistic muscles is termed *reciprocal inhibition*, which is actually a special case of inhibition. Through such mechanisms, timed movement is produced with smoothness. This balance of interplay between various sets of antagonistic muscles is achieved in large measure by the extrapyramidal system.

The *first series* of the extrapyramidal tract fibers is rather involved. The reader should refer frequently to the schematic illustration (Fig. 16-12) for clarity in following the pathway through the description following. (1) Large cell bodies in the gyri anterior to the precentral gyrus (4 and 6), as well as certain parts of the temporal and parietal lobes, supply axons

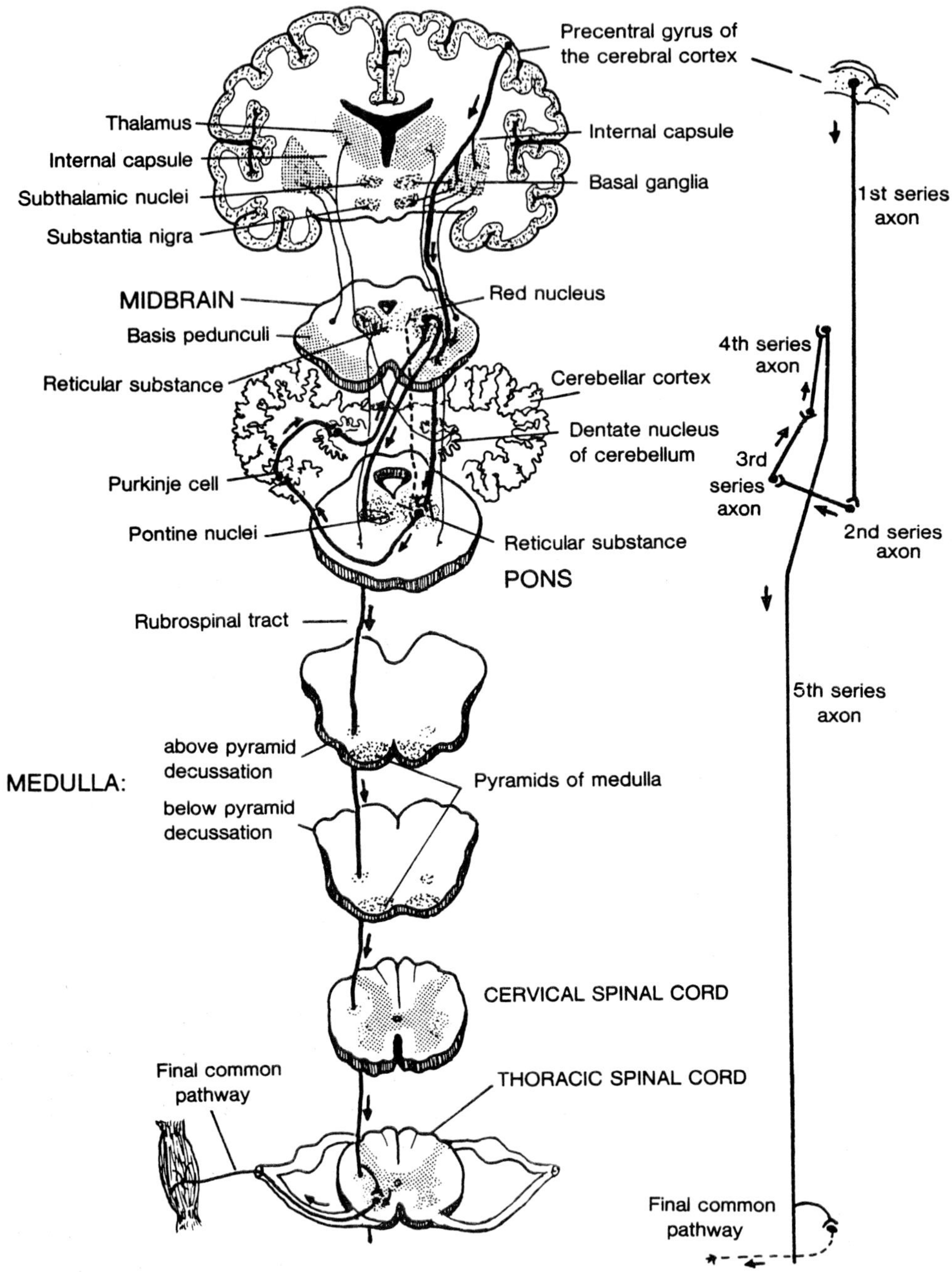

Figure 16-12. Efferent pathway — the extrapyramidal tract. Notice that the extrapyramidal tract is less well defined than the pyramidal pathways. A fifth-series axon is required to reach the final common pathway. This system is associated with phylogenetically older portions of the brain as indicated by the course of the axons. (From Crouch, J.E.: *Functional Human Anatomy*, 3rd ed., 1978. Courtesy of Lea & Febiger, publisher.)

coursing downward by means of the (2) corona radiata, the internal capsule, cerebral peduncle, to the (3) pons, where the axons terminate. Here, scattered collections of cell bodies, (4) the pontine nucleus, are found providing a *second series* of axons that pass through the (5) middle cerebral peduncle to the (6) contralateral cerebellar cortex. It is this group of fibers that makes up the transverse-course bundles of fibers and characteristic feature of the pons. In the cerebellar cortex, the axons end at the (7) Purkinje cells. These large cell neurons transmit *third-series* fibers to the (8) dentate nucleus within the cerebellum, where the *fourth series* of neurons is introduced. These axons reenter the brain stem through the (9) superior cerebellar peduncles, rise to the (10) midbrain, decussate, and enter the (11) red (ruber) nucleus. The rubrospinal tract, *fifth-series axons*, leaves the red nucleus, decussates, and travels downward in the (12) anterior column cells with lower motor neurons and influences the anterior nerve roots controlling skeletal muscles. For novelty, one might apply a descriptive name to this tract: "cerebropontocerebellodentaterubrospinal tract." Unconscious muscle coordination in response to feedback from proprioceptive impulses is provided by the rubrospinal tract — the lowest division of the meandering tract described.

Other pathways of the extrapyramidal system representing still further variety in the routes taken by fibers may be depicted as the *reticulospinal* and *vestibulospinal tracts*. These paths originate mainly in the precentral motor cortex, as do the pyramidal tracts. They descend through the internal capsule like the pyramidal and other extrapyramidal fibers described previously.

But these present fibers are primarily found traveling by way of the basal ganglia. From there, some fibers travel to the brain stem, namely, the vestibular nuclei of the medulla for the vestibulospinal tract, which sends axons down the anterior funiculus of the spinal cord, terminating in the anterior columns around motor cells. This tract serves to assist in the maintenance of muscle tone and equilibrium by translating impulses received from the semicircular canals into automatic postural adjustments.

The other tract, the reticulospinal tract, also descends the cord but mainly in the lateral funiculus. A network of neurons and fibers in the brain stem, known as the reticular formation, provides the principal origin for the fibers of the reticulospinal tract. Neither the reticulospinal, rubrospinal, or vestibulospinal tract enters the pyramids of the medulla; hence, they are termed members of the extrapyramidal system.

THE PERIPHERAL NERVOUS SYSTEM

The PNS is made up of two major groups of nerves, all paired, which relate the brain and spinal cord to sensory receptors and muscles remotely located from the central nervous system. The cranial nerves and spinal nerves are the two principal groups of the peripheral nervous system. Each group contributes fibers to a third group of nerves termed the autonomic portion of the peripheral nervous system.

Cranial nerves arise from the brain and, with only one exception, distribute their fibers to the head and neck. These nerves may transmit *both* sensory and motor impulses, *only* sensory impulses, or *only* motor impulses. All spinal nerves carry *both* sensory and motor impulses. The *spinal nerves* distribute their fibers to the body walls and limbs. The *autonomic nervous system* (ANS) is also a division of the PNS and may be termed a visceral motor system; it is associated with control of involuntary muscles and glands.

Typically, the peripheral nervous system is thought of only as groups of nerve fibers. Collections of cell bodies are included, too. Cell bodies found in the primary sensory organs are associated with the PNS. These cell bodies are also found in the cerebrospinal ganglia located on the dorsal roots of spinal nerves and the ganglia of the autonomic nervous system (see Figs. 16-5 & 16-24).

A review of the concept of a *nerve* is appropriate before becoming submerged in the description of the numerous major nerves of the peripheral nervous system. A nerve is composed of a collection of microscopic nerve fibers, so that the amassed bundle is visible to the naked eye. Several hundred or thousand nerve fibers are required to make up a nerve. Nerve fibers are axonal processes of neurons. Their role is to conduct impulses between nonnervous tissue and the central nervous system. The fibers may be sensory (afferent), to the central nervous system, or motor (efferent), away from the central nervous system. A nerve transmitting both sensory and motor fibers is known as a *mixed nerve*.

Cranial Nerves

Attached to the brain, the twelve pairs of symmetrically arranged cranial nerves are distributed primarily to the head and neck structures. Part of the vagus nerve winds inside of the trunk, serving the viscera. Each cranial nerve is specially named, often after the specific structure served. Additionally, they are numbered as are the spinal nerves; it is appropriate to use Roman numerals instead of Arabic numerals in designating the cranial nerves.

A number of clever sentences using the first letter of each nerve have been generated by anatomy students (and professors) seeking recall clues. The letters OOOTTTAFVGVAH represent a real challenge to the speech and hearing student who attempts to use only structures and terms closely associated with his or her field. For example, *Oral occlusions optionally tie tongues anteriorly; fauces vary grossly; velums attenuate hypernasality*. As can be seen from Table 16-I, many of these nerves are vitally important to communicative functions including aural, visual, and oral modalities.

The cranial nerves follow very complicated routes to and from the peripheral structures served. Most of the twelve pairs of cranial nerves have their nuclei in the hindbrain (Fig. 15-46). On occasion, they combine to form a *plexus* to serve a given area or structure. They do not follow segmentation as in the spinal cord. Hence, they are said to be functional units rather than segmental units.

Olfactory, I

This purely sensory nerve of smell originates in the upper part of the nasal cavity. Dendritic filaments in the olfactory mucous membrane send their axons through the perforations of the cribriform plate of the ethmoid bone into the olfactory bulb lying on the floor of the anterior cranial fossa. The tract courses backward from the bulb and terminates at the base of the brain near the optic chiasma (Fig. 16-13). Ultimately, the fibers reach the hippocampus of the cerebrum, the location of the olfactory sense areas.

Movement of air through the nose carries particles that stimulate the recep-

Table 16-I
THE CRANIAL NERVES

Name	Divisions	Distribution	Functions
I Olfactory	Afferent	Olfactory mucous membrane	Smell
II Optic	Afferent Efferent	Retina To retina	Vision
III Oculomotor	Efferent		
	i. Somatic	Superior, medial, & inferior recti and inferior oblique muscles of eyeball; levator palpebrae superior muscle	Movement of eyeball and elevation of upper eyelid
	ii. Parasympathetic Preganglionic	Ciliary ganglion; ciliary and sphincter pupillae muscles of eyeball	Focusing; control of pupillary size
	Afferent	The muscles above	Proprioceptive (via ophthalmic part of trigeminal)
IV Trochlear	Efferent somatic	Superior oblique muscle of eyeball	Depression of the medially rotated cornea
	Afferent	Superior oblique muscle of eyeball	Proprioceptive (via the ophthalmic part of trigeminal)
V Trigeminal	Efferent branchial	Muscles of mastication Digastric (anterior belly) and mylohyoid muscles	Mastication and swallowing
		Tensor veli palatini	Tenses anterior part of soft palate
		Tensor tympani	Modifies movements of tympanic membrane and malleus
	Afferent	Skin of face and anterior scalp Mucous membrane of mouth including teeth, gums, and tongue (anterior 2/3) Mucous membrane of nasal cavities and paranasal sinuses Meninges (dura mater) and skull	General sensibility
		Sensory to muscles of mastication and facial expression, and of tongue and eyeball	Proprioceptive
VI Abducent	Efferent somatic Afferent	Lateral rectus muscle of eyeball Lateral rectus muscle of eyeball	Turns cornea laterally Proprioceptive
VII Facial	Efferent		
	i. Branchial	Muscles of facial expression and scalp Digastric (posterior belly), stylohyoid and stapedius muscle	Facial expression, closure of eye, movements of lips Elevation of hyoid bone in swallowing Control of stapes
	ii. Parasympathetic Preganglionic	Pterygopalatine ganglion; glands of nasal cavity, of hard and soft palates, and lacrimal gland Submandibular ganglion; submandibular and sublingual glands and glands of tongue (anterior 2/3)	Secretomotor
	Afferent	Mucous membrane of tongue (anterior 2/3) (excluding vallate papillae)	Taste

Table 16-I (*continued*)

Name	*Divisions*	*Distribution*	*Functions*
VIII Vestibulo-cochlear	Afferent	From duct of cochlea	Hearing
		From maculae of utricle and saccule From ampullae of semicircular ducts	Equilibration (measurement of position of head relative to gravity and to changes in angular velocity)
	Efferent	To organ of Corti and vestibular apparatus	
IX Glossopharyngeal	Efferent		
	i. Branchial	Stylopharyngeus muscle Middle constrictor	Pharyngeal movements and elevation of larynx
	ii. Parasympathetic preganglionic	Otic ganglion Lingual ganglia	Secretomotor to parotid gland, and glands in posterior 1/3 of tongue
	Afferent	Mucous membrane of pharynx and tongue (posterior 1/3) including vallate papillae, and tonsillar region	General sensibility and taste
		Mucous membrane of auditory tube, tympanic cavity and antrum, and mastoid air cells	General sensibility
X Vagus (and Cranial root of Accessory	Efferent		
	i. Branchial	Levator veli palatini muscle, palatoglossus	Lifts soft palate
		Pharyngeal muscles	Swallowing and other pharyngeal movements
		Laryngeal muscles	Control of larynx in respiration and phonation
	ii. Parasympathetic preganglionic	Plain muscles of esophagus, stomach, intestine (as far as splenic flexure), and gall-bladder	Movements of these viscera
		Pancreas and gastric glands	Secretomotor
		Heart (nodal tissue & cardiac muscle)	Cardiac depressor
		Lungs (plain muscle & glands of bronchi and bronchioles)	Bronchoconstriction
	Afferent	Mucous membrane of alimentary and respiratory passages from epiglottis to transverse colon	General and visceral sensibility, and some taste
		Part of auricle and external acoustic meatus	General sensibility
		Lungs	Respiratory reflexes
		Glomus aorticum and wall of aorta	Vasosensory
XI Accessory (Spinal Root)	Efferent branchial	Trapezius and sternocleidomastoid muscles	Movements of head and shoulder
XII Hypoglossal	Efferent somatic	Extrinsic and intrinsic muscles of tongue, except palatoglossus	Movements of tongue

tors through a chemical reaction upon contact. The sense of smell compliments the sense of taste. Laryngectomized individuals often complain about food tasting flat due to loss of ability to direct the breath stream through the nose; they breathe through a stoma in the neck.

Optic, II

Concerned with sight, the optic nerve and retina are sometimes described as an extension of the brain. The reason is that the optic nerve is formed by an outpouching (evagination) of the diencephalon and is surrounded by the meninges and cerebrospinal fluid. The rods and cones of the retina transduce light stimuli into neural impulses, which are relayed via the optic nerve.

Both optic nerves fuse at the optic chiasma, where the fibers from the nasal half of the eyeball of each eye cross over. Leaving the chiasma, the nerve bundles become known as the *optic tracts*. After encircling the midbrain, the majority of the fibers pass to the lateral geniculate body in the thalamic complex. The fibers leave the diencephalon as optic radiations, carrying visual impulses to the calcarine cortex of the occipital lobes. Other fibers of the optic tract arrive at the superior colliculi of the midbrain. Synapses with

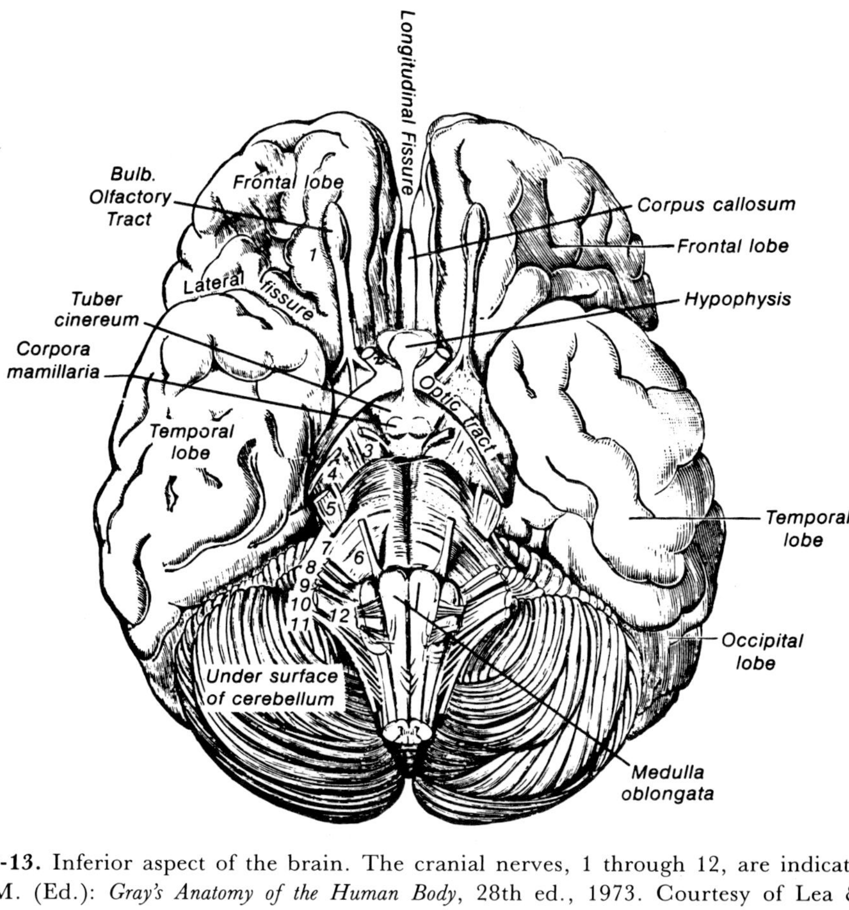

Figure 16-13. Inferior aspect of the brain. The cranial nerves, 1 through 12, are indicated. (From Goss, C.M. (Ed.): *Gray's Anatomy of the Human Body*, 28th ed., 1973. Courtesy of Lea & Febiger, publisher.)

other tracts permit capability of widespread motor responses from light stimuli (Figs. 16-13, 16-15).

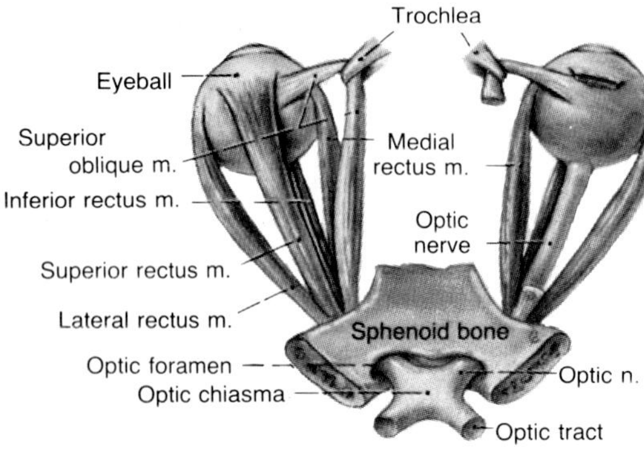

Figure 16-14. The optic tract (cut) viewed from behind, forming the chiasma and sending optic nerves into foramen of the sphenoid and to the eyeball. (From Jacob, S.W., and Francone, C.A.: *Structure and Function in Man*, 4th ed., 1978. Courtesy of W.B. Saunders Co., Philadelphia.)

Lesions associated with the visual pathways or visual cortex result in different types of deficits. Interruptions along the visual pathway, between the retina and visual cortex, result in various patterns of blindness of the two eyes. In one given eye, a lesion of the visual pathway may result in total blindness, half blindness (hemianopsia), one-fourth blindness or quadrant deficit, or no functional deficit.

Destruction of area 17 of the cerebral cortex, if bilateral, results in total blindness. Partial blindness may be expected if only one side is destroyed. The term for this condition is *homonymous hemianopsia*. The central area of the blinded visual field is spared without explanation except for the possibility that there is duplicate cortical representation for the central visual region in each side of the brain. This would posit the probability that some fibers from

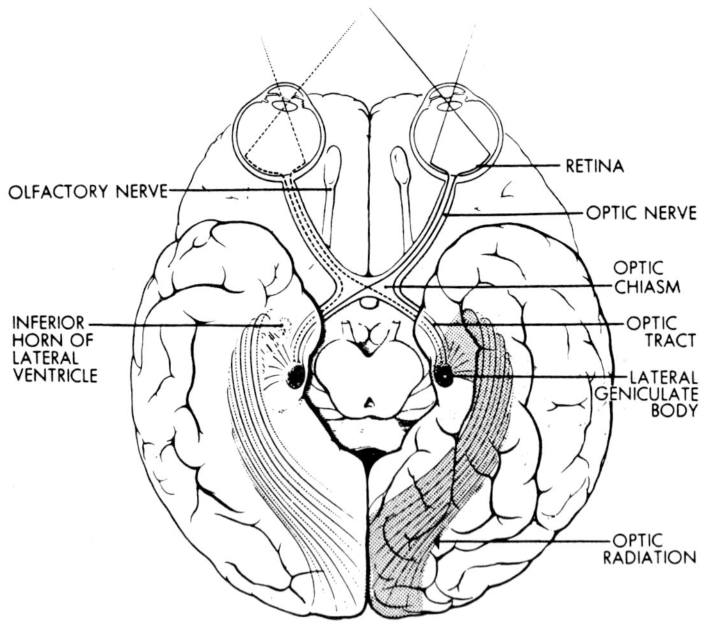

Figure 16-15. The visual pathways from the eyes to the visual cortex of the brain. (From *Man, Microbes, and Matter* by B.C. Block. Copyright © 1975. Used with the permission of McGraw-Hill Book Company.)

the central part of the retina and posterior to the optic chiasma cross over before reaching the visual cortex.

Areas 18 and 19 of the occipital lobe serve to coordinate visual sensation with other functions. Tumors in these regions may produce visual hallucinations, sometimes quite vivid and complex. Impairment of written communication may be present. A patient can see the word but is unable to associate its meaning; however, the patient may use the word in speech correctly without effortful recall. The term *word blindness* has been used to describe this observation. Coordination between visual and motor activity may also be subnormal from such interference of these two visual association areas.

Oculomotor, III

This nerve has a motor function for five of the extrinsic muscles of the eye: medial, superior, and inferior rectus muscles, inferior oblique, and levator palpebrae. These enable eye movement in the nasalward, upward, and downward directions plus raising the superior tarsal plate (upper eyelid). Additionally, the oculomotor nerve provides autonomic fibers to the ciliary and sphincteric muscles of the iris. These efferent fibers are concerned with pupillary light reflex and accommodation of the lens.

An abnormality of the eye, *strabismus*, which means deviation of either eye or both, generally is the fault of the extrinsic eye muscles. The image does not fall on corresponding points of the two retinas, so two images result. This condition of twin vision is known as *diplopia*. Correction may involve surgical alteration of the extrinsic ocular muscles or of the lens to

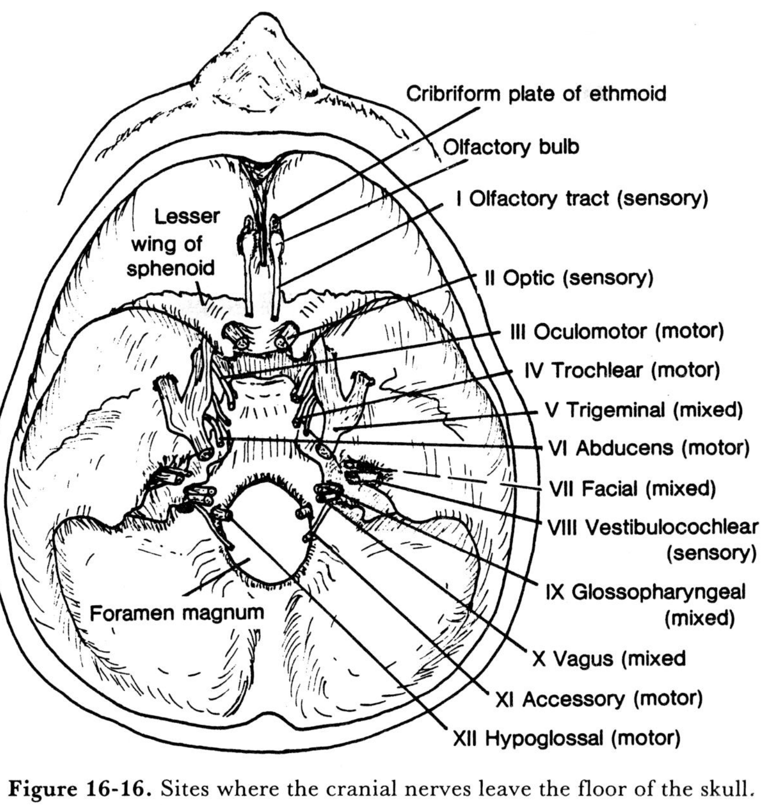

Figure 16-16. Sites where the cranial nerves leave the floor of the skull.

move the image onto corresponding points on the two retinas.

Trochlear, IV

The trochlear is a motor nerve serving only one muscle, the superior oblique muscle of the eye. It moves the eyeball.

Trigeminal, V

The fifth nerve, the trigeminal (*three twins*), is the largest of the cranial nerves. It passes sensory information to the central nervous system from the skin of the face, the mucous membrane lining the cavities of the mouth, nose, paranasal sinuses, the greater part of the tongue, and teeth. Three divisions, *ophthalmic, maxillary*, and *mandibular*, branch extensively from the semilunar (gasserian) ganglion to send sensory fibers to the face, head, and jaws. The *sensory portion* of this nerve is by far the largest (*see* Figs. 16-16 & 16-17).

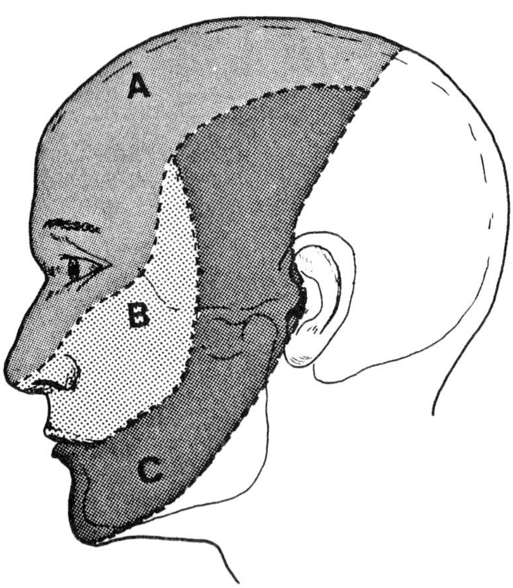

Figure 16-17. Distribution of sensory fibers by the three branches of the trigeminal nerve: (A) ophthalmic division, (B) maxillary division, (C) mandibular division.

Nuclei of this nerve are found in the medulla and pons. Connections between the sensory nuclei of the trigeminal in the lower medulla and the spinal tract are associated with pain. Relief from excruciating facial pain may be had by making an incision in the posterior part of the medulla and interrupting the pain fibers transmitting this sensation to consciousness.

Trigeminal neuralgia (pain) is a well-known anomaly. Seldom bilateral, this condition may arise from numerous causes, and it usually appears more often between the ages of forty and sixty, perhaps due to vascular deficiency. Two conditions involving the sensory division of the trigeminal are loss of sensation and paresthesia (abnormal sensation as burning, prickling, crawling, etc.). Sometimes there is a loss of proprioceptive data, which are responsible for maintaining the occlusal stability necessary for normal mastication and also, to some extent, speech sound production.

The *motor root* of the trigeminal shares the route of the maxillary division of this nerve. Every dentist and speech pathologist should be able to recite the names of the eight muscles innervated by the motor division of the trigeminal nerve. It supplies four muscles of mastication: *lateral* and *medial pterygoids, masseter*, and *temporalis*. The other muscles supplied are the *tensor tympani, tensor palatini, mylohyoid*, and the anterior belly of the *digastric*. Deficiencies influencing speech may result from paralysis of the motor division of the trigeminal, but *no clear documentation has been shown to indicate that soft palate function is impaired by paralysis of the tensor palatini*. The tensor tympani, which may be an anatomical continuation of tensor palatini, maintains tension on the eardrum. It may act to increase the efficiency of the drum or protect the ear from damaging, loud, sustained sound. Lack of tonus in the tensor tympani could result in excessive oscillations of the eardrum, with potential damage to the hair cells in the cochlea. As will be seen later, the action of the tensor tympani is not effective as a protective *reflex* to loud sounds (see Figs. 17-6 & 17-7).

Abducens, VI

This nerve is solely motor and innervates only one muscle, the lateral rectus of the eyeball. It turns the eyeball outward upon contraction.

Cranial nerves III, IV, and VI are tested together since all supply the muscles that control eye movements. A convenient mnemonic is LR6,SO4,3. This reminds the clinician that the lateral rectus muscle is supplied by the sixth (abducens) and the superior oblique by the fourth (trochlear). All other ocular muscles receive innervation by the third (oculomotor) nerve.

Facial, VII

The facial nerve, like the trigeminal, serves important functions in communication. Although primarily motor, it transmits both sensory and motor fibers. On its course from the brain stem, it joins the eighth nerve through the internal acoustic meatus, passes through the facial canal of the temporal bone, and emerges from the skull at the stylomastoid foramen. Initial branches supply the stylohyoid and posterior belly of the digastric muscles, the auricularis muscles, and the occipitalis. Further branchings provide fibers for all other muscles of facial expression. The facial muscles below the forehead receive contralateral innervation via crossed corticobulbar tracts, while the forehead receives bilateral innervation.

Other efferent fibers are derived from the *chorda tympani* branch, which travels through the middle-ear cavity to the sublingual and submaxillary salivary glands and the lacrimal gland (Fig. 16-18). The fibers are part of the autonomic system serving the viscera of the body. The stapedial branch leaves the facial nerve as it transverses the facial canal to supply the tiny stapedius muscle of the middle-ear cavity. This small muscle responds reflexively to loud sounds to prevent inner-ear damage (see Figs. 17-6 & 17-7).

The sensory division carries taste fibers from the anterior two-thirds of the tongue and some muscle proprioception (Fig. 16-19). The division is often termed *nervus intermedius* and is sometimes described as a small mixed cranial nerve accompanying the facial nerve. Such a dichotomy is not followed here.

Lesions of the motor division of the facial nerve often present clinical cases of asymmetry in facial expression if only one side is affected. For example, if the lesion is unilateral and involves but a branch of the nerve, only one muscle may be involved. The individual, however, upon contracting the frontalis, would have function on both sides, since the forehead has bilateral innervation and is not paralyzed by unilateral lesions. Interruptions in branches serving muscles around the mouth render the person unable to smile or use the lips effectively in articulation.

Bell's palsy is one of the more common types of facial paralysis, perhaps in response to exposure and chill or to a mild infection. In most cases, the functions return gradually, but involuntary facial spasms may occur for years afterward (see Chap. 22).

Lesions occurring after the nerve exits the stylomastoid foramen of the skull include the following: inability to whistle, inability to wrinkle the forehead, drooping of the angle of the mouth, collecting of food between cheeks and gingiva, and a loss of deep sensation in the facial mus-

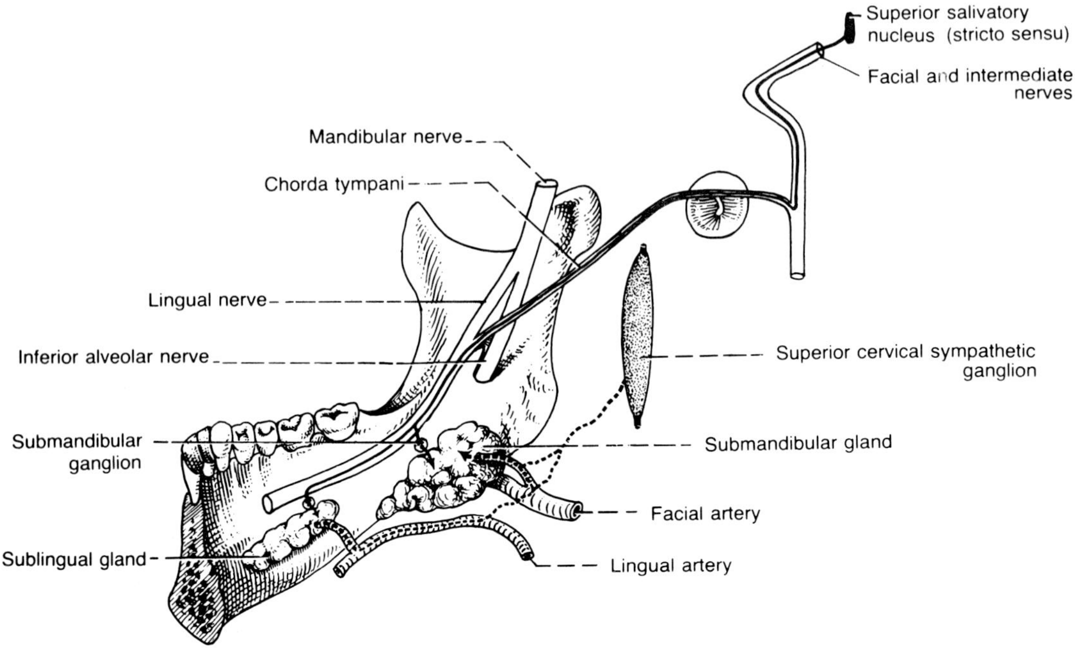

Figure 16-18. Branches of the trigeminal and facial cranial nerves. The mandibular division of the trigeminal is both sensory and motor. Notable are the lingual (illustrated) and muscular branches. The chorda tympani is from the facial nerve and innervates structures in the middle ear before joining the lingual nerve. (From Bossy, Jean: *Atlas of Neuroanatomy and Special Sense Organs*, 1970. Courtesy of W.B. Saunders Co., Philadelphia.)

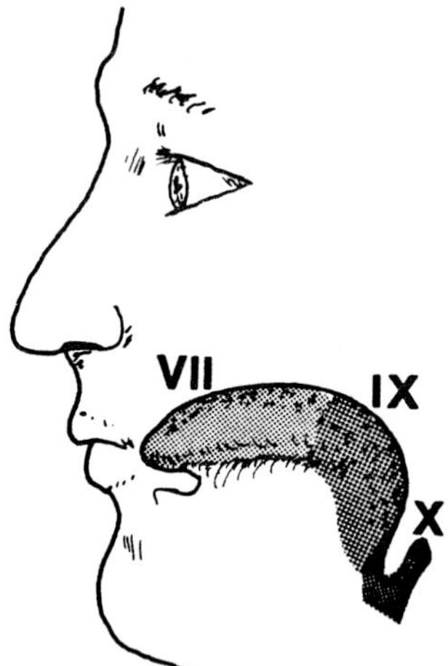

Figure 16-19. Cranial nerves innervating areas of the tongue associated with taste: (VII) facial, anterior two-thirds; (IX) glossopharyngeal, posterior one-third; (X) vagus, area adjacent the epiglottis.

cles. If the lesion is in the facial canal or nearer the nucleus of the facial nerve in the pons, additional symptoms may be added: loss of taste at the anterior two-thirds of the tongue, reduced salivation, and hyperacusis from failure of the stapedius muscle to check excessive movement of the stapes in response to loud sound, characterized both by acute pain in the ear and subsequent noise-induced hearing loss.

Flaccid paralysis of one side of the face is not rare as an aftermath of a stroke. The pattern and deficits of facial muscle movement may indicate the site of a lesion if the courses of facial nerve branches are well understood. The cosmetic problem is sometimes of more concern to the patient than such lost functions as labial sound deficiencies.

Vestibulocochlear, VIII

There are two divisions of the vestibulocochlear nerve; vestibular (balance) and cochlear (hearing); both are sensory pathways. The *cochlear division* contains cell bodies in the spiral ganglion of the cochlea. Small fibers travel to the organ of Corti in the cochlea where they pick up auditory impulses. (This mechanism of the peripheral auditory system is described in the next chapter).

The cochlear nerve fibers join the vestibular fibers and comprise the eighth nerve as it courses out of the temporal bone through the internal acoustic meatus. The nerve enters the medulla at the inferior peduncle, where the cochlear fibers terminate in the ventral and dorsal cochlear nuclei in the pons. The auditory pathway continues, with tracts carrying impulses to the inferior colliculus of the midbrain and to the medial geniculate body near the posterior part of the thalamus. From here, other fibers send the impulses on to the auditory center in the temporal cortex (Fig. 16-20).

While there are some 25,000 to 30,000 *afferent* nerve fibers in the cochlear division of the VIIIth cranial nerve, there are an additional 500 *efferent* nerve fibers. Known as the olivocochlear bundle, these efferent fibers arise in the medulla and terminate on the hair cells of the cochlea or on afferent fibers, innervating them. The function of the efferent fibers is not yet understood, but they could provide a basis for selective attention to particular sounds through inhibitory and facilitory influences on the hair cells and afferent cochlear fibers.

Disorders of the auditory apparatus and pathways may involve anomalies ranging from deafness to mild hearing losses at selective frequencies.

The *vestibular division* provides impulses to the cerebellum and vestibular nuclei for maintenance of equilibrium. The receptors are located in contiguous structures of the inner ear, namely, the semicircular canals, saccule, and utricle. The cell bodies are found in the vestibular ganglion at the inner end of the internal auditory meatus. Peripheral processes of the cells pass to the components of the balance receptor, while central processes of the neurons join fibers of the cochlear division. These are the primary neurons.

The vestibular fibers terminate in the four vestibular nuclei of the medulla. Secondary neurons convey impulses on to the spinal cord for reflex postural adjustments. Other fibers convey impulses to the nuclei of the third, fourth, and sixth cranial nerves for reflexive eye movements associated with body movements. Still others connect to salivatory and vomiting centers in the brain stem and bring about the condition of nausea in response to dizziness. Some go to the cerebellum.

Vertigo is a condition characterized by hallucinations in which the victim feels that the world around him is in motion or that he is moving in space. The condition may involve severe disturbances, with the patient complaining of dizzy spells. Vertigo and dizziness are not synonymous. Vertigo is an hallucination of *motion*; dizziness involves sensations of being *unsteady, confused, light-headed*, or *faint*.

Glossopharyngeal, IX

The ninth cranial nerve transmits both afferent and efferent fibers. The *sensory division*, its largest, serves to pass general sensation from the soft palate, tonsils and pharynx, auditory tube, and tympanic cavity. Other afferent fibers transmit taste sensation from the posterior third of the tongue (Fig. 16-19). Participation in reflexive control of blood pressure, as well as taste and swallowing, are

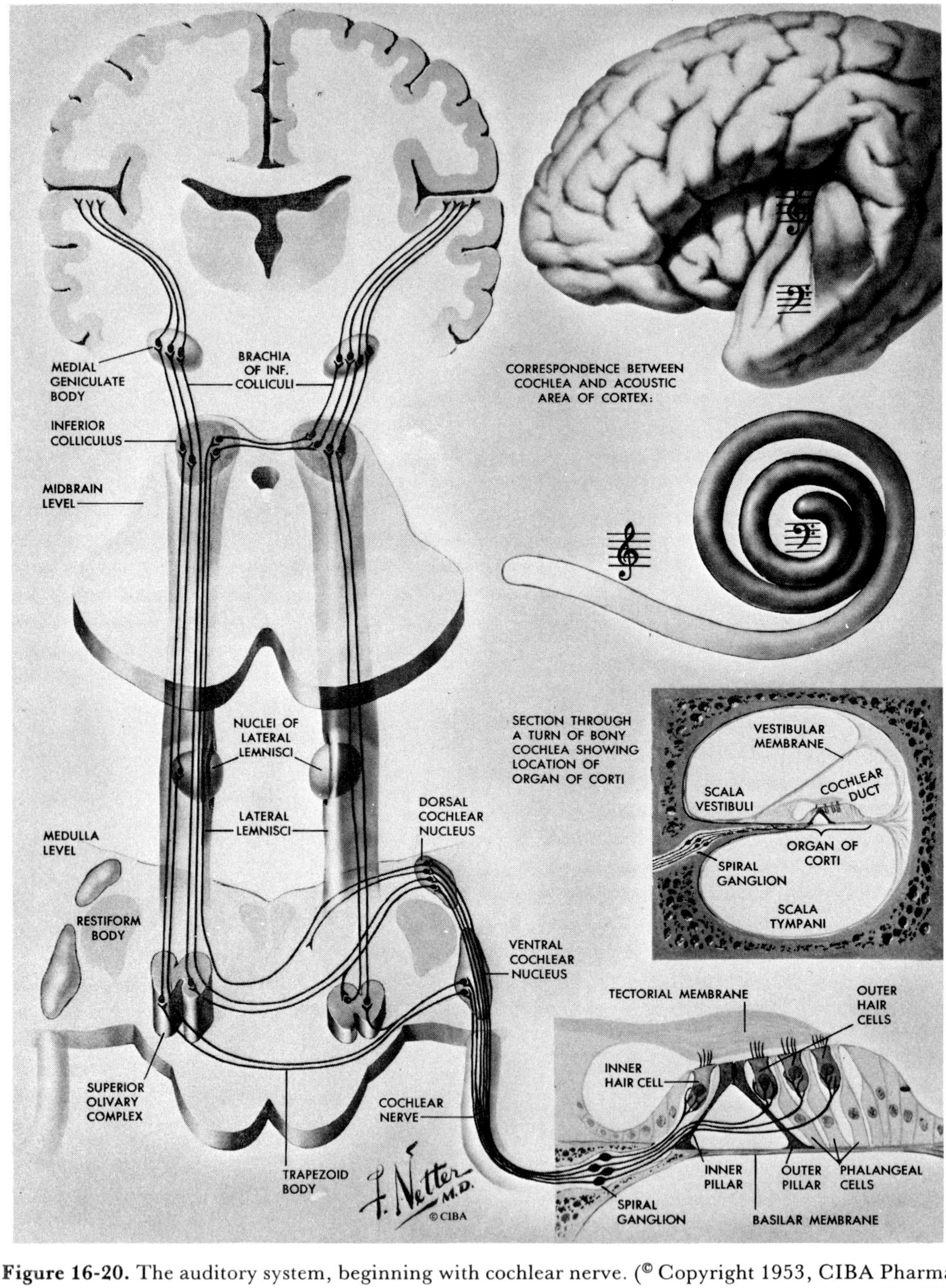

Figure 16-20. The auditory system, beginning with cochlear nerve. (© Copyright 1953, CIBA Pharmaceutical Company, Division of CIBA-GEIGY Corporation. Reprinted with permission from *The CIBA Collection of Medical Illustrations* illustrated by Frank H. Netter, M.D. All rights reserved.)

provided by the sensory division.

The *efferent division* innervates the stylopharyngeus, but this is of little functional significance since it is not possible to demonstrate damage in this muscle by clinical examination. A few fibers combine with the vagus to form the *pharyngeal plexus*. Branches from this plexus innervate pharyngeal muscles and its mucous membrane. Other efferent fibers stimulate the parotid salivary gland to pour saliva through the cheek wall onto the molars during mastication.

When interrupted by injury, the paralysis affects taste and swallowing activity and possibly causes insufficient saliva for masticatory purposes or for the articulators to move smoothly (avoiding the dry feeling during speech). Also, there may be a loss of the gag response from disturbance of the sensory stimuli for gagging, loss of taste in the posterior one-third of the tongue, and loss of sensation in the soft palate, fauces, and posterior one-third of the tongue; the uvula may deviate to the normal side, and there is inability to constrict the posterior pharyngeal wall toward the soft palate when asked to say *ah*. Another condition, glossopharyngeal neuralgia, is marked by intense pain in the area of the auditory tube orifices and behind the ear. It may be initiated by sneezing, coughing, or throat clearing.

Vagus, X

Known as the wanderer, the vagus is the most widely distributed of the cranial nerves. It is a sensory and motor nerve. The *sensory division* serves the heart (reflexive inhibition), lungs (stretch reflexes for respiratory control), mucous membrane of the bronchi, trachea, larynx, pharynx, the skin of the external ear, posterior portion of the external auditory meatus, and abdominal viscera. Thus, the vagus provides regulation for circulation, heart rate, respiration, digestion, and several other sensation-type functions, including the larynx.

Anesthesia of the upper larynx and epiglottis results from lesions involving the superior laryngeal branch of the vagus. Normal sensation in this area tends to rule out involvement of the branch known as the *superior laryngeal nerve*, which serves an important tensor muscle of the vocal folds, the cricothyroid (Fig. 16-21).

Another condition with possible connection to the sensory division of the vagus and involving the larynx concerns hyperactivity of the mucosa in the larynx. With excess mucus, the folds tend to stick together and tend to be uneven in weight. Thus, the normal synchronous vibratory movement required for acceptable tone production during phonation is upset. Vocal nodules can result from such a situation, since abusive efforts from "throat clearing" and extra strain during phonation may result. Also, accumulations of mucus may act as secondary vibrators, producing interference with normal phonation as is sometimes experienced by all with the "morning voice."

The *motor division* of the vagus is significant to structures serving generation and control of speech sounds. The *superior laryngeal nerve*, a branch of the vagus, was mentioned above relative to its afferent fibers. It innervates the cricothyroid muscle of the larynx, the tensor muscle of the vocal folds. If interrupted, phonation is still possible, but the folds become slack and the voice fatigues quickly under normal use.

Demyelination associated with the superior laryngeal branch in individuals with multiple sclerosis leads to the characteristic scanning speech pattern in these patients because of altered conduction velocities in the nerve.

The other laryngeal nerve, the *recurrent*, branches from the vagus below the

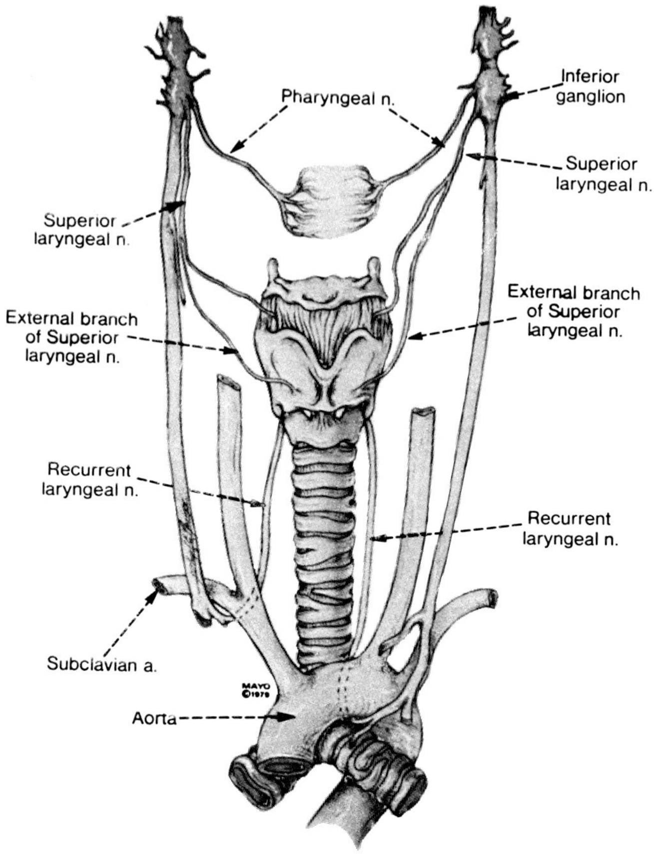

Figure 16-21. The relation of the vagus nerve to the larynx. (From Aronson: *Clinical Voice Disorders*. New York, Thieme-Stratton, Inc., 1980.)

larynx at the level of the pulmonary plexus and arch of the aorta and courses cranialward to enter the lower aspect of the larynx. All intrinsic muscles of the larynx, the vocalis, thyroarytenoid, and the several arytenoid muscles, are innervated by the recurrent; only the cricothyroid and inferior pharyngeal constrictor gain innervation from another branch of the vagus.

Special concern is warranted if the recurrent nerve supply to the posterior arytenoid muscle is lost, for this muscle is the safety muscle of the larynx. It widens the laryngeal opening after the folds have closed, making it possible to breathe. The other muscles provide for adduction and tension of the folds for fine modulation of the airstream in the production of voice.

Two pathologic conditions are present with paralysis of the recurrent nerve. First, there is adduction paralysis, in which one or both folds are unable to move to the midline for closure during phonation. The other type, already described, is abductor paralysis. Accidents, particularly cuts or blows on the neck, pressure from adjacent structures such as an aneurysm of the aortic arch, around which the left recurrent loops, or of the subclavian artery, around which the right recurrent loops (Fig. 16-21), surgical damage, thyroid gland disease (such as goiter), and other similarly localized conditions may impair or prevent impulses arriving at the muscles of the vocal folds and related glottal adjusting muscles. Too, lesions in the medulla at the site of the tenth nerve nuclei may paralyze the folds.

Continuing with the efferent division of the vagus, the following structures are also innervated:

A. levator palatini and palatoglossus
B. uvulus and pharyngeal muscles (innervation of the pharyngeal muscles is shared with the glossopharyngeal through the pharyngeal plexus)
C. the esophagus (swallowing)
D. heart muscle (slowing heartbeat)
E. muscles of the bronchial tree (constriction)
F. glands of the stomach and pancreas
G. muscles of the intestines (peristalsis)

Dysphagia, any impairment in swallowing, may relate to a dysfunction to those vagal fibers associated with the autonomic nervous system and contained in the series of structures listed here.

Due to the use of the oral and pharyngeal muscles alluded to in articulatory-resonant functions for speech, lesions involving these branches may directly influence speech sound production and cause inaccurate articulation and nasalized speech (Fig. 16-22).

Spinal Accessory, XI

This nerve is frequently listed as the accessory cranial nerve. Its two divisions are not according to sensory and motor fibers; rather, the classification follows the origin of the fibers, the lateral aspect of the upper spinal cord (cervical segment), and the medulla.

Both divisions of the eleventh nerve carry motor fibers. The spinal fibers innervate the upper part of the sternocleidomastoid and trapezius muscles. The cranial portion provides efferent fibers to the pharynx, upper larynx, uvula, soft palate, and larynx. These fibers, serving the listed voluntary muscles, belong to the vagus, since upon joining the vagal motor fibers, the accessory fibers become indistinguishable.

Lesions of the cranial root of the spinal accessory may be expected to result in the same type of disabilities as those pertaining to faulty articulatory-resonance that were mentioned in connection with the vagus.

Hypoglossal, XII

The hypoglossal contains only motor fibers. It is the essential nerve of tongue movement, supplying both intrinsic and extrinsic muscles. It is important to note that the motor supply of the tongue musculature is dense in comparison with the number of muscle fibers served by each nerve. This enables more precise control of the tongue and is necessary for the high diadochokinetic rate that the muscles of the tongue require for rapid articulation of speech sounds. Added to the list are the thyrohyoid and geniohyoid muscles, but the hypoglossal cooperatively innervates these muscles, along with branches from the upper cervical spinal nerves.

Several disorders of the nerve supply to the tongue may result in disabilities of articulation (see Chap. 22). Unilateral peripheral paralysis, or hemiplegia, causes the tongue to deviate to the affected side upon protrusion. If a supranuclear lesion (a lesion occurring above the level of the nucleus of the nerve) is involved, the tongue, upon protrusion, deviates to the side opposite the lesion. If branches supplying the intrinsic muscles are disturbed, precise shaping of the tongue for proper modulation of the airstream is impaired

PATHWAYS OF PHONATION

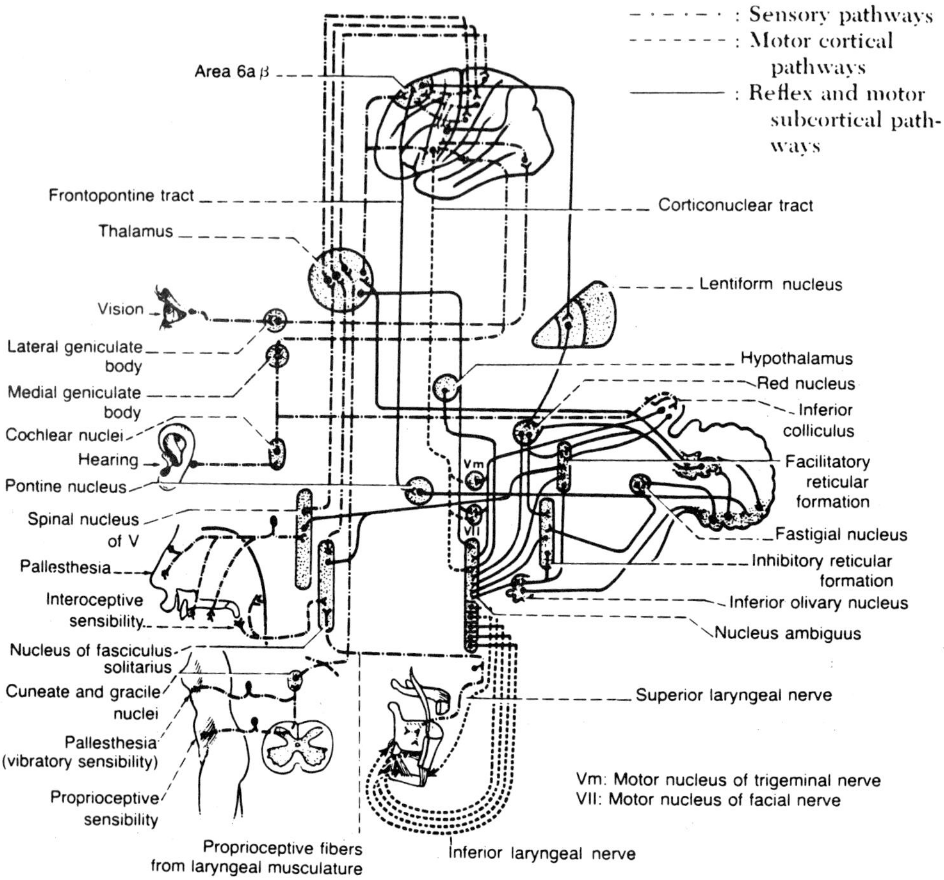

Figure 16-22. Schematic detail of the nerve tracts associated with phonation. (From Bossy, Jean: *Atlas of Neuroanatomy and Special Sense Organs*, 1970. Courtesy of W.B. Saunders Co., Philadelphia.)

or lost. The result is sluggish, inaccurate articulatory patterns.

Spinal Nerves

Earlier in this chapter, reference was made to the spinal nerves, a part of the peripheral nervous system. They arise according to thirty-one segments of the spinal cord and may be grouped as follows: eight cervical, twelve thoracic, five lumbar, five sacral, and one coccygeal (see Fig. 16-4).

Reviewing the origin of the spinal nerves, it should be recalled that each nerve arises by a dorsal and ventral root in the spinal cord (see Figs. 16-6 & 16-9). Also, the segmentation of the spinal cord is represented by the respective pairs of spinal nerves originating in each of the thirty-one segments and serving a given segment of the body. The fibers course laterally and are combined as one bundle, a spinal nerve, as they pass through the intervertebral foramina.

In the upper portion of the spinal cord, the nerves leave the cord almost horizontally; but progressively downward, the respective foramina through which the spinal nerves must reach to exit the vertebral canal become lower in relation to the cord itself (Fig. 16-23). Thus, as the spinal cord ends at the level of the second lumbar vertebra, some ten pairs of spinal nerves are found in the vertebral canal below the cord. These comprise the *cauda equina* (resembling a horse's tail) and are bathed in cerebrospinal fluid. In this area, lumbar punctures are made for clinical purposes; there is little danger to the spinal nerves, since they are slippery in the fluid suspension (see Fig. 15-21).

The *ventral root* transmits only efferent fibers. It serves as the final common pathway for all motor impulses leaving the spinal cord. The cell bodies of these fibers are located in the spinal cord.

The *dorsal root* contains only sensory fibers. Characteristic of each root is an ovoid swelling near each intervertebral foramen. This is the spinal ganglion (see Fig. 16-6). The spinal or dorsal root ganglion holds the cell bodies of primary sensory neurons. Most of the spinal ganglia are found within the cavities of the intervertebral foramina, encased by extensions of the dura mater. Just lateral to the ganglion, the dorsal root combines with the ventral root.

As the ventral and dorsal roots merge during the exit of the intervertebral foramina, the spinal nerve is formed. It travels but a fraction of an inch before dividing into anterior and posterior divisions, or *rami* (see Fig. 16-3). The anterior division or ventral ramus comprises the main part of the spinal nerve, serving the muscles and skin of the extremities and areas of the trunk. The posterior division or dorsal ramus supplies the muscles attached to the vertebral column and the skin of this area.

Plexuses

Nerve networks termed *plexuses* (braids) are formed by the ventral rami of the spinal nerves. These are combinations of more than one spinal nerve to form a definite nerve, for example, the phrenic nerve. Such interlacing of two or more spinal nerves to form these networks is commonplace in all regions of the spinal nerves except the thorax. Most nerves originating in these plexuses transmit fibers of neurons from more than one segment of the spinal cord (Fig. 16-23).

The *cervical plexus*, formed by the ventral rami of the lower cervical spinal nerves and the first thoracic spinal nerve, give rise to some nerves important in control of speech structures. The ansa cervicalis is one such nerve; it innervates the sternohyoid, sternothyroid, thyrohyoid, geniohyoid, and omohyoid muscles. Another, the phrenic nerve, serves the diaphragm.

The *brachial* (arm) *plexus* is a complex nerve network. It is also formed by fibers from the lower cervical nerves and the first thoracic nerve. Nerves arising from this plexus serve all the muscles of the upper extremity, back, and chest. This includes muscles of the rib cage participating in respiratory movement except for the costals, which are innervated by the intercostal branches of the thoracic spinal nerves; however, there are no thoracic plexuses.

The *lumbosacral plexus* is formed by the last thoracic nerve plus lumbar and sacral nerves. Frequently, these are divided into lumbar and sacral plexuses. From the *lumbar plexus*, the iliohypogastric nerve arises to assist in supplying respiratory muscles of the anterolateral abdominal wall, particularly the external, internal, and transverse abdominal muscles. Other

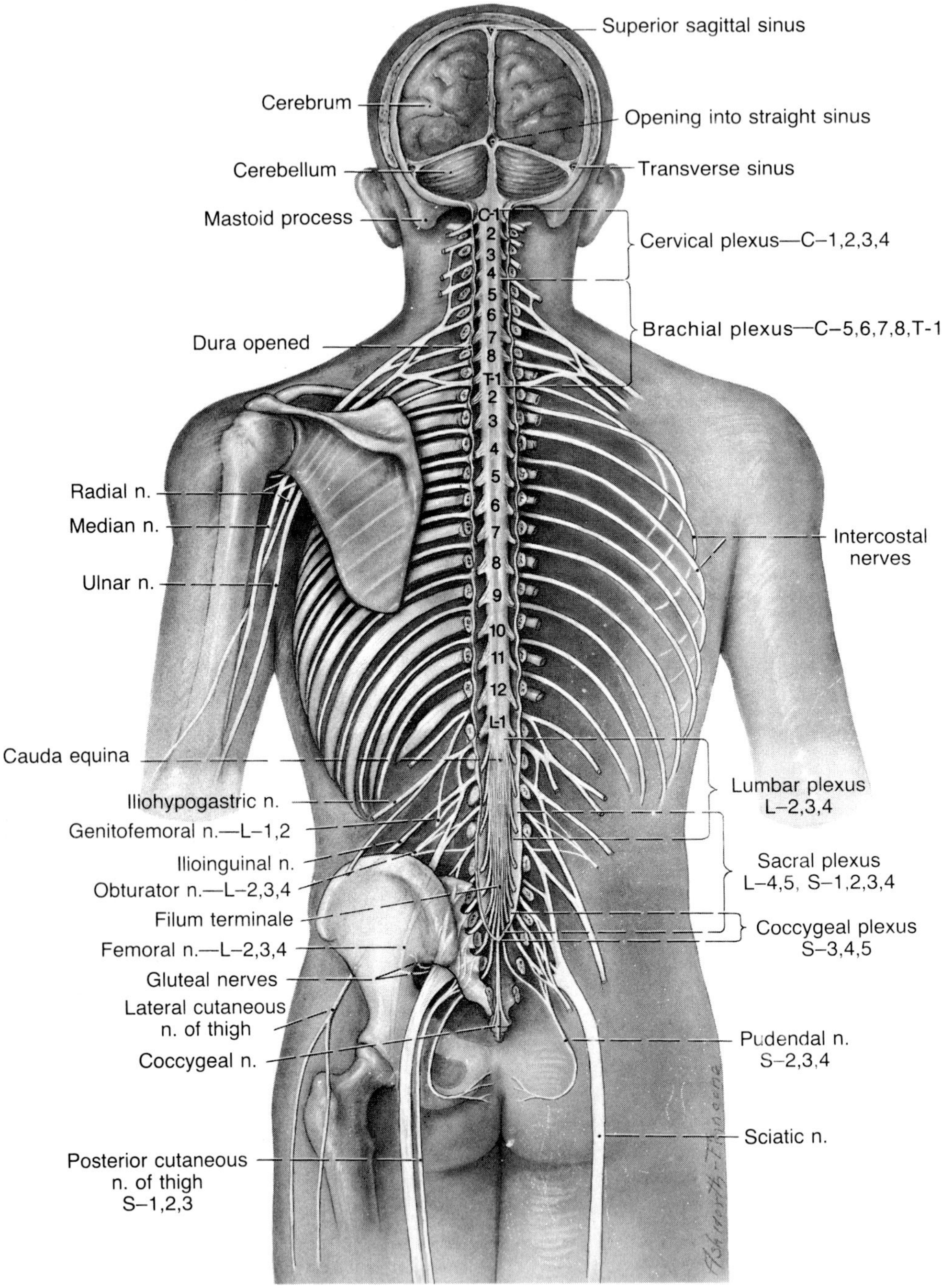

Figure 16-23. The spinal cord and its nerves. (From Jacob, S.W., and Francone, C.A.: *Structure and Function in Man*, 4th ed., 1978. Courtesy of W.B. Saunders Co., Philadelphia.)

nerves arise to serve muscles and the skin of the lower portions of the trunk and the lower limbs.

As mentioned earlier, there are no plexuses formed by the thoracic nerves. Instead, the ventral rami of the first eleven thoracic nerves give rise to eleven intercostal nerves and a twelfth, the subcostal. These nerves course in the subcostal groove on the inferior border of each rib and correspond numerically to the sheltering rib. They innervate muscles of the anterior thoracic wall, namely, the intercostal, subcostal, and transversus thoracis muscles. In addition, an important muscle of the anterior abdominal wall, rectus abdominis, receives efferent fibers from the intercostal and subcostal nerves. These nerves also provide fibers to the other three abdominal muscles. The respiratory functions of these muscles were outlined earlier.

The cutaneous branches of the nerves discussed here were not explained. These fibers, a part of the plexuses and spinal nerves referred to previously, provide for sensations from the skin according to a *dermal segmentation* pattern. Most anatomy texts illustrate the areas of the skin over the body served by the cutaneous branches of the various spinal nerves. Only the facial area of the head is not included; this area is served by the trigeminal cranial nerve. Loss of sensation in a given segment of the skin may provide important diagnostic clues relative to a paralytic disability in a portion of the body.

Autonomic System

There is but one nervous system. The subdivisions being presented should be thought of as arbitrary. Furthermore, it should be understood that there is functional continuity of nervous tissue from one part to another. The autonomic system is a vital extension of the nervous system, since it is concerned with actions beyond ordinary willful control.

All motor nerves supplying muscles of the visceral organs, glands, heart, and blood vessels, and peripheral involuntary muscles such as the intrinsic eye muscles and hair follicles are included in the autonomic nervous system. In essence, then, the autonomic system is an *efferent system*, supplying structures of the body not under voluntary control. There are afferent fibers in the system, although the efferent fibers predominate. While the autonomic system is viewed as a distinct system, it is not something anatomically apart from the whole of the nervous system as might be implied above. Without question, it is inextricably tied in with other components of the nervous system of the body.

It is common to specify the nomenclature used for such a system before describing its detail, since there are variations in the way some writers view the autonomic nervous system. It is taken here to include visceral efferent fibers and visceral afferent fibers. The term *autonomic* implies self-control without dependence on outside influences. The fibers convey (1) afferent impulses from the viscera (not reaching conscious levels), (2) organic sensations such as hunger, distention, sexual sensations, nausea, and (3) pain (see Fig. 16-1).

Divisions of the Autonomic Nervous System

The major feature is that the autonomic system is a two-motor-neuron system. Most organs served by fibers of this system have a dual nerve supply with opposing function. One set of fibers is the activator of the organ or gland, as the case may be. These fibers provide stimulation for muscle contraction or for glands to se-

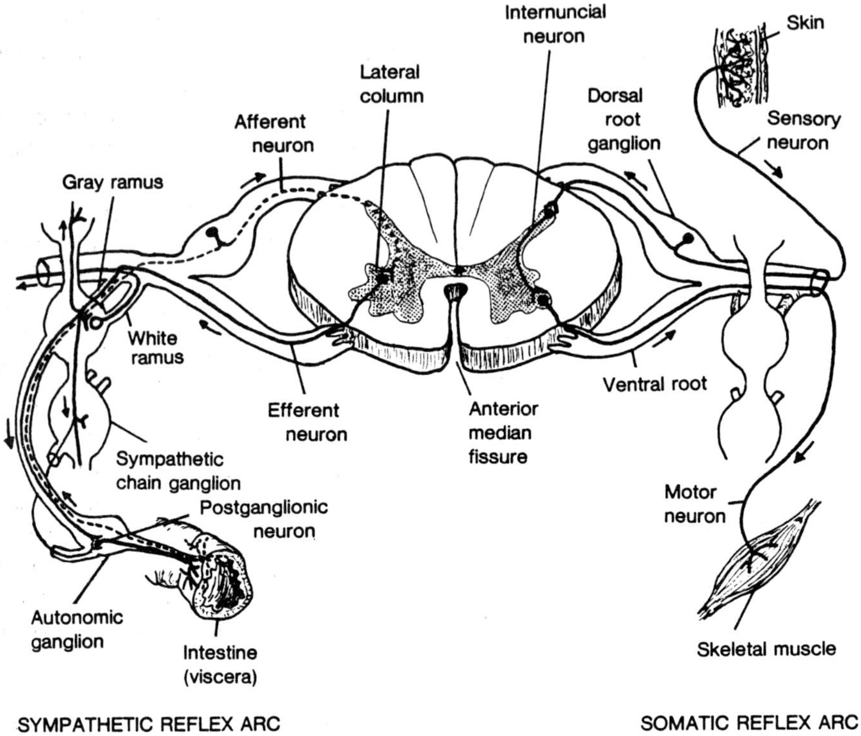

Figure 16-24. A comparison of the general arrangement of the autonomic and somatic pathways at a level of the thoracic cord. The simple reflex arc of both systems is schematized.

crete. This is the *sympathetic division* of the autonomic nervous system. Another set of fibers usually inhibits these activities; this is the *parasympathetic division*.

Both divisions require two neurons for the conduction of an efferent impulse. This is distinctive, since the somatic part of the nervous system requires only one neuron (Fig. 16-24). Each division is characterized by a number of ganglia (collections of nerve cell bodies). These ganglia contain motor neuron cell bodies as contrasted with the posterior (dorsal) root ganglion of the spinal cord, which contains sensory neuron cell bodies.

The sympathetic and parasympathetic divisions contain efferent and afferent fibers. Branches in the spinal nerves serve the sympathetic division, supplying both efferent and afferent fibers.

Cranial nerves III, VII, IX, and X contribute cell bodies to the parasympathetic division along with cell bodies in the sacral region of the spinal cord.

In the head region, the third cranial nerve (oculomotor) supplies the involuntary muscles of the eye for altering pupil size and shape of the lens. A branch of the seventh cranial nerve (facial) innervates glands in the mucous membrane of the mouth, palate, nose, nasopharynx, and lacrimal gland. Another branch of the seventh nerve, the chorda tympani, innervates the submandibular and sublingual salivary glands. The ninth cranial nerve (glossopharyngeal) innervates the parotid salivary gland and contributes to the tympanic plexus by a branch, the tympanic nerve. All of the above carry ganglionic synapses.

The tenth nerve (vagus) wanders about in the thorax and abdomen supplying the heart, lungs, digestive tract and its related organs, and the kidneys. The most extensively distributed of the parasympathetic fibers are those of the vagus.

Leaving the spinal cord in the ventral nerve roots of the corresponding spinal nerves of the second, third, and fourth sacral segments of the cord, the parasympathetic fibers of the sacral region go directly to the structure supplied. Sacral parasympathetic fibers innervate the terminal part of the gastrointestinal tract, the reproductive organs in the pelvic cavity, and genital structures.

The sensory fibers of the parasympathetic division follow much the same pattern as those of the sympathetic division. They travel from the viscera to reach their cell bodies, which are located either in the sensory ganglia of the cranial nerves or in the dorsal root ganglia of the sacrospinal nerves. The ventral axons enter the spinal cord to contribute to the formation of a local reflex arc or pass cranialward to higher centers of the autonomic nervous system (Fig. 16-25).

Autonomic System Functions

The role of the autonomic nervous system is to maintain constancy of composition of the body's internal environment or to serve as a *homeostatic mechanism.* This vital responsibility is indispensable. As mentioned earlier, much of the viscera is innervated by both types of fibers of the autonomic system. When fibers of one division are stimulated, activating a given

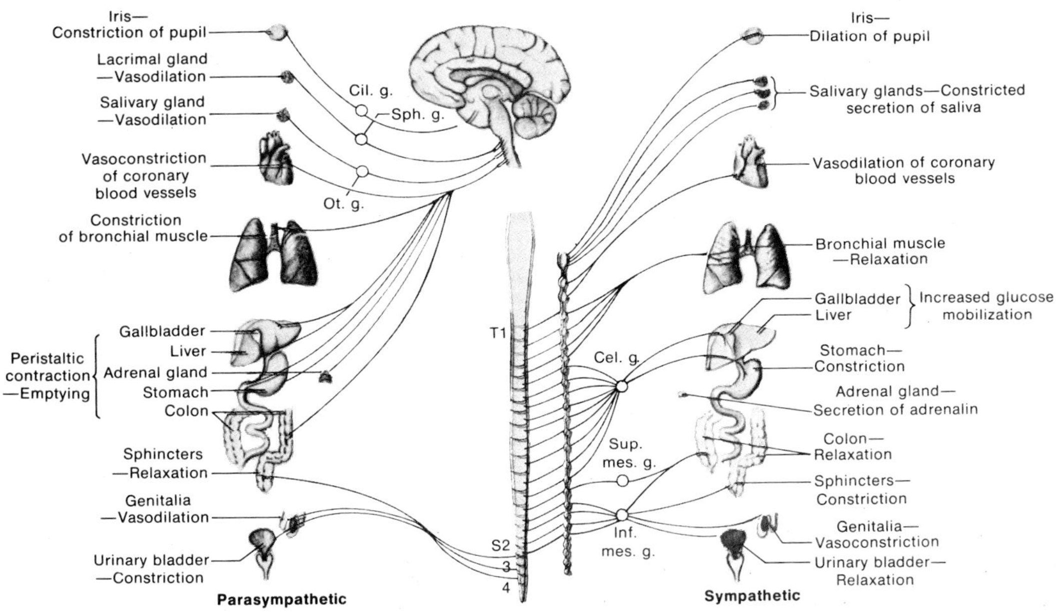

Figure 16-25. Diagram of the autonomic nervous system showing the chain of parasympathetic ganglia on the left and the sympathetic ganglia and their nerves on the right. Abbreviations: (Cil. g.) ciliary ganglion, (Sph. g.) sphenopalatine ganglion, (Ot. g.) otic ganglion, (Cel. g.) celiac ganglion, (Sup. mes. g.) superior mesenteric ganglion, (Inf. mes. g.) inferior mesenteric ganglion. The outer columns indicate functions of the sympathetic and parasympathetic divisions. (From Anderson, P.D.: *Clinical Anatomy and Physiology for Allied Health Sciences*, 1976. Courtesy of W.B. Saunders Co., Philadelphia.)

structure, just the opposite effect may be expected from the parallel fibers of the other divisions. For example, activities that operate when no expected emergency looms are handled by the parasympathetic division, while the sympathetic division is in abeyance.

On the other hand, in a case of sudden fright, the sympathetic division takes charge. To better understand this function, the reader might imagine himself as being in a sleeping bag on a dark night and suddenly hearing the sound of something walking through bushes and breathing heavily a short distance away. The autonomic system's efferent impulses of the sympathetic division would respond to the stimulus as follows:

1. Pupils dilate for better vision.
2. Hair stands up, or at least "gooseflesh" occurs.
3. Heartbeat accelerates for better blood supply.
4. Blood is directed to skeletal muscles for flight and fight.
5. Muscles tense for action.
6. Salivation ceases.
7. Respiration rate is increased for added oxygen and energy sources.
8. Digestion is slowed.
9. Activity of sexual organs is inhibited.
10. Bladder wall musculature is relatively relaxed, etc.

These vegetative functions of the two divisions of the autonomic nervous system have been studied in relation to stuttering more so than any other communicative disorder. There is little evidence to date to warrant a conclusion that much stuttering has its onset in frightening situations accompanied by a vigorous response of the autonomic system. Perhaps shocks or intense scares cause an existing dysfluency problem to manifest itself clearly, and the individual reads the many autonomic system reactions as a conditioned stimulus that sets off the stress on the communicative act interfering with speech fluency. In other words, the awareness to the individual that these physiological behaviors are taking place becomes associated with adversive, fluency-fracturing stress.

The phenomenon of stage fright, a universal problem, has also caused specialists to examine the autonomic nervous system to attempt to better understand the mechanism involved. Few students in a basic public speaking course in college are without the gnawing problem of stage fright during each speech before the class. Typically, the autonomic system gets a good workout by the students of each such class. Certainly this negative feedback (the physiological changes triggered by fear of the audience) could act as a strong reinforcer that things just are not going well, which in turn causes a more intense response by the autonomic system and so on in a vicious cycle. This is physiological speculation, but it seems somewhat plausible in view of the autonomic system's structure and known functional areas.

Other systems that no doubt are intricately related to the autonomic system and each other are the *limbic system* and the *reticular system*. These latter systems are profiled in Chapter 22, which pertains to "Applied Clinical Anatomy and Physiology." The reader is encouraged to compare the ANS with the limbic and reticular systems for possible evidence of integrative, control, and coordinating activities.

BIBLIOGRAPHY

Aronson, A.E.: *Clinical Voice Disorders*. New York, Thieme-Stratton, 1980.

Bossy, J.: *Atlas of Neuroanatomy*. Philadelphia, Saunders, 1970.

Chusid, J.G.: *Correlative Neuroanatomy and Functional Neurology*, 14th ed. Los Altos, CA: Lange, 1970.

Crouch, J.E.: *Functional Human Anatomy*, 3rd ed. Philadelphia, Lea and Febiger, 1978.

Darley, F.L., Aronson, A.E., and Brown, J.R.: *Motor Speech Disorders*. Philadelphia, Saunders, 1975.

Gardner, E.: *Fundamentals of Neurology: A Psychophysiological Approach*, 6th ed. Philadelphia, Saunders, 1975.

Hausman, L.: *Illustrations of the Nervous System*, Atlas III. Springfield, IL: Thomas, 1961.

Matzke, H.A., and Foltz, F.M.: *Synopsis of Neuroanatomy*, 3rd ed. New York, Oxford U Pr, 1979.

Netter, F.H.: *Nervous System*. Summit, NJ: CIBA, 1962, vol. I.

Truex, R.C., and Carpenter, M.B.: *Human Neuroanatomy*, 6th ed. Baltimore, Williams and Wilkins, 1981.

Chapter 17

THE AUDITORY SYSTEM

INTRODUCTION

THE ear, like the eye, is classified as a distance receptor. In other words, direct contact with a stimulus-object is not necessary to elicit stimulation of the receptor. Touch, pain, pressure, etc. are direct sensors. The two ears are only narrowly separated insofar as the actual organ of hearing is concerned. They are located within the temporal bones, the most intricate bones of the human body. But the ear is not always found in the head of all creatures. It is located on the thorax of some grasshoppers and the legs of some crickets.

Conventionally, the ear is divided into three major units for anatomic and clinical descriptions. Representing three cavities, with distinct boundaries for each, the cavities are arranged in a peripheral-central order. These are the *outer, middle* and *inner* ears. It seems a misnomer to designate these parts of the total ear structure as ears, but tradition has firmly seated this approach. Perhaps it is not misleading if one remembers that only the innermost aspect, the inner ear, is that portion housing the mechanism for transducing mechanical vibrations into neural impulses. In this chapter, the structure of these portions of the ear will be treated. But the auditory system must include the neural pathways also. A separate division traces the route of the neural tract to the cortex of the brain, and another section is concerned with the sound serving as the stimulus for hearing.

THE OUTER EAR

Two principal structures make up the outer ear. These are an exposed appendage and a canal piercing the side of the head. Under the earlier discussion of superficial anatomy of the head, these structures were first described. In analogy, the outer ear has the shape and function of a megaphone, funneling sound into the head where *hearing* may take place. Its parts are the auricle and external auditory canal.

Auricle

In sound reception, the outermost part of the external ear, the auricle or pinna, is not of real acoustic significance in humans. Man must turn his entire head to focus sound energy from a particular direction. Most animals are able to turn the ears, usually independently of each other. Cats and dogs may be observed sleeping, yet both ears appear alert. When faint sounds are detected, they turn the ears until maximum reception is obtained. Man and most mammals are not able to close the canal to prevent water or dirt from entering. Both seals and moles possess special valvular mobility for this purpose.

Therefore, for man, the auricle is an expanded appendage with quite limited functions. Its three relatively weak muscles, the anterior, superior, and posterior auricularis (see Fig. 13-12), are capable of moving the auricle only slightly. One may cup a hand behind an ear, reflexively in attempting to understand a speaker located at some distance away, but this may not be much more effective than placing a hand over the sternum to aid in throat clearing. (The psychological benefit is probably more important.) To a questionable degree, though, the auricle is a sound gatherer. It is also a sound deflector, shading one ear from a posteriorly located sound source and enabling the auditory mechanism to note the characteristic delay in arrival time of the waves. This permits directional perception of sound.

Elastic cartilage provides the framework for the auricle. The resiliency serves to prevent damage when the auricle is deformed by wrestling, headwear, sleeping on the side, or other reasons. It springs back to its characteristic shape because of the flexible cartilage. A rupture of the cartilage, however, from boxing or accidents may heal by an overgrowth of tissue, which may resemble a cauliflower. The auricle may then be described as a cauliflower ear. A seldom noted feature is that one ear may be lower than the other; this often goes completely unnoticed until

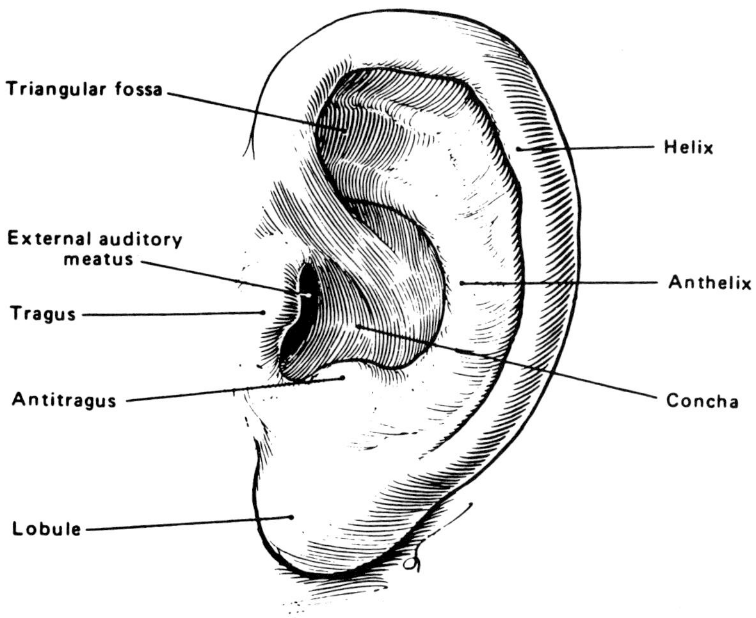

Figure 17-1. The auricle. (From *Dynamic Anatomy and Physiology*, 3rd ed., by L.L. Langley, I.R. Telford, and J.B. Christensen. Copyright © 1969. Used with the permission of McGraw-Hill Book Company.)

the individual begins to wear glasses. Adjustments of the frames must then be made to enable the lenses to sit level.

A number of landmarks may be readily recognized on the auricle. The folded rim forming the auricle is known as the *helix*; it begins above the canal orifice and extends around most of the edge. The helix stops at the lower part at the *lobule* or earlobe, the site for affixing decorative fixtures, earrings. Piercing of the lobule is not uncommon in some societies for both males and females. This landmark is fleshy and has relatively few pain endings, but it is vascular and may be used to take small blood samples. There is also a sebaceous gland in the earlobe that should be avoided when piercing the ear. An infection in the sebaceous gland can spread quickly and necessitate the use of antibiotics in the affected individual.

The small shield anterior to the canal orifice is the *tragus*. It may be depressed by the finger to occlude the canal from sound.

The spacious cavity leading into the canal orifice is the *concha* (shell-like). It is bounded by the tragus anteriorly, the *antitragus* below, and the *anthelix* behind. It is the anthelix that is usually absent or undifferentiated as a fold in individuals with lop ears. Surgical correction for lop ears essentially involves the construction of an anthelix from existing cartilage in the ear.

The *triangular fossa* is positioned in the upper front corner of the ear. It is bounded superiorly by the helix and inferiorly by the anthelix.

External Auditory Canal

Leading from the concha of the auricle

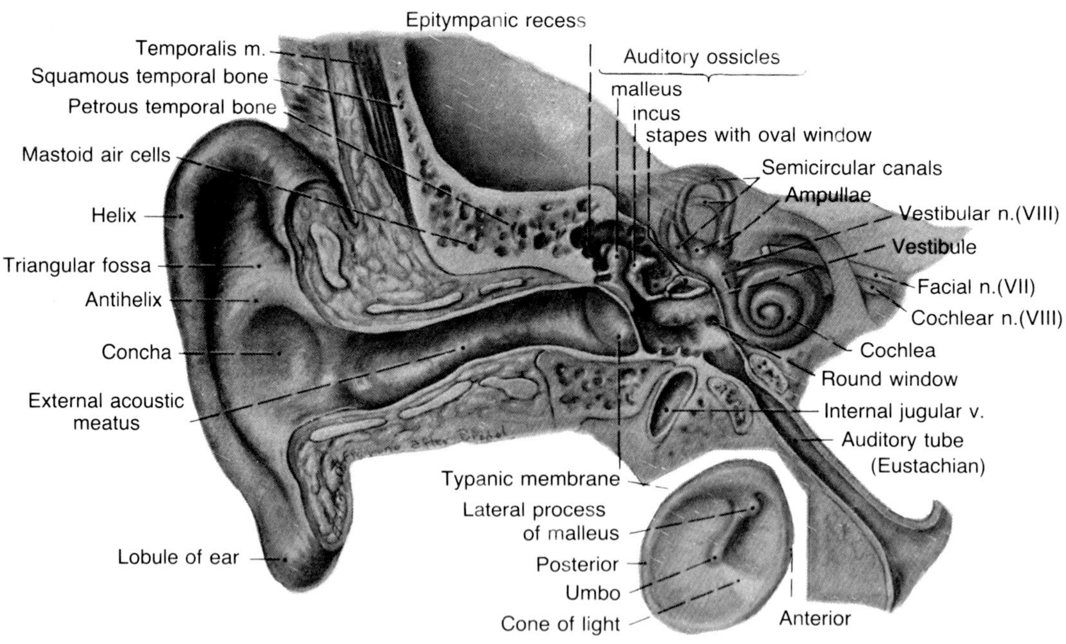

Figure 17-2. Frontal section through the outer, middle, and inner ear. Notice the bony portion of the external auditory meatus begins near the location of the tympanic membrane. (From Jacob, S.W., and Francone, C.A.: *Structure and Function in Man*, 4th ed., 1978. Courtesy of W.B. Saunders Co., Philadelphia.)

into the skull is the external auditory meatus or ear canal. Its length is about 2.5 cm ending medially at the tympanic membrane or eardrum. From a cross section, the ear canal is oval shaped and about a little over 0.5 cm in diameter, or less than the diameter of a pencil (Fig. 17-2).

In youth, the canal is relatively straight but becomes curved with progressive age to adulthood. To see the eardrum at the end of the canal, one must pull the auricle upward and backward to straighten the canal. Or, a speculum, a funnel-shaped instrument, may be introduced for efficient viewing of the tympanic membrane. This is the device found on the otoscope (a specialized ear flashlight) used by physicians and audiologists for inspecting the ear canal and tympanic membrane.

The outer aspect of the canal contains a cartilaginous framework as the concha leads into the canal. Medially, the temporal bone serves this purpose, since the canal size is governed by the size of the opening into the temporal bone in this part of the canal.

Along the internal surface of the canal are short, stiff hairs and wax-producing glands. These are protective agents. The wax glands secrete a dark substance known as *cerumen*. It has a bitter taste and helps discourage insects from entering the canal. Additionally, the cerumen prevents the canal from drying out. Temperature and humidity controls that protect the eardrum are also functions of the external meatus.

It is the responsibility of the external auditory canal to transmit sound to the drum membrane. In the process, it actually amplifies slightly those frequencies that correspond to the *resonance frequency* of the external auditory tube. Hence, the tube serves as a resonator. This special range of frequencies is from 2000 Hz (cycles per second) to 5500 Hz. Consequently, human hearing is more acute for these frequencies than if the drum membrane were at the opening of the canal on the surface of the head. This range, as might be expected, includes those frequencies of greatest acuity for human reception.

Obstruction of the external auditory meatus interferes with sound transmission to the eardrum. Occasionally, excessive wax secretions will occlude the canal, necessitating removal. Children and a few adults have had numerous foreign objects removed from their ears; the canal almost seems to invite some persons to insert everything from rocks and beans to safety pins (opened, no less). Insects in the canal may irritate more than impede sound transmission. Some otologists have photographed maggots in the canal. Also, allergies and infections cause inflammation of the lining, which, along with secretions, may occlude the canal.

At the medial end, the canal is capped by the tympanic membrane or drum membrane, as indicated previously. This membrane represents the beginning of the middle ear; therefore, the tympanic membrane is not considered a part of the outer ear.

THE MIDDLE EAR

Tympanic Membrane

The tympanic membrane seals the inner end of the external auditory meatus. It appears as a pearl-gray membrane, oval in shape. It is composed of an outer layer, middle layer, and inner layer. The outer skin layer of the epithelium covering the auricle is continuous in the external canal and forms the outer layer of the tympanic membrane. Specifically, it is squamous (pavement-like) epithelium. The medial layer is connective tissue. The inner layer is mucous membrane, which lines the mouth, nose, and pharynx. It is continuous with the linings of the walls of the middle-ear cavity and the nasopharyngeal cavity through the auditory tube.

An *annulus*, a strong ring of fibrous tissue, anchors the margins of the tympanic membrane to the bony wall of the canal in the *tympanic groove*. Sometimes this attachment point is described as a ring of bone. Around the margins of the membrane, blood vessels enter and are quite responsive indicators of infectious conditions in the area. In such cases, they appear prominent and alter the coloring of the drum.

A very thin membrane, the tympanic membrane is semitransparent (Fig. 17-4). When viewed from the side, it has the shape of a loudspeaker cone, facing the external auditory meatus. Faint radiating lines in the eardrum are formed by its middle fibrous tissue layer. In addition to the radial fibers, there are two sets of circular fibers. They are arranged near the margins of the drum and near its center. Interestingly, they exhibit the same design that one might find in the cone of a typical loudspeaker, which has tiny accordion wrinkles near its periphery and at its center in order to respond more efficiently over a wide frequency range.

The major portion of the eardrum is known as the *pars tensa*, which is the tense area of the drum. A second portion, the

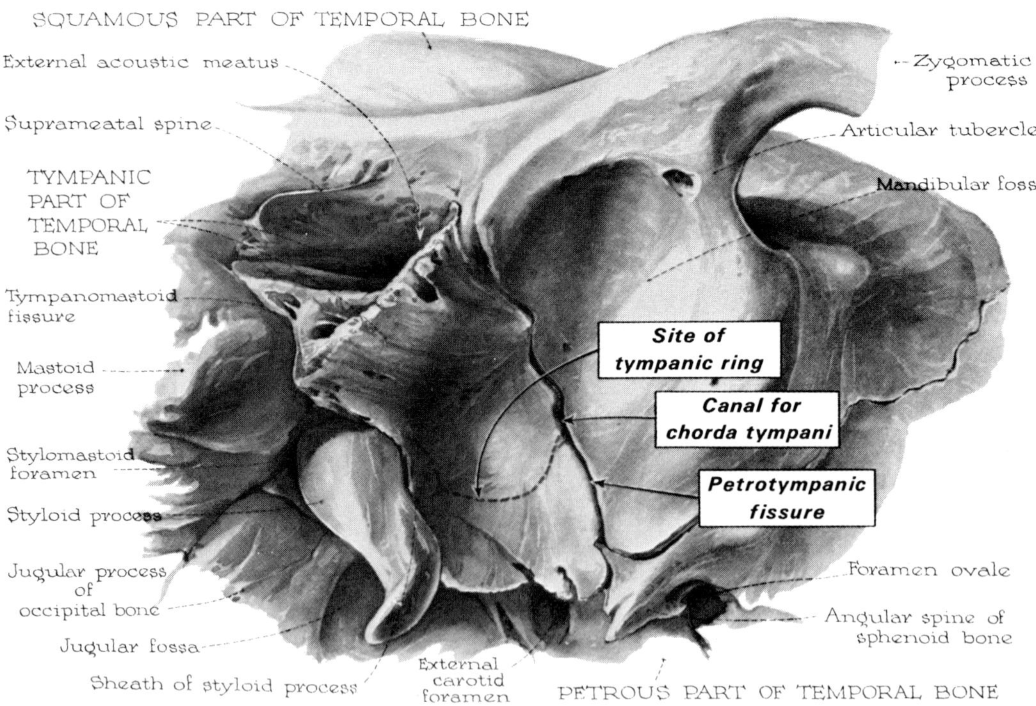

Figure 17-3. Inferior surface of the temporal bone showing the site of the annulus (dotted line) in the external auditory meatus. (From Anson, B.J., and Donaldson, J.A.: *Surgical Anatomy of the Temporal Bone and Ear*, 3rd ed., 1981. Courtesy of W.B. Saunders Co., Philadelphia.)

pars flaccida, is but a small section at the top of the drum. The middle layer of the tympanic membrane is found in the pars tensa but not in the pars flaccida. On the lower aspect of the pars tensa is a well-recognized landmark, a *reflected cone of light*. This is present due to the angular surface of the eardrum reflecting the examiner's light source.

Another feature that may be observed on the drum is the outline of a portion of the *malleus*, the first bone of the ossicular chain. A process of the malleus, the *manubrium* (handle), extends to the middle area

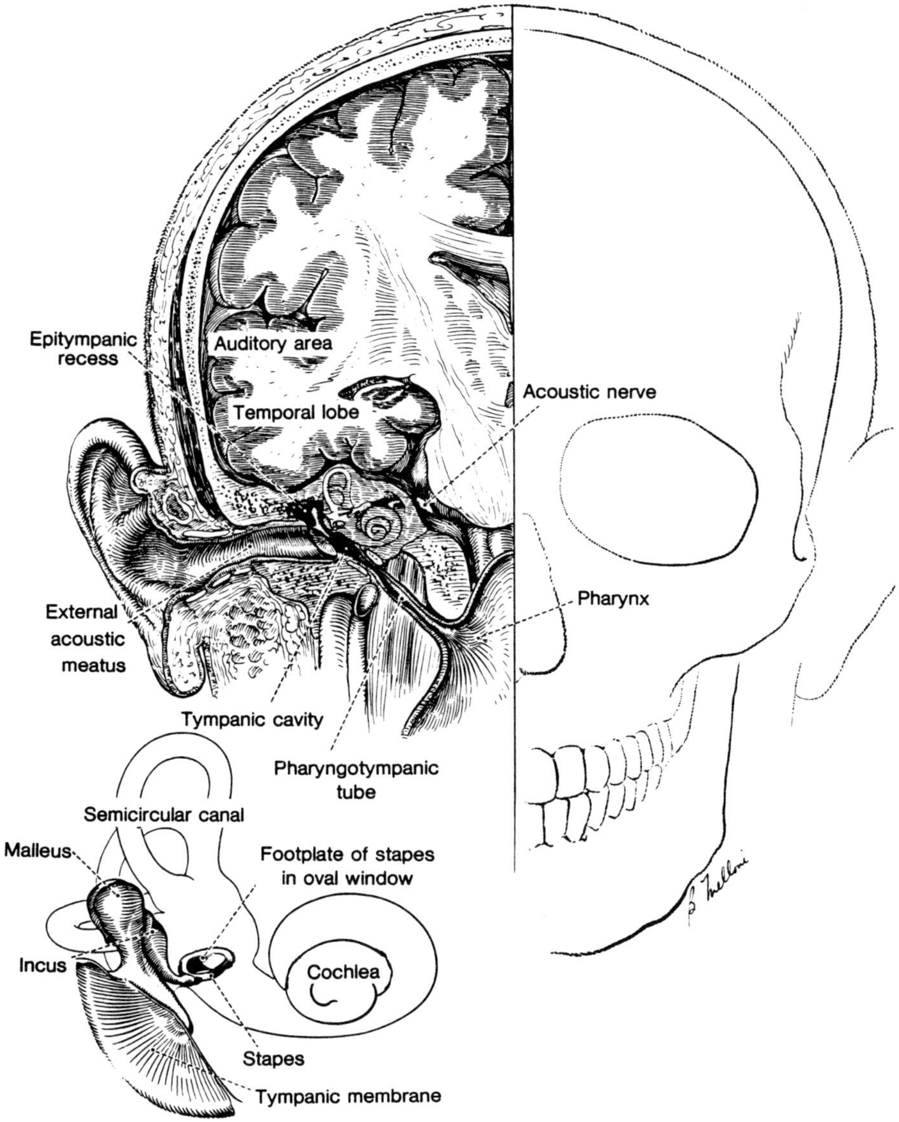

Figure 17-4. Coronal section showing the divisions of the ear in relation to adjacent structures. The tympanic membrane, ossicles, and cochlea are enlarged at the lower left. (Reproduced from: "Some Pathological Conditions of the Eye, Ear and Throat." Courtesy of Abbott Laboratories.)

of the eardrum in contact with the drum. The tip of the manubrium ends at the central area of the drum. This area represents the peak of the drum's depression and is known as the *umbo*. Fibrous tissue attaches the eardrum to the manubrium as well as to the *lateral process* of the malleus. Thus, movement of the tympanic membrane results in movement of the malleus. The outline of the malleolar tympanic membrane process appears like the hour hand of a clock pointing to the eleven-o'clock position in the left ear and in the one-o'clock position in the right ear. These malleolar processes of the two ears point anterosuperiorly in the same plane as horns might be positioned if humans had such.

The Auditory Ossicles

The three ossicles of the middle ear are the smallest bones in the body. Together with the tympanic membrane, they effectively comprise a mechanism for transducing sound waves in the air into physical vibrations (at the eardrum), which are transmitted to the inner ear. The vibrations are again transduced, this time into neural impulses. Specifically, the ossicular chain transfers the mechanical vibrations of the drum across the air-filled cavity of the middle ear to the oval window of the inner ear.

Malleus

The malleus, the first bone behind the eardrum, is in direct contact with the eardrum as described above. Its process, the manubrium, is connected through a neck to an enlarged portion, the *head* of the malleus. From a certain view, the malleus does have the appearance of a crude hammer (Fig. 17-5).

Incus

Through the head of the malleus, movement is transmitted to the incus, the second ossicle in the chain. Its triangular shape has given it an alternate name, the anvil, to coincide with the hammer. An articular facet on its major part or body at the root of its long process meets the head of the malleus. A short process extends off at a 90-degree angle for a limited dis-

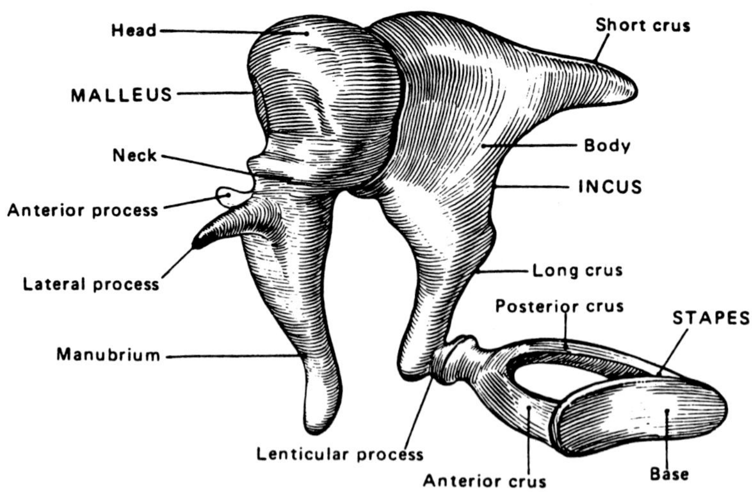

Figure 17-5. The ossicles. (From *Dynamic Anatomy and Physiology*, 3rd ed., by L.L. Langley, I.R. Telford, and J.B. Christensen. Copyright © 1969. Used with the permission of McGraw-Hill Book Company.)

tance. The *long process* or *crus* (neck) of the incus ends in a sharp turn termed the *lenticular process*. The short lenticular process points toward the inner wall of the middle cavity, ending in an articular facet.

Stapes

The third auditory ossicle, the stapes, the most delicate of the ossicular chain, has the distinct appearance of a stirrup. Its head articulates with the lenticular process of the incus. An *anterior crus* and a longer *posterior crus* end on an oval-shaped *footplate*. The two crura add strength to the linkage. The footplate or base fits within an oval-shaped opening, the *oval window* in the inner wall of the tympanic cavity and is held in place by a ring of ligamentous fibers called the *annular ligament*. By means of the stapes, vibrations are passed from the incus through the oval window into the fluid-filled cavity of the inner ear.

Suspension of the Ossicles

At least five ligaments are responsible for holding the ossicles in their respective positions within the middle-ear cavity. The stapes and incus each have one ligament, while the malleus has a total of three ligamentous suspensions. Sometimes the incus contains two ligaments, however. All of these serve to anchor the ossicular chain. The ligament of the stapes differs somewhat in that it is a ring, the annular ligament described above (Fig. 17-6).

Ossicular Movement

In response to frequencies under 800 Hz, the ossicles vibrate as a unit to transfer vibrations from the eardrum to the oval window of the inner ear. For frequencies below 800 Hz, the leverage afforded by the way the ossicles are put together may help in providing a more suitable stimulus in combination with intensity adjustments to the fluid of the inner ear. But intensity transmission of the sound is another story. The actual displacement of the eardrum is extremely small in response to sounds of conversational loudness. This displacement has been equated

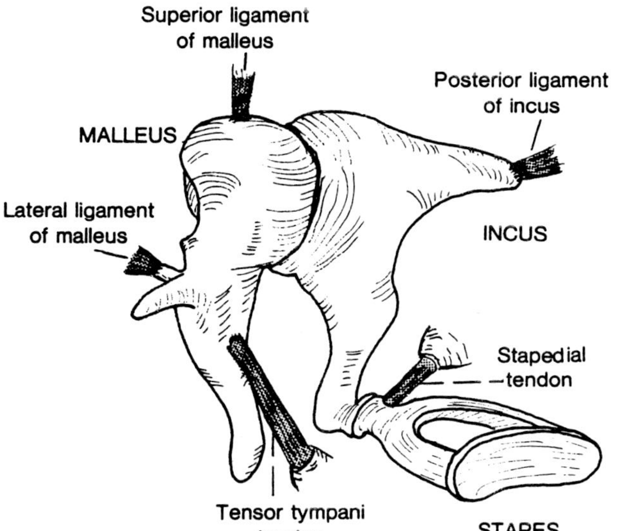

Figure 17-6. The ossicles. The suspensatory ligaments and tendons attaching to the ossicles are shown. (Modified from *Dynamic Anatomy and Physiology*, 3rd ed., by L.L. Langley, I.R. Telford, and J.B. Christensen. Copyright © 1969. Used with the permission of McGraw-Hill Book Company.)

to about 1/100 millionth cm, the diameter of a hydrogen molecule. Whispering would cause the eardrum to move to a much less degree.

Due to the combination pivot and lever connections between the ossicles, an efficient system for amplifying the power of the vibrations is present. This is necessary since the air medium carrying the vibration energy to the drum is much less dense than the fluid medium receiving the energy. Oscillatory movements at the oval window are less in amplitude but greater in power by about 2:1. Thus, the drum and the ossicles make up what is known as an *impedance-matching mechanism*. The fluid of the inner ear offers much more resistance to vibratory motion than to air, necessitating some effective coupling device to change wide but relatively weak airborne oscillations to sufficiently strong oscillations in fluid. The ossicular chain along with the eardrum is the necessary impedance-matching mechanism.

Ossicular Muscles

Two muscles act on the ossicular chain. The outer two ossicles receive the insertion of these two small muscles.

Tensor Tympani

The tensor tympani arises in a bony semicanal above the auditory tube. It travels toward the ossicular chain on the medial side of the tympanic cavity and ends in a slender tendon, which emerges into the cavity to insert on the malleus (Figs. 17-6 & 17-7). It makes a bend to insert into the malleus at the root of the manubrium. The fifth cranial nerve, the *trigeminal*, innervates the tensor tympani muscle.

Stapedius

The stapedius is a very small muscle. It is also contained in a bony cavity but is counterpositioned to the tensor's canal. A bony pyramid marks the location of the cavity for the stapedius. From the hidden muscle, a tendon enters the cavity of the middle ear and inserts into the posterior crus and/or the neck of the stapes (Figs. 17-6 & 17-7). A branch of the *facial* cranial nerve innervates the stapedius muscle.

Role of the Middle-Ear Muscles

The tensor tympani and stapedius are usually described as protectors of the inner ear from excessively intense sounds that impinge upon the eardrum. It is true that the muscles can stiffen the positions of the respective ossicles and check the extent of their oscillations during sound transmission from the eardrum to the inner ear. In response to a loud sound such as a jackhammer, the stapedius may reflexively contract and pull the head of the stapes posteriorly to reduce its effective coupling with the inner ear.

Some persons can voluntarily contract the tensor tympani, pulling the eardrum medially to stiffen it. This may be accomplished also in association with contraction of muscles around the orbital cavity. Nonetheless, the ear's sensitivity is reduced as the middle-ear muscles fixate the structures to which they are attached. However, the tensor tympani is not active in reflexive contraction to sounds softer than those producing a startle effect.

The action of the stapedius muscle in guarding the inner ear may be compared to some extent with the reflexive response of the iris to intense light. But sudden, high-level sounds such as gun blasts or explosions are able to reach the ear before either of the two middle-ear muscles have time to react. These intense sounds may permanently damage the inner ear's sensitivity at certain frequencies.

A second role of the middle-ear muscles conventionally described in textbooks is to increase the sensitivity of the ear to faint sounds, making the eardrum more responsive to airborne vibrations and the ossicular chain more efficient for transmitting vibrations to the oval window. In essence, it is purported that they take the slack out of the chain and tense the tympanic membrane for improved function in the conduction of sound. This role is not well supported by the literature. In fact, the topic of improved auditory activity based on middle-ear muscle function is sparsely treated. There is evidence that the tensor tympani possesses a proprioceptive circuit for maintaining tonus, but the benefit to acuity is not clear.

The Middle-Ear Cavities

The major cavities of the middle ear are, in reality, composed of the middle-ear cavity proper and two extensions of this principal cavity. The space containing the ossicular chain is the central cavity and is sometimes called the *tympanum*. It is very small, however, contrary to the impression given by drawings in books. Its height is its greatest dimension, and this comes to just over 1 cm. The ossicles almost fill the middle cavity so that the air volume surrounding the ossicles is small; the capacity is about seven drops of water.

The lining of the middle-ear cavity is mucous membrane. The cavity contains a roof, floor, and four walls. Five openings or membranous windows permit the middle-ear cavity to relate to other cavities. Direct openings are provided to the auditory tube and to the air cells of the mastoid bone. Through the oval and round windows on the bony inner wall, communication to the inner ear is allowed, and the tympanic membrane permits contact with the external auditory meatus.

It should be pointed out that there are two parts of the tympanic cavity, the tym-

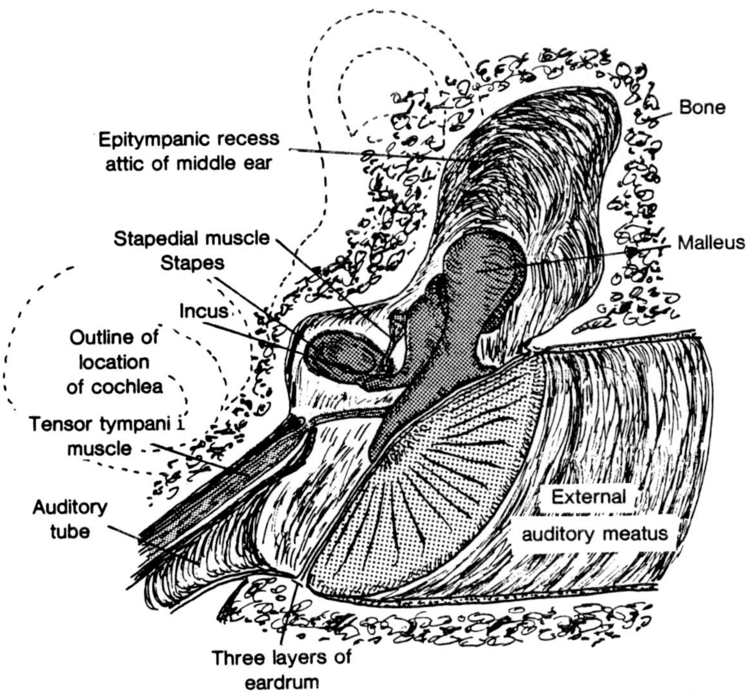

Figure 17-7. The middle-ear cavity showing the location of the two muscles associated with the ossicles.

panic cavity proper and the *attic* or *epitympanic recess*, which extends above the eardrum and houses the upper parts of the malleus and incus (Fig. 17-7).

The posterior wall of the middle-ear cavity contains a small conical eminence, the *pyramid*, which encapsulates the stapedius muscle. The tendon of the stapedius emerges into the middle-ear cavity to attach onto the stapes. An opening from the epitympanic recess onto the posterior wall communicates with the mastoid air cells in the temporal bone; this communication route is the *tympanic antrum* or aditus.

The roof of the middle-ear cavity is separated from the cranial cavity by a thin layer of bone known as the *tegmen tympani*. The floor of the cavity is sometimes designated as the jugular wall, since the jugular vein forms a bulb as it turns just under the thin bony floor of the middle-ear cavity.

The lateral wall is made up principally by the tympanic membrane but also includes a ring of bone at the medial end of the external auditory meatus, attaching the margins of the eardrum. This ring is interrupted by a notch, the notch of Rivinus, at its superior part.

The medial wall is a thin bony boundary between the middle and inner ears. It is marked by two openings and two notable prominences. The pair of openings is comprised of the *oval* and *round windows*. As described earlier, the footplate of the stapes fills the oval window for transmitting eardrum movement to the fluid under the footplate. The round window is named for its shape. A membrane closes this defect in the inner wall, preventing fluid from escaping the inner ear. The

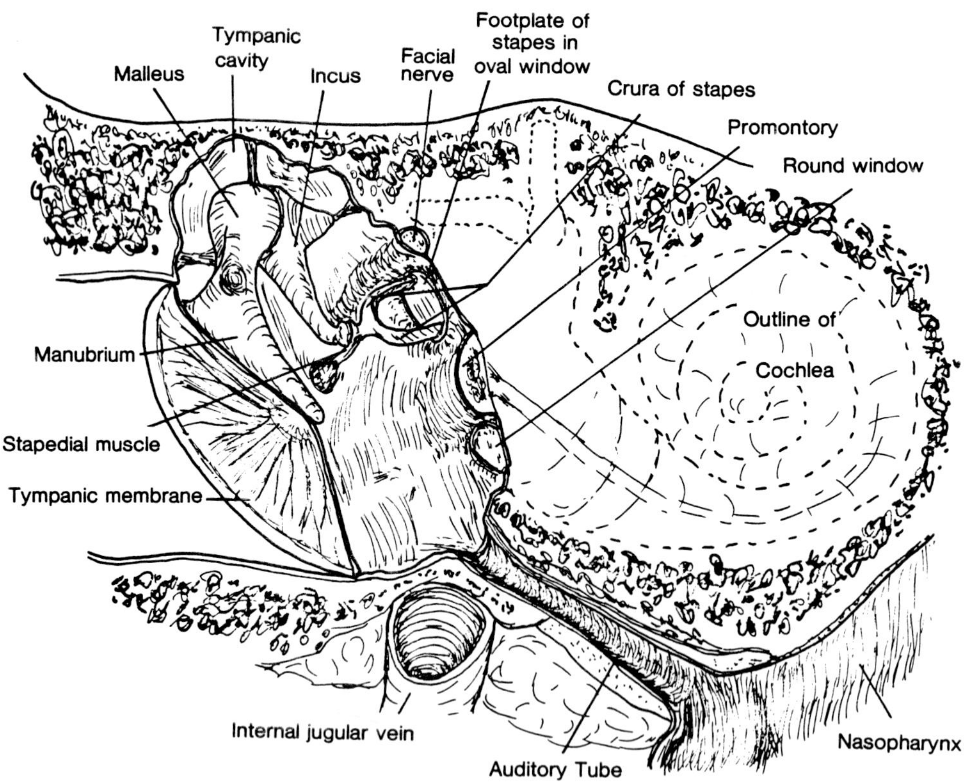

Figure 17-8. The middle-ear cavity emphasizing aspects of the medial wall.

promontory, a rounded prominence, separates the superiorly placed oval window from the inferiorly placed round window. A second prominence is formed by the canal housing the facial nerve. The path of the facial canal prominence enters above the level of the tensor tympani canal and arches across the inner wall of the middle-ear cavity, passing just above the oval window. It continues on to the posterior wall, curving nearly vertically and inferiorly to leave the cavity along the mastoid wall. A branch arises out of the facial nerve as it descends into the temporal bone. This is the chorda tympani nerve, which rises back into the middle-ear cavity in its own canal and transverses the cavity across the medial surface of the tympanic membrane on top of the manubrium of the malleus before exiting on the anterior wall on its way to the mouth (Fig. 17-8).

The final wall of the middle-ear cavity is the anterior wall. This wall is also termed the carotid wall, since it separates the cavity from the internal carotid artery, which travels in its own canal on the other side of this wall. On the upper aspect of the wall, the opening of the canal encapsulating the tensor tympani muscle is located. The orifice of the auditory tube is found breaking the anterior wall at a point slightly above the level of the floor of the cavity.

The Auditory Tube

Through the auditory tube, also known as pharyngotympanic and eustachian tube, the middle-ear cavity communicates with the nasopharynx. The mucosal lining of the middle-ear cavity and pharynx are continuous through this tube. It is about 3.75 cm in length, traveling from the ear to the nasopharynx. On its way, it drops about 35 degrees while angling medialward about 45 degrees. As it leaves the middle-ear cavity, its walls are composed of bone, but after about 1.25 cm along the way, the bone gives out and cartilage continues on to the nasopharynx. Ending under the mucous membrane of the pharynx, the cartilage creates a bulge caused by an elevation of the mucosa, the *torus tubarius*, at the termination of the cartilaginous portion of the auditory tube. Here, the diameter of the tube is greatest. Normally, the tube is closed so that sounds of breathing and speech are not piped into the middle ear.

Upon swallowing or yawning, the torus tubarius is distorted to open the tube. This is accomplished in part by the salpingopharyngeus and the dilator tubae (a small group of short muscle fibers at the orifice of the tube). Other muscles associated with opening the tube are those connected with palatal action, the tensor and levator muscles.

The major function of the auditory tube is to permit air to enter and leave the middle-ear cavity. This serves to equalize air pressure on either side of the eardrum. The importance of this function is demonstrated each time one takes a plane ride, which results in changes in atmospheric pressures on the eardrum. These changes are balanced by air entering or leaving the middle-ear cavity via the auditory tube. A second use of air in the middle-ear cavity is for an oxygen supply to the mucous membrane of the cavity. If blocked, the membrane will absorb the oxygen trapped in the cavity, decreasing the pressure on the medial aspect of the eardrum. The drum is then put under pressures from the atmosphere, which causes drum retraction and interferes with its movement. Pain may arise from this condition in some cases. If unchecked, the mucous membrane will secrete excess fluid, further complicating the condition. It is thought that the tube is not an efficient drainage system for the cavity, although in this role, it is able to evacuate a portion of the middle-ear cavity to some extent.

In unusual cases, the orifice of the auditory tube will remain open. Reactions to certain drug therapy or sudden losses of weight may be the primary causes. The tube is chronically patent (*patulous*), or open, and transmits breathing and speech sounds directly into the ear cavity, a highly annoying condition. By contrast, the tube that does not open normally is known as a nonpatent tube and may require mechanical inflation to avoid pain and fluid collection.

The auditory tube may be blamed for many middle-ear infections. Bacteria may enter the middle-ear cavity through this tube from the nasopharynx. Due to the limited volume of the middle-ear cavity, the space may fill rapidly with fluid from mucous membrane secretions during an infectious invasion. The condition, *otitis media*, may cause both pain and hearing loss. If the fluid is made up of pus, it is termed *purulent otitis media*. If the fluid contains no bacteria, the condition is non-purulent otitis media. A fluid line or meniscus may be seen through the tympanic membrane if the middle cavity is only partially filled. An earache is the common symptom of middle-ear infection. The drum may be distended from pressure created by accumulation of fluid in the cavity. It is possible for the drum to rupture if the pressure continues to build and the auditory tube remains closed. An additional danger is that the tiny ossicles may be damaged or destroyed if the condition is recurring or chronic.

THE INNER EAR

The cavity forming the inner portion of the hearing organ occupies one cavity compartment, the balance organ occupies another compartment, and the two are joined by a vestibule. These two cavity systems comprising the hearing and equilibrium senses are fluid filled. The cavity system forming the end-organ for hearing is known as the *cochlea*. The other cavity system concerned with the sense of balance is termed the *vestibular apparatus*. Together they are referred to as the inner ear.

The cochlea and vestibular apparatus may be thought of as being adjoining cavity systems; they are housed in the temporal bone. The cavities are complex, winding as a series of encircling tunnels and forming labyrinthine bony canals. The size of the hollowed-out spaces in the temporal bone is actually quite small; they are typically illustrated in textbooks highly enlarged, giving the impression of rather large structures. The size of the cochlea is about the size of a pea (Fig. 17-9).

The *bony labyrinth* is housed in the petrous portion of the temporal bone. Encased within the bony labyrinth is the *membranous labyrinth*, which includes several membranous sacs and ducts with interconnections. This latter labyrinth may be thought of as a separate cavity, suspended in the middle of the bony labyrinth. It contains its own fluid system. There is both a cochlear and vestibular membranous labyrinth. The two portions communicate by way of the *ductus reuniens*. Each will be described in separate divisions. Surrounding the fluid-filled membranous tube forming the delicate membranous labyrinth is the *perilymph*, another fluid system. The perilymph (*peri* = outside) is analogous to the cerebrospinal fluid, which acts as a shock absorber for the central nervous system.

The Vestibular Division

The vestibular portion or equilibrium sense organ of the internal ear is sometimes referred to as the posterior labyrinth, since it is situated posteriorly to the cochlea. Additionally, the vestibular division of the internal ear lies in the lateral portion, the cochlea being more medially positioned in the petrous portion of the temporal bone. There are three circular-shaped canals, the *semicircular canals*, plus the *utricle* and *saccule* (two small sacs) included in the vestibular apparatus.

That portion of the bony labyrinth forming the vestibular apparatus is marked by a relatively enlarged entrance, the *vestibule*, which is a common entrance to both the vestibular and cochlear ducts. From the vestibule, the three semicircular canals arch around at various angles. At one end of each of the semicircular canals is a swelling termed an *ampulla*.

Each semicircular canal is named after the plane or position in which it lies. They are named the *superior, lateral*, and *posterior canals*. The arrangement permits them to be placed at right angles to each other. Both the superior and posterior semicir-

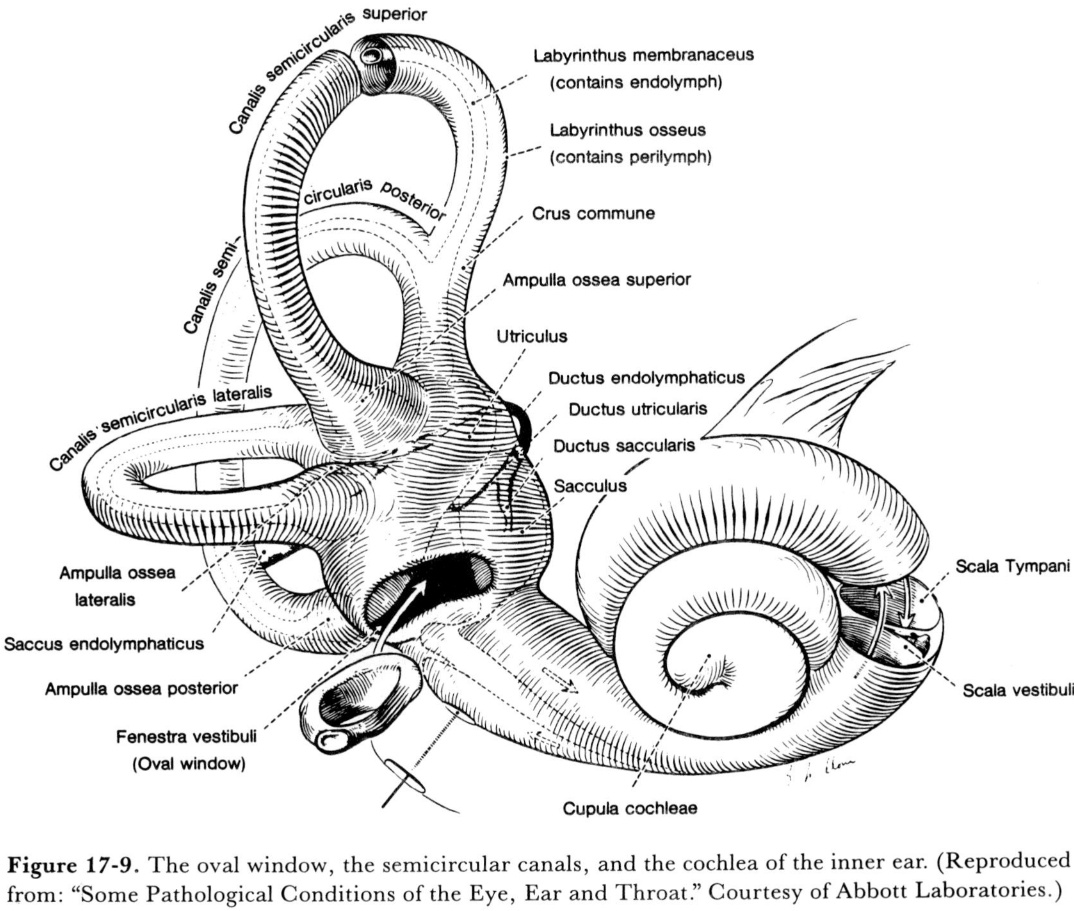

Figure 17-9. The oval window, the semicircular canals, and the cochlea of the inner ear. (Reproduced from: "Some Pathological Conditions of the Eye, Ear and Throat." Courtesy of Abbott Laboratories.)

cular canals are vertical or near vertical and provide sensation from movements of the head related to the vertical plane, as in nodding the head to and fro or from side to side. The lateral or horizontal canal is activated when the head is bent forward about 30 degrees and the head is rotated. The superior and posterior canals unite near the vestibule so that a total of five orifices are needed to join the bony semicircular canals with the vestibule. Each canal is lined with membranous ducts of the same shape but only one-fourth the diameter of the bony canal. These ducts are part of the membranous labyrinth (Fig. 17-10).

The membranous ducts open into a membranous sac within the vestibule, the utricle. Through a small duct, the utriculosaccular canal, the utricle communicates with another smaller membranous sac within the vestibule, the saccule. Two passages lead from the saccule. One is the *ductus reuniens* (Hensen's duct), referred to earlier as the communication between the membranous labyrinth of the vestibular apparatus and the cochlear duct. The other is the *endolymphatic duct*, which ends in a blind pouch or sac on the posterior surface of the petrous portion of the temporal bone. Here, the *endolymphatic sac* lies in contact with the dura mater of the brain case.

As mentioned previously, this entire

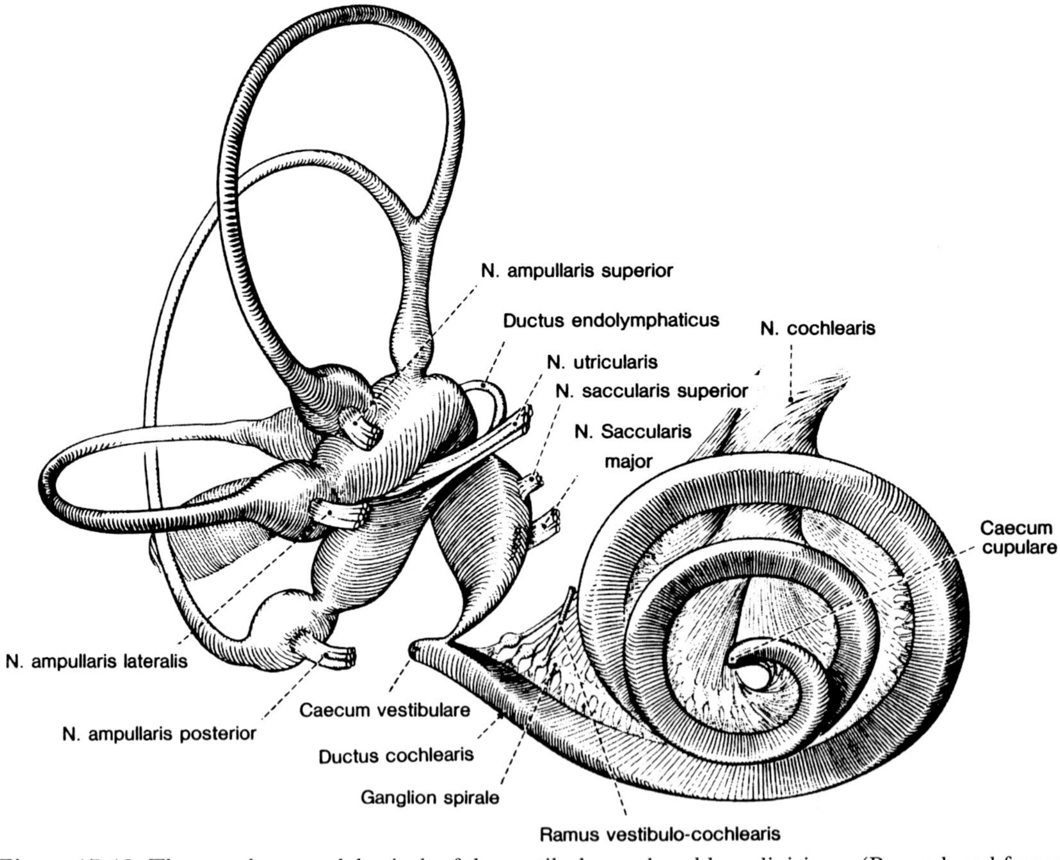

Figure 17-10. The membranous labyrinth of the vestibular and cochlear divisions. (Reproduced from: "Some Pathological Conditions of the Eye, Ear and Throat." Courtesy of Abbott Laboratories.)

membranous labyrinth is filled with endolymph (*endo* = inside) and is surrounded by perilymph. It is the responsibility of the perilymph to separate the membranous labyrinth from the bony labyrinthine walls. To keep the membranous labyrinth from swinging against the bony walls, small strands of connective tissue anchor the membranous canals within the semicircular canals. A similar provision is made to keep the utricle and saccule in place. Perilymph is produced by the epithelium lining the walls of the bony labyrinth (Fig. 17-11).

The utricle functions in response to the pull of gravity and acceleration. It is thought that the saccule also shares this responsibility. In contrast, the semicircular canals react to all movement involving turning in space. Body position seems to be an important part of the utricle's role in the vestibular apparatus.

Neuroceptors

A total of five sites in the vestibular system are specialized for detecting movement of the fluid contained in the vestibular labyrinth. These are the neuroceptors. In the utricle, a special chamber is pro-

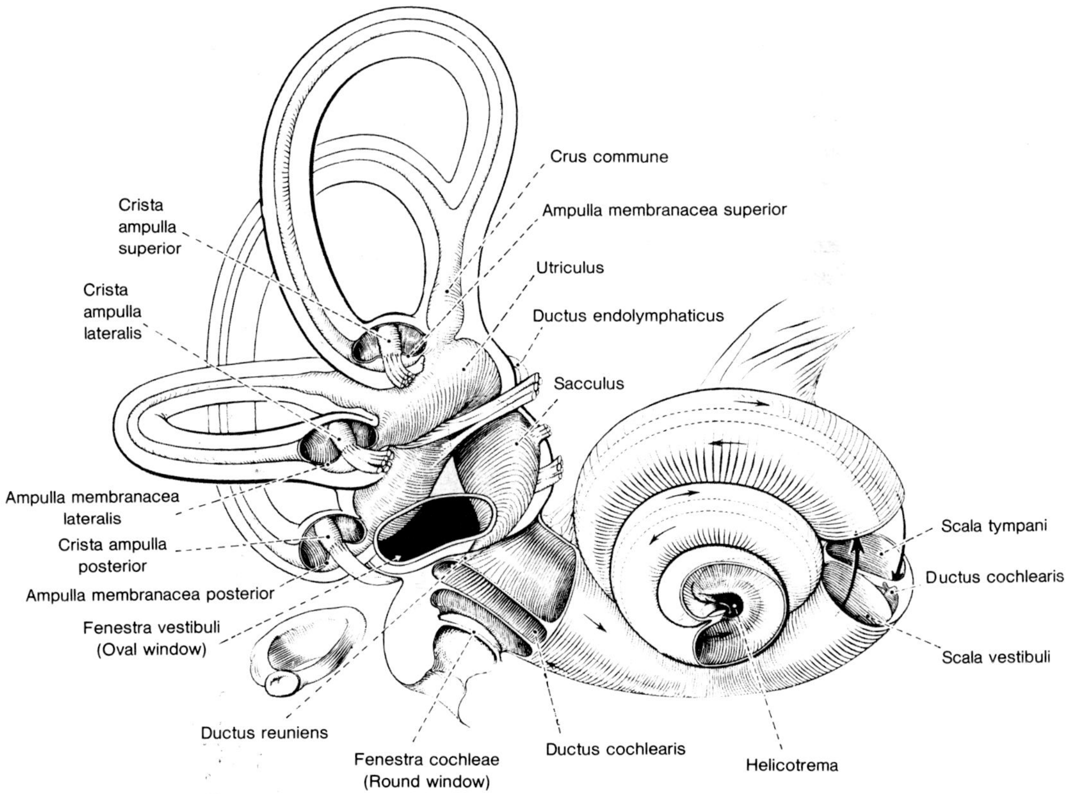

Figure 17-11. The vestibular apparatus and cavities of the cochlea. Notice that the round window is associated with the scala tympani. (Reproduced from: "Some Pathological Conditions of the Eye, Ear and Throat." Courtesy of Abbott Laboratories.)

vided for housing the *macula acustica utriculi* or one of the neuroceptors. Likewise, the saccule contains a neuroceptor, the *macula acustica sacculi*. A single neuroceptor, termed a *crista*, is found in each of the three ampullae of the semicircular canals.

The *crista ampullaris* is responsible for detecting movement or pressure changes in the endolymph of the semicircular canals. The cristae present a mound of gelatinous material covering it. Movement of the endolymph in the semicircular canal stimulates the hair cells since each crista blocks the endolymph to intercept flow changes. A quick turn of the head causes the fluid to lag behind due to inertia, resulting in pressure changes from movement of the endolymph. To visualize this principle of inertia, one can cover the bottom of a large, flat pan with water, set it on a table and give it a cautious, quick push. The reaction of the water illustrates the fluid motion of the endolymph (Fig. 17-12).

Nervous filaments transmit the neural impulses from the hair cells of the crista ampullaris and maculae to the brain stem via the eighth cranial nerve, the vestibulocochlear nerve. A vestibular ganglion on the eighth nerve is located at the inner end of the internal auditory meatus. Here are contained the cell bodies of the *primary neurons* serving the vestibular neuroceptors. These are bipolar cells, the peripheral process bringing the impulse from the hair cell and the central process transmitting the impulse to the vestibular nuclei of the medulla. *Secondary neurons* provide impulses (1) to the spinal cord, for reflexive adjustments of posture and movements; (2) to the nuclei of the third, fourth, and sixth cranial nerves, enabling reflexive eye movements that correspond with body movements; (3) to centers of salivation and vomiting in the brain stem associated with these actions, as a response to dizziness; (4) to the cerebellum, for coordination of movement and equilibrium based on vestibular sensations; and (5) probably to the sensory cortex via the thalamus, for awareness of body position and velocity changes.

Functional Considerations

The consequences of the two vestibular systems operating in disharmony may present a grave condition in extreme cases. Such a condition may arise if one vestibular system feeds motion data unlike that coming from the contralateral system. It seems hardly feasible that one system should be stimulated as a result of movement and leave the other system idle. The entire reflex mechanism for control of equilibrium is upset, to say nothing of the disturbance to the patient. The term *vertigo* is used to describe such conditions in which there is a hallucination of motion. (Dizziness is used to describe the patient who feels light-headed, faint, or confused and refers to symptomatology that is circulatory or psychoneurotic, not vestibular, in origin.) Surgical destruction of one offending vestibular apparatus can eliminate the malady; individuals can function satisfactorily with only one labyrinthine apparatus.

It should be remembered that the vestibular system is but a part of the mechanism enabling a person to perceive his position in space. Instant integration of neural impulses from the eyes and proprioception organs as well as from the vestibular apparatus establish a relation to orientation in space. How can it be determined whether an elevator is going up or down without looking at the floor indicator? Pressure changes on the bottom of the feet and a response by the semicircular canals provide adequate data to the brain in the absence of visual clues.

Almost every child has tested the reaction of his equilibratory mechanism by

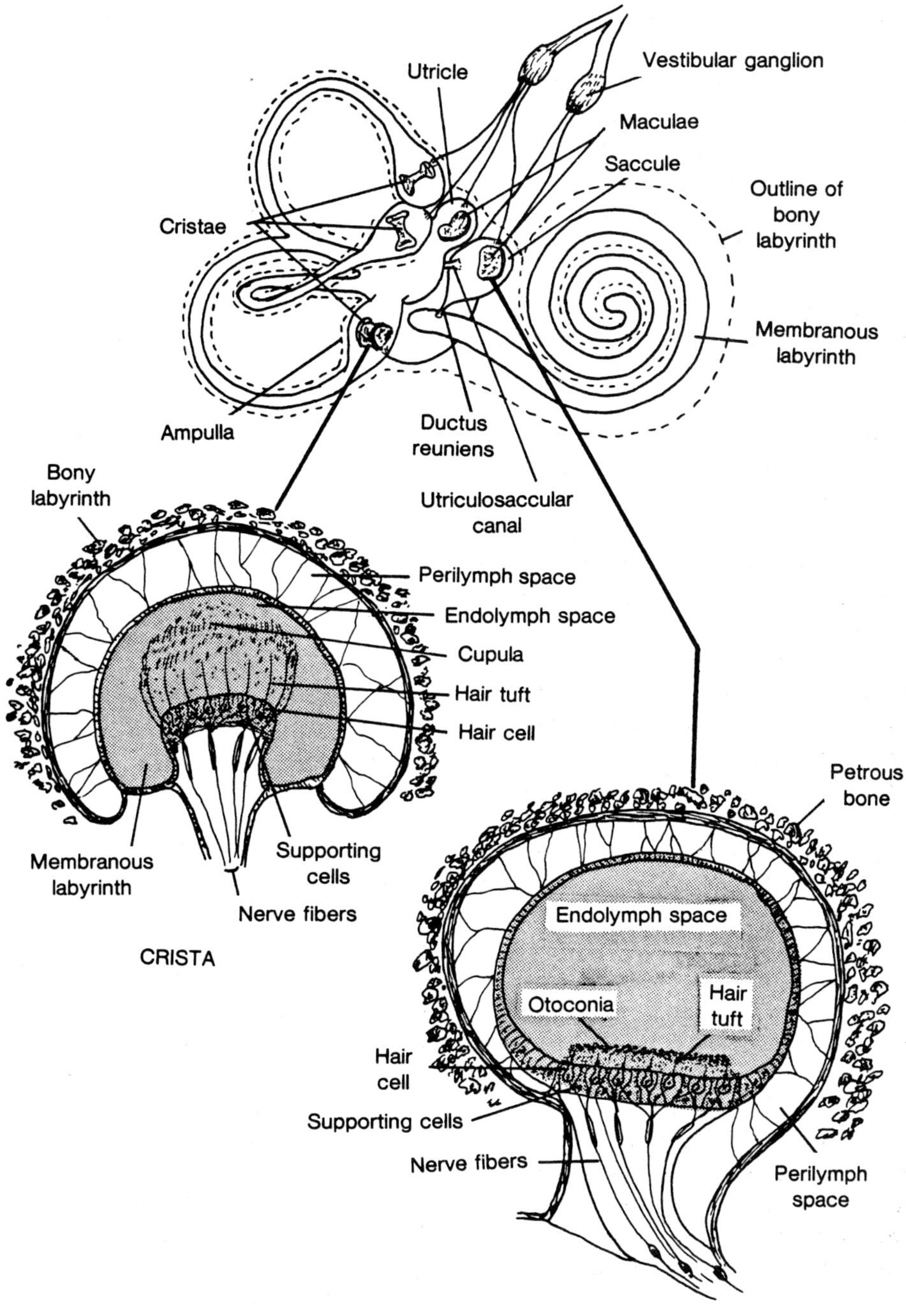

Figure 17-12. Diagrammatic representation of the neuroceptors of the vestibular apparatus. Notice that the membranous labyrinth is held suspended within the perilymph space by anchoring strands of connective tissue, termed *trabeculae*. Redrawn from Minckler, Jeff: Vestibular system. (In Minckler, Jeff (Ed.): *Introduction to Neuroscience*, St. Louis, 1972, The C.V. Mosby Co.)

spinning around in place and then trying to stand still. The child will, without fail, fall to one side. It is a judgment problem, for the subject is under the impression of falling to one side when in reality he is not. Such disturbances of the vestibular apparatus may be termed postrotational errors of movement. Pilots who erroneously rely on the vestibular organ rather than on instruments during combat-type maneuvers may be left to fatal judgment errors.

Acceleration and deceleration in rotating a subject produces a semicircular canal reaction, causing the eyes to rhythmically oscillate in the opposite direction of the acceleration; the reverse is true during deceleration. This oscillation is termed *nystagmus*. Movement of the endolymph, stimulating the cristae of the canals, occurs only during velocity *changes* and is not determined by velocity as such. Thus, steady movement would cause nystagmus to disappear.

If the endolymph accumulates in excess quantities so as to displace most of the perilymphatic space, the condition is known as *Meniere's disease*. The chief symptom appears to be vertigo, with nystagmus, from vestibular dysfunction. But other concomitant symptoms include hyperexcitability involving tinnitus (head noises or ringing in the ears) and elevated thresholds of hearing (loss of hearing), the latter two being from cochlear dysfunction. It is not surprising to find the cochlea involved, since the endolymph is continuous between the vestibular and cochlear portions of the inner ear. The cause of Meniere's disease is not known.

Cochlear Division

The internal ear, as mentioned previously, houses both the vestibular apparatus and the organ of hearing. In terms of relative position, the vestibular division is located behind and slightly lateral to the cochlear division, which is more medially placed in the petrous portion of the temporal bone (Fig. 17-13).

The cochlea is part of the bony labyrinth of the internal ear. Inside, there is a membranous labyrinth, paralleling the structure of the vestibular labyrinths. Furthermore, the cochlea and vestibular apparatus are united through a medial connection, the vestibule. The coupling does not end with the vestibule, since the membranous labyrinth is continuous from one division of the internal ear to the other. As described earlier, this passage is known as the ductus reuniens; it permits the endolymph of the saccule of the vestibular apparatus to communicate with the endolymph of the cochlear duct. It should be remembered that the endolymph is the fluid of the membranous labyrinth and is surrounded by perilymph.

The cochlea is coiled like the shell of a common snail (*cochlea* = snail shell). If one should inject a hardening liquid into the cochlea, let it set, and then break the bone away, the cast would emerge as a spiraling structure with about two-and-one-half turns. At the basal end where it joins the vestibule, the cast would appear greater in diameter than in the medial or apical (tip) sections. The cochlea exhibits spirals stacked in a conical shape. In the living subject, these coils are wrapped around a central, matching conical-shaped bone known as the *modiolus*. The diameter of the basal coil is about 9 mm, a little more than the diameter of a pencil.

The modiolus has a thin bony ridge projecting from its central mass into the space of the cochlear canal or the spiral space hollowed out in the petrous portion of the temporal bone. Following the cone shape of the coils, this bony ridge gives the modiolus the appearance of a short,

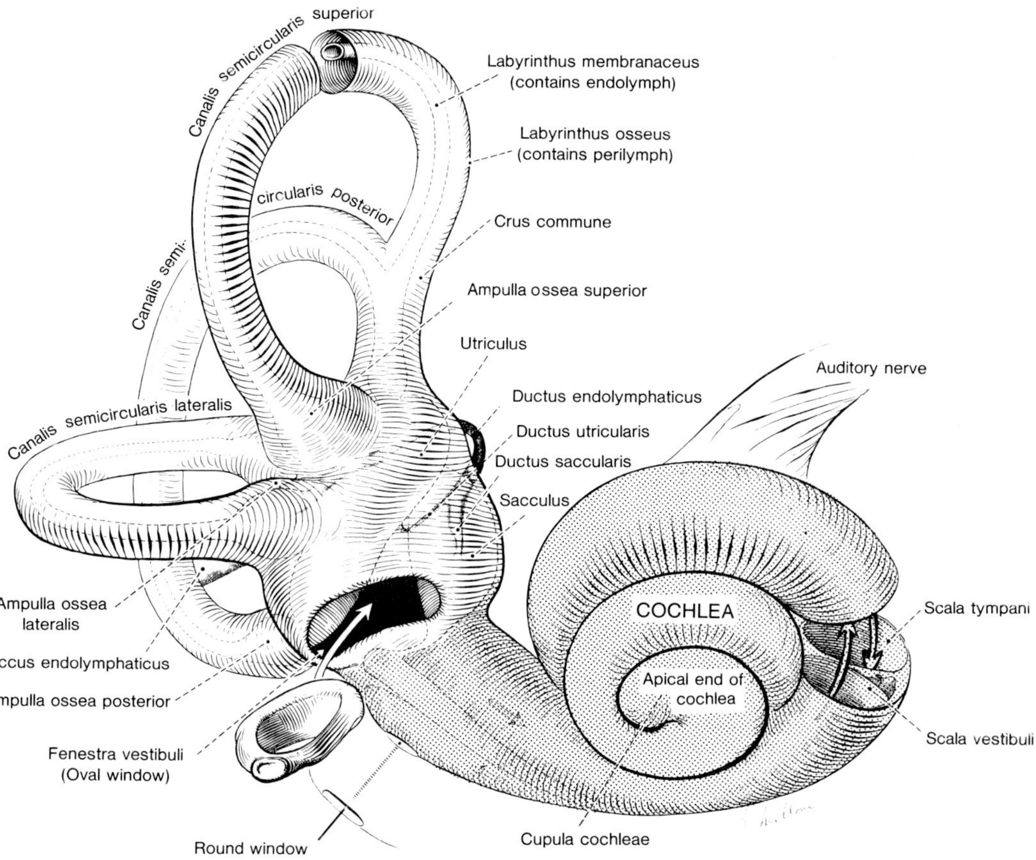

Figure 17-13. The cochlear division of the inner ear. (Reproduced from: "Some Pathological Conditions of the Eye, Ear and Throat." Courtesy of Abbott Laboratories.)

blunt screw, with the ridge of bone serving as the threads. This spiral structure is known as the *spiral lamina* or *bony spiral lamina*.

Extending from the spiral lamina across the space of the cochlea and attaching to the oposite wall is a fibrous membrane termed the *basilar membrane*. This fibrous membrane is anchored on the wall opposite the spiral lamina by the *spiral ligament*. But it should be clearly understood that the basilar membrane is not under tension, a feature vital in understanding the role of this membrane in hearing physiology. Another membrane, the *vestibular* or *Reissner's membrane*, rises from the area of the spiral lamina at about a 45-degree angle (relative to the basilar membrane) to attach to the bony wall in the region above the spiral ligament.

Therefore, the cochlea is divided into three canals by two partitions. There is one central duct and a canal running on either side. The vestibular space or *scala vestibuli* lies next to the vestibular membrane and the bony wall of the cochlea. On the opposite side of the central duct is the tympanic space or *scala tympani*, also extending to the bony wall of the cochlea. These two canals are filled with perilymph. They spiral to the apex of the cochlea, where they communicate with each other through an opening called the *helicotrema* (see Fig. 17-11). Obviously, then, any vibrations introduced into one of the two spaces, vestibuli or tympani, would be transmitted to the other, setting the fluid of the two canals into motion.

The vestibular space communicates with the middle-ear cavity through the vestibule to the oval window. This window is covered with a membrane to which is attached the footplate of the stapes. The tympanic space also communicates with the middle-ear cavity via the round window. The membrane covering the round window has been referred to as the secondary tympanic membrane (Fig. 17-14).

That canal formed between scala vestibuli and scala tympani is the *cochlear duct* or *scala media*. It is a part of the membranous labyrinth and, as previously mentioned, is filled with endolymph. In contrast to the other two canals, this duct is closed at the apex. As the cochlear duct uncoils coming out in the basal turn, it passes between the openings, forming the oval and round windows. It ends in the vestibule but sends up a small tube, *ductus reuniens*, to the saccule to make the membranous labyrinth continuous between the vestibular and cochlear divisions.

Organ of Corti

The essential structure relating to the physical support and function of the organ of Corti is the basilar membrane. Spiraling from the basal portion at the vestibule to the apical end, the basilar membrane gradually widens. Laid out straight, it is a variably tapered structure, being much wider at the apical end than

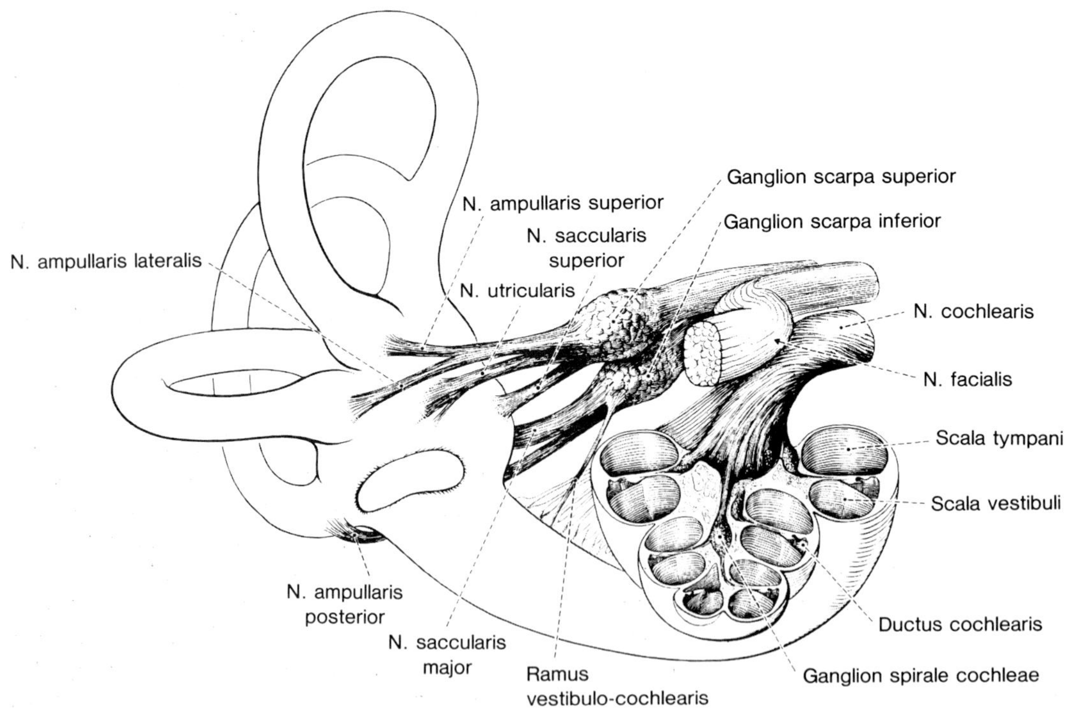

Figure 17-14. Cross section of the coils of the cochlea showing the three fluid-filled spaces. Notice that the facial nerve travels along with the two divisions of the vestibulocochlear nerve. (Reproduced from: "Some Pathological Conditions of the Eye, Ear and Throat." Courtesy of Abbott Laboratories.

at the vestibular end (see Fig. 18-14). Although there is a remarkable dimension variation from subject to subject, the total length is about 32 mm and is about 0.36 mm at the apex and 0.21 mm at the basal turn.

Lying on the basilar membrane and contained within the cochlear duct is the organ of hearing. It is known as the *organ of Corti*. This complex structure appears in cross section to be something like an extension of the bony spiral lamina. Tiny canals in the spiral lamina perforate it from basal to apical sections to provide passage of the nerve filaments innervating the receptor cells in the organ of Corti.

The receptor units in the organ of Corti are specialized sensory hair cells. Supporting cells and structures of the organ of Corti serve the complicated operation of this organ's essential receptors. A mound of cells and related structures on the basilar membrane seem to be clustered around the hair cells. Indeed, they represent the most prominent placement in the complex mass. A striking feature is the *tectorial membrane* above the hair cells. This gelatinous structure runs along the top of the hair cells as a delicate ribbon from one end of the cochlea to the other. The tips of the hair cells are embedded in the tectorial membrane (Fig. 17-15).

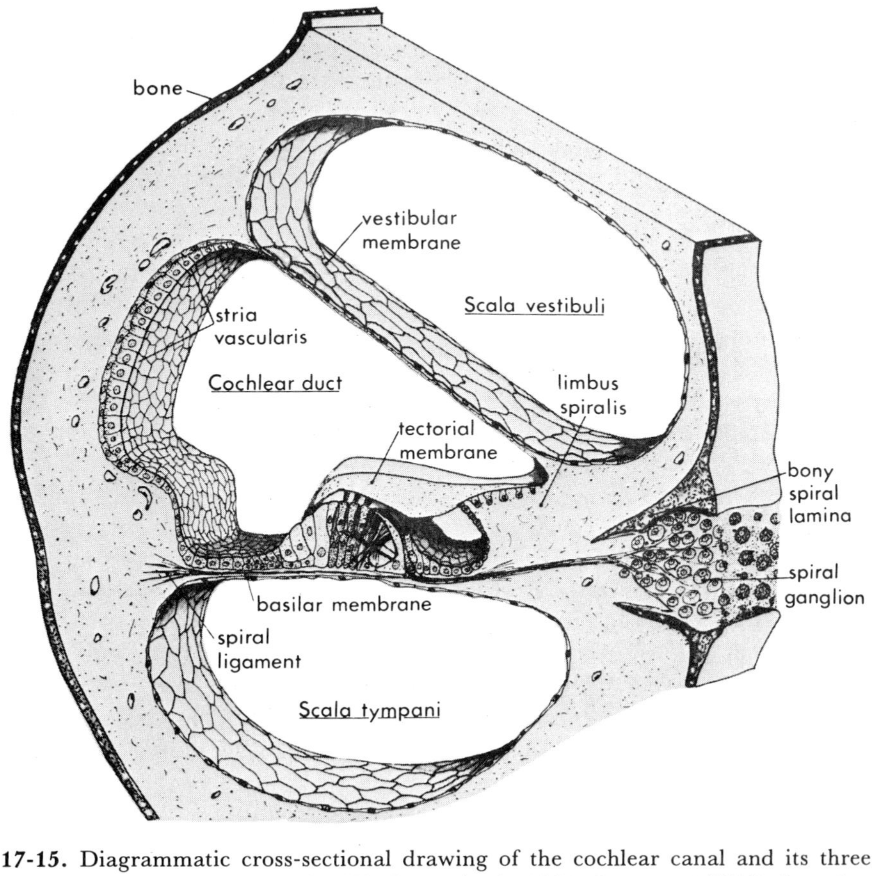

Figure 17-15. Diagrammatic cross-sectional drawing of the cochlear canal and its three divisions. (From Leeson, C.R., and Leeson, T.S.: *Histology*, 4th ed., 1981. Courtesy of W.B. Saunders Co., Philadelphia.)

To keep the tectorial membrane in place, it is fastened by the *vestibular lip*, which is a part of the *spiral limbus*, at times just referred to as the limbus. This amassed tissue is found on the bony spiral lamina ending at the foot of the vestibular membrane. The other end of the tectorial membrane is attached to the *cells of Hensen*. In older publications, illustrations depict this end of the tectorial membrane floating above the outer row of hair cells, with the tip sometimes retroflexed. Investigators were unaware that the preparation process was shrinking the membrane significantly and pulling its distal end loose.

Adjacent to the border cells near the origin of the basilar membrane are the *inner phalangeal cells*, also known as *Deiters' cells*. The designation as phalangeal cells stems from a phalangeal process associated with Deiters' cells. The phalangeal cell body itself rests upon the upper shelf of the bony spiral lamina as a support for the inner hair cells. Rising up from the base of the phalangeal cell is a process that terminates at the level of the top of the inner row of hair cells. As this rather rigid process reaches its upper limit, it broadens to form, with similar adjoining processes, a net-like lamina known as the *reticular lamina*. The hair cells on the inner row are supported at their bases by the phalangeal cells and at their apexes by the lamina formed by the phalangeal processes. It is this membrane (the reticular lamina) that contains a net-like arrangement through which the inner and outer rows of hair cells may protrude their hair-like processes into the tectorial membrane above.

The *rods of Corti* are placed on the basilar membrane near its medial attachment. There are two prominent, heavy phalanges that emanate from basilar cells just on the outside of the inner phalangeal cells. Known also as the *pillars of Corti*, these two structures slant toward each other until they meet near their apexes. At the upper limit, the processes expand to join the headplates of the phalanges and form the reticular lamina. This row of rods extends throughout the length of the spiraling basilar membrane. There are nearly 6000 inner rods and nearly 4000 outer rods (Fig. 17-16).

On the outer wall in the cochlear duct is the *area vascularis* or *stria vascularis*, so named for the rich supply of tiny blood vessels formed on this wall. The stria vascularis lies over the upper portion of the spiral ligament. A special group of cells in this upper region of the stria vascularis serves as secretory cells to produce endotic fluid of the cochlear duct. The endolymph provides a nutritive transport function, transmits mechanical vibrations for hearing purposes, and probably serves as a protective hydraulic shock absorber to protect the delicate complex comprising the organ of Corti.

HAIR CELLS. The hair cells in the organ of Corti are special mechanoreceptors. They respond to stimuli created by pressure waves traveling through the cochlear fluid. In essence, movement by the tectorial membrane results in bending of the hair-like projections located at the top of the cells and embedded in the tectorial membrane. This bending action is sufficient stimulus to trigger the hair cell into discharging a neural impulse through a nerve fiber connected to the cell.

There are about 17,000 hair cells per ear, though this varies widely from subject to subject. They spread across one inner row nearest the bony spiral lamina and three to five outer rows. The inner row ranges from about 3000 to 4000 hair cells. Collectively, the outer rows range from about 12,000 to 18,000 hair cells.

The inner row is but a single row of hair cells from basal to apical extremes.

However, the outer rows, those distal to the rods of Corti, present a variable number of rows: three rows in the basal cochlear turn, to four rows in the middle turn, and up to five rows in the apical turn (the turn farthest from the oval window).

Each cell contains a number of hair-like projections or *stereocilia*, pointed verti-

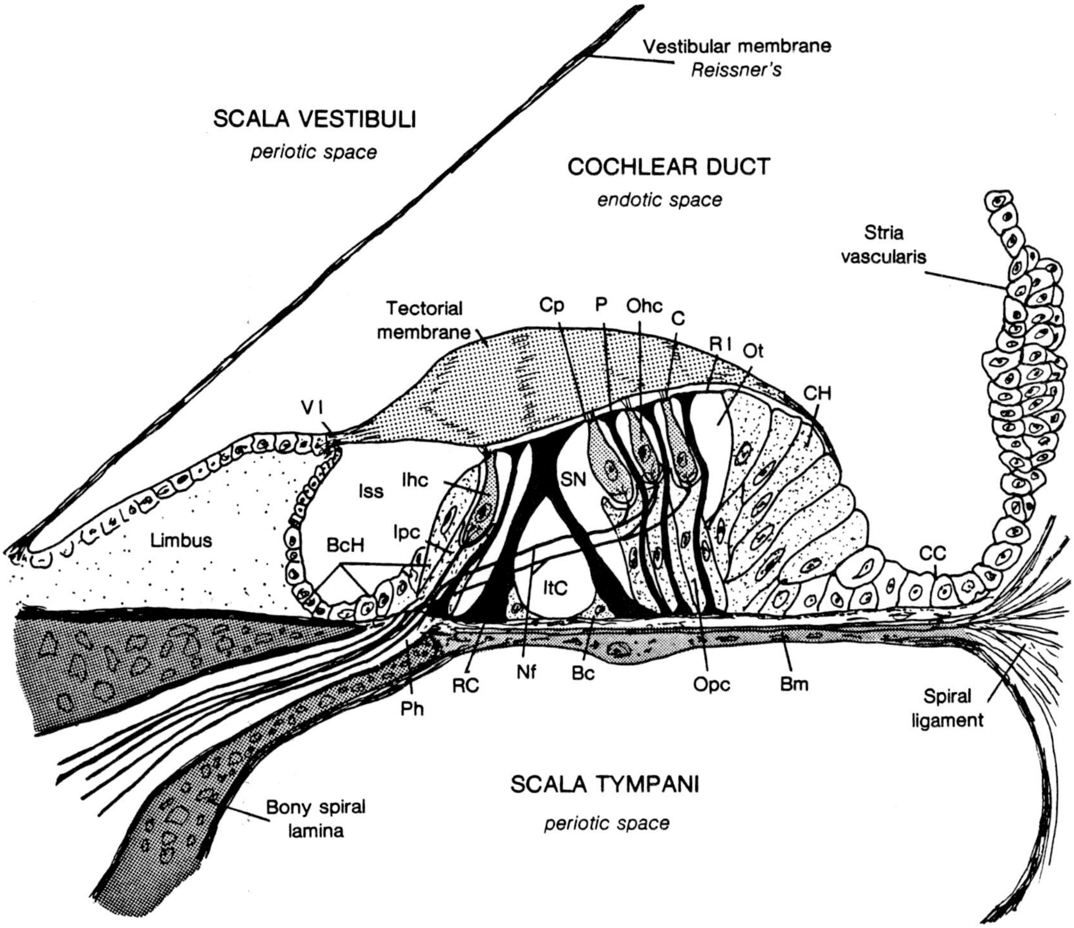

Bc Basilar cells
BcH Border cells *of Held*
Bm Basilar membrane
C Cilia
CC Cells of Claudius
CH Cells of Hensen
Cp Cuticular plate
Ihc Inner hair cell
Ipc Inner phalangeal cell
Iss Internal spiral sulcus
ItC Inner tunnel of Corti
Nf Nerve fibers
Ohc Outer hair cell
Opc Outer phalangeal cell
Ot Outer tunnel
P Phalanges
Ph Perforate habenula
RC Rods of Corti
R l Reticular lamina
SN Space of Nuel
V l Vestibular lip

Figure 17-16. Histology of the organ of Corti as seen in cross section. The single inner and three outer hair cells are distinguished by shading. The phalangeal cells, also known as Dieters' cells, send phalanges from the basilar membrane to the reticular lamina.

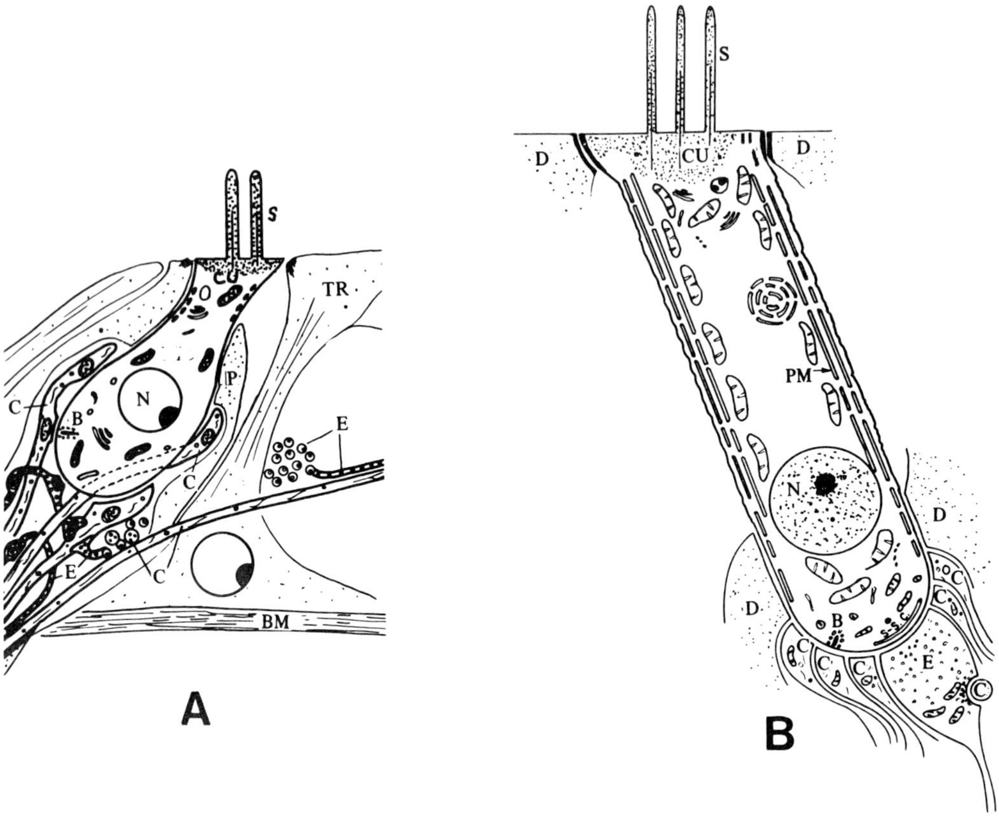

Key:

BM — Basilar membrane
C — Afferent cochlear nerve fibers
E — Efferent nerve fibers of the olivocochlear bundle
P — Phalangeal supporting cell
TR — Tunnel rod
B — Synaptic bar
N — Nucleus
D — Dieters' cells
CU — Cuticular plate
S — Stereocilia

Figure 17-17. (A) Diagram of an inner hair cell, and its innervation. (B) Diagram of an outer hair cell and its innervation (drawn at a higher magnification than A). (Courtesy of C.A. Smith. Reproduced by permission from *Advancement of Science*, Vol. 24, No. 122, 1968).

cally from a *cuticular plate* at the top of each cell. These hairs vary in number according to the location of the hair cell. Outer hair cells exhibit about 40 to 115 hairs per cell; inner cells have about 35 to 60 per cell. Further, the hairs appear arranged in a W pattern on the outer hair cells, but the W pattern flattens out to a straight-line pattern on the inner hair cells.

Nerve fibers leave the base of the hair cells and embark upon a path directed toward the bony spiral lamina, where the fibers enter the bony structure through openings beginning with the *perforate habenula*. Afterwards, the fibers reach into the modiolus and form a spiral ganglion on the way, the *spiral ganglion of Corti*. Continuing onward, the fibers twist together from all sections of the cochlea into a central nerve bundle known as the *auditory* or *cochlear nerve*. The cochlear nerve is then joined by the vestibular branch to form the vestibulocochlear cranial nerve, which passes to the brain stem, entering the central nervous system on either side of the brain stem.

BIBLIOGRAPHY

Afifi, A.K., and Bergman, R.A.: *Basic Neuroscience*. Baltimore, Urban and S, 1980.

Anson, B.J., and Donaldson, J.A.: *Surgical Anatomy of the Temporal Bone and Ear*, 2nd ed. Philadelphia, Saunders, 1973.

Bloom, W., and Fawcett, D.W.: *A Textbook of Histology*, 10th ed. Philadelphia, Saunders, 1975.

Gerber, S.E.: *Introductory Hearing Science: Physical and Psychological Concepts*. Philadelphia, Saunders, 1974.

Harris, J.D.: *Anatomy and Physiology of the Peripheral Hearing Mechanism*. Indianapolis, Bobbs, 1974.

Jerger, J. (Ed.): *Handbook of Clinical Impedance Audiometry*. Dobbs Ferry, NJ: American Electromedic Corp., 1975.

Leeson, C.R., and Leeson, T.S.: *Histology*, 3rd ed. Philadelphia, Saunders, 1976.

Lim, David J.: Cochlear anatomy related to cochlear micromechanics, a review. *JASA, 67*:1688, 1980.

Myers, D., Schlosser, W.D., and Winchester, R.A.: *Otologic Diagnosis and the Treatment of Deafness*. Summit, NJ: CIBA, 1962.

Newby, H.A.: *Audiology*, 4th ed. Englewood Cliffs, NJ: P-H, 1979.

Chapter 18

BASIC HEARING PHYSIOLOGY

THE role of the ear is to act as a receptor for sound vibrations. For the most part, the ear is concerned with those sounds within a certain range of frequencies. Very low frequency vibrations and very high frequency vibrations are not easily perceived by the ear. If the sound vibrations should be slower or faster than that for which the ear is capable of transmitting, no sound would be heard, of course, for the sound would be infrasonic or ultrasonic.

Most sound vibrations arise in the environment at a distance from the ear. These are airborne sounds traveling as waves through the medium of air. With the head under water, the sound waves travel through the water to the ears. Water is an even more efficient medium than air. Pools sometimes have speakers installed below the water line for the underwater swimmers' pleasure.

The ears monitor airborne sound waves during speaking, but sounds also reach the ear via bone and tissue. Due to this second channel for monitoring one's own speech, individuals are surprised upon first hearing themselves from a recording, since the recording is feeding the sound back to the ear only by way of waves through the air, without the benefit of direct coupling from the vocal tract to the ear.

One is also able to hear many biological sounds in the body though seldom is any attention paid to these. In a very quiet room such as an anechoic chamber, breathing and swallowing noises are the loudest sounds in the room. Even the arterial pulse may be heard by some. A stomach growl emits embarrassingly clear, raucous tones without the usual competition from environmental sounds, which often mask biological sounds from the ears.

In the previous chapter, the structures comprising the ear were treated. References to the function of the various structures were frequently made but most often without physiological continuity, since the purpose was to describe the makeup of the parts of the ear and its neural pathways. Presently, the discussion will concern itself with the path of the sound vibration pattern into the ear and through it, examining how the ear structures handle it along the way. The nature of sound energy impinging upon the brain will also be considered initially.

SOUND ENERGY

The Stimulus for Hearing

Although the function of the vocal tract is speech sound production, one hears many more sounds than he is able to produce. An introduction to acoustics, with emphasis on the nature of sound serving as a stimulus for hearing, will be examined first. It is an elementary treatment of the anatomy and physiology of sound.

Sound is so common that it is taken for granted, and most of it (that which fails to catch one's attention) is ignored. Almost any sort of motion generates sound, even the tree falling in the forest with no one present (if one follows the physicist's point of view). Some of the energy from moving bodies travels through the air in all directions, striking everything in the vicinity, including human ears, if present. Examples might be the bells on the telephone, the strings of a guitar, the splatter of water running into a glass, the vocal folds, the emission of a strong thrust of air from a jet engine, and thousands of other common sources. All of these sounds illustrate airborne sounds and are the type that are of interest here, but sound energy may be found in mechanical, electrical, and hydraulic forms.

Acoustics is the study of the science of sound and is a branch of physics. Like light, electricity, and heat, sound is a form of energy. Whenever sound is made extremely intense, it can overheat and kill small furry animals due to heat absorption by the fur. When dissipated, sound, like all other forms of energy, may then be expected to be eventually converted into heat. But the usual amount of heat present is negligible because the energy level of most sounds is so minute.

From the physicist's view, sound is comprised of physical vibrations of a generator or of disturbances in some medium. The psychologist is concerned about the effects of sound as a stimulus.

While most would probably adopt the psychologist's view most readily, it must be realized that the physical components that determine the attributes of sound as a stimulus are indeed the shapers of the stimulus itself and thereby mark the fashion in which the sound serves as a stimulus and carries information.

Of course, a driving force such as wind or a hammer is necessary to set a given vibrator into motion, which in turn will set the surrounding air particles into a wave of vibratory motion. Examples of sound generators and their energy sources may be of a wide variety: vocal folds operated by expiratory breath stream; reeds, as in woodwind instruments, also operated by the breath stream; piano strings struck by hammers, bells rung by a striker; guitar strings that are plucked; columns of air in an organ pipe operated by wind pressure across the mouth of a pipe; and speaker cones driven by electromagnetic fields, to name but a few.

In each case, each sound generator is capable of passing its own vibratory motion on to the surrounding air medium. And in each case, the type of wave motion propagated meets the requirements of the ear for sound perception.

Harmonic Motion

The nature of vibratory motion relative to sound waves is basic to an understanding of sound. *Simple harmonic motion* is the most elementary pattern; it produces a pure tone. When traced out on paper, simple harmonic motion is a symmetrical wavy line. This vibratory pattern may be illustrated with a pendulum having a marker attached. While moving from side to side, a sheet of paper is pulled forward under the swinging marker. The to-and-fro marks represent a sinusoidal pattern, since it has mathematical properties and can be described in sine and cosine functions. It will be observed that the pattern of the pendulum is a simple continuous line going back and forth on the page, but it meets all the requirements for simple harmonic motion. The periodicity of simple harmonic motion makes it possible to discuss the wave pattern in cycles (Fig. 18-1).

Frequency

One attribute of a sound wave present in the sine waves described above is *frequency*. How often the wave completes a full 360-degree (one to and one from motion, and back to the starting point) cycle determines the frequency of the wave pattern per unit of time. This is expressed in cycles per second, abbreviated *cps*. The present abbreviation of Hz (after Hertz, in recognition of his contributions in electromagnetic waves) is now preferred by international agreement. If the frequency of vibration is below 16 Hz or above 20,000 Hz, the sound will not be audible, except to only a very few persons at most (less than 1% of the population) (see Fig. 18-6).

Frequency is a purely physical phenomenon. Sound waves pass by at a rate of about 1,130 feet per second under standard conditions. Successive parts of the wave, the waves of *compression* and *rarefaction*, will pass at rates dependent upon the frequency of the sound. The wavelength for a given frequency may be measured in feet, calculated by the formula

$$\text{Wavelength} = \frac{\text{velocity of wave}}{\text{frequency}}$$

To find the wavelength of middle C on the piano, the known values are inserted as follows (Fig. 18-2):

$$\text{Wavelength} = \frac{1130 \text{ ft.sec.}}{261.6 \text{ Hz}} = 4.32 \text{ feet}$$

Thus, from the crest to crest of the specified frequency of middle C, it will require a distance of almost 4½ feet for the wave to develop completely or finish one cycle. In the process, 261.6 wavefronts will pass by per second at this frequency. The ear, then, is receiving 261.6 minute pressure wavefronts striking it per second (Table 18-I). (This frequency for middle C on the piano is based on A = 440 Hz, which is termed *concert pitch* or the pitch reference used in America and some

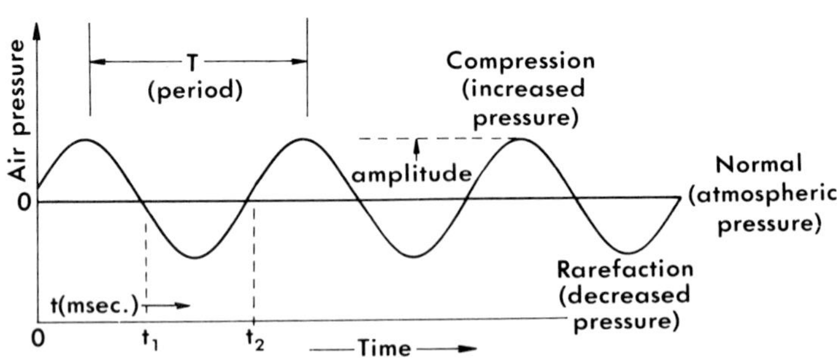

Figure 18-1. Sinusoidal wave. The waveform changes show how the air pressure changes with *time* when it is measured at a given point. (From Gerber, S.E.: *Introductory Hearing Science*, 1974. Courtesy of W.B. Saunders Co., Philadelphia.)

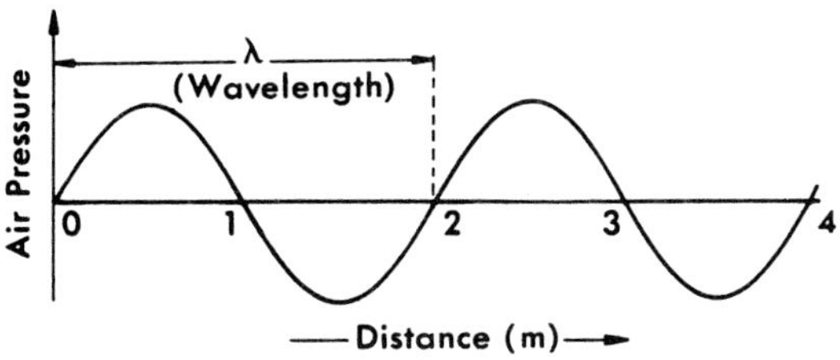

Figure 18-2. Sinusoidal wave. The waveform shows how the air pressure changes with *distance* from a given point at a given time. (From Gerber, S.E.: *Introductory Hearing Science*, 1974. Courtesy of W.B. Saunders Co., Philadelphia.)

Table 18-I

FREQUENCY AND WAVELENGTHS OF THE MUSICAL NOTE *C*

Octave Note	*Frequency*	*Wavelength in Feet*
C_1	32.75	37.50
C_2	65.5	17.25
C_3	131	8.62
C_4	262	4.32
C_5	524	2.15
C_6	1048	1.08
C_7	1096	.54
C_8	4192	.27

other countries. A *C* tuned to 256 Hz is based on another pitch reference, particularly the older, French standard.)

It may not be extremely clear now, but one can discover that the long wavelengths, associated with low frequencies, would require more time before they completed a cycle in comparison with short wavelengths associated with high frequencies. The longer wavelengths are able to bend around objects and corners much more readily than high-frequency sounds, which have the shorter wavelengths. Walking into another room in some houses while a full-range stereo is playing will illustrate this difference in frequency. The bass will continue to be heard, while there may be a noticeable attenuation of the high frequencies.

BASIC FREQUENCY AND OVERTONES. Few sounds in the environment approach the simple wave form of the sine wave. Musical sounds as well as noise are complex sounds. Any time two or more different frequencies are combined, a complex sound results. All speech sounds are comprised of complex sound waves (Fig. 18-3).

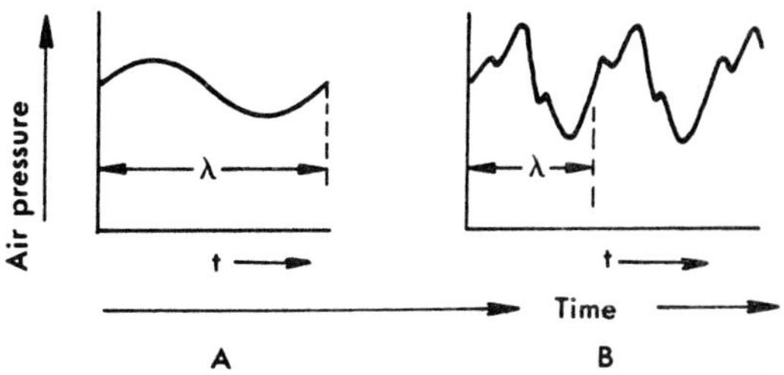

Figure 18-3. Wavelength for a pure tone or single frequency (A) and for a complex, multi-frequency wave (B). (From Gerber, S.E.: *Introductory Hearing Science*, 1974. Courtesy of W.B. Saunders Co., Philadelphia.)

FREQUENCY AND PITCH. Frequency is a physical attribute of sound. Its psychological correlate is pitch. Related to and dependent on frequency, pitch does not have a one-to-one ratio with frequency. However, pitch rises as frequency rises. On a musical scale, each note has a frequency 1.059463 times that of the note just below it, yet as the scale progresses to the octave above middle C, the musical intervals become subjectively larger as frequency increases.

At high and low frequencies, pitch may be significantly altered by the loudness of the tone. This is true primarily with pure tones and not with complex tones; otherwise, the poor musician would never be certain of the accuracy of his pitches. To demonstrate the relation between pitch and loudness, one can strike a tuning fork of a 150 Hz and gradually move it toward and then away from the ear. Its pitch will lower as it is brought

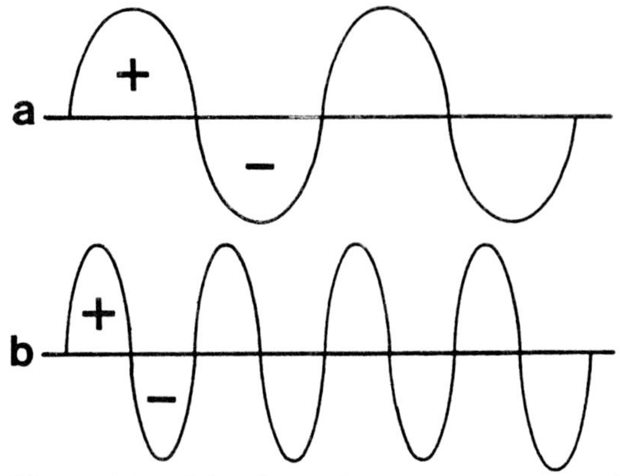

Figure 18-4. Waveforms for two pure tones of equal *amplitude* but different *frequencies*. Tone *b* is twice the frequency of *a*.

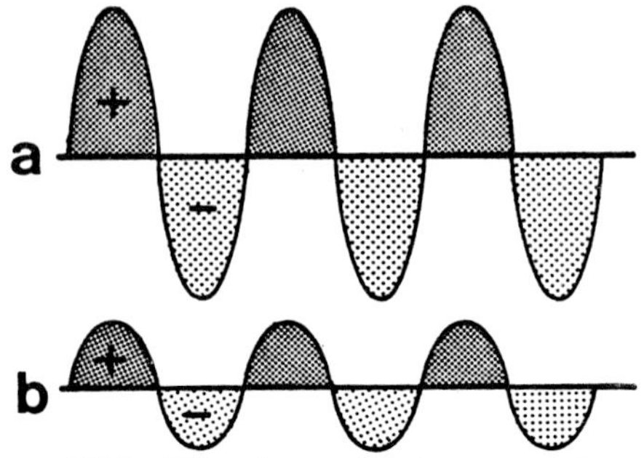

Figure 18-5. Waveforms for two pure tones of equal *frequency* but different *amplitude*. Tone *a* is approximately twice the amplitude of *b*.

closer and rise as it is moved away (Fig. 18-4).

Intensity

Amplitude denotes the maximum excursion from the rest position as illustrated with a pendulum. It will be seen that the sine waves presented illustrate that the vibrator moved in such a pattern so that two excursions were traced in each cycle. But this aspect of a sound wave is independent of the frequency of oscillations. Wide excursions, for example, do not result in lower or higher vibration rates. This dimension of a wave represents the amount of *energy* contained within the sound waves (Fig. 18-5).

Intensity is the energy density of a sound wave. It is a reference to the compactness of the air molecules and their energy potential during the compressional part of the sound wave. Graphically, intensity is represented by the area produced during the positive and negative excursions of the vibrator. From such a plot, with time on the abscissa, it can be seen that intensity involves energy flow per unit of time, like frequency, which is the number of cycles per unit of time. In terms of the manner that sound waves are propagated in ever-enlarging concentric shells away from the sound source, intensity would be described as the energy flow per unit of time through a unit area perpendicular to the direction of energy flow. Energy flow is measured in *watts*.

For the speech and hearing clinician, a more practical approach is to deal with the sound *pressure* rather than watts or the energy level since the ear, microphones, earphones, and speakers are constructed to sense or produce minute pressure variants in the air. Sound pressure is the force per unit area.

The *dyne* is the unit of force most frequently used in measurement of sound pressure. A one-gram weight is about one-thirtieth of an ounce (most letters weigh less than an ounce, the base reference for postage). A gram is equal to about 1000 dynes. In terms of sound pressure reference, the amount of pressure per square centimeter of area is used. This means that the amount of force acting over a given area of surface is taken as the standard method for determining pressure for sound waves. Atmospheric pressure at standard conditions exerts about one million dynes per square centimeter of surface area.

A football may require inflation to ten

pounds per square inch, which would be equivalent to around 700,000 dynes/cm^2. This example is given to show that larger units of measurement are more practical for higher pressures. But the minute pressures required to move air particles so that the measurement of a barely audible sound is on the order of a fraction of a dyne, about 0.0002 dynes/cm^2. Expressed in pounds per square inch, this page would probably be too narrow to hold on one line all the zeroes necessary to show the value. On the other hand, the ear cannot tolerate sound pressures of 2000 dynes/cm^2 and above.

THE dB SCALE. The decibel (dB) is the most frequently used unit of measurement of sound pressures. Named in honor of Alexander Graham Bell, the decibel is one-tenth of a Bel and is the logarithm of the ratio between two sound intensities or sound pressures. By international agreement, the standard reference pressure of a value of 0 dB is 0.0002 dynes/cm^2, the threshold of hearing at 1000 Hz for an average young, healthy adult without a history of ear pathology. This 0 dB level may also be expressed as 0 dB sound pressure level (SPL) at the standard reference given. Therefore, alone, the decibel contains no specific quantity of sound; it must be used in reference to a standard sound to have a physical meaning (Fig. 18-6).

Even though the dB of a sound is often expressed without the reference actually being given, it is understood to be the sound pressure level (SPL) of 0.0002 dynes/cm^2. Conversational speech averages about 60 dB from a distance of three feet; this implies that the sound is 60 dB above the SPL above.

If two sounds of 70 dB, identical in frequency and phase, were combined, it might seem that the result would be a sound of 140 dB. In reality, such a 2:1 ratio yields only 6 dB increase, or 76 dB. Increasing the sound by a ratio of 10:1 means a 20 dB increase in the sound, and a 100:1 increase gives a 40 dB increase. Obviously, there is no one-to-one ratio between increases in the ratio of the two sounds and the corresponding dB in pressure values. The relation is logarithmic as indicated previously. In fact, increasing the ratio of the sound to 1,000,000:1 raises the SPL to only 120 dB. At a 60 dB conversational speech level, one is 1000 times above the reference of 0.0002 dynes/cm^2, the threshold of hearing (Tables 18-II & 18-III).

By charting the SPL as a function of frequency, one can quickly determine the level required for minimum hearing as well as the maximum (tolerance) hearing limits for sound pressures. As is true for frequency sensitivity, the ear is also not equally sensitive at all intensity levels. Those frequencies near the resonant frequency of the external auditory tube are loudest. Still, frequency values and dB levels are not related to one another as physical phenomena, but they are inextricably related as psychological attributes of sound. Also, the ear is more sensitive to small increases in sound pressure at low levels than at high levels of intensity. Some ears are able to detect less than a 1 dB increment in sound pressure at certain frequencies and intensity levels.

INTENSITY AND LOUDNESS. The psychological correlate of intensity or SPL is loudness. Occasionally, the term *volume* or *level* is used to denote loudness, especially on radios and phonographic equipment. The loudness of a tone, as indicated earlier, is dependent on its intensity but also on its pitch.

Spectra

Breaking down a complex tone into its component sine waves is an operation called Fourier analysis. Each sine wave

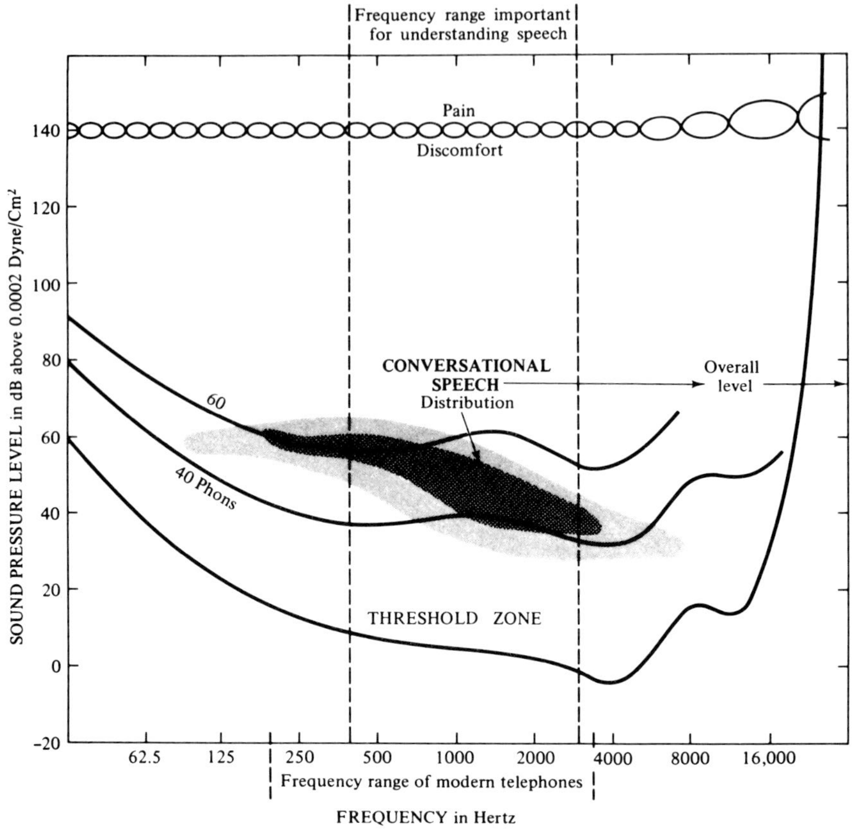

Figure 18-6. The speech area. Speech is a mixture of complex tones, wide-band noise, and transients. Both the intensities and the frequencies of speech sounds change continually and rapidly. It is difficult to measure them and logically impossible to plot them precisely in terms of sound pressure levels at particular frequencies. This figure shows the approximate distribution of sound-pressure levels with respect to frequency that would occur if brief but characteristic bits of phonemes of conversational speech were actually sustained like pure tones. The density of the shaded area represents roughly the probability of finding in a sample of speech the particular combinations of intensity and frequency. Individual voices differ, however, and obviously the boundaries of the speech area are not sharp.
The overall sound-pressure level of the stronger vowels in conversational speech at 1 meter is about 72 dB. The fundamental frequency of a deep bass voice is about 100 Hz, but the fundamental frequencies of most women's and children's voices are about 250 Hz. The strongest individual sounds in the frequency range below Hz are louder than 60 phons, and the weakest significant elements are about 30 dB below the strongest. The weaker elements are very often masked by background noises. For good understanding of everyday speech, the range from 400 to 3000 Hz is sufficient. This range includes almost all of the "formant" frequency bands of speech, which distinguish the vowels and many consonants. (From *Hearing and Deafness*, 3rd edition, by Hallowell Davis and S. Richard Silverman. Copyright © 1947, 1960, 1970 by Holt, Rinehart and Winston. Reprinted by permission of Holt, Rinehart and Winston, CBS College Publishing.)

Table 18-II

THE DECIBEL SCALE
WITH A PRESSURE REFERENCE

dB	*Pressure (dynes/cm²)*
0	0.0002
20	0.002
40	0.02
60	0.2
80	2.0
100	20.0
120	200.0
140	2000.0

Table 18-III

EXAMPLES OF SOUND LEVELS
RE: 0.0002 Dynes/cm² SOUND PRESSURE LEVEL

Nature of Sound	*dB*
Low whisper, four feet from speaker's lips	20
Barely intelligible speech in a quiet room	40
In an audiometric test suite	45
Conversational speech at three feet away	65
Trim room in a sawmill	90
Shout, one foot from the ear	100
Jet aircraft on takeoff (outside and behind engines)	115

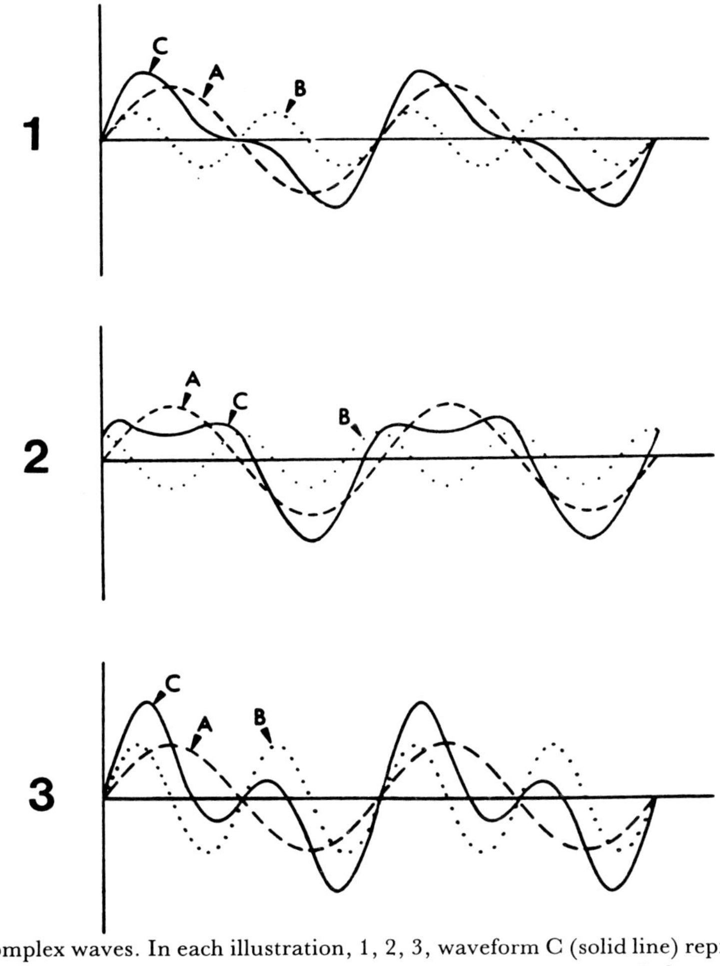

Figure 18-7. Complex waves. In each illustration, 1, 2, 3, waveform C (solid line) represents a composite of waves A and B. The variations in summing the two primary waves result from shifting the phase or amplitude or frequency of B. (From Gerber, S.E.: *Introductory Hearing Science*, 1974. Courtesy of W.B. Saunders Co., Philadelphia.)

contributing to a complex tone has two attributes for which an accounting must be made. These are frequency and intensity. An especially effective way of representing the distribution of these two attributes for a given complex sound is known as a *sound spectrum* (Fig. 18-7).

Primarily, the frequency components of a complex wave are being identified through a spectral or frequency analysis. The most graphic perhaps is the *line spectrum*, in which a line is drawn up from the abscissa for each frequency present in the complex wave being analyzed. The height of each line is dependent on the amount of sound energy present, so that amplitude is shown on the ordinate. In one sense, a line spectrum is like looking at the *end* of a group of sine waves, for no cyclic patterns are included, only frequency and amplitude without a time base. All sounds may be plotted by such a line spectrum (Fig. 18-8).

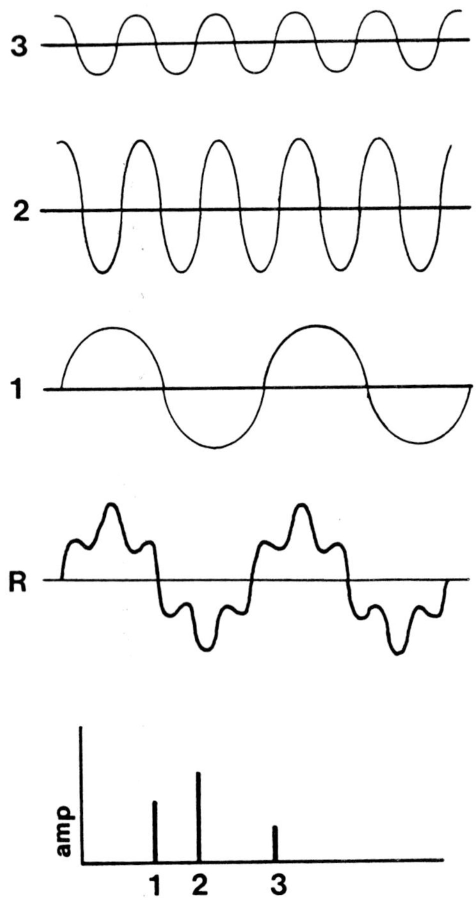

Figure 18-8. Line spectrum (bottom). R is the hypothetical sum of waves 1, 2, and 3. The amplitude for each primary wave is shown on the ordinate of the line spectrum, and the frequency would be designated on the abscissa.

If the first line on the line spectrum represents the fundamental frequency, the other lines, which are whole-number multiples of the fundamental, would represent the overtone structure of the sound. If every frequency is represented with some sound energy, then the ear will probably interpret the sound as noise. Such a spectrum would become a *continuous spectrum*, showing an unbroken line over the frequencies at the given amplitudes (Fig. 18-9).

SPECTRUM AND QUALITY. The spectral composition of a sound is important to the listener. The distribution of the frequencies and the amount of energy present for each is related to the *quality* of the sound. Quality, then, is the psychological correlate to the physical phenomenon, which has been discussed as spectrum. For example, if a clarinet and singer are heard producing the same pitch in alternating fashion, even the nonmusician can easily determine the difference in the sound source. The two sounds are distinct in quality because of the distribution of the overtone structure.

Spectral analysis is useful in determining the behavior of the vocal tract as a resonator. The vocal tract is around 17 cm from the glottis to the lips. If a tube of this length, uniform in structure, is considered, the resonance frequency would be

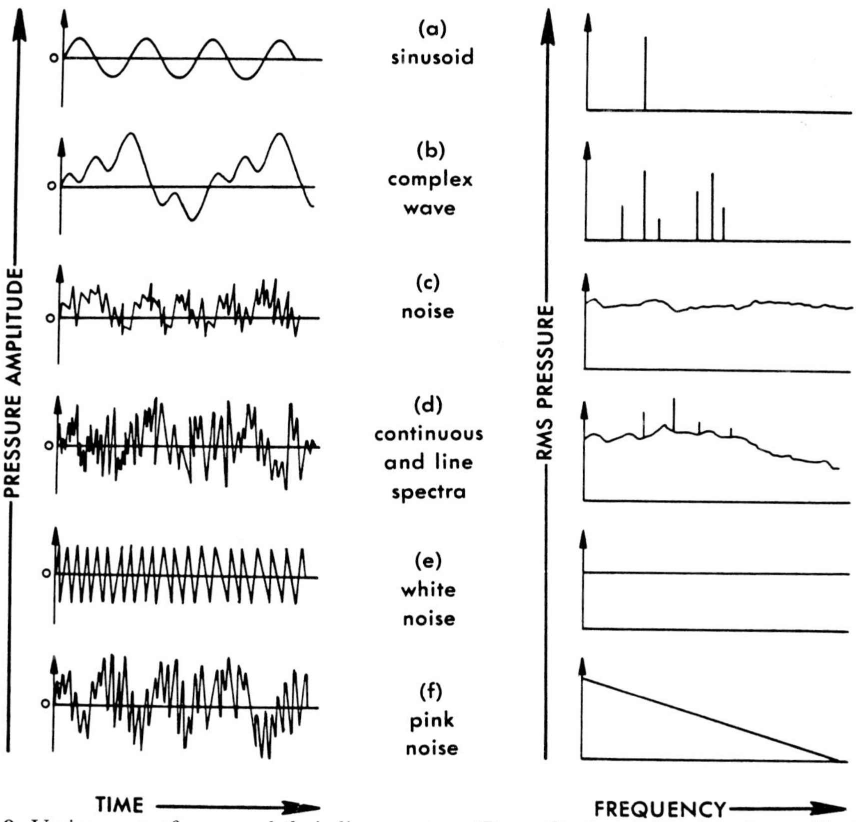

Figure 18-9. Various waveforms and their line spectra. (From Gerber, S.E.: *Introductory Hearing Science*, 1974. Courtesy of W.B. Saunders Co., Philadelphia.)

about 500 Hz, since one end of the tube, the glottis in this analogy, would be closed. The overtones, like those previously described with the tube resonators and segmental vibrations of a string, would be primarily 1500 Hz, 2500 Hz, 3500 Hz, and 4500 Hz.

The buzz-like phonatory sound at the glottis contains a number of frequencies that excite or ring the natural frequencies and, to some extent, their overtones of the vocal tract. But due to the irregularities of the tract, the overtones will be anything but evenly distributed. There will be a peak for each of the tract's natural frequencies. Changing the shape of the tract, e.g. moving the tongue, changes the amplitudes of the natural frequency as the natural frequency is altered. Those points of peak resonance of the vocal tract are termed *formants*. The formant frequencies are those frequencies present at the peaks. These are named according to position. The lowest formant frequency comprises the first formant. In successive order, the next highest frequency is the second formant and so on.

It appears that the ear is tuned well to vowel formants. For example, if one ear hears one formant of a vowel through earphones and the other ear hears a second formant, the auditory system can merge the two and permit the brain to recognize the vowel. In fact, the ear is phenomenal in handling complex acoustic energy.

OUTER-EAR PHYSIOLOGY

Auricle and Head

The intention of the auricle is to collect, concentrate, and conduct sound waves into the external auditory meatus. The auricle's function is probably most responsive to higher frequencies. However, its importance is not considered as great as for many animals that possess the ability to focus the ear in various directions with efficient auricular muscles. Man's outer ear muscles are capable of but weak movements of the auricle.

Turning the ear to concentrate sound waves into the ear canal enables man to gain an advantage in determining the directionality of a sound source. Of course, he must move his entire head to accomplish this. Then, the shadow effect of sounds striking the spherical-shaped head must be considered, for an obstacle in the sound path alters the timing of the wave arriving at the two ears.

With two ears to correlate differences in the time that the sound waves arrive at each ear, stereophonic listening is possible. Sounds coming from one's back are shadowed by the auricle so that if the head is turned even slightly to give advantage to one ear, the deflecting action of the auricle of the contralateral ear will delay the time of arrival of the sound. Thus, one will perceive the location of the sound source. In such circumstances, the auricle appears to aid in determining the sound source by providing time delays. This is thought to be highly personal since each person learns the manner by which his own head and auricles create sound shadows. After all, no two heads and/or auricles are identical in size and shape. Therefore, it is unreasonable to expect all persons to listen alike in directionality tasks, since each has custom-developed equipment.

With a little practice, one can begin to appreciate the slight asset afforded by cupping the hands at the rim of the auricles. When listening to widely separated stereo speakers at low listening levels, the cupping enhances the collection of some sounds not heard without the artificial extension of the auricle. These added bits of sound are limited to fine nuances, usually, and may involve more psychological attending than enhancement of hearing acuity.

The External Auditory Canal

Sound waves focused by the auricle enter the external auditory meatus. After traveling down this canal, the waves are blocked at the medial end of the canal by a membrane, the tympanic membrane or eardrum. The canal itself affects the sound stimulus, since it acts as a resonator in the same way that some tube-shaped musical instruments act upon the articulation of musical sounds.

As a resonator, the auditory canal reinforces frequencies between 2000 and 5000 cycles per second. These frequencies will be perceived as being louder than frequencies outside this range. Tones between 3000 and 4000 cycles per second are especially amplified, to a point of two to four times greater than that at the entrance to the outer ear. This more specific range is due to the fact that cavities possess a specific frequency at which they are maximally efficient in most rapidly radiating sound energy ringing the resonator. The canal's inherent resonant frequency lies somewhere in this range for most ears. In essence, the sound at the tympanic membrane has greater intensity than the sound at the auricle because the canal, as a resonator, enables the energy of the sound to be utilized more rapidly.

Without moving parts, the outer ear and the canal are probably the most trouble-free structures of the auditory mechanism, especially when considered over a lifetime. Some individuals do experience occasional problems with excessive accumulation of cerumen, which blocks sound passage to the tympanic membrane, but this is not a serious problem (Fig. 18-10).

MIDDLE-EAR PHYSIOLOGY

Six structures of the middle ear are involved in the transmission of sound to the inner ear. These mechanical structures are concerned with transducing sound from air to a mechanical mode for exciting the fluid in the cochlea. This function of the middle ear is necessary to transfer acoustic energy from a medium of very low density (air) to a medium of relatively high density (perilymph). Termed *impedance matching*, such an operation is required if there is efficient transfer from a medium having a low resistance to movement by a vibrator to a medium having a high resistance to the same vibrator. If sound waves in the air were put into direct contact with the fluid of the cochlea, the greater part of the sound energy would be reflected back through the air as rejected sound energy because of the stiff medium.

The tympanic membrane and ossicles provide an efficient link between acoustic energy and cochlear fluid through a special mechanical linkage. The design is not to transmit the actual acoustic energy but to transduce it into mechanical energy, an oscillatory movement. Stated another way, when introduced into the inner ear by the middle-ear structures, the stimulus is effectively altered for favorable reception by the fluid medium.

The six structures considered vital in the impedance matching responsibilities outlined above are the tympanic membrane, three ossicles, and two middle-ear muscles (Fig. 18-11).

The Tympanic Membrane

If two loudspeakers are placed so as to face each other, the first connected to a live program source and the second unconnected, the second will move sympathetically, to a limited extent, to the movement of the first. The principle enabling the passive speaker to respond to the first is the same for the eardrum in a basic sense. Electrical energy causes the first speaker's cone to oscillate, which in turn sets up waves of oscillation through the air. Upon striking the second speaker's cone, the waves find an elastic obstacle and, rather than merely bouncing off the obstacle, the acoustic energy sets the mechanical structure into an oscillatory movement. Hence, acoustic energy is transduced into mechanical energy and is now classified as a new mode of energy.

Just as the cones of loudspeakers are able to respond to a wide range of frequencies, the eardrum is likewise able to do so. The conical shape of the eardrum is an asset for its response to high tones. The circular fibers at the umbo and near the circumference give it a particular advantage in responding to a wide range of frequencies. The off-center and extended attachment of the manubrium of the malleus enables the eardrum to pass onto the manubrium a considerably greater amount of sound energy than if the load

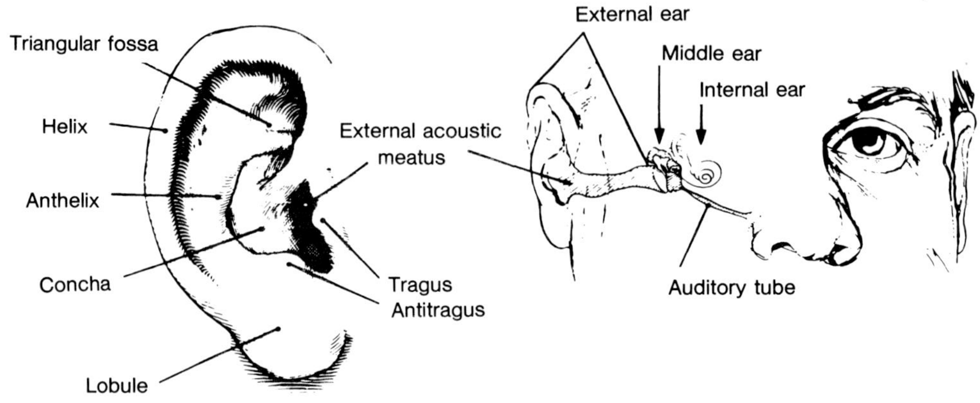

Figure 18-10. The auditory apparatus. The auricle is attached at the opening of the external acoustic meatus. Notice the auditory tube provides pressure equalization to the middle-ear cavity. (From *Dorland's Illustrated Medical Dictionary*, 26th ed., 1981. Courtesy of W.B. Saunders Co., Philadelphia.)

of the ossicles were added by a true central attachment. Unlike the speaker cone, though, the whole of the eardrum does not respond to acoustic stimuli. The effective area is about two-thirds of the total area. Even with this limitation, the area of the drum is considerably greater than that of the footplate of the stapes. (see Fig. 17-4).

It will be remembered that the task of the middle ear is not to transpose unchanged the same energy in the outer ear to the cochlear fluid, for the two mediums offer differing resistances or impedances. The eardrum provides the initial step in what may be described as the introduction of nonlinearity between the outer and inner ear insofar as the transfer of sound is concerned. Low-intensity sounds cause the eardrum to act as a levered surface to set the ossicular chain into action. During high intensities, the membrane behaves differently, for the mode of vibration changes to reduce the excursion usually required by the manubrium. Such nonlinearity between the acoustic stimulus and the middle-ear structures becomes progressive through the ossicular chain.

The tympanic membrane is stiff enough to hold its conical shape, even after being dissected from its peripheral attachment. Yet it displays elastic properties sufficient for a sensitive response to acoustic energy impinging upon it. The amount of displacement, however, is not directly proportional to the displacing force. At very high intensities, the eardrum shows appreciable distortion due to nonlinearity. At low frequencies, the eardrum moves more as a single unit and changes to vibrate in sections for higher frequencies.

Excursions of the tympanic membrane are miniscule, even during response to conversational-level sounds. The distance traveled by the drum while receiving typical speech sounds is on the order of the diameter of a large molecule.

Perforations in the eardrum may interfere with the drum's ability to respond to sound; however, from a functional view, the drum is less vulnerable to such anomalies as rupture, thickening, or retraction than is generally thought. Hearing problems are sometimes associated with the listed anomalies, but the condition may not be the real cause of a hearing complaint. The temptation is great to stop upon discovery of such a clinical feature as listed and not pursue the problem further. The tympanic membrane is more functional under atypical conditions than it was once thought.

The Ossicles

Attached to the tympanic membrane, the ossicles faithfully follow the movement of the eardrum. Their responsibility is to transmit the mechanical energy of the eardrum to the fluid of the inner ear (Fig. 18-11).

A mechanical advantage is built into the ossicular chain. The leverage gained from the structural makeup of the three middle-ear bones allows a power gain ratio of about 1.3:1. This means that vibrations at the oval window will have less displacement than vibrations being put into the chain by the eardrum, but there will be a trade-off of amplitude of the excursion for increased power.

The arrangement of the ossicles is somewhat like that of an inverted V, with the pivot near the point of the V (Fig. 18-12). When the manubrium of the malleus, which is attached to the eardrum, moves either in or out, the malleus head and the major part of the incus move in the opposite direction, due to the ligaments that suspend the ossicles and govern their movements. The stapes, attached to the long process of the incus, moves synchronously (in the same direction) with the eardrum. The ossicles thereby comprise a compound lever apparatus. That process of the malleus serving as a lever, the manubrium, is about 1.3 times larger than the lever of the incus, the long process. This provides an increase in power by the same amount, 1.3:1. The purpose in this design is to change acoustic energy with long excursions and low pressure into mechanical energy having short excursions but higher pressure. The product of this design is to apply appropriate pressure waves to the perilymph of the cochlea, since it is more difficult to set up pres-

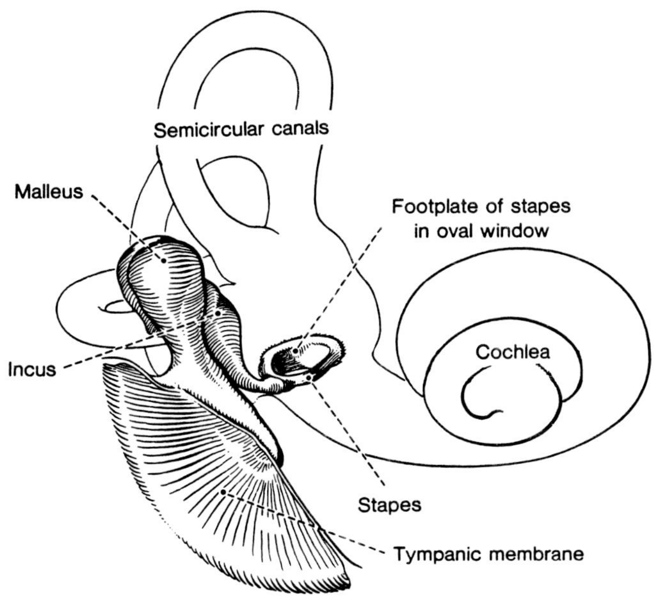

Figure 18-11. The tympanic membrane (cut) and ossicles. (Reproduced from: "Some Pathological Conditions of the Eye, Ear and Throat." Courtesy of Abbott Laboratories.)

sure waves in the fluid of the inner ear than in lighter mediums such as air.

In addition to the power advantage in the ossicular chain, the eardrum (being much greater in area than the oval window) introduces an even greater increase in power. This ratio is about 14:1 when the power of the drum vibrations are passed to the much smaller oval window. This ratio represents the differences in areas of the two structures; that is, the effective area of the eardrum (about 43 mm^2) divided by the area of the footplate of the stapes (about 3 mm^2) equals about 14. Hence, the eardrum is about 14 times larger in surface area than the footplate.

By calculating the total power advantage contributed by both the ossicles and the eardrum area, the pressure transformation ratio comes to about 18:1 (1.3 × 14–18). These figures vary from one source to another and from one specimen and species to another, but the 18:1 ratio appears to be a good ratio for humans, even though it is perhaps on the conservative side. This represents about a 25 decibel (dB) increase or hearing advantage (a whisper from 5 feet away registers about 20 dB).

The footplate of the stapes is responsible for transferring the mechanical energy of the eardrum and ossicles into the cochlear fluid. Its pattern of movement is not purely that of an in-and-out motion like a piston. Instead, there is a hinge action, with the stapes moving back and forth on a vertical axis, like a door opening and closing. The point of greatest fixation is the posterior part of the annular ligament. This mode of movement is used for moderate sound intensities. For high intensities, the mode changes to a hinge action along a horizontal axis. The effect is to diminish the intensity of the mechanical sound energy. A good analogy of such a shift in hinging is that feature provided by the versatile rear door on some auto stationwagons. It may be opened out like a car door, or it can be made to swing down by flipping four latches to change the hinging. Shifts in axes of rotation of the stapes are associated with a protective mechanism to prevent damaging sound vibrations from entering the cochlear fluid.

A well-known cause of hearing impairment resulting from transmission loss due to fixation of the footplate of the stapes is otosclerosis. Resorbed bone is replaced by sclerotic bone of irregular design. The pathologic bone progressively invades the oval window and immobilizes the stapes. This condition is peculiar to the otic capsule in man.

Tympanic Muscles

Behavior of the tympanic membrane and the ossicles is affected by the two small middle-ear muscles. Both are connected to the ossicles, the *tensor tympani* to the manubrium of the malleus and the *stapedius* to the head of the stapes.

The tympanic muscles differ in their influence on the ossicles. The tensor tympani appears not to be active in initiating a protective reflex to sounds of high intensity. Stimulation to the skin of the periorbital region of the face or contractions of the muscles in this region appears to be one way to elicit tensor tympani reflexive contraction other than by sounds that elicit a startle response.

Contraction of the tensor tympani puts the eardrum under tension, as its name implies. While this action may raise the natural and resonant frequency of the drumhead and the ossicles, the result is a decrease in sensitivity due to transmission loss. Through clinical studies, low and middle audiometric frequencies have been found to improve up to 15 dB after

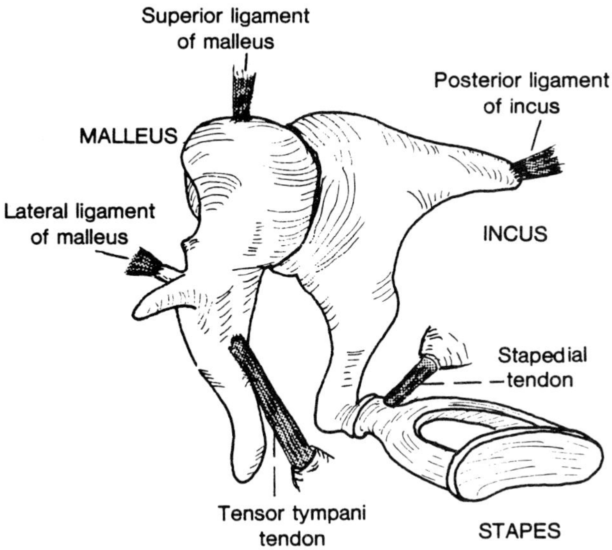

Figure 18-12. Ligaments and muscles associated with the ossicles. (Modified from *Dynamic Anatomy and Physiology*, 3rd ed., by L.L. Langley, I.R. Telford, and J.B. Christensen. Copyright© 1969. Used with the permission of McGraw-Hill Book Company.)

release of a prolonged contraction of the tensor tympani, which had kept the eardrum markedly retracted (drawn inward). The release was accomplished through an anesthesia block to the sphenopalatine ganglion in the back of the nose or by cutting the muscle's tendon. In addition to improved hearing acuity in such clinical cases, distortion of speech was reduced.

Conditions involving a prolonged contraction of the tensor tympani appear at first to be a needed protective function; nonetheless, the tensor tympani is unable to contract reflexively with sufficient speed to serve as an efficient protector against sudden loud sounds. The stapedius muscle is at least eight to ten times faster, and even this muscle has a brief lag time.

Perhaps the tensor tympani could be thought of as a protective agent under preparatory set situations. This could involve squinting the orbital musculature, which in turn may cause the tensor tympani to contract as the individual anticipates loud sounds such as watching the fuse of a firecracker burn to the explosive powder. This is conjecture, of course.

The stapedius muscle is credited with the true acoustic reflex, also called the stapedial reflex. Although the stapedius may contract reflexively in response to moderately loud sounds, its response comes about 10 milli-seconds after the sound has reached the eardrum. The action of this muscle displaces the stapes and decreases its coupling to the inner ear, which increases the stiffness of the ossicles. This resistance may be studied by measuring the proportion of the energy from a test tone that is reflected from the eardrum. Such a procedure depicts the principle of impedance measurements of the middle ear. Acoustic impedance-admittance measurements are routine in audiometric evaluations in almost all clinics. The instrumentation includes an impedance bridge along with an audiometer and graphic recording equipment.

These audiometric procedures are termed *tympanometry*.

The stapedial reflex is dependent on the individual's threshold for signal detection. Usually, this means that the sound must be around 75 to 80 dB above his threshold or in the vicinity of 90 dB, which is well above the level of normal conversational speech. These figures vary among authors and among subjects as well as frequency and type of sound, i.e. pure tone versus noise.

INNER-EAR PHYSIOLOGY

Cochlear Fluid

The usual path of sound leads into the inner ear through the oval window, referred to by some as the second tympanic membrane. Fluid fills the inner-ear structure. Vibrations are thus transmitted directly from the membrane over the oval window to the perilymph, since the fluid is in contact with the medial wall of this membrane. Liquids are incompressible; hence, inward excursions by the membrane at the oval window must be compensated by an outward excursion at the membrane covering the round window, which also opens into the tympanic cavity. Likewise, outward excursions at the oval window result in inward movement at the round window. The structure may be compared to a U-shaped glass tube filled with a liquid and having membranes on either end (Fig. 18-13).

In the action-reaction cycle by the two membranes at the oval and round windows, waves of vibration are sent through the spirals of the cochlea. These waves deform the cochlea partition and its associated parts, namely, the fluid in the duct

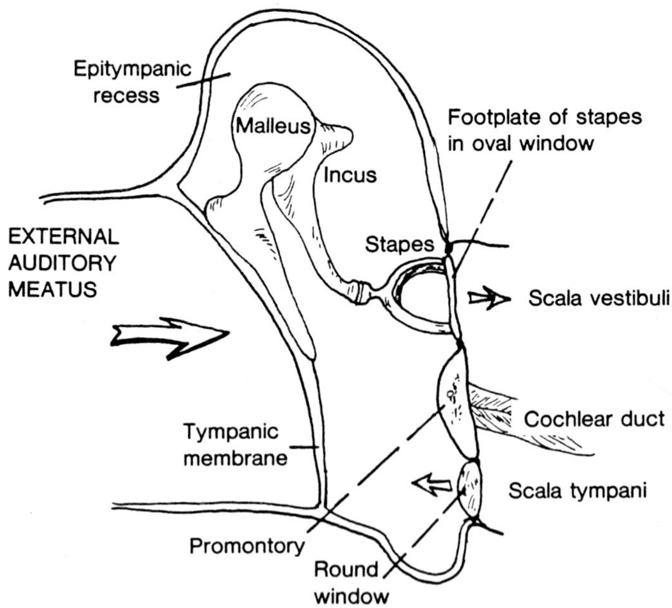

Figure 18-13. Schematic representation of the action of the tympanic membrane, footplate, and round window in response to a wave of pressure (compression phase) entering the ear.

known as scala media and the basilar membrane. It is these mechanical deformations that cause the hair cells of the organ of Corti to be stimulated. They in turn send a neural impulse to the brain over the auditory nerve.

To follow the path of sound to the hair cells of the organ of Corti requires a more detailed account of what happens in the cochlea than was implied in the sketch of the preceding paragraph. Initially, as stated above, the footplate of the stapes at the oval window acts upon the perilymph of the scala vestibuli. During an inward movement of the footplate, the fluid is displaced toward the apical opening of the scala vestibuli, the helicotrema. After passing through this opening, the wave of displacement travels through the scala tympani toward the basal end of the cochlea and toward the tympanic cavity. At the basal end is the round window, which moves to accommodate the fluid displaced by the footplate of the stapes. To account for a complete cycle of movement, the movement of the footplate and round window reverse in direction of excursion and move outward and inward, respectively.

In following a wave of displacement throughout the cochlea in the manner above, one very important route of the displacement was not considered. This pertains to the short-cuts between the two windows. The partition separating the windows is the cochlear partition and is filled with endolymph fluid, which does not communicate with either of the scalae (vestibuli or tympani) on its two sides. This is a flexible partition, with Reissner's membrane on the scala vestibuli wall and the basilar membrane on the scala tympani side.

A sharp wave of sound pressure is heard as a click. The pressure wave, introduced at the oval window, imposes a quick, strong wave of pressure in the perilymph of the scala vestibuli. There is little time for the pressure wave to displace fluid around the helicotrema and back to the round window. Instead, the sudden wave of pressure pushes on the cochlear partition. In turn, this partition bends in the direction of the scala tympani and the round window, where the relief from pressure is obtained by a reciprocal movement of the membrane covering the round window (Fig. 18-14).

The initial displacement of the cochlear partition occurs at the basal portion of the cochlea in the area between the windows for impulsive, click-like sounds. The bulge spreads toward the helicotrema in a wave pattern. It requires only about three milliseconds for the wave to travel to the helicotrema, where fluid will flow through this apical opening if the footplate remains displaced inward for the duration of the three milliseconds. This mechanical deformation of the cochlear partition, particularly of the basilar membrane, is an important process in analysis of sound by the inner ear (Fig. 18-15).

Slower increases in pressure, such as the sensation of a thump, permit some flow of perilymph around the helicotrema in addition to the bulge of the cochlear partition. However, the bulge is broader and begins more toward the helicotrema in comparison with short-lasting click-like sounds. In either case, and with all sounds, the propagation of a wave of deformation along the cochlear partition, and specifically the basilar membrane, travels toward the apical end. This direction of travel is due to the structure of the basilar membrane; it is narrowest at the basal end and widest at the apical end, but the taper is not entirely uniform. It is also of variable stiffness because of the fibers within the membrane, yet it is a flexible structure. As explained in an earlier chapter, the basilar membrane is not under

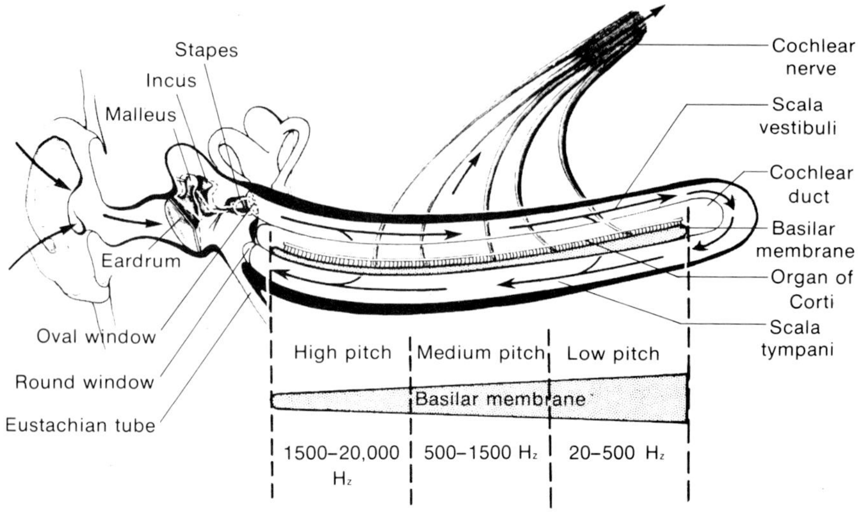

Figure 18-14. Diagram of an "uncoiled" cochlea showing the transmission of sound through the canals. Notice the relative widths of the basilar membrane at the basal (high Hz) and apical (low Hz) ends. (From Anderson, P.D.: *Clinical Anatomy and Physiology for Allied Health Sciences*, 1976. Courtesy of W.B. Saunders, Philadelphia.)

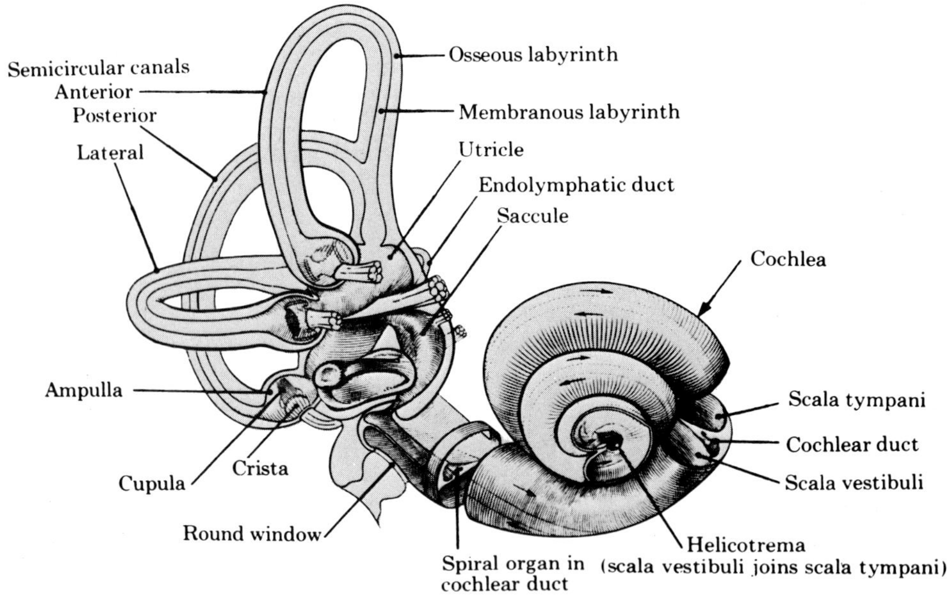

Figure 18-15. The cochlear canals showing the helicotrema. (From *Dorland's Illustrated Medical Dictionary*, 26th ed., 1981. Courtesy of W.B. Saunders Co., Philadelphia.)

tension; if slit, there is no gaping of the margins of the wound.

Along the cochlear partition are areas particularly sensitive to certain frequencies. A layout of these frequencies according to place on the basilar membrane has been established through experimental studies. The portion of the cochlear partition near the windows responds to very high frequencies. Moving along the partition toward the apical end, the frequency map shows progressively lower frequencies. By the time the first half of the cochlea has been considered, the frequency has dropped to about 2500 cycles per second and continues to drop until the apical portion is reached. It can be seen, then that the cochlea is ordered progressively from high to low frequency and that stimulation to regions along the membrane would explain in part how the ear handles frequency analysis of sound.

The other membrane of the cochlear partition, Reissner's membrane, is without stiffness, so that vibrations in the scala vestibuli are transmitted directly to the endolymph within the cochlear partition. Its function apparently keeps the perilymph and endolymph separated. Whether this is significant in the process of hearing is not understood. The pliability of Reissner's membrane does permit

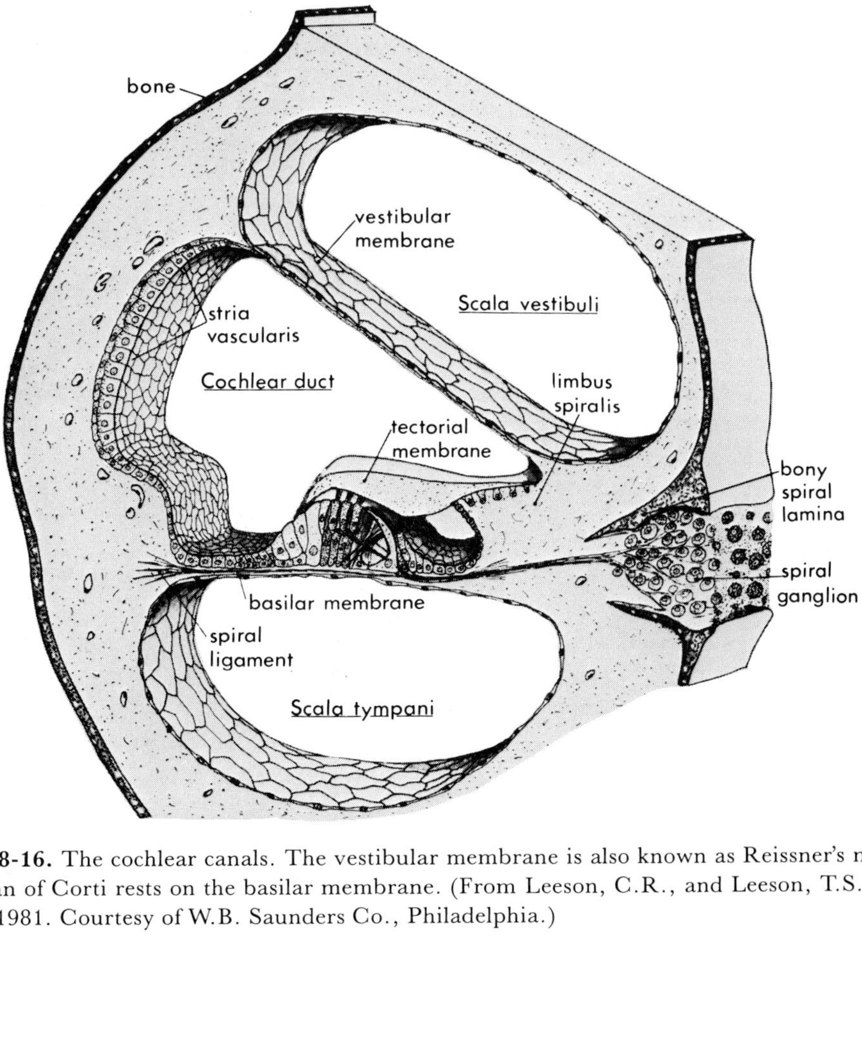

Figure 18-16. The cochlear canals. The vestibular membrane is also known as Reissner's membrane. The organ of Corti rests on the basilar membrane. (From Leeson, C.R., and Leeson, T.S.: *Histology*, 4th ed., 1981. Courtesy of W.B. Saunders Co., Philadelphia.)

the basilar membrane to be deformed readily as waves of pressure are introduced into the perilymph (Fig. 18-16).

HEARING BY BONE CONDUCTION. Up to this point, vibrations in the cochlear fluid have been considered as having been introduced by the stapes at the oval window or transmission of airborne sound introduced into the inner ear via the ossicles. Another route by which the cochlear fluid may be stimulated is by bone conduction. Setting the bones of the skull into vibration will, in turn, set the cochlear fluid into sympathetic motion, since the fluid is encased by bone. Sound above 40 dB in intensity may be heard even in the absence of the auditory ossicles.

As pressure waves strike the skull, they set the cranial bones into vibration, which activates the inner-ear fluid. Bone conduction oscillators (a type of contact speaker) are used to shake the skull bones very slightly during certain audiometric studies. The purpose is to assess hearing acuity by the mode of bone conduction for comparison with hearing acuity by air conduction. In short, this procedure is a study of the efficiency of the middle-ear transmission structures.

Organ of Corti

Described earlier as the organ of hearing proper, the organ of Corti contains the sensory cells. Its structural stiffness resists any deformation that might trigger a response of the sensory cells, except at the precise location of the upper structure of the cells where the cells are very sensitive to even the most minute movement (Fig. 18-17).

The gelatinous, ribbon-like tectorial membrane is fastened at the limbus and to the cells of Hensen; the latter attachment is made possible through a more flexible and thinner lip than the medial lip (at the modiolus). In response to up-and-down movements of the basilar membrane, the tectorial membrane is able to swing up and down more like the hard cover of a book but within the restriction of its double attachments. The tectorial membrane may also slide across the reticular lamina as the two structures rise and fall together in response to the introduction of sound vibration into the cochlear fluid.

The sliding action of the tectorial membrane is present principally due to differences in points of attachment of the

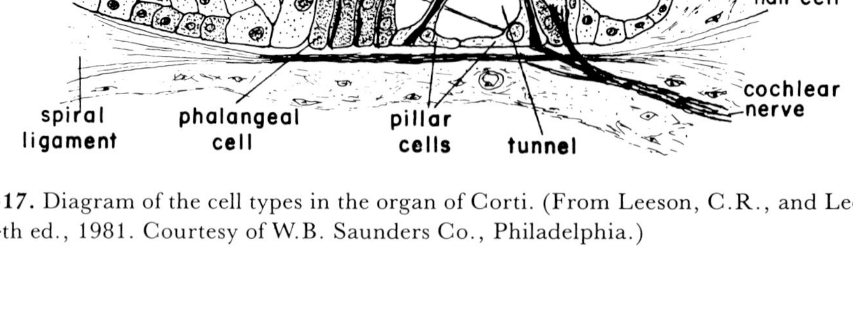

Figure 18-17. Diagram of the cell types in the organ of Corti. (From Leeson, C.R., and Leeson, T.S.: *Histology*, 4th ed., 1981. Courtesy of W.B. Saunders Co., Philadelphia.)

tectorial and basilar membranes. Even as the two membranes move up and down in parallel fashion to one another, the difference in hinging results in an important sliding and shearing action. The more medially hinged tectorial membrane causes it to slide in this direction (medially) as the basilar membrane moves up (Fig. 18-18).

Thus, deformation of the organ of

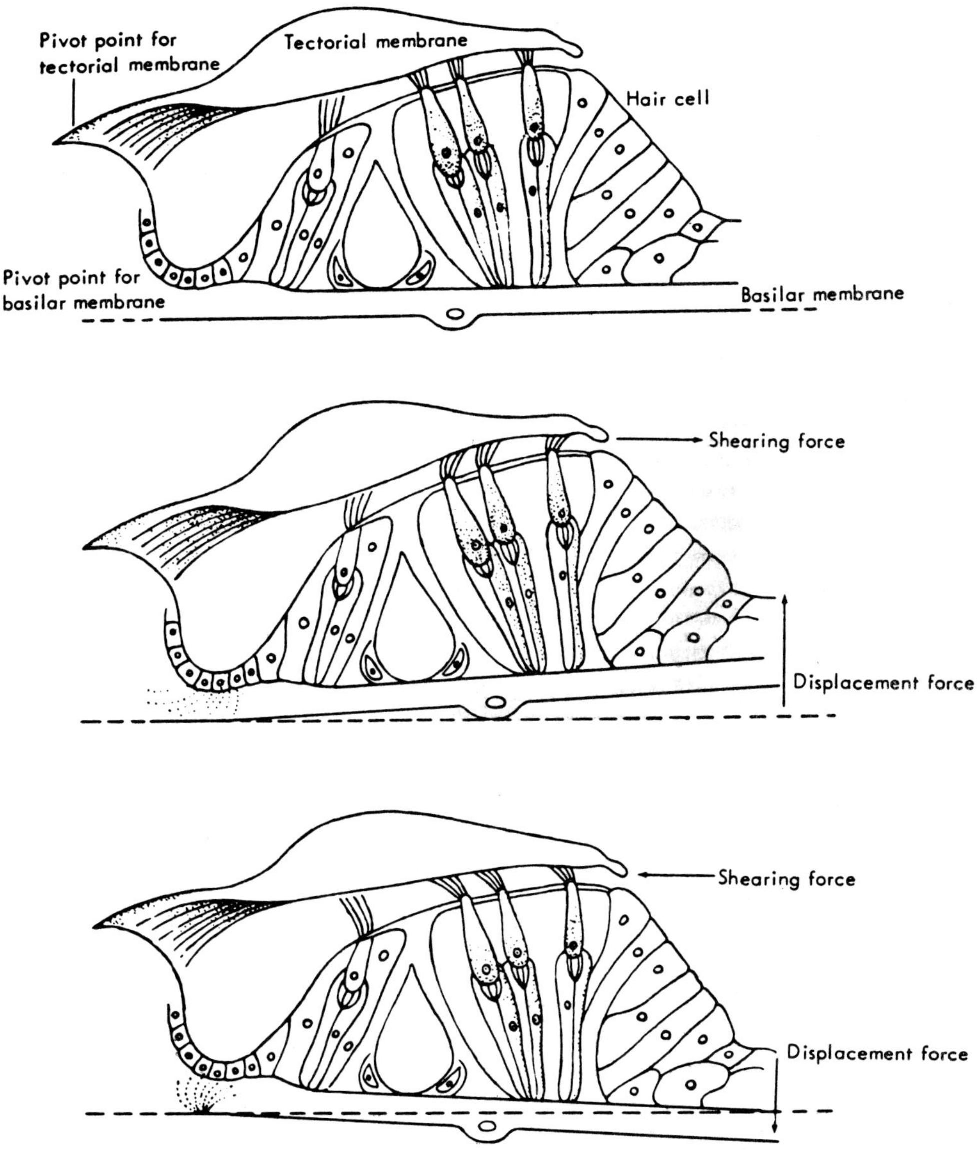

Figure 18-18. Schematic illustration of the shearing force between the tectorial membrane and the cilia of the hair cells — caused by separate hinging of tectorial membrane and the basilar membrane. (From Daniloff, R., Schuckers, G., and Feth, L.: *The Physiology of Speech and Hearing: An Introduction.*, ©1980, p. 402. Reprinted by permission of Prentice-Hall, Inc., Englewood Cliffs, N.J.)

Corti through fluid pressure on the Reissner's membrane or on the basilar membrane moves the tectorial membrane in a shear between the two structures. The stereocilia of the hair cells are bent according to the direction of the shear, since they are embedded in or attached to the tectorial membrane.

Bending of the stereocilia away from the modiolus (toward the double-pointed base of the W pattern of the stereocilia) triggers an impulse in the nerve fiber connecting with the hair cell. If the stereocilia are bent toward the modiolus, as may be the case when the basilar membrane rises, the nerve impulses are inhibited. Bending the stereocilia sideways results in no effect relative to nerve impulses being triggered (Fig. 18-19).

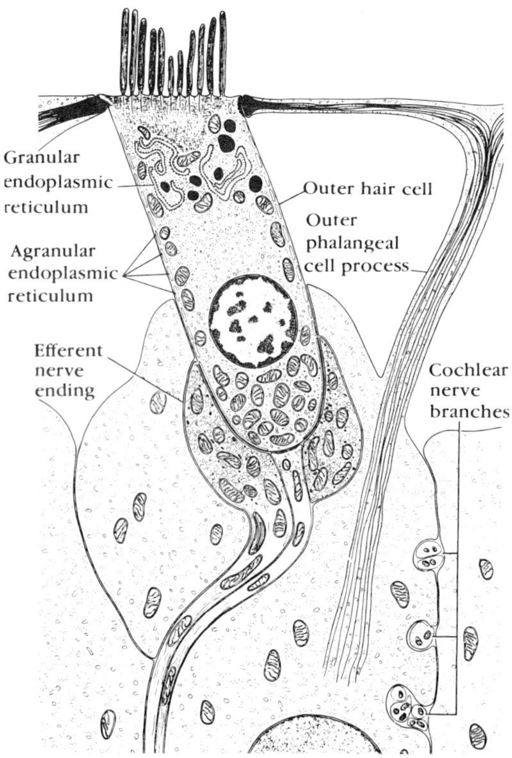

Figure 18-19. Microscopic detail of an outer hair cell. Notice the efferent nerve ending at the base of the cell adjacent to the afferent nerve ending at the midbase of the cell. (From Bloom, W., and Fawcett, D.W.: *A Textbook of Histology*, 10th ed., 1975. Courtesy of W.B. Saunders Co., Philadelphia.)

It is not understood how the bending of the stereocilia triggers the release of a chemical transmitter substance at the opposite end of the hair cell. The hair cell represents a mechanoreceptor like some other major sense organs, but the ear is the most sensitive. The sensitivity of mechanical deformation is placed at atomic dimensions and at energy levels fairly close to that of molecular movements, which are due to heat.

COCHLEAR MICROPHONICS. By some process, the hair cells are able to respond to deformations with an electrical output. This produced energy or potential is in proportion to the amplitude of mechanical displacement of the hair cell. The electrical energy pattern can be readily demonstrated by implanting two electrodes within the cochlea, one on each side of the cochlear partition. The resulting recording of the electrical potential is known as the *cochlear microphonic*. The term *microphonic* is used to denote the transduction of the acoustic signal into an electrical form after the fashion of a microphone. The resemblance is more closely associated with a resistance microphone, in which the mechanical movement of a plate or diaphragm in the microphone modulates the flow of electrical current from a battery. The hair cell acts as a battery, maintaining an electrical potential between the inner and outer surface of the cell itself. It is able to recharge itself.

When the stereocilia are bent by the shearing motions between the tectorial membrane and the hair cells, the cells convert this motion into an electrical voltage and amplify it at the same time. This voltage then excites the nerve endings, which lead via the auditory pathway to the brain. But this is not a description based on hard evidence and therefore must be accepted with caution.

Theories of Hearing

Describing the shearing force on the hair cells and their discharge of an electrical potential leaves untouched yet a more involved phenomenon. Thus far, no account for pitch discrimination has been mentioned. How the organ of Corti and the brain operate in the coding and perception of pitch introduces the topic concerning theories of hearing. Only a brief introduction is within the scope of this introductory text, however.

PLACE THEORY. A popular theory of cochlear action involves a place principle of frequency analysis. It has been given credence by Georg von Békésy and others who have mapped the mechanical activity of the basilar membrane and described how it acts as a frequency analyzer. Driving sounds of various frequencies and intensities were fed into the ear by von Békésy. By using a microscope, he actually observed the amplitudes of movement at different *places* along the membrane in response to the different frequencies. Others have measured the electrical output of the hair cells and confirmed this mapping of the basilar membrane. Frequency sensitivity of the basilar membrane progresses from high to low, beginning at the basal or window end. Thus, the relationship between place of maximum amplitude of the membrane corresponds well to the perceived frequency. Since the membrane is tapered from narrow to wide, the mapping of frequency would be expected to occur from high to low as the findings have indicated (Fig. 18-20).

Stated in another way, the place theory may also be discussed as a type of *traveling wave theory*. This appears proper, since waves in the cochlea are moving in a progressive pattern respective to the two windows. Waves traveling along the basilar membrane resulting from a given frequency have a wave formant or crest producing a greater displacement at one place along the membrane. This place relates to the relative stiffness of the basilar membrane because of its taper. The shearing force is greatest at this point, causing the hair cells to become active electrically. Frequency is therefore depicted by the place of greatest stimulation along the basilar membrane.

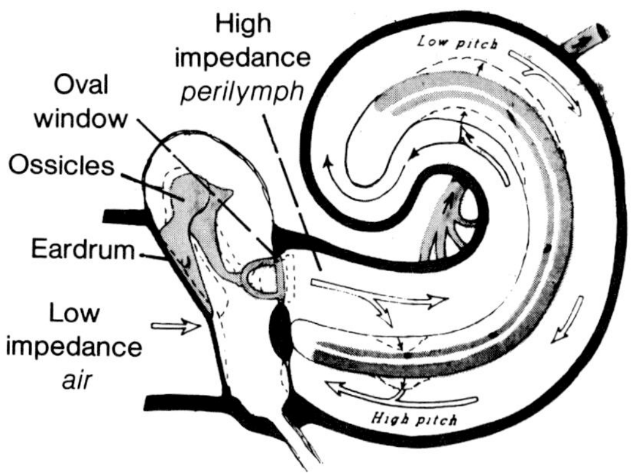

Figure 18-20. Schematic illustration of the path of an energy wave traveling through the ear. (From Jacob, S.W., and Francone, C.A.: *Laboratory Manual of Structure and Function in Man*, 5th ed., 1982. Courtesy of W.B. Saunders Co., Philadelphia.)

While much evidence may be cited in support of a place principle of hearing, there are definite deficiencies or unanswered questions. For example, the keenness of the ear in detecting frequency changes cannot be explained by localization of the vibration on the basilar membrane. Man's sense of pitch is based on more than just the point of maximum deformation along the membrane. Another question concerns the *missing fundamental*. In a musical tone having a number of overtones but whose fundamental pitch component or dominant frequency has been removed, the ear subjectively hears the missing sound, which acts to set the pitch of the tone. Hence, one hears a pitch whose *place* is not excited.

A further problem with the place theory is its difficulty in coping with the fact that pitch perception is altered by a change of intensity.

VOLLEY THEORY. This theory has been offered to account for frequency coding. It is known also as a neural-rate theory. It does not require that there be discrete analysis of sound so that all frequency components are spatially analyzed by the organ of Corti along the basilar membrane. Instead, all timing information of the stimulus sound is transmitted to the nervous system. It is then the responsibility of the nervous system to code the auditory data, including frequency, through patterns of discharge.

In response to a stimulus, the nerve impulses coming from the hair cells are grouped together in successive bursts or volleys. Each axon fires in a type of timing pattern for every cycle or most every cycle of the volley of impulses being sent out. This means that a given axon will not fire at exactly the same phase of every cycle; further, it may also skip one or more cycles between discharges. This is sometimes a physiological necessity, since a neuron may discharge an impulse only so many times per unit of time; it must have recharging time between firings. In recording the neural impulses from the auditory nerve of animals, the grouping of impulses is distinct for 500 Hz. Such groupings or volleys are also present up to 4000 Hz, though with decreasing clarity of the volleys. Above 4000 Hz, the timing of the impulses breaks down into a random pattern. Without question, the grouping of neural impulses below 1000 Hz is an important mechanism for coding frequency and transmitting this information to the brain. From this point, it is the task of the brain mechanisms to interpret the coded auditory data.

In reflecting on the two theories of hearing presented above, one may realize that both theories (the place principle and the frequency principle) have validity and that both could be considered interacting without nullifying each other. Measures of neural activity in the auditory tracts of the brain have not strongly supported or negated either theory. Thus, place and neural-rate information may be important components of the coding process whereby the brain receives all characteristics of a given sound stimulus, including pitch and loudness. Apparently, some central system is sensitive to the manner in which these components are coded so that, as the information progresses through brain levels phylogenetically, nerve structures utilize an even different type of analysis of the auditory impulses.

The two major theories presented above are representative of two classes of auditory theories, although the presentation here is sketchy and of an introductory nature. Even if all the available literature had been included (this would have required several volumes alone), the results would still leave the student without adequate information to explain satisfactorily how the ear transforms sound into neural impulses, preserving pitch and loudness data. Typically, this is considered in greater detail at graduate level studies. With further research into neural mechanisms and elaboration of present laboratory investigations of the ear, the gaps concerning auditory theory are sure to narrow.

BIBLIOGRAPHY

Denes, P.B., and Pinson, E.N.: *The Speech Chain: The Physics and Biology of Spoken Language*. Garden City, NY: Anch Doubleday, 1973.

Harris, J.D.: *Anatomy and Physiology of the Peripheral Hearing Mechanism*. Indianapolis, Bobbs, 1974.

Myers, D., Schlosser, W.D., and Winchester, R.A.: *Otologic Diagnosis and the Treatment of Deafness*. Summit, NJ: CIBA, 1962.

Newby, H.: *Audiology*, 3rd ed. Englewood Cliffs, NJ: P-H, 1972.

Whitfield, I.C.: *The Auditory Pathway*. Baltimore, Williams and Wilkins, 1967.

Wolfson, R.J., Schlosser, W.D., and Winchester, R.A.: *Vertigo*. Summit, NJ: CIBA, 1965.

PART FIVE: APPLIED CLINICAL ANATOMY AND PHYSIOLOGY OF THE HEAD AND NECK

Chapter 19

POSTNATAL GROWTH AND DEVELOPMENT

THE growing child is a dynamically changing organism that shows remarkable learning capabilities in the presence of dramatic body change. The miracle of normal speech is well appreciated, in part, by consideration of the growth timetable for various parts of the speech machinery. That speech is developed and habituated in spite of dramatic changes in facial, oral, and pharyngeal morphology is a unique capacity of the human organism.

In this chapter, selected components of the speech apparatus will be reviewed from a growth and development perspective. The postnatal growth of the face, oral cavity, and pharynx will be discussed and some clinical applications will be considered.

POSTNATAL PATTERN OF GROWTH

All living organisms exhibit a basic order or pattern in their growth. Vertebrate embryos express a pattern whereby the head develops first. A longitudinal, or head-to-toe, axis is established initially. The organism develops in this order: the head first, next the trunk, and then the front and hind limbs.

There is a great deal of variation within patterns of human development. The head of the newborn child is out of proportion with the rest of the body, as compared with the proportions of the adult. This disproportion is eventually eliminated as the postnatal growth of the head progresses at a slower rate than other parts of the body.

Each component of the organism's body exhibits its own growth timetable for *rate, direction*, and *magnitude*. The variations observed in each component account for the fact that each individual is slightly different from others. Some identical twins provide interesting exceptions to this generality.

The differential growth of the human head causes changes in *size* to be observed but not in the *pattern* of growth; that is, the individual continues to be recognized as the same individual in spite of size changes. Measurements of the human head indicate that, postnatally, the *head grows most in height, less in the anteroposterior dimension, and least in width*. The growth in height continues longest.

Two of the last components of the face to reach full growth are the chin and the nose. The chin is especially influenced by vertical growth in the late-teen years. The development of a "chin button" is associated with apposition (laying on) of bone on the chin that impacts on the horizontal and vertical dimension of the chin. Many adults, especially females, recall the early teen years as including dissatisfaction with their nose. The nose may appear somewhat large during this period when the lower face has not achieved its full growth potential. As the mandible and chin continue to develop (vertically and horizontally in the late teens), the balance of the face is improved and the nose takes on a more normal-size configuration. Even in the adult, it is not uncommon for a patient to seek out the care of a plastic surgeon for a nose reduction and end up having a chin implant instead. Some of the criteria for evaluating individual proportions of the face are presented in Chapter 21.

An appreciation for the growth of the facial area can be gained by briefly considering the growth rates for various other body tissues, which grow differentially in such a way that they become distinguishable from each other.

In 1930, Scammon reported findings for the differential growth rate for different body tissues. These data are included in the training of every medical and dental student, as it is an important data base for understanding postnatal growth and development. Scammon classified the body tissues into four categories: *general, neural, lymphoid*, and *genital* tissue types. The *general* or body-type growth, as shown in the curves that Scammon constructed from his research, reflects the pattern of growth of the body as a whole. This includes muscular, skeletal, circulatory, respiratory, and excretory systems. Height and weight are good examples of the product of general body growth. As shown in Figure 19-1, the curve for general body tissues tends to be roughly sigmoidal (S-shaped). This suggests a fairly high rate of growth during the early years of life. From ages six to ten years, there is a period of more moderate growth. From ages ten to sixteen years, there is a period of accelerated growth, the *prepubertal* and *pubertal* growth period. Following this, there is another period of leisurely growth until adult proportions have been achieved at around age twenty years.

The general body curve includes most muscles and bones. The central nervous system, by contrast, grows at a different rate, following the *neural* curve. The rapid postnatal growth of the brain, spinal cord, and special senses tends to be completed before puberty. The rapid acceleration of the growth of neural tissues of the body results in the characteristic curve shown in Figure 19-1.

The growth curve that Scammon portrays for *lymphoid* tissues of the body is characterized by exceedingly rapid growth up to puberty. At puberty, the amount of lymphoid tissue in the body is about 200 percent of that found at full maturity (around age 20 years). At puberty, the lymphoid tissues begin to involute or atrophy and eventually reach adult proportion (or 100% of adult growth as per Scammon's data). It is noteworthy that Scammon was not evaluating the growth of the tonsils and adenoids in his original research. His data for lymphoid tissue were derived from samples taken in the intestinal organs, the thymus, and other internal organs. More recently, it has been established that the growth and development of the tonsils and adenoids mirrors what Scammon reported for other lymphoid tissues, especially in the absence of recurrent bouts of tonsillitis or

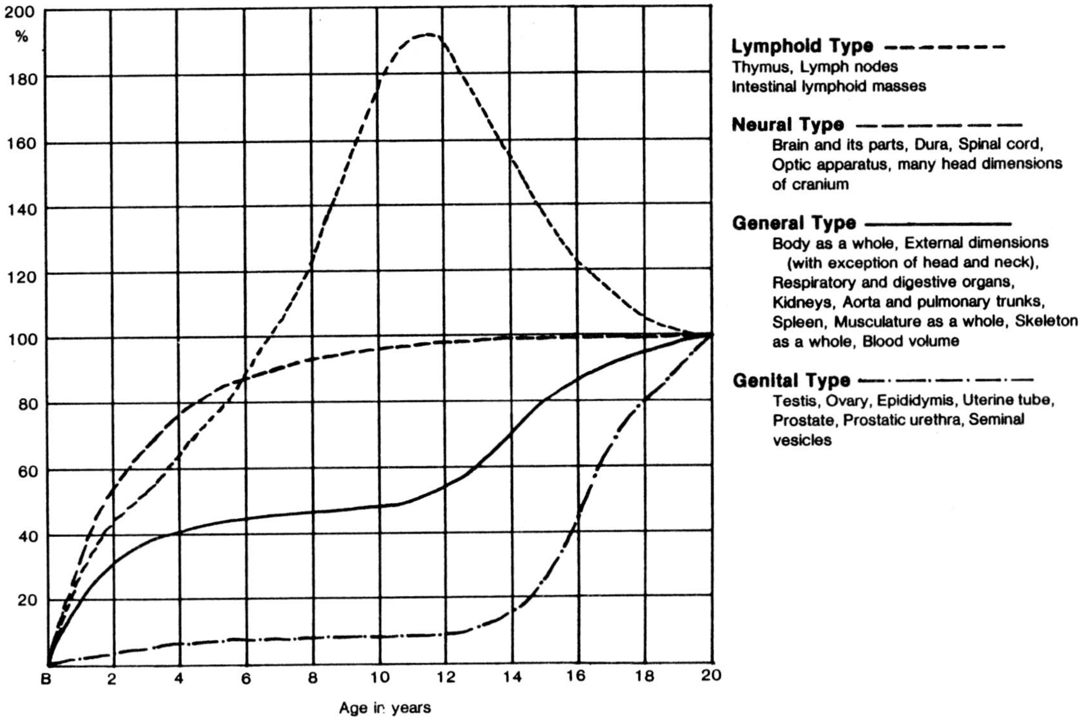

Figure 19-1. Tissue and organ growth curves from the research of Scammon (1930). The solid line shows general or body type, which includes most muscular, skeletal, circulatory, respiratory and excretory systems. The long-dashed line shows neural-type tissues, and the short-dashed line shows lymphoid-type tissues, such as in the thymus, lymph nodes and intestinal lymphoid masses. The combination dashed-dotted line depicts genital-type tissues.

adenitis.

The curve for *genital* tissues shows acceleration just prior to puberty when the sex organs undergo rapid maturation. It is interesting that the curve for genital tissue acceleration has moved over to the left more than one full year since the 1930s. That means that the inception of puberty is currently a year sooner than it was in the 1930s.

Consideration of the Scammon curves for tissue and organ growth should stimulate interesting speculations by speech clinicians as to some possible links to speech development. It is interesting that in the early speech-formative years, neural tissue growth and maturation far exceed muscle and bone growth. It is possible that the disparity in the growth and maturation of neural tissues and the muscles that they supply accounts in part for the sequence of events that lead to a speech sound-developmental pattern. Those sounds that involve a fairly rudimentary neuromotor gesture pattern for production, as well as a fairly stable contactual surface, may have a greater chance of being learned at an early age. Conversely, those sounds that require a more advanced neuromotor gesture pattern and lack a stable contactual surface would logically be habituated later in the developmental sequence of speech learning.

An example of a rudimentary neuromotor gesture pattern is coaptation (contact or approximation) of the lips. The ability to open and close the lips is an obvously important neuromotor skill for the infant in sucking. In addition to the early development of this neuromotor pattern, it is also interesting that the sensory innervation of the lips is well developed.

The sensory receptors in the lips are highly developed, providing a high level of innervation (or a low innervation ratio between nerve fibers and muscle fibers served). The high level of sensory innervation of the lips is compatible with the early development and mastery of speech sounds involving lip contacts, such as /p/, /b/, and /m/. From a neurological design and function standpoint, as per motor and sensory innervation, the emergence of bilabial sounds early in the speech-learning sequence is logical and would be somewhat predictable from laboratory study of the nervous system.

Another example of a rudimentary neuromotor gesture is a protrusion or retraction pattern of the tongue. Even the neurologically impaired individual can protrude the tongue. The sensory distribution of nerve endings in the tongue is extremely rich in the tongue tip, while there are few receptors in the dorsum and blade of the tongue. As a consequence, contacts of the tongue tip against other structures provide a much higher level of sensory feedback from the tongue than a like contact from other areas of the tongue. From a neurological standpoint, it is not surprising that sounds that involve some contactual relationship of the tongue tip are matured earlier than those sounds that do not depend on tactile feedback.

A contrast to early learning of speech sounds is the /s/ sound. Any public school speech clinician will testify to the high incidence of /s/ misarticulations as characteristic of the school speech therapy case load. The /s/, for most individuals, is produced in the absence of tongue-tip contact with other structures. The tongue is usually raised in the oral cavity to some arbitrary position. The child learning speech has to depend more on kinesthetic or proprioceptive feedback to master production of /s/ than for sounds that have tactile components. On this basis, it is not surprising that so many children exhibit difficulty in the mastery of the /s/ sound. Often, a sound with tactile feedback is substituted for /s/, such as "th." As Scammon's curves indicate, the growth patterns for neural and general body tissues plateau and grow more leisurely at around age six years. It is at this time that most speech sounds are expected to be matured.

GROWTH AND DEVELOPMENT OF THE MAXILLA

The maxilla grows by adding new bone (apposition) onto some of its surfaces, while in other areas, bone is resorbed as a means of maintaining the pattern of growth. Generally, apposition occurs at the sutures, the posterior tuberosity area, and at the alveolar margins. Apposition of bone at the alveolus is, of course, associated with dental development (Fig. 19-2).

In a literal sense, the maxilla and nasomaxillary complex of structures grow posteriorly and superiorly if one considers the areas where new bone is laid down. The effect of this growth, however, is an anterior and somewhat inferior displace-

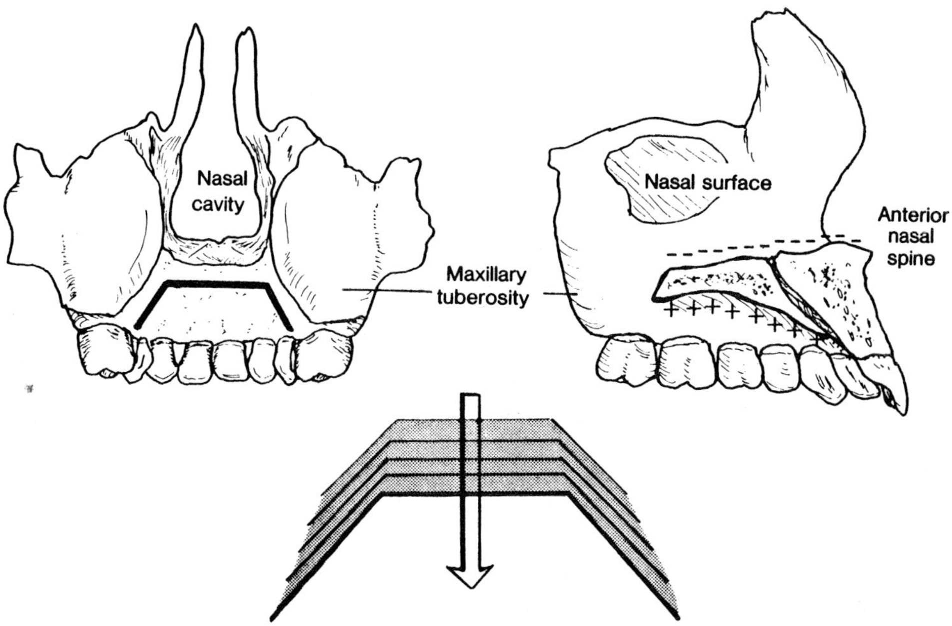

Figure 19-2. Growth of the maxilla showing areas of bone apposition (+) and resorption (-), and the "V" principle of palatal growth, with apposition on the oral surface and resorption on the nasal surface of the hard palate. Posterior aspect (left); sagittal aspect (right). (Modified from Enlow, D.H.: *Handbook of Facial Growth*, 2nd ed., 1982. Courtesy of W.B. Saunders Co., Philadelphia.)

ment of the maxilla. The reason for this is that the maxilla abuts the pterygoid plates, body of the sphenoid, cribriform plate of the ethmoid, and orbital plates of the frontal bones superiorly and posteriorly. New bone laid down on the maxilla at these abutments would result in displacement of the maxilla in the opposite direction, since these abutments are fixed areas. Accordingly, even though the maxilla grows in a posterior and superior direction, the clinical result is an anterior and inferior repositioning of the maxilla and nasomaxillary complex. Most clinicians simply say that the maxilla grows downward and forward in expressing the result of the processes of apposition and resorption that account for maxillary growth.

The rate of growth of the maxilla is more rapid than the mandible in the prepubertal period. As shown in Figure 19-3, the maxilla has expressed most of its growth by around age twelve, although some growth of the maxilla does occur variably between ages twelve and twenty.

The hard palate grows along with the maxilla since the majority of the bony palate is maxillary in origin. The appositional growth of bone on the maxillary alveolus and the oral surface of the hard palate accounts in large measure for the size increments in these structures. With bone being laid down on the oral side of the hard palate as the alveolar ridge grows and with resorption on the nasal (superior) surface, the palate tends to widen over time. This has been referred to as the "V" principle of growth, as shown in Figure 19-2. The shape of the palatal vault is somewhat related to the width of the dental arch. The tongue apparently plays a variable role in influencing palatal shape by a molding influence in a rest position. Another possible influence on the vault of the hard palate is the amount of restraint on the vertical descent of the hard palate exhibited by the midline nasal structures, i.e. the bony septum. A septum that lags in vertical development may encourage the vault of the palate to be increased. The maxillary dental arch may also narrow as it develops in the presence of restricted vertical development of the bony septum. The exact mechanisms by which the palatal vault is observed to be excessively vaulted or the dental arch is narrow remain highly speculative. It would be safe to assume that the reasons for the configuration of an individual hard palate are multifactorial.

GROWTH AND DEVELOPMENT OF THE MANDIBLE

Like the maxilla, the mandible is seen clinically to be displaced over time in a downward and forward direction. Hence, most people summarize that mandibular growth is in a downward and forward direction. Like the maxilla, mandibular growth is actually in a posterior and somewhat superior direction from the standpoint of the addition of new bone, which accounts for growth.

Bone is added in many places on the mandible, especially onto the condyles and posterior border of the ramus. These are major areas of apposition, while the anterior border of the ramus is a resorptive area. Apposition of bone also occurs along the alveolar ridge of the mandible, on the inner surface of the symphysis area (at the lingual tuberosity), and on the outer surface of the mandible at the lower border of the chin area. The area above the chin, or above the mental depression, is resorptive. This is why most of us develop a prominent "chin button," especially between the time of puberty and adulthood when development in this area is most active (Fig. 19-4).

The condyles of the mandible grow upward and backward. Since they abut with the glenoid (mandibular) fossae of the temporal bone, the growth pattern is a downward and somewhat forward response. As the ramus grows vertically

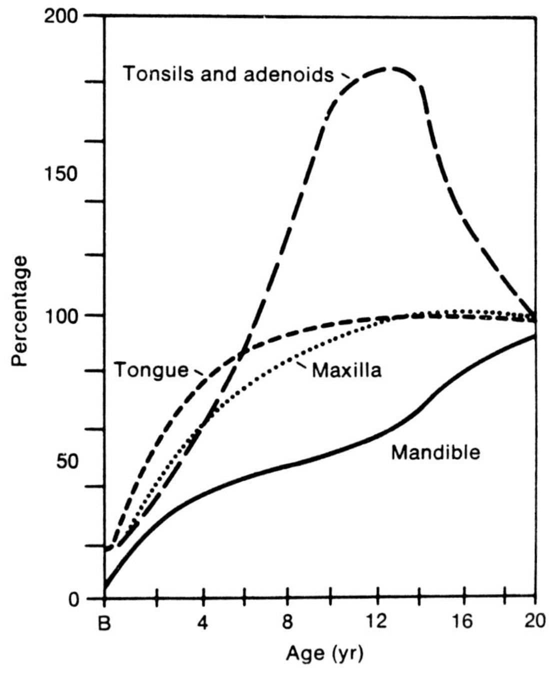

Figure 19-3. Summary of growth patterns for the tonsils and adenoids, tongue, maxilla, and mandible following the format developed by Scammon. (From Mason, Robert M., and Glass, Lillian: Speech and language. Part I: Normal speech and language development. In Stewart, Ray E., Barber, Thomas K., Troutman, Kenneth C., and Wei, Stephen H.Y., (Eds.): *Pediatric Dentistry* St. Louis, 1983, the C.V. Mosby Co.)

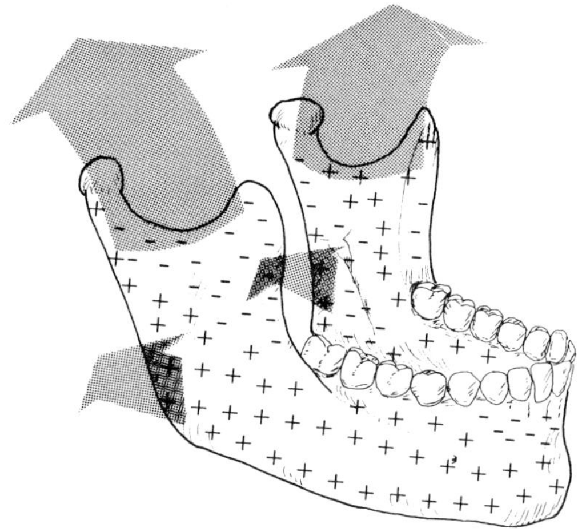

Figure 19-4. Growth of the mandible showing areas of bone apposition (+) and resorption (–) that account for mandibular growth and the maintenance of form with growth.

along with the condyles, the tongue is able to assume a more inferiorly positioned posture in the oral cavity.

The neck of the condyle is involved primarily in remodeling changes and area relocation. The body, or corpus of the mandible, lengthens as a function of displacement and remodeling activity of the ramus.

The role of the mandible in contributing to facial form is important. Figure 19-3 shows an overall growth cycle of the mandible in comparison with adjacent structures. The mandible grows slowly and steadily in a child, tapering off somewhat between ages of eight and twelve years, and then undergoing a pubertal and post-pubertal growth spurt up to around eighteen years. Additional increments in mandibular growth are seen into and past the twenties in some patients. The vertical portion of the mandible — the ramus — grows more slowly than the mandibular corpus during the elementary school years. In spite of growth differences within the two major parts of the mandible at various ages, the ramus is generally about one-third shorter than the corpus length at any age. This can be estimated by measuring from the mandibular condyle to the angle of the mandible and comparing this measurement with the distance recorded from the mandibular angle to the chin point.

The prolonged growth potential of the mandible influences the progression of facial development changes observed from childhood into the twenties. Interruption of the growth pattern during any stage of development can be responsible for anomalous facial form. An example is the child who fractures a mandibular condyle with resultant ankylosis of the affected joint area. If left untreated, a marked lower facial asymmetry will develop. The restricted side of the mandible will appear shortened, while ramus growth on the normal side appears elongated.

GROWTH AND DEVELOPMENT OF THE TONGUE

The tongue is one of the most adaptable organs of the body. Its adaptive capacity in respiratory and speaking functions is well known. Due to the strategic location of the tongue in the oral cavity and anterior boundary of the oropharynx, it is influenced by adjacent structures while also exerting influence on the growth and development of surrounding tissues.

Because of the adaptive nature of the tongue, it is extremely difficult to measure the size of the tongue in living specimens. Hence, tongue size and developmental status is a highly subjective clinical assessment.

It is generally held that the tongue exhibits a growth velocity similar to the neural tissues of the body, as shown in Figures 19-1 and 19-3. The tongue appears to grow steadily and approach maximum size at or near eight years of age. Since the tongue is attached to the mandible, in part, and since the tongue grows more rapidly than the mandible, the clinical implication of the growth differential is a natural tendency for the tongue to be positioned relatively high and forward in the oral cavity in the early growth years. The subjective impression of tongue size and resting tongue posture, at any point in time, is linked to the size and shape of adjacent structures as well as the respiratory demands placed upon the tongue. That is, in order to maintain an adequate size of the airway, the tongue may alter position as an adaptive response to some form of airway restriction. For this reason, the clinical procedure in evaluating the tongue should include an assessment of the relationship of oral and pharyngeal structures and the configuration of the airway (Fig. 19-5).

The adaptive nature of apparent tongue size and resting position is well demonstrated by the patient with mandibular prognathism. A prominent tongue in size and anterior resting posture prior to corrective jaw surgery most often appears to be smaller and rests more posteriorly following surgical setback of the mandible. Radiographic studies indicate that hyoid position changes after surgery (the hyoid drops inferiorly), which is an indication of how the tongue is able to adapt to the changes in the size and shape of the oral cavity created by surgery.

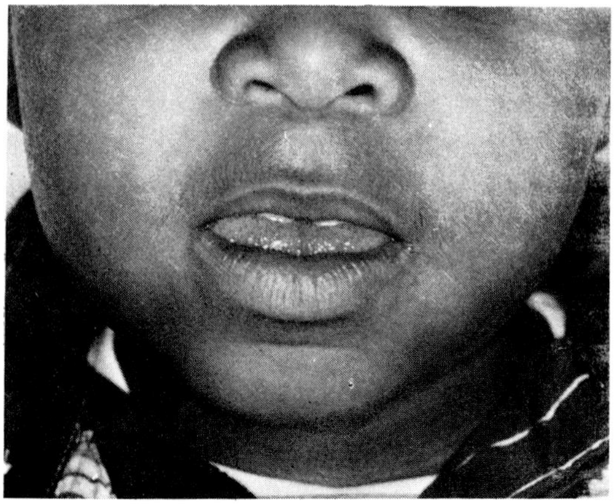

Figure 19-5. Patient with a prominent tongue with the "appearance" of macroglossia. However, this is a normal-sized tongue positioned forward in the oral cavity, with mandible hinged open, as a response to a small posterior airway. The tongue has adapted to respiratory demands and is postured forward.

Overall, tongue size remains an elusive component to document in individual patients. The clinician should be advised to resist labeling a tongue as exhibiting macroglossia unless there is some known mass-occupying lesion involved such as a hemangioma or neurofibroma. A macroglossia from natural hypertrophy of the tongue is a rare observation in clinical examination. As a case in point, the child with mongolism who has a mouth-open posture and forward-resting tongue posture appears to have a large tongue but in actuality possesses a small oral cavity. The maxilla is often retruded in a patient with mongolism, and the oral isthmus size is small. This necessitates a forward tongue carriage with the mandible hinged open to maintain the airway.

GROWTH AND DEVELOPMENT OF THE LIPS

The upper and lower lips are soft tissue masses that grow along with the facial bones that they overlap. The vertical growth of the lips is a variable phenomenon that does not exactly mirror the vertical growth of the facial skeleton.

There are marked growth changes in the lips between ages four and twenty years. The upper lip grows more steadily than the lower lip, reaching full growth expression by ages seventeen to nineteen years. The lower lip grows more than the upper lip and possesses a more varied pattern of growth. For the lower lip, there is an accelerated period of growth between ages nine and thirteen years. Growth is completed by around age eighteen to nineteen.

During the process of facial growth, it is normal for children to possess a lips-apart, mouth-open posture. This is related, in part, to the incomplete expression of lip growth in the vertical plane of space. The *lips*, as discussed here and in the supporting studies that have documented lip-lengthening over time, refer to the area between the anterior nasal spine and the margin of the upper lip, and the base of the chin and the margin of the lower lip (Vig and Cohen, 1979) (Fig. 19-6).

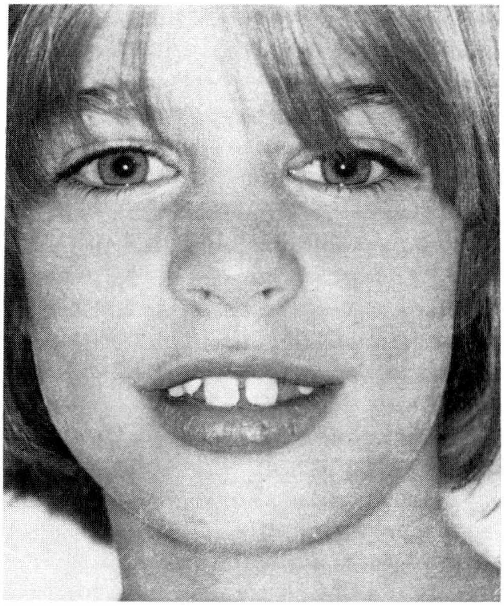

Figure 19-6. A ten-year-old child with incomplete lip growth showing *lip incompetence*. This is *not* mouthbreathing or the so-called adenoid facies (see text for discussion).

In those children whose lip growth does not equal vertical growth of the facial skeleton, it is often difficult to rest the lips together in a comfortable manner. A child or adult who has a lips-apart posture, showing several millimeters of tooth structure, but who can close the lips only with noticeable muscle effort in the mentalis area is said to have *lip incompetence*. From a growth and development standpoint, lip incompetence in children is a normal developmental stage, not to be mislabeled as "mouthbreathing." As stated previously in this text, mouthbreathing is a label that should only be used following a carefully controlled aerodynamic study of the airway using appropriate instrumentation. Studies indicate that most individuals with a lips-apart posture are able to drive an adequate amount of air intranasally (Fig. 19-7).

Lip incompetence spontaneously disappears in most children as the full potential of lip growth is experienced. That is, children with short lips at age twelve, with accompanying lip incompetence, usually show a lips-together resting posture at age eighteen or nineteen.

GROWTH AND DEVELOPMENT OF TONSILS AND ADENOIDS

The faucial tonsils, or "tonsils," and the pharyngeal tonsils, or the "adenoids" (or adenoid mass), have traditionally been discussed together. This is an unfortunate grouping, since the growth cycle of these two tissue organs may differ markedly in the same patient. They will be discussed in the same category in this text to point

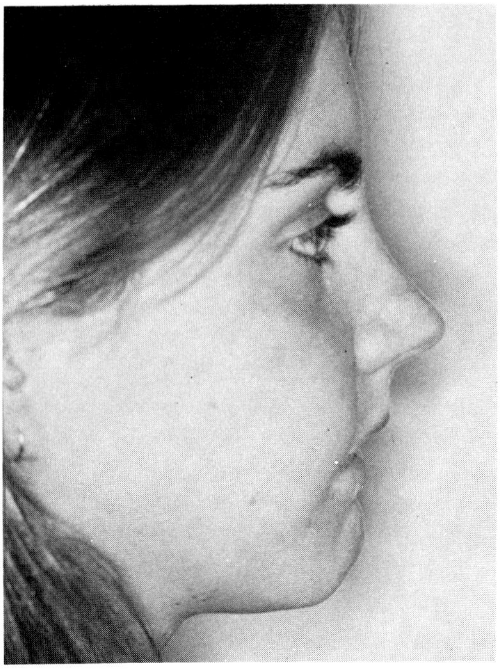

Figure 19-7. Lip incompetence in an adult. The lips are typically apart in repose. In this view, the patient closes the lips and shows the "mentalis strain" typical of lip incompetence.

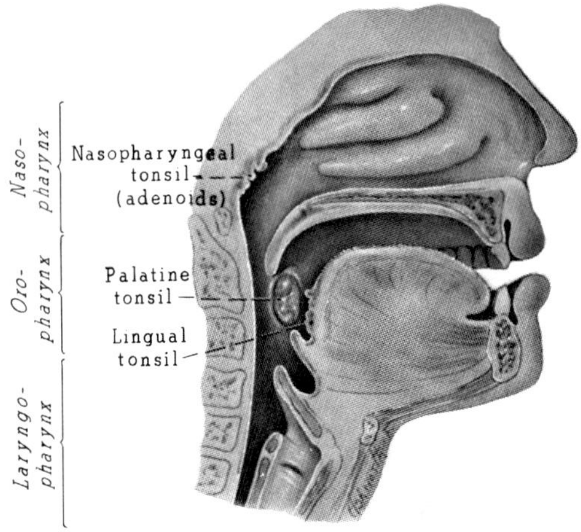

Figure 19-8. Lateral view of tonsillar tissues constituting Waldeyer's ring, including the pharyngeal tonsils (adenoids), faucial tonsils, and lingual tonsillar masses. (From Jacob, S.W., and Francone, C.A.: *Elements of Anatomy and Physiology*, 1976. Courtesy of W.B. Saunders Co., Philadelphia.)

out the differences between these masses, especially as related to speech function (Fig. 19-8).

The faucial tonsils have little to do with speech. It is a rare clinical finding that the tonsils prevent the tongue or other oral structures from performing their appointed duties for speech. By virtue of their strategic location at the oral isthmus, however, the tonsils can compete for the same space as the posterior portion of the tongue. In such instances, the tongue adapts by moving forward in the oral cavity to maintain airway adequacy.

If the faucial tonsils are exceptionally large, the surrounding faucial pillars can be displaced a sufficient distance from one another to effectively impede the soft palate from its full range of posterosuperior movement. That is, faucial pillars under great tension from large tonsils can bind the velum to some extent and prohibit full elevation of the velum. This situation, however, is thought to occur infrequently, as most children possess large tonsillar masses and normal velopharyngeal closure.

The faucial tonsils, and the adenoids, can be seen to begin proliferating at about age six weeks. These tissue masses grow continuously to the extent that they occupy a large area of the nasopharyngeal or oropharyngeal space in the growing child. The growth rate for tonsils and adenoids is generally depicted in Figure 19-3, but there is considerable variation between and among individuals. The clinician should *not* be tempted to presume about the size of the adenoid mass by intraoral observation of the size of the faucial tonsils. While these tissue organs are related in tissue type and general growth, they should be treated as separate entities for clinical purposes.

Faucial tonsils play a minor role in contributing to speech disability due to the normal adaptive capabilities of the tongue and surrounding structures. Removal of the tonsils should never be recommended by a speech clinician. The considerations for a tonsillectomy (or an adenoidectomy) should be for *medically* related factors.

The speech clinician is well aware that the *pharyngeal tonsil*, or *adenoid*, is a diffuse aggregation of lymphatic tissue found on the posterior and lateral walls of the nasopharynx. The adenoid mass reaches a growth peak around age twelve, with considerable variation among individuals. After puberty, the adenoid mass (and the faucial tonsils) are seen to involute or atrophy (waste away) spontaneously. By age twenty, little adenoid tissue remains (Fig. 19-9).

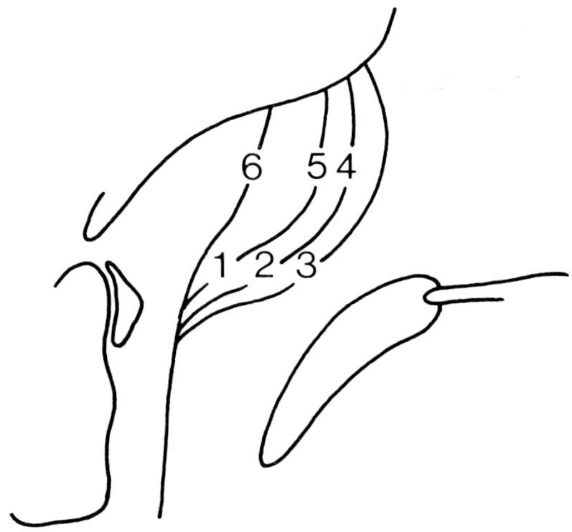

Figure 19-9. Lateral x-ray film tracing composite of growth and involution of the adenoids: (1) age 3, (2) age 8, (3) age 12, (4) age 15, (5) age 18, and (6) age 20. This is a normal cycle of adenoid growth and involution, although there is considerable variation in a normal population.

An enlarged adenoid may occlude the auditory (eustachian) tube's opening, resulting in a lowering of the air pressure inside the middle-ear cavity as the oxygen contained in the trapped air is absorbed by the membranous lining. Such a situation often calls for surgical removal of the adenoid tissue (adenoidectomy). The presence of large adenoids in the absence of pathology, however, does not signal a need for adenoidectomy. Adenoids are expected to be large in children.

Surgical removal of the adenoid creates a dramatic change in the dimensions of the nasopharyngeal space. Most of the time, velar activities adjust to this change and velopharyngeal closure is maintained. In about 5 percent of adenoidectomies, speech becomes persistently hypernasal following adenoid removal. This is associated with an inability for palatopharyngeal closure to take place in the new, surgically created environment of the nasopharynx. Factors that have been linked to speech deterioration following total adenoidectomy include defects in the anatomy and kinesiology of the soft palate, an anatomically wide nasopharyngeal port that was masked by the presence of a large adenoid mass, and an irregular (obtuse) configuration of the cranial base.

It should be obvious that the speech clinician can play an important role in identifying those individuals with normal nasal voice quality who are likely to become hypernasal if a complete adenoidectomy is performed. Such identifications represent one of the few opportunities in clinical work to *prevent* a speech problem. For this reason, such activity is particularly exciting and important. Clinical guidelines for identifying potential cases of hypernasality from adenoid removal are found in Mason (1973).

GROWTH AND DEVELOPMENT OF THE SOFT PALATE

The soft palate, during speaking and swallowing, serves to separate the nasal cavity from the oral cavity as needed. It participates in closure of the nasopharyngeal port with other structures in the pharynx in a sphincteric activity referred to as *velopharyngeal closure*.

The soft palate grows steadily from birth up to around age 18 years. During function, the soft palate normally buckles or dimples at about 80 percent of its overall length, as measured from the hard palate to the tip of the uvula. The *velar dimple*, as it is known clinically, represents the major attachment location in the soft palate of the levator palatini fibers. In individuals with a repaired cleft palate or a submucous cleft palate, the muscle fibers of the levator palatini are often found to be positioned farther forward in the velum. In these instances, the velar dimple occurs less than 70–80 percent of the distance from hard palate to uvula (Fig. 19-10).

The location of the velar dimple is an important diagnostic observation, in that the velar tissue beyond the dimple is not useful in the processes of velopharyngeal closure. That is, there is a difference in *overall* velar length (from hard palate to tip of uvula) and *effective* velar length (from hard palate to velar dimple). In velopharyngeal closure, the elevated soft palate at the area of the velar dimple makes a contact with the posterior wall of the pharynx, as seen in the lateral dimension. The tissue from velar dimple to the uvula hangs pendant.

The clinical importance of the estimation of velar dimple location in the soft palate is especially significant in the patient who is a candidate for adenoid removal. Some patients, with an anterior position of the velar dimple in the soft palate, are rendered hypernasal following a total adenoidectomy. In such instances, the adenoid mass was acting as a target for contacts of the velum in speech. Following adenoidectomy, the short effective length of the velum was insufficient to adjust to the changes in nasopharyngeal port opening (Fig. 19-11).

Examining for the velar dimple is a subjective evaluation in most clinical settings. The clinician should follow the guidelines offered in Chapter 23 in gaining proficiency in conducting an intraoral examination.

It is not common for a child to develop a hypernasal speech pattern with spontaneous involution of the adenoid mass. The apparent reason for the normal nasal resonance balance with adenoid involution is the adaptive capacity of the velum to accommodate to changes in the nasopharyngeal space. These adjustments have been described as the *velar stretch* capacity.

Velar stretch is associated with the ability of the levator palatini muscles to adjust to the gradual changes in nasopharyngeal architecture. This phenomenon is apparently diminished in the patient with an anterior location of the velar dimple or in a repaired cleft patient or an individual with a submucous cleft. These patients may experience deterioration of nasal voice quality over time with adenoid involution. In such patients, surgical removal of the adenoid mass can create a severe level of hypernasality due to the inability of the velum to adjust to the dramatic changes in nasopharyngeal dimensions.

Any patient with normal speech and preoperative intraoral findings that create suspicion about adequacy of velopharyngeal closure following total adenoid removal should be carefully evaluated by radiographic means. Alternative surgical

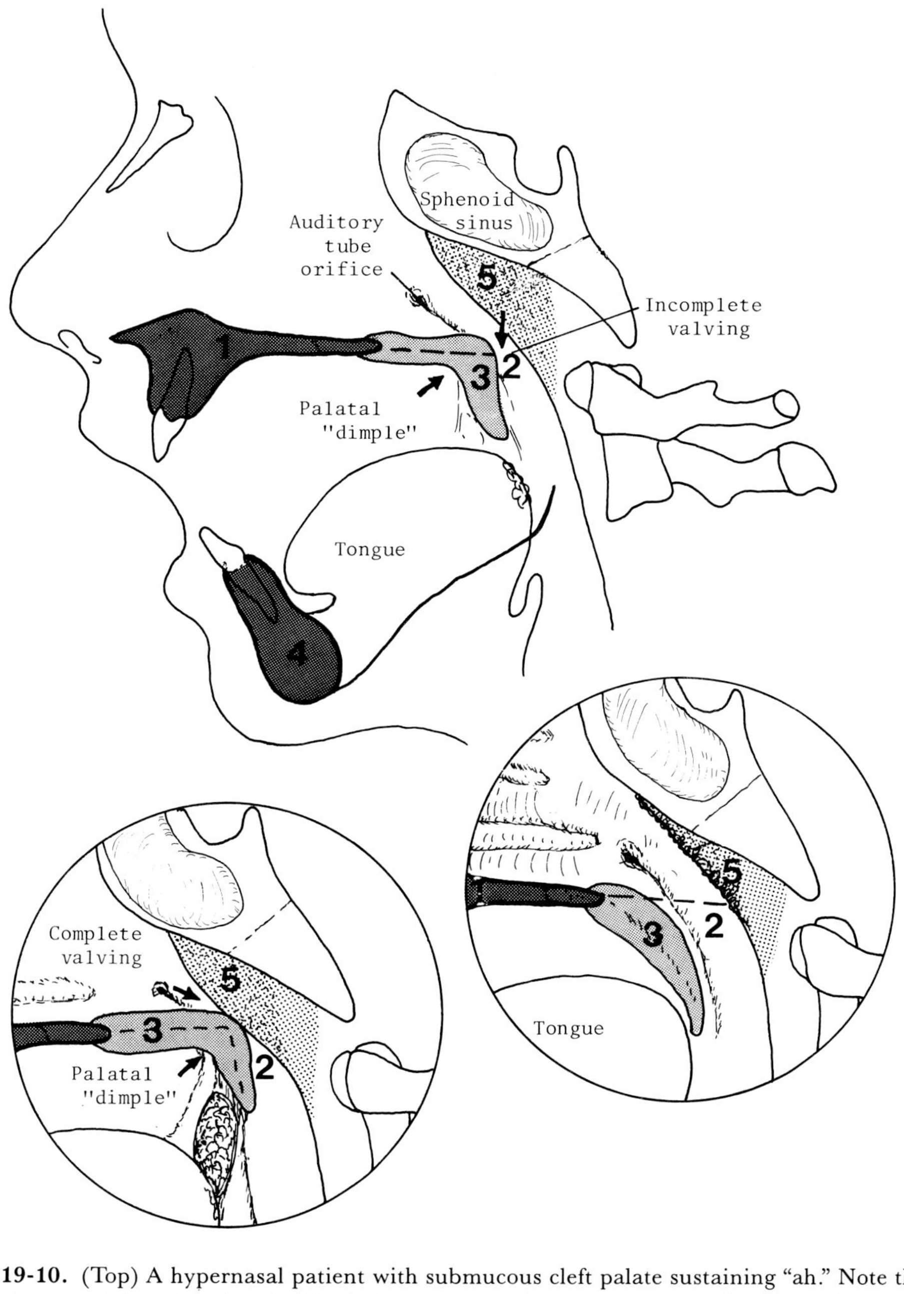

Figure 19-10. (Top) A hypernasal patient with submucous cleft palate sustaining "ah." Note the short effective length of the velum (dashed line) due to an anterior point of velar dimpling. Nasal resonance was normal prior to adenoid removal, although some adenoid tissue remains. (1) hard palate, (2) velopharyngeal gap, (3) velum, (4) mandible, (5) adenoid. (Right inset) The velum at rest, showing relationship to posterior pharyngeal wall. Note the torus tubarius and eustachian orifice on lateral wall of pharynx. (Left inset) Patient with normal point of velar dimple at the 80 percent mark exhibiting velopharyngeal closure during "ah" phonation. This patient is at risk for developing hypernasality if an adenoidectomy is performed due to a deep pharynx that is masked by the adenoid mass.

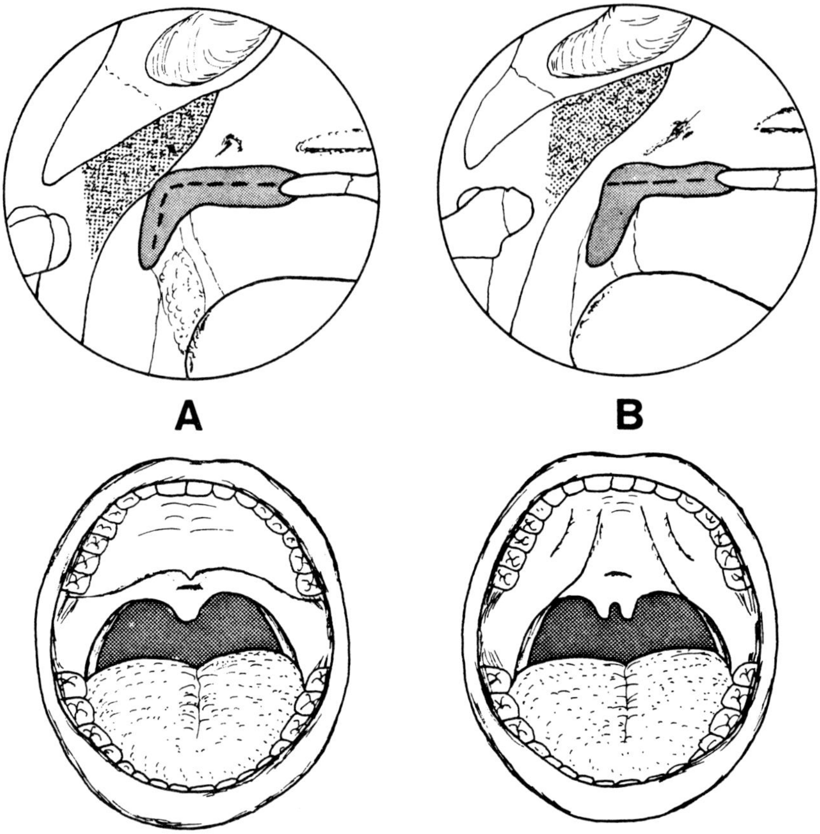

Figure 19-11. Intraoral and radiographic findings in patients with a normal velum (A) and a submucous cleft palate with bifid uvula (B) during sustained phonation of "ah". Both patients have a deep pharynx. Patient A will likely become hypernasal if the adenoid pad is removed even though the velar dimple is in a normal location. Patient B is hypernasal, with an anterior velar dimple and a deep pharynx. These patients demonstrate that evaluation of the palates is only one of several considerations in velopharyngeal function.

procedures, such as a lateral or peritubal (around the auditory tube orifice) adenoidectomy, may be recommended as a means of attempting to retain normal speech functions and yet eliminate adenoid encroachment from around the auditory or eustachian orifice.

GROWTH AND DEVELOPMENT OF THE NASOPHARYNX

The pharynx grows in response to the tissues that serve as its boundaries. As the facial skeleton descends away from the cranial base in growth, the nasopharynx widens to some extent. Growth of the face is accompanied by growth of the adenoid mass, and the dimensions of the pharynx are influenced by the amount of adenoid at any level, especially in the horizontal dimension.

The vertical dimension of the nasopharynx is determined by several considerations, including the angle of the cranial base and the height and contour of the cervical vertebrae. Some of these considerations are discussed in Chapter 24.

In the growing child, the nasopharynx tends to remain rather small in comparison with the configuration in the adult. This is largely associated with the presence of a large adenoid mass. In the pubertal period and beyond, the nasopharynx elongates as the individual grows taller. Increments in the vertical dimensions of the bodies of cervical vertebrae account for the dimensional changes in the vertical size of the nasopharynx. At the same time, the adenoid is involuting spontaneously to increase the horizontal diameter of the nasopharynx.

In the anteroposterior dimension, the hard palatal plane and the atlas of the first cervical vertebra maintain a fairly stable horizontal relationship over time. This may add some stability to the target area for velopharyngeal closure through the major growth and development stages, although no one knows for sure.

Overall, the nasopharyngeal space undergoes some dramatic changes in soft tissue mass and dimensionality from infancy to adulthood. The soft palate appears to be the major resource in adjusting to the changes in nasopharyngeal architecture as a means of maintaining normal patterns of speech through velopharyngeal closure.

BIBLIOGRAPHY

Bjork, A.: Variations in the growth pattern of the human mandible: longitudinal radiographic study by the implant method. *J Dent Res, 42*:400, 1963.

Brodie, A.J.: Consideration of musculature in diagnosis, treatment, and retention. *Am J Orthod, 38*:823, 1952.

Brodie, A.J.: The Relation of Glossopharyngeal Complex to Orthodontic Therapy: Growth Aspects. Presented at the meeting of the American Association of Orthodontists, Los Angeles, May, 1962.

Enlow, D.H.: *The Human Face*. New York, Harper and Row, 1968.

Handelman, C., Pruzansky, S., and Mason, R.: Hypoplastic Adenoids and Early Involution of Adenoids in Congenital Palatopharyngeal Incompetence. Presented at the meeting of the American Cleft Palate Association, Mexico City, May 1966.

Mason, R.: Preventing speech disorders following adenoidectomy by preoperative examination. *Clin Pediatr, 12*:405-414, 1973.

Mason, R., and Glass, L.: Speech and language. In Stewart, R., Barber, T., Troutman, K., and Wei, S. (Eds.): *Pediatric Dentistry: Scientific Foundations and Clinical Practice*. St. Louis, C.V. Mosby, 1982.

Mason, R., and Serafin, D.: The tongue: multidisciplinary considerations. In Serafin, D., and Georgiade, N. (Eds.): *Pediatric Plastic Surgery*, St. Louis, C.V. Mosby, 1984.

Moyers, R.E.: *Handbook of Orthodontics*, 3rd ed. Chicago, Year Book, 1973.

Pruzansky, S., and Mason, R.: The "stretch factor" in soft palate function. *J Dent Res, 48*:972, 1969.

Scammon, R.: The measurement of the body in childhood. In Harris, J. (Ed.): *The Measurement of Man*. Minneapolis, U of Minneapolis Pr, 1930.

Vig, P., and Cohen, A.: Vertical growth of the lips: a serial cephalometric study. *Am J Orthod, 75*: 405-425, 1979.

Chapter 20

FACIAL OSTEOTOMIES

GENERAL CONSIDERATIONS

THE modern oral and maxillofacial surgeon has the capability of surgically placing the jaws in virtually any new position. Individuals who suffer functionally or cosmetically from a *skeletal dysplasia* (abnormality of development) can benefit from surgery on the upper or lower jaw or both. Surgery is usually designed to provide a more functional dental arch relationship while also improving facial profile (cosmesis) (Fig. 20-1).

The general name for surgical repositioning of the jaws is an *osteotomy* (surgical cutting of a bone). In most facial osteotomies, some preoperative orthodontic treatment and postoperative orthodontic stabilizing treatment is required. Because of the collaborative contributions of the orthodontist and oral and maxillofacial surgeon, as well as a goal of surgery being the creation of a normal, straight (orthognathic) profile, the performance of facial osteotomies is usually referred to as *orthognathic surgery*.

The teeth usually adapt in some way to a dysplastic position of the jaws. Overgrowth of the mandible, or *prognathism*, for example, is usually accompanied by changes in the position of the upper and lower incisor teeth. As the lower jaw grows forward, the lower incisors adapt by tipping the incisal edges backward (lingually) to maintain contact with the upper incisors. Likewise, the incisal edges of the maxillary incisors usually flare facially to some degree. As a result of this, orthodontic work preoperatively involves *decompensating* the dentition – that is, removing the natural compensations to the dentition that have occurred over time as the dysplastic jaw growth was experienced. The removal of the dental compensations prior to surgery permits the teeth and jaws to be arranged in a more stable position at the time of surgery. Most often, orthodontic treatment preoperatively increases the magnitude of the problem. In the case of mandibular prognathism, the lower anterior teeth are flared forward while the upper incisors are tucked back lingually. The surgical procedure, then, is designed to align the teeth appropriately, one arch to the other, by changing the position of the jaws (Fig. 20-2).

Orthognathic surgery has become a common and predictable endeavor. Most often, the planning for surgery is done

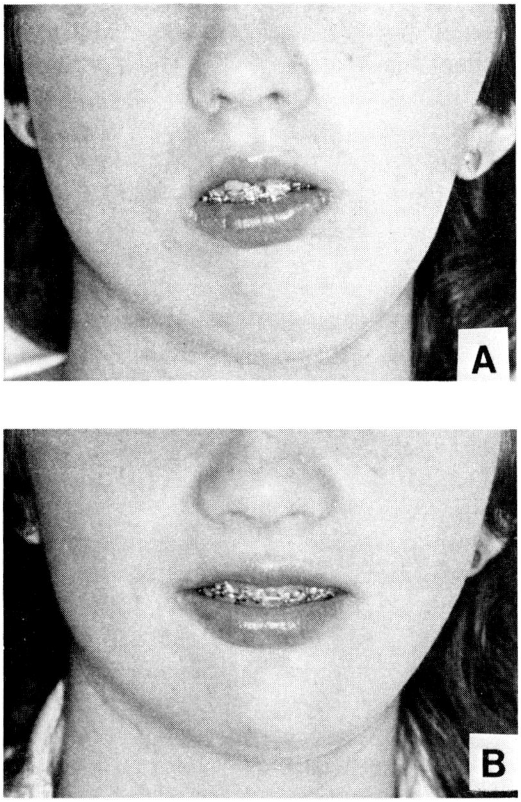

Figure 20-1. (A) 16-year-old patient with left hemifacial microsomia. The jaws were hypoplastic on the left side and the dental arches and supporting bone were malpositioned. The lower jaw deviated to the left. Both dental arches were "canted"; that is, tilted up on the left and down on the right. The patient is lip incompetent and shows 80 percent of upper tooth structure at rest. (B) Following maxillary impaction (8 mm up on the right) and rotation and uprighting of the mandible—combined maxillary and mandibular osteotomies.

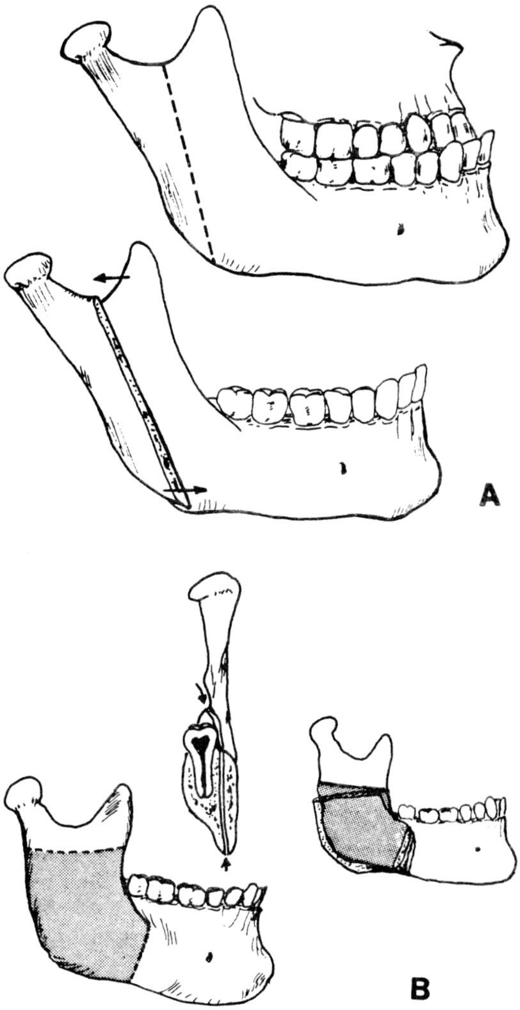

Figure 20-2. Mandibular osteotomies for prognathism. (A) Vertical oblique osteotomy, which can be done with an extraoral or intraoral approach. (B) A sagittal-split osteotomy, where cuts are made on the inside of the ramus and the outer surface of the body of the mandible.

mutually by an orthodontist and an oral and maxillofacial surgeon. In many instances, the orthodontist constructs the stabilizing acrylic splint utilized in surgery. Frequently, the orthodontist assists in surgery by placing the splint and wiring the jaws together. The splint is wired to the teeth by means of either orthodontic bands or archbars (metal bars wired to the teeth in surgery). Most surgical procedures in orthognathic surgery necessitate fixating the jaws in an immovable position for about six weeks to facilitate healing of the surgically modified jaw(s) position. Wiring the jaws closed as a part of surgery is called *intermaxillary fixation*. Such fixation is always utilized when one or both jaws have been moved surgically, while a partial or segmental osteotomy of a jaw may not require intermaxillary fixation.

MAXILLARY OSTEOTOMIES

An osteotomy to reposition the upper jaw, in most cases, referred to as a *LeFort osteotomy*, named after a French surgeon-anatomist, Rene LeFort. In 1900 and 1901, LeFort reported on experiments designed to illustrate the sites of fractures of the facial skeleton. These reports are now known as LeFort's classification of maxillary fractures. It is of general interest that LeFort's experiments involved whole cadavers. LeFort subjected the cadaver heads to a variety of traumatic blows, while varying the direction and force of the application. Afterward, the lines that the fracture patterns followed were carefully determined.

It is somewhat ironic that these early experiments have provided the modern surgeon with guidelines for the selection of surgical incisions for orthognathic procedures. The three basic patterns of facial fractures of the maxilla are now referred to as LeFort I, II, and III osteotomies.

A *LeFort I* osteotomy involves surgery to the alveolar process of the maxillae. The position of the teeth is changed, and there are also changes expected in the base of the nose, the upper lip, and in the nasal cavity. It might be expected that as the underlying bony framework is modified, adjacent soft tissues also change variably. The incisions for a LeFort I osteotomy are in the upper vestibule of the oral cavity, crossing the nasal floor, through the septum of the nose, and often at or near the anterior nasal spine (Fig. 20-3). The entire surgery is done intraorally.

The LeFort I osteotomy is the procedure of choice for advancing the lower portion of the maxilla (maxillary advancement), such as in an individual with a repaired cleft lip and palate who shows midfacial retrusion but who has acceptable orbits and cheek contours or projections (Fig. 20-4). This operation is also suggested for the patient who has vertical maxillary excess with a toothy smile (vertically). A strip of bone is removed on both sides and the maxilla is repositioned superiorly in a *maxillary impaction* LeFort I osteotomy (Figs. 20-5, and 20-6A–D).

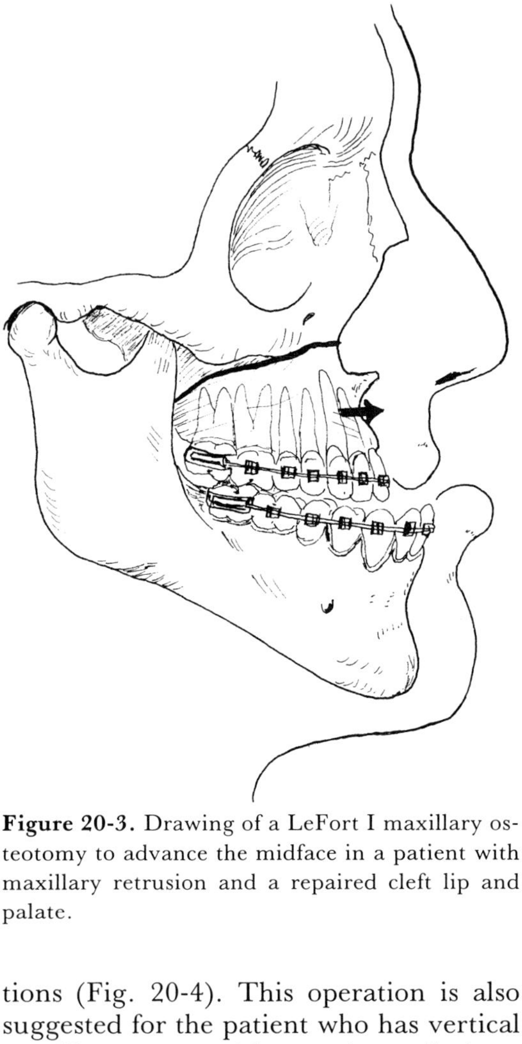

Figure 20-3. Drawing of a LeFort I maxillary osteotomy to advance the midface in a patient with maxillary retrusion and a repaired cleft lip and palate.

A partial or segmental osteotomy can be done to a portion of the dental arch, such as the surgical removal of first bicuspids and supporting bone at that area, and followed by retraction of the anterior dental and skeletal segment into the space created. Even a single tooth osteotomy

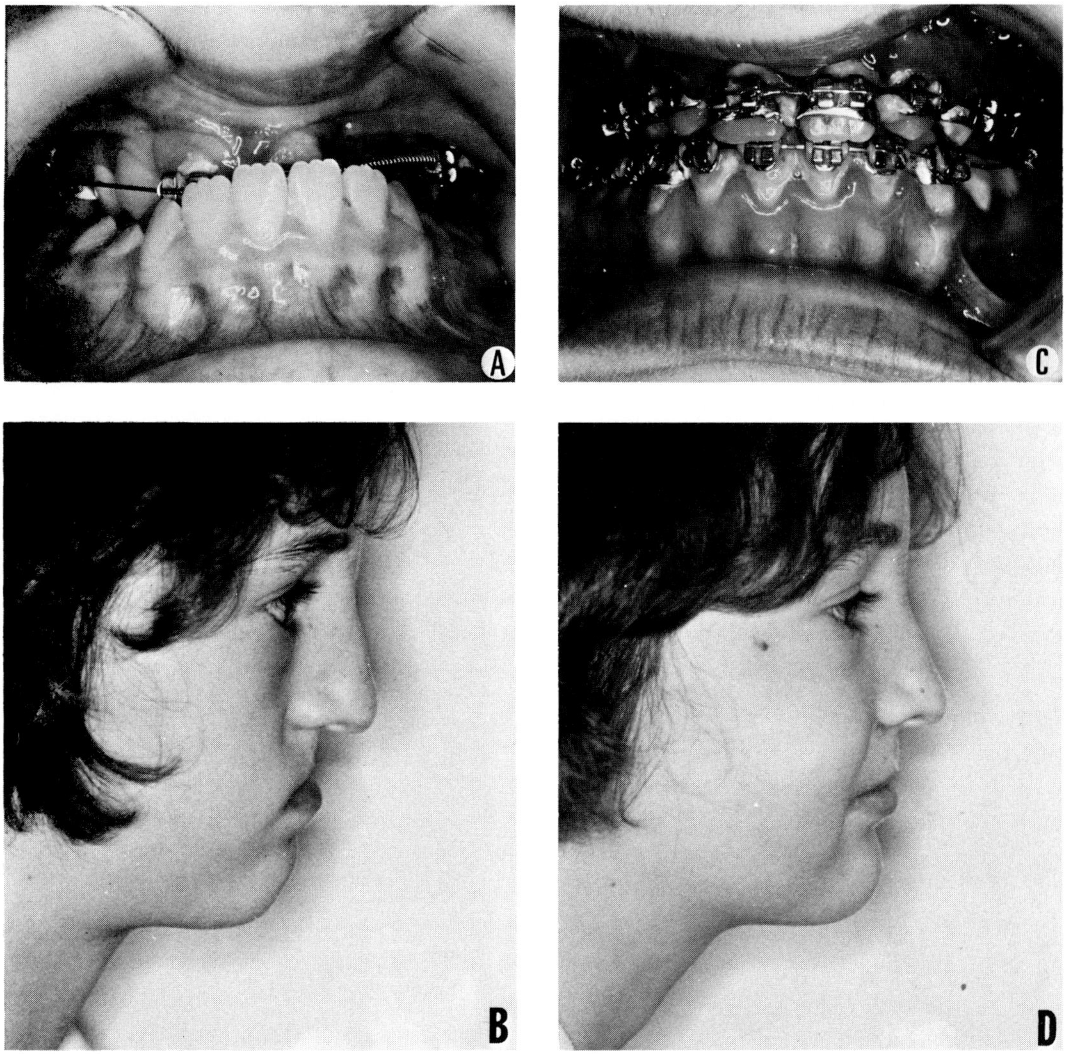

Figure 20-4. Dental occlusion and facial profile views of thirteen-year-old patient with a repaired cleft lip and palate who underwent a maxillary advancement osteotomy for midfacial retrusion.

can be done, although this would not be described as a LeFort osteotomy.

A *LeFort II* osteotomy is more extensive in its scope and in the vertical dimensions of the operation (Fig. 20-7). Incisions extend across the nasal bridge and across the orbital floor and lateral walls of the orbits. This procedure is indicated especially for a patient with midfacial retrusion who also has flattening of the cheeks and orbital floor area. Hence, a LeFort II maxillary advancement can correct midfacial retrusion while simultaneously improving a flattened upper midface. Most of the surgery is done intraorally on LeFort II osteotomies.

A *LeFort III* osteotomy is an extensive procedure used primarily with craniofacial

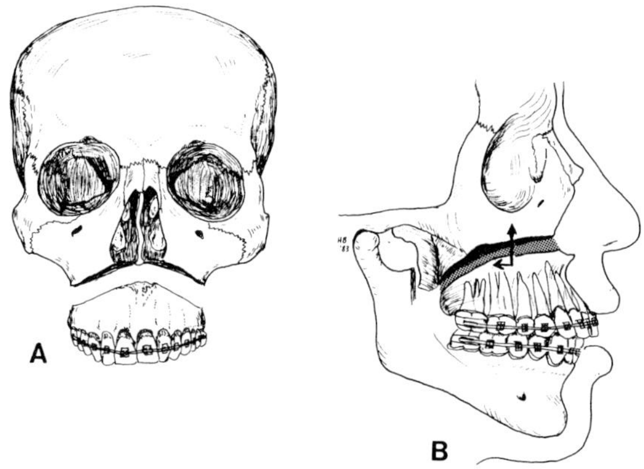

Figure 20-5. A LeFort I maxillary impaction for a patient with vertical maxillary excess. The stippled area of bone in B is removed by surgery (intraorally), and the maxilla is repositioned in a superior position. (A) Coronal view of the area where the maxilla is cut. (B) Lateral view.

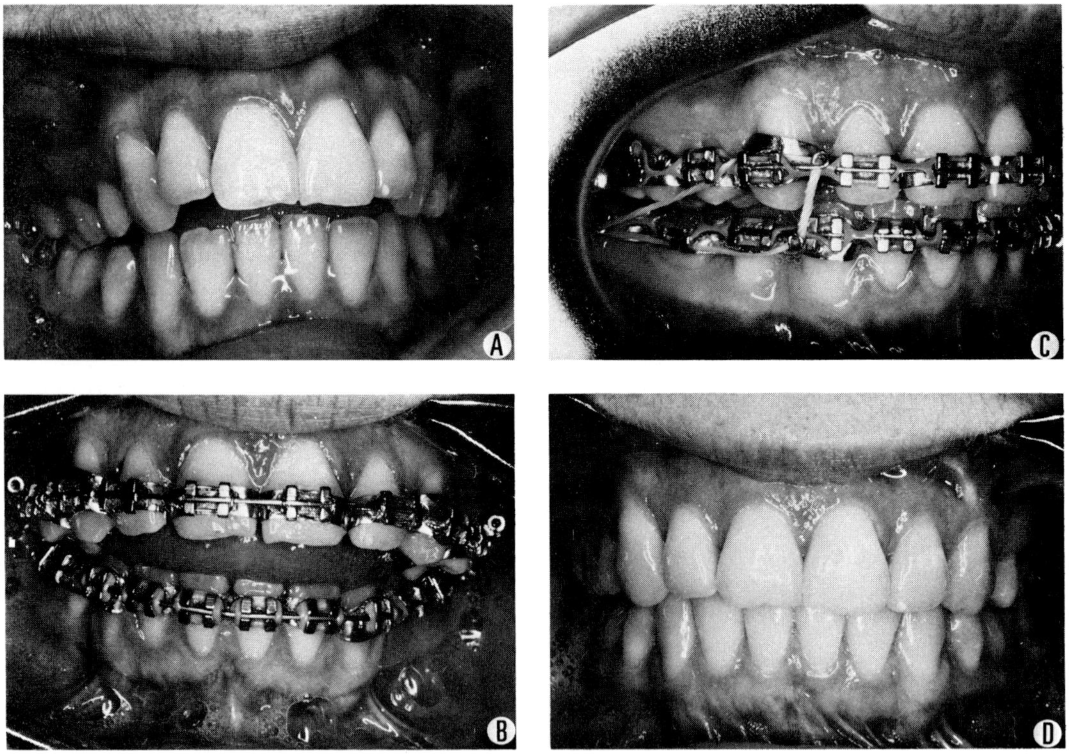

Figure 20-6. (A) Preoperative dental appearance of a patient with vertical maxillary excess. Due to the anterior openbite, the posterior portion of the maxilla needs to be surgically impacted more than the anterior part.

(B) Preoperative orthodontic treatment. The openbite increases as the arches are leveled, revealing the true extent of the deformity.

(C) Following surgery and release from intermaxillary fixation (jaws wired shut), light elastic forces are used to aid in stabilization and healing.

(D) The dental interdigitation following the completion of post-operative orthodontics.

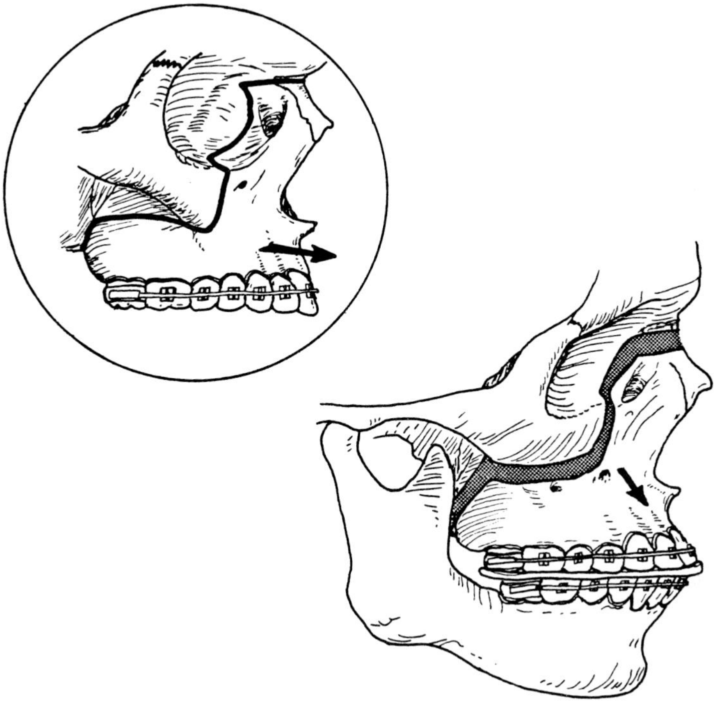

Figure 20-7. Drawing of a LeFort II osteotomy, which is used for midfacial advancement. Bone grafts are used to fill in the area where the advancement cuts are made.

syndrome patients such as Crouzon's, Apert's, and Pfeiffer's syndromes. Such patients usually have severe midface retrusion and other craniofacial variations. The LeFort III osteotomy mobilizes the entire maxillary-orbital structures and involves incisions across the vertex of the skull as part of the surgery (Fig. 20-8). Most of the surgery on the maxilla can be done intraorally.

The results of surgery can be extremely dramatic, especially since it is no longer rare to operate on both jaws and the chin (genioplasty) in the same procedure. Where two-jaw surgery is done in the same operation, it is common to employ two acrylic splints in surgery: (1) an intermediate splint to position the newly mobilized jaw via the unoperated jaw and (2) the final splint used in intermaxillary fixation. Needless to say, any osteotomy, especially involving the maxilla, necessitates careful planning and coordinated efforts between surgeon and orthodontist.

Speech Implications from LeFort Osteotomies

The speech clinician should have some rudimentary knowledge of osteotomies, in that patients of mutual concern to the speech clinician and a variety of other specialists will undergo such procedures. Surgical advancement of the maxilla can potentially compromise the velopharyngeal mechanism and create hypernasality, especially in patients with repaired cleft palate where the adaptive capacity of the velum is reduced (Mason, Turvey, and Warren, 1980; Mason and Warren, 1981). In other patients, such as those with mid-

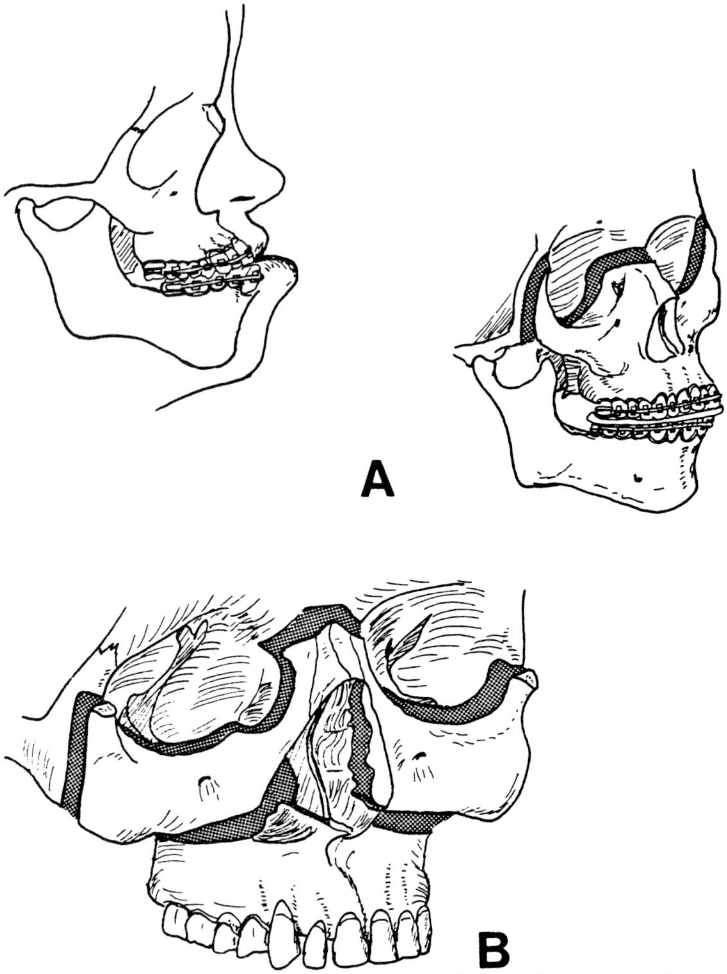

Figure 20-8. A LeFort III osteotomy, which is used for midfacial advancement in patients with severe midfacial retrusion. The entire maxillary-orbital structures are mobilized.

face retrusion syndromes (Crouzon's, Apert's, and Pfeiffer's), the nasopharyngeal area is very small prior to surgery, with a hyponasal voice quality. Following midfacial advancement surgery, speech is usually seen to improve in nasal resonance balance. Thus, speech may improve spontaneously with surgical correction of a skeletal and dental malocclusion.

MANDIBULAR OSTEOTOMIES

Orthognathic surgery for the mandible may involve positioning the mandible up, down, back, forward, or any combination of positional changes. A genioplasty to augment or decrease the chin prominence may be done in combination with a mandibular osteotomy.

A mandibular osteotomy for *prognath-*

ism involves separation of the two sides of the mandible at the angle and sliding the corpus of the mandible posteriorly to overlap the distal segment (ramus) as a means of reducing the prognathism. There are a variety of surgical approaches and procedures available to the surgeon. The operations can be done intraorally by making incisions in the facial portion of the oral vestibule. An extraoral approach may be utilized, with a small incision made below the angle of the mandible. An operative procedure to retract the mandible may be done with a vertical oblique osteotomy (see Fig. 20-2A), either by extraoral or intraoral approach, or a sagittal-split osteotomy done intraorally. While other surgical variations are also utilized by some, the procedures noted above are the procedures employed most often.

Advancement of the mandible involves a sagittal-split osteotomy as the procedure of choice (see Fig. 20-2B). The ramus and corpus are split and slid along one another to affect the lengthening of the mandible. It is not common for fixating wires to be required at the surgical site at the ramus-corpus level. The fixation of upper and lower teeth into the acrylic splint provide sufficient stability to permit the area of surgery to heal and solidify in the position created.

When a mandibular or maxilary osteotomy is done, it is often necessary to modify the position of the chin. A genioplasty to either augment or reduce the lower face height and horizontal position of the chin is a common operation, either in isolation or in combination with other procedures. This work can be done intraorally, using a patient's own bone to either slide upon or reduce, or use a graft from another area of the body as a wedge. It can also be done using an extraoral incision below the chin with the implantation of silicone or some other inert implantable material.

Speech Implications from Mandibular Osteotomies

The surgical modification of the position of the lower jaw should also affect a change in the resting posture of the tongue as well as functions of the tongue. Generally, it has been found that an individual's speech adapts quickly and effectively to an osteotomy. On the basis of available clinical evidence, it is not expected that speech should deteriorate following a mandibular osteotomy (or a maxillary procedure either). Individuals who undergo osteotomies, who previous to the operation had normal speech, appear to adapt quite well to a surgically modified oral environment. The patient with abnormal speech patterning preoperatively may spontaneously adapt to a normal speech physiologic pattern but may also require therapy postoperatively.

A case in point is the patient with a skeletal anterior openbite malocclusion. Such a patient may be expected to exhibit a tongue thrust swallow pattern and a cosmetic tongue thrust in speaking. Sibilants may be produced normally or abnormally from an acoustic standpoint prior to surgery. Such a patient has a fairly good prognosis for achieving a nonthrusting pattern following surgery.

The individual with a skeletal openbite is also a good example of biological considerations in the choice of surgical procedure. It is a general rule in orthognathic surgery that anterior skeletal openbites are *not* closed by mandibular osteotomies. Of all the types of osteotomies, a mandibular osteotomy for anterior openbite is subject to more postoperative relapse than almost any other procedure. This is associated with the strong pull of the strap muscles of the neck in fighting the rotational repositioning of the mandible. In most cases, the posterior portion of the maxilla is *impacted* as a means of closing an anterior openbite. Impaction of the maxilla, either in one piece or with posterior bilateral segmental impactions, permits the mandible to autorotate closed as the upper molar teeth are repositioned superiorly. Autorotation of the mandible is a natural adjustment of the facial skeleton, while rotating the corpus of the mandible relative to the ramus, especially with an elevation of the front of the mandible, is evidently not a normal rotation. Most important, however, is the principle that an operation should be performed in the jaw area where the cause of the problem is located. Most patients with an anterior skeletal openbite have posterior maxillary vertical excess, which is why the maxilla is surgically impacted posteriorly. Of course, many other problems may be evident in a given patient, such as prognathism with an openbite, where the mandible may be implicated as the dysplastic area and, therefore, will be the focus of surgical planning.

While speech is not expected to change significantly in the typical osteotomy patient, positive and negative changes may be seen individually. Some patients may require adaptive speech therapy for a period of time postsurgically, while others who show adaptive problems immediately after release from intermaxillary fixation may spontaneously improve over time (Figs. 20-9A&B).

One area of potential importance to the speech clinician is the patient with lip incompetence preoperatively. Lip incompetence present over a period of many years may leave the lips rather flaccid following surgery. While the jaws are arranged in a manner compatible with the ability for a natural lip seal, the lip incompetence may be retained in the presence of poor lip tonicity. Some supportive therapy may be indicated for such patients. Currently, measures of lip tonicity, as well as therapeutic measures to provide lip competence, are fairly subjective. Nonetheless, this potential problem is mentioned here as a possible consideration for the clinician working conjointly with a patient who undergoes a mandibular (or maxillary) osteotomy.

While current research has not strongly implicated the speech of the osteotomy patient as significantly deteriorating following any osteotomy, there are a host of considerations yet to be exploited by the speech clinician as to the impact of surgery on speech.

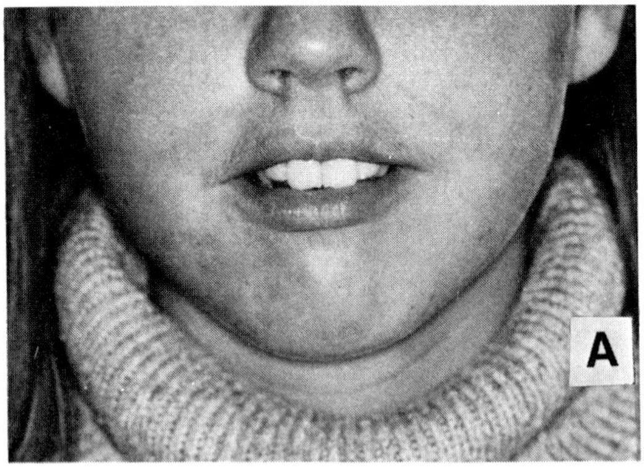

Figure 20-9A. Patient with vertical maxillary excess and lip incompetence pre-operative to a LeFort I maxillary impaction osteotomy.

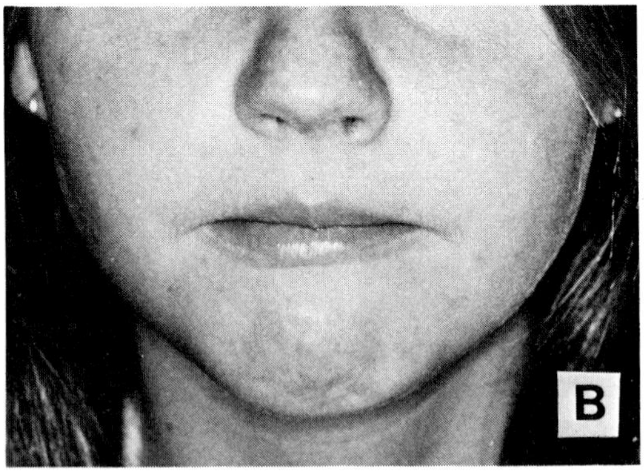

Figure 20-9B. Patient following maxillary impaction. Some lip incompetence remains.

BIBLIOGRAPHY

Bell, W.H., Proffit, W.R., and White, R.P.: *Surgical Correction of Dentofacial Deformities*, Philadelphia, Saunders, 1980, Vols. I and II.

Epker, B.N., and Wolford, L.M.: *Dentofacial Deformities: Surgical-Orthodontic Correction*. St. Louis, C.V. Mosby, 1980.

Epker, B.N., and Wolford, L.M.: Middle third face osteotomies. *J Oral Surg, 33*:491-514, 1975.

Mason, R.M., Turvey, T.A., and Warren, D.W.: Speech considerations with maxillary advancement procedures. *J Oral Surg, 38*:752-758, 1980.

Walker, G.F.: A new approach to the analysis of craniofacial morphology and growth. *Am J Orthod, 61*:221-230, 1972.

West, R., and Epker, B.N.: Posterior maxillary surgery: its place in the correction of dentofacial deformities. *J Oral Surg, 30*:562-575, 1972.

Chapter 21

FACIAL ESTHETICS

AN appreciation of the anatomy and physiology of the facial area, and facial esthetics, is a necessary requisite for treatment planning in orthodontics, oral and maxillofacial surgery, and plastic surgery. Some of the delimitations between tooth and lip position, for example, may require collaborative planning between surgeon and orthodontist. The speech clinician may become involved in adjusting speech patterns to dental or surgical changes in the oral environment.

Consideration will be given in this chapter to some selected soft-tissue components of the facial area that have multidisciplinary value. The philosophy expressed here is that the speech clinician can participate more fully on interdisciplinary teams by appreciating the diagnostic guidelines and data base utilized in other disciplines.

FULL FACE EVALUATION

When the face is separated into two equal halves at the midline, symmetry of facial parts can be determined. There is a wide range of asymmetry in facial halves that falls within the normal range. In the adult, deviations in the nasal septum occur in about 85 percent of the population. It is also common for one eye or ear to be positioned a little higher or lower than its corresponding mate (Fig. 21-1).

The Horizontal Dimension of the Face

The face has been described as being five eyes wide; that is, the face can be conveniently separated into five horizontal parts. One eye width is usually found to occupy the space of the bony tissues that separate the two eyes. The wider the space, the flatter the nasal bridge. Pathologic spacing between the eyes is *hypertelorism* and is a feature of some craniofacial disorders (dysostoses).

The *medial canthus* is the end of the fissure at the medial corner of the eye. The intercanthal width, or the distance between the medial slits of the eye, is usually found to correlate in the normal face with the alar base width of the nose (Fig. 21-2), while the width of the mouth at the commissures usually corresponds roughly with the medial limbus of the eye (the medial rim of the iris).

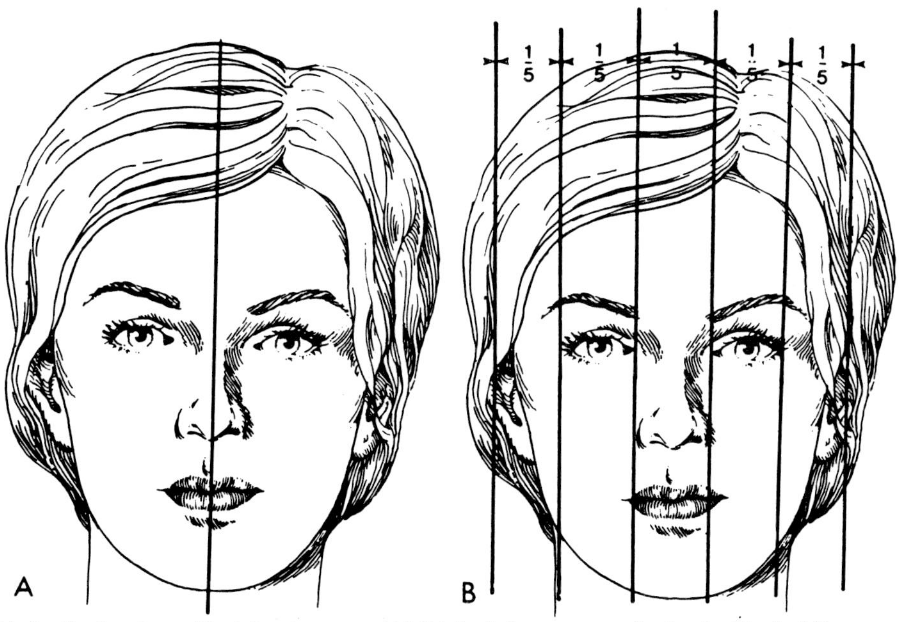

Figure 21-1. Evaluation of facial symmetry. (A) Right-left symmetry is checked by holding an object so it bisects the glabella (of the frontal bone), nasal tip, upper lip, and chin. (B) The face is divided approximately into fifths. (From Bell, W.H., Proffit, W.R., and White, R.P.: *Surgical Correction of Dentofacial Deformities,* Vols. I and II, 1980. Courtesy of W.B. Saunders Co., Philadelphia.

The Vertical Dimensions of the Face

Normal soft-tissue vertical proportions of the face can be determined by separating the face into three parts (see Fig. 21-2). The *upper third* of the face extends from the midline hairline to the eyebrow area, while the *middle third* extends from eyebrows to the base of the nose. The *lower third* of the face can be further separated into thirds, with the lips positioned in the upper third of the lower face (Fig. 21-2B).

If the vertical dimension of the face is considered from the standpoint of the bony skeleton, the face extends from the nasal bridge to the chin. The area above the eyes is considered part of the neurocranium rather than the facial area from a skeletal standpoint. This distinction between *soft-tissue face* and *facial skeleton* should be noted, as references and terminology in clinical reports of soft tissues will differ from cephalometric reports of the facial skeleton.

THE FACIAL PROFILE

In profile, the relationship between the two jaws is most evident. The underlying skeletal framework of the face influences how the soft tissues of the face will relate and be positioned. There are three basic facial patterns into which all *facial skeletal combinations* can be placed. Regardless of the superimposition of hard and soft-tissue variations, the analysis of mandibular and maxillary relationships to the cranial base and to each other prevails in a basic facial pattern. The basic reference is

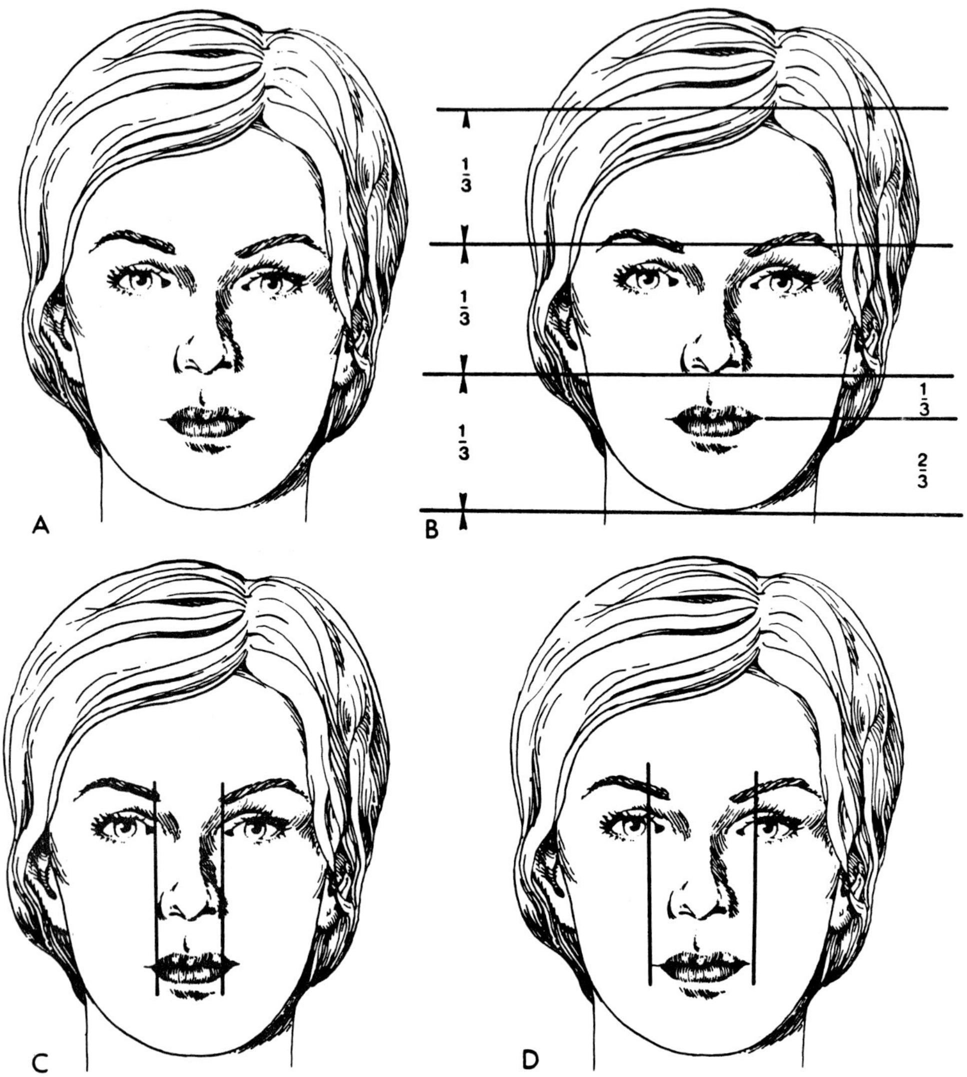

Figure 21-2. Normal facial proportions. (A) Patient in "natural head position" facing examiner. (B) Thirds of the face are roughly equal in vertical dimension. (C) Inter-inner canthal distance is about equal to alar nasal base width. (D) Mouth is as wide as distance from right to left medial limbus. (From Bell, W.H., Proffit, W.R., and White, R.P.: *Surgical Correction of Dentofacial Deformities,* Vols. I and II, 1980. Courtesy of W.B. Saunders Co., Philadelphia.)

the horizontal plane of space.

The *Class I* facial pattern is considered normal and refers to a good balance of skeletal and soft-tissue components. The components of a Class I pattern are shown in block diagrams in Figure 21-3. One should always remember that ethnic characteristics alter concepts of profile evaluation (Peck and Peck, 1970).

The *Class II* facial pattern (Fig. 21-4) shows midface protrusion. This may be due either to maxillary protrusion or mandibular retrusion or a combination of both. Such considerations relate to the

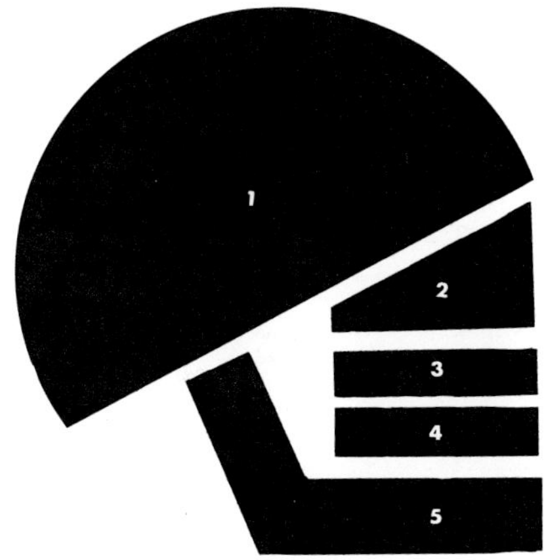

Figure 21-3. Class-one facial pattern. Block 1 represents the neurocranium. Block 2 is the maxilla. Block 3 represents the maxillary dental arch, while Block 4 is the mandibular dentition. Block 5 represents the mandible. (From Khouw, F.E., Proffit, W.R., and White, R.P.: Cephalometric evaluation of patients with dentofacial disharmonies requiring surgical correction. *J Oral Surgery*, Vol. 29, 1970.

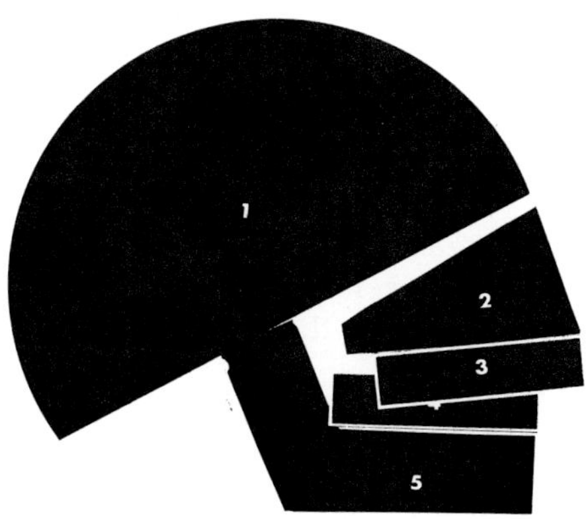

Figure 21-4. Class-two facial pattern (see Fig. 21-3 legend). (From Khouw, F.E., Proffit, W.R., and White, R.P.: Cephalometric evaluation of patients with dentofacial disharmonies requiring surgical correction. *J Oral Surgery*, Vol. 29, 1970.)

denture bases (skeletal bases for the dentition, as shown in blocks 2 and 5 of Figure 21-4), not the teeth per se (blocks 3 and 4). Using the reference points shown in Figure 21-4, the clinician can identify

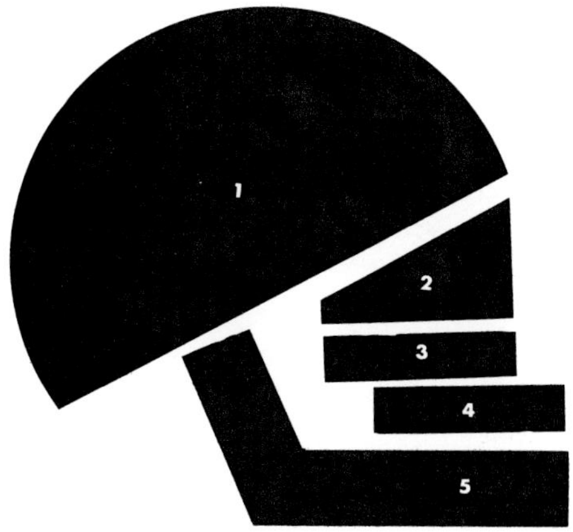

Figure 21-5. Class-three facial pattern. Notice the dental compensations: The upper front teeth are positioned forward and the lower dental arch is positioned somewhat distal to the mandible (see Fig. 21-3 legend). (From Khouw, F.E., Proffit, W.R., and White, R.P.: Cephalometric evaluation of patients with dentofacial disharmonies requiring surgical correction. *J Oral Surgery*, Vol. 29, 1970.)

midfacial protrusion by lateral facial analysis.

The *Class III* profile is characterized by protrusion of the lower face (Fig. 21-5). Again, the relative position of the maxilla (deficient or retrusive) or mandible (protrusive), or a combination of each, may contribute to the Class III pattern. It is interesting that although bony profiles may look alike, their etiology can be entirely different.

The clinical evaluation of the profile usually involves inspection of the soft tissues of the face. The usual guideline for the clinical evaluation of the *soft-tissue profile* is a slightly convex line that is connected in the examiner's imagination between the bridge of the nose, base of the nose, and most prominent point of the chin. The straight, or orthognathic, profile seen in Figure 21-6 is the standard reference for profile analysis, but the slightly convex, or anteriorly divergent, profile is considered as the most esthetically pleasing facial configuration.

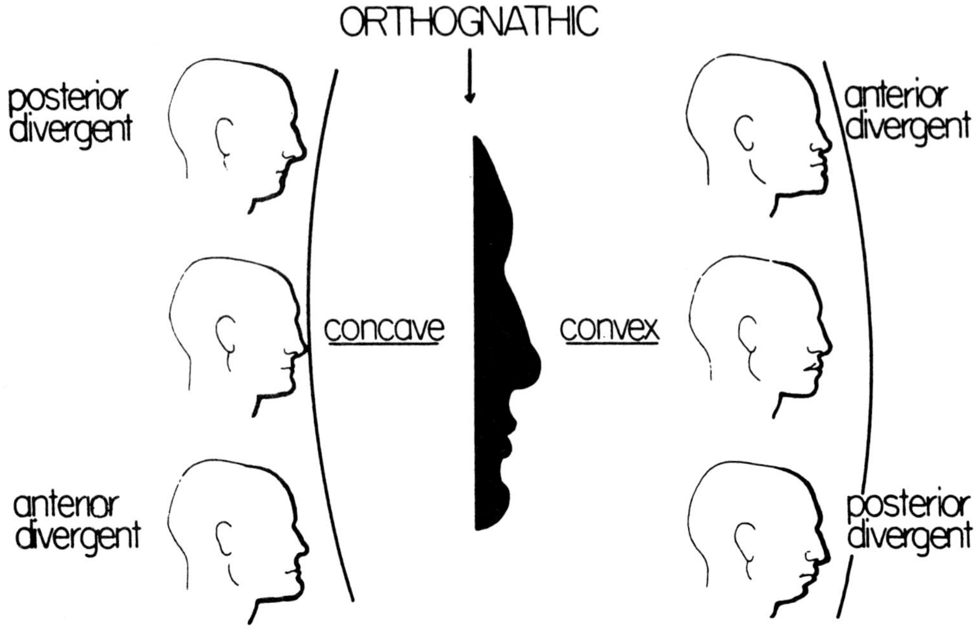

Figure 21-6. Convexity/concavity of lips and anterior or posterior divergence of the overall profile interact to produce several profile combinations. An individual with a straight or orthognathic profile has neither convexity/concavity nor divergence, but in other persons these conditions may exist singly or in combination. (From Graber, T.W., and Swain, B.F.: *Current Orthodontic Concepts and Techniques,* 2nd Ed., 1975. Courtesy of W.B. Saunders Co., Philadelphia.)

THE NOSE IN PROFILE

The angle contributed by the confluence of the nose and lips (the nasolabial angle) should normally be 90 to 110 degrees, as shown in Figure 21-7. An excessive or *obtuse nasolabial angle* is often seen with excessive vertical maxillary growth. It is as if the face grew vertically and the soft tissues were distorted in the vertical plane as an adaptation to this growth. Individuals with long faces also tend to show an excessive amount of tooth structure at rest and may have difficulty in achieving a natural lip seal at rest (lip competence). Surgical treatment of vertical maxillary excess, as with a LeFort I maxillary impaction, would create a more esthetic relationship between resting lips and teeth, as well as restoring the nasolabial angle to the normal range.

An *acute nasolabial angle* is seen in some patients with midfacial retrusion, such as a patient with a repaired cleft lip and palate who experienced some horizontal and vertical collapse of the maxilla. A nasal mode of respiration may be somewhat impaired in such patients as the angle of the liminal valve of the nose (see Chap. 2) acts to impede the flow of air through the nares.

Figure 21-7. (Left) The nasolabial angle is esthetic when it is 90 degrees to 110 degrees. (Right) Soft-tissue nasion, lips, and chin are most esthetic, in whites, when arranged as seen. (From Bell, W.H., Proffit, W.R., and White, R.P.: *Surgical Correction of Dentofacial Deformities*, Vols. I and II, 1980. Courtesy of W.B. Saunders Co., Philadelphia.)

THE LIPS IN PROFILE

One of the most useful guidelines in facial esthetics deals with the relationship of teeth to lips as seen in the lateral profile view. A line extended from the tip of the nose to the most prominent point of the chin should find the lower lip of a child on or just behind this reference line. In an adult, with added growth of the nose and chin, the lips are most likely positioned a few mm behind this line (Fig. 21-8). The *esthetic line*, as it is referred to in orthodontics, is subject to some modification in different racial groups.

An orthodontist finds the esthetic line to be an especially useful diagnostic aid in determining whether extraction of teeth will be indicated in treatment. The patient whose lips extend beyond the esthetic line is a likely candidate for extraction of teeth, while a patient whose lips rest several mm behind the line would be considered for a nonextraction approach to orthodontics, if possible.

Tooth-Lip Relationships

The teeth provide lip support and are important contributors to facial esthetics. With the mouth at rest and the lips slightly parted, about 2 mm of upper tooth structure should be visible for a pleasing tooth-to-lip cosmetic effect. With lips together, the lower lip normally covers 2 mm of the upper incisor on its facial surface. If an orthodontist can achieve this resting relationship of lower lip and upper incisor in a completed case, where there had been protrusion of upper teeth prior to orthodontics, relapse of the teeth forward is highly unlikely.

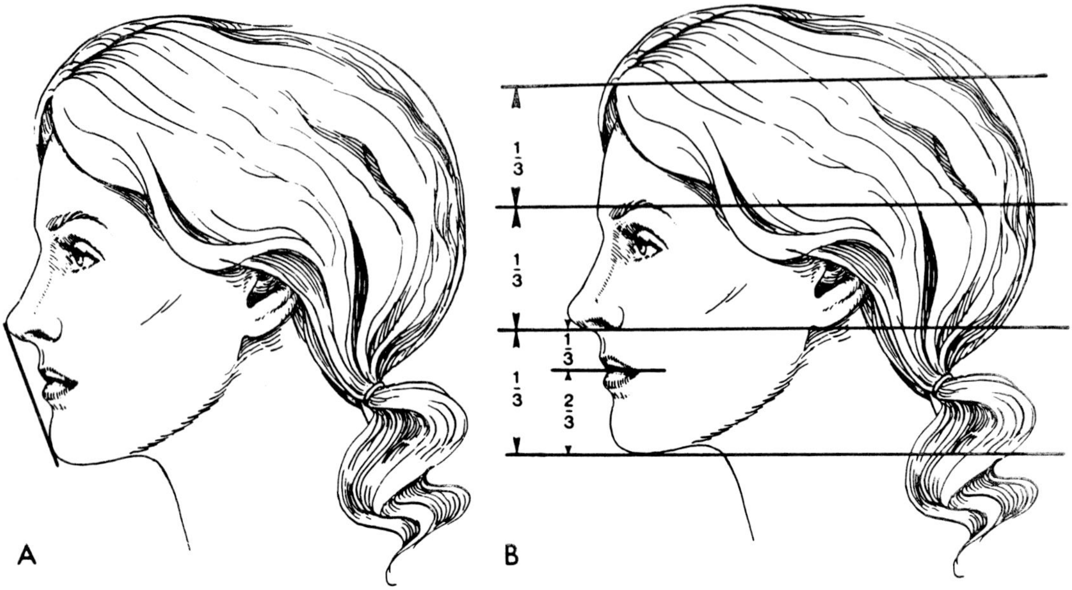

Figure 21-8. Normal facial relationships. (A) Head in "natural head position" with visual axis parallel to floor. The esthetic line is shown. (B) The thirds of the face are equal, and the lower third is broken into one-third and two-thirds. (From Bell, W.H., Proffit, W.R., and White, R.P.: *Surgical Correction of Dentofacial Deformities*, Vols. I and II, 1980. Courtesy of W.B. Saunders Co., Philadelphia.)

SUMMARY

Some selected topics of facial esthetic evaluation have been presented. Consideration should be given to the vertical and horizontal planes of space. In addition, the important contribution of the nose and lips to facial esthetics should be recognized.

The orthodontist and surgeon utilize concepts related to facial esthetics in treatment planning for patients with skeletal and dental dysplasias. As a member of the multidisciplinary team for such patients, the speech clinician should be knowledgeable of basic considerations in facial esthetics.

BIBLIOGRAPHY

Bjork, A., and Skieller, V.: Facial development and tooth eruption. *Am J Orthod, 62*:330-383, 1972.

Burstone, C.J.: Lip posture and its significance in treatment planning. *Am J Orthod, 53*:262, 1967.

Enlow, D.R.: *Handbook of Facial Growth*, 2nd ed. Philadelphia, Saunders, 1982.

Khouw, F.E., Proffit, W.R., and White, R.P.: Cephalometric evaluation of patients with dentofacial disharmonies requiring surgical correction. *J Oral Surg, 29*:789-798, 1970.

Peck, H., and Peck, S.: A concept of facial esthetics. *Angle Orthod, 40*:284-318, 1970.

Proffit, W.R., Epker, B.N., and Ackerman, J.L.: Systematic description of dentofacial deformities: the data base. In Bell, W.H., Proffit, W.R., and White, R.P.: *Surgical Correction of Dentofacial Deformities*, Vol. I. Philadelphia, Saunders, 1980.

Sassouni, V.: Analysis of dentofacial vertical proportions. *Am J Orthod, 50*:801-823, 1964.

Sassouni, V.: *The Face in Five Dimensions*, 2nd ed. Morgantown, WV: West Virginia U Pr, 1962.

Vig, P.S., and Cohen, A.M.: Vertical growth of the lips: a serial cephalometric study. *Am J Orthod, 75*:405-415, 1979.

Williams, R.: The diagnostic line. *Am J Orthod, 55*:458-476, 1969.

Chapter 22

NEUROLOGICAL ASPECTS AND SOME CNS DISTURBANCES

INTRODUCTION

THREE functional parts of the brain have been proposed in an attempt to explain some of the basic problems in understanding aspects of neurolinguistics (Luria, 1973). The **first** part is responsible for providing adequate *tone* of the cortex and is responsible for alertness. The apparatus for this first functional part includes groups of cells in the upper brain stem and in the cerebral cortex. These are structures of the limbic system.

The **second** functional part is responsible for reception, elaboration, and storage of exteroceptive (external environment) information. Additionally, the basic cerebral mechanism of cognition is included in this functional part. Structurally, the apparatus for the second part involves the posterior aspects of the brain hemispheres.

The **third** functional part is responsible for programming, regulating, and controlling our behaviors. Simultaneously, the third part is involved with descending fibers that provide a regulation of vigilance and attention. This servomechanism provides feedback circuits to ensure that our behavior coincides with our conscious goals and motives. The anterior parts of the cerebral hemispheres contain the apparatus for the third functional part of the brain.

Psychobiologists are especially interested in the brain's ability to organize itself into functional parts for the various complex operations of alertness, information processing, and action. These functions are not mere compartments or territories on the surface of the brain with well-defined boundaries. Instead, they are diffuse, "systems" often with mysterious interconnections. The intrigue of these complicated and puzzling cooperative operations has motivated specialists in medicine, psychology, biology, anatomy, and speech to work diligently to explain the brain's functional secrets.

LIMBIC SYSTEM

A number of anatomists object to the use of terms like *limbic system* or *limbic lobe*. They feel such designations might lead students to think of the limbic system as a distinct anatomic and functional entity. It is a functionally integrated system involving a number of concentrations or aggregations of nerve cells; therefore, the limbic system is really not a structural unit.

While the limbic system cannot be alluded to as an isolated anatomic and functional entity, it is nonetheless an essential portion of the complex neural circuitry governing learning experiences, integrating present experiences with past ones, setting into action plans, and modulating behavior and emotional reactions. Medical students learn to think of the limbic system as controlling the four F's: feeding, fighting, fright, and sexual behavior. Its role, then, is to mediate over fundamental body functions such as breathing, digestion, heartbeat, biological rhythms, and sexual reflexes. Since the limbic system participates in various aspects of emotional behavior and in basic body functions, it is sometimes aligned with the "visceral" brain.

Parts of Limbic System

The term *limbic* means border or hem, which identifies the location of this system to some extent. It is comprised of two rings of medially positioned cerebral cortex — literally, a ring around the upper aspect of the brain stem. The precise connections of these structures are not understood, but they appear to be interconnected in both direct and indirect pathways.

Hypothalamus

While the limbic system occupies only the lower one-fifth of the brain, its parts are tied bidirectionally to a mass of cell bodies known as the *hypothalamus*. There are two connecting outposts to the hypothalamus: the septal areas in front and the amygdala along the sides. The hypothalamus receives information from almost all parts of the brain. It can also send information to these parts. Because of the extensive interlinks with areas of the brain that mediate basic functions, the influence of the limbic system on behavior is unbelievably extensive. Functional attributes of the hypothalamus alone imply far-reaching influences on biological activities. The most well-known functions of the hypothalamus are as follows:

1. Regulation of heart rate and arterial blood pressure.
2. Control of hunger and regulation of body weight.
3. Regulation of water and electrolyte balance.
4. Regulation of body temperature.
5. Control of movements and glandular secretions of the stomach and intestines.
6. Production of agents that in turn stimulate the pituitary gland to release various hormones.
7. Regulation of sleep and wakefulness.

Components of the limbic system are not easy to pinpoint. Some writers list only a few structures while others make a rather extensive listing. Isaacson (1974) qualifies his description by naming the inner core of the limbic structures based on the premise that all have strong interconnections with the hypothalamus and with all other regions. He identifies the inner-core members as the hypothalamus, the amygdala, the hippocampus, and the septal area, with consideration also being given to the cingulate cortex. A descriptive comparison of the detail makeup of the limbic system is beyond the scope of this text.

Amygdaloid Complex and Hippocampus

The *amygdala* is a subcortical structure located at the dorsomedial tip of the temporal lobe. Lesions involving the amygdaloid bodies suggest a relation to mechanisms concerned with feeding and abnormal sexual behavior (Fig. 22-1).

The *hippocampus* is associated geographically with the olfactory centers but has only very indirect functional relations with them, in contrast to earlier viewpoints. The hippocampal formation and its connections via the fornix seem to constitute a mechanism of fundamental importance for learning and memory. Some patients with bilateral hippocampal lesions can retain in memory for as long as fifteen minutes a short sentence or a short sequence of numbers, provided they are allowed to keep their attention on it (e.g. no one speaks to them and there is no other interference). When these patients shift their attention to something else, even for a moment, the sentence or numbers are completely forgotten. However, the patients can remember skills in their premorbid occupations and events of their earlier lives. It seems, then, that the hippocampus is involved in selecting, categorizing, and organizing stimuli for storage; it may also be involved in temporary storage (Peele, 1977).

The Limbic System and Frontal Lobe

More than 45 years ago, Papez (1937) proposed a model of the neural basis of emotional behavior. Current ideas do not differ much from those of this early neurobiologist. Many continue to think of the hypothalamus — the pivotal component of the limbic system — as functioning to integrate the motor and endocrine responses that produce appropriate emotional behavior. Coming into play is the *forebrain*, which suppresses emotional responses to trivial and inconsequential stimuli. Hence, it is the forebrain that connects the hypothalamus to the outer world in such a manner that appropriate anatomic (muscle expression and movement) and glandular (endocrine) concomitants of emotion are expressed in response to external conditions. These activities are an integral part of the autonomic nervous system (Fig. 22-2).

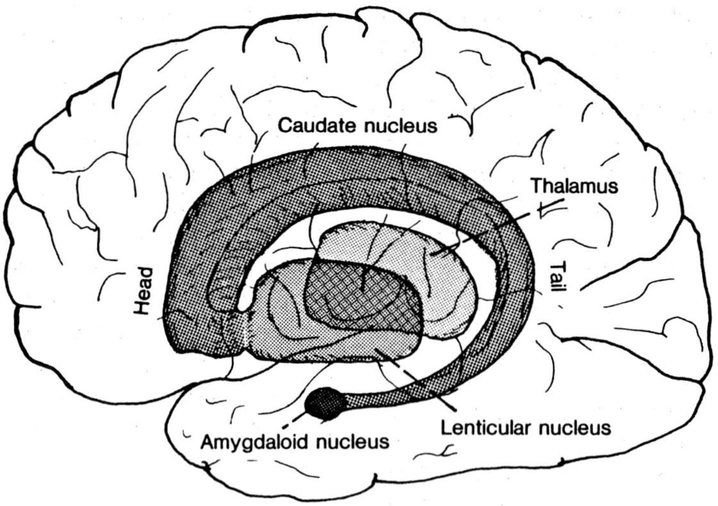

Figure 22-1. Basal ganglia. The caudate nucleus is an arching structure within the cerebral hemisphere. Its tail swings laterally into the temporal lobe to end in a knob-shaped mass known as the amygdaloid nucleus. The head of the caudate nucleus connects a lens-shaped body, the lenticular nucleus. These structures plus the claustrum make up the basal ganglia of each hemisphere of the brain.

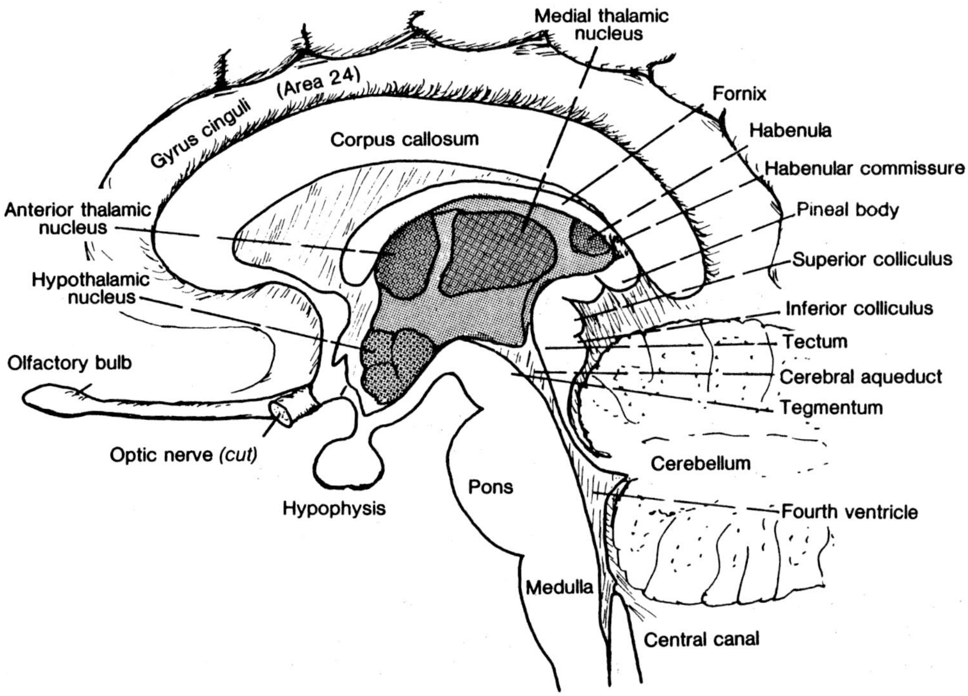

Figure 22-2. The hypothalamic and thalamic nuclei, the cingulate gyrus, and habenula. These structures are central in providing important integrative mechanisms for autonomic activity. The frontal lobe is also viewed as important.

It is the forebrain that is crucial for conscious experience of emotions and memory. During learning, there is a clear focus of attention which appears to wander progressively downward to the limbic system where muscles may perform the action without attention to every movement.

Papez, in attempting to answer the question of how higher cortical centers communicate with the hypothalamus, suggested that cortical influences were funneled to the *hypothalamus* by way of the *cingulate gyrus* to the *hippocampal area*. The hippocampal formation processed this information and projected it (by way of the *fornix*) to the *mammillary body* of the hypothalamus. The hypothalamus in turn provided information to the *cingulate gyrus* by a pathway that coursed through the anterior *thalamic nuclei* thus, a feedback system between the cortical areas and the limbic structures is established (Fig. 22-3).

Hypothalamus and the Autonomic Nervous System

Since the hypothalamus sends fibers into the pituitary (hypophysis), this master gland links the endocrine system with the central nervous system. The hypothalamus also mediates conventional reflexes involving simple neural inputs and outputs. Hess (1954) found that autonomic responses evoked by hypothalamic stimulation did not occur in isolation but in characteristic constellations that gave the appearance of being organized behavior. Not only is the hypothalamus thought to be a motor nucleus for the autonomic nervous systems, it is also a coordinating center that integrates various inputs to ensure

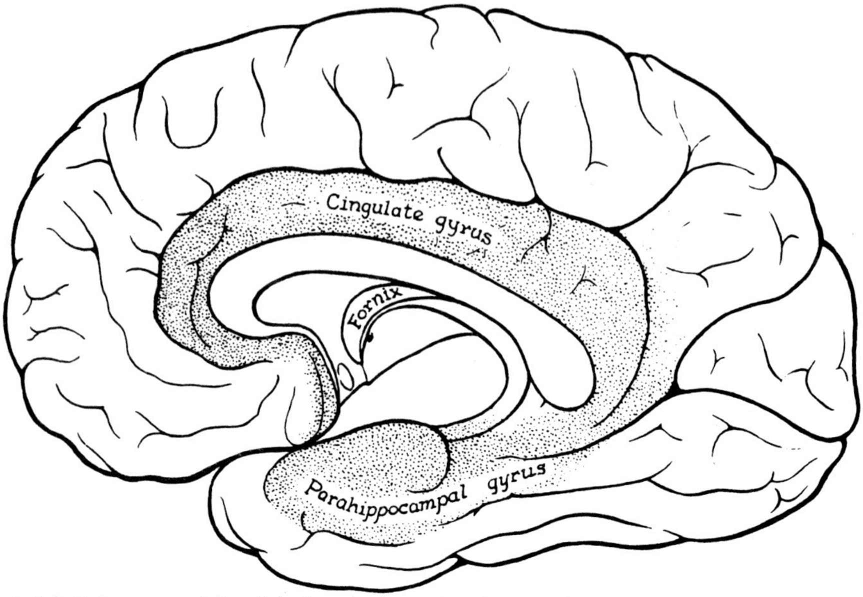

Figure 22-3. Medial aspect of the right hemisphere showing the fornix, which arches back and down to join the hippocampus. Notice the parahippocampal gyrus. (From Truex, R.C. and Carpenter, M.B.: *Human Neuroanatomy*, 6th ed. Copyright 1969, The Williams and Wilkins Co. Reproduced by permission.)

a well-organized, coherent, and appropriate set of autonomic and somatic responses.

Summary of the Limbic System

The ring of cortical structures in the lower part of the brain is termed the *limbic system*. Neurobiologists and psychobiologists have linked this group of structures to hormones, drives, temperature control, and reward and punishment centers.

The limbic system complex functions to produce emotional feelings of anger, fear, pleasure, and sorrow, which can alter the behavior of individuals. This system appears to *recognize* disturbances in a person's physical or psychological condition that might threaten the individual. By causing pleasant or unpleasant feelings about experiences, the limbic system directs the individual into behaviors that are likely to enable survival.

One part of the system, the hippocampus, has been linked to *memory* formation. A reverberating neural circuit may cycle messages full circle through five limbic structures: hippocampus, mammillary body of the hypothalamus, anterior thalamus, cingulate cortex, and hypothalamus. The role of the fontal lobe is vital in suppressing some emotional stimuli from the limbic structures. Because of strong interconnections between the hypothalamus and other structures of the limbic system, as well as with other cortical, autonomic, and endocrine links, the hypothalamus appears to play a central role in the physiological and behavioral regulatory mechanisms that make life possible in all of us.

RETICULAR FORMATION

A steady state of mild contraction of muscles, even resting muscles, is known as *muscle tone*. Without some degree of muscle tone, muscles become weak, flabby, and eventually paralyzed. Wakefulness may be described as the maintenance of a state of optimal "cortical tone." This neural tone of the brain is necessary for the carrying out of all organized mental functions. When cortical tone is lost, we pass into a state of sleep or coma. The loss of cortical tone reduces our ability to respond to our environment; there is also a considerable decrease in the brain's functional capacity, and oral language processing is gravely disturbed.

To locate the wakefulness center of the brain, we go to the brain stem. A special formation extending vertically in the brain stem and interconnecting widely different structures in the brain is the locus of this wakefulness center. Known as the *reticular formation*, the organization of the cell networks comprising this formation is designed for the rapid spread of excitation throughout the whole brain. The reticular formation once was designated by a functional title, the reticular activating system, but this has fallen into disuse.

Impulses can go both ways in the reticular formation. Ideas may originate in the cerebral cortex, descend as a wave of excitability *down* to the reticular formation in the brain stem, and activate muscles to cause movement. We go to the refrigerator for a canned drink, for example. We are then able to walk back to our favorite spot in front of the television, whereupon we resume our attentiveness to the program. All the while we need not devote attention to movement or to drinking because the reticular formation maintains the needed *ascending* excitement of muscle groups to accomplish these feats. Obviously, throughout this whole episode, the attention is centered on the television. In this "preoccupied" state, we may even respond to a simple question without confounding our attention on the program.

Wakefulness, therefore, depends on a *vertically* arranged, two-way system capable of conducting information from outside our bodies inward, as well as from the cortex downward to the nerves and muscles controlling the body.

Structural Aspects

Before moving into other functional areas of the reticular formation, we should outline its structural aspects. This network of *nerve fibers* and tiny *islands of gray matter* (cell bodies) scattered throughout the medulla oblongata, pons, and midbrain extends from the upper spinal cord to the diencephalon. Its fibers interconnect centers of the cerebellum, basal ganglia, hypothalamus, and cerebrum with fibers in all the major ascending and descending pathways (Fig. 22-4).

The unique feature of reticular neurons, regardless to which island of cell bodies they belong, is the far-flung distribution of their axons. These extensive axons course in both rostral and caudal directions in the brain stem. Neurons with this characteristic structural property are ideal for governing the level of activity of the nervous system as a whole.

Psychobiologists concentrated on the cerebral cortex up until the discovery of the reticular formation. They viewed the cerebral cortex as the essential apparatus of wakefulness. With the concept of a reticular network, a new principle of brain organization was introduced. These investigators could no longer ignore the so-called primitive divisions of the brain, namely, the lower brain stem. The reticular formation of the lower brain stem

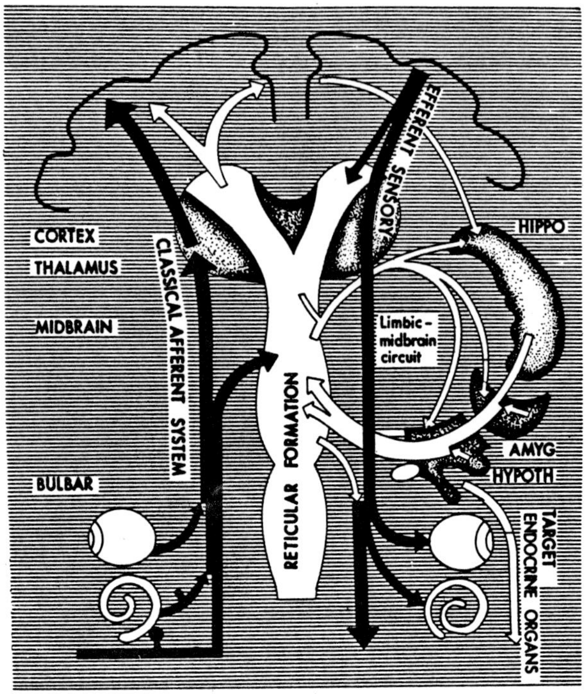

Figure 22-4. The reticular formation and its relations to other parts of the brain systems involved in learning. (From *The Central Nervous System and Behavior*, page 289. Josiah Macy, Jr., Foundation, 1959.)

ushered in the concept of brain organization along a *vertical dimension.*

Nonetheless, even though the reticular formation is nestled in the brain stem, the information capacity of the brain is located principally in the cerebral cortex. Instead of a nerve net like the reticular formation, the cortex contains isolated nerve cells with numerous short and long interconnecting fibers (Fig. 22-5).

Functional Aspects

When sensory impulses arrive in the reticular formation, it reacts by signaling the cerebral cortex. This is the essence of the *cortical tone* alluded to earlier, since the cortex is aroused into a state of wakefulness. Otherwise, the cerebral cortex remains unaware of the stimulus, nor can it interpret, analyze, or store sensory information.

Disabling the reticular formation, then, means the individual will remain unaware of incoming information. A blood clot or brain tumor no larger than the size of a match head, if located near the reticular formation in the brain stem, can bring about a state of irreversible coma *even when the cerebral cortex remains completely normal.* The vital speech receptive and expressive centers are as nonfunctional as a fine automobile without fuel.

It should be pointed out that a comatose state is not a form of sleep. Sleep is a highly active neurophysiological state with levels of oxygen uptake by the cerebrum equal to waking levels. Oxygen uptake by comatose patients has been shown in every study to fall below normal resting levels. Behaviorally, the arousal thresholds between sleep and coma are the distinguishing features.

Of course, not all sensory data reaches

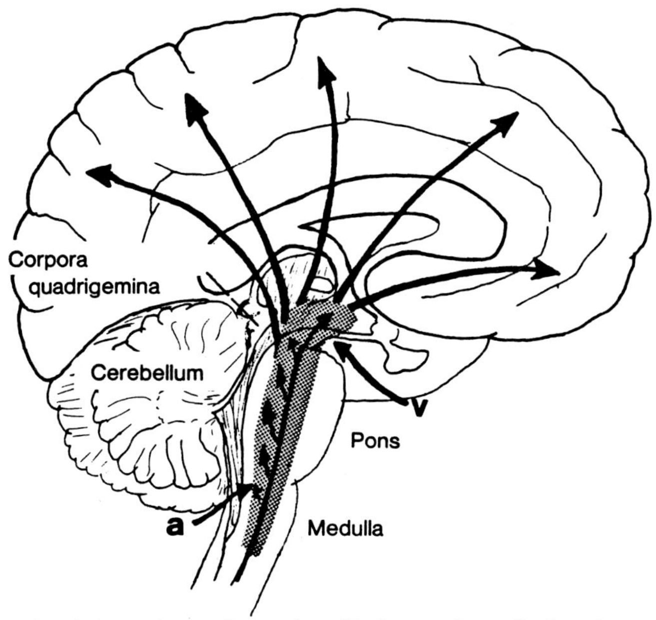

Figure 22-5. Schematic of the reticular formation. Pathways from the basal ganglia, the cerebellum, brain stem nuclei, and the spinal cord feed impulses into the mass of gray matter in the brain stem known as the reticular formation (RF). Two other important input channels are vision (v) and auditory (a). Diffuse projections are shown emanating from the RF to the cerebral cortex.

the brain. It appears that the reticular formation acts as a filter for incoming sensory impulses. Those impulses that are interpreted as meaningful and useful are passed on to the cerebral cortex; others are disregarded. The selective screening action by the reticular system frees the cerebral cortex from what would otherwise be a continual barrage of sensory stimulation and permits the cortex to focus upon more significant information.

It is apparent from the foregoing that the reticular formation is involved in numerous varied functions. In the area of *alertness*, it is concerned with states of sleep and various levels of awareness. Its importance to speech development is obvious. With motor activities, the reticular core regulates the proper levels of muscle tone and organizes somatic and autonomic motor activities that produce respiration, vasomotor (varying the diameter of blood vessels), and cardiac responses as well as ordinary, voluntary motor performance such as speech production and general movement (Fig. 22-7).

Reticular Formation and Attention

By means of its neuronal axons projecting to the diencephalon and the cortex, the reticular formation is important in the effecting and maintenance of arousal and alerting states. Because this formation is a vital component to the *attentional mechanism*, any disturbance might impair or, in certain cases, destroy attention.

Attention appears to be necessary for the transfer of briefly retained material to the motor system or for storing it in a more permanent form. If attention is not directed to the trace within a few seconds of being laid down, there is little or no recall of the trace.

Maintenance of attention to a particu-

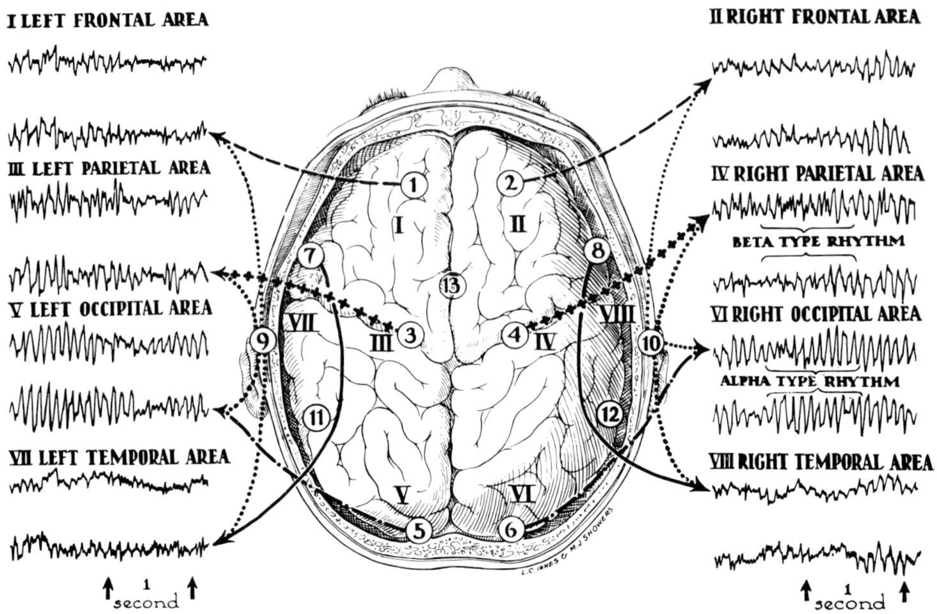

Figure 22-6. Diagram of electroencephalogram: The numbers 1, 2, 3, etc, are corresponding locations on the brain to electrodes placed on the subject's scalp. These may be used in several combinations; one combination is shown above where 1, 2, 3, etc., are attached to 9 and 10. The region of the brain represented by records of electrical waves are indicated by Roman numerals. Typical alpha rhythm of 8–12 waves per second and beta rhythm of 18–32 waves per second are shown. Electroencephalograms can detect the electrical activity of the brain to provide clinical data associated with brain disorders. (From King, B.G., and Showers, M.J.: *Human Anatomy and Physiology*, 1969. Courtesy of W.B. Saunders Co., Philadelphia.)

lar stimulus or continuous stream of stimuli is probably *less* difficult to explain than attempting to explain sudden *changes* of attention. If we are attending to one stimulus, e.g. studying fish swimming about in an aquarium, some "cell assembly" in the perceptual networks generates expectancies and facilitates related neural activities. At the same time, this cell assembly provides inhibition to deprive other cell assemblies, which are also entertaining stimuli, from gaining access to the motor system — to turn the head and eyes away from the aquarium — or from gaining access to those premotor and memory processes that appear to be related to conscious awareness. So long as one cell assembly is the preferred apparatus for a given theme of attention, a "set" is maintained to ensure continuity of attention. It may be a function of the reticular formation in attention to break up the dominance of a particular cell assembly and to permit a new "set" to be established.

A number of investigators, but especially Eysenck (1968) and Broadbent (1962), have posited that the attention mechanism is tied in with the personality traits of introversion-extroversion. Breaking up a dominance of a cell assembly to shift attention is much easier in extroverts than introverts. By nature, introverts are supposed to thrive on a narrow field of arousal stimuli and are able to focus their attention on one theme for much longer periods than extroverts.

Some linguistically handicapped children exhibit behaviors that may be the result of a disturbance of attention. A disturbance in the reticular formation may result in a lack of set-to-attend or the ability *to attend or respond selectively*. It could also be postulated in reverse, *learning not to respond* (discriminative responses). Finally, students should remember that the makeup and functional roles of the reticular formation are not well understood. A reason is that all known major pathways pour collateral fibers into the reticular core, permitting a vast, complicated interacting system. It is not organized with deft specificity, which means that it is largely precluded from fine movement coordination. However, the reticular formation probably plays a part in coordination of gross movements by large groups of muscles.

SPEECH DISORDERS: DYSARTHRIA, APRAXIA, AND APHASIA

Nowhere is localization of higher functions of the cerebral cortex more dramatically demonstrated than in disturbances of language. If, for example, the disturbance involves the inability to program muscles to execute complex learned or skilled movement, such as using the articulators to generate appropriate speech signals, the disorder is termed *apraxia*. But apraxias are unassociated with disability resulting from paralysis, sensory loss, or dementia, so the disturbance is a different kind of disintegration of function than is usually encountered with an interrupted nerve supply to a muscle. Important insights into the processing of information by the cerebral cortex are available from clinical studies and observation of changes in behavior produced by insults on the CNS from trauma and disease. Some of these that are pertinent to applied neuroanatomy and neurophysiology are described below.

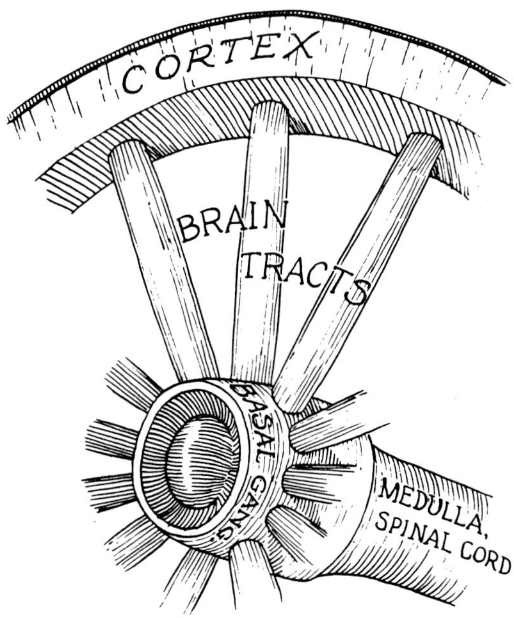

Figure 22-7. Schematic illustration of the relations between the basal ganglia, the cortex and medulla. (From Silverman, S.I.: *Oral Physiology*, 1961. Reproduced here with the permission of Doctor Silverman. Originally published by The C.V. Mosby Co., St. Louis.)

Before addressing each of the selected speech disorders independently, a brief overview of the compass of each clinical term will be used to differentiate among them.

Dysarthria (disturbance in articulation) refers to imprecise productions of consonant sounds resulting from damage to the central or peripheral nervous system. It is characterized by some weakness, slowness, incoordination, or altered muscle tone associated with the speech mechanism. When there is faulty programming or articulatory movements by the brain circuits specifically responsible for this activity, the resulting articulatory disorder is apraxia, not dysarthria. *Aphasia* may also result in faulty speech production, but the etiology (origin of the problem) is the *language perceptual system*, which is a reduction in the capacity to decode (interpret) and encode (formulate) language, including semantic and grammar elements. Different neuroanatomical and neurophysiological components are associated with these clinical entities (Darley, Aronson, and Brown, 1975) (Fig. 22-8).

Dysarthria

It was indicated above that *dysarthria* refers to a speech impairment due to disturbances in muscular control of the speech mechanism. Following Darley, Aronson, and Brown, the term is best restricted to neurogenic speech disturbances, excluding apraxia and aphasia, and includes coexisting motor disorders of respiration, phonation, articulation, resonance, and prosody. Clinicians should remember that dysarthria is a *motor speech disorder*.

The neuroanatomy of dysarthria focuses on lesions in four areas. These may be termed *classes* of dysarthria after Darley, Aronson, and Brown. The reader should be aware that various names are used to refer to a given area of motor activity carried out by the nervous system. Anatomical, physiological, and clinical terms are employed with an anatomical term used at one level of motor organization and a clinical term at another.

Flaccid Dysarthria

At the lowest level of motor function is the *lower motor neuron*. Large neurons located in the anterior horns of the spinal cord and in the motor nuclei of cranial nerves with axons projecting from the central nervous system to muscles comprise the lower motor neuron. It functions reflexively. These neurons serve the speech musculature as well as other muscles of the general somatic motor system. Regardless of the origin of an impulse on its way to a muscle served by the lower motor neuron, it must use this neuron as the final pathway. Hence, the lower motor neuron is also termed the *final common pathway* (Fig. 22-9).

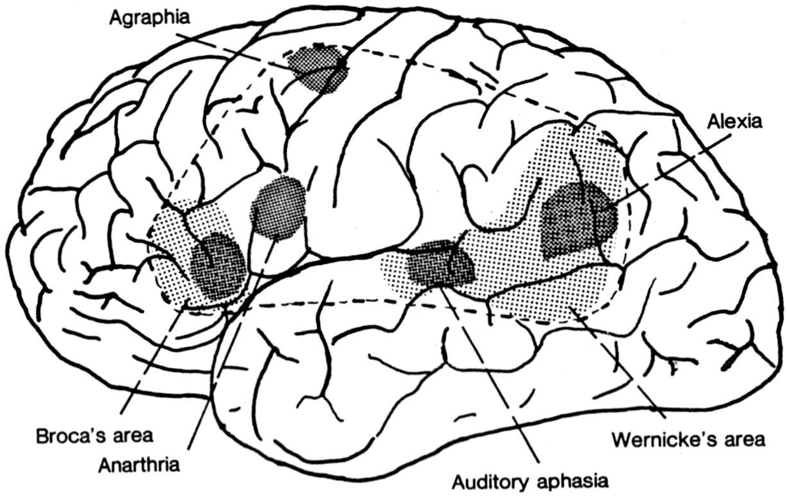

Figure 22-8. Sites on the left cerebral cortex in which lesions may produce disturbances in one or more forms of language.

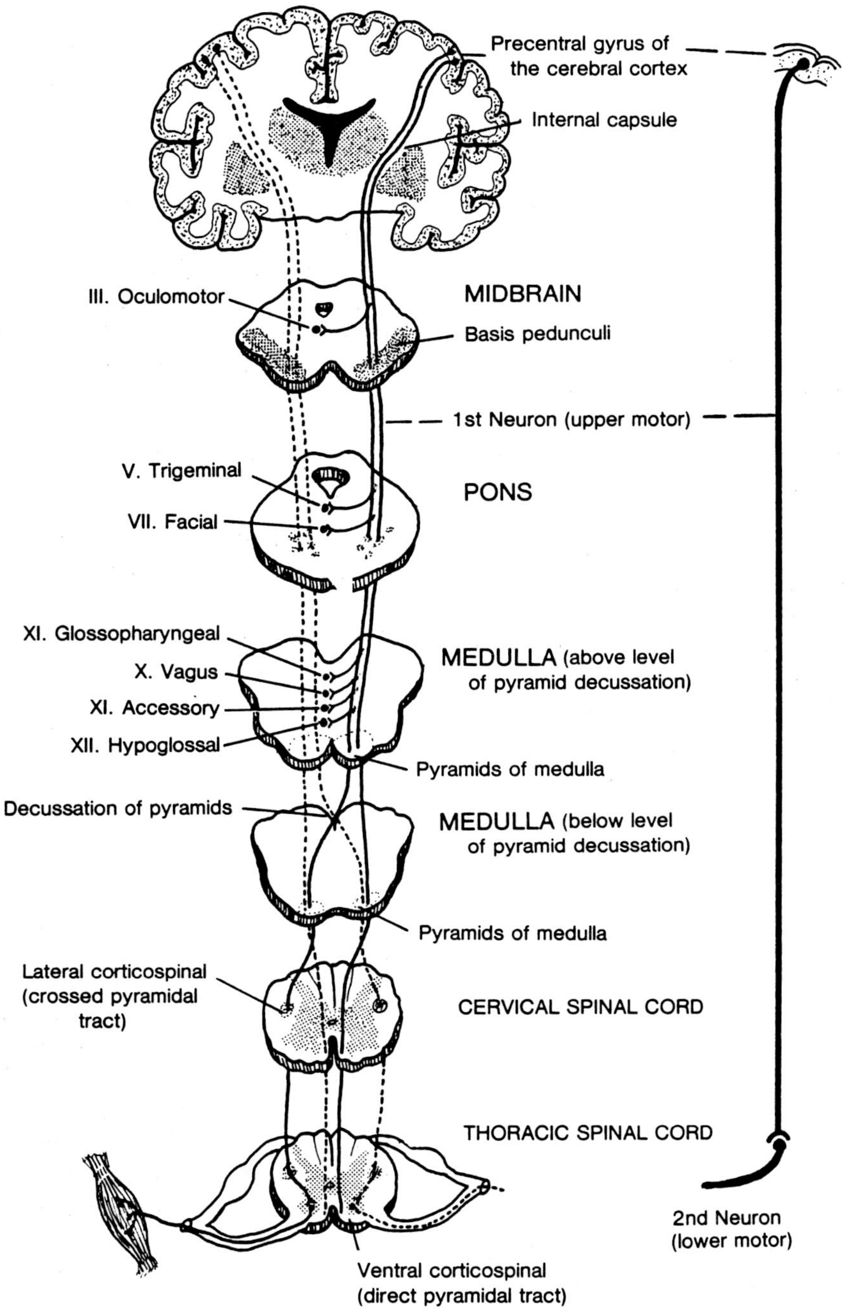

Figure 22-9. The pyramidal motor pathway. The upper motor neuron has its nucleus imbedded in the cerebral cortex with its long axon coursing down through the brain to the brain stem (corticobulbar) or spinal cord (corticospinal). The second principal neuron, which may be located in the brain stem or spinal cord, sends its axon directly to a muscle and is termed the final common pathway or lower motor neuron. All motor neurons of cranial and spinal nerves are lower motor neurons. (From Crouch, J.E.: *Functional Human Anatomy*, 3rd ed., 1978. Courtesy of Lea & Febiger, publisher.)

Clinically, the term *bulbar* level is used to designate the lowest level of motor organization for speech. (Bulbar refers to the medulla oblongata, or brain stem, where lower motor neuron nuclei are found.) Lower motor neuron lesions are associated with loss of muscle tone (hypotonicity) leading to *flaccid dysarthria*.

Spastic Dysarthria

A second class of disturbance in the control of motor speech concerns elevated muscle tone (hypertonicity). This locus of disturbance is the *upper motor neuron* and is a rather high level in the hierarchy of motor organization. The neurons are located in the cerebral cortex. Upper motor neuron is a term denoting the specific motor tracts contributed by the cerebral cortex, particularly motor tracts concerned with voluntary movement. Both the pyramidal and extrapyramidal tracts are associated with the upper motor neuronal level (see Chap. 16) (Fig. 22-9).

The upper level motor neuron is designated as a *corticobulbar tract* for those neurons projecting their axons to the cranial nerve motor nuclei in the brain stem. If the axons reach the spinal cord, *corticospinal* is used.

Bilateral upper motor neuron lesions result in hypertonic conditions of the muscles associated with speech, among other things. A characteristic *spastic dysarthria* is attributed to bilateral lesions. Termed *pseudobulbar palsy*, bilateral lesions of the corticobulbar tracts result in profound dysfunction. Specifically, this disorder causes the patient to experience difficulty with facial expression, tongue mobility, chewing, swallowing, articulation, phonation, and breathing. Cranial nerves V, VIII, X, and XII are critically involved (see Chap. 26). Inappropriate spells of laughing or crying are sometimes present in pseudobulbar palsy.

Unilateral damage to the corticobulbar tract does not result in massive dysfunction, because each tract connects *bilaterally* with most of the cranial nerve motor nuclei. Clinicians should remember this point. Therefore, if one tract is damaged, the other tract carries on the function. But cranial nerves VII (facial) and XII (hypoglossal) are exceptions, since corticobulbar tract lesions result in contralateral lower facial paralysis and, in some cases, contralateral weakness of the tongue.

Cerebellar Dysarthria

The third neuroanatomic class of dysarthria can be termed clinically as *ataxic dysarthria*. Overall, there is clumsiness of posture and gait, poor coordination of movement, and an inability to perform rapid alternating movements. Speech clinicians will note the presence of *scanning speech*, which is the awkward use of the speech musculature. There will be unnatural prolongation of sounds, even stress on the syllable within words, and a mechanical evenness in the utterance of words.

Lesions in the cerebellum or its pathways produce ataxic dysarthria. *Intentional movements*, such as speaking or walking, are affected by cerebellar disorders and are to be distinguished from basal ganglia disorders, which result in meaningless *unintentional movements* occurring unexpectedly. The cerebellum does not initiate motor activity but is vital in the control of the accuracy of muscle responses initiated at other levels of motor organization in the central nervous system (Fig. 22-10).

Extrapyramidal Dysarthria

This is a mixed class of dysarthrias found in association with several disease processes. These disorders involve the extrapyramidal system — a diffuse or in-

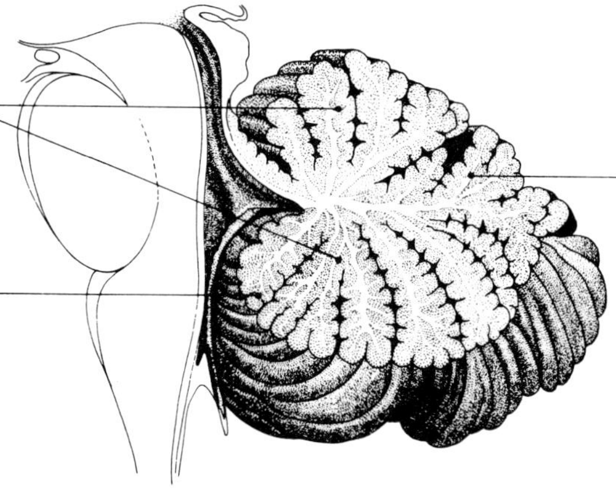

Figure 22-10. Sagittal section through the cerebellum illustrating areas contributing to smooth coordination of movement and posture. (From *Man, Microbes, and Matter* by B.C. Block. Copyright© 1975. Used with the permission of McGraw-Hill Book Company.)

direct motor pathway controlled by the basal ganglia. This system appears to control muscle tone rather than movement in normal subjects. (See Chapter 16 for a detailed description of the interconnections of this wandering nerve tract.)

Two basal ganglia disorders influence the extrapyramidal system importantly and have notable association with dysarthria. The reader will recall that the basal ganglia are islands of gray masses (cell bodies) situated within the white matter of the cerebral hemispheres. There are four large subcortical groups of nuclei that comprise the basal ganglia: caudate nucleus, lenticular nucleus which is made up of the putamen and globus pallidus, claustrum, and amygdala. The basal ganglia and several associated subthalamic and midbrain structures are referred to as the extrapyramidal system.

Parkinson's disease is marked by rigidity, so-called pill-rolling movements, a resting tremor, mask-like facies, and a shuffling gait. The primary anatomical locus is the substantia nigra, a nucleus in the midbrain. Other structures involved are the globus pallidus and, to a lesser extent, the putamen and caudate. Also known as paralysis agitans, this anomaly is actually a disease of the extrapyramidal system. Dysarthria associated with parkinsonism is characterized by a diminution of movement and is termed *hypokinetic dysarthria*. Patients with lesions that are widespread in the corpus striatum (*striped body*) experience uncontrolled movements that occur at random (Fig. 22-11). (The corpus striatum is made up of the caudate and lenticular nuclei combined with the internal capsule in between, which give it stripes of gray and white matter.) In addition to striatum lesions, another primary anatomic site of lesion is the *cortex*. These patients take on the appearance of someone attempting to execute a peculiar dance. In fact, this characteristic dance-like movement is known as *chorea* (Greek for dance). There are several forms. Muscle groups may be characterized by quick involuntary movements, a slowness of movement, and hypertonicity, which affects speech and yields a hyperkinetic dysarthria.

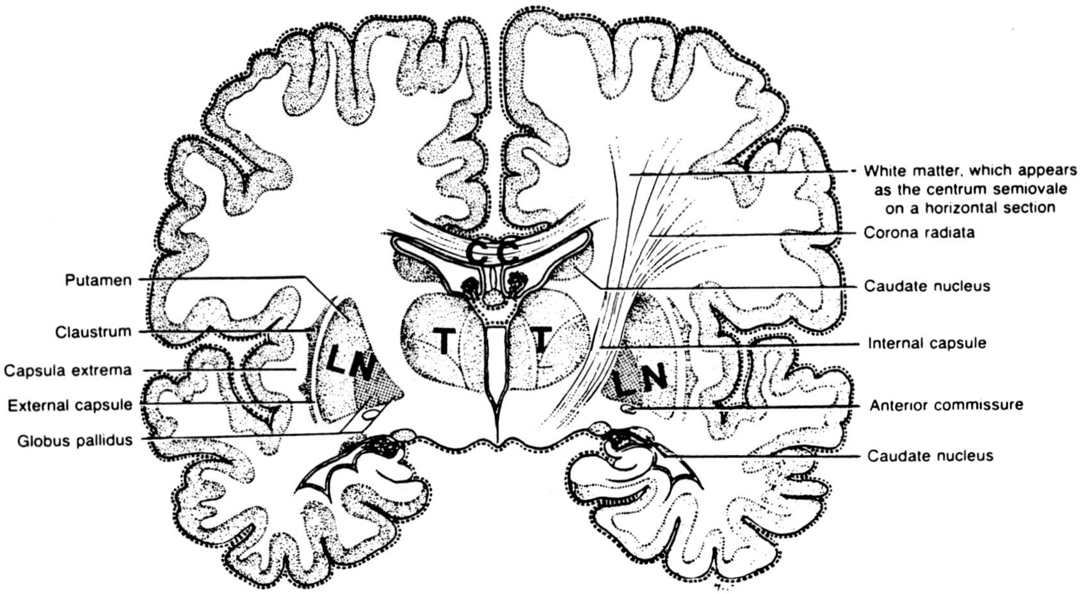

Figure 22-11. Coronal section of the brain showing gray (cell masses) and white (myelinated fibers) matter that may be involved in parkinsonism. Deep within the white matter are the basal ganglia. Members of the basal ganglia: lenticular nucleus (LN), which is made up of the putamen and globus pallidus, claustrum, amygdaloid nucleus, and the caudate nucleus (see also Fig. 22-1). (From Bossy, Jean: *Atlas of Neuroanatomy and Special Sense Organs*, 1970. Courtesy of W.B. Saunders Co., Philadelphia.)

Summary

Dysarthria is a variable disorder associated with the motor system responsible for speech production. This speech disorder is, then, directly related to muscle dysfunction, with its chief symptoms being slowness, weakness, incoordination, or change in tone of the speech musculature. While the most salient error made by patients with dysarthria is imprecise production of consonants, all the basic motor processes — respiration, phonation, resonation, articulation, and prosody (melody of the speech as determined by pitch, quality, loudness, duration variations) — are variably involved.

Apraxia

The inability to carry out complex learned movement is *apraxia* if the disorder cannot be explained by weakness, incoordination (ataxia), sensory loss, poor verbal comprehension, or inattentiveness to instructions; yet, apraxia is not synonymous with aphasia. It is unlike dysarthria because there is no motor impairment causing a breakdown in purposive behavior. Apraxia is a disorder of the *cerebral cortex*.

To understand the neuroanatomy of apraxia, we need to trace the path of speech processing through a communication loop in the brain. The anatomical model found later in this chapter is useful (Fig. 22-14). From an *input source*, such as a verbal command, the signal is fed into Wernicke's area since this area is the *auditory language comprehension center*. Termed also the *posterior speech cortex*, this area is located primarily on the left temporal lobe, its posterior aspect. After appopriate analysis and associations are made, neurons send the appropriate words and proper sequences of words to transform meaningful internal content into language for external output. This is accomplished by converting word sequences into a neural code of instructions for the *motor speech programmer*.

The pathway is from the language processing area to the programmer via a tract of association fibers that interconnects Wernicke's area with Broca's area. This tract is the *arcuate fasciculus* (bundle). It leaves the temporal lobe by curving backward, upward, and inward to negotiate the deep lateral fissure and the insular lobe (Isle of Reil) (Fig. 22-12). Next, the arcuate fasciculus enters the parietal lobes beneath the cortex and turns forward to join another bundle of association fibers from the parietal and occipital lobes. Together, these fused bundles form the superior longitudinal fasciculus. This new, enlarged bundle courses anteriorly just below the cerebral cortex to reach the frontal motor association areas anterior to the motor cortex (Benson, 1973). Here lies *Broca's area*, the anterior speech area.

After the wave of inpulses from Wernicke's area travels over the arcuate fasciculus, it engages the motor speech programmer in Broca's area. Named after Paul Broca, a nineteenth century French neurologist, Broca's area is the *motor speech area* in the inferior frontal lobe gyrus of the left hemisphere and is concerned with the organization of motor speech patterns. In Broca's area, the next operation involves *programming* more than nine dozen muscles to act in a precise, selective sequence and with precision of duration and strength of contraction. As reported by Darley, Aronson, and Brown, this entails 140,000 neuromuscular events per second. Obviously, understanding and appreciating this programming of the speech muscles is more than a day's work! Somehow, though, speech is produced and a response to the original message is made.

We should not overlook the "lesser" hemisphere in tracing the path of speech in the brain from a heard command to spoken response. The posterior and anterior language areas of the dominant hemisphere are connected with their corresponding areas in the nondominant hemisphere. Fiber bundles of the *corpus callosum* provide extensive communication between the two halves of the brain. Still, it is the role of the posterior area of the left hemisphere that serves as the analyzer for speech reception and as an integrating processor for all language modalities. It is the anterior speech area (Broca's area) of the dominant hemisphere that contains the motor speech programmer.

A lesion associated with the motor speech programmer will be manifest by one or more of the following clinical symptoms.

1. *Articulation is often off target*; the patient attempts with considerable effort to correct the error, but errors persist with extensive variations.
2. *Speech production slows*; the words are spaced, and the patient may stress syllables and words evenly.
3. Patients *make few substitutions, additions, repetitions*, and *prolongations* in articulation.
4. There is *no difficulty with word finding or managing meaning-bearing units* of language; the problem is confined to processing.
5. Patients *visibly and audibly grope as they effortfully attempt to produce correct articulatory postures* and to execute a sequence of postures of the articulators in forming words.
6. If a *paralysis* is present, it *is insufficient to explain the articulation errors* present, although no impairment of muscle function is usually noted (Darley, Aronson, Brown, 1975).

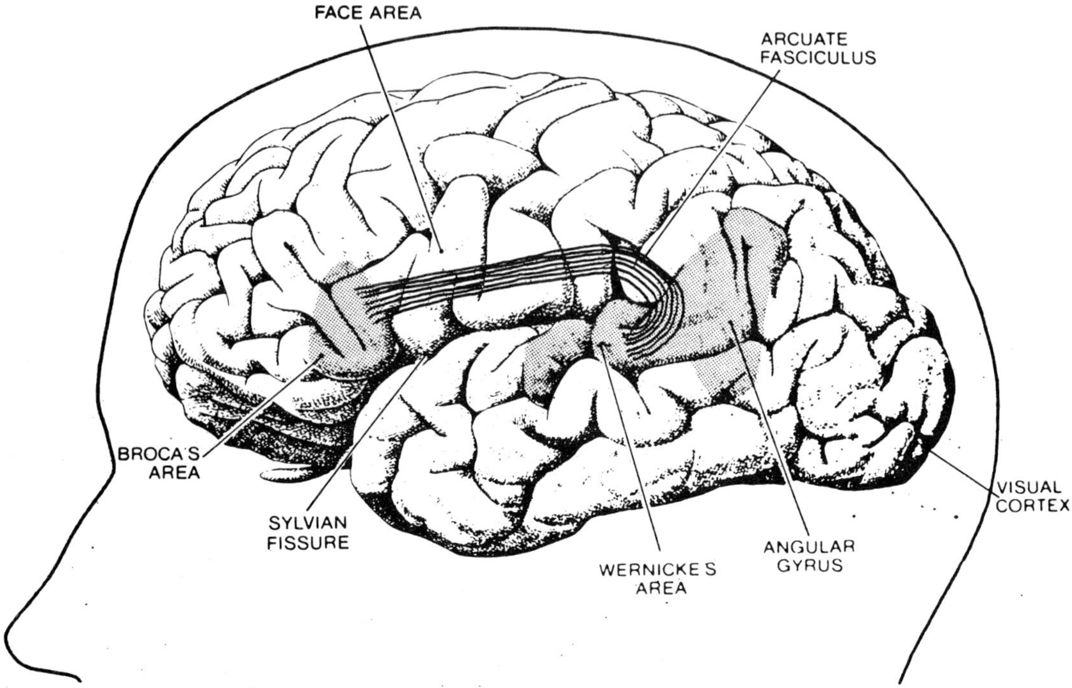

Figure 22-12. Language perception, formulation, and expression centers, left cerebral cortex. The arcuate fasciculus is a bundle of fibers interconnecting the posterior speech cortex (Wernicke's area) with the anterior speech cortex (Broca's area). (From Language and the Brain by Norman Geschwind. Copyright© 1972 by *Scientific American, Inc.* All rights reserved.)

Aphasia

Often encountered in the medical literature is the clinical entity known as *Broca's aphasia*. Synonymous with Broca's aphasia is the term *motor aphasia*. In the preceding section, disturbances in Broca's area were depicted as *not* involving disturbances of language. Here is a good case of confusion of clinical terminology and neurophysiology pertaining to speech and language functions. Actually, Broca proposed in 1861 the term *aphemia* to describe impaired motor speech programming associated with expressing language externally. Later, others substituted the term *aphasia* to refer to many language disturbances, ignoring the specific symptomatology and neuroanatomic basis Broca had described. Wernicke in 1874 described aphasia as belonging to the left hemisphere and, in particular, the temporal lobe and acknowledged Broca's aphemia. Nonetheless, use of aphasia in reference to the frontal-lobe speech center has persisted.

Aphasia is impaired *comprehension, formulation*, and *expression* of language. (Perhaps the student should habitually think of *internal expression of language* to be the programming area of Broca as related to this context to avoid confusion with externalizing of language — articulation.) Numerous aphasias are listed, with an especially good summary of these being found in the dictionary by Nicolosi (1982). The disorder is unexplained by any impairments, if present, with primary sensory deficits, dementia, or psychological problems.

So, the contrast between the effects on speech from lesions in Broca's and Wernicke's areas are distinct. Broca's area is responsible for the conversion of formulated ideas into smoothly articulated sounds. The speech area of Wernicke is more involved in the process of conveying meaning. A patient with Broca-area problems might utter stumbling, painfully slow telegraphic sentences, yet they would be appropriate ideas. Those with destructive lesions in Wernicke's area speak with perfect articulation but make little sense.

CEREBRAL DOMINANCE

Speech clinicians find the topic of cerebral dominance an intriguing phenomenon. How the brain develops functional laterality for various complex activities has far-reaching implications in understanding aphasias as well as in understanding general neurophysiological concepts pertinent to topics encountered by speech clinicians.

Linguistic dominance in the brain has been studied with considerable intensity but much remains to be learned. An example of the scope of this problem concerning language lateralization is shown in Table 22-I. The data were gathered by Branch, Milner, and Rasmussen (1964) through sodium amytal tests. This technique involves injection of this fast-acting barbiturate into the left or right internal carotid artery located in the neck. The patient counts aloud as the drug is injected. That hemisphere directly above the injection dysfunctions briefly as the drug reaches the hemisphere supplied by the injected carotid artery. When the hemisphere dominant for speech is suppressed by the drug, the patient ceases counting and cannot respond to commands to continue until the drug wears off. As can be seen in Table 22-I, a number of left-handed individuals show weakness or absence of lateralization.

Specialization of function in one hemisphere seems to result in more efficient performance. It reduces the possibility of an interference phenomenon — to prevent a neurological log jam.

Table 22-I

Lateralization and Hemisphere Dominance

Handedness	*Dominant hemisphere (%)* Left	Right	Both
Left handed	64	20	16
Right handed	90	10	0
Ambidextrous	60	10	30

It is known that hemisphere dominance develops gradually. In fact, either side of the brain can develop into full linguistic competence in young children. Clinical impressions hold that the right hemisphere in girls is not specialized for a particular function. In boys, the right hemisphere seems to be activated consistently during spatial tasks; girls experience more difficulty and do not rely heavily on one hemisphere. But language appears to transfer more readily to the right hemisphere in females than males following damage to the left hemisphere during early childhood. This extended *plasticity* in young girls suggests that they may have a lower incidence of developmental disorders associated with left-hemisphere dysfunction. Males do display a higher incidence of developmental dyslexia, developmental aphasia, and infantile autism. Each of these impairments has language deficits as a dominant theme (Kelly, 1981).

For some time, it was thought that the triggering mechanism for specialized brain lateralization was the development of language. In essence, we had to await language development before a hemisphere could emerge dominant. Then, more critical investigation of the asymmetries of the brain anatomy focused on cortical areas that correspond to the speech centers. Not only were these areas found to be larger, but the inference was made that the left hemisphere, in most instances, seemed to be *preset* and *genetically programmed* for dominance (Fig. 22-13).

It is interesting to observe that earlier conclusions about the mechanism for brain lateralization are attributed to language developed and were derived from studies of children suffering left-hemisphere injury in infancy and early childhood. Speech had not yet developed in these patients. The trend reported was that an insult to the left hemisphere at 2–3 years of age may result in mild impairment, increasing in severity according to the age of the child at the time of the injury. Children 9–10 years old experience permanent language impairment related to left-hemisphere disturbances (Kupferman, 1981).

In Chapter 15, it was pointed out that the two cerebral hemispheres are not mirror halves. However, they have bilateral representation for sensory and motor functions. Through the corpus callosum — that mass of interconnecting fibers (200 million) between the hemispheres — the language and speech programming centers of the left hemisphere tell the other half what is going on. Further, the right hemisphere may even be capable of primitive understanding of language.

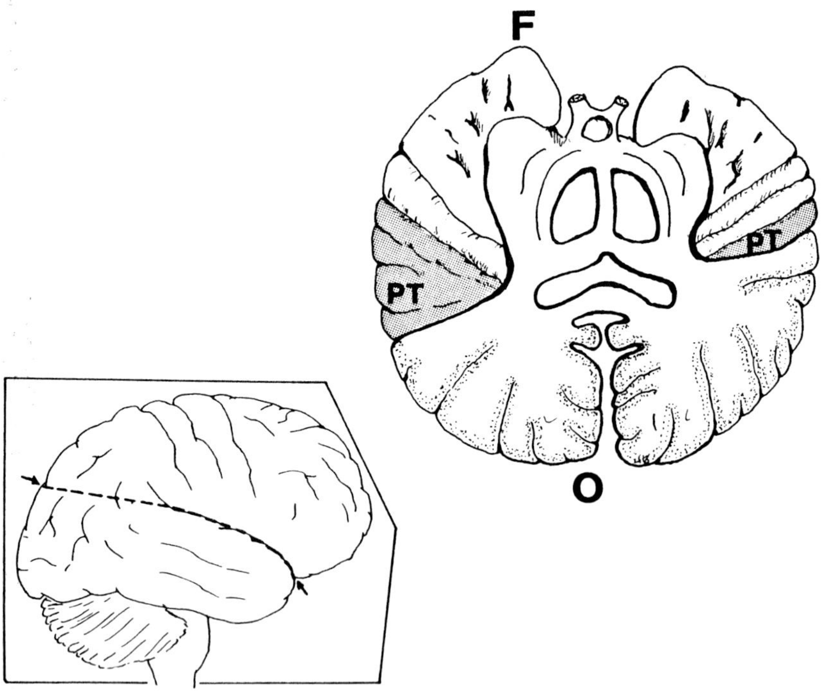

Figure 22-13. Section through the brain following path of dotted line (inset) to show asymmetry of the superior temporal lobe. The planum temporale (shaded) occupies a greater area on the left temporal lobe in a typical specium. The left front (F) and occipital (O) lobes are also indicated. (From Geschwind, N. and Levitsky, W.: Human brain: left-right asymmetries in temporal speech region, *Science, 161*:186-187, 1968. Copyright 1968 by the American Association for the Advancement of Science.)

Studies involving commissurotomies (cutting the corpus callosum) have been reported. Subjects were asked to identify a stimulus verbally or nonverbally. Words projected to the right hemisphere were read and understood. The more complicated the stimuli, the less accurate the response. It appears that the right hemisphere is almost completely incapable of language output even though it may have some capacity to process inputs (Peele, 1977).

Although the right hemisphere may lack verbal abilities, it is superior to its mate in certain perceptual tasks. Another interesting point derived from studies of commissurotomized subjects is that the two hemispheres may function independently. At the great risk of oversimplification, we may say that the normal brain is a composite of a left hemisphere that tends to be an *intellectual, rational, verbal*, and *analytical thinker* and a right hemisphere that tends to be a *perceiver* and an *emotional, nonverbal*, and *intuitive* thinker. In principle at least, each hemisphere has the capacity of independent function, but normally, integration of function is maintained by means of extensive commissural connections.

ANATOMICAL MODEL OF THE ORAL COMMUNICATION SYSTEM

It should be rather obvious from the introduction to the speech process that dynamic oral communication must depend on the interaction of a series of intriguing automatic and reciprocal relationships within the speaker. A problem may reflect itself in any one of several operations of the communicative process and result in some type of oral communicative disorder, a point described earlier in this chapter. But malfunctions may arise from speech generation or feedforward sources, as well as from speech monitoring or feedback sources.

Receptor Unit

Figure 22-14 is a theoretical model to illustrate the functional design of the oral communication process. A simplistic anatomical schema is used after Mysak (1971). First, this model differs in the manner it represents the origins of an idea found in the communication model in Figure 1-1 (see Chap. 1). Since ideas (intentive behaviors of the speaker) must be sparked from some stimulus, sensations through one or more sensory channels serve to gain the attention of our consciousness. These stimulus channels (the eye, ear, tactile, and, in some cases, proprioception) (the awareness of muscle position and spatial relation), make up the *receptor unit*.

The eye and ear are the dominant speech receivers to prompt an idea for communicating to a listener. Techniques emphasizing tactile and proprioception clues have been employed with deaf-blind patients, with cerebral palsied children, and with the retarded to teach sensing of the articulatory movements, concepts of speech-sound generation, and understanding of language. It should be noted that we usually receive multisensory external speech stimuli, probably utilizing two or more channels simultaneously (Fig. 22-15).

Integrator Unit

The *integrator unit* is represented by two small illustrations of the brain with markings on each to illustrate the proposed functional area contributing to this second component of the model. The sensory

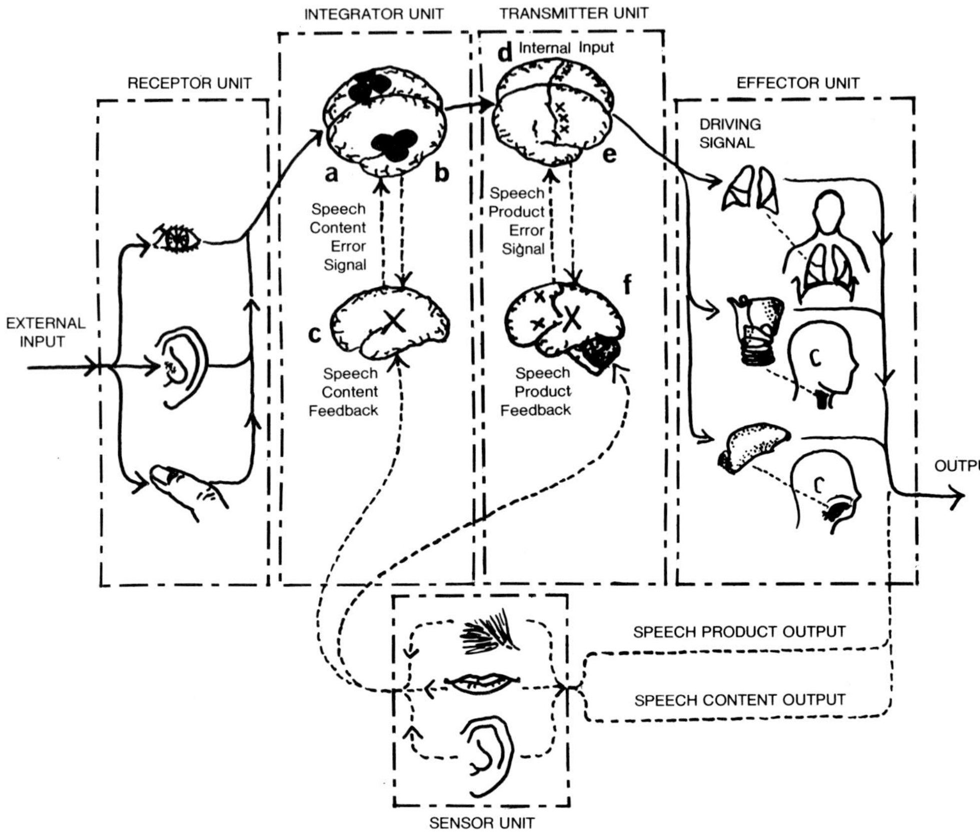

Figure 22-14. Anatomical schema of the speech system. (Modified from Mysak, E.D.: *Speech Pathology and Feedback Theory*, 2nd ed., 1971. Courtesy of Charles C Thomas, Publisher, Springfield, Illinois.)

areas of the brain that recognize and attach significance to incoming neural patterns of auditory, visual, and tactile stimuli are included in phase *a*. Phase *a*, then, represents the notion of forming perceptions in the brain.

Phase *b* in the integrator unit involves the interpreting and the elaborating of incoming information and forms verbal and nonverbal response attitudes. A storage component is also included that retains or releases held information on command.

The speech content error-measuring loop, phase *c*, compares the *actual* speech content with the *intended* speech content and determines if any errors are present and if they warrant corrective action. Thus, a speech content *comparator* and a speech content *corrector* serve together as the integrator unit's corrector device. Therefore, while speech is being sent out, the word neuronal-patterns (organized barrages of nerve impulses), representative of the information being sent, persist simultaneously somewhere in the left side of the brain for comparator and corrector circuits to examine and execute changes as desired. If the comparator determines that an error is present and a change is needed, the corrector device will launch a scan for different neuronal-concept patterns that lead to different neuronal word-patterns and a different speech output. Thus, the semantics of the

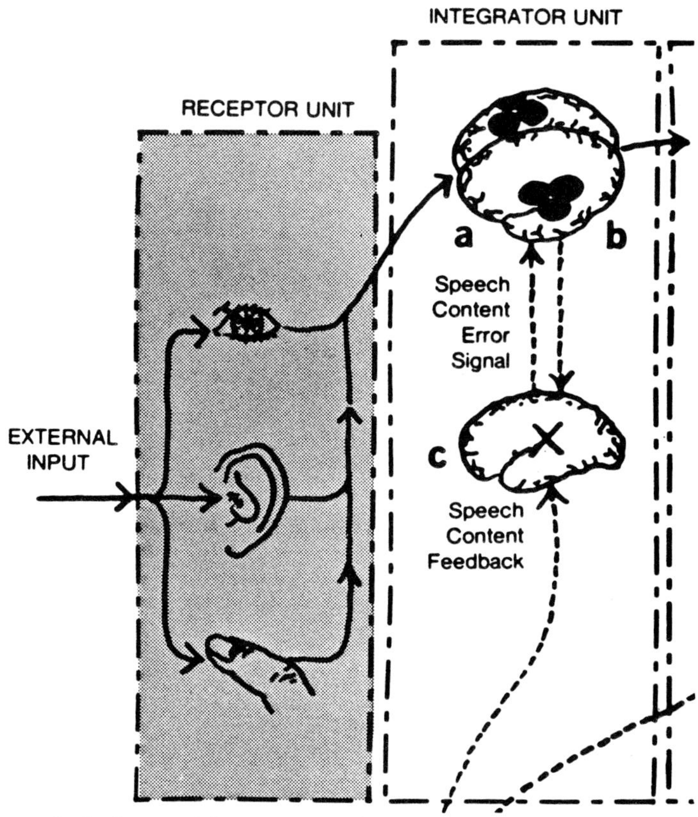

Figure 22-15. Anatomical schema of the receptor unit and its relation to the integrator unit. (From Mysak, E.D.: *Speech Pathology and Feedback Theory*, 2nd ed., 1971. Courtesy of Charles C Thomas, Publisher, Springfield, Illinois.)

output system are watched over.

Once the integrator has selected a particular response, ideas or speech intentions issuing from the integrator unit automatically excite word patterns in the transmitter unit. Even after discharging the neuronal word-picture pattern to the next unit, this signal also has the potential for keeping the neuronal word-pattern active for later checking.

Transmitter Unit

Again, figures of the brain appear in our model for yet another unit. This time, however, the activity moves to the actual production of audible speech. It is apparent that the brain is carrying on a considerable amount of communication with itself during speech production. Speech intentions flowing from the integrator unit automatically excite word patterns in phase *e* of the *transmitter unit*. Phase *d* is charged with sending operating signals for programming the muscular systems of the expiratory, vibratory, and articulatory components that are charged with producing oral communication (Fig. 22-16).

A speech *product* comparator, parallel to the *content* comparator of the integrator unit, receives signals from within the brain circuitry as well as from output feedback signals. If the comparator unit judges the two signals to differ, a message

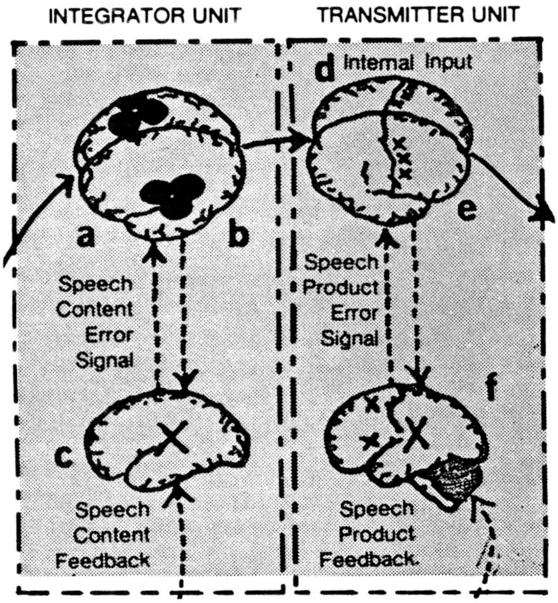

Figure 22-16. Anatomical schema of the integrator and transmitter unit. (From Mysak, E.D.: *Speech Pathology and Feedback Theory*, 2nd ed., 1971. Courtesy of Charles C Thomas, Publisher, Springfield, Illinois.)

is sent to the speech product corrector, which then combines the error signal and input signal into a new corrected driving signal. Interestingly, there is a "predictor potential" existing within the speech product comparator. Command signals are permitted to progress rapidly without feedback monitoring when error-free speech units are anticipated. A general analogy may be drawn between this speeded-up speech output mode and the expert typist who zips along on, "Now is the time for all good men to come to the aid of their country." The typist assumes quite a different monitoring precision when the typing material consists of unfamiliar professional jargon.

Note that different areas of the brain are marked in the transmitter unit. Still, the left side is more crucial. Phases *c, d,* and *f* represent principal involvement of the left-brain hemispheres, while phases *a, b,* and *e* represent involvement of both halves of the brain.

Effector Unit

Three components comprise the *effector unit*: an energy source, a vibrator, and a modulator. The *energy source* is the expired lung air forced through the vocal tract by the expiratory musculature. This is the motor unit in the first aspect of speech production. In the larynx, the vocal folds become the *vibrator* to generate sound from the outgoing air (Fig. 22-17). The lips, tongue, and soft palate are the articulators that constrict, stop, and alter the resonance of the sound stream. The articulators are, therefore, the *modulators* for transducing the sound stream into meaningful chunks of sound called *words*. Implied within these three components is the act of resonation, which is automatically present in association with the latter two components.

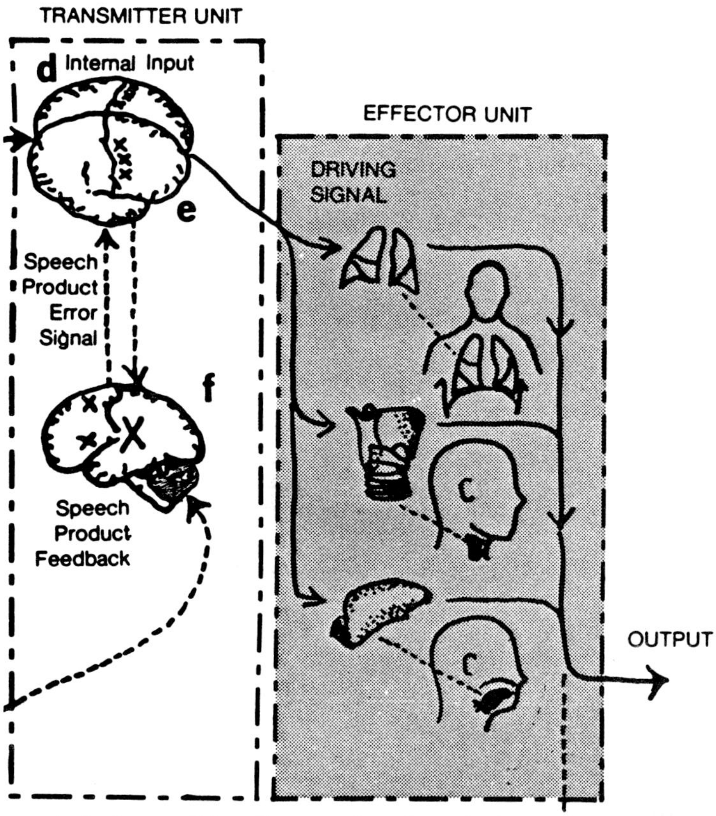

Figure 22-17. Anatomical schema of the effector unit and its relation to the transmitter unit. (From Mysak, E.D.: *Speech Pathology and Feedback Theory*, 2nd ed., 1971. Courtesy of Charles C Thomas, Publisher, Springfield, Illinois.)

The effector unit is driven by the transmitter unit, which sends neural impulses to the respective muscles of each component. The musculature contracts in an intricate and precise operation, following a computer-like programming from the speech integrator and transmitter units.

All of the speech mechanism is interrelated even though the aspects of speech are frequently illustrated and discussed as if they were rather independent actions in the oral communication process. The student is again reminded that the reason for breaking down these complex interactions is to simplify the structures and functions, since the intent of this text is only introductory.

Sensor Unit

Notice the location of the *sensor unit* centered directly under the integrator and transmitter units. The three components illustrated are an *auditory sensor*, a *tactile sensor*, and a *proprioceptive sensor*. If the speaker should watch himself/herself in a mirror while speaking, a *visual sensor* would be added (Fig. 22-18).

The sensors are responsible for providing feedback of the speech content and speech product. The ear represents the auditory system, the lips are in contact to represent tactile sensors of the speech articulators, and the muscle represents the proprioceptive sensors of the speech mus-

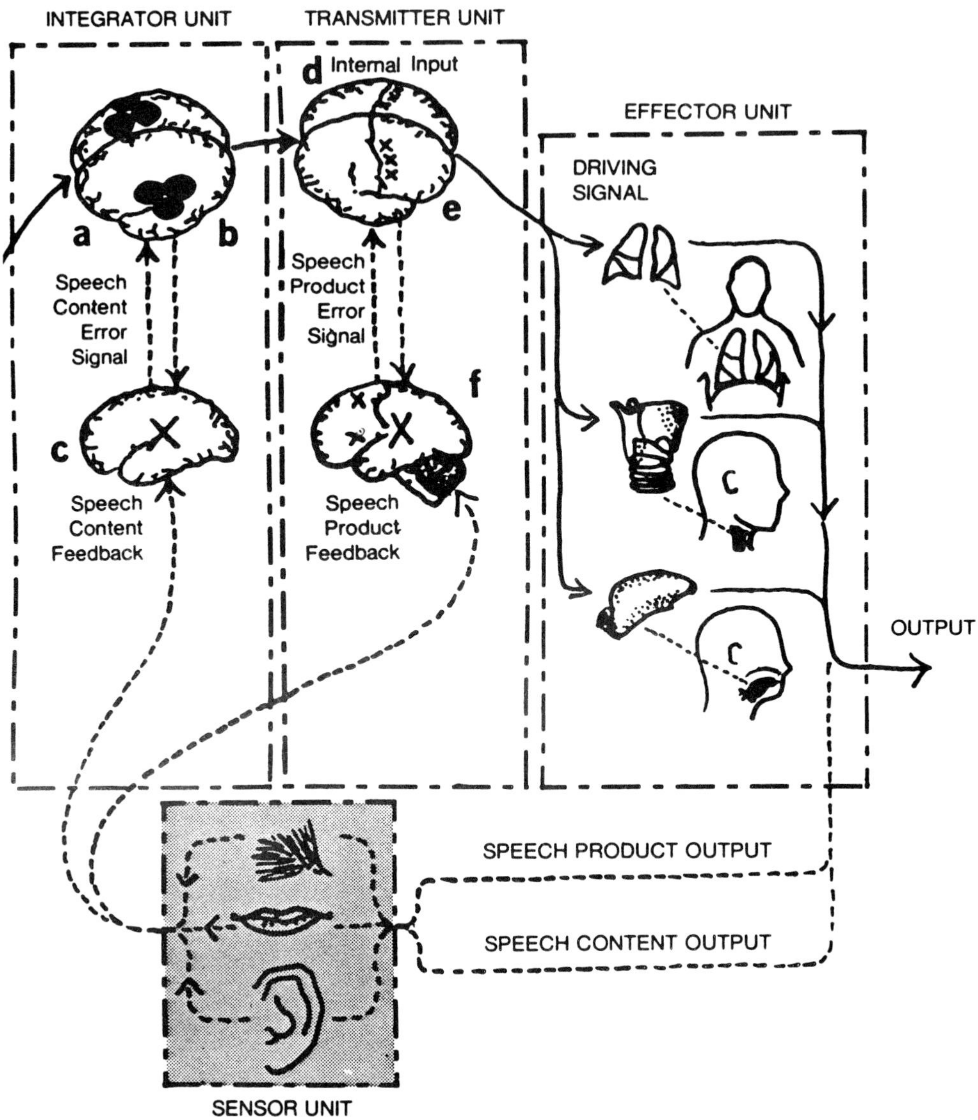

Figure 22-18. Anatomical schema of the sensor unit and its relation to other units. The three components shown are responsible for feeding back speech product and content data to the brain. Sensor 1 feeds back a proprioceptive (muscle sense) dimension; sensor 2, the tactile dimension; and sensor 3, the auditory dimension. This unit may also include sensor 4, the visual dimension. (From Mysak, E.D.: *Speech Pathology and Feedback Theory*, 2nd ed., 1971. Courtesy of Charles C Thomas, Publisher, Springfield, Illinois.)

culature. Of course, the eye would be added when this channel is operative.

Our sensor unit sends two signals, the speech content and speech product feedback signals to the integrator and transmitter units, respectively, where comparisons are made between the ongoing processing by the brain of the signal and the actual product articulated. External output is monitored for correct meaning (semantics) and correct speech sound pattern (articulation).

BIBLIOGRAPHY

Arbib, M.A., Caplan, D., and Marshall, J.C. (Eds.): *Neural Models of Language Processes*. New York, Academic Pr, 1982.

Benson, D.F., and Geschwind, N.: The aphasias and related disturbances. In Baker, A.B., and Baker, L.A. (Eds.): *Clinical Neurology*, Vol. I, 4th ed. Hagerstown, MD: Har-Row, 1973.

Branch, C., Milner, B., and Rasmussen, T.: Intracarotid sodium amytal for the lateralization of cerebral speech dominance. *J Neurosurg, 21*:399-405, 1964.

Broadbent, D.E.: Attention and the perception of speech. *Sci Am, 206*:143-151, 1962.

Cote, L.: Basal ganglia, the extrapyramidal motor system, and diseases of transmitter metabolism. In Kandel, E.R., and Schwartz, J.H. (Eds.): *Principles of Neural Science*. New York, Elsevier/North-Holland, 1981.

Darley, F.L., Aronson, A.E., and Brown, J.R.: Clusters of deviant speech dimensions in the dysarthrias. *J Speech Hear Res, 12*:462-496, 1969.

Darley, F.L., Aronson, A.E., and Brown, J.R.: Differential diagnostic patterns of dysarthria. *J Speech Hear Res, 12*:246-269, 1969.

Dworkin, J.: Tongue force in normal and in dysarthric patients with amyotrophic lateral sclerosis, *J Speech Hear Res, 23*:828-837, 1980.

Eysenck, H.J., and Eysenck, S.B.G.: *Personality Structure and Measurement*. San Diego, CA: Knapp, 1968.

Hess, W.R.: *Diencephalon: Autonomic and Extrapyramidal Functions*. New York, Grune and Stratton, 1954.

Isaacson, R.L.: *The Limbic System*. New York, Plenum Pr, 1974.

Jackson, F.E.: *The Pathophysiology of Head Injuries*. Summit, NJ: CIBA, 1966.

Kandel, E.R.: Brain and behavior. In Kandel, E.R., and Schwartz, J.H. (Eds.): *Principles of Neural Science*. New York, Elsevier/North-Holland, 1981.

Kelly, D.D.: Sexual differentiation of the nervous system. In Kandel, E.R., and Schwartz, J.H. (Eds.): *Principles of Neural Science*. New York, Elsevier/North-Holland, 1981.

Kupferman, I.: Localization of higher functions. In Kandel, E.R., and Schwartz, J.H. (Eds.): *Principles of Neural Science*. New York, Elsevier/North-Holland, 1981.

Logemann, J.A., and Fisher, H.B.: Vocal tract control in Parkinson's disease. *J Speech Hear Disord, 46*:348-352, 1981.

Lapointe, L., and Horner, J.: Palalia. *J Speech Hear Disord, 46*:34-38, 1981.

Luria, A.R.: *The Working Brain*. New York, Basic, 1973.

Mayeux, R.: Disorders of speech and learned movements. In Kandel, E.R., and Schwartz, J.H. (Eds.): *Principles of Neural Science*. New York, Elsevier/North-Holland, 1981.

Milner, P.M.: *Physiological Psychology*. New York, H, R & W, 1970.

Mysak, E.D.: *Speech Pathology and Feedback Theory*. Springfield, IL: Thomas, 1971.

Nicolosi, L.: *Terminology of Communication Disorders*, 2nd ed. Baltimore, Williams and Wilkins, 1982.

Papez, J.W.: A proposed mechanism of emotion. *Arch Neurol Psychiatry, 38*:725-743, 1937.

Peele, T.L.: *The Neuroanatomic Basis for Clinical Neurology*, 3rd ed. New York, McGraw, 1977.

Penfield, W., and Roberts, L.: *Speech and Brain-Mechanisms*. Princeton, NJ: Princeton U Pr, 1959.

Sussman, H.: Evidence for left hemisphere superiority in processing movement related tonal signals. *J Speech Hear Res, 22*:224-234, 1980.

Whitaker, H.A. (Ed.): *Brain and Language*. New York, Academic Pr, 1974.

Chapter 23

OROFACIAL EXAMINATION*

THE examination of a patient in the typical clinical setting usually includes a description of the status and function of the facial and intraoral structures. Assessment of orofacial form and function provides a basis for understanding the physical system that the patient is using for speech.

Since the examination findings may be sent to a physician or dentist, the use of "orofacial examination" rather than "oral-peripheral examination" is recommended in clinical description and report writing. The use of oral-peripheral begs the question: "Peripheral to what?" Clarity of descriptions should be one goal of examination reporting.

EXAMINATION OF THE FACE

General Considerations

In conducting an examination of the face, the clinician should remember that facial expression, coloration, skin conditions, and other findings help to assess the general health and personality of the patient. General appearance of expression (facies) is important and often characteristic. The child with a congenital heart defect usually presents a facies characterized by a blue or purple tint to the tissues of the lips, under the eyes, and around the cuticles of the fingers. These signs are important to recognize, since children with heart problems are not good risks for surgery and associated general anesthesia. In such patients, speech problems of hypernasality, when present, often assume a secondary role to more important problems. Accordingly, the clinician should exercise caution in recommending surgery to correct hypernasality in patients with heart defects. This situation arises with some frequency in cleft palate patients, who have a 10 percent incidence of

*This chapter is based on an article that was published in *Communique*, the journal of the North Carolina Speech, Hearing, and Language Association (*8*:2-25, 1978), and the *International Journal of Oral Myology* (*6*:3-16, 1980). Portions have been reprinted here with the permission of the editors. Examination forms that parallel the information in this chapter were published in *Language, Speech, and Hearing Services in Schools* (*8*:161-163, 1977) by Mason and Simon.

congenital heart defects (Gorlin and Pindborg, 1964).

Not all facies are associated with clearly observed physical symptoms as found in patients with heart defects. Some children possess indefinite characteristics that confuse the clinician, resulting in descriptions such as the "Funny-Looking Kid Syndrome" (FLK). Rather than representing a definite constellation of findings, the FLK syndrome denotes that the clinician recognized that vague facial variations are present but was unable to describe them accurately.

The so-called *adenoid facies* is another example of associations of cause and effect being made inappropriately. The adenoid facies is supposedly produced by adenoidal enlargement and characterized by mouth hanging open, pinched nares, and somewhat retracted chin (see Fig. 10-25). In recent years it has been found that this label has been used inappropriately, in that there is not a characteristic facial configuration or growth pattern that follows from adenoidal enlargement. Further, there is an incorrectly presumed association between adenoid size, breathing pattern, and mandible and tongue posture. Most clinicians have ceased to use the term adenoid facies, as it is an inaccurate description of a physical and functional situation.

Lateral Observation of the Face

The Facial Profile

Examination of facial features in the lateral dimension permits evaluation of the adequacy of the facial profile. There are three basic points to identify as referents for facial profile adequacy. These are (1) the bridge of the nose, (2) the base of the nose, and (3) the most prominent point on the chin. (4) The base of the chin is useful in assessing vertical facial height (Fig. 23-1).

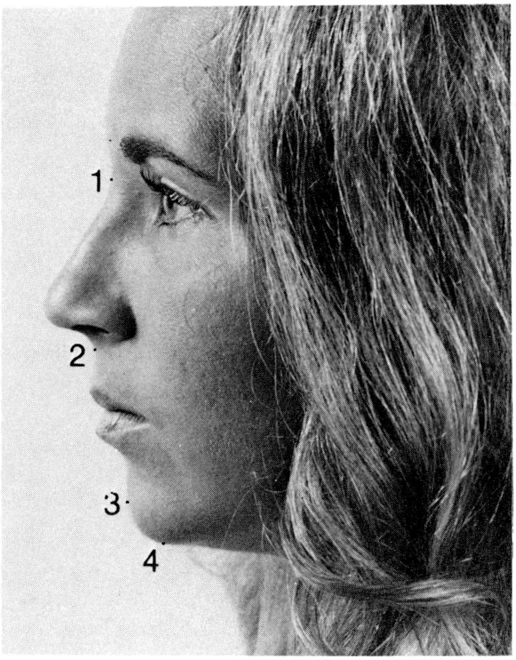

Figure 23-1. Lateral view of the face showing the points used to evaluate the facial profile: (1) the bridge of the nose; (2) the base of the nose; (3) the most prominent point on the chin; and (4) the base of the chin (helpful to use in assessing facial height).

In a normal profile, the examiner should be able to connect points 1 and 3 with either a straight line or slightly convex line. Deviations of the lower or middle face from this imaginary reference can be described as either a *retrusion* (deviation posterior from the normal reference) or a *protrusion* (deviation in an anterior direction). The use of these terms permits descriptions of profile in the horizontal planes of space.

There are three basic facial patterns into which all facial skeletal combinations can be placed, as explained in Chapter 21. The influence of an anomalous condition (such as a forward-resting tongue posture) on the expression of the growth pattern over time is the special challenge for the clinician to understand.

Facial Height

For practical purposes, clinicians measure facial length within the confines of the midface and lower face. That is, the upper face area (from midline hairline to eyebrows) is most often not included in facial measurements. All measurements and clinical guidelines to follow will adhere to this format, considering the face as extending from nasal bridge to base of chin, but referring to this area as midface and lower face area.

In the vertical direction, the midface accounts for about 40 percent of overall facial height in adults, while the lower face accounts for approximately 60 percent. While these figures are quite variable in children, these dimensions are of general use for all age groups. In the clinical situation, it is convenient to remember that lower facial height should generally be 20 percent greater than midfacial height (Fig. 23-2).

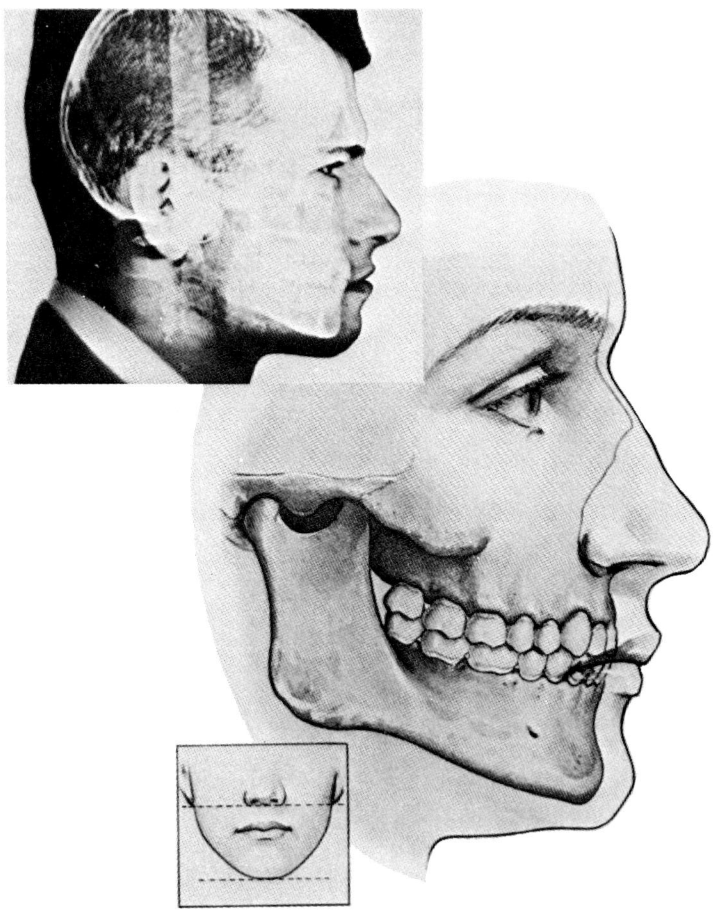

Figure 23-2. The face in profile, demonstrating good proportions in the horizontal and vertical planes. Lower face height is highlighted in the inset. (From Massler, M., and Schour, I.: *Atlas of the Mouth*, 2nd ed., 1958. Copyright by the American Dental Association. Reprinted by permission.)

Midface Height

Clinicians can measure midfacial height by placing a finger on the patient's nasal bridge (the upper limit of midface height) with the thumb positioned at the base of the nose. It is important to be sure to keep this dimension constant; next, the clinician removes his/her fingers from these points and repositions the index finger at the base of the nose with thumb now resting wherever it may touch on the chin. In the normal, the thumb fits into the soft-tissue groove above the chin prominence leaving approximately 20 percent of additional facial height uncovered. There is, of course, considerable variation in a normal population in the vertical dimension and, apparently, more variation among female adults.

Lower Face Height

In instances where lower facial height is reduced, it will probably be determined that the mandible is also retruded and reduced in overall size (Fig. 23-3). Where lower facial height appears increased, excessive growth increments in the corpus of the mandible are expected to have occurred. These skeletal variations suggest accompanying dental problems, since the bony framework for occlusal relationships of the dental arch is modified. It should be noted, however, that a patient with a Class II or III skeletal pattern need not possess Class II or III dental characteristics. Nonetheless, dental malocclusions of Class II and III variety are typically attributed to skeletal form changes. Many can be treated successfully by orthodontic procedures, but severe problems most often require combined orthodontic and oral surgical (orthognathic) management for a stable result.

From a lateral view of the face, the examiner can study the configuration of each side of the mandible, with particular reference to the *mandibular plane*. The clinician is encouraged to palpate along the lower border of the corpus of the mandible. The mandibular plane should be roughly parallel with a plane established by a line connecting the upper margin of the ear to the nostril, in the lateral view (Fig. 23-2).

A *steep mandibular plane* is a clinical finding in which there is an increase in lower anterior face height (see Figs. 23-4 and 23-5). Often, this condition is seen with a short ramus and a higher posterior tongue position in the oral cavity. This can lead to a *skeletal* openbite malocclusion where the tongue adapts to this situation by resting forward against or between the anterior teeth. The mandible hinges open a few millimeters and the posterior teeth are seen to supraerupt, while the anterior teeth are discouraged from erupting by the resting anterior tongue position. If this posture persists for an extended period, such as months or years, a skeletal openbite can be created (see Chap. 10, "The Dentition").

Clinicians will find in their clinical work that *skeletal* openbite malocclusions, almost without exception, will be accompanied by a tongue-thrust swallow. There is a temptation to unjustly blame the openbite on the swallow. The tongue-thrust swallow is simply filling up a space that was already there. The resting anterior tongue posture, however, can be etiologic to the problem by creating the differential dental eruption pattern that causes supraerupted molars to contact prematurely and hinge the mandible open somewhat, as explained in Chapter 10.

The Ears

In addition to assessing the profile from the lateral view, the clinician is interested in observing the status of the ears. When an external ear is malformed,

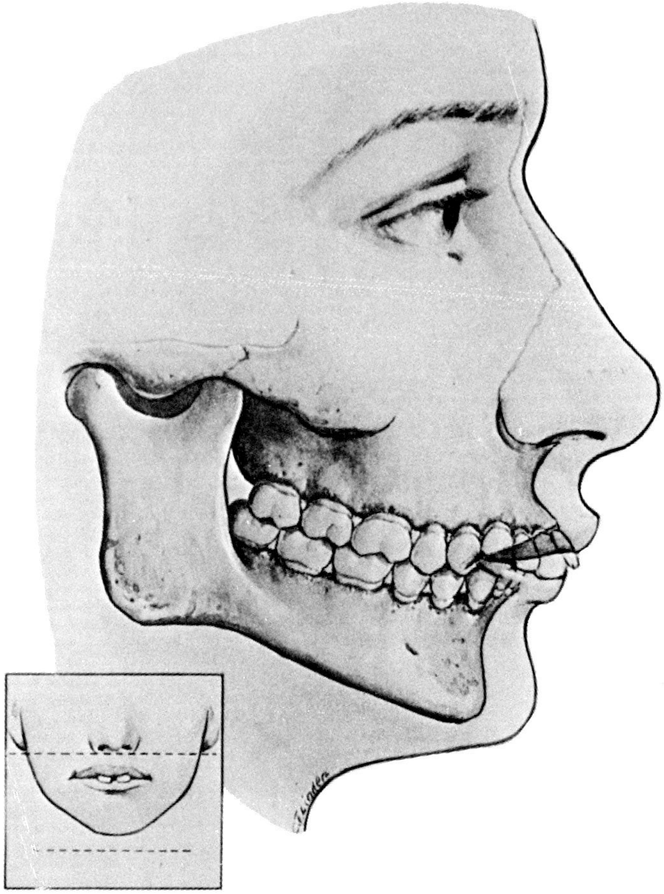

Figure 23-3. Patient with reduced lower face height associated with a retrusive mandible. This patient also has some maxillary dental and skeletal protrusion. Notice the acute angle of the mandible. The reduced lower face height is highlighted in the inset. (From Massler, M., and Schour, I.: *Atlas of the Mouth*, 2nd ed., 1958. Copyright by the American Dental Association. Reprinted by permission.)

the general term used most frequently to describe this is *microtia*. A microtic ear may or may not be accompanied by closure (stenosis) or absence (atresia) of the external auditory canal (Fig. 23-6).

External ear deformities are of clinical interest, in that the outer ear is an embryonic derivative of the first branchial arch. Therefore, the presence of outer-ear malformation should alert the examiner to the possibility of generalized variations in the maxilla and mandible. A poor correlation exists between the severity of the external-ear malformation and the status of the middle-ear cavity. Accordingly, it is important that air conduction audiometry be attempted in the presence of microtia and atresia.

Frontal Observation of the Face

Normal facial structures are not exactly symmetrical on the two sides of the face, and differences should be recorded. The midline is a handy reference for the examiner to use in observing symmetry and function of the face.

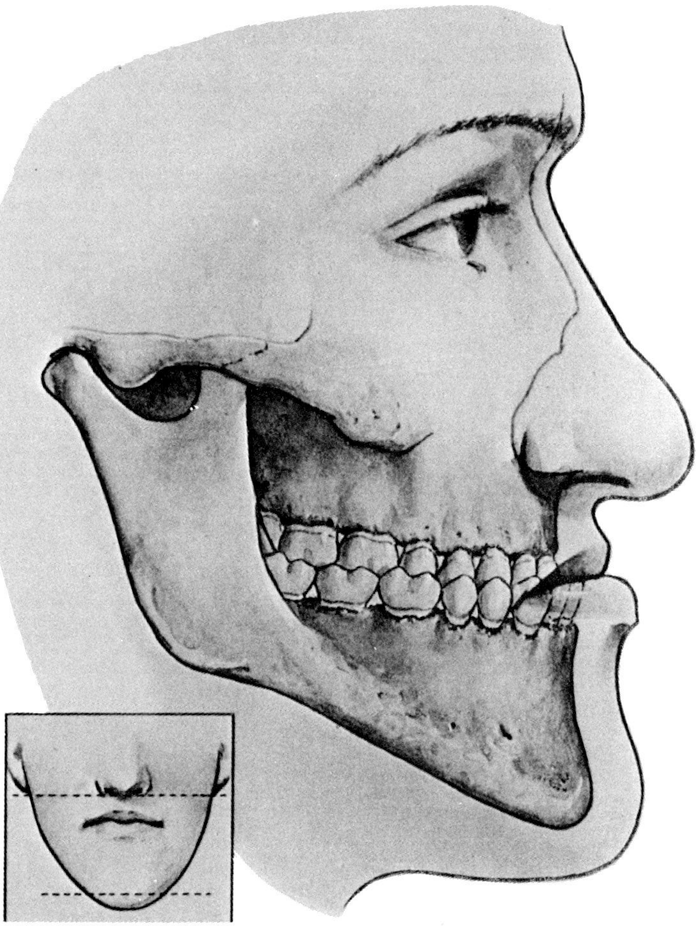

Figure 23-4. Patient with a steep mandibular plane, increased lower face height, and a severe crossbite condition as a consequence of hypertrophy of the mandible. (From Massler, M., and Schour, I.: *Atlas of the Mouth*, 2nd ed., 1958. Copyright by the American Dental Association. Reprinted by permission.)

The Upper Midface Area

The Eyes

The eyes and adjoining areas constitute an important part of facial form. The many variations in spacing, degree of protrusiveness, and bordering bony- and soft-tissue support account for some individual differences in facial form.

It is not unusual to find the orbits positioned on different planes in the *vertical* dimension. This is especially marked in children with hemifacial hypertrophy. In comparing the relative horizontal position of the eyes, the generalization can be followed that the *face is five eyes wide*, as explained in Chapter 21 (see Fig. 21-1). That is, there should be room for one eyewidth in the space occupied by bony tissues that separate the eyes, as shown in Figure 23-8. This observation is helpful in determining whether the orbital globes are separated excessively. Such spacing between the eyes (hypertelorism) is a common finding in mental retardation and other syndromes. The individual

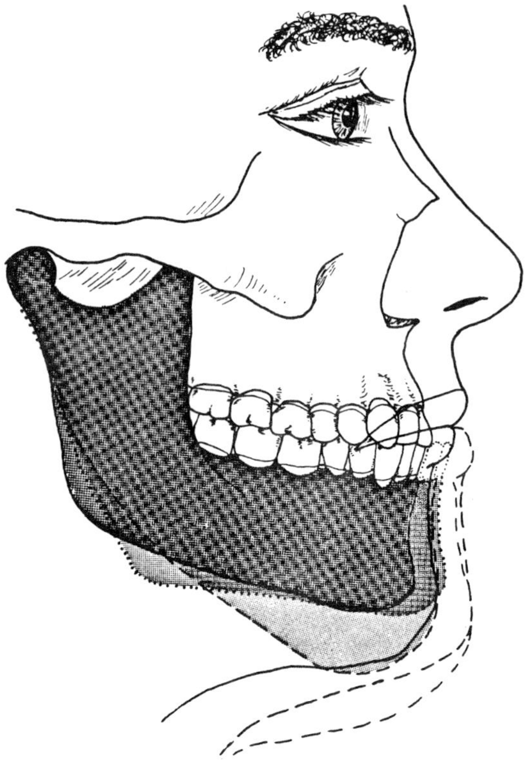

Figure 23-5. Superimposition of mandibular plane types showing the relationship of mandibular angle configuration and lower anterior face height. Generally, the more open the mandibular angle, the longer the lower face height.

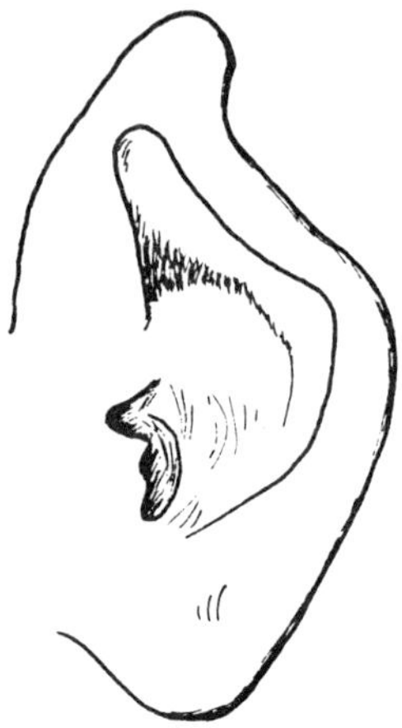

Figure 23-6. Microtic ear in a child that is characterized by a principal deformity in the helix.

with hypertelorism should *not*, however, be arbitrarily catalogued as mentally retarded. The interested reader can find a table of normal bony interorbital distances by age in the research of Currarino and Silverman (1960).

Individuals vary as to size of the bony orbits (Fig. 23-9). This accounts for the observation of deep-set eyes (enophthalmos) or protrusiveness of the eyes (exorbita, proptosis, exophthalmos). Enophthalmos may also be related to orbital tissues that lose moisture or fat causing the eyes to sink into the sockets. Exorbita may be associated with hyperthyroidism or other metabolic factors.

An unusual facies may include the presence of *proptosis* (bulging eyes) and *ptosis* (drooping of the upper eyelid). The resultant appearance most often dominates the facial features. *Ptosis* of the eyelid need not accompany proptosis, however. As noted in Chapter 13, the levator palpebrae superioris muscle is responsible for elevating the superior tarsal plate (upper eyelid). When the eyelid is ptotic, the examiner should suspect damage either to Cranial Nerve III or to the cervical sympathetic nerve supply (Fig. 23-10).

The inability to close the eyelids is seen with weakness or paralysis of Cranial Nerve VII. This can be evaluated by having the patient attempt to close the eyes tightly against efforts of the examiner to open the eyelid by pressing upward under the eyebrow. Evaluating the strength of the sphincteric response of the ocular muscles and the stability of muscular resistance provides a basis for commenting on the neuromotor status of the upper face in all patients. In the normal, considerable ocular muscle resistance should be noted during attempts to force the eyelid open.

In examining the orbit area, the sup-

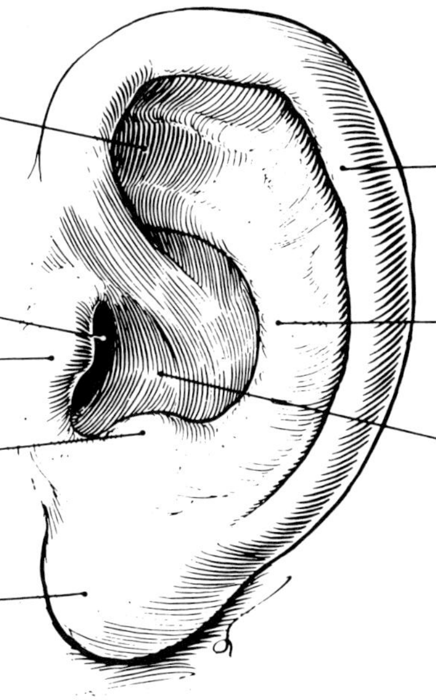

Figure 23-7. A normal external ear. (From *Dynamic Anatomy and Physiology* by L.L. Langley, I.R. Telford, and J.B. Christensen. Copyright© 1969. Used with the permission of McGraw-Hill Book Company.)

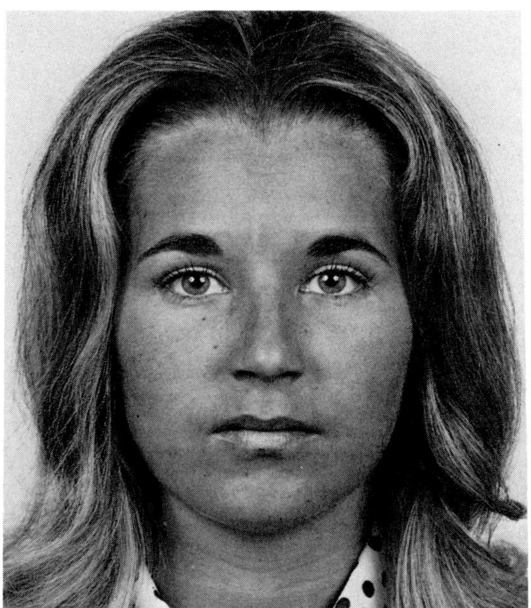

Figure 23-8. Frontal view of the face demonstrating the normal spacing between the eyes.

porting areas below the orbits also contribute much to the appearance of the face. Underdevelopment of bony or soft tissues in the maxillary and zygomatic areas can dramatically modify facial appearance — the face can be made to appear either more horizontal or vertical. When tissues are *hypoplastic*, the eyes may seem more prominent. From a frontal view, hypoplasia can be assessed by palpation of suspicious areas to distinguish whether the reduction is because of deficiency in soft tissue or underlying bony support.

The Lower Midface

The Nose

Examination of the external nose quickly reveals that deviations in the nasal septum are commonplace in the normal population (see Figs. 23-11 and 2-18). The variety of noses seen in clinical work is very impressive! The most important

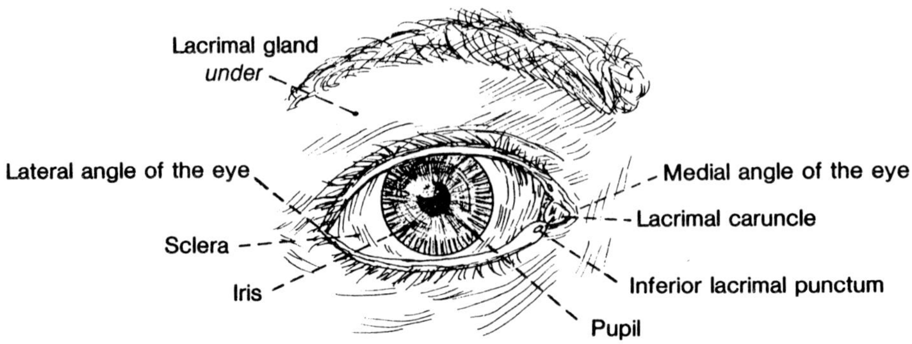

Figure 23-9. Anterior view of a normal eye.

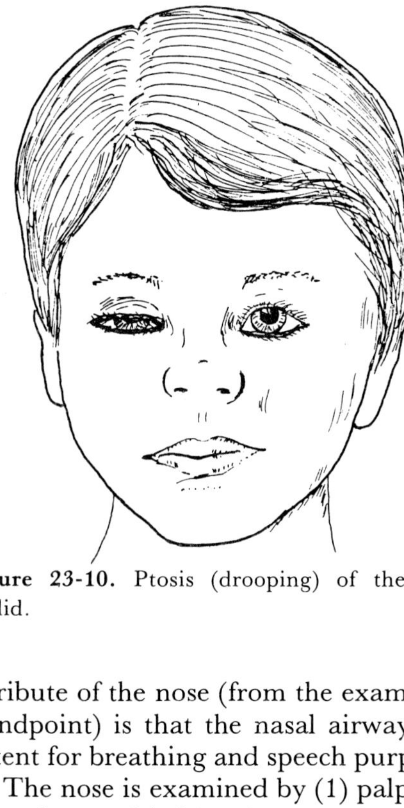

Figure 23-10. Ptosis (drooping) of the right eyelid.

attribute of the nose (from the examiner's standpoint) is that the nasal airways are patent for breathing and speech purposes.

The nose is examined by (1) palpating from the nasal bridge downward to identify any anterior septal deviation, (2) observing the nares, and (3) examining the nasal cavity through the nostrils. The external nares may show variations in symmetry and width of alar cartilages. A useful "rule of thumb" in evaluating alar width is that this dimension should be equivalent with intercanthal width (space between the medial corners of the eyes) for normal lateral balance of the face (Fig. 23-11). In individuals with repaired complete cleft lip deformities, the nasal ala may be flared laterally and inferiorly on the side of the cleft. This is related to the disturbance of the nasal floor, as in Figure 23-14. Additionally, there may be some reduction in the length of the column of tissue — the columella nasi — that separates the nostrils and joins the nasal tip with the base of the nose. Shortening of the columella, and related lateral flaring of the nasal alae, is characteristic of the condition of bilateral complete cleft lip (Fig. 23-12).

In examining the nasal cavity, the nostrils should be spread individually using either a speculum or, perhaps, part of a tongue blade. On the lateral wall of the nasal cavity, the clinician examines for the presence and color of the inferior nasal concha (or turbinate) (Fig. 23-13). The characteristic color of a normal turbinate is deep red. When the turbinate appears pale or gray, this suggests the possibility of allergic rhinitis (as with hay fever). At or near the midline of the nasal cavity is the septum. The examiner looks at the septum in relation to the turbinates. In instances where there is contact of the septum with the inferior turbinate, as shown

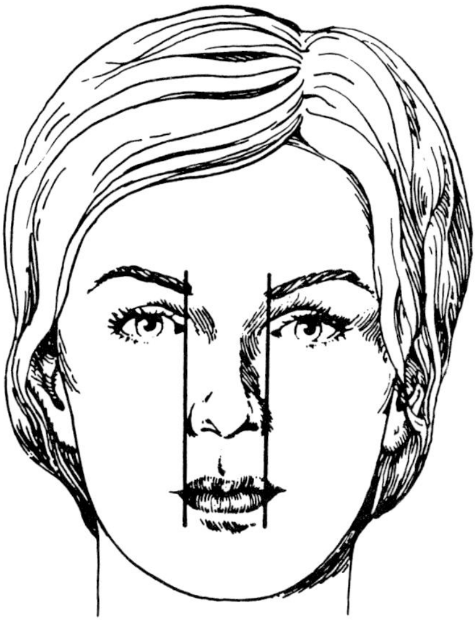

Figure 23-11. The alar base width of the nose coincides normally with the intercanthal width of the eyes.

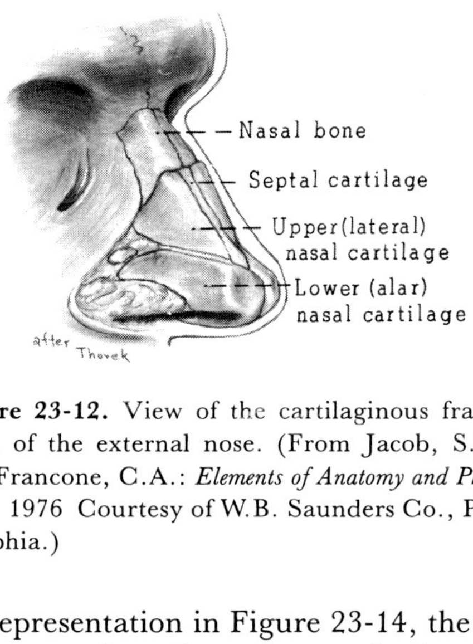

Figure 23-12. View of the cartilaginous framework of the external nose. (From Jacob, S.W., and Francone, C.A.: *Elements of Anatomy and Physiology*, 1976 Courtesy of W.B. Saunders Co., Philadelphia.)

by representation in Figure 23-14, the resulting nasal configuration can lead to hypernasality (anterior blockage on the affected side and an excessively large opening in the opposite nasal chamber). The nasal configuration for the individual in Figure 23-14 is typical of many repaired unilateral cleft lip conditions and may account for residual hypernasality subsequent to pharyngoplasty in some cases. With air being recirculated on the side of the cleft (excessive nasal resonance) and increased nasal turbulence on the noncleft side (excessive nasal emission), the preoperative hypernasal speech complaints are maintained by the unresolved anatomical problems in the nasal cavity. Such patients sometimes require nasal reconstruction to eliminate the speech and breathing problems. It is encouraging that these individuals can be identified initially by orofacial examination, followed by aerodynamic and other studies.

The Lower Face

The Lips

The lower face may show variations in amount of available tissue between the base of the nose and upper lip. This *philtrum* area is highly variable among individuals. This is often related to connective tissue variations in the upper oral vestibule (sulcus) or to the configuration of the dental arch. Absence of a vestibule, as in some repaired cleft lip and palate cases, may require surgical undermining to lengthen the philtrum and free the upper lip for more efficient function in sphincteric activities of the mouth.

In the noncleft individual, the length of the upper and lower lip is of considerable diagnostic importance to the speech clinician. Children can be inappropriately identified as mouthbreathers who show adequate nasal respiration by aerodynamic studies. Such children are usually found to be lip incompetent rather than mouthbreathing, as pointed out in Chapters 19 and 21 (Fig. 23-15).

The lower face and mouth area can be assessed efficiently by having the patient exaggerate productions of /i/ and /u/. These tasks involve symmetrical excur-

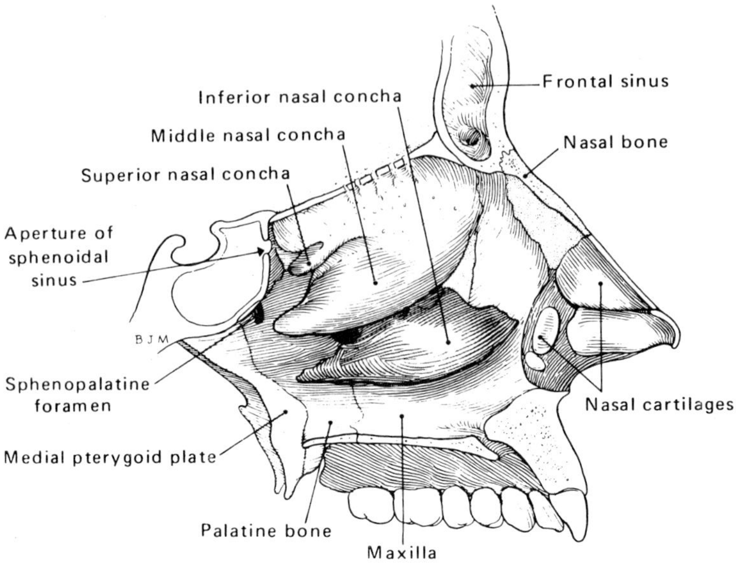

Figure 23-13. The nasal conchae (turbinates). (From *Dynamic Anatomy and Physiology*, 3rd ed., by L.L. Langley, I.R. Telford, and J.B. Christensen. Copyright© 1969. Used with the permission of McGraw-Hill Book Company.)

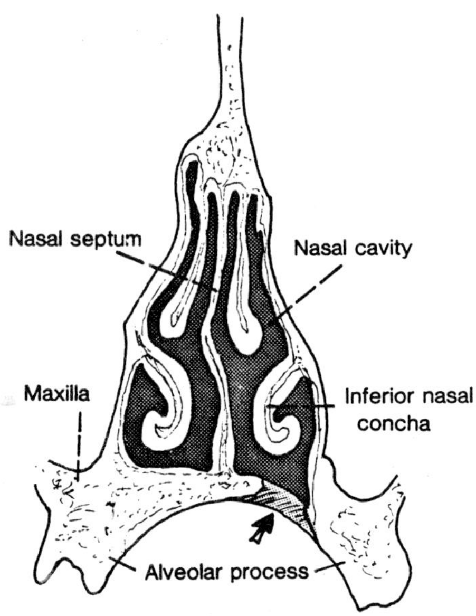

Figure 23-14. Frontal view of the nasal cavity in a patient with a repaired unilateral complete cleft lip and palate.

sions of the facial musculature, permitting evaluation of the relationship of the mimetic muscles (muscles of facial expression) with the orbital ring around the mouth. The /i/ and /u/ tasks also constitute a basic neurological assessment of the integrity of the facial nerve (Cranial Nerve VII). This should be supplemented by diadochokinetic testing of the lips on /p/. The rapid sequential responses of the lips on repeated utterances of /p/ may provide the examiner with information concerning the neuromotor readiness of the lips for speech production purposes.

The upper lip-line should conform to "Cupid's bow" (the anatomical name for this line), the derivative of which is an example of anatomy as a descriptive science. If there is a deficiency in upper lip-line from cleft lip repair, there may be an underlying muscular deficiency. On exaggerated /u/, it would be expected to observe asymmetrical pursing of the oral sphincter in such cases. This observation may be appropriately followed by consultation with a plastic surgeon regarding eventual revision of the upper lip and

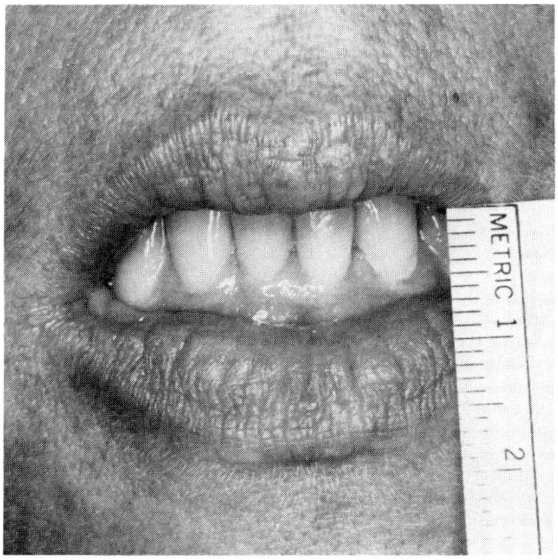

Figure 23-15. Lip incompetence. This patient has a normal pattern of nasal respiration, as demonstrated by clinical and aerodynamic (air flow) studies. This posture does not automatically signal "mouthbreathing."

underlying musculature (Fig. 23-16).

If the upper-lip musculature is not joined properly to form adequate muscular union, a typical related problem is drooping or protrusion of the lower lip. Since the lip area is a sphincter of musculature, when one part of the sphincter is defective, sagging or weakness can occur at the opposite side of the sphincter.

Oral habit behaviors involving the lips can be detected in the form of severe chapping and/or indents in the vermilion border caused by lip sucking and chewing.

Examining Mandibular Movement

From a frontal view of the face, the mandible can be assessed in comparison to overall facial form. The mandibular condyles are isolated by placing the entire surface of two fingers vertically on each side of the head in front of the ear tragus. As the patient's mouth opens upon command, the condyles can be identified as they shift in and out of the mandibular (glenoid) fossa.

In addition to isolating the condyles for purposes of measuring ramus height, the examiner has an opportunity to evaluate the symmetry of jaw opening on the two sides. As the patient opens the jaw, the examiner focuses visually on the central incisors of the mandibular arch to look for deviation of jaw movement off of the midline. If the mandible deviates to one side while the condyles are felt to

Figure 23-16. Unoperated clefts of the lip. The bilateral cleft patient on the right has no muscle fibers in the prolabium tissue below the nose. The orbicularis oris musculature is displaced along the lateral margins of the nose. Achieving good muscle continuity across the upper lip area from surgical closure of a cleft lip is not always an easy or feasible task.

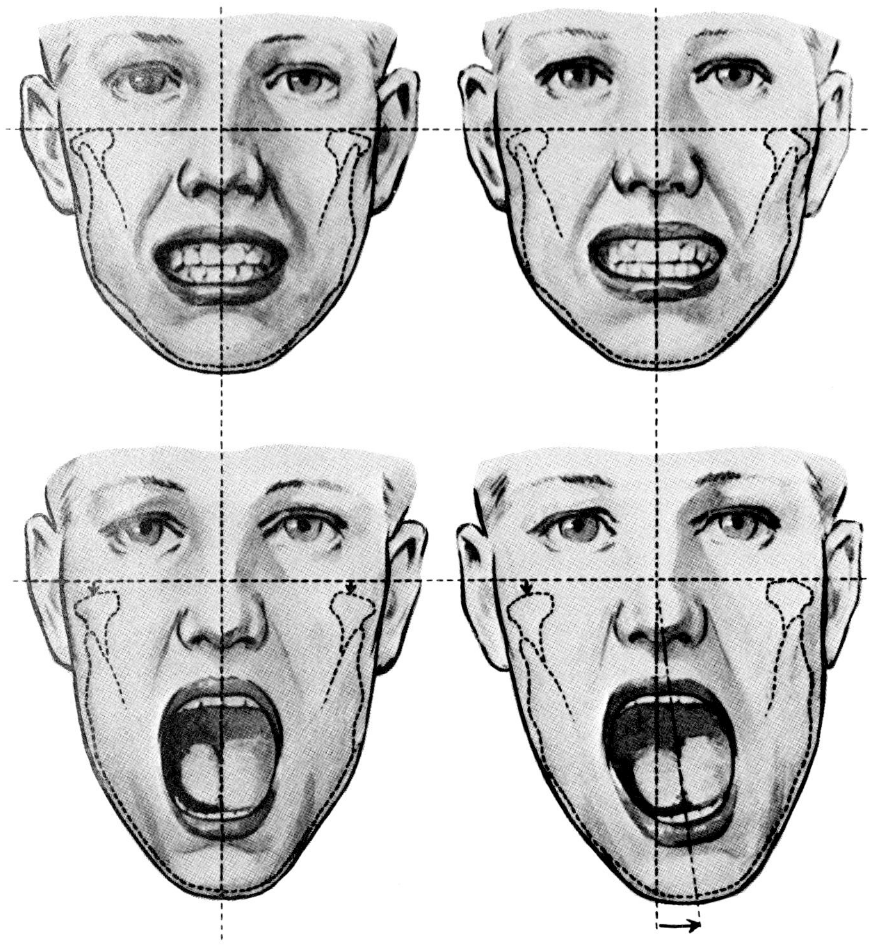

Figure 23-17. Normal (left) and asymmetrical (right) jaw opening. (From Massler, M., and Schour, I.: *Atlas of the Mouth*, 2nd ed., 1958. Copyright by the American Dental Association. Reprinted by permission.)

move equally and symmetrically out of the glenoid fossa, it would appear that the mandible is smaller on the side where the deviation is seen. If, however, mouth opening is vertical (does not deviate laterally) at the front of the face while one condyle moves farther out of the fossa than the other, the side with greatest excursion out of the fossa is revealed as the smaller side of the mandible (Fig. 23-17).

EXAMINATION OF THE INTRAORAL MECHANISMS

General Considerations

The process of intraoral examination involves making observations and decisions about form and function, normal and abnormal, and deficit and capability.

The experience is therefore an exciting and difficult one and is especially challenging for the speech clinician. There is a natural tendency for clinicians to generate cause-effect relationships following superficial evaluation of few variables in the oral cavity. This tendency can be avoided by developing an appreciation for the *range* of variation seen in the normal and by assimilating signs and symptoms related to the abnormal.

Examination of the oral mechanisms proceeds logically if a systematic progression is followed. Observations should start at the front of the mouth and progress posteriorly in the oral cavity. The tongue should be examined last, however. Tongue posture and functions should always be evaluated in light of contributing influences from adjacent anatomy.

The Oral Vestibule

As the upper or lower lip is pulled forward, the examiner can investigate the oral vestibule. The vestibule is lined with a pale mucosa that covers the inner surface of the lips and cheeks and the periosteum of the jaws. Dentists prefer the term *gingiva*, rather than gums, to describe the dental periosteum.

Pyorrhea is a general term used to indicate most forms of inflammation of the gingiva. Where there is resorption of the gingival margins around the roots of teeth, the gingival disease may involve the presence of pus, which can be squeezed out from the gingival margins. It is important to identify instances of gingival (gums) and periodontal (tooth attachment tissues) disease and make appropriate referral to a dentist, since an individual can lose his/her teeth from periodontal disease in the absence of dental caries (cavities).

In addition to examining the oral vestibule to obtain a quick estimate of the general oral health of the individual, the vestibule should be examined for quantity and location of connective tissues. A midline frenum or frenulum is normal, as are small frenulae at the level of the cuspid teeth (canines). Multiple buccal frenulae in the maxillary vestibule are a finding more prevalent among individuals with oropharyngeal malformations. Therefore, identification of extra connective bands in the buccal vestibule should

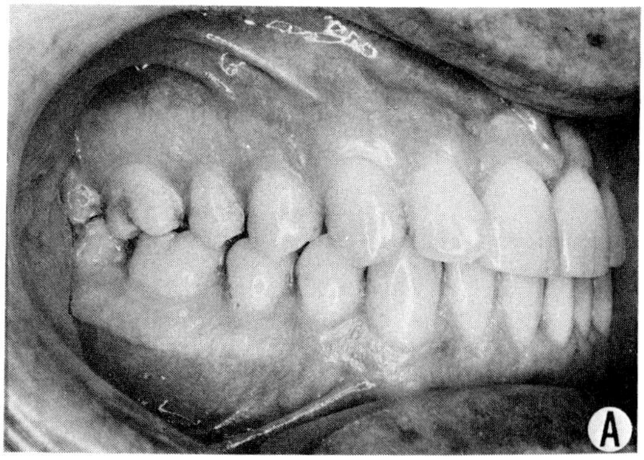

Figure 23-18. A. Appearance of healthy gingiva in an adult.

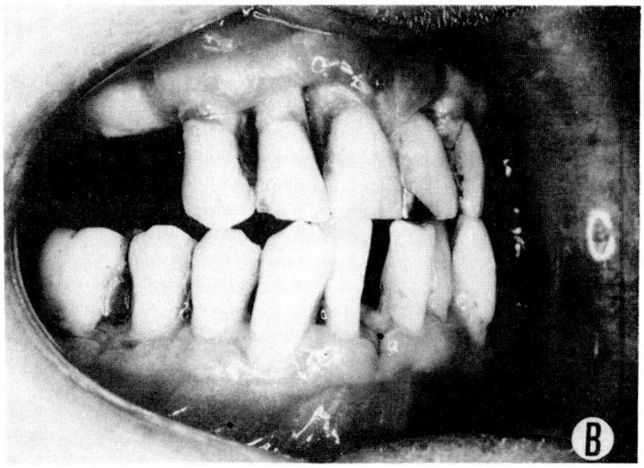

Figure 23-18. B. Severe periodontitis in an adult. The teeth are mobile due to destruction of some of the attachment apparatus (bone loss and ligamentous damage). The tongue is prominent due to the spacing between teeth. The loss of interdental gingiva can serve to create a "cumulative open-bite."

alert the clinician to the possibility of additional variations in the oral cavity and pharynx regions, especially cleft palate and submucous cleft palate.

The Dentition

Background Information

The child presents a unique dentition that is representative of his particular dental age. The dentition is time-linked to biological age and as such is expected to exhibit constant change up through the final eruption of the permanent teeth. Therefore, the nondentist may experience considerable difficulty in determining whether the state of the dentition represents a normal biologic change or whether an anomalous condition has also extended to the teeth and supporting bone, or if there is a combination of effects (Fig. 23-19).

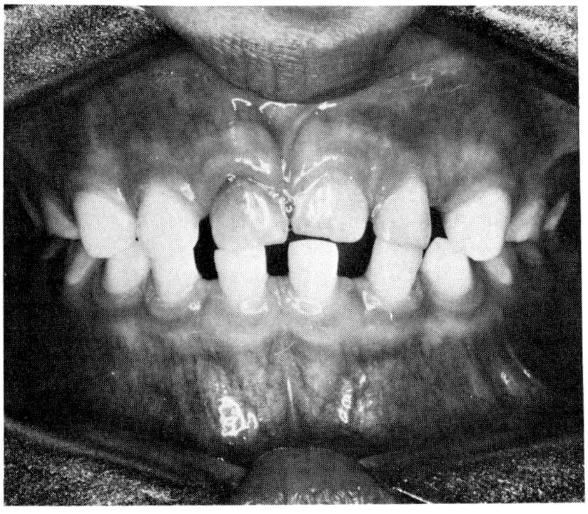

Figure 23-19. The dentition in a child showing typical anterior spacing.

A tooth of primary importance in the dental arch is the cuspid because of its strategic location at the point of greatest curve in the arch. Removal of a single maxillary cuspid in the permanent dentition results almost universally in a midline shift toward the missing spot.

When the two dental arches are considered together in a contactual relationship, the most important teeth in each arch become the molar teeth. Their importance in occlusion is based upon the fact that they have large contacting surfaces for stability. Since the first molars erupt early and tend to be retained in the permanent dentition, occlusal relationships of the upper and lower dental arches are expressed with reference to first molar contacts.

When the teeth are in normal occlusion, the mandibular first molar is positioned one-half tooth ahead of the maxillary first molar (see Fig. 23-2). In describing this situation by intraoral inspection by spreading the cheek laterally to afford inspection of the molars, it is seen that the cusp farthest forward on the outside of the maxillary molar fills in the groove that is in the middle of the first mandibular molar. In more appropriate terminology, the mesial-buccal cusp of the maxillary first molar occludes in the buccal groove of the mandibular first molar.

When occlusion is abnormal, the biting relationships of the jaws are modified. Occlusion and malocclusions are currently described by an old system devised by Edward Angle (1907) and based upon horizontal relationships between the dental arches. This system of classification has some inherent shortcomings, and newer systems of classification have been developed, such as by Ackerman and Proffit (1969) (see also Proffit and Ackerman, 1973). The Angle system prevails at present and is a convenient way of communicating information to individuals in dentistry and speech pathology. It should be noted that dentists prefer to use the designations of occlusal classes used by Angle, rather than the descriptive terms such as *mesi-occlusion* and *distocclusion*.

Specific Dental Conditions

Of the many occlusal conditions seen in children that are within normal limits at various ages (such as end-on-end molar contacts between ages 5 and 8 years), some conditions are noteworthy and merit referral to a dentist. These include: *crossbites* (especially with a functional shift) anterior or posterior openbites, and Class II and III malocclusions that appear to be developing skeletal problems (see Chapter 10 for explanations).

Special Techniques for Dental Examination

To evaluate occlusal relationships, the clinician should instruct the patient to "bite on the back teeth" and separate the lips. This procedure is recommended, since it is common for a patient to protrude the mandible and contact the mandibular and maxillary incisors end-on-end when attention is focused to the front of the mouth. The recommended procedure usually eliminates the tendency to extend the mandible.

With teeth in contact, the anteroposterior, transverse, and vertical relationships of the teeth are noted. The clinician should first note the dental midlines, as compared with the skeletal midlines. Where growth deviations have occurred, asymmetries in the midlines are expected findings.

Next, the clinician should count the number of teeth on each side of the midline in each jaw. Missing or extra teeth are common findings in patients with facial anomalies.

Third, the clinician should evaluate posterior occlusal relationships as per the occlusal classes discussed in Chapter 10.

If the patient demonstrates a Class II Division One malocclusion, examination of the anterior dentition in occlusion should reveal excessive overjet. Such patients may also have problems with anterior fricative productions. These individuals should be instructed to produce /s/ with teeth in occlusion and, again for comparison, with incisors in an end-to-end relationship. This is accomplished by rotating the mandible forward until the incisal edges contact. It is not uncommon for /s/ to be produced correctly in such examination testing when the anterior airway dimension is constricted by approximating the front teeth end to end. In selected instances, the clinician may recommend teaching /s/ to this patient with mandible rotated forward as a means of adapting to the problem created by excessive anterior overjet. If the patient also has an openbite malocclusion anteriorly, this procedure may not be successful without added reduction of airspace by the tongue tip.

Finally, the patient should be instructed to open the mouth as wide as possible while the examiner notes the general state of each tooth. Decayed teeth, evidence of rampant caries, broken cusps on molars, and excessive crowding during eruption stages should be brought to the attention of a dentist.

Although dental problems are often linked with speech problems in individual cases, there is no basis for projecting a one-to-one correlation between the presence of dental abnormality and speech problems. Too many children with missing lateral incisors have normal /s/ productions to discourage the speech clinician from working on /s/ with those frontal or lateral lispers where incisor or canine teeth may be missing. The adaptation of speech function to oral-form variations is a naturally occurring phenomenon that may, however, require some therapeutic encouragement in selected cases.

The Hard Palate

Since the hard palate is a bony platform onto which the bulk of the soft palate finds attachment and support, it is logical to assess the hard palate before the velum. Inspection of the hard palate is most easily accomplished with the patient extending the head backward. This position is recommended because it provides exposure of the entire hard palate. Unlike the soft palate, there are no anatomical characteristics of the hard palate that would interfere with the examination as a function of head posture (Fig. 23-20).

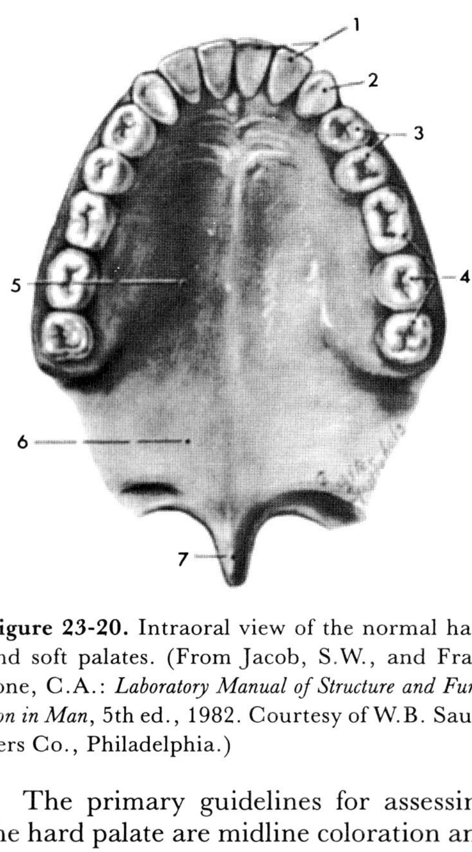

Figure 23-20. Intraoral view of the normal hard and soft palates. (From Jacob, S.W., and Francone, C.A.: *Laboratory Manual of Structure and Function in Man*, 5th ed., 1982. Courtesy of W.B. Saunders Co., Philadelphia.)

The primary guidelines for assessing the hard palate are midline coloration and location of the posterior border. Normal midline colors are pink and white, whereas a blue tint on the midline should generate suspicion about the adequacy of the bony framework of the palate. Individuals with submucous cleft palate deformities often have a bluish tint at the midline of the hard palate and soft palate, denoting that the blood supply is abnormally close to the mucosal surface.

When a blue tint is seen lateral to the midline of the hard palate and taking the configuration of an oval, this is most likely associated with a developing *torus palatinus*. This is a bony variation that develops in about one in five individuals (Schaumann et al., 1970). A torus palatinus is an extra growth of bone (a benign exostosis) that develops over the mid-palate suture line of the hard palate. It begins to express itself in the child at around age six years, sometimes first as an oval coloration, and growing at varying rates until a prominent ridge of bone is clearly seen covering the midline. The blue coloration associated with a torus in some children is thought to be the outline of the displaced blood supply at the lateral edges of the tissue proliferation. At adulthood, the coloration distinction often disappears. A palatal torus only poses a problem in some adults who undergo fitting of a denture in the maxillary arch.

The location of the posterior border of the hard palate can usually be identified by the color differential between hard palate and soft palate. The characteristic scalloping of the posterior border of the hard palate aids in this identification in the normal as seen in Figure 23-21. The clinical guideline for evaluating the length of the hard palate is to connect an imaginary line between the posterior borders of the maxillae (tuberosities of the maxilla). The posterior nasal spine of the hard palate should fall within this reference line. Palpation along the midline for the posterior nasal spine should also be performed when the coloration differences are not clear or in those instances where a submucous cleft of the hard palate is suspected. The intraoral and radiographic signs of the expected consequences of a normal and submucous cleft of the hard palate are shown in Figure 23-21.

The contour or vault of the hard palate has been a subject of much discussion in speech pathology, dentistry, and medicine. The height of the hard palatal vault is largely related to the width of the maxillary arch. Accordingly, those individuals with wide maxillary arches tend to have flat palatal vaults, and vice versa. Although the contour of the hard palate may influence lingual-alveolar contacts during speech, most individuals with unusual (especially, constricted) vaults learn to adapt by moving those tongue placements to a lower level against the gingival margins of the incisors. Individuals who have not adapted and whose problematic speech is related to the anterior alveolus should be taught a different location for tongue contacts in speech therapy.

A determinant of maxillary-arch-form variations has been linked by many to oral habits such as thumb sucking. In most instances, speech behaviors are expected to adapt to form variations of the hard palate in spite of the unusual nature of the maxillary arch in many thumb suckers.

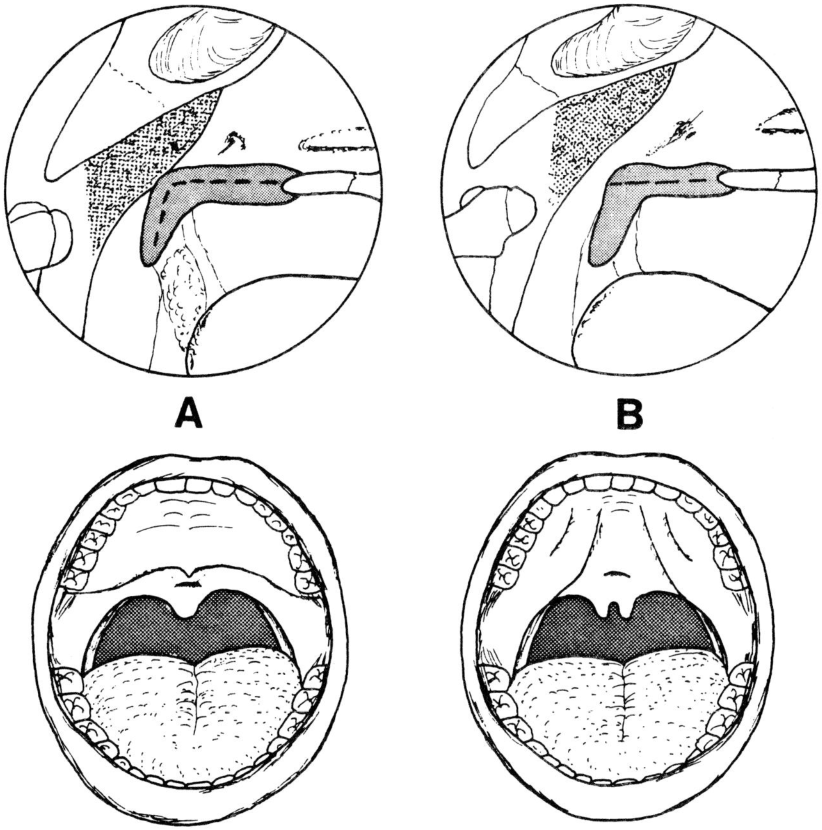

Figure 23-21. Intraoral and radiographic findings in two individuals sustaining a vowel phonation. Normal findings (A) and anomalous findings (B) are portrayed in the hard palate, soft palate, and uvula. Both individuals have a deep pharynx, although the person on the left has velopharyngeal competence with the aid of an adenoid pad.

The Soft Palate

Controlling Examination Artifacts

The ideal position for examining the soft palate is one which reflects the typical intraoral relationships used by the individual in speaking. For the majority of patients, this will entail sitting erect in a chair with eyes focused straight ahead. Accordingly, the examiner should resist the temptation to have the patient extend the head backward to afford easier inspection of the soft palate. Modification of the "natural" head posture can distort muscle relationships in the head and neck. These changes produce physiological differences that do not reflect the normal function of the velum. Hence, intraoral inspection of typical muscular function of the soft palate should be accomplished with head in a natural, upright position. The authors recommend seating patient and examiner in a relationship so that the patient's mouth, with head erect, can be aligned in a *horizontal plane* with the eye level of the examiner, as shown in Figure 23-22.

Assessment of velar elevation should not be attempted while the patient protrudes the tongue. The anterior faucial pillars attach inferiorly into the lateral borders of the tongue and are therefore pulled forward with the tongue during lingual protrusion. If the palatoglossus muscles are contracted excessively, the range of velar elevation can be restricted.

As the patient depresses the mandible to permit examination of the oral cavity, notation should be made of the position of the mandible for maximum mouth opening. This position is unfortunately selected by some individuals in response to instructions to open the mouth widely. It is emphasized that this is *not* a recommended patient posture for examination. With mouth locked open, the musculature of the pharynx is also immobilized. This is due to the functional relationship between the oral and pharyngeal sphincters, as determined by anatomical communications between these areas. The oral ring of musculature connects with the pharyngeal ring at the pterygomandibular raphe, where fibers of the buccinator join with those of the superior constrictor. Upon dissection of the muscle bundles in this area, one is impressed with the continuity of many muscle bundles joining the buccinator and superior constrictor. As a consequence of the reciprocity between these oral and pharyngeal valves, when the mouth is open *maximally*, velar activity is *minimal*. Conversely, the individual who exhibits passive anterior oral activity during speaking also tends to possess hypernasal voice characteristics.

Examination of the function of the intraoral structures is recommended with the mouth open three-fourths of maximum distance. The comfort of the patient in maintaining this mouth opening can be checked by moving the mandible laterally to insure that facial muscles are not rigid.

CONSIDERATIONS AND GUIDELINES IN VELAR EXAMINATION

Evaluation of the soft palate requires integration of a number of anatomical and kinesiological factors. A primary factor relating to the velum is adequacy of muscle union at the midline. The velum is, therefore, best examined in phonation when contraction of velar muscles is outlined under the mucosal surface. The examiner will be especially interested in the midline color of the velum, which should normally be pink and white, similar to that found in the hard palate. When the

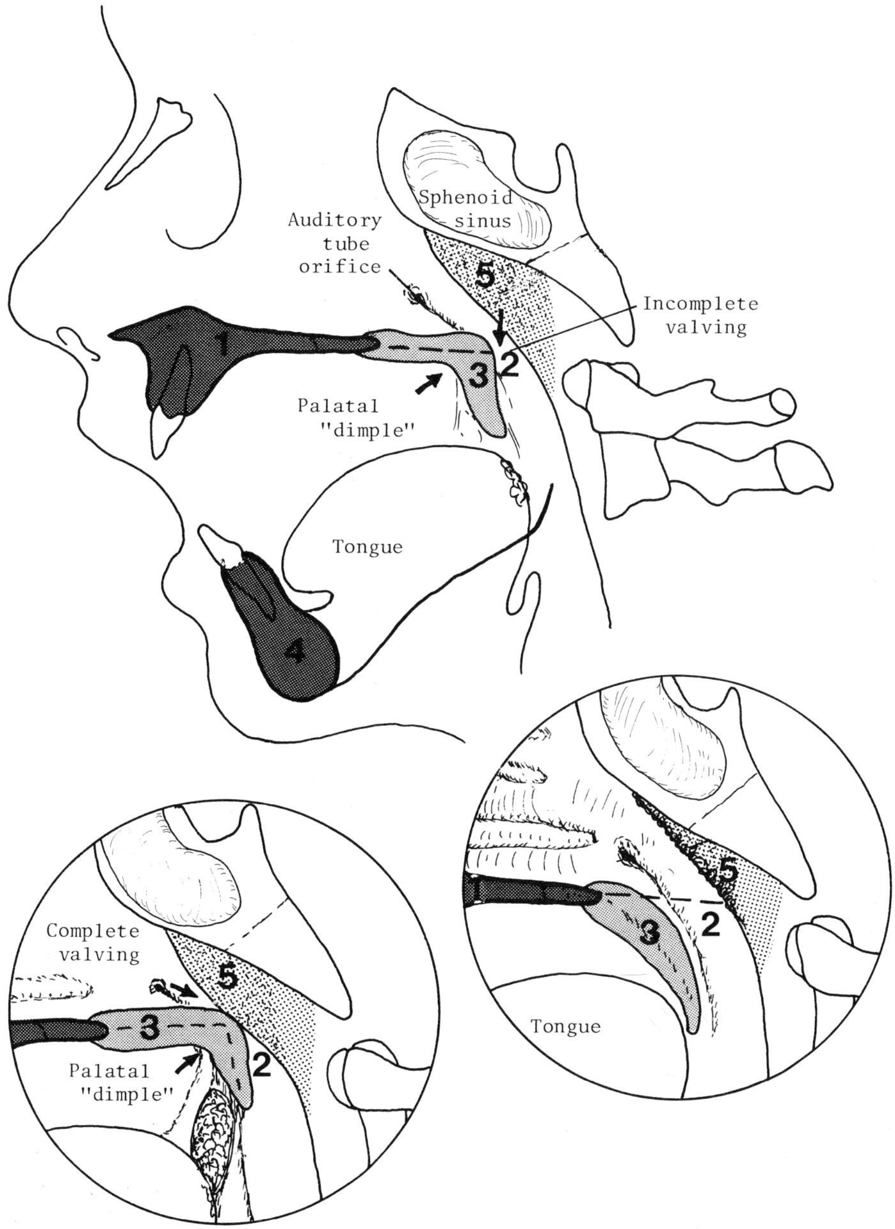

Figure 23-22. (Top) Recommended patient head posture for examination of the soft palate. The velar dimple location can be compared to the depth of the pharynx. This patient is hypernasal. (Lower left) Patient sustaining "ah" who achieves velopharyngeal closure but who is observed intraorally to have a deep pharynx. Adenoid removal would present a hazard to the maintenance of normal nasal resonance balance. (Lower right) The soft palate at rest. Overall velar length at rest provides insufficient information about the potential for velopharyngeal closure.

midline appears blue or purple, the presence of a submucous cleft of the velum is signalled. This indicates muscle deficiency at and across the midline of the velum and is usually accompanied by an A-shaped, abnormal, anterior coursing of velar musculature into the area of the hard palate. The normal communications of soft palate and hard palate are by attachments of velar musculature onto the aponeurosis at the posterior edges of the hard palate. In a submucous cleft of the hard palate, the aponeurosis forms along the margins of the bony cleft underlying the mucosa. A submucous cleft of the hard palate influences the soft palate by shortening the functional and overall lengths of the velum, as shown in Figures 23-21 & 22.

Some individuals with submucous clefts have normal nasal resonance balance. This is often due to the presence of adenoidal tissue, which aids in obturating the excessive nasopharyngeal space created from anterior velar displacement. It is critical that the adenoid is maintained in these patients to insure adequate velopharyngeal function for speech. In such cases, the normal involution process of the adenoids does not usually render the individual hypernasal because the soft palate accommodates to the increased pharyngeal depth by stretching in phonation (Pruzansky and Mason, 1969; Mason and Warren, 1980).

For diagnostic purposes, the *effective length* of the velum is a more critical designation than overall velar length. The effective length of the velum represents that portion of the elevated soft palate that obturates the nasopharyngeal port. The effective velar length is defined as the amount of tissue filling the space between the posterior border of the hard palate and the posterior wall of the pharynx, measured in a line with the plane of the hard palate, as shown in Figures 23-21 and 23-22 (Calnan, 1961; Mason and Grandstaff, 1971).

The effective length of the velum can be extrapolated from the intraoral identification of the point where the soft palate *buckles* or *dimples* during phonation, in relation to the anteroposterior diameter of the pharynx. The velar dimple on the oral surface of the soft palate is an indication of the point of maximum velar elevation on the nasal surface of the velum. The speech clinician can estimate effective velar length by intraoral identification of the velar dimple, as depicted by the x-ray tracings in Figure 23-22.

The dimple of a normal elevated soft palate is seen to occur about 3 to 5 mm in front of the posterior border of the soft palate. This represents a dimpling point at approximately 80 percent of the total length of the velum, including the uvula. In Figure 23-21, it is demonstrated that although the overall resting length of the velum is similar in these cases, the soft palate on the right buckles at approximately 50 percent of its overall length during phonation. This indicates a short effective length of the velum. In this situation, the abnormal muscle architecture and attachment cannot be remedied by speech therapy exercises.

In addition to identifying the location of the velar dimple, it is also important to assess the amount of elevation of the soft palate. A normal velum elevates to the plane of the hard palate and can also increase in overall length by stretching in phonation. The judgment of whether the velum moves superoposteriorly with a "minimum," "moderate," or "excellent" range of excursion is valuable for comparing the function of the soft palate with activity in other parts of the velopharyngeal sphincter.

The Uvula

The uvula is often inappropriately regarded as a benign and innocuous extension of the soft palate. In spite of the fact that the uvula plays no important role in speech, it is treated as a separate category from the velum in this section because of the diagnostic importance of uvular variations. It is known that about one in seventy-five individuals has a bifid uvula (Meskin et al., 1964). Bifidity of the uvula can occur as an unrelated and isolated finding. However, the coexistence of bifidity of the uvula with submucous clefts of the hard and soft palates and excessive nasopharyngeal dimensions should be recognized (see Fig. 23-21). A bifid uvula often goes unnoticed until the cleft uvular tags are separated by a conscientious examiner.

A bifid uvula that is held together by mucous can be identified initially by observing the heart-shaped scallops at the base of a bifid uvula. Any suspicion about the integrity of the uvula should be followed by attempts to separate it at the midline by gently stimulating the uvula with a tongue blade.

When a bifid uvula is confirmed, the examiner should check carefully in the areas of the hard palate, soft palate, and pharynx for other variations. Those individuals with bifid uvula who are candidates for adenoidectomy should be reviewed by lateral x-ray examination if the pharyngeal depth appears to be questionable by intraoral examination. An x-ray evaluation technique for such patients has been proposed by Mason (1973*a*).

Although bifidity of the uvula may not be a remarkable finding in a selected case, when bifid uvula is present, the clinician should be suspicious about the anatomy of the oral cavity and pharynx until these areas are closely inspected.

Other variations in the uvula suggesting the same concern about adjacent areas as bifid uvula include a short uvula, thin uvula, and missing uvula. These variations have been found in accompaniment with submucous clefts and other abnormal findings.

The Fauces

The speech clinician should evaluate the area of the faucial isthmus as a part of the intraoral examination. Of special interest is the determination of the degree to which the faucial (or palatine) tonsils, if present, reduce the diameter of the faucial isthmus. Reduction in the faucial airspace can impede the normal flow of air into the oral cavity.

It is usually difficult to observe the bulk of the tonsillar masses in children due to the high position of the tongue in the oral cavity. In such situations, diagnostic information about the presence and size of the faucial tonsils can be obtained by observing the uvula. If the uvula appears to be positioned in front of the posterior faucial pillar, as shown in Figure 23-23, this signals a history of large tonsils. Enlarged tonsillar masses displace the faucial pillars one from the other. The posterior faucial pillar is pushed away from the tonsils, and the uvula often appears to have moved to an anterior position (Fig. 23-23).

While much has been speculated about the role of the faucial tonsils in speech production, in most instances they are not a factor. The tongue usually adapts to the presence of enlarged tonsils by moving to a forward or lower position in the oral cavity. However, some tongue thrusting in children has been linked to reduced faucial airway size from enlarged tonsils (Ricketts, 1968).

The Pharynx

The clinician is afforded a view of the oropharynx during intraoral examination, while velopharyngeal closure efforts

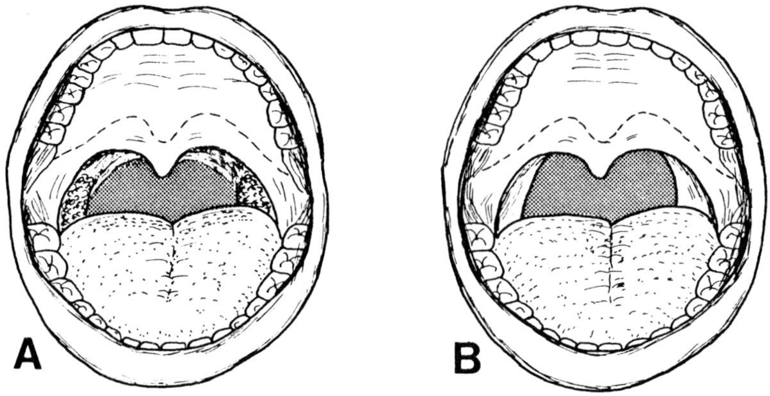

Figure 23-23. Intraoral views comparing the configuration of the fauces with tonsils (left) and without tonsils (right). Notice the anterior appearance of the position of the uvula, often seen when large tonsillar masses are present, in the view on the left.

take place in the nasopharynx above the level observed perorally. Nonetheless, valuable information about velopharyngeal activity can be obtained by viewing the oropharyngeal area. A specific technique for evaluating velopharyngeal closure has been reported by Mason and Grandstaff (1971).

The anteroposterior depth of the nasopharynx is difficult to estimate by intraoral inspection and becomes a highly subjective judgment. Nonetheless, the clinician should evaluate this dimension in relation to the effective length of the velum during sustained phonation of "ah." Notation should be made of any movement of the posterior wall of the pharynx. Movements may be seen to occur in almost any direction (including posteriorward), and it is also common for no pharyngeal wall movement to be observed during phonation. The maximum amount that the posterior wall is expected to move anteriorly is 3 mm (Hagerty et al., 1958).

In some individuals, a Passavant's cushion is observed on the posterior wall of the pharynx. This is a bunching up of the superior constrictor musculature as a horizontally directed tissue prominence seen at the approximate level of the hard palatal plane. It should be noted that, when present, the appearance of Passavant's bar is related *only* to functional activities of the velum and is not seen at rest. It appears in approximately one-third of the cleft palate cases and occasionally in normal speakers. The presence of a Passavant's pad should lead to a suspicion about the integrity of the velopharyngeal mechanism. Contrary to popular belief, Passavant's cushion is *not* regarded as a typical clinical finding.

The adenoid mass, when present, may be regarded as the most anterior projection of the posterior pharyngeal wall. Adenoidal tissue cannot be observed in its entirety by unaided intraoral inspection, in that the bulk of the adenoid is located in the nasopharynx. Ideally, the clinician should examine the velopharyngeal port and adenoid mass using a dental or gutteral mirror, but this procedure is not feasible in most clinical settings (Fig. 23-24).

One of the most obvious ways of determining whether adenoids are intact is to ask the patient. In addition, the clinician can equate the subjective appearance of the nasopharyngeal airspace with what is heard as the patient phonates. An individ-

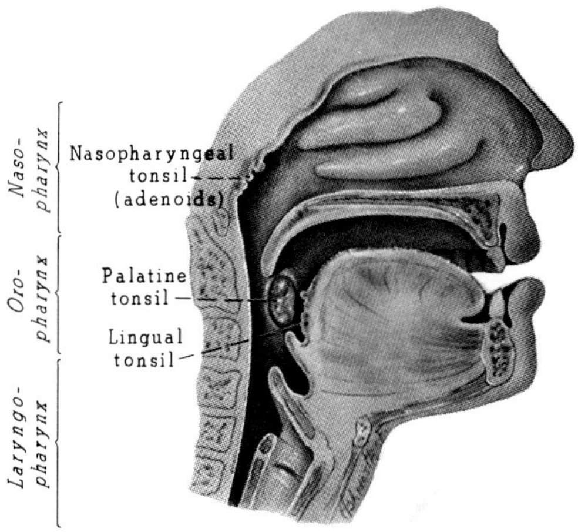

Figure 23-24. Lateral view of tonsillar tissue showing the convoluted surface of the adenoid. (From Jacob, S.W., and Francone, C.A.: *Structure and Function in Man*, 4th ed., 1978. Courtesy of W.B. Saunders Co., Philadelphia.)

ual judged to have a deep pharynx, but with normal nasal resonance balance, is probably contacting the velum against the adenoid pad during phonation to achieve a velopharyngeal seal.

Considerable diagnostic attention has been directed to the medial movement patterns of the lateral pharyngeal musculature (Podvinec, 1952; Chase, 1960; Kelsey et al., 1972). Where velar elevation is minimal or deficient, patients are often recommended for speech therapy because of compensatory medial movement of lateral pharyngeal musculature. Frontal cinefluoroscopy reveals that medial activity of the lateral pharyngeal walls, seen at the level of intraoral inspection, is not highly predictive of pharyngeal wall movements at the nasopharyngeal sphincter (Young, 1970; Mason, Young and Stallworth, 1973). Accordingly, the clinician should avoid being too encouraged about a hypernasal individual's potential to respond to speech therapy *solely* on the basis of observed medial activity of the lateral pharyngeal walls. However, active medial movements of lateral pharyngeal muscles during phonation substantiate that the nervous supply to the pharynx is intact.

The pharyngeal gag response, which is not a true reflex, can occasionally be a useful diagnostic sign. This activity may take a variety of forms. In the neurologically impaired individual, it can be either heightened or absent depending on whether damage is in the central or peripheral nervous system. Some individuals with intact nervous systems, however, will not respond to gagging stimuli.

Gagging is accomplished by either pressing firmly on the base of the tongue with a tongue blade or by stimulating the velum with the blade. Individuals who fail to demonstrate a gag with these procedures are often successfully gagged by stimulating the posterior pharyngeal wall with a tongue depressor.

Gagging the patient will result in maximum excursion of the velum in an upward and posterior direction and constriction of the faucial pillars and pharynx. This activity represents the potential of the velum to function in a nonspeech context. Also, marked medial displacement of the lateral pharyngeal muscles is expected.

The contraction of pharyngeal muscles produced by gagging provides a comparison with the pharyngeal muscles' responses seen during phonation. While such differences may be of general curiosity, they probably offer little of value in treatment planning considerations. Altogether, gagging an individual is most likely an interesting inclusion into the oropharyngeal examination sequence, but it is of guarded value in treatment planning.

The Tongue

The tongue is the most adaptable

organ of articulation. It has the capability to adjust to many of the structural variations found in the oral cavity. The postural and movement adaptations of the tongue are apparently made in response to structural variations and spaces available in the oral cavity. Due to its adaptive nature, it is appropriate to evaluate the tongue last in the sequence of orofacial examination, after variations in the structures and spaces contiguous to the tongue are identified.

Size of Tongue

The size of the tongue is impossible to determine accurately in living specimens due to its adaptive nature. Nonetheless, some notion as to the relative size of the tongue should be estimated in selected patients. Deviation in the overall size of the tongue is not a common finding in the case load of any clinician (Tulley, 1969). Due to the growth discrepancy between tongue and jaws, especially in the elementary school years, a child may appear to have a tongue too large for the oral cavity. The situation is usually not one of *macroglossia* but of disproportionate tongue size in relation to jaw and oral cavity size (Subtelny, 1970) (Fig. 23-25). (Macroglossia can be seen in patients with hemangioma, hematoma, or neurofibroma of the tongue.) The clinical test for macroglossia involves having the patient bite down. If the teeth can approximate without biting the tongue, a case cannot be made for a diagnosis of macroglossia, in the view of many physicians.

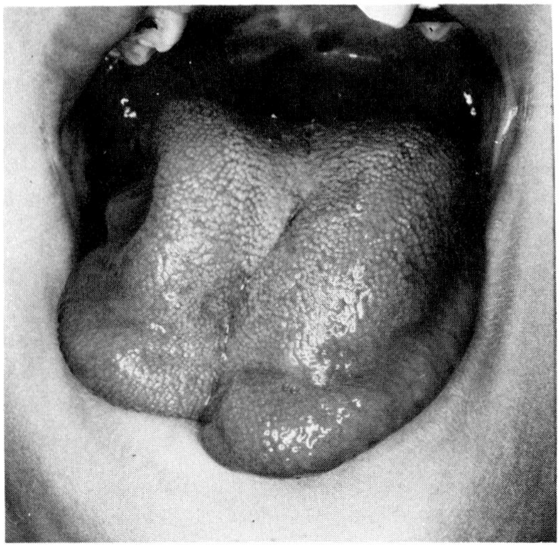

Figure 23-25. Patient protruding tongue. Notice the redundant tissue on the lateral borders of the tongue. Even so, this is not considered to be a macroglossia.

Tongue Adaptation to the Oral Cavity

There are many instances in children where the oral cavity is restricted in size and the posterior oral airspace dimension is reduced. This situation requires adaptive positioning and movements of the tongue. Such children have sometimes been subjected arbitrarily to tongue-retraining exercises without appropriate regard for the presenting morphology. Subtelny (1970) points out that tongue adaptation is not always feasible (either naturally or with training) when morphological limitations are present in the oral cavity and pharynx. In spite of this, for some patients myofunctional procedures have demonstrated improvement in tongue posture and functions (Barrett and Hanson, 1978).

Tongue Movements

There is evidence to indicate that tongue movements for speech and nonspeech activities show little learning reciprocity, as these movements are mediated by different control mechanisms in the central nervous system (Hixon and Hardy, 1964; Shelton et al., 1966). Consequently, there is no logical rationale for testing nonspeech tongue movements (such as swallowing) to assess the potential for tongue function in speech activities.

Protrusion and lateralization activities of the tongue provide basic information about the integrity of the hypoglossal nerve and are suggested for use in screening cranial nerve XII function (see Chap. 26). Deviations of the tongue at rest and upon protrusion and lateralization should be correlated with information obtained from testing the pharyngeal gag response and from oral diadochokinetic testing. Suspicions about central or peripheral nerve damage should be based on several observations.

The degree of neuromotor maturation of the tongue for speech can be estimated in children from lingual diadochokinetic testing. In testing lingual function of rapid repetitions of *tuh* and *luh*, a child under five years of age may show an anteriorly directed tongue-tip pattern in place of lingual elevation. This is a normal, expected pattern for a child of this age.

An alternate pattern seen in children below age seven-and-one-half years on rapid *tuh* and *luh* repetitions is positioning the tongue tip interdentally and producing the sequence with vertical movements of the mandible as the primary gesture. By age seven or seven-and-one-half years, a child should have experienced sufficient neuromotor maturation to produce rapid movements of the tongue independent of *mandibular assist*. Consequently, the child who retains a mandibular assist on diadochokinetic tasks at age seven-and-one-half or older is suspected of having a neuromotor delay for speech, especially if exhibiting difficulty in elevating the tongue tip in speech activities.

In diadochokinetic testing with *tuh* and *luh*, it is recommended that the examiner focus on the *pattern* of tongue movement and the *consistency* of contacts made rather than counting repetitions per second. On testing with *kuh* (which is probably of limited value for assessing the posterior tongue), the gestures of the tongue are expected to occur more slowly. This relates to the added tongue mass posteriorly and the sparse motor and sensory innervation, in comparison with the front of the mouth.

Evaluating Tongue Function

In evaluating tongue function by intraoral inspection in tongue thrusters, it is suggested that diagnostic emphasis be placed on resting tongue posture, airway size, and skeletal growth variations. Observed variations may be especially related to growth discrepancies involving the jaws or disproportionate size between tongue and jaws that results in oral-form changes and an anterior resting-tongue posture. This diagnostic emphasis on anatomy contiguous to the tongue appears to be a logical and conservative approach for relating lingual function variations to the orofacial milieu.

The clinician will also be interested in the relationship of the lingual frenum to speech production. Many individuals are able to speak normally in spite of restricted anterior tongue movements from a short lingual frenum. Too often, a short frenum is inappropriately isolated as the factor that accounts for delayed speech development.

The frenum can be evaluated by having the patient attempt to contact the tongue tip against the maxillary alveolus while biting on a tongue blade placed between the molar teeth. Many children who cannot elevate the tongue tip this high, however, demonstrate normal speech. If the child can contact the maxillary alveolus with the tongue tip, the length of the frenum is probably adequate for speech purposes. Such individuals may produce linguo-alveolar sounds by protruding the tongue tip in a compensatory movement against or between the teeth. This is a natural adaptation for a frenum which restricts the full range of anterior tongue movements (Fig. 23-26).

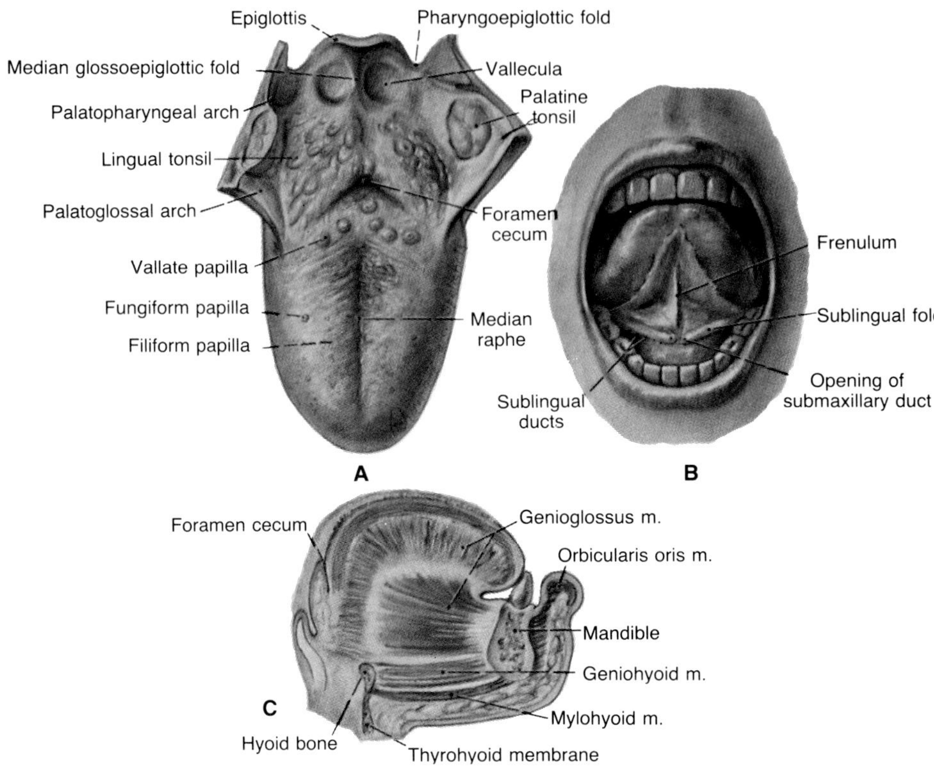

Figure 23-26. Intraoral view of the lingual frenum area. (From Jacob, S.W., and Francone, C.A.: *Structure and Function in Man*, 4th ed., 1978. Courtesy of W.B. Saunders Co., Philadelphia.)

SUMMARY

The principles and guidelines presented above represent some important factors of orofacial examination. The procedures of examination follow a logical progression of observations. Delineations should be made between form and function and their interactions in the facial area, oral cavity, and pharynx.

Orofacial Examination Checklist — Short Form

I. Facial characteristics
   A. Frontal view — eye spacing: zygomatic bones, nasal area, vertical dimensions, lips
   B. Profile — straight/convex, excessively convex, concave
II. Intraoral Characteristics
   A. Dentition — general
      1. occlusal relationships — anterior and posterior
      2. bite-normal: excessive overbite or overjet, openbite, crossbite
      3. sibilant production/teeth relationships: /s/, /z/, /f/, /v/
   B. Hard palate — bony shelf for-

mation
1. coloration – midline, lateral
2. bony midline (palpate): normal, submucous cleft (short)
3. contour – vault/maxillary arch relationship

C. Soft palate or velum
1. midline muscle union – complete, submucous cleft
2. length at rest and effective length during phonation
3. location of dimpling on velum from nasal spine to uvula tip
4. velar elevation and range of velar excursion
5. acoustic correlates: amount of nasality during counting in 60s, 70s, 80s, and 90s

D. Uvula – shape and position
E. Fauces – isthmus opening and coloration, tonsils intact?
F. Pharynx
1. depth between velar dimple and pharyngeal wall on "ah"
2. observation of Passavant's pad during function
3. ask patient if adenoids removed and subsequent speech status
4. gag response

G. Tongue
1. size related to oral environment
2. diadochokinetic rate on: tuh, luh, kuh, puh-tuh-kuh
3. lingual frenum

III. General Observations and Other Findings

BIBLIOGRAPHY

Ackerman, J.L., and Proffit, W.R.: The characteristics of malocclusion: a modern approach to classification and diagnosis. *Am J Orthod, 56*:443-454,1969.

Angle, E.H.: Malocclusion of the Teeth, 7th ed. Philadelphia, S.S. White Dental Mfg. Co., 1907.

Barrett, R.H., and Hanson, M.L.: *Oral Myofunctional Disorders*, 2nd ed. St. Louis, C.V. Mosby, 1978.

Bjork, A.: *The Face in Profile*. Lund, Sweden: Berlingska Boktryckreriet, 1947.

Bjork, A.: Variations in the growth pattern of the human mandible: longitudinal radiographic study by the implant method. *J Dent Res, 42*:400-411, 1963.

Bloomer, H.H.: Speech defects associated with dental malocclusions and related abnormalities. In Travis, L. (Ed.): *Handbook of Speech Pathology and Audiology*. New York, Appleton, 1971.

Brodie, A.J.: Comparative anatomy of the pharynx. In Pruzansky, S. (Ed.): *Congenital Anomalies of the Face and Associated Structures*. Springfield, IL: Thomas, 1961.

Brodie, A.J.: Consideration of musculature in diagnosis, treatment, and retention. *Am J Orthod, 38*:823-835, 1952.

Calnan, J.: Modern views on Passavant's ridge. *Br J Plast Surg, 10*:89-113, 1957.

Calnan, J.: Palatopharyngeal incompetence in speech. In Pruzansky, S. (Ed.): *Congenital Anomalies of the Face and Associated Structures*. Springfield, IL: Thomas, 1961.

Chase, R.: An objective evaluation of palatopharyngeal competence. *Plast Reconstr Surg, 26*:23-29, 1960.

Currarino, G., and Silverman, F.N.: Orbital hypertelorism, arhinencephaly, and trigonocephaly. *Radiology, 74*:206-217, 1960.

Gorlin, R.J., and Pindborg, J.J.: *Syndromes of the Head and Neck*. New York, McGraw, 1964.

Graber, T.M.: *Orthodontics: Principles and Practice*. Philadelphia, Saunders, 1966.

Hagerty, R., Hill, M., Pettit, H., and Kane, J.: Posterior pharyngeal movements in normals. *J Speech Hear Res, 1*:203-210, 1958.

Hixon, T.J., and Hardy, J.S.: Restricted motility of the speech articulators in cerebral palsy. *J Speech Hear Disord, 29*:293-306, 1964.

Kelsey, C.A., Ewanoski, S.J., Crummy, A.B., and Bless, D.M.: Lateral pharyngeal-wall motion as a predictor of surgical success in velo-pharyngeal insufficiency. *N England J Med, 287*:64-68, 1972.

Khouw, F.E., White, R.P., and Proffit, W.R.: Cephalometric evaluation of patients with dentofacial disharmonies requiring surgical correction. *J Oral Surg, 29*:789-798, 1970.

Mason, R.M.: Orofacial anomalies and speech. In Wingo, J., and Holloway, G.: *An Appraisal of Speech Pathology and Audiology: A symposium.* Springfield, IL: Thomas, 1973.

Mason, R.M.: Preventing speech disorders following adenoidectomy by preoperative evaluation. *Clin Pediatr, 12*:405-414, 1973.

Mason, R., and Grandstaff, H.L.: Evaluating the velopharyngeal mechanism in hypernasal speakers. *Ohio J Speech Hear, 4*:23-28, 1970, and reprinted in *Lang Speech Hear Serv in Schools, 4*:53-61, 1971.

Mason, R.M., and Proffit, W.R.: The tongue-thrust controversy: background and recommendations. *J Speech Hear Disord, 39*:115-132, 1974.

Mason, R.M., and Pruzansky, S.: Clinical, Radiographic, and Genetic Study of 110 Cases of Congenital Palatopharyngeal Incompetence. Paper presented at meeting of the American Cleft Palate Association, New York City, May, 1965.

Mason, R.M., and Simon, C.: An orofacial examination checklist. *Lang Speech Hear Serv in Schools, 8*:155-163, 1977.

Mason, R.M., and Warren, D.W.: Adenoid involution and developing hypernasality in cleft palate. *J Speech Hear Disord, 45*:469-480, 1980.

Mason, R.M., Young, E.C., and Stallworth, W.: The Predictive Value of Lateral Pharyngeal Wall Movement in Speech Diagnostics. Paper presented at American Speech Hearing Association Convention, Detroit, 1973.

McWilliams, B.J., Musgrave, R., and Crozier, P.: The influence of head position upon velopharyngeal closure. *Cleft Palate J, 5*:117-124, 1968.

Meskin, L.H., Gorlin, R.J., and Isaacson, R.J.: Abnormal morphology of the soft palate: I. The prevalence of cleft uvula. *Cleft Palate J, 3*:342, 346, 1964.

Podvinec, S.: The physiology and pathology of the soft palate. *J Laryng Otology, 66*:452-461, 1952.

Proffit, W.R.: Muscle pressure and tooth position: a review of current research. *Australian Orthodont, 3*:104-108, 1973.

Proffit, W.R., and Ackerman, J.L.: Rating the characteristics of malocclusion: a systematic approach for planning treatment. *Am J Orthod, 64*:258-269, 1973.

Pruzansky, S.: The contribution of roentgenographic cephalometry to the study of congenital and acquired malformations of the face. *Dent Clin North Am, 7*:395-417, 1960.

Pruzansky, S., and Mason, R.M.: The "stretch factor" in soft palate function. *J Dent Res, 48*:972, 1969.

Ricketts, R.M.: Respiratory obstruction syndrome. *Am J Orthod, 54*:495-507, 1968.

Schaumann, B.F., Peagler, F.D., and Gorlin, R.J.: Minor craniofacial anomalies among a Negro population. *J Oral Surg, 29*:566-575, 1970.

Selder, A.: Hemifacial Microsomia: Audiometric Correlations. Paper presented at the First International Conference on Craniofacial Malformations, New York, 1971.

Shelton, R.L., Arndt, W.B., Krueger, A.L., and Huffman, E.: Identification of persons with articulation errors from observation of non-speech movements. *Am J Phys Med, 45*:143-150, 1966.

Smith, S.: 100 years since Passavant. *Folia Phoniatr, 16*:139-154, 1964.

Speidel, T.M., Isaacson, R.J., and Worms, F.W.: Tongue-thrust therapy and anterior dental openbite. *Am J Orthod, 62*:287-295, 1972.

Subtelny, J.D.: Malocclusions, orthodontic corrections, and orofacial muscle adaptation. *Angle Orthod, 40*:170-201, 1970.

Tulley, W.J.: A critical appraisal of tongue-thrusting. *Am J Orthod, 55*:640-650, 1969.

Vig, P.S., and Cohen, A.M.: Vertical growth of the lips: a serial cephalometric study. *Am J Orthod, 75*:405-415, 1979.

Young, E.C.: A Cinefluorographic Study of Movement Patterns of Lateral Pharyngeal Musculature. Unpublished Master's Thesis, University of Tennessee, 1970.

Chapter 24

CEPHALOMETRIC X-RAY FILM ANALYSIS

OVERVIEW

X-RAY films of the head and neck (roentgencephalograms) have been used for over fifty years in clinical practice in medicine, dentistry, and to some extent in speech pathology. Cephalometric x-ray films provide a means of evaluating the interrelationships of cranial, facial, and pharyngeal structures either on a longitudinal or serial basis.

Speech clinicians have utilized lateral single-exposure (static) x-ray head films to gain information about functions and dysfunctions of the velopharyngeal mechanism. Such films have been obtained while the patient sustains a phonation of /a/, /i/, or /s/. These tasks are thought to assess the *functional potential* of the velopharyngeal apparatus to achieve velopharyngeal closure, as seen in the two dimensions of the lateral cephalometric x-ray projection.

The dynamics of speech are better appreciated by the adaptation of motion picture x-ray techniques (Fig. 24-1). Cinefluoroscopy is a technique that uses an electronic image-intensifier as a means of providing continuous motion with a relatively low level of radiation exposure. The fluoroscopic tube, used as an image intensifier, provides much less radiation exposure than cineradiography (without fluoroscopy). Within the past ten years, the videofluoroscope has been utilized with an x-ray source to provide dynamic studies of the speech mechanism. A multiview technique (lateral, frontal, and basilar views) has provided important information about the dynamics of the speech process and the interrelationships of activity at various valves in the speech mechanism.

Radiographic techniques are no longer acceptable for use in *human* research studies. The potential hazards of excessive radiation exposure and the potential damage to various tissue organs such as the thyroid and cornea preclude x-ray use on a routine, research basis. Currently, prudent use of radiographic data is practiced by most clinicians. As a consequence, the static, lateral radiograph persists as a useful *clinical* means of describing presenting morphology and function or dysfunction in the speech mechanism.

Instrumental techniques other than radiography are being perfected as alternates to radiographic study. Nasendoscopy and air-flow instrumentation are examples of equipment that is commonly used in many multidisciplinary centers to

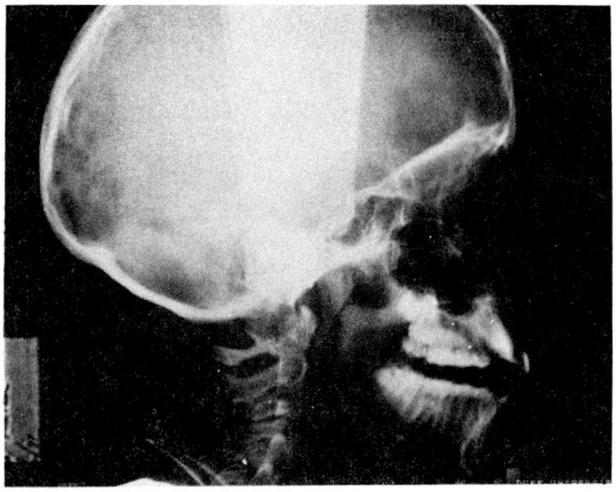

Figure 24-1A. Lateral roentgencephalometric x-ray film of an individual sustaining the vowel /i/.

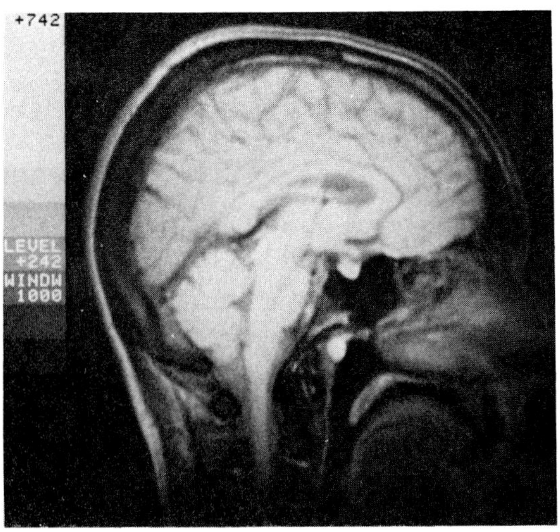

Figure 24-1B. A midsagittal view of the skull obtained with a nuclear magnetic resonance imaging technique. Courtesy of Duke University Medical Center, Division of Imaging, Department of Radiology.

assess velopharyngeal function and other components of the speech mechanism. Results of examinations using this equipment have revealed important insights as to the status of various parts of the speech apparatus while also providing predictive information about the development of future problems.

Overall, the speech clinician has contributed much to the understanding of the characteristics of velopharyngeal closure, coarticulation components of speaking, and speech sound characteristics as viewed radiographically.

In this chapter, selected information about the evaluation of lateral cephalometric radiographic films will be presented. The purpose is to equip the speech clinician with background information utilized in dentistry and medicine, especially orthodontics, relative to the morphologic description of a patient. Such information should be especially useful to the speech clinician working in a multidisciplinary setting, as well as being useful for evaluating selected components of the speech mechanism. The lateral, static x-ray film is used for discussion, since this is the most common radiographic data obtained in orthodontic analysis of the head and neck. Also, more normative data has been developed in relationship to the lateral film than any other radiographic view.

BASIC EQUIPMENT

An oriented lateral cephalometric x-ray film, as obtained in an orthodontist's office, utilizes a cephalostat or head-stabilizing device. The cephalostat and the x-ray source are positioned in a fixed relationship so that the anode of the x-ray source is five feet from the midsagittal plane of the head, or center of the cephalometer. The cephalostat immobilizes the head by ear rods. Often, a fixation arm

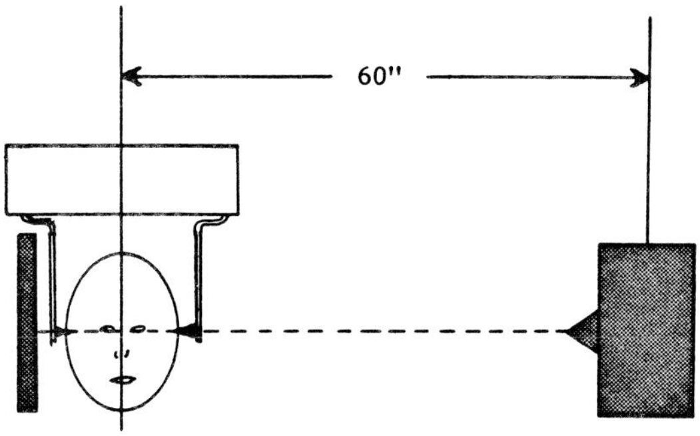

Figure 24-2. Schematic representation of the five-foot distance between the anode of the x-ray source and the midsagittal plane of the head in an orthodontic cephalometric x-ray setup. The patient's head is stabilized in a cephalostat, and the x-ray film cassette is positioned close to the patient's head for lateral x-ray view.

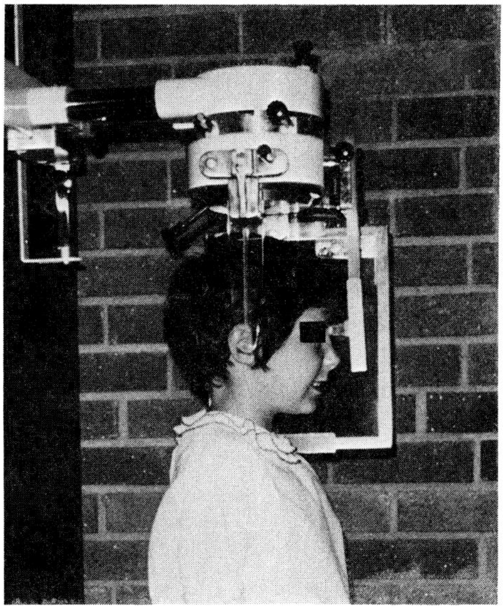

Figure 24-3. Patient positioned in the cephalostat ready for a lateral x-ray film to be made of the head and neck area. The view is from the anode of the x-ray tube. Note the square, black x-ray film holder (cassette) beyond the patient.

resting against the nasal bridge is also used to stabilize the head in a position in which the patient is upright, looking toward "infinity." In the absence of nasal bridge fixation, the cephalogram obtained is said to utilize a *natural* head position (Fig. 24-2).

Most hospital radiology units do not have a cephalostat with a five-foot anode to mid-cephalometer dimension (Fig. 24-3). All orthodontic cephalometers, however, are based upon the standardized relationships described above. A complete description of the instrumentation is reported in a text by Broadbent, Broadbent, and Golden (1975).

TRACING CEPHALOMETRIC RADIOGRAPHS

Although "eyeballing" a radiograph is a convenient way of assessing orofacial form and some functional relationships, most often the presenting morphology of the cranial and facial area is subjected to comparative judgments. The application of normative data to an individual lateral cephalometric x-ray film is best done if the x-ray film is traced on matte acetate paper. The paper is placed over the x-ray film, and various landmarks and anatomical structures are identified or outlined with a number 2 pencil. Linear and angular measurements can be made from the tracing to assess the structures and areas under consideration.

EVALUATION OF THE CRANIAL BASE

Since the facial skeleton attaches onto the cranial base and is influenced in growth by the configuration of the base of the cranium, this is a logical place to begin a lateral cephalometric x-ray film analysis. The landmarks *nasion* (N) and *basion* (Ba) serve to separate the neurocranium from the facial skeleton. Nasion is the frontonasal suture, while basion is the most anterior projection of the foramen magnum. Unfortunately, a straight line drawn from Nasion to Basion does not reveal much of value about the configuration of the cranial base. The addition of another landmark, *sella* (S), permits an angular measure to be made as an expression of the cranial base area and its relationship with the facial skeleton (Fig. 24-4).

The *cranial base angle* is formed by connecting lines between nasion, the middle of sella turcica, and basion. Normal cranial base angle (N-S-BA) is 130 degrees, plus or minus 10 degrees. That means, of

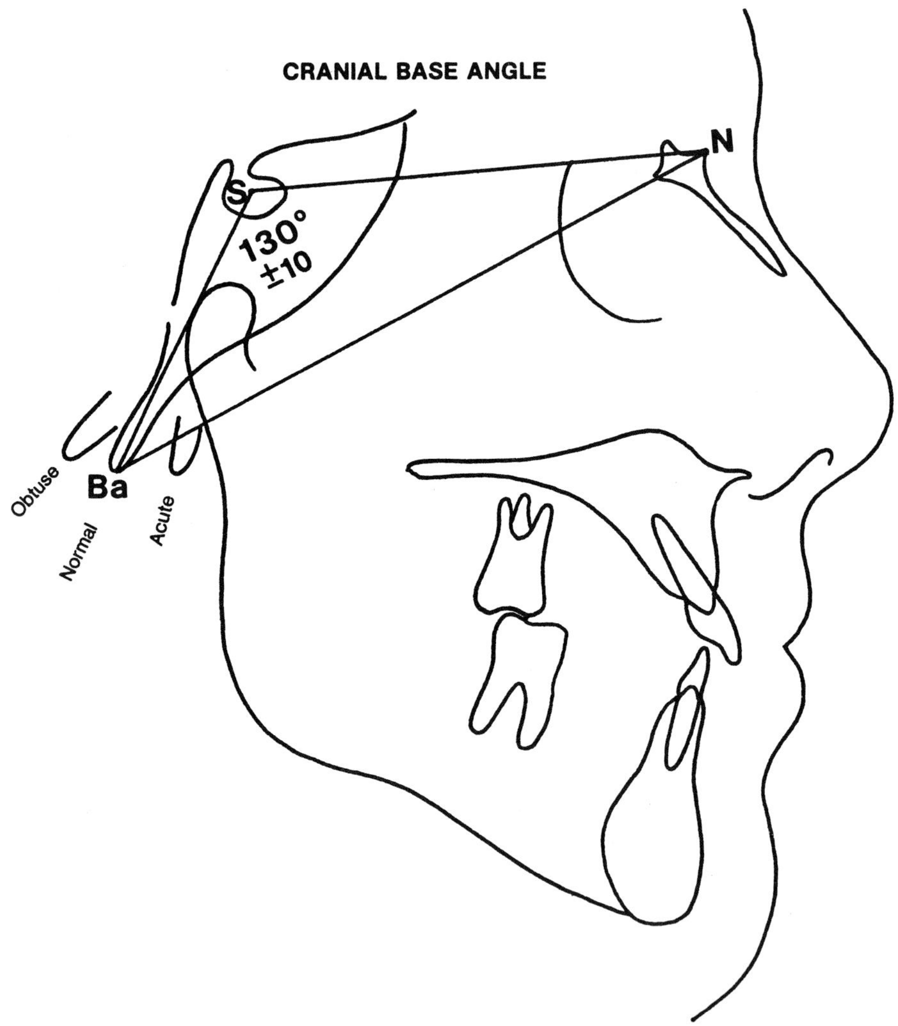

Figure 24-4. X-ray film tracing of the cranial base area from nasion (N) to basion (Ba). The configuration of the cranial base is expressed by cranial base angle from nasion (N) to sella turcica (S) to basion (Ba). Variations in cranial base angle are shown.

course, that an individual is within the normal range who demonstrates a cranial base angle from 120 to 140 degrees. This is usually written $130° \pm 10°$. Consequently, a cranial base angle less than 120 degrees would indicate an acute cranial base, while one greater than 140 degrees would be considered an obtuse cranial base.

Cranial base angle does *not* change greatly from childhood to adult form. Some fluctuation in the S-Ba arm of the angle does change, however. Nonetheless, an individual with an acute cranial base early on maintains this pattern over time.

An acute cranial base could contribute to reduction of the nasopharyngeal space while also encouraging a *forward* growth tendency of the face. An obtuse cranial base angle, by contrast, would encourage a *downward* growth pattern of the face. This is because the temporomandibular joint (TMJ) area is positioned more posteriorly, as are other cranial structures.

There are obviously exceptions to the tendencies cited here. With respect to the velopharyngeal mechanism, for example, the amount of soft tissue in the nasopharynx (the adenoid) is an important variable contributing to the soft-tissue depth of the pharynx where the bony or osseous pharyngeal depth may be influenced by the cranial base. Accordingly, observation of the cranial base area provides a useful description of the primary attachment area for the facial skeleton and upper portion of the speech mechanism.

Finding the landmarks N and S on a radiograph are not usually too difficult. Nasion represents the most involuted point at the frontonasal area. The middle of sella turcica is most often readily seen. The most difficult of the cranial landmarks to identify is basion.

Two methods of identifying Ba are suggested. The first is to find the odontoid process of the second cervical vertebra (C2) and look for a radiopaque (light) area about 3 mm above the tip of the odontoid. Ba will appear light, or radiopaque, since it is a bony landmark and will not permit the x-ray energy to pass through. This area, like other bony areas of increased density, will therefore appear *white* or underexposed on the x-ray film. Spaces and soft tissues, by contrast, will appear radiolucent or *dark* on the x-ray film, since the x-ray energy passes more readily through these areas. In many individuals, Ba will appear as if it is resting right on top of the tip of the odontoid. This is rarely the case. In such individuals where this appears the case, it is usually calcification of the ligaments that join basion with the odontoid that accounts for this perception. Close inspection usually finds a rounded, wider area 3 mm above the odontoid. This is the true location of basion (Fig. 24-5).

The second method of locating basion is to identify the rounded configuration of the occipital condyle. If this curve is followed forward from posterior to anterior, the upswing anteriorly is usually 2 mm ahead of basion. If one looks for an opaque area just posterior to the anterior margin of the occipital condyle, the location of basion is usually confirmed.

THE CERVICAL SPINE

The morphology of the cervical spine area is an important component of a lateral radiographic assessment. The upper vertebral column serves as an attachment for the muscles and soft tissues of the nasopharynx. If the cranial base

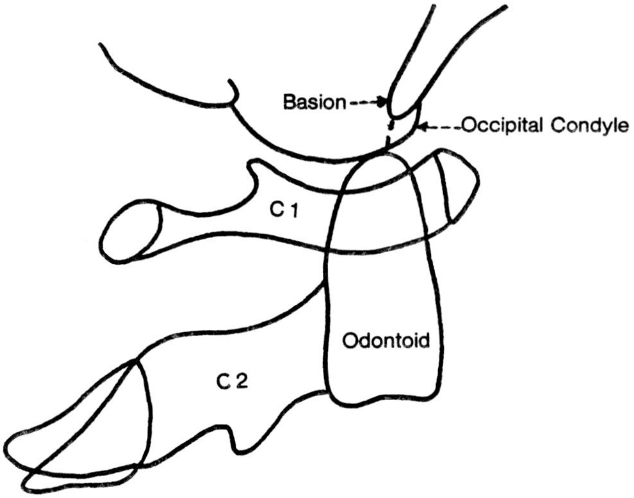

Figure 24-5. The location of basion on a lateral x-ray film is shown in this tracing as being about 3 mm above the tip of the odontoid process of the second cervical vertebra (C2), or about 2 mm behind the anterior upswing of the occipital condyle.

angle is obtuse, for example, and the cervical spine is obliged to be positioned more posterior than normal because of the flexion of the cranial base, the bony or osseous depth of the nasopharynx is increased. This observation may be an especially important one in the patient who is a candidate for an adenoidectomy. Removal of the adenoid mass may lead to the development of excessive, hypernasal speech following the adenoidectomy because of the increased osseous pharyngeal depth. In such situations, the adenoid mass is said to have *masked* a morphologic problem that was there all along (Figs. 24-6 & 24-7).

The upper three cervical vertebrae are most appropriate for examination, since they are located in the area where velopharyngeal and lingual activity is prominent. It is quite common that the area of velopharyngeal closure is adjacent to the first cervical vertebra, the atlas, or C1. The normal atlas, as seen in the lateral cephalometric projection, has an anterior tubercle that extends about 2 to 3 mm anterior to the plane of the other cervical vertebrae (see Fig. 24-6). This is true whether the configuration of the cervical spine is straight (a "military" spine) or slightly curved (a *lordotic* cervical spine). In most adults, some normal lordosis is observed.

It is well known from cephalometric studies that a velopharyngeal gap of 3 mm is sufficient to create a hypernasal condition. Consequently, small variations in the configuration of the anterior tubercle of the atlas can potentially contribute in a significant way to the increased depth of the nasopharynx (Fig. 24-8). The most common variations in the anterior tubercle of the atlas include flattening and superoposterior rotation. These conditions are developmental variations that are *not* associated with any neurological deficit. Consequently, a radiologist may not attach any special significance to these findings. While they are morphological variations, they are common in the cleft palate population but not in a normal sample. Their appearance should alert the clinician to the possible deleterious consequences to normal nasal resonance bal-

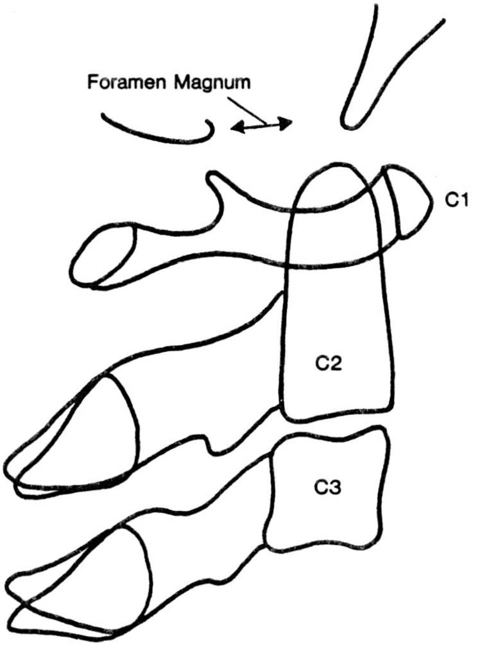

NORMAL CERVICAL SPINE

Figure 24-6. Right lateral view of a normal cervical spine showing C1, C2, and C3 in relationship to the foramen magnum.

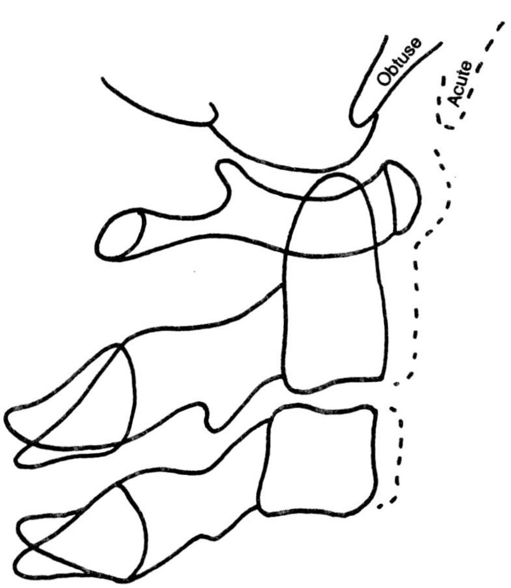

Figure 24-7. Right lateral view of the effect of cranial base flexion on the position of the cervical vertebrae position. The solid line depicts an obtuse cranial base angle, and the dashed line shows an acute cranial base and forward position of the vertebral column.

ance from a total adenoidectomy.

The atlas can be positioned in a superior position, above the usual location for velopharyngeal closure, if the occipital condyles are flattened or hypoplastic. The epitome of this situation is where the atlas is fused to the base of the skull. *Occipitalization of the atlas*, as this condition is called, can be potentially dangerous to the patient's health, since the second cervical vertebra is also positioned superiorly. If the odontoid process of C2 is positioned within the confines of the foramen magnum, the space for the spinal cord is compromised. This situation is referred to as *basilar invagination* (of C2 or the *axis* into the foramen magnum). Any suspicious instances of the odontoid appearing to invade the space of the foramen magnum should be referred to a radiologist for a definitive evaluation. Such a patient needs to be positioned very carefully on the operating table for intubation if surgery is undertaken. Hyperextension of the head, as is done routinely for intubation, serves to bring the *opisthion* (Greek for *rear*) — the central, posterior margin of the foramen magnum — forward. The spinal cord could potentially be damaged between the back of the odontoid and opisthion (Fig. 24-9).

Individuals with submucous cleft deformities have a relatively high incidence of cervical spine variations. In addition to the conditions mentioned above, fusion of the bodies or spinous processes of the second (C2) and third (C3) cervical vertebrae are common variations. Fusion of C1 and C2, however, is very rare. The observation of fusion of cervical vertebrae, especially C2 and C3, should alert the clinician to look carefully for other variations, such as clefting. While cervical spine variations are more prominent in cleft than noncleft individuals, it appears that the frequency of their occurrence increases as the severity of clefts decreases.

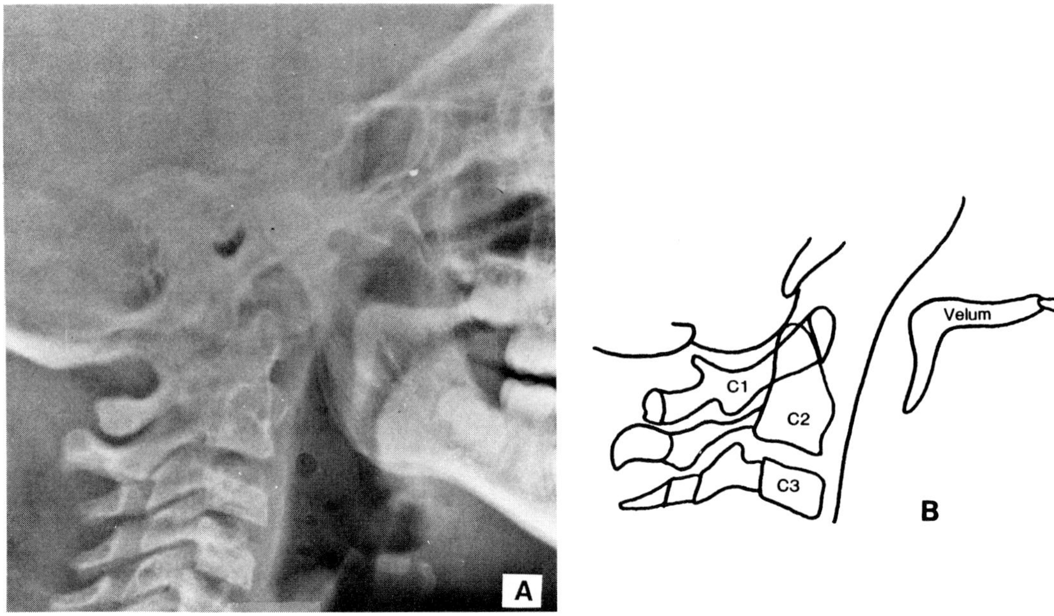

Figure 24-8. Flattening and superoposterior rotation of the anterior tubercle of the atlas. This minor variation can contribute to an increased depth of the pharynx.

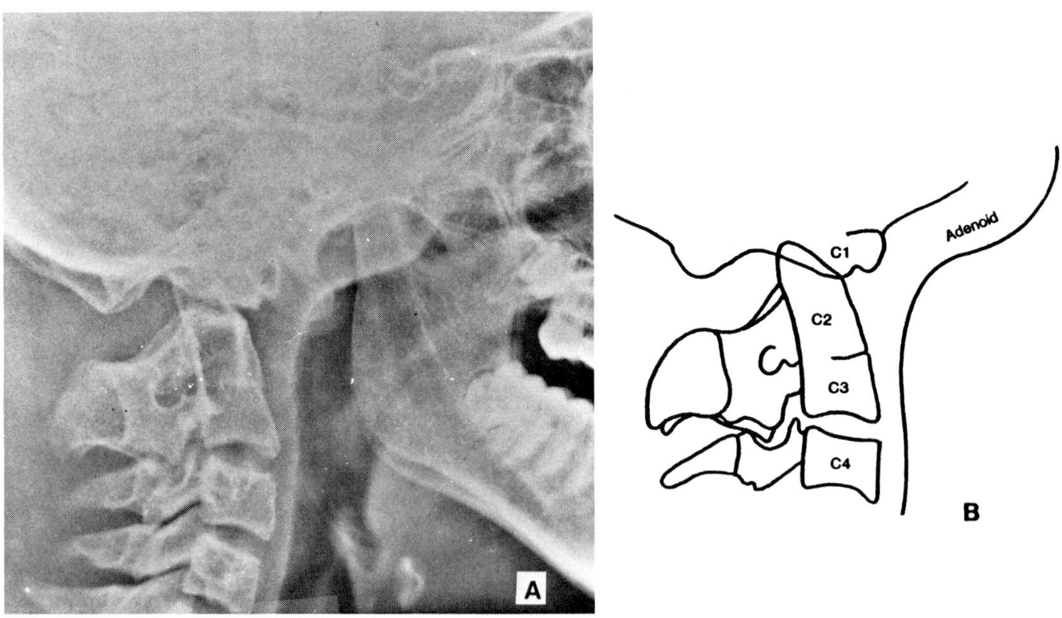

Figure 24-9. An anomalous cervical spine in a patient with a submucous cleft palate deformity. The atlas is fused to the base of the skull (occipitalization of the atlas), and the spinous processes of C2 and C3 are fused. The odontoid process of C2 has invaginated into the foramen magnum creating a potential hazard (see text for discussion).

Consequently, they are seen to occur more readily in submucous clefts and isolated palatal clefts. As such, these variations serve as clues to the examiner to look carefully intraorally and in the family history for other variations. Of course, cervical spine anomalies are also frequent in many syndromes, such as Klippel-Feil (short neck, low hairline, reduction in number of cervical vertebrae or fusion of several vertebrae).

Altogether, the cervical spine area comprises one of the components of the osseous pharyngeal depth. Its configuration also contributes to the framework for the velopharyngeal portal. The anteroposterior location of the cervical spine is determined in large part by the configuration of the cranial base.

THE UPPER FACE

The most common method of assessing the relative position of the maxilla is in relation to the cranial base. In the anteroposterior dimension, the angle formed by landmarks S-N-A provides this assessment. The S-N arm of the angle contributes the cranial component, while the line drawn from nasion to point A expresses the position of the maxilla. Point A is the most involuted portion of the anterior maxilla. This point is less variable than the anterior nasal spine (ANS), which is buried under the soft tissues of the nose and, accordingly, would not be useful for this purpose. By like token, the point where the upper teeth and alveolar bone meet (supradentale) is not useful, in that this point expresses the position of the upper incisors rather than bony maxilla. It is well known that incisors can protrude (along with supradentale) without commensurate protrusion of the maxilla (Fig. 24-10).

A normal SNA is 82° ± 4°. Therefore, an SNA of 70° would indicate midface retrusion, while an SNA of 90° would indicate maxillary protrusion. There are, of course, exceptions to such findings. A low position of sella would affect this angle and would have to be corrected to obtain an accurate portrayal of the SNA significance.

There are no well-established cephalometric norms for the vertical component of the maxillae. Overall facial height, as described in Chapter 21 on "Facial Esthetics," is a suitable clinical measure. Upper facial height is usually measured from nasion to ANS, while lower facial height from ANS to *Menton* (lowest point on the chin). The normative data for upper facial height (N-ANS) in adult males is 60 mm ± 4, and for females, 56 ± 3. Caution is recommended in the use of these norms, however, as upper facial height should be viewed as a proportion of total facial height rather than an isolated finding, as pointed out in Chapter 21.

Useful information about facial height can also be obtained from the lateral radiograph by identifying the location of ANS and PNS. The palatal plane, as is determined by a line extending from ANS to PNS, is evaluated in relationship with the vertical location of the atlas shown in Figure 24-10. Usually, the palatal plane and the atlas fall on the same horizontal. If ANS is higher or lower than PNS, this may signal a vertical discrepancy in the growth of the maxilla. A "canted" maxillary plane, as this situation would be described, is often the sign of a skeletal dysplasia of the maxilla, such as posterior vertical maxillary excess where the palatal

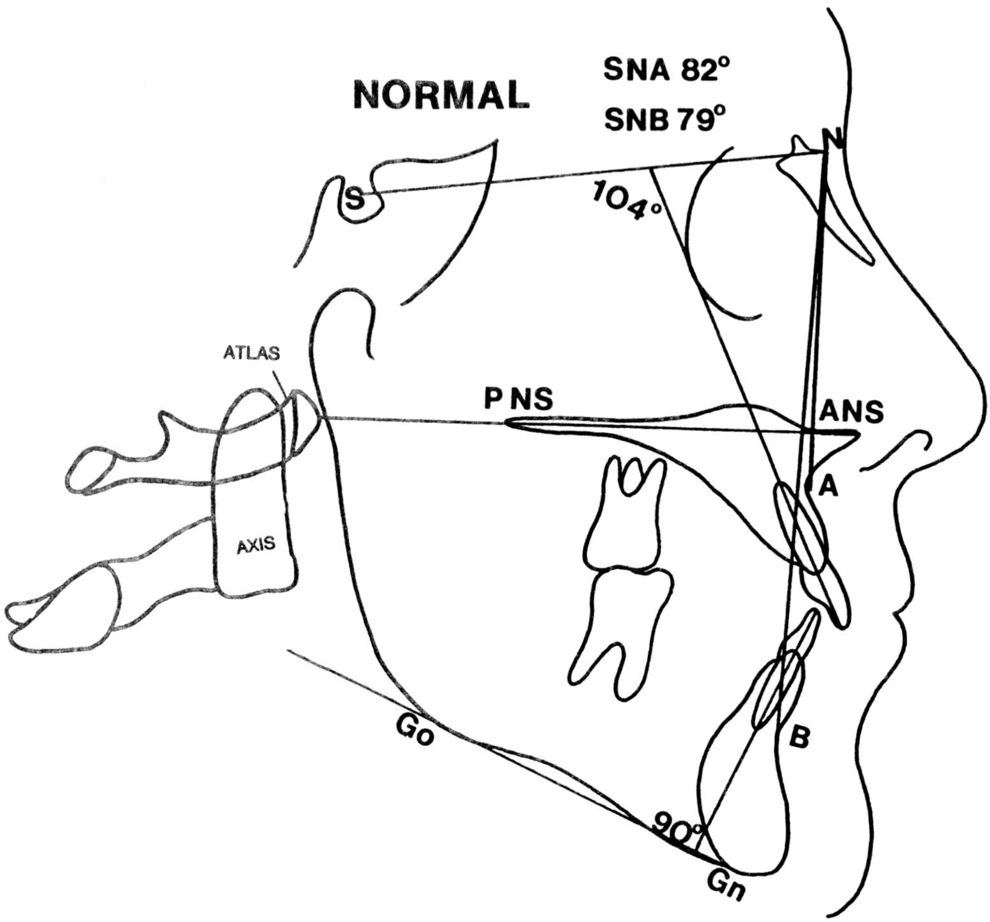

Figure 24-10. Selected radiographic landmarks and angles to express the position of maxilla and mandible to the base of the skull. SNA expresses the horizontal position of the maxilla. SNB expresses the horizontal position of the mandible. ANS-PNS shows the palatal plane, which is typically at the same vertical level as the atlas. The upper incisor to the SN plane is shown, and the lower incisor to the mandibular plane (GoGn) is demonstrated.

plane is canted from PNS being displaced inferiorly. This would, incidentally, also create an anterior openbite malocclusion.

THE LOWER FACE

The mandible is certainly an important structure to the speech clinician and others, since it interrelates with the tongue, hyoid bone, and the oropharyngeal area. Its development and position is closely associated with the functions of breathing, eating, swallowing, and speaking.

The most frequently used measure of the horizontal position of the mandible is

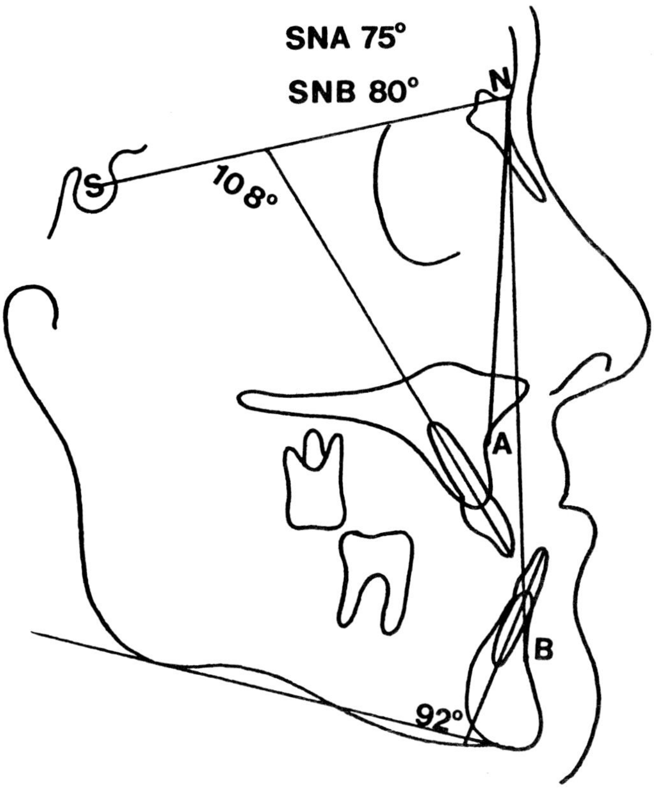

Figure 24-11. Lateral x-ray film tracing of a patient with pseudoprognathism. The cephalometric measures indicate a normal mandible and a retruded maxilla in spite of the clinical appearance of a large mandible.

the angle S-N-B, or SNB. Point B is the most involuted portion of the anterior mandible. This area is above the chin and below the alveolus and is known as *supramentale*. This measurement expresses the horizontal position of the mandible with reference to the cranial base. The norm for SNB is 79° ± 3°, as seen in Figure 24-10. The SNB measurement, in comparison with SNA, provides the clinician with a basis for determining whether the patient has a true prognathism or a pseudoprognathism. As one might expect, a normal SNB in combination with a diminished SNA might give the appearance of mandibular prognathism rather than the midfacial retrusion that is often the real cause of the jaw discrepancy (Fig. 24-11).

While SNA and SNB relate the maxilla and mandible to the cranial base area, the measurement ANB relates maxilla to mandible. An ANB of 3° is normal, with a range from 0° to 5°.

THE DENTITION

Of the many dental measurements utilized in orthodontics and other areas of dentistry, the most useful for describing the position of the teeth relative to the bones to which they are attached involves the incisors. The angulation of the upper incisors is measured by extending a line from upper incisor tip, through the root of the incisor, and extending to intersect a line from sella to nasion. The norm for upper incisor to NS is $104° \pm 6°$ (see Fig. 24-10). For the lower incisor, the line from incisor tip through the root is extended to intersect a line that follows the mandibular plane. This is constructed by connecting a line from the mandible angle (*gonion*) to the most anteroinferior point of the chin (*gnathion*). The norm for lower incisor to Go-Gn (the mandibular plane) is $95° \pm 7°$.

When the upper and lower incisor positions are considered, especially in comparison with the maxillary and mandibular positions (SNA and SNB), the examiner is provided with sufficient information to evaluate whether a dental or skeletal or a combination of variations is present.

THE PALATES

The speech clinician will be especially interested in the cephalometric appearance of the hard and soft palates. The hard palate is, of course, a bony framework for the attachment of the soft palate, as mentioned in other chapters. In the lateral cephalometric x-ray projection, the hard palate normally ends at the pterygomaxillary fissure (PTM) — a tear-shaped fissure that is seen as a radiolucent shadow on a radiograph. The front part of the teardrop is the posterior border of the maxillary tuberosity, while the posterior part is the anterior border of the pterygoid plates. The posterior nasal spine of the hard palate will never be longer than the PTM teardrop. Finding the PTM and following the contour of the maxillary tuberosity inferiorly until it intersects the palatal plane is the usual method of identifying the location of the posterior nasal spine (PNS) in the normal. In a repaired cleft palate or submucous cleft palate patient, the posterior nasal spine will be absent. Often, the end of the hard palate is identified anterior to the PTM in such patients (Fig. 24-12).

The posterior nasal spine is an important landmark in assessing the length of the soft palate, either at rest or in function. Normative data for resting length of the velum are found for age and sex in the research of Subtelny (1957) and are too numerous to include in this text. While the resting length of the velum is of some interest, especially in relationship to the soft-tissue depth of the pharynx (Subtelny, 1957), the *functional* length of the velum is of special importance.

Normally, the soft palate increases in length during functional activities such as speaking. The ability of the velum to increase overall length during function has been termed the "*stretch factor*" in soft palate function (Pruzansky and Mason, 1969). The most important radiographic finding of the elevated velum, as seen on a phonation lateral x-ray film, is the *effective* length of the velum (the amount of velar tissue used in attempts to obturate the nasopharyngeal space). Research has determined generally that the normal soft palate elevates up to the level of the hard palatal plane. Furthermore, the point of greatest flexion of the velum is located about 80 percent of the total length of the velum, as measured from the PNS to the tip of the uvula (Fig. 24-13). The point of greatest velar elevation is known as the *velar dimple* and is created by the attachment scheme of the levator palatini muscles in the velum. The shorter the effective velar length (as measured from PNS to the velar dimple), the narrower the nasopharyngeal depth must be to maintain a velopharyngeal seal (Fig. 24-14).

The patient with an anterior (short) point of levator attachment in the velum, but who is normal in nasal resonance balance due to a velopharyngeal seal involving the adenoid mass, is not a good candidate for total adenoidectomy. This surgery could potentially create a persistent hypernasality following the operation.

An individual with a normal palate who undergoes the normal process of adenoidal involution over time is not expected to develop hypernasality with adenoid involution. This is associated with palatal stretch, which is apparently a compensatory adaptation to permit the maintenance of normal nasal resonance balance with changes in the architecture of the pharynx (Mason and Warren, 1980).

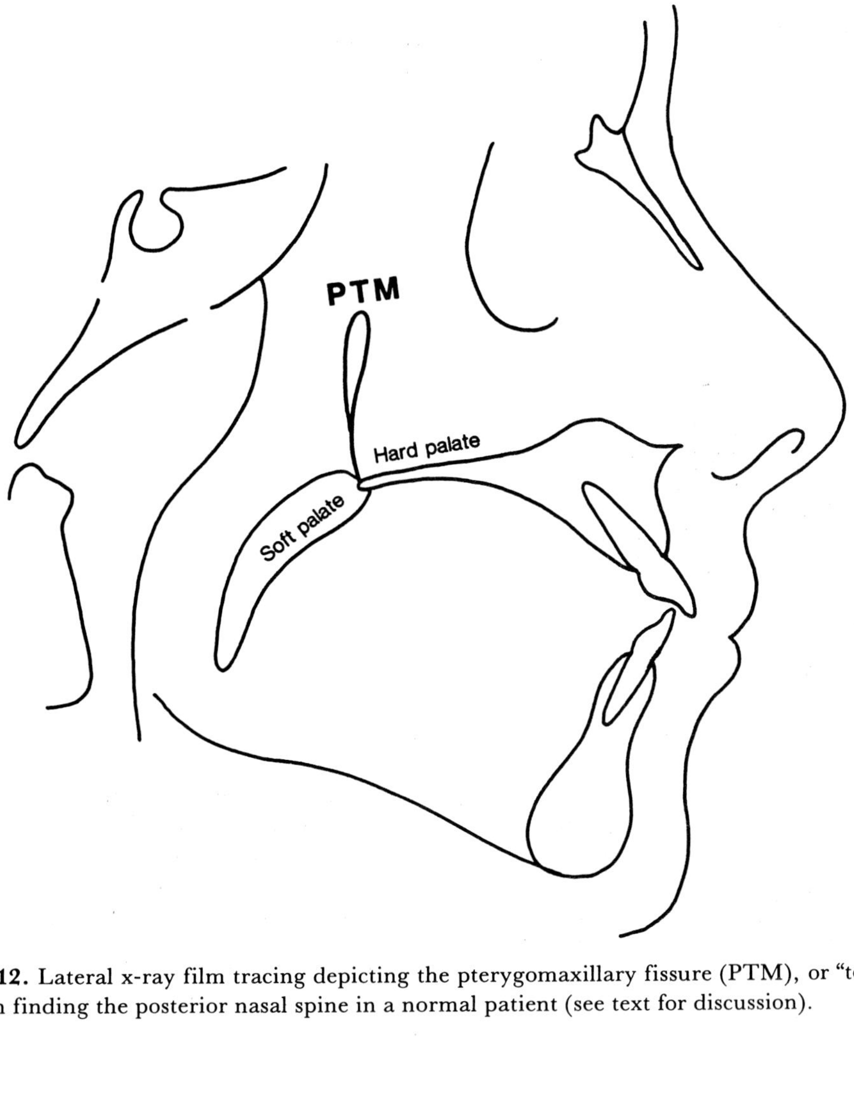

Figure 24-12. Lateral x-ray film tracing depicting the pterygomaxillary fissure (PTM), or "teardrop." This aids in finding the posterior nasal spine in a normal patient (see text for discussion).

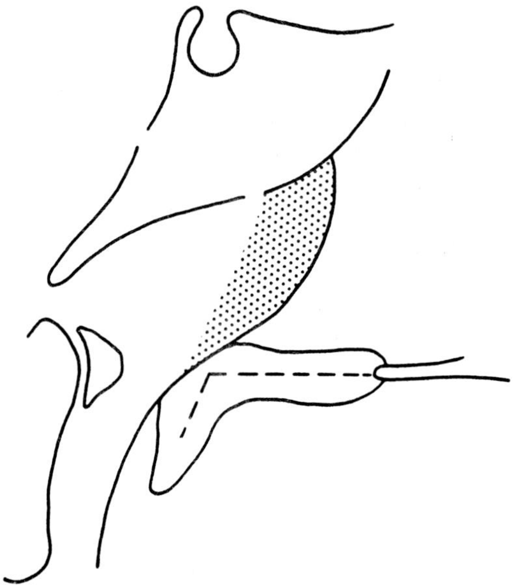

Figure 24-13. Lateral cephalometric x-ray film tracing during sustained /i/ showing a broad velopharyngeal seal and the soft palate "dimple" or flexion. Notice that the velum has elevated to the level of the hard palate plane. The adenoid mass is shown as the stippled area. This patient would maintain a velopharyngeal seal if a total adenoidectomy were performed.

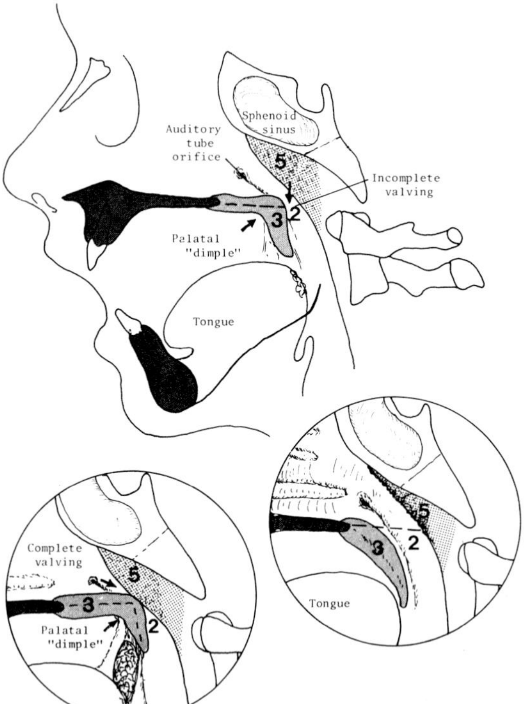

Figure 24-14. Demonstration of the velopharyngeal apparatus through x-ray film analysis. Effective length of the soft palate is measured from the posterior nasal spine (PNS) (which is located at the limit of the hard palate) to the velar dimple. Incomplete valving is shown (top) as well as normal valving (lower left). (Lower right) The palate is at rest, with a dashed line added to indicate the position for velopharyngeal valving. Notice the important role of the adenoid mass in the tracings presented.

THE ADENOID MASS

Information about the adenoid mass has been presented in many other sections of this text. By now, the reader should be quite conversant about the significance of the adenoid mass in velopharyngeal closure.

From a diagnostic standpoint, the adenoid should be evaluated, in part, with regard to the contiguous anatomy. That is, the cranial base angle, status of cervical vertebrae, and configuration of hard and soft palate should be closely observed.

It is tempting to look at the overall contour of the adenoid mass and incorrectly presume about the size of the adenoid. As pointed out in Chapter 19, "Postnatal Growth and Development," the adenoid is expected to be large in size from around age three to twelve years. While the portion of the adenoid at the level of the palatal plane is important as a contact site for velopharyngeal function, it is the portion of the adenoid at or near the posterior choanae that determines whether the airway has been compromised. Blockage of the posterior choanae by adenoidal tissue certainly precludes normal respiration intranasally. Hyponasality associated with adenoidal hypertrophy occurs when the posterior choanae is blocked with adenoid. It is not the amount of adenoid at the velopharyngeal hiatus that causes a hyponasal voice quality, except in rare instances. The posterior choanae, or the posterior entrance into the nasal cavity, is observed on the lateral x-ray film in two ways. One is to find the posterior border of the vomer, which is seen to project cranialward in an oblique direction from the PNS area. The other method is to find the posterior margin of the PTM, which can be considered to be a reasonable representation of the posterior opening of the nasal cavity. The adenoid mass at or near the PNS in the vertical dimension is the area in which the soft-tissue mass of adenoid should be evaluated for airway patency or blockage.

Overall, the adenoid mass as observed on the two dimensions of the lateral x-ray film represents the midline contour of the adenoid. This shadow does not reveal the amount of adenoid on the lateral walls of the pharynx. For speech purposes, nonetheless, it is the midline convexity of the adenoid that is important to identify and evaluate with regard to the presence, absence, maintenance, or deterioration of velopharyngeal closure.

SUMMARY

The lateral cephalometric x-ray film is a standard part of an orthodontic data base. It is also a potentially important tool of examination to the speech clinician. For those clinicians working in concert with an orthodontist, a sustained film of phonation can reveal several important characteristics of velopharyngeal function, current status, and predictions for subsequent history.

The lateral x-ray film provides an oriented look at the morphology of the head and neck area. Normative data are available with which to evaluate the adequacy of the facial structures for speech and other functions. Some selected orthodontic norms, as well as a discussion of palatal and cervical spine variations, are provided for the speech clinician to consider. The normative data selected for inclusion in this chapter represents only the most common observations of interest to the speech clinician.

BIBLIOGRAPHY

Bean, J., Kramer, J.R., and Khouw, F.E.: A simplified method of taking radiographs for cephalometric analysis. *J Oral Surg, 28*:675-678, 1970.

Broadbent, H.B., Broadbent, H.B., Jr., and Golden, W.H.: *Bolton Standards of Dentofacial Developmental Growth.* St. Louis, C.V. Mosby, 1975.

Johnson, N.A.: Xeroradiography for cephalometric analysis. *Am J Orthod, 69*:524-526, 1976.

Moorees, C.F.A., and Kean, M.R.: Natural head position: a basic consideration for analysis of cephalometric radiographs. *Am J Phys Anthropol, 16*:213-234, 1958.

Riolo, M.L., Moyers, R.E., McNamara, J.A., Jr., and Hunter, W.S.: *An Atlas of Craniofacial Growth.* Monograph No. 2, Craniofacial Growth Series, Center for Human Growth and Development, U of Michigan, 1974.

Steiner, C.C.: Cephalometrics in clinical practice. *Angle Orthod, 29*:8-29, 1959.

Wylie, W.L.: The assessment of antero-posterior dysplasia. *Angle Orthod, 17*:97-109, 1947.

Chapter 25

INDIRECT LARYNGOSCOPY AND PITCH ASSESSMENT

TECHNIQUE FOR VIEWING THE LARYNX IN INDIRECT LARYNGOSCOPY

ONE of the most exciting skills that a speech clinician can learn is indirect laryngoscopy. It opens up more insights into the structure and function of the larynx than any other approach. While speech clinicians must strictly avoid rendering medical diagnoses, following a patient's physical (laryngeal) response to therapy for voice problems such as vocal nodules can be quite useful (Fig. 25-1).

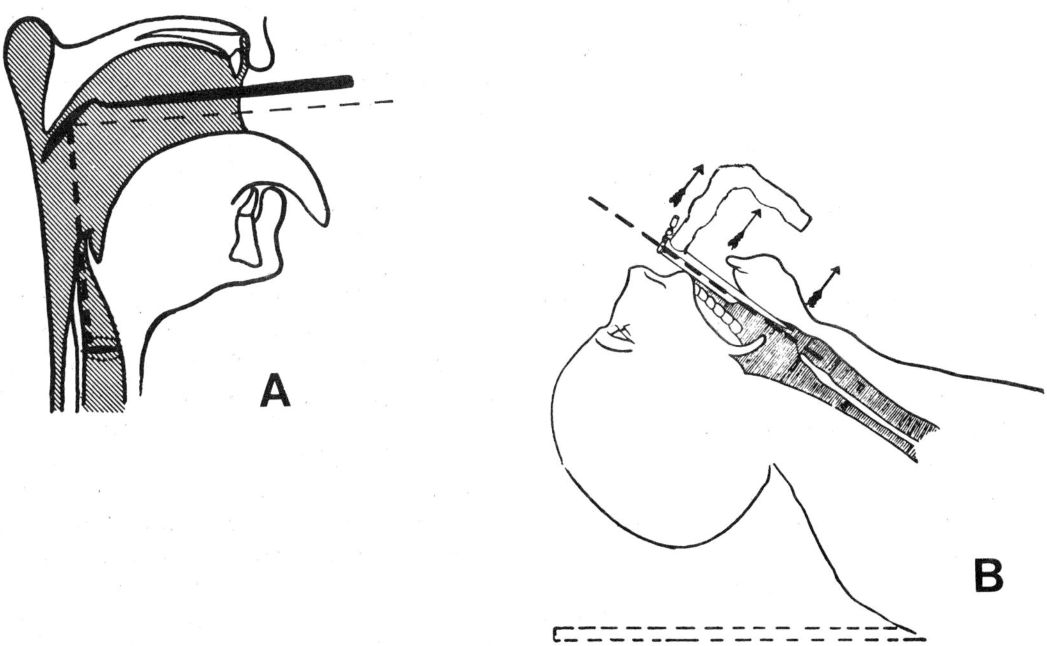

Figure 25-1. (A) Indirect laryngoscopy (mirror examination of the larynx). (From Gardner, E., Gray, D.J., O'Rahilly, R.: *Anatomy*, 4th ed., 1975. Courtesy of W.B. Saunders Co., Philadelphia.) (B) Direct laryngoscopy. The dotted line indicates the line of sight through the laryngoscope to the larynx.

The first step in learning the technique of mirror examination by students is to adopt a serious disposition and an all-business approach. Learning to manipulate the mirror in one's own mouth in front of a well-lighted mirror is the next step. The third step is to find a cooperative friend, especially a fellow student in speech and hearing, to serve as a practice subject. Together you should review the anatomy of the structures associated with the examination. The fourth step is to become intimately familiar with the technique as described next while at the same time reviewing mirror illustrations of the vocal folds (Fig. 25-2).

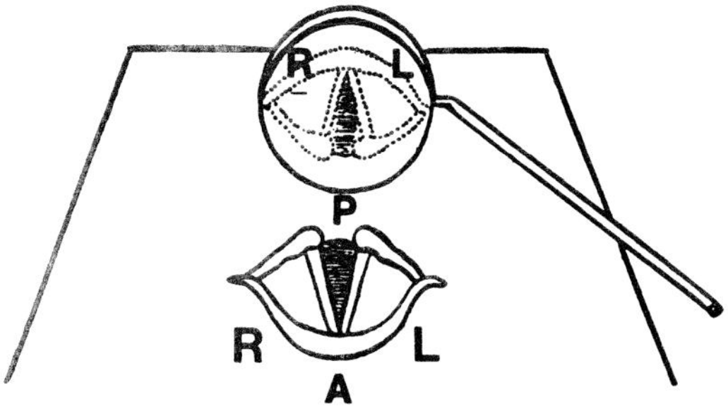

Figure 25-2. Schematic diagram illustrating proper designation of left-right, anterior-posterior of the vocal folds. The apex of the V formed by the folds is at the anterior aspect of the glottal opening. In the mirror, the V seems to point in an opposite direction, but left-right positions are unchanged. (From Gallander: Surgical Anatomy.)

The equipment necessary to perform an indirect laryngoscopic examination includes a headlamp or head mirror with a 25 to 35 cm focal length, a laryngeal mirror or dental mirror, gauze squares, and an alcohol lamp or suitable warmer. Additionally, if the head mirror is used, a light source, such as a clear 100- or 150-watt electric bulb is required.

POSITION OF THE PATIENT. The patient should be seated upright in a chair with

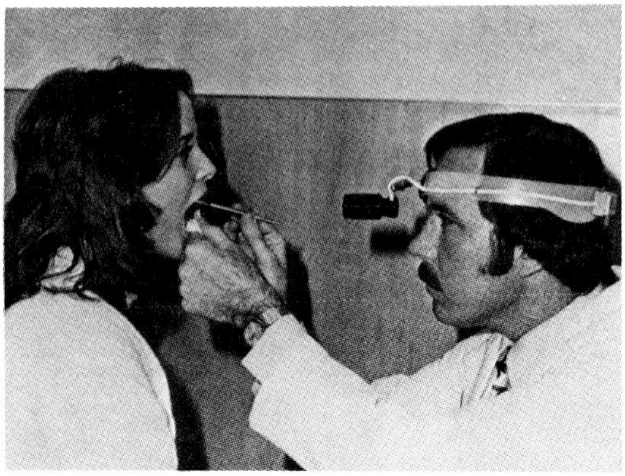

Figure 25-3. Indirect laryngoscopy — position of the patient and examiner.

the back arched forward, hands placed on the knees for stability, and neck hyperextended; the knees are together.

POSITION OF THE EXAMINER. The examiner positions himself for examination so that his eye level is at the mouth level of the patient. This requires adjustment of the patient's chair or that of the examiner. If the head mirror is used, a position on the same side as the light source may be more practical, in which case the light source would be placed behind and to one side of the patient's head. With the headlamp, the examiner may straddle the patient's knees.

ADJUSTMENT OF THE ILLUMINATION SOURCE. The headlamp is positioned as close to the eye level of the examiner as possible. An interorbital location of the headlamp beam is not distracting to the examiner. With the head mirror, the examiner should move his/her head forward and backward, focusing the light to produce a small, brilliant spot of illumination. The head mirror should be as close to the face or glasses of the examiner as possible, with the aperture directly in front of the left eye (Fig. 25-3).

ORIENTING THE PATIENT. The patient is instructed to look straight ahead and to keep the head in a stable position, as if looking "through" the examiner. The patient should be told to "pant slowly." This activity serves to elevate and spread the palate toward the posterior pharyngeal wall, providing space for mirror placement in the oropharynx. Furthermore, this activity reduces the tendency to gag. As an alternative with the patient who has difficulty with the panting technique, the task of maximum or forced inhalation is often effective. Topical anesthetics should be used only under the supervision of laryngologists or dentists.

POSITION OF THE TONGUE. The clinician who is right-handed will use his left hand to stabilize the patient's tongue and head position. The patient is asked to protrude the tongue; the examiner wraps a gauze square around the protruded area and grasps the tongue by placing the flat part of his thumb over the surface of the tongue and the middle finger on the bottom. The third finger rests against the groove above the mental protuberance area, and the index finger is braced against the upper lip or upper teeth to increase steadiness while holding the tongue. The tongue must not be pulled too tightly for risk of cutting the inferior surface of the tongue on the lower incisors. However, the examiner should pull the tongue far enough forward to obtain a good open oropharyngeal space. In the protruded tongue position, the epiglottis is pulled forward from the field of vision for laryngeal observation (Fig. 25-4).

THE LARYNGEAL MIRROR. A number 5 or 6 laryngeal mirror or dental mirror is used. An anti-fog cream is applied to prevent fogging of the mirror during the examination. Warming the mirror over an alcohol flame or in warm water is an alternative to prevent fogging. Before the mirror is introduced into the patient's mouth, it is tested on the back of the examiner's hand to determine that it is not too hot.

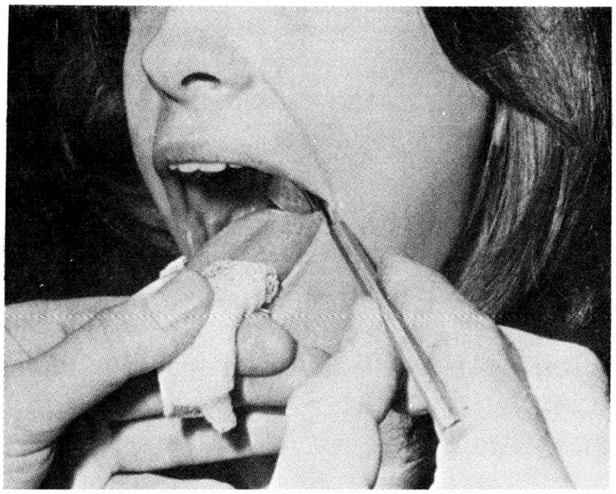

Figure 25-4. Position of the tongue and laryngeal mirror during indirect laryngoscopy.

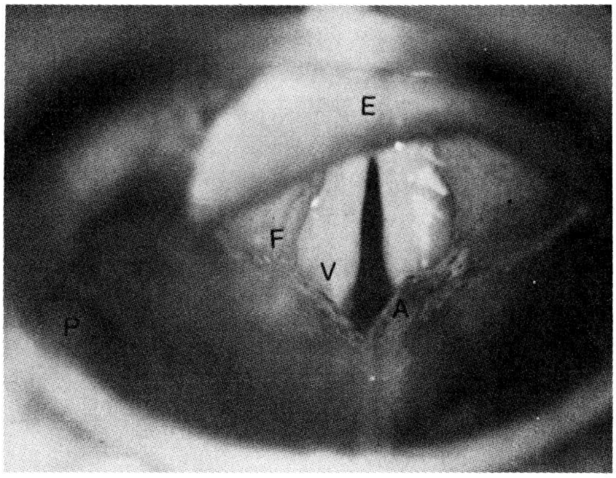

Figure 25-5. Mirror view of the larynx showing the epiglottis (E), arytenoids (A), piriform recess (P), ventricular or false folds (F), and vocal folds (V). The glottis is evident between the folds.

The mirror is cradled between the thumb and index finger only, with the shaft of the mirror located on the right side at the three-o'clock position. As the mirror is introduced into the oral cavity, the shaft is placed against the corner of the patient's mouth. The mirror is passed to the back of the mouth, with the glass surface parallel to the tongue and the shaft sliding over the corner of the lip. The mirror should not be permitted to touch any surfaces on the way to the oropharynx. The light source is adjusted as needed, and the patient is reminded to pant slowly. In some individuals, the mirror can be positioned in the oropharynx so that the mirror head is not in contact with any structures. In the typical examining situation, the mirror is pressed gently against the uvula and velum and the mirror tilted as needed by rotating the mirror shaft to facilitate an optimal view of the larynx (Figs. 25-6 & 7).

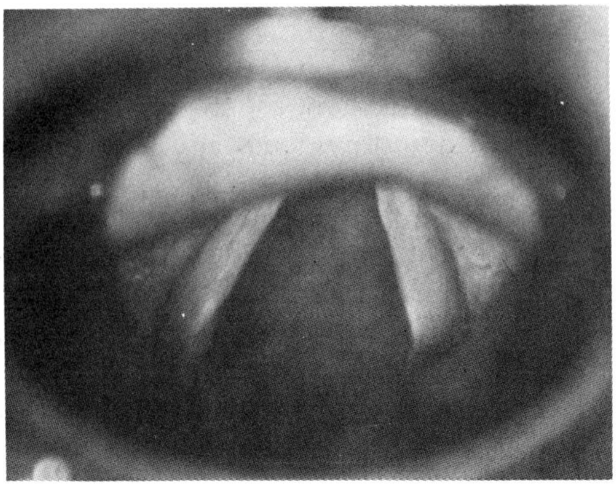

Figure 25-7. The glottis during deep inspiration.

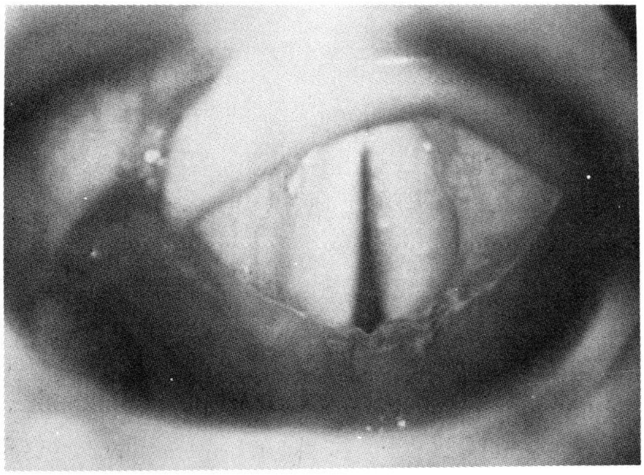

Figure 25-6. Position of the folds during phonation.

APPEARANCE OF THE LARYNX. The larynx can be examined efficiently during a structured breathing activity. The folds will move with each phase of breathing. To view the folds in adduction, the patient should be asked to sustain a phonation of *ee* during mirror examination. Although an *ah* sound will be produced, having the patient think of the *ee* sound will aid in keeping the tongue forward and lying low in the oral cavity. The epiglottis will thereby be kept forward in order to provide an unimpeded view of the larynx.

PITCH ASSESSMENT

The rationale for selecting only one clinical procedure for this section of applied anatomy stems from the belief that this area pervades the gamut of voice problems and relates often to clinical techniques used in the habilitation or rehabilitation of patients with voice problems. Pitch assessment is also germane to the concept of applied clinical anatomy. Caution is advised so that the reader does not assume that pitch assessment is more important than any other dimension of voice evaluation.

Since the purpose of this text is not to supplant diagnostic texts or other clinical books covering voice, only an introduction to pitch assessment is provided. The student is encouraged to utilize the bibliography at the end of this chapter.

Concepts of Pitch Levels

Appropriateness of pitch levels for individuals is a variable factor. Age, sex, personality, and even body-build influence decisions of appropriateness for pitch usage. But there is no single best pitch for any speaker; rather, there are a range of pitches that represent for a given voice efficient phonation.

Optimum Pitch

That range of pitches (closely clustered for most) is the result of efficient phonation and is known as the *optimum pitch range*. These pitches are produced with ideal coupling of the larynx with the resonators so that they complement one another.

To demonstrate the optimal pitch level used by a patient, two methods may be used. The approach suggested by Cooper (1977) is particularly easy to implement.

> A very effective method of finding the correct pitch level and proper tone focus is to say "um-hum" spontaneously, sincerely, and easy with an upward inflection as though agreeing with someone. If correctly done, the individual should experience a tingle or buzz about the lips or around the bridge of the nose, an area known as the *mask*. Another method is to say *hello*, again spontaneously, sincerely, and easy, pretending that the other person is a little hard of hearing. Normally when the incorrect pitch is altered to the natural level, the tone focus is also changed to a properly-balanced oro-naso-laryngo-pharyngeal tone focus. It must be emphasized and understood that when the correct tone focus is used, the low pitches in the voice may be comfortably and easily used as part of the normal range.

Another method that has been described in the literature for many years is attributed to Fairbanks (1960). This approach implies that there is one musical note that will represent the most efficient pitch level. As noted above, the implication should be expanded so that even as Fairbanks' method is used, a range of pitches around the focal musical note is included among the optimal pitches.

Using a piano, the person, say a tenor, is asked to sing from middle C to the lowest note that he can reach. He returns to middle C and sings to the highest note that he can reach, including falsetto. The total number of notes, counting either half steps (counting from any given key on the piano to its *closest* neighboring key) or whole steps (counted from any given key on the piano to the *second closest* key) is determined. Next, the total is divided by 8 (half-step method) or 4 (whole-step method). The quotient yields a note somewhere close to the optimum pitch for the tenor. (See Figure 25-8 for an example of the whole-step procedure.)

Yet, a third method has been proposed by Hollien and others (1971) for assessing pitch. Termed the *phonational frequency range* (PFR), or just *phonational range*, this useful technique, like the Fairbanks' approach, yields a measure of the range of the voice that may be compared with norms established by a number of investigators. However, unlike Fairbanks, no attempt is made to compute an optimal pitch, only the range of the subject's phonational pitch. The PFR is based on *physiological limits* of the larynx rather than on the average or ideal speaking fundamental frequency.

To obtain the phonational frequency range, the limits of the PFR are first determined. The subject matches his or her voice to a series of tones, using either a piano or a calibrated audio oscillator as the source for reference tones. Adult females would begin with a note around E_4 (330 Hz), and males would begin one octave lower, or E_3 (165 Hz). The subject is asked to phonate an /a/ on pitch with the reference tone as it descends in whole steps. The *lowest frequency s/he can sustain* becomes the lower limit of the phonational frequency range. Next, the subject returns to the starting note and follows the

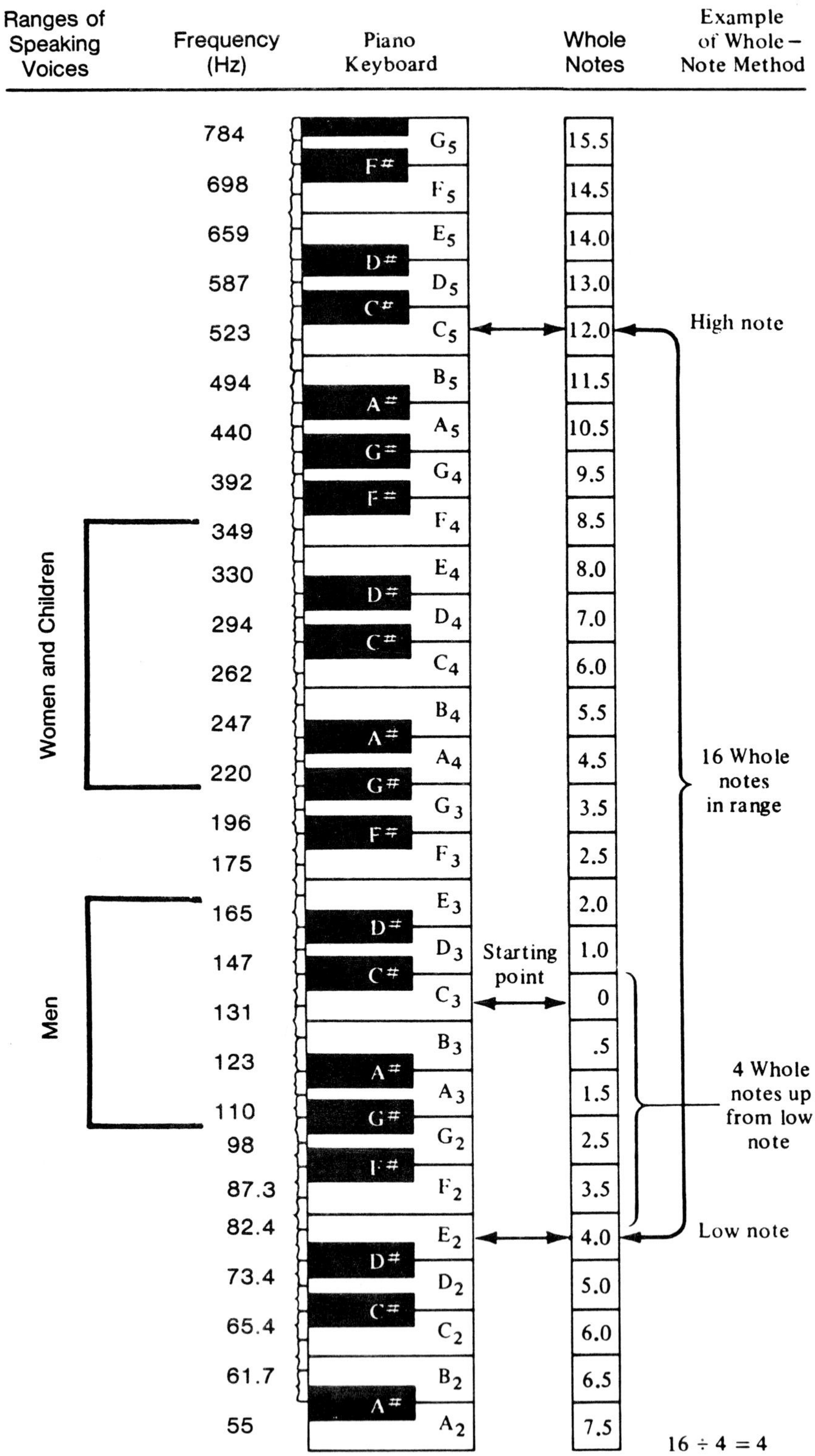
Ranges of Speaking Voices
Frequency (Hz)
Piano Keyboard
Whole Notes
Example of Whole-Note Method
Women and Children
Men
784
698
659
587
523
494
440
392
349
330
294
262
247
220
196
175
165
147
131
123
110
98
87.3
82.4
73.4
65.4
61.7
55
G5
F#
F5
E5
D#
D5
C#
C5
B5
A#
A5
G#
G4
F#
F4
E4
D#
D4
C#
C4
B4
A#
A4
G#
G3
F#
F3
E3
D#
D3
C#
C3
B3
A#
A3
G#
G2
F#
F2
E2
D#
D2
C#
C2
B2
A#
A2
Starting point
15.5
14.5
14.0
13.0
12.0
11.5
10.5
9.5
8.5
8.0
7.0
6.0
5.5
4.5
3.5
2.5
2.0
1.0
0
.5
1.5
2.5
3.5
4.0
5.0
6.0
6.5
7.5
High note
16 Whole notes in range
4 Whole notes up from low note
Low note
16 ÷ 4 = 4

reference tone up the scale until the *highest note the subject can sustain*, including falsetto for males. Each descending and ascending procedure should be repeated to ensure that the note at each limit can be sustained briefly.

The difference between the two frequencies, counting by *half steps* on a keyboard of the notes sustained, is the PFR. Clinicians can compare the norms reported by Hollien et al. (1971) to determine if a given patient's PFR is impaired. Table 25-I presents the mean lowest and highest frequencies for males and females as well as the phonational range.

Clinicians can determine if a patient's phonational range is too extensive or too restrictive. Subjects with normal vocal mechanisms will usually exhibit a PFR of at least two octaves (i.e. 24 half steps/semitones or 16 whole steps/notes). Therefore, clinicians may apply this data and conclude that any patient who is unable to generate a PFR of only one octave

Table 25-I

MEAN PHONATIONAL RANGES BY SEX

Group	*N*	*Age in Years*	*Lowest Frequency (Hz)*	*Highest Frequency (Hz)*	*Phonational Range in Half Steps (semitones)*
Males	332				
Mean		21.4	73.4 (D_2)	659.28 (E_5)	37.9
Range		18 – 36	58.97 – 116.54 $A_2^{\#} - A_3^{\#}$	207.65 – 1400.0 $G_3^{\#} - F_6^{\#}$	13 – 55
Standard Deviation		3.5	2.3 semitones	4.4 semitones	5.0
Females	202				
Mean		22.8	130.81 (C_3)	1046.5 (C_6)	37
Range		18 – 36	92.50 – 185.0 $F_2^{\#} - F_3^{\#}$	554.37 – 1975.5 $C_5^{\#} - B_7^{\#}$	23 – 50
Standard Deviation		4.8	2.5 semitones	4.6 semitones	5.3

SOURCE: Hollien, 1977.

Figure 25-8. Chart for locating optimum pitch based on the Fairbanks method. The total number of notes vocalized, including falsetto, taken each time from a comfortable starting pitch (16 whole notes in the example on the chart). The sum (16) is divided by 4. Next, count up from the lowest note (E_3) 2 whole notes (16 ÷ 4 = 4) to find the estimated optimal pitch (C_3). Notice that any two adjacent keys (white-black/white-white) are *one-half* note or step apart. The numbers in the blocks under the column *Whole Notes* illustrate the proper counting for whole notes in the example. Also shown are the ranges of typical speaking voices. (Modified from Fig. 6-1 (p.67) in *Modifying Vocal Behavior* by John P. Moncur and Isaac P. Brackett. Copyright© by John P. Moncur and Isaac P. Brackett. By permission of Harper and Row, Publishers, Inc.)

←

undoubtedly is probably experiencing a clinical voice problem.

With older subjects, the PFR would be expected to drop, although no definitive norms are available for older subjects. Good data for children are also lacking. It is estimated, based on limited information from studies, that adolescents produce a higher mean phonational range than adults. Contrary to commonly held expectations, preadolescents exhibit the narrowest range, while children progressing through puberty, as well as postadolescents, have the most extensive phonational range. Voice clinicians may expect normal phonational ranges even during the period of voice change.

Age and Pitch

In the Fairbanks' method of determining optimum pitch above, the musical pitch computed for the subject may be designated as the *fundamental frequency*, assuming that the optimum and habitual pitches coincide. Usually, the mean fundamental frequency is assessed by electronic equipment in speech science laboratories. Electronic counters and computers provide quick and accurate mean fundamental frequency measurements.

The fundamental frequency of the voice lowers with growth to adulthood. The descent of the fundamental frequency accompanies the descent of the larynx in the neck between birth and puberty. Table 25-II shows a comparison of the mean fundamental frequency for several age ranges. It should be noted that differing methodologies among the investigators were used, so the interpretation of results must be treated accordingly.

Table 25-II

MEAN FUNDAMENTAL FREQUENCY (Hz)
AS A FUNCTION OF AGE AND SEX

Age in Years	*N*	*Sex*	*Mean Fundamental Frequency*	*Investigators*
7	15	M	294	Fairbanks, Wiley, Lassman (1949)
7	15	F	281	Fairbanks, Herbert, Hammond (1949)
8	15	M	297	Fairbanks, Wiley, Lassman (1949)
8	15	F	288	Fairbanks, Herbert, Hammond (1949)
10	6	M	263.5	Curry (1940)
11	6	F	266	Duffy (1970)
14	6	M	232.7	Curry (1940)
15	6	F	237	Duffy (1970)
17	193	F	211.5	Hollien & Paul (1969)
18	6	M	115.9	Hollien, Malcik, Hollien (1965)
18	6	M	133.1	Curry (1940)
20 – 29	175	M	120	Hollien & Shipp (1972)
30 – 39	175	M	112	Hollien & Shipp (1972)
40 – 49	175	M	107	Hollien & Shipp (1972)
50 – 59	175	M	118	Hollien & Shipp (1972)
60 – 69	175	M	112	Hollien & Shipp (1972)
70 – 79	175	M	132	Hollien & Shipp (1972)
80 – 89	175	M	146	Hollien & Shipp (1972)

SOURCE: Aronson, A.E.: *Clinical Voice Disorders* (New York, Thieme-Stratton, Inc., 1980).

Habitual Pitch

Sometimes, the average optimal and average habitual pitch ranges are several notes apart and may indicate abuse of the vocal mechanism. However, comparing optimal and habitual pitches for individuals with active vocal fold pathologies may yield no useful clinical information, since abnormal changes in mass, elasticity, and contour of the vocal fold edges may prohibit the patient from producing good approximation of an optimal pitch.

BIBLIOGRAPHY

Aronson, A.E.: *Clinical Voice Disorders*. New York, Thieme-Stratton, 1980.

Boone, D.R.: *The Voice and Voice Therapy*, 2nd ed. Englewood Cliffs, NJ: P-H, 1977.

Cooper, M., and Cooper, M.H. (Eds.): *Approaches to Vocal Rehabilitation*, Springfield, IL: Thomas, 1977.

Fairbanks, G.: *Voice and Articulation Drillbook*. New York, Har-Row, 1960.

Moncur, J.P., and Brackett, I.P.: *Modifying Vocal Behavior*. New York, Har-Row, 1974.

Chapter 26

EXAMINATION OF THE CRANIAL NERVES

GENERAL CONSIDERATIONS

NEUROLOGICAL assessment, as presented in this chapter, provides clinical implications followed by assessment for each of the cranial nerves. Where the examination of the integrity of a portion of the neurological system involves procedures beyond the scope of the speech clinician, the technique given is designed to acquaint the student with the neurologist's objective and to afford insight into *clinical neurophysiology*. Speech clinicians will find a number of useful procedures outlined that may serve as neurological screening techniques to understand better the nature of communication impairments. From the profile of the patient's assets and impairments, appropriate management strategies can be drawn and implemented.

The *cranial nerves* are not just peripheral structures. It will be recalled that their nuclei lie principally in the brain stem and are tied to the major tracts and systems. Therefore, proper evaluation of the cranial nerves will of necessity require consideration of the central nervous system components affecting cranial nerve function.

From Chapter 16, we learned that the cranial nerves supply important information from outlying sensory receptors to brain stem neurons. They are also responsible for bringing about the necessary changes in the peripheral *effector apparatus*. This apparatus includes glands, smooth (involuntary) and striated (voluntary) muscles, or inhibition to peripheral sensory components.

A number of the cranial nerves were earlier described as essential to receptive or expressive language. These nerves will command most of our attention in presenting a clinical anatomy approach to the cranial nerves. The reader is encouraged to refer frequently to Chapters 15 and 16 while progressing through the present chapter for a refresher review of the cranial nerves and their nuclei.

PRINCIPLES OF NEUROMUSCULAR FUNCTION

Only with normal sensory and motor systems can we expect speech production to be on target. The sensory system provides instantaneous feedback to correct errors and to guide in the execution of fine motor coordination during speech articulation. The major sensory channels for receptive language are also a function of the cranial nerves.

The proper production of speech depends on a high precision of *timing* and *strength* of muscle contraction, the **first** principle of neuromuscular function. Consider the stutterer whose initiation of speech is painfully effortful due to tonic blocks. The victim's speech musculature appears frozen, overcontracted. By contrast, a flaccid paralysis will result in the slurred speech associated with weak muscles. If the defect focuses on an impairment of nerve impulse conduction at the level of the muscle fibers, repetitive use of the speech muscles will result in fatigue. As would be expected, hypertonic or hypotonic muscles, whatever the etiology, cannot be regulated for the great precision of timing needed for competent speech functions.

A **second** principle of neuromuscular function underlying normal motor performance is *range of movement*. Clinicians may be miscued if the patient is able to execute single or repetitive movements. A diadochokinetic task may reveal a limited range of excursion or movement. On occasion, the range may vary unpredictably. Patients with ataxia or parkinsonism, for example, display impairment in range of movement.

To generate normal speech, a series of rapid, discrete, and unsustained muscle contractions are essential. These *phasic* movements account for most of the delicacy and accuracy needed for good speech production. Thus, observation of *speed of movement* is the **third** principle in motor speech. This principle has greatest influence on the prosody of speech when the impairment is not severe.

The **fourth** feature of motor speech concerns the *accuracy of movement*. It is a composite of a muscle's timing, strength, range, and speed. More difficult speech sounds are associated with greater frequency of misarticulations, for example. There may also be signs of inappropriate pitch and loudness fluctuation due to inaccurate control of phonation in coordination with expiration.

In Chapters 16 and 22, *muscle tone* was described as a gentle, steady-state contraction. For normal tone, healthy muscles at rest exhibit mild resistance to displacement or passive stretching. Variations in tone may affect the phonatory apparatus more than other parts of the speech mechanism. Tremors — a breakdown in muscle steadiness — interfere with normal tone. If not severe, however, tremors generally do not interfere with articulation, resonance, or the prosody of speech.

CRANIAL NERVE I — THE OLFACTORY NERVE

The olfactory nerve is actually a tract structurally associated with the brain (see Fig. 15-42). It is more of a process of the central nervous system than a cranial nerve, as the nerve does not actually leave the cranial cavity. In humans, the olfactory system plays a relatively unimportant part in the life of the individual. If rendered *anosmic* (loss of the sense of smell) as a result of disease or injury, there is little handicap. Most individuals depend only slightly on this cranial nerve for recognition of potential danger, for the sorting out of food, for determining our proper territory, and for searching out a suitable companion. Still, the olfactory system

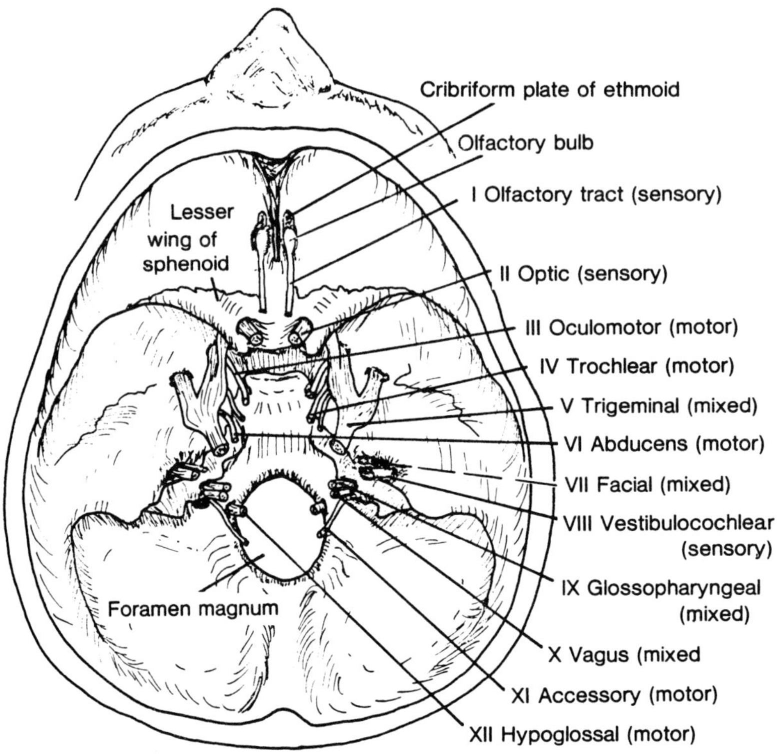

Figure 26-1. Floor of the brain case showing the location of the cranial nerves.

provides the ability to discern certain potentially harmful agents (a gas leak) and the ability to anticipate pleasurably the joys of certain foods and companions and to reject others.

Clinical Assessment

Ordinarily, speech and hearing clinicians will not test the olfactory nerve. Neurologists will first examine the nasal passages for possible obstructions and then assess the integrity of this nerve by asking the patient to identify familiar odors with eyes closed. Odors such as cloves, coffee, and oil of wintergreen are used. Each nostril is tested separately.

CRANIAL NERVE II — THE OPTIC NERVE

Clinical Implications

Like the olfactory nerve, the optic may stretch the classification of a separate cranial nerve, since it appears to be structurally more similar to a tract than a nerve. The optic nerve is covered by the meninges.

The center of the retina, the fovea, is the area of most vision. Fibers from the fovea project to the tip of the occipital lobe. A strong blow to the back of the head may cause the patient to experience bilateral central visual field defects if both occipital poles are damaged. Optic nerve fibers temporal to the fovea connect with the brain ipsilaterally (same side). Fibers nasal to the fovea cross over to the oppo-

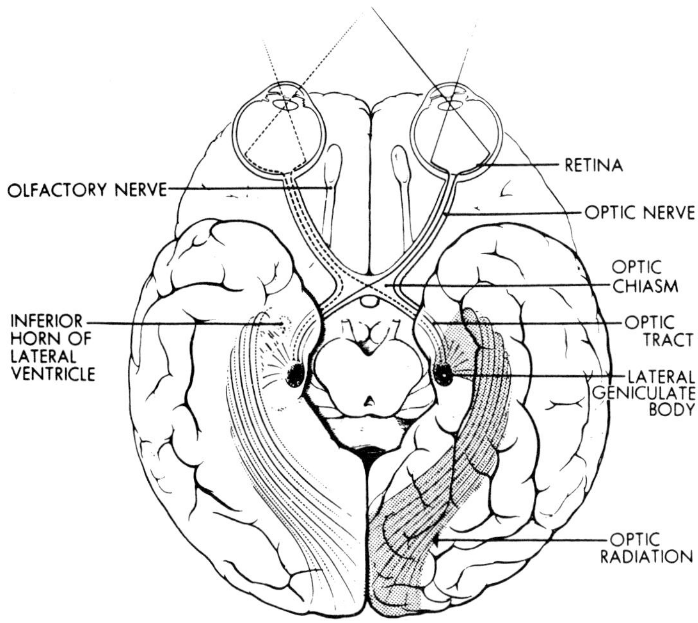

Figure 26-2. Cranial nerve II and the optic tract. Notice the relation between the olfactory nerve (I) and the optic. (From *Man, Microbes and Matter* by B.C. Block. Copyright© 1975. Used with the permission of McGraw-Hill Book Company.)

site side (contralateral) at the optic chiasma. A lesion of the optic tract on the right, for example, results in loss of the left visual field in *each* eye. Patients with a *central scotoma* (visual field defect) will see a doctor sooner than one with a gradually constricting peripheral visual field, because the central impairment causes a profound loss of visual acuity, owing to fovea involvement (Fig. 26-2).

Patients with suspected optic chiasma lesions should be tested for olfactory nerve involvement, since the olfactory tracts pass close to the optic chiasma. A tumor pressing upon the optic tract, for example, might be discovered, since it would be causing a deterioration in olfactory nerve function.

Clinical Assessment

Testing of optic nerve integrity will not usually be attempted by the speech and hearing clinician. Tests include the measuring of visual acuity using Snellen charts, perimetry assessments of visual fields, color blindness tests, and funduscopic examination. A visual screening machine (such as that produced by *Tracor*), tests for visual acuity, phoria (any abnormal convergence of the visual axis), and hyperopia (a refractive error in which the focus of the image falls behind the retina, i.e. farsightedness), automatically changing the test field after each subject response.

CRANIAL NERVES III, IV, VI — OCULOMOTOR, TROCHLEAR, ABDUCENS

Clinical Implications

Three cranial nerves supply the six pairs of muscles responsible for eye movements. The motor fibers of the oculomotor nerve are of two types: they provide voluntary and involuntary functions. It is the parasympathetic components of cra-

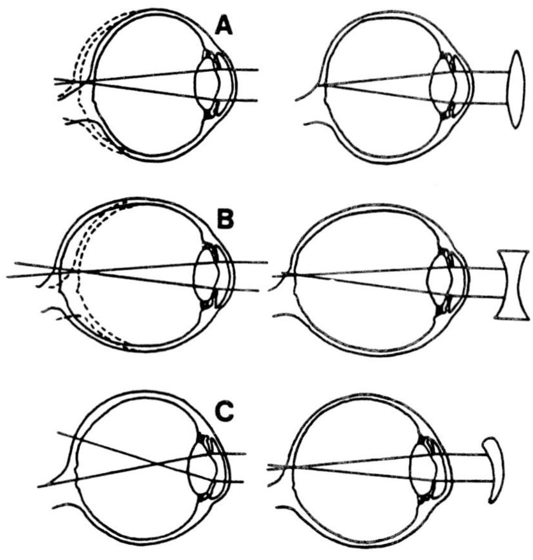

Figure 26-3. (A) Hyperopia. The focus of the image extends beyond the retina (left) until corrective lens are added (right). (B) Myopia. The focus of the image is in front of the retina (left) with corrective lens added (right). (C) Astigmatism. Uneven focusing of the image resulting from distortion of the curvature of the lens (left) with corrective lens added (right).

nial nerve III that innervate the smooth (involuntary) muscles of the eye to cause *accommodation* of the eye, i.e. focusing and pupillary constriction. Accommodation involves a neural circuit first via the optic nerve to the visual cortex so that it can tell the eye on what to focus. Then, assuming the eyeball is pointed in the proper direction, the involuntary fibers of the oculomotor nerve stimulate the muscles that reshape the lens to focus on the appropriate object (Fig. 26-3).

In a total lesion of one of the third cranial nerves, two prominent symptoms are evident. One is *ptosis* resulting from paralysis of the superior levator palpebrae muscle of the involved eye (see Fig. 23-10). When the examiner raises the drooping eyelid, the pupil will be dilated due to paralysis of its constrictor muscle fibers. Also, the affected eye will show deviation outward and slightly downward through action of the other two ocular muscles. If only the superior levator palpebrae is impaired, tarsorrhaphy is performed (operation to keep eyelid open).

If there is damage to the left motor area of the cerebral cortex visuomotor area (area 8, see Fig. 15-31), the patient cannot look to the right. The eyes tend to deviate to the left as the patient "looks at the lesion." This pattern occurs because the pathway from the left hemisphere innervates the right lateral rectus muscle (right cranial nerve VI) and left medial rectus muscle (left cranial nerve III). Speech clinicians will readily recognize such *conjugate gaze* problems in patients, although they are not expected to differentiate the site of the lesion. This is the neurologist's domain (Fig. 26-4).

Clinical Assessment

Ocular movements are tested by having the patient follow the examiner's finger in all directions of gaze. If the oculomotor nerve is damaged, the affected eye will not be able to move in a superior, inferior, or medial direction. Also, dilation of the pupil and ptosis (drooping) of the upper lid will be present. If the trochlear nerve is damaged, the affected eye will be unable to move inferiorly and laterally. If the abducens nerve is involved, the affected eye will not be able to move laterally.

Other tests include pupillary light reflex, accommodation, and *consensual light reflex* (shining a light into one eye and noting the change in the opposite pupil).

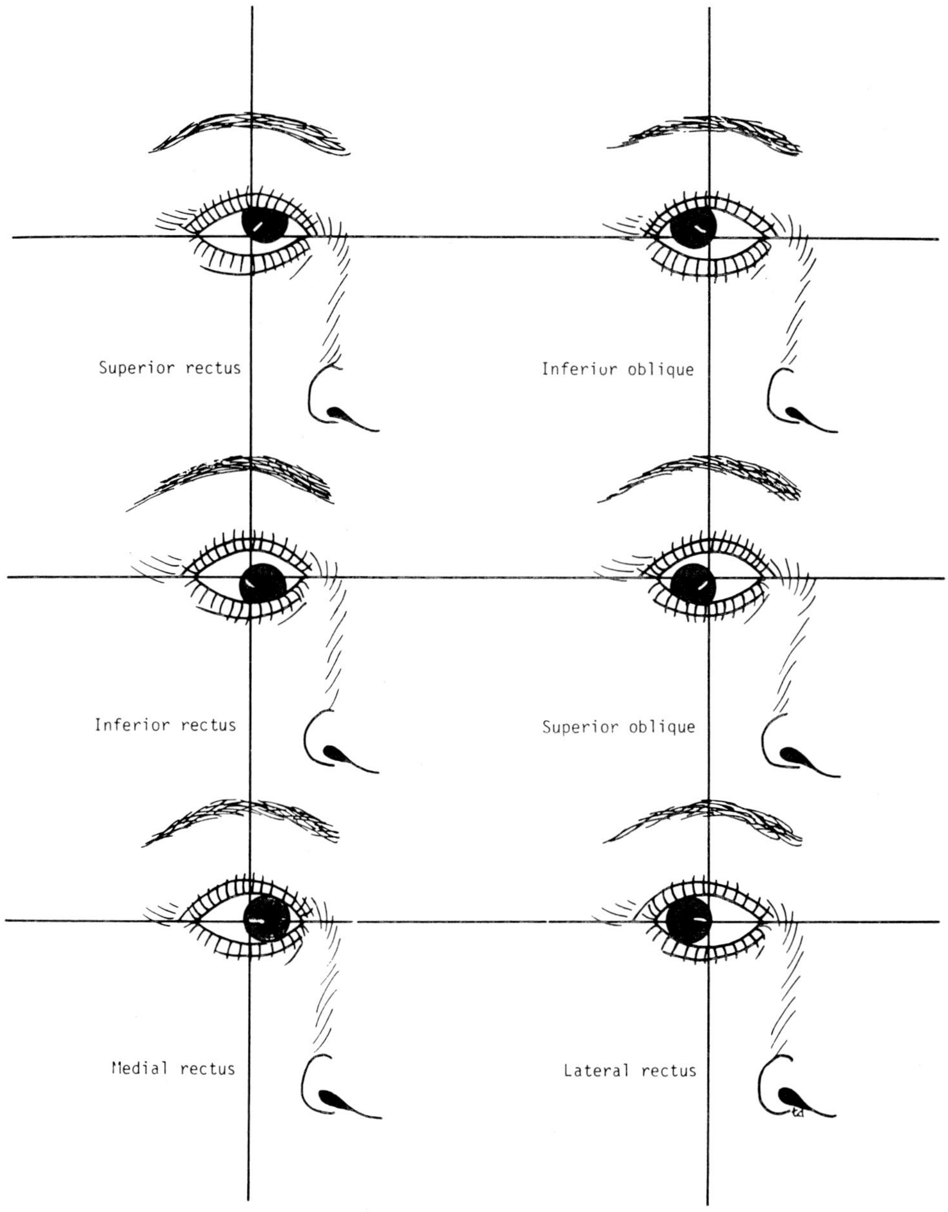

Figure 26-4. Actions of four recti and two oblique muscles of right orbit. Position of pupil in relation to vertical and horizontal planes should be noted in each one. (From Snell, R.S.: *Clinical Anatomy for Medical Students*, 2nd ed., 1981. Courtesy of Little, Brown and Co., Boston.)

CRANIAL NERVE V — TRIGEMINAL

Clinical Implications

SENSORY COMPONENTS (or Division). The student should bear in mind that the trigeminal nerve has a remarkable number of different modalities of sensation that are carried from the periphery to the central nervous system over but a single nerve and relayed to different sites in the brain. This fifth cranial nerve is known as trigeminal because it has three major peripheral branches: ophthalmic, maxillary, and mandibular. Respectively, they leave the cranium through the *superior orbital fissure,* the *foramen rotundum,* and the *foramen ovale*.

All three branches of the trigeminal contain sensory fibers, and each branch serves an area of the face with less overlap than is found among the spinal dermatomes of the trunk. Sensory fibers from the ophthalmic division innervate the upper area of the face and scalp. Division-two fibers, the maxillary nerve, serve the midface. Division three, the mandibular nerve, serves the chin and parts of the cheek and temporal area (Fig. 26-5).

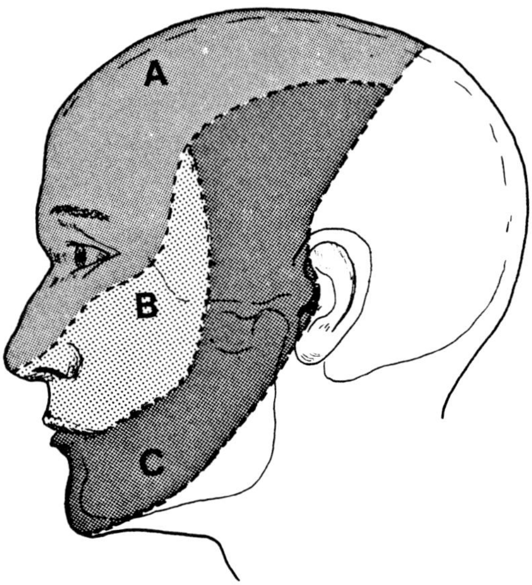

Figure 26-5. Distribution of sensory fibers by the three branches of the trigeminal nerve: (A) ophthalmic division, (B) maxillary division, (C) mandibular division.

Three physiological classes of receptors have been identified in the regions of the face described above: (1) mechanoreceptors for sensitivity to light touch, and pressure, (2) thermoreceptors for detecting changes in skin temperature, and (3) nociceptors (pain receptors), which are sensitive to stimuli that damage the skin.

Within the mouth, the trigeminal nerve provides sensory innervation to most of the oral mucosa over the *hard palate* to the anterior two-thirds of the *tongue*. Sensation from these surfaces enables *tactile* information to be fed back to the brain during speech articulation. Taste sensations from the anterior two-thirds of the tongue are carried by the facial nerve, not the trigeminal.

The *corneal blink reflex* is a function of sensory division of the trigeminal and the facial nerve. A wisp of cotton applied to the cornea causes afferent impulses to travel the ophthalmic branch of the trigeminal to its nucleus in the medulla. Through internuclear fibers, the motor nucleus of the facial nerve is activated bilaterally to bring about contraction of the orbicularis oculi muscles.

Other pertinent regions innervated by the Vth nerve are the opening to the *eustachian tube*, the mucous membrane within the cheek, the *external auditory meatus*, and part of the *tympanic membrane*. Toothaches and pain in response to dental procedures are transmitted by this nerve. At the alveolus, the soft tissue, *gingiva* and *periodontal membrane*, as well as *tooth pulp*, are innervated by trigeminal sensory fibers.

Of particular relevance to speech clinicians is the sensory innervation by the trigeminal to the muscles of *mastication* (see Figs. 13-13–13-15). By means of their proprioception (position sense) from stretch receptors in the jaw muscles, feedback is provided to one of the trigeminal nuclei located in the midbrain. Collateral fibers between this nucleus and the motor nucleus of the trigeminal nerve establish a monosynaptic reflex arc to the motor neurons — the jaw-jerk reflex.

MOTOR COMPONENTS (or Division). Through the mandibular nerve of the trigeminal, the masseter, temporal, and two pterygoid muscles receive motor innervation. Additionally, the anterior belly of the digastric muscle, the mylohyoid, the tensor tympani, and the tensor palatini are supplied by axons from the motor nucleus of the trigeminal.

Peripheral trigeminal motor paralysis is noted from disturbances of masticatory function. During opening, the jaw *deviates toward the side of the paralyzed muscles* because of the contralateral, normal functioning pterygoid muscles. In pseudobulbar palsy (see Chap. 22), the corticobulbar fibers on both sides are interrupted by areas of softening of the brain tissue, usually from cerebral vascular disease. With this type of upper motor neuron disease, *volitional movements are impaired*, but the jaw-jerk reflex is usually increased. This reflex, normally absent in health, is tested by tapping the chin gently as the patient holds the lower jaw slightly open. Normally, a reflexive closure occurs, but this reflex is absent in bilateral, lower motor neuron disease, such as bulbar palsy, because the motor neurons in the nuclei are diseased. Other symptoms of lower motor neuron disease include (1) paralysis of the muscles of mastication for all movements, (2) a lower jaw that hangs open, and (3) atrophy (wasting away) of the muscles.

Clinical Assessment

The sensory and motor divisions of the trigeminal nerve are tested separately. In those occasional instances where a clinician may wish to assess sensory integrity of the trigeminal, the key observation is symmetry of sensations on both sides of the face. Wisps of cotton, pinpricks, and cold and warm objects applied to the forehead, cheeks, and jaws indicate sensitivity to light touch, pain, and temperature, respectively. In all likelihood, the speech and hearing clinician will not actually test trigeminal reflexes, which include corneal blink, jaw-jerk, and sneezing. Clinicians should appreciate the functional divisions of this nerve, however.

The motor division of the trigeminal is assessed by palpation of muscles during opening and tight closure of the jaws. The temporal and masseter muscles are evaluated by external palpation. The anterior and posterior heads of the temporalis can be palpated in front of and directly above the ear helix. A portion of the pterygoid muscles can be palpated by placing a finger into the maxillary oral vestibule on each side (in separate examinations) and applying pressure in a superior direction just beyond the maxillary tuberosity, and again more posteriorward. Fibers of the medial and lateral pterygoids can be identified with this approach.

Deviation of the jaw to one side when the mouth is opened is a sign of motor damage to the trigeminal. Pain or sensitivity in the muscles of mastication or temporomandibular joint may signal a problem of dental origin. Such patient complaints merit referral to a dentist for evaluation.

The mylohyoid and anterior belly of the digastric can be evaluated generally by applying external resistance of the fingers under the mandible as the individual lowers the jaw against this resistance. The platysma muscle is also being assessed in such activity, and this is appropriate since the platysma plays a role in depressing the mandible.

The tensor palatini and tensor tympani are impractical to assess, and no specific test is indicated since there are no reliable means of evaluating functional responses to testing stimuli.

CRANIAL NERVE VII—FACIAL

Clinical Implications

Unfortunately, disturbances of the *seventh nerve* in its nucleus or along its course in the bony canals are relatively common. Tumors or vascular impairment may damage the facial nuclei, which are located in the posterior aspect of the pons close to the fourth ventricle, or damage the fibers. Neighboring nuclei and fibers may also be affected. The classic clinical symptom is facial paralysis with other neurologic signs if there is involvement of adjacent cranial nerves.

Bells's palsy is the most common disorder of the facial nerve. The site of injury is often along the nerve as it travels the facial canal. Possibly caused by some local swelling, the patient will awaken to discover one side of the face paralyzed. The face is usually asymmetrical, the nasolabial fold is flat, the cheeks sag, and the corner of the mouth frequently droops. The eye on the affected side cannot be closed voluntarily or reflexively, as in the corneal blink reaction. The patient cannot distend the cheek or purse the lips tightly. Saliva may drool from the paralyzed side of the mouth in some patients. Return of function is gradual. In fact, in a majority of the cases of Bell's palsy, the facial muscles regain their function (Fig. 26-6).

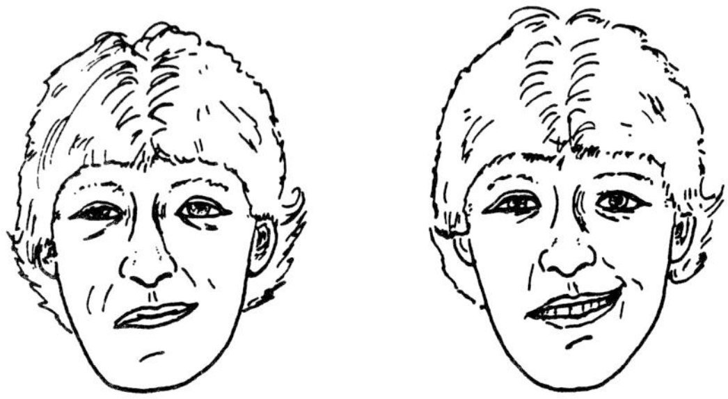

Figure 26-6. Right unilateral facial weakness. The affected side droops at rest (left). During smiling (right), the normal side retracts and rises.

Since the cranial nerve accompanies the vestibulocochlear nerve (VIII) as they travel into the temporal bone, tumors or other insults to the VIIIth nerve may affect the facial nerve. Even middle-ear infections are hazardous to the facial nerve, which courses through this region of the temporal bone. Depending on the site, certain peripheral facial disorders may result: changes in lacrimal secretions, salivary secretions, and taste over the anterior two-thirds of the tongue.

Since a branch of the facial nerve innervates the stapedial muscle, protection of the ear from loud sounds may be diminished. The stapedial muscle contracts reflexively to loud sounds and fixates the ossicles to prevent overstimulation of the organ of Corti. The facial motor nucleus receives fibers from the superior olive of the medulla for the completion of the *acousticostapedial reflex arc*. Through this arc the tension of the stapedius is regulated relative to certain sound intensity levels.

Interruptions in the corticobulbar tracts on one side coming to a facial nerve nucleus, which frequently occur in cerebral vascular accidents (CVA) or strokes, result in an obvious loss of movement in muscles over the lower half of the face contralateral to the lesion. This is an *upper motor neuron* or *central lesion* (see also Chap. 22). That portion of the facial nucleus supplying the frontalis, orbicularis oculi, and corrugator (muscles of the upper face) receives corticobulbar fibers from the precentral gyrus of each cortex. If only one corticobulbar tract is damaged, the upper facial muscles are only *partially paralyzed*. The muscles of the lower face suffer more impairment. *This is an important diagnostic point*. They are innervated by a dorsally located cell group in the facial nucleus that is connected not with a direct tract (corticobulbar) but with a wandering cortical tract. These aberrant corticobulbar fibers follow an indirect route and give the impression of not having dual origins in the precentral gyrus.

When disease destroys the facial nucleus or interrupts the complete facial nerve in its peripheral course, the entire half of the face is usually paralyzed. Such a peripheral facial paralysis would be indicative of a *lower motor neuron type*. With corticofacial paralysis, this would be an upper motor neuron type and indicative of central facial paralysis.

Interestingly, in one type of central facial paralysis, damage to corticobulbar fibers precludes voluntary contraction of the lower facial muscles; however, the patient may have good function in spontaneous smiling or laughing. The explanation for this functional disparity is that movements pertaining to emotional reaction may be initiated through the basal ganglia and diencephalic centers and their projections to the facial nerve nuclei. These fibers are not usually damaged with the corticobulbar tracts. Disturbances in the extrapyramidal centers that feed the facial nerve nuclei produce another type of central facial paralysis. It is the *reverse* of the first type, i.e. there is a loss of emotional movement with unimpaired voluntary movement.

Central connections of the facial nuclei with the oculomotor (III) nucleus provide interconnections to correlate eyelid and eye movement. The seventh cranial nerve *closes* the eye, whereas the third *opens* the eye. This is a vital clinical point in differential diagnosis.

The facial motor nucleus also cooperates with the trigeminal (V) nucleus. Fibers from the sensory trigeminal nuclei project to the facial motor nucleus to complete the central arcs for the corneal blink reflex (see under "Trigeminal Nerve" presented earlier). Tectobulbar (*tectum* = roof area of mesencephalon comprising the colliculi) fibers also end in the facial nucleus for completion of other visuofacial and auditory reflexes. Blinking of the eyes in response to a sudden bright light is an example of this reflex.

Clinical Assessment

Testing the facial nerve by the speech clinician involves assessment of motor status. This involves, simply, evaluations at the orbicular sphincters — oral and ocular. The patient is asked to exaggerate productions of /i/ and /u/. The examiner can attempt to pull the corners of the lips forward on /i/ and back on /u/. This is done by placing the fingers on the cheeks. Symmetry of response on the sides of the face is noted during these tasks and during smiling.

The upper face is evaluated by having the patient close the eyes tightly while the examiner attempts to open the lids by finger pressure on the lids in a superior direction. Normally, muscle pulsations are found in resistance to the pressure applied by the examiner, but these pulsations should be bilaterally symmetrical. The symmetry of the face is also evaluated at rest and during voluntary facial movements.

The sensory status of the facial nerve is

assessed by having the patient protrude the tongue and identify the taste of sugar (sweet), citric acid (sour), salt (salty), or quinine (bitter) applied to the anterior portion of the tongue.

Reflexes of the face, cornea, conjunctiva, and eyelid are also examined in a complete test of the facial nerve.

CRANIAL NERVE VIII—VESTIBULOCOCHLEAR

Clinical Implications

The *eighth cranial nerve* and its central connections were described in Chapter 16. Since there are two distinct divisions of this nerve, each will be examined separately.

COCHLEAR DIVISION. Disorders arising from impairment of the hearing apparatus, namely, the organ of Corti, usually have a *gradual* onset of hearing loss. The high frequencies become progressively restricted; there is comparatively good acuity for middle frequencies; and there is normal hearing for low frequencies. Prolonged noise exposure is often blamed, but there are a number of other etiologies. As acuity for the upper frequencies diminishes, the speech may be noticeably affected by imprecise sibilant production, especially /s/, since these are high-frequency sounds occurring often in conversational speech. The patient hears a distorted acoustic stimulus decreasing his/her listening intelligibility.

Cochlear nerve disease may be charac-

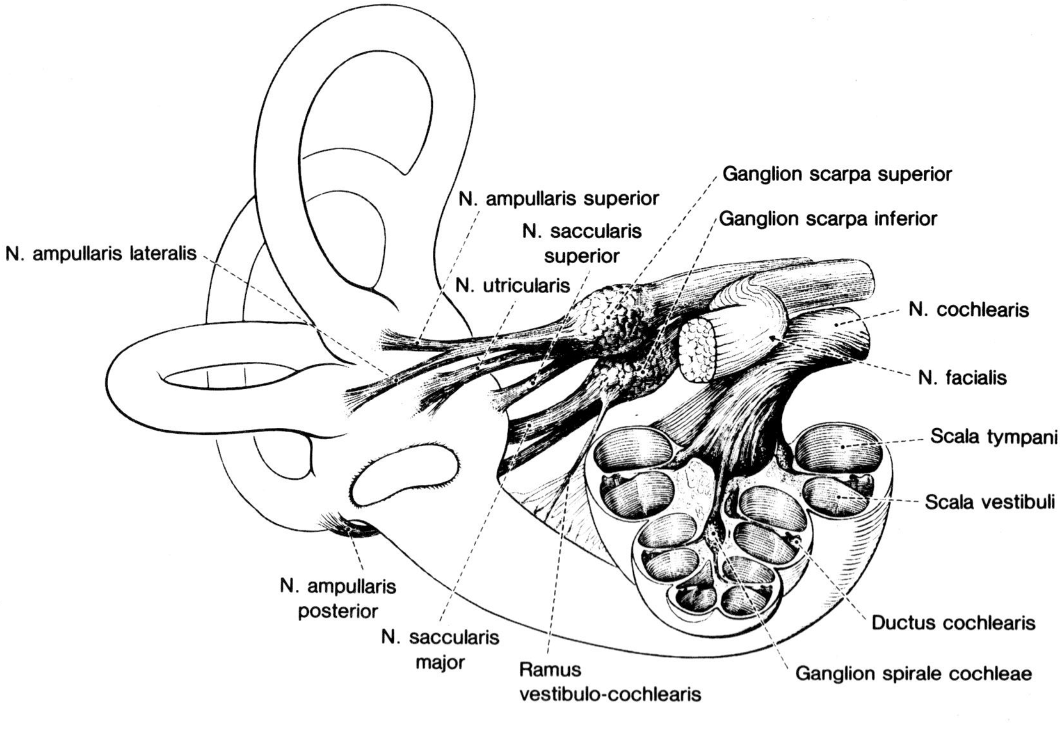

Figure 26-7. Cross section of the coils of the cochlea showing the three fluid-filled spaces. Notice the facial nerve travels along with the two divisions of the vestibulocochlear nerve. (Reproduced from: "Some Pathological Conditions of the Eye, Ear and Throat." Courtesy of Abbott Laboratories.)

terized by more *rapid* onset of hearing loss, restriction in low-frequency hearing, and normal or near-normal perception of upper tones. Whether the site of lesion is in the inner ear or retrocochlear (located behind the cochlear — eighth nerve and beyond), the characteristic hearing loss is termed *sensori-neural*. Conductive loss pertains to a problem in transmitting acoustic energy through the external auditory meatus and across the auditory ossicles of the middle ear. The mechanized vibrations impinging on the organ of Corti in the inner ear are transduced by the hair cells of the inner ear to neural discharges that are sent from the cochlea to the brain stem via the cochlear division of the eighth cranial nerve.

Hearing deficits, then, that are attributable to the inner ear or retrocochlear structures are sensori-neural in nature. A common complaint by patients suffering from sensori-neural loss is *tinnitus*. Characterized by subjective head noises in the absence of known stimulation, tinnitus is usually reported as a high-pitched ringing in the ear. These sounds may be more bothersome to the patient than the hearing impairment.

Otologists (ear specialists) and hearing clinicians are particularly concerned with the site of pathology when a sensori-neural loss is found. Tumors associated with the eighth nerve are not exceedingly rare. They may develop from the Schwann cell sheath, growing slowly to create an insidious hearing loss that takes considerable time to be noticed. Acoustic neurinoma, although a benign tumor of the eighth nerve, is nevertheless life-threatening. A unilateral sensori-neural loss, according to Ginsberg and White (1978), should be considered a potential acoustic neurinoma until proven otherwise.

As eighth nerve tumors enlarge, the vestibular division of the vestibulocochlear nerve is also compressed or stretched. Later, the sensory root of the trigeminal and facial nerves may be involved. While hearing loss may be the earliest symptom noticed in cases of acoustic nerve tumors, other symptoms may overshadow the hearing problem and other cranial nerve involvements. This course of clinical symptomatology occurs when a tumor develops on the nerve nearest the brain stem rather than the cochlea. Patients develop increased cranial pressure, the brain stem is slowly displaced, and the blood system in the area is constricted. There may also be mental symptoms, disturbances of consciousness, epilepsy, and corticospinal motor tract signs (Fig. 26-8).

VESTIBULAR DIVISION. The labyrinth of the vestibular apparatus, in conjunction with retinal and proprioceptive neural impulses, serves to establish and maintain equilibrium. The position of the head in space is known to the brain because of the vestibular system. The system is functional during active and passive movement of the head and also with the head at rest. In response to the need for postural adjustments, muscle-tone alterations are provided by the system.

Disorders in equilibrium result if the labyrinths, retinas, and proprioceptive receptors (the sensory endings in muscles and around joints) do not provide corroborative information to higher neural centers. In essence, one vestibular apparatus could send contrasting information to the brain from its mate, or the eyes may project a strange image concerning equilibrium that counters proprioceptive data. Probably the most common symptom of malfunction of the postural-maintenance system is *vertigo*, the subjective sensation of movement. Although this symptom arises most commonly from the labyrinths, it may also result from cardiac, gastric, or ocular disorders, or several other sources. Vertigo is a hallucination of

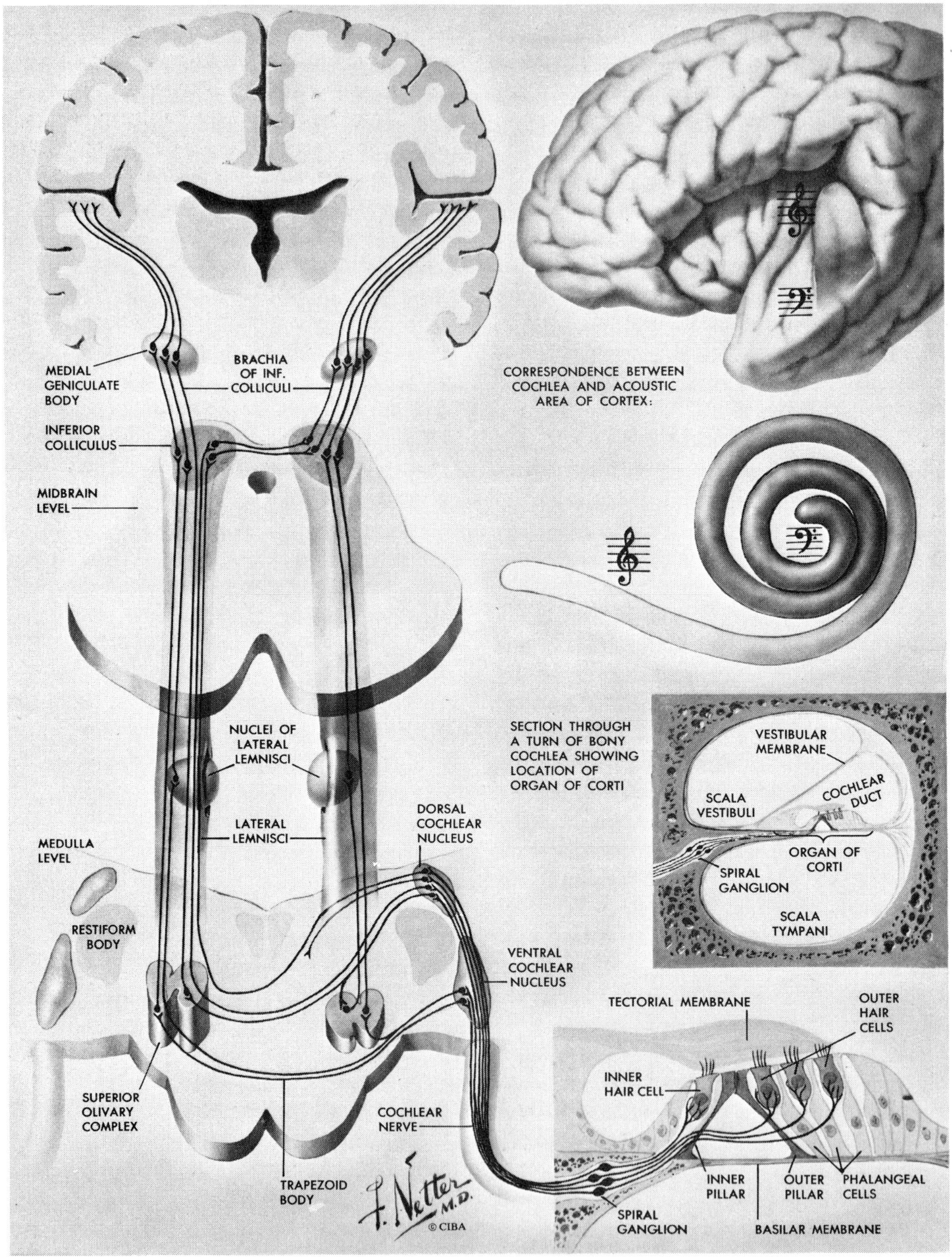

Figure 26-8. The auditory nerve and central auditory pathways. (Copyright© 1953, CIBA Pharmaceutical Company, Division of CIBA-GEIGY Corporation. Reprinted with permission from THE CIBA COLLECTION OF MEDICAL ILLUSTRATIONS. Illustrated by Frank H. Netter, M.D. All rights reserved.)

movement, a sensation as if the external world were revolving around the patient or as if the patient were revolving about in space. Vertigo and dizziness are not synonymous. Patients with labyrinthine pathology usually have accompanying nausea, vomiting, paleness, heart palpitation, and sweating.

Oscillation of the eyes is frequently associated with labyrinthine disorders as well as with a number of central disturbances. If the head, held erect, is turned about a vertical axis, the turning produces a deviation of the eyes in the direction *opposite* to that of the turning. The subject seems to seek to maintain the image of a fixed object on the retinas. Shortly after a certain point of deviation, the eyes are quickly pulled *toward* the direction of turning. These two movements of the eyes, when repeated successively, are known as *nystagmus*. By varying the position of the head and the direction of turning, it is possible to bring specific pairs of semicircular canals into the plane of rotation. When nystagmus is induced through clinical tests (the patient is rotated 10 turns in 20 seconds and suddenly stopped), it usually persists from 14 to 40 seconds in normal subjects. Errors in pointing (past pointing) with outstretched arms, tendency to fall, and vertigo may linger as long as the rhythmic oscillations of the eyeballs do. Exaggerated responses are found in diseased labyrinths.

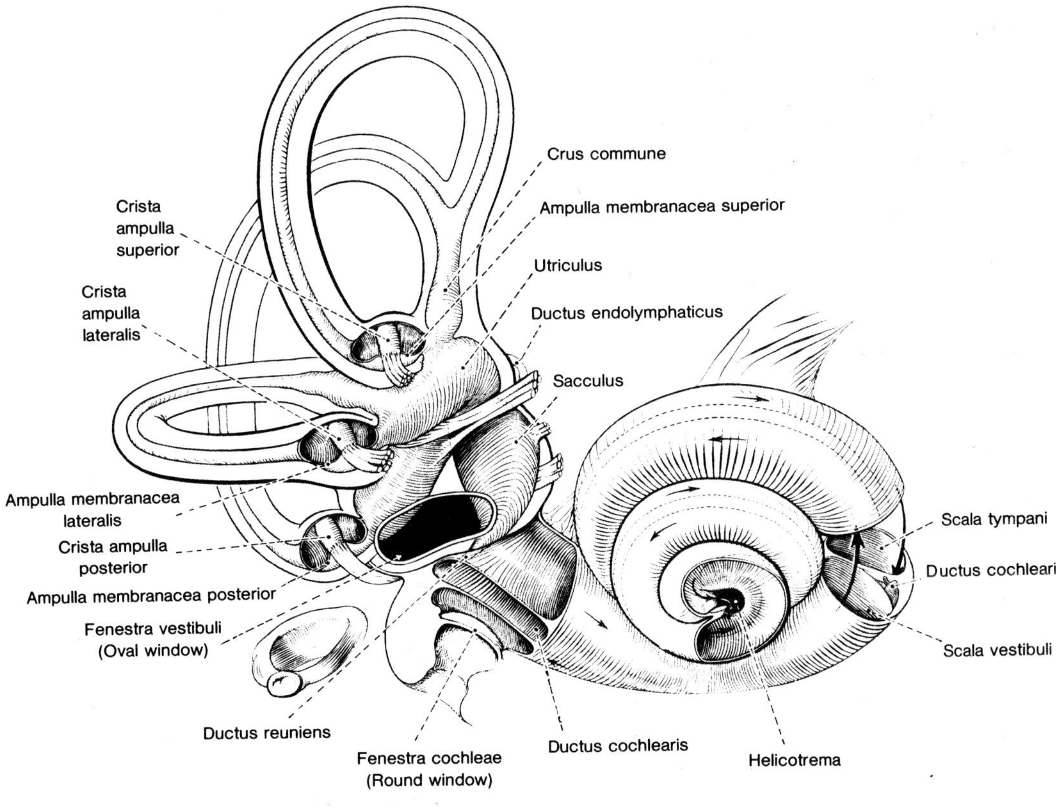

Figure 26-9. The vestibular apparatus. (Reproduced from: "Some Pathological Conditions of the Eye, Ear and Throat." Courtesy of Abbott Laboratories.)

Each of the five divisions of the vestibular component — three semicircular canals, a utricle, and a saccule — send fibers to the vestibular nucleus in the medulla. Neurons in this nucleus provide information to the visual and somatosensory systems. From that portion of the nuclei connecting the semicircular canals, fibers travel to the oculomotor, trochlear, and abducens nuclei. These oculomotor nuclei, as described earlier, innervate the muscle moving the eyeball. The intriguing arrangement of the fibers of semicircular canal receptors and the pattern of connections with the nuclei between the oculomotor nuclei is such that it permits the receptors to influence motion in the plane of the canal in which the receptor resides.

Cold water injected into the ear canal will produce reflex ocular responses in the form of nystagmus. These cold caloric responses are mediated by brain stem nuclei. There is no response in comatose patients. Caloric stimulation creates activation of a single labyrinthine mechanism, while rotation provides bilateral stimulation. The caloric stimulus is introduced into the external auditory meatus and produces a reaction by the meatus, eardrum, and middle ear to the temperature change. The semicircular canal, situated very close to the middle ear, also undergoes a temperature change. The fluid of the semicircular canal will change in density and *rise* or *fall*, depending on whether a warm or cold caloric stimulus was used. This causes fluid movement and nystagmus, which corresponds to testing by head rotation.

Medical clinics with otology-audiology teams utilize electronystagmography (ENG), which includes caloric testing, in evaluating patients with hearing and vestibular complaints when there is a suspected peripheral (vestibular and eighth nerve) impairment or a central neurological disorder.

Bilateral labyrinthine destruction does not produce vertigo or nystagmus but a disturbance of equilibrium. However, with time, this disturbance is compensated to a great extent by the optical and proprioceptive mechanisms.

CRANIAL NERVE IX—GLOSSOPHARYNGEAL

Clinical Implications

Isolating functions of the *ninth cranial nerve* is sometimes impossible since it shares central nuclei with the vagus nerve. However, it is the sole supplier of motor fibers to the stylopharyngeus muscle. With the vagus, the glossopharyngeal sends motor fibers to the pharyngeal constrictors. It is efferent to the parotid salivary gland and afferent for general skin sensation from the auricle. Other sensory fibers (general visceral sensation) come from the posterior third of the tongue, the pharynx, and the middle ear. Taste sensation from the posterior third of the tongue is also the responsibility of the ninth nerve; however, the vagus nerve seems to share this duty. In selected, chronic-pain patients, severing the glossopharyngeal intracranially (to reduce pain) has not resulted in loss of taste or other sensory responses except for pain.

Sectioning (cutting) the ninth nerve intracranially does produce paralysis of the stylopharyngeus, but this produces little disability. There may also be some motor disturbances of the pharyngeal constrictors on the side of the damaged nerve, but due to overlap of innervation by the vagus, the impairment may not be evident. When both the ninth and tenth nerves are together impaired, with respect to those fibers each sends to the posterior third of the tongue, the deficit is pronounced. The same degree of deficit is found if both ninth nerves are at fault. Possible disturbances inlcude loss of taste over the posterior aspect of the tongue, anesthesia (loss of sensation) over the posterior pharynx and posterior one-third of the tongue, and loss of the gag response (Fig. 26-10).

Damage to the glossopharyngeal is sustained with corticobulbar tract lesions, but the damage is certain to be associated with injury to similar fibers of other close-by cranial nerve motor nuclei. *Pseudobulbar palsy*, an upper motor neuron disorder, is a resulting clinical condition, described earlier, which will be alluded to later in this chapter.

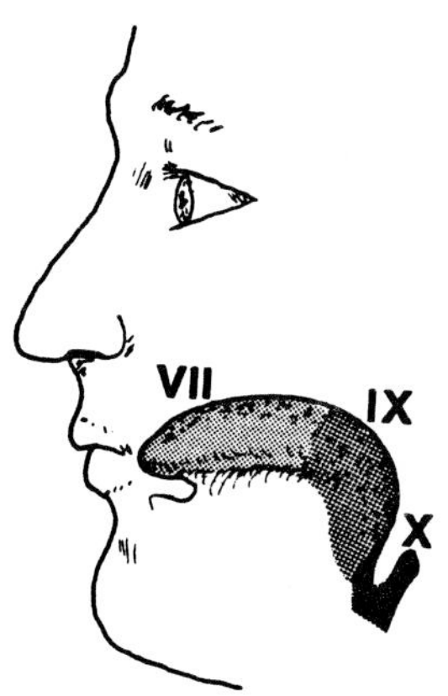

Figure 26-10. Cranial nerves innervating areas of the tongue associated with taste: (VII) facial, anterior two-thirds; (IX) glossopharyngeal, posterior one-third; (X) vagus, area adjacent the epiglottis.

Clinical Assessment

When testing for integrity of the glossopharyngeal, the examiner must remember that the failure to elicit a gag response is not conclusive of a pathologic ninth nerve, since gagging is not a true reflex. Some patients have become conditioned not to respond to a gag stimulus or have never developed any appreciable gag response. Cleft palate patients as a group have reduced gag responses and may tolerate well the introduction of a tongue blade for manipulation of the posterior wall of the pharynx and the faucial pillars without initiating a gag. Other diagnostic clues such as loss of touch, sensation, and taste are important when studying the glossopharyngeal nerve. The examiner will wish to be alert for clinical symptoms relating to other cranial nerves, since there is overlap among some of the cranial nerves.

CRANIAL NERVE X—VAGUS

Clinical Considerations

The *tenth cranial nerve* has the most components and innervates more individual structures than any single nerve in the body. It supplies the major parasympathetic innervation of the autonomic system and is vital for speech production.

The major signs and symptoms that may be observed in patients with disorders affecting the vagus nerve fall into three groups. First, the motor disorders include impairments in phonation, swallowing, and velopharyngeal valving. Second, the sensory disturbances include laryngeal and lower pharyngeal pain or paresthesia, cough and respiratory irregularities, and salivation. A third group includes cardiac and gastric dysfunctions. These symptoms should be attributed to both the vagus and glossopharyngeal, since the two nerves are generally involved jointly in most of the disorders listed.

The vagus begins with nuclei deep within the reticular formation of the medulla. Actually, the nucleus ambiguus is an elongated aggregation of neurons serving the glossopharyngeal, vagus, and accessory cranial nerves. The vagus exits the skull through the jugular foramen. At least eight branches are recognized, but only those closely associated with aspects of speech production will be described here (Fig. 26-11).

The pharyngeal branch contains sensory and motor fibers. Upon joining branches of the glossopharyngeal, the fibers form the *pharyngeal plexus*. From this plexus, fibers project to the pharyngeal constrictors. Vagus fibers alone innervate the glossopalatinus, salpingopharyngeus from the pharyngeal plexus, uvulus, and levator palatini. (The other muscle of the soft palate — the tensor palatini — is served by the trigeminal.)

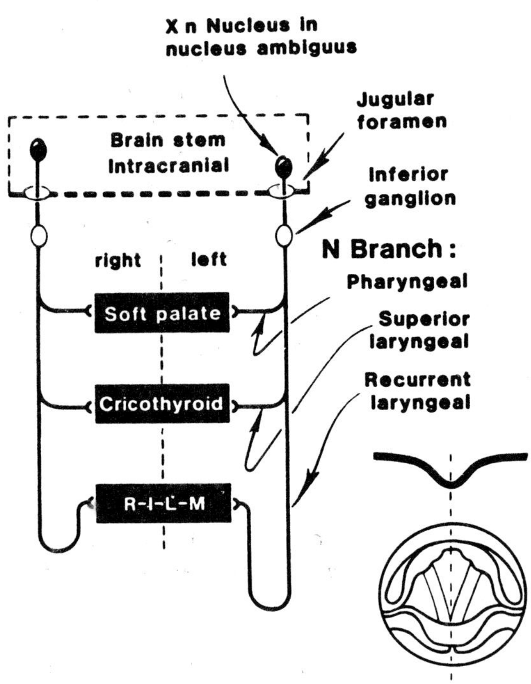

Figure 26-11. Diagrammatic illustration of the pharyngeal and laryngeal branches of the vagus. At the lower right, the soft palate and larynx are shown. Notice the differences in lengths of the left-right recurrent laryngeal branches. (From Aronson: *Clinical Voice Disorders*. New York, Thieme-Stratton, Inc., 1980.)

When the pharyngeal branch of the vagus is interrupted, the soft palate fails to elevate on the involved side. In bilateral paralysis, both sides of the palate hang low and there is decrease or loss of mobility. There is hypernasality and nasal emission. The patient may be unable to cough, has difficulty swallowing, experiences nasal regurgitation of food, and has pharyngeal paralysis (Fig. 26-12).

A superior laryngeal nerve branch, also carrying sensor and motor fibers, travels from the inferior ganglion of the vagus, located near the base of the skull, divides and reaches the larynx. An external branch innervates the cricothyroid muscle, the *tensor* of the vocal folds, and the inferior pharyngeal constrictor. An internal branch enters the larynx through the thyrohyoid membrane to provide sensory fibers from (1) the mucous membrane of the larynx in the upper laryngeal cavity, (2) the muscle spindles and muscle stretch receptors in the larynx, and (3) the mucous membrane of the aryepiglottic folds and the arytenoid cartilages. This branch may provide important data during clinical examinations of laryngeal dysfunction (Fig. 26-13).

Paralysis of the superior laryngeal nerve is uncommon. Only the cricothyroid muscle and sensation to the upper part of the larynx are lost if this branch of the vagus is interrupted. However, if there is unilateral paralysis of the vocal folds with anesthesia on the same side in the laryngeal vestibule and accompanying cricothyroid paralysis, then it may be surmised that the site of pathology is cranialward from the inferior ganglion of the vagus or is a central lesion.

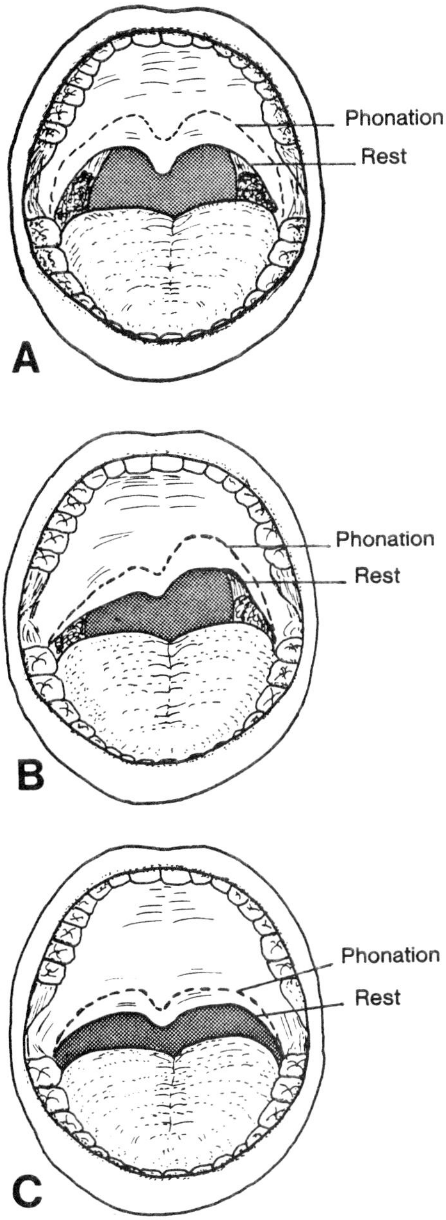

Figure 26-12. Positions of the soft palate, normal and abnormal. (A) Normal appearance of the soft palate at rest and during phonation (*dotted line*). (B) Right unilateral soft palate paralysis showing a lowered right side at rest and during phonation (*dotted line*). (C) Bilateral soft palate paralysis at rest and during phonation (*dotted line*). (After Darley, Aronson, and Brown: *Motor Speech Disorders*, 1975. W.B. Saunders.)

The recurrent laryngeal nerve, as described in Chapter 16, has an unusual course (Fig. 26-13). It branches from the trunk of the vagus at a point below the larynx and ascends to enter the larynx inferiorly. It is not symmetrical, since the left recurrent is longer. Together, these recurrent nerves supply all of the intrinsic muscles except the cricothyroid. Like the other branches of the vagus above, sensory fibers are found in the recurrent laryngeal nerve. They carry sensory fibers from the mucous membrane below the vocal folds and from muscle receptors of the intrinsic muscles (Aronson, 1980).

Laryngoscopic examination of a patient with unilateral recurrent laryngeal nerve paralysis will show the impaired fold to be in a *paramedian position*. As the patient phonates for the examiner, the normal fold moves across the midline to achieve adduction. It will appear shorter than the healthy fold. A noticeable disturbance in the quality of the vocal fold sound may be evident, characterized by a breathy hoarseness. If the patient has bilateral paralysis, the examiner will be unable to view the full length of the folds due to excessive overhang of the epiglottis; the folds will also appear bowed. Phonatory quality is breathy-hoarse with reduced loudness and pitch range (Fig. 26-14 & 15).

Peripheral disturbances of the vagus nerve may be difficult to recognize unless they are bilateral *or* affect speech and respiration. This seems to hold true also for upper motor neuron (central) pathology. While a *unilateral* disturbance of the vagus will be manifest by moderate dysphonia, hypernasality, nasal regurgitation, and a swallowing disorder, among other things, a *bilateral* lesion of the vagus may result in grave impairment of phonation, deglutition, and respiration. Bilateral lesions can cause death due to asphyxia unless a tracheostomy is performed promptly. In complete laryngeal paralysis, the vocal

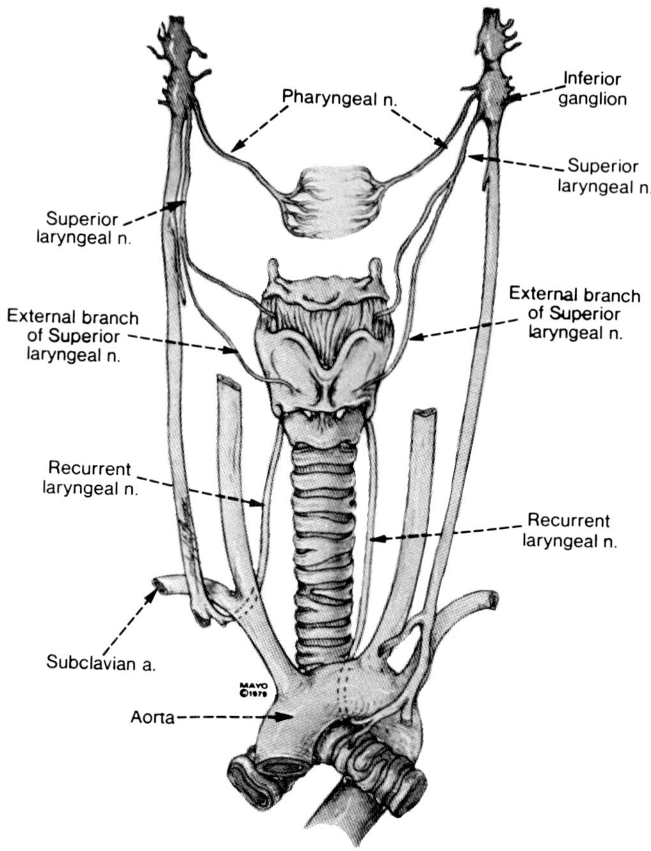

Figure 26-13. Course of the vagus nerve branches innervating the soft palate (pharyngeal), the cricothyroid muscle and mucous membrane in the aditus of the larynx (external laryngeal), and the remaining intrinsic muscles (recurrent laryngeal). (From Aronson: *Clinical Voice Disorders*. New York, Thieme-Stratton, Inc., 1980.)

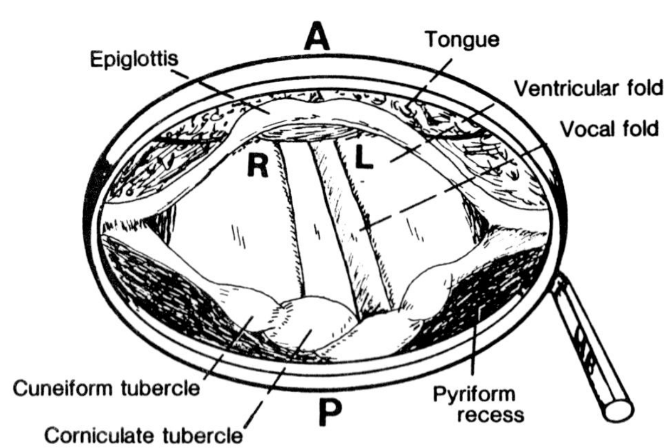

Figure 26-14. An oblique glottis due to unilateral paralysis of the cricothyroideus (left side) as viewed through a laryngeal mirror. The posterior end of the glottis points *away* from the healthy side and toward the paralyzed side. The paretic fold is also lower than the unaffected fold.

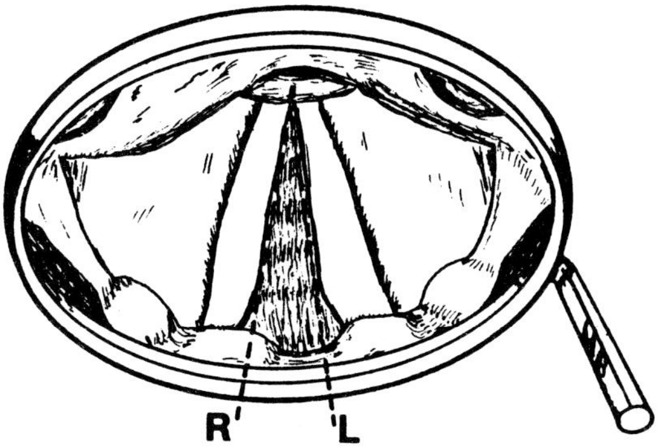

Figure 26-15. Mirror view of the larynx showing the right fold in a paramedial position with the left fold at midline during initial adduction of the folds. The healthy fold (LK) then crosses the midline to contact the paralyzed fold (R) (see also Figure 26-14).

folds lie too close to the midline for heavy use of the voice or physical exertion; they cannot be abducted for adequate inhalation. However, this latter situation is not typical because the folds usually lie in a paramedian position causing the patient to be aphonic. Many types of paralyses may be temporary; voice clinicians should be familiar with the medical literature on the course of the disease for patients referred to them for treatment.

Clinical Assessment

As noted under the ninth cranial nerve, the vagus and glossopharyngeal nerves are functionally related and are tested together. This is accomplished by evaluating the pharyngeal gag reflex or response. The examiner can usually elicit a gag by touching the sides of the pharynx with a tongue depressor or by touching the soft palate.

Other tests of the vagus include swallowing and phonation. Swallowing is a highly individualized behavior, and the examiner should be conservative about diagnosing swallowing disorders on the basis of asymmetry or tongue thrusting. The size of the posterior airway has a great deal of influence on swallowing patterns. If the patient reports coughing following swallows in eating, a dysphagia (swallowing difficulty) is suspected. In known or suspected cases of neurologic damage to the vagus nerve, the speech clinician should be discouraged from having an individual swallow food or liquid during testing. Aspiration of food into the trachea should not be attributed to speech testing.

Normal function of the vagus is indicated by a clear laryngeal tone at various pitch and loudness levels. When hoarseness is present in the voice, indirect laryngoscopic examination of the vocal folds is indicated.

CRANIAL NERVE XI—ACCESSORY

Clinical Implications

Many consider the *accessory* or *spinal accessory cranial nerve* to be purely motor. It has two divisions. One arises from cells in the medulla and one from cells in the anterior gray column of cells in the upper cervical part of the spinal cord. The portion arising from the medulla is in close association with the nucleus of the vagus nerve.

Disturbances involving the dorsal motor nucleus of the accessory nerve or its axons may be difficult to verify. There is too close an anatomic and physiologic relationship with the vagus nuclei and nerves. Thus, the two nerves can mask each other's functions or malfunctions.

Disorders involving the nucleus ambiguus or the accessory nerve fibers arising in this nucleus cause some impairment in swallowing, speech, and respiration, depending on the extent of the pathology.

Paralysis of the sternocleidomastoid and trapezius muscles on one side would represent a unilateral lesion of the spinal nucleus. A peripheral paralysis may occur if the nerve fibers innervating these two muscles are damaged as they leave the jugular foramen at the base of the skull. The patient has a drooping shoulder and restrictive arm elevation with a paralysis of the trapezius. Sternocleidomastoid paralysis results in weakness of head rotation to the opposite side of the impaired muscle. If both sternocleidomastoids are paralyzed, the patient is unable to hold the head erect or to raise it from a recumbent position.

Clinical Assessment

Since at least a portion of the accessory nerve is almost indistinct in origin and the course of travel of the nerve fibers, only that division pertaining to the spinal (cervical) root is assessed. The examiner will observe the patient for ability to (1) raise each arm from the side until it is pointing straight up, (2) turn the head to the left and right, and (3) shrug each shoulder.

Besides this testing, the examiner should palpate the sternocleidomastoid while the muscles are contracted against resistance and observe them for weakness or hypertonicity.

CRANIAL NERVE XII – HYPOGLOSSAL

The *hypoglossal nerve* makes articulate speech possible. This nerve innervates both the intrinsic and extrinsic muscles of the tongue. It also transmits proprioceptive fibers from the muscle spindles of the tongue muscles.

Divided into cell groups corresponding to the tongue muscles they supply, the nucleus of the hypoglossal nerve extends throughout much of the medulla. Corticobulbar fibers from each motor cortex go to each hypoglossal, thus putting the tongue under *bilateral* cortical control. Extrapyramidal nuclei are also interconnected with the hypoglossal nucleus.

Lesions of the hypoglossal nerve will result in lower motor neuron paralysis of the tongue musculature on the same side (homolateral) to the lesion. There is loss of movement and tone, muscular atrophy, and deviation of the protruded tongue to the atrophic side (due to action of the normal contralateral genioglossus). Small fascicular quiverings appear before atrophy. With the tongue at rest in the oral cavity, the tongue may deviate toward the normal side through the pull of the normal styloglossus (Fig. 26-16).

Figure 26-16. Unilateral hypoglossal paralysis (right). Upon protrusion, the tongue deviated to the affected side. Atrophy of the tongue on the impaired side is also evident.

With upper motor neuron damage (pseudobulbar palsy), the voluntary movements of the tongue are greatly impaired and speech is dysarthric. Muscle fibrillations and atrophy do not usually occur. Usually, other cranial nerve nuclei will be involved with upper motor neuron paralysis.

Clinical Assessment

The first aspect of an examination of the tongue is a careful study of the tongue at *rest*. The examiner should seek answers to these questions: Does the tongue appear smaller than normal? Shrunken or furrowed? Both halves equal in size and shape? Are there any fasciculations? Does it have a tendency to refuse to remain at rest? Are there spontaneous gross rotations, elevations, protrusions, lateralizations, retractions?

Assessment during movement should include having the patient protrude the tongue maximally to examine strength and range of motion of the tongue. Any deviation to one side should be carefully noted. The examiner evaluates the ability of the patient to elevate the tongue to touch the upper lip while the examiner holds the mandible in a moderately depressed position. Lateral tongue stretch can be assessed by having the patient place the tongue against the cheek, pressing against the examiner's finger and resisting his/her efforts to force the tongue inward. Moving the tongue from side to side as rapidly as possible, the patient can demonstrate for the examiner speed, regularity, and range of motion.

BIBLIOGRAPHY

Aronson, A.E.: *Clinical Voice Disorders*. New York, Thieme-Stratton, 1980.

Aronson, A., and Hartman, D.: Abductor spastic dysphonia as a sign of essential voice tremor. *J Speech Hear Disord, 46*:52-58, 1981.

Carpenter, M.B.: *Core Text of Neuroanatomy*, 2nd ed. Baltimore, Williams and Wilkins, 1978.

Cooper, M.: Direct vocal rehabilitation. In Cooper, M., and Cooper, M.H. (Eds.): *Approaches to Vocal Rehabilitation*. Springfield, IL: Thomas, 1977.

Curry, E.T.: The pitch characteristics of the adolescent male voice. *Speech Monographs, 7*:48-62, 1940.

Daniel, B., and Guitar, B.: EMG feedback and recovery of facial and speech gestures following neural anastomosis. *J Speech Hear Disord, 43*:9-20, 1978.

Duffy, R.J.: Fundamental frequency characteristics of adolescent females. *Lang Speech, 13*:14-24, 1970.

Fairbanks, G.: *Voice and Articulation Drillbook*, 2nd ed. New York, Har-Row, 1960.

Fairbanks, G., Herbert, E., and Hammond, E.: An acoustical study of vocal pitch in seven and eight-year-old girls. *Child Dev, 20*: 71-74, 1949.

Ginsberg, I.A., and White, T.P.: Otological considerations in audiology. In Katz, J. (Ed.): *Handbook of Clinical Audiology*, 2nd ed. Baltimore, Williams and Wilkins, 1978.

Glattke, T.J.: Electronystagmography. In Katz, J. (Ed.): *Handbook of Clinical Audiology*, 2nd ed. Baltimore, Williams and Wilkins, 1978.

Hollien, H.: The registers and ranges of the voice. In Cooper, M., and Cooper, M.H. (Eds.): *Approaches to Vocal Rehabilitation*. Springfield, IL: Thomas, 1977.

Hollien, H., and Paul, P.: A second evaluation of the speaking fundamental frequency characteristics of post-adolescent girls. *Lang Speech, 12*:119-124, 1969.

Hollien, H., and Shipp, T.: Speaking fundamental frequency and chronologic age in males. *J Speech Hear Res, 15*:155-159, 1972.

McGlone, R.E., and McGlone, J.: Speaking fundamental frequency of eight-year-old girls. *Folia Phoniatr, 24*:313-317, 1972.

Minckler, J.: (Ed.): *Introduction to Neuroscience.* St. Louis, C.V. Mosby, 1972.

Rontal, M., and Rontal, E.: Lesions of the vagus nerve: diagnosis, treatment, and rehabilitation. *Laryngoscope, 87*:72-86, 1977.

Saunders, W.H.: *The Larynx.* Summit, NJ: CIBA, 1964.

Wilson, K.: *Voice Problems of Children*, 2nd ed. Baltimore, Williams and Wilkins, 1979.

Zwitman, D.H.: Bilateral cord dysfunctions. *J Speech Hear Disord, 44*:373-378, 1979.

INDEX

T

W

Y

Z